ENCYCLOPEDIA OF
HURRICANES, TYPHOONS, AND CYCLONES

NEW EDITION

DAVID LONGSHORE

☑Checkmark Books®
An imprint of Infobase Publishing

ENCYCLOPEDIA OF HURRICANES, TYPHOONS, AND CYCLONES, New Edition

Checkmark Books
An imprint of Infobase Publishing
132 West 31st Street
New York NY 10001

Library of Congress Cataloging-in-Publication Data

Longshore, David.
Encyclopedia of hurricanes, typhoons, and cyclones / David Longshore.—New ed.
p. cm.
Includes bibliographical references and index.
ISBN-13: 978-0-8160-6295-9 (hc)
ISBN-10: 0-8160-6295-1 (hc)
ISBN-13: 978-0-8160-7409-9 (pbk)
ISBN-10: 0-8160-7409-7 (pbk)
1. Hurricanes—Encyclopedias. 2. Typhoons—Encyclopedias.
3. Cyclones—Encyclopedias. I. Title.
QC944.L66 2008
551.5503—dc22
2007032336

Checkmark Books are available at special discounts when purchased in bulk quantities for businesses, associations, institutions, or sales promotions. Please call our Special Sales Department in New York at (212) 967-8800 or (800) 322-8755.

You can find Facts On File on the World Wide Web at http://www.factsonfile.com

Text design by Annie O'Donnell
Illustrations by Bob Cronan
Photo research by Tobi Zausner, Ph.D.

Printed in the United States of America

VB MSRF 10 9 8 7 6 5 4 3 2 1

This book is printed on acid-free paper.

For the late Raymond Edward Moore, M.D.

Friend, Patron, and Survivor of the
Great New England Hurricane of 1938

CONTENTS

ACKNOWLEDGMENTS

I am indebted to numerous people and scores of fine institutions for their kind and invaluable contributions to the creation of this revised edition of the *Encyclopedia of Hurricanes, Typhoons, and Cyclones*. Some gave their assistance as part of their daily work, while others spent precious hours of their own conducting research, reading copy, and making the coffee. I am also appreciative of all the correspondence I received from the first edition's readers, and for their thoughtful suggestions and notes of enthusiasm. In many respects, these people and organizations operated not unlike the rain bands of a tropical cyclone, all spinning inward together for the greater cohesiveness of the whole.

They are: Mr. and Mrs. John Alamo; Mr. Thomas Alamo; Amherst College; Dr. George F. Bailey; Professor Christopher Bellavita; the Boston Public Library; Capt. C. Raymond Calhoun, USN (Ret.); Mr. and Mrs. Raymond Dee and family; the Dudley Knox Library at the Naval Postgraduate School; Dr. Kerry Emanual; Mr. and Mrs. Robert Enrione; Professor Fredric L. Cheyette; Señor Jorge Farinas; Mr. Gregory James Flail; the Robert Frost Library at Amherst College; Dr. Jason Gonsky, M.D.; Mr. J. J. Haines; Mr. Christopher Hetherington; Ms. Jeanette Iafolla; Mr. Yuki Karakawa; the Los Angeles Public Library; Mr. Francis E. McCarton; Mr. James McConnell; Mr. and Mrs. Harry Maxon, IV; Metropolitan College of New York (MCNY); the Museum of the City of New York; the Miami (FL) Public Library; the Center for Homeland Defense and Security (CHDS) at the Naval Postgraduate School; Northern Essex Community College; the New York Public Library; the New York City Mayor's Office of Emergency Management (OEM); the New York City Department of Transportation (NYCDOT); Mr. John T. Odermatt; Mr. James Peralta; Mr. Caryl Phillips; the Saltmarsh Library at Pinkerton Academy; Ms. Linda Quick; Mr. Francis M. Reed; Mr. Andy Siegel; Mr. Richard J. Sheirer; Capt. Robert Simeral, USN (Ret.); the Savannah (Ga.) Public Library; the Truro (Mass.) Historical Society; the U.S. Department of Homeland Security; Mr. Sebastian Windgassen; and the Zanger family.

I would like to further extend a special note of appreciation to my editor, Mr. Frank Darmstadt, a true gentleman who skillfully and patiently guided the creation of this work without encountering any major storms and keeping us all off the reef. I would also like to recognize the tenacious efforts of the book's photographic researcher, Tobi Zausner, Ph.D., for her discriminating eye and delicious scones; and Ms. Melissa Cullen-DuPont and Ms. Alana Braithwaite for their painstakingly detailed editorial assistance. Mr. Michael J. Alamo contributed a great deal of original research to this project, as well as his enthusiastic support for its completion, and I will forever be indebted to his longtime kindness.

Spiraling inward toward the eye where the winds blow the strongest, special recognition must be given to my beloved parents, to my sisters and brothers-in-law, and to my two nephews, for weathering with incredible forbearance so many hurricanes, typhoons, and cyclones.

INTRODUCTION

It has been a decade since the first edition of the *Encyclopedia of Hurricanes, Typhoons, and Cyclones* was published. During that time the science, history, and culture of tropical cyclones around the globe has remained ever evolving, providing the meteorological world with improved forecasting techniques, altered naming systems, new intensity and duration records, and all-too-many fresh horrors. Since this work was published, the naming lists for those tropical cyclones that tear across the western North Pacific Ocean every year, for example, were entirely changed in the year 2000. Instead of purely western male and female names, they now feature identifiers that are unique to the countries and cultures most vulnerable to typhoon activity. Cambodia, for instance, submitted Sarika, which translates as "singing bird." Typhoon Singing Bird . . . as a naming system, it certainly does catch the public's attention. Since this work was published, 1992's Hurricane Andrew was officially upgraded to a Category 5 hurricane; and in 2005, no fewer than 31 tropical systems originated in the North Atlantic Ocean. Among them was the most destructive and deadly tropical cyclone witnessed in the United States during the last half a century, Hurricane Katrina. Hurricane Gilbert's 17-year reign as the most intense tropical cyclone yet observed in the North Atlantic basin was deposed by a late-season Hurricane Wilma, which mounted an astonishingly low central barometric pressure reading of 26.05 inches (882 mb), while improvements in forecasting and warning systems have made the various weather services' landfall and intensity predictions that much more accurate and timely, thereby reducing economic costs while enhancing life safety. In Australia in March 1999, Cyclone Vance generated the strongest wind gust yet observed on that continent (166 MPH [267 km/h]); and in April 2006, Cyclone Monica produced a central pressure reading of 25.96 inches (879 mb) as it passed just to the north of the Australian continent, the most intense cyclone yet seen in that

region of the globe. Typhoon activity in the western North Pacific Ocean has also set new records: In 1998, Super Typhoon Zeb generated a pressure reading of 25.75 inches (872 mb) as it blasted across the northern Pacific, while the 2002 season saw Cyclone Zoe traverse the South Pacific Ocean with a frighteningly low central pressure of 25.75 inches (879 mb). We have even seen the formation of a mature stage tropical cyclone in the very inactive *South* Atlantic Ocean: Caterina of March 2004.

Just as the study of tropical cyclones has evolved and changed, books like the *Encyclopedia of Hurricanes, Typhoons, and Cyclones* have changed with it. In addition to dozens of new scientific, historical, and cultural entries, I have added to this revised edition new photographs, maps, and drawings that illustrate the always freshening changes in the world of tropical cyclones. I have also retained the philosophical underpinnings of the first edition, and have sought to preserve the strength of the prose found in the first edition. In large part, this is a book about the wind, and it can often prove an aesthetic and informational challenge to find new ways to describe the screeching winds, the pelting rains, the rolling surges, and the sheer drama that remains one of nature's most awesome workings. I have further worked to correct any deficiencies or inaccuracies that might have occurred in the previous edition, and remain grateful to those readers who have shared their personal experiences, their views and opinions, to bring a renewed perspective to this endeavor.

As with the first edition, I have used a number of sources—many more than could possibly be accommodated in the bibliography—to compile this revised edition. This was, in of itself, an enormous logistical and research challenge, and there were many times when I felt as though a metaphorical Category 5 hurricane of paper and information was bearing down on the project. I was determined to read every scrap of information, sift through every archive, relive every moment in order to compile the most comprehensive one-volume work on the subject I could. There are hundreds of secondary sources available on all aspects of tropical cyclone activity, from scholarly works to narrative histories, and many of them are excellent and enjoyable to read. Media reports, for instance, continue to serve as excellent primary source material for tropical cyclone activity; and yet, there is often contradictory information contained within them. Even the best of the official Web sites and reference sources provided by meteorological agencies around the globe have their space and informational limitations, and it is much too time-consuming and expensive to cull every archive in every nation for every rumor, every story, every anecdote, and every logbook of data for every last detail. Moreover, were all that data to be included in a *book*, it would have as many volumes as a tropical cyclone has water vapor particles, and could not be taken to the beach for some seaside informative reading—especially when a hurricane or typhoon or cyclone is bearing down on that very beach.

And yet amid the shifting winds of science, history, and culture, some things about the tropical cyclone remain unchanged. Tropical cyclones continue to be among nature's most efficient delivery systems; when they reach their final destination, they deliver their waterlogged cargo by simply dissolving. For all their painful destructiveness, their slicing capacity to kill, we are learning to live with the tropical cyclone's double edges. We have come to understand the exact nature of their atmospheric balancing act, and of how difficult—and necessary—it is to shift heat and moisture in an ever changing (and perhaps, increasingly challenging) atmosphere. We have come to quantify and qualify and grieve over the ravages of the tropical cyclone's winds, rains, and seas, while at the same time recognizing that special way in which they can relieve even more destructive and life-threatening droughts and famines. After Tropical Storm Kammuri struck southeastern China in August 2002, heavy rains alleviated a persistent drought condition over Guangdong Province. When Hurricane Debby passed near Cuba in 2000, it, too, broke a long-standing drought condition across the island. And Tropical Storm Floyd, as it passed northward up the eastern United States in September 1999, became the then-deadliest landfalling hurricane in the United States since Agnes in 1972. But its tremendous downpours (which measured 20 inches [50.8 cm] in some places) did end a long drought that had resulted in stringent water usage restrictions for many months and may have precipitated an even larger series of emergencies. There is even evidence to suggest that the tropical cyclone helps feed the very oceans that

in turn feed it. According to a report from the Associated Press dated December 7, 2002, scientists in Taiwan and the United States "used a trio of NASA satellites to observe how the passage of even moderate typhoons over the South China Sea can generate upwellings of nutrient-rich waster from deeper in the ocean and spark massive blooms of phytoplankton."

Whatever the changes, whatever the stasis, one fact remains of paramount importance; that we continue to understand and respect the power and potential that even a weak tropical storm has to cause loss of life, loss of property, and a loss of security. As of this writing, the historic and vibrant city of New Orleans still faces many years of reconstruction and rebirth in order to return to what it was before Hurricane Katrina undermined the city's levees and swept away centuries of hard work. National meteorological agencies and research institutions continue to suffer from the deadly effects of under-funding and political neglect. Annual financing for the Hurricane Research Division of the National Oceanic and Atmospheric Administration (NOAA), for example, has never exceeded $5.1 million, while its staffing numbers have declined by 30 percent since 1996. In September 2006, the U.S. National Science Board renewed the call for the federal government to invest more money—to the tune of about $300 million—in tropical cyclone research and engineering science. Between 2005 and 2006, hurricane-related damage in the United States totaled $168 billion, and nearly 1,500 lives were lost; grim tallies that certainly lend credence and urgency to the call for a centrally organized, multiagency response to tropical cyclone activity in the United States, and indeed, around the globe.

Further enhancements also remain to be achieved in the fields of emergency management and tropical cyclone preparedness. While local, state, federal, and private agencies can and do make a crucial difference (and with limited resources) in saving lives and saving systems during periods of tropical cyclone activity, it ultimately remains up to individuals (particularly those who live in storm-prone areas) to educate themselves, their families, and friends on tropical cyclone preparedness and on those steps that can be taken to save your own life, your own systems, when a hurricane, typhoon, or cyclone threatens. Prepare a hurricane evacuation plan well in advance. Think of what will need to be done with pets in the event of an emergency. Prepare a "Go Bag," a portable safety kit that contains critical documents, contact information, food, water, medicine, a flash light or light stick, some form of communications, and other material such as a first aid kit—and keep the Go Bag in good condition and readily available in the unhappy event that it is needed. Mariners should always pay close attention to meteorological conditions before getting under way, while surfers and surf-watchers should always remember that an on-coming tropical cyclone is like an atmospheric octopus—once the riptide or high wave catches, it is almost impossible to break free. And, of course, if one does live in a hurricane, typhoon, or cyclone-zone, and an evacuation order is issued, heed it and go.

It is my hope that you will enjoy reading and studying the new edition of the *Encyclopedia of Hurricanes, Typhoons, and Cyclones*. It is a work that is designed to be read as well as referenced, and as the workings of hurricane preparedness and storm safety are experiential in nature, it does contain long and hard-learned scientific, historic, and cultural lessons that one day may prove of life-saving use to the stormy and ever-changing world around us.

Accumulated Cyclone Energy Index (ACE) Developed by the Tropical Prediction Center (TPC) and regularly employed by the National Oceanic and Atmospheric Administration (NOAA) and the NATIONAL HURRICANE CENTER (NHC) since the latter part of the 1990s, the Accumulated Cyclone Energy (ACE) index assists meteorologists and forecasters in analyzing, quantifying, and in some cases, predicting, the total seasonal activity of TROPICAL CYCLONES within each of the Earth's tropical cyclone generation zones. Determined through the use of mathematical formulas and models, the ACE index—which according to the NHC is a "wind energy index, defined as the sum of the squares of the maximum sustained surface wind speed (knots)"—is calculated every six hours for existing NORTH ATLANTIC tropical systems of tropical storm intensity and higher. By including additional data such as the number of tropical storms and hurricanes, the number of major HURRICANES, and geographical location, NOAA can use the ACE to categorize hurricane, typhoon, and cyclone seasons as being *above-normal, near-normal,* and *below-normal*. Since implementing the ACE Index, meteorologists at the NHC have (retroactively) applied the technique to North Atlantic tropical cyclones dating from 1950–2005, and used the ACE Index to determine a base period from 1951–2000. These studies indicate that the median annual index for this base period was 87.5.

In terms of definitional clarity, an *above-normal* season generally possesses an ACE Index value above 103, or 117 percent of the median, and includes either 10 TROPICAL STORMS, six hurricanes, or two major hurricanes. A *near-normal* season carries an ACE Index value of between 66 and 103, and is generally representative of 75 percent to 117 percent of the median. Additionally, a near-normal season is characterized by having less than the long-term average number of tropical storms, hurricanes, or major hurricanes. And a *below-normal* season has an ACE Index value of 65 and below, which generally translates to 75 percent of the median as determined by the base period data.

According to records maintained by NOAA, the top five North Atlantic tropical cyclone seasons as determined by the ACE Index were: 2005, 1950, 1995, 2004, and 1961. In the case of the extraordinarily active 2005 North Atlantic HURRICANE SEASON, an ACE Index of 248 was derived. In addition to the severity of the storms produced during that season, there were a total of 27 named tropical systems, with seven of them reaching major hurricane status. Although the 1950 season produced less than half the number of named storms observed during the 2005 season (a total of 13 tropical systems), it generated an above-average number of major hurricanes (eight), which gave it an ACE Index value of 243.

Adrian, Hurricane *Eastern North Pacific Ocean–Central America, May 15–23, 2005* The first TROPICAL CYCLONE of the 2005 eastern NORTH PACIFIC OCEAN HURRICANE SEASON, and the first tropical cyclone to have made a recorded landfall on the western coast of Honduras, Hurricane Adrian destroyed roads, spawned flash floods and landslides, uprooted trees, downed power lines, and forced the evacuation of some 14,000 people as it trundled ashore near the Gulf of Fonseca on May 19, 2005. A weak Category 1 hurricane with a CENTRAL BARO-

METRIC PRESSURE of 28.99 inches (982 mb) that had been downgraded to a TROPICAL DEPRESSION by the time it made landfall, Adrian nevertheless delivered sustained winds of 40 MPH (65 km/h) to the coastal town of Acajutla, located approximately 35 miles (55 km) west of the El Salvadoran capital of San Salvador. Adrian's high winds churned across Central America at 12 MPH (19 km/h), killing an El Salvadoran military pilot whose aircraft crashed as it was being moved from Adrian's path, while another two people in neighboring Guatemala perished in a mudslide caused by Adrian's drenching rains. On May 20, Adrian disintegrated over Central America before its remnants reached the Caribbean Sea.

On June 21, 1999, the western Mexican states of Coahuila and Colima were inundated by the outer RAIN BANDS associated with an earlier Hurricane Adrian. Formed near the Gulf of Tehuantepec, approximately 240 miles (386 km) southeast of Acapulco, MEXICO, during the early morning hours of June 18, the system was upgraded to TROPICAL STORM intensity later the same day, and to hurricane status on June 20. Then located approximately 430 miles (692 km) south-southeast of Baja, California, Adrian's minimum central pressure of 28.73 inches (973 mb) produced sustained winds of 98 MPH (158 km/h) and heavy seas. Upon interacting with lower sea-surface temperatures and southeasterly WIND SHEAR, Adrian quickly weakened, returning to tropical storm status on June 21, and almost completely dissipating by the early evening hours on June 22. Although Adrian remained an offshore system for its entire lifespan, it was responsible for at least six deaths, including four people who drowned after being swept from a beach in Chiapas by a large wave generated by Tropical Depression Adrian on June 18; and another three people who died elsewhere in Mexico from riverbank flooding associated with Adrian's large rainfalls.

The name Adrian has been retained on the rotating naming lists for eastern North Pacific tropical cyclones.

advection In the term's strictest sense, advection is the transfer of any atmospheric property—be it heat, moisture, or motion—through horizontal movement. In a more general sense, advection refers to one of two meteorological processes (the other being CONVECTION) that creates wind.

In a HURRICANE, TYPHOON, or CYCLONE, advection results in the destructive winds that scour the surface of the land or sea—the flow of air that results when the cool, dense air that is descending toward Earth's surface on one side of a convective cell is suddenly moved *horizontally* across the system's bottom, only to be lifted up on the cell's opposite side by rapidly *rising* or by convective currents of warm, light air.

This cyclical motion, which results from the uneven heating of the air by unequal surface temperatures, in turn serves to power the convective cell and will remain in continuous motion as long as the two contrasting surface temperatures persist. In tandem with the convective process, advection forms one of the principal mechanisms by which all atmospheric circulations, from global weather systems, to TROPICAL CYCLONES, to localized supercell thunderstorms, are sustained.

Agnes, Hurricane *Southern–Eastern United States, June 15–25, 1972* A relatively weak Category 1 HURRICANE, its intense precipitation nevertheless initiated one of the most extensive flood emergencies in U.S. history. During a five-day sojourn up the eastern seaboard in June 1972, Agnes forced the displacement of more than 100,000 people, caused nearly $2 billion in flood damage, and claimed 134 lives in FLORIDA, GEORGIA, Pennsylvania, NEW YORK, VIRGINIA, and Maryland.

An early season TROPICAL CYCLONE that originated off the northeast coast of CUBA on June 15, 1972, Agnes was just barely of hurricane strength when it first made landfall at Apalachicola, Florida, on June 19. Its 75-MPH (121-km/h) winds knocked out power lines, downed several trees, and caused an estimated $10 million in property damage; it otherwise spared Florida's Panhandle the full promise of its later fury. Densely laden with more than 20 cubic miles of PRECIPITATION, Agnes slowly moved northeast, spawning no less than 17 TORNADOES in central Georgia. Twelve people in Florida and Georgia were killed.

After briefly reintensifying off Georgia's east coast, Agnes quickly swept across SOUTH and NORTH CAROLINA and struck Virginia with hurricane-force gusts and torrential rains. In the state capital, Richmond, severe flooding caused the James River to crest at nearly 36 feet (12 m) above mean low water, breaking a record that had stood since 1771. Richmond's reservoirs were polluted with silt and storm-swept debris, while several multistory buildings in the business district were almost completely submerged. Close to 10,000 people were left homeless. Touring the area by helicopter, then President Richard M. Nixon quickly declared 63 of Virginia's 96 counties federal disaster areas on June 26.

Slowly progressing up the coast, Agnes then battered significant portions of eastern Maryland. Fierce rains and lingering winds forced more than 2,000 people to flee their homes and caused an estimated

$50 million in damage. Extensive flash flooding completely immersed communication networks across the state, making it impossible to restore service for several weeks thereafter. Fifteen lives were lost.

During the night of June 22, Agnes thudded into Pennsylvania. Swiftly rising flood waters washed away numerous highways and bridges and completely isolated the state capital of Harrisburg. The governor's mansion was severely damaged when eight feet (3 m) of water burst through its lower story, sweeping it clean of its luxurious furnishings and historical artifacts. More than 100 miles (175 km) of the Erie-Lackawanna Railway's main line was deluged, forcing the financially strapped railroad to file for bankruptcy protection on June 26. In Lock Haven, Pennsylvania, the Piper Aircraft Corporation suffered the near total loss of its principal manufacturing facility. The sprawling residential community of Wilkes-Barre was completely inundated when an astonishing 19 inches (48.3 cm) of rain forced the low-lying Susquehanna River to breach its 38-foot (13 m) dikes. Scientists estimated that by the morning of June 24, the Susquehanna's thunderous flow was exceeding an astronomical one billion cubic feet of water per second. Agnes further ravaged 14 counties in southern New York before blowing itself out over southern NEW ENGLAND on June 25. The identifier Agnes has also been used in the western NORTH PACIFIC OCEAN for two systems that were of tropical storm intensity at landfall. Between September 25 and 28, 1965, Tropical Storm Agnes killed five people and caused several injuries as it passed ashore in Hong Kong. This large Chinese city was again affected by Severe Tropical Storm Agnes between July 24 and 30, 1978, in which three people were killed and 134 injured.

The name Agnes has been retired from the rotating list of North Atlantic tropical cyclone names.

air mass By definition, air masses are vast individual bodies of air within which the horizontal distribution of TEMPERATURE and moisture is fairly uniform and stable. Covering expansive tracts of Earth's surface in a single stretch, they are formed when the air remains stationary over a particular geographical location for an extended period of time, thus deriving a uniform temperature from that of the land or the sea below.

In practice, however, air masses are the purveyors of the world's varied weather systems, including those unique elements required for TROPICAL CYCLONE generation. Because the atmosphere is in constant motion, an air mass will not remain stationary indefinitely; in time, other air masses will move to displace it, creating in the convergence of different temperatures an atmospheric collision that yields the highs and lows associated with clement and stormy weather, respectively.

A *front* is the line where two air masses come together. The line where a warm air mass overtakes a cold air mass is a *warm front*; a cold air mass displacing a warm one is a *cold front,* and the situation where, because of similarities in temperature, neither air mass is able to overcome the other is a *stationary front.* Either way, the result is generally vicious weather, be it rain, snow, sleet, or thunderstorms.

Although HURRICANES, TYPHOONS, and CYCLONES are not EXTRATROPICAL CYCLONES and so do not contain the sort of frontal systems generally associated with stormy weather, they are nevertheless subject to the influence of air masses in motion. First, tropical cyclones begin as a revolving collection of thunderstorms over warm ocean waters—thunderstorms that are in themselves the direct result of a collision between air masses of contrasting temperature and moisture content.

Second, a tropical system develops as its CENTRAL BAROMETRIC PRESSURE begins to fall and curving cloud bands start to gather around what will eventually become its EYE. Rising air masses within the cyclonic system establish a crucial temperature difference between the storm's core and its external environment. Following the laws of ADVECTION and CONVECTION, this temperature differential allows cooler air masses to descend toward the ocean's surface and warmer air masses to ascend to what is known as an OUTFLOW LAYER. In time this relentless mechanical process determines the central pressure of the tropical cyclone and the intensity of its winds.

Furthermore, air masses often determine what course, or track, a particular hurricane will take across Earth's surface. Because most tropical cyclones contain weak internal STEERING CURRENTS and so are not always able to choose where or when—if ever—they will make landfall, the presence of powerful external air masses becomes the primary means of locomotion for these storms. By deciding in what direction and at what speed a tropical cyclone will travel, air masses determine whether that system will make a potentially destructive landfall or will harmlessly pass out to sea.

Alabama Although its 53-mile (85-km) coastline is the shortest of the five southern states—FLORIDA, MISSISSIPPI, LOUISIANA, and TEXAS—that border on the hurricane-prone GULF OF MEXICO, Alabama has had a long and often violent history of TROPICAL CYCLONE activity. Between 1559 and 2006, the state was affected by no less than 51 documented hurricanes, 27 of these being direct landfalls in the picturesque Dauphin Island–Mobile Bay-Perdido Key areas. Since 1711, 17 major hurricanes—storms with BAROMETRIC PRESSURES of less than 28.47 inches (964 mb) and nine-to-12-foot (3–4 m) STORM SURGES—have come ashore in Alabama, causing thousands of deaths and billions of dollars in property damage. On average, the state's low-lying coast, treasured for its white beaches and genteel winter resorts, is threatened by a hurricane every two-and-a-half years and is directly affected by one every five years. At times, it has even received several hurricanes in succession—as happened in 1740 when the Twin Mobile Hurricanes came ashore within one week of each other, and in 1860 when two storms struck within a month's time. Indeed, as recently as 1985, Alabama sustained damage from three of that year's major storms: Hurricane Danny (August 15, 1985), Hurricane ELENA (August 31, 1985), and Hurricane JUAN (October 26, 1985).

Alabama's location at the northern crest of the Gulf of Mexico affords it a very long and vulnerable HURRICANE SEASON. Since 1711, Alabama has been struck by at least four June hurricanes, although none were of major intensity. In July, the probability of a hurricane strike increases slightly, with five reported landfalls between 1711 and 1994; one of these July storms, the Mobile Hurricane of 1916, was judged by survivors to have been the most destructive storm on record—it killed four and caused more than $3 million in damage. However, in August and September, when the first CAPE VERDE STORM arrives in the gulf, the number of strikes jumps to an ominous 23 and 33, respectively. Memorable midseason hurricanes in Alabama include the BAY ST. LOUIS HURRICANE that passed from Louisiana into Mobile Bay on the morning of August 26, 1819, killing an estimated 200 people; the Great Mobile Hurricane of August 27, 1852, that claimed the lives of 300; the disastrous storm of August 30–31, 1856, that brought great destruction to the state's heavily wooded interior; and the hurricane of August 11–12, 1860, that flooded the settlements around Mobile Bay with more than four inches (102 mm) of rain. In September 1906, the mighty Pascagoula-Mobile Hurricane left 134 dead in Mississippi and Alabama and caused several million dollars in damage to shipping in Mobile Bay. More recently, Hurricane FREDERIC slammed into Mobile on September 12, 1979, with sustained winds of 132 MPH (213 km/h) and a staggering 14-foot (5-m) storm surge. Eight people were killed, and $2.3 billion dollars in damage were assessed. The lingering effects of a powerful September hurricane initially struck New Orleans in 1711, forcing the relocation of the fledgling city of Mobile to its present position at the sheltered head of the bay that now bears the same name.

During the 1997 North Atlantic hurricane season, Alabama was affected by Hurricane DANNY, which rolled ashore in southwestern Alabama on the night of July 19 and 20. A Category 1 hurricane at landfall, Danny's central pressure of 29.32 inches (993 mb), 75 MPH (121 km/h) sustained winds, and 30 inch (762 mm) rainfalls downed trees and power lines, but caused no deaths or serious injuries in the state.

In September of 2002, a relatively weak Tropical Storm Hanna bundled ashore near Mobile, Alabama, on the 14th of the month. Sustained winds of 52 MPH (84 km/h) and a pressure reading of 29.58 inches (1,002 mb) battered trees and other structures, but caused no serious damage. Tracking to the northeast over Alabama, Georgia, and the Carolinas, Hanna spent the next two days dropping torrential amounts of rain, causing widespread flooding, particularly in Georgia.

On September 16, 2004, a very erratic Hurricane IVAN lurched ashore in Gulf Shores, Alabama, as a powerful Category 3 system with a central pressure reading of 27.55 inches (933 mb) at landfall. With sustained winds of 132 MPH (212 km/h) at its center, Ivan moved along the eastern shores of Mobile Bay before tracking north-northeast across the state, causing moderate damage to foliage and smaller structures. The barrier island of Gulf Shores was inundated with some eight feet (2.5 m) of seawater, while several TORNADOES touched down in the neighboring Florida Panhandle. Two people in Alabama indirectly lost their lives to Ivan's trek through the state.

The remarkable 2005 North Atlantic hurricane season saw the state of Alabama directly and indirectly affected by no fewer than four tropical cyclones, including a downgraded Tropical Storm KATRINA. On June 11, 2005, the former Hurricane ARLENE—now downgraded to tropical storm intensity—made landfall in southwestern Alabama. In terms of barometric pressure more intense at landfall than 2002's Tropical Storm HANNA, (29.26 inches [991 mb] as opposed to 29.58 inches [1,002 mb]), Arlene's sustained, 52-MPH (84-km/h) winds overturned small structures and downed tree limbs, but claimed no lives.

Heavy rains, a minimum pressure reading of 29.50 inches (999 milibars), and 35-MPH (56-km/h) winds generated by Hurricane CINDY's visit to Alabama on July 5, 2005, swamped portions of Mobile. Numerous transportation arteries—including the Mobile Bay Causeway—experienced spot flooding. Recurving to the northeast over Alabama as a tropical depression, Cindy delivered heavy rains to the northern reaches of the state.

Downgraded to tropical storm status with a pressure reading of 28.93 inches (980 mb) at landfall, the former Hurricane DENNIS menaced much of Alabama with 52-MPH (84-km/h) winds and heavy rains on July 11, 2005. No deaths or serious injuries resulted.

The now legendary Hurricane KATRINA brought nearly unprecedented meteorological conditions to southern and western Alabama between August 29 and 30, 2005. In the coastal town of Bayou La Batre, near the Alabama-Mississippi border, a storm surge gauged at between 12 and 14 feet (4–5 m) in height destroyed several houses and caused extensive shoreline erosion. To the east of Bayou La Batre, in Mobile, Hurricane Katrina generated the largest storm surge recorded at that location since July 1916, nearly 11.5 feet (4 m). As sustained winds of 104 MPH (167 km/h) battered Mobile, much of the downtown area was flooded, and a dusk-to-dawn curfew was imposed. Downgraded to a tropical storm by the late-evening hours of August 30, Katrina's steadily rising barometric pressure of 28.93 inches (980 mb) generated 52-MPH (84-km/h) winds over Alabama's northwestern quadrant. Two people died in Alabama as a result of Katrina, and property damage estimates across the state ranged as high as $1 billion.

Such relentless hurricane activity has also reshaped many of the principal geographical features of the Alabama coastline. A number of its low-lying peninsulas, spits, and lagoons have been repeatedly scoured by the torrential winds and inundating tides of passing hurricanes, causing incalculable damage to the area's fragile ecosystem. The popular resort island of Dauphin, which lies across the entrance to Mobile Bay, has been raked by so many hurricanes in the past century that it has literally doubled its length and moved several hundred feet to the west. On several stormy occasions, it has also been breached and cut into pieces by the furious onslaught of a hurricane's storm surge. Nearby Sand Island, once home to a towering lighthouse, almost completely disappeared in the 1947 hurricane, while Fort Morgan Peninsula, which serves to narrow the entrance to Mobile Bay, was seriously eroded along both its gulf and bay shorelines. It has been estimated that during Hurricane Frederic, most of Fort Morgan Peninsula was submerged beneath 15 feet (5 m) of water. Although a multitude of steps—including the construction of seawalls and the legal preservation of sand dunes—have been taken to better protect Alabama's coastal integrity, it is unlikely any of them will prove an adequate match for the ferocious cycle of hurricanes that continues to blight the Cotton State.

Alberto, Tropical Storm *Southern United States, July 3–7, 1994* Between 1982 and 2006, five North Atlantic TROPICAL CYCLONES have been identified with the name Alberto. Of these systems, two were of HURRICANE intensity, while the remaining three were classified as TROPICAL STORMS.

Less than three days after the official start of the 1982 North Atlantic HURRICANE SEASON, the first Hurricane Alberto originated over the warming waters of the extreme western GULF OF MEXICO on June 2, 1982. At its peak a powerful Category 1 hurricane, a meandering Alberto's central pressure reading of 29.08 inches (985 mb) produced sustained winds of 86 MPH (138 km/h) and heavy rains over MEXICO's Yucatán Peninsula, western CUBA and southern FLORIDA, but caused no deaths or serious injuries.

On August 8, 1988, a weak Tropical Storm Alberto chugged ashore in Newfoundland, CANADA, delivering 40-MPH (64-km/h) winds and a central pressure of 29.58 inches (1,002 mb) to the province's rocky but picturesque coastline. Formed off the coast of Georgia on July 5, Tropical Storm Alberto tracked to the northeast, brushing past NORTH CAROLINA's Outer Banks as it followed the nourishing waters of the Gulf Stream, northward. No deaths or injuries were reported in Tropical Storm Alberto's wake.

Even though its 60-MPH (97-km/h) winds were not of HURRICANE strength when it first came ashore at Destin, FLORIDA, on July 3, 1994, TROPICAL STORM Alberto was later responsible for some of the worst flooding to strike neighboring GEORGIA in almost a century.

Originating off the northwest coast of CUBA on June 30, Alberto steadily moved north across the GULF OF MEXICO on a course that closely paralleled that of a 1919 unnamed tropical storm that began in roughly the same place and at about the same time in the HURRICANE SEASON. As was the case in the earlier storm, Alberto, with a CENTRAL BAROMETRIC PRESSURE of 29.32 inches (993 mb), made a midmorning landfall in the vicinity of Pensacola with winds of less than hurricane intensity but bearing immense quantities of tropical PRECIPITATION. At Fort Walton, Florida, Alberto's 66-MPH (106-km/h) gusts created isolated power outages while whipping longleaf pine trees and road signs, but it caused no fatalities.

While coastal flooding ruined that year's valuable oyster harvest in Apalachicola Bay, it was not until Alberto moved inland, where it collided with an entrenched high-pressure system over the southern United States, that the storm's torrential rains were finally unleashed on July 5. Americus, Georgia, received as much as 21.1 inches (533 mm) of rain in a single 24-hour period, causing the Flint and Ocmulgee

Rivers to overrun their banks quickly. Forty thousand people in Albany, Georgia, were forced to seek higher ground when steadily rising floodwaters inundated their homes. Cemeteries burped up their caskets, and budding cotton fields were washed clean of crop and topsoil alike. Thirty-one people were killed, and hundreds of millions of dollars in property damage were assessed in Tropical Storm Alberto's wake.

Between August 4 and 23, 2000, Hurricane Alberto tracked a long and convoluted course across the North Atlantic basin. Originating near the western African coast, Alberto tracked to the northwest, executed a large loop to northeast of BERMUDA, then moved northward into the North Atlantic. At its peak a hurricane of Category 3 intensity, Alberto's central pressure of 28.05 inches and sustained winds of 127 MPH (204 km/h) remained well offshore for its entire existence. As of the end of the 2005 North Atlantic hurricane season, Alberto (2000) was the second longest-lived tropical cyclone on record in the Atlantic basin, with a total distance traveled of 6,500 miles (10,500 km). Alberto's marathon existence was topped only by 1966's Hurricane Faith, which traveled some 7,500 miles (12,500 km).

The first named tropical system of the 2006 North Atlantic hurricane season, Alberto, formed off the western coast of CUBA on June 10. A disorganized system with a weak convective center hampered by an unfavorable WIND SHEAR environment, Alberto remained a TROPICAL DEPRESSION (with a central pressure of 29.61 inches [1,003 mb]) for its entire existence in Cuban waters before recurving to the northeast and eventually making landfall near Adams Beach, FLORIDA, as a powerful tropical storm. Like most tropical depressions and tropical storms, Alberto was a strenuous rainmaker, dropping significant amounts of precipitation on western Cuba and the southeastern U.S. While some Cuban reports indicated up to 16 inches (406 mm) of precipitation from Alberto, the system produced smaller counts during its passage over Florida and the mid-Atlantic states. Ruskin, Florida, received 6.71 inches (170 mm) of precipitation on June 13, while Raleigh, NORTH CAROLINA, observed 7.16 inches (182 mm) on June 14, as the system was undergoing extra-tropical deepening. When the system trundled ashore on the Florida Panhandle on June 13, its central barometric pressure of 29.44 inches (997 mb) generated sustained wind speeds of 69 MPH (111 km/h). Meteorologists attributed Alberto's abrupt strengthening configuration prior to landfall to its interaction with the fueling waters of the loop current. Not surprisingly, Alberto's constituent thunderstorms generated scores of TORNADOES, including four in Florida. Although there were many reports of STORM SURGE flooding in Florida and flash floods in North Carolina, Alberto's passage was not considered particularly destructive. Two people indirectly lost their lives to Alberto, while damage estimates of $50 million were tallied.

The identifier Alberto has been retained on the list of North Atlantic tropical cyclone names, and will be given to the first named tropical system of the 2012 season.

Alex, Hurricane *Southern–Eastern United States– North Atlantic Ocean, July 31–August 6, 2004* The first TROPICAL CYCLONE of the 2004 North Atlantic hurricane season, and one of only two known North Atlantic tropical cyclones to have reached major hurricane status (Category 3 on the SAFFIR-SIMPSON SCALE and above) in the cool ocean waters north of 38 degrees, Alex was born off the NORTH CAROLINA coast on July 31, 2004. Spawned by a weak tropical depression (TD 1) located some 175 miles (282 km) south-southeast of Charleston, SOUTH CAROLINA, Alex slowly drifted to the north-northeast before intensifying into a Category 2 hurricane (with sustained winds of 100 MPH [161 km/h]) on the afternoon of August 3. Passing very close to North Carolina's picturesque but vulnerable Outer Banks, Alex delivered Category 1 winds and rains to the landmark lighthouse at Cape Hatteras, but remained an offshore system that caused no deaths or injuries in the Tarheel State. An extreme gust of 102 MPH (164 km/h) was observed on the Outer Banks on the afternoon of August 3.

On August 4, 2004, while situated over the leading expanses of the Gulf Stream some 800 miles (1,285 km) southwest of Newfoundland, CANADA, Alex became the most powerful tropical cyclone yet observed north of 38 degrees. Driven by an estimated CENTRAL BAROMETRIC PRESSURE of 28.26 inches (957 mb), sustained winds of 120 MPH (195 km/h) lashed the sea surface, generating enormous waves and breaking Hurricane Ellen's 1973 record of sustained 115-MPH (185-km/h) winds in roughly the same area. For four British rowers who had come within 300 miles (483 km) of breaking a 108-year old record for rowing across the Atlantic (55 days) from west to east, however, Alex's record-setting passage proved a degree of competition they could have done without. A huge wave produced by the hurricane smashed their small, lightweight craft and cast them into the raging sea, where they remained for six terrifying hours before being rescued by a Danish cargo vessel. Alex remained at Category 3 intensity through August 5, before steadily weakening over the chill expanses of the western NORTH ATLANTIC OCEAN.

Between July 27 and August 3, 1998, an earlier Tropical Storm Alex—the first North Atlantic tropical cyclone to be dubbed Alex—produced a minimum pressure reading of 29.52 inches (1,000 mb) and sustained 52-MPH (84-km/h) winds, but remained to the east of the Leeward Islands, over the open Atlantic.

The name Alex has been retained on the rotating list of North Atlantic tropical cyclone names and is scheduled to reappear in 2010.

Alice, Hurricane *Caribbean Sea, December 30, 1954–January 6, 1955* The first North Atlantic TROPICAL CYCLONE identified with a name taken from the female list of hurricane identifiers, preseason Tropical Storm Alice carved a bizarre trajectory through the southwestern CARIBBEAN SEA and GULF OF MEXICO between May 25 and June 6, 1953. A powerful TROPICAL STORM whose CENTRAL BAROMETRIC PRESSURE of 29.44 inches (997 mb) produced sustained winds of 69 MPH (111 km/h), Alice affected Central America and northwestern CUBA before bounding ashore in the FLORIDA Panhandle on June 6. No deaths or significant property losses were recorded.

The 1954 North Atlantic hurricane season also opened with a tropical cyclone named Alice—one of hurricane intensity. An early season system, Hurricane Alice delivered 81-MPH (130-km/h) winds and torrential rains to northern MEXICO and southern TEXAS between June 24 and 26, 1954. No deaths or injuries were tallied.

Ironically, the second Hurricane Alice of 1954, commonly known as Alice 2, became one of the few tropical cyclones to originate outside of the standard NORTH ATLANTIC HURRICANE SEASON, which extends from June 1 to November 30 of each year. Originating on December 30, 1954, from an extratropical cyclone that had acquired tropical characteristics, and lasting until January 6, 1955, Alice rang in the New Year by delivering a central barometric pressure of 29.73 inches (1,007 mb) and sustained 81-MPH (131-km/h) winds to the island of Grenada, in the Leeward Islands. Although Hurricane Alice caused tens of thousands of dollars worth of property damage to the Leeward Islands, it also delivered much needed precipitation to the island chain, including PUERTO RICO.

The last Hurricane Alice observed in the North Atlantic basin occurred between July 1 and 7, 1973. A strong Category 2 system, 1973's Alice produced a pressure reading of 29.11 inches (986 mb) and sustained, 92-MPH (148-km/h) winds as it tracked along the U.S. eastern seaboard. It remained an offshore system until July 6, when it ground ashore in eastern Newfoundland, CANADA, as a powerful tropical storm. A pressure reading of 29.26 inches (991 mb) produced 69-MPH (111-km/h) winds, but caused no deaths or significant damage in the province.

Alice has also been used as an identifier for tropical cyclones in the western North Pacific Ocean. Between May 17 and 21, 1961, four people died as Typhoon Alice twirled ashore at Hong Kong.

The name Alice has been retired from the list of North Atlantic tropical cyclone names.

Alicia, Hurricane *Southern United States, August 14–19, 1983* This moderately powerful Category 3 HURRICANE's 115-MPH (185-km/h) winds and 12-foot (4-m) STORM TIDE lashed large portions of south TEXAS on August 18, 1983. With a central pressure at landfall of 28.41 inches (962 mb), Alicia was the first storm of hurricane intensity to strike mainland United States since Hurricane ALLEN came ashore near Brownsville, Texas, on nearly the same date in August 1980.

A midseason hurricane that formed over the warm waters of the eastern GULF OF MEXICO on August 14, Alicia leisurely intensified before striking the port city of Galveston with diligent fury in the early morning hours of August 18. Damage in the hurricane-prone city of 60,000 was severe. Alicia's six-foot (2 m) STORM SURGE, coupled with above-average tides, broke over Galveston's famed seawall, flooding several low-lying areas of the island city. All electricity and telephone service was cut, and downpours made driving exceptionally hazardous. There were several incidents of looting. Six people were killed, and another 30 were injured in Galveston, but Alicia could have been much worse: The ferocious GREAT GALVESTON HURRICANE of 1900 killed between 6,000 and 12,000 people, while a lesser hurricane in 1915 breached the newly built seawall and drowned 275 people. The lifesaving benefits of modern construction techniques and hurricane awareness programs were made readily apparent by Alicia's spirited assault on the one-time "New York of the South."

Its wind and rain virtually undiminished, Alicia spiraled inland, hammering Houston and its surrounding communities just before midday on August 18. Already reeling from an economic recession brought on by a collapse in oil prices, Houston sustained close to $1 billion in property damage as Alicia's winds knocked out power lines, uprooted trees, and overturned cars. In downtown Houston, Alicia spawned small TORNADOES that pried glass curtain walls away from some of the city's tallest buildings. Deadly shards of glass and torn aluminum crashed to the street as the wind and rain shattered the glittering facades of the 71-story Allied Bank Plaza building (now the Wells Fargo Bank Plaza) and the 33-story Hyatt Regency Hotel. Forty-two thousand people fled as flash floods invaded their homes; nearly 50 people were arrested for looting. Even though Alicia was downgraded to a TROPICAL STORM by three o'clock that afternoon, its damage was so extensive that several days passed before telephone and water services was fully restored. Fifteen people in the Houston area were killed.

On August 19, then President Ronald Reagan declared both Galveston and Houston disaster areas and made available several hundred million dollars in federal funds for emergency relief. The American Insurance Association later set the total insured property damage from Alicia at $675 million, although this figure was eventually adjusted higher. At the time, Alicia's $1 billion price tag made it the third-most-costly hurricane in U.S. history. Owing to the extensive damage wrought by the hurricane, the name *Alicia* was retired from the revolving list of HURRICANE NAMES the following year.

Allen, Hurricane *Eastern Caribbean–Southern United States, August 4–10, 1980* One of the most intense Category 3 HURRICANES on record, Allen spent August 4 through August 10, 1980, beating a trail of destruction through the eastern CARIBBEAN SEA, HISPANIOLA, CUBA, MEXICO, and the gulf coast of TEXAS. Allen intensified to Category 5 status on the SAFFIR-SIMPSON SCALE three times during its 1,200-mile (1,920-km) trek from the Cape Verde Islands. Its lowest central pressure, 26.55 inches (899 mb), was recorded off the northeast tip of Mexico's Yucatán Peninsula on the evening of August 7. Before dissipating into a string of TORNADOES over south-central Texas on August 10, Allen caused close to $1 billion damage to six countries and left 272 people dead.

A classic CAPE VERDE STORM, Allen was born as a low-pressure TROPICAL DISTURBANCE off the northwest coast of Africa on August 1. Carried westward by the equatorial trade winds, Allen rapidly intensified over the mid-Atlantic's warm tropical waters. By August 3, as the hurricane bore down on the islands of BARBADOS, St. Lucia, and DOMINICA, weather reconnaissance flights through its eye indicated BAROMETRIC PRESSURES as low as 27.23 inches (922 mb) and intermittent winds upward of 200 MPH (320 km/h). Touted as one of the twentieth century's most potentially dangerous hurricanes, Allen whisked across Barbados and St. Lucia on August 4. Winds of 130 MPH (209 km/h) tore into St. Lucia's vital banana plantations, uprooting trees, leveling houses, and killing 16 people. On the resort island of Barbados, Allen's 11-foot (4-m) STORM SURGE pounded empty beachfront hotels, smashing windows, flooding swimming pools with seawater, and washing away yacht piers.

On August 5, Allen careened across southern Haiti. Sustained winds of 120 MPH (193 km/h) dev-

astated a large portion of the country's coffee crop, while torrential rains spawned flash floods that killed an estimated 220 people. Brushing past the island of JAMAICA on the morning of August 6, Allen's slightly diminished 100-MPH (161-km/h) winds battered northern cities Port Maria and Port Antonio and unleashed deadly rains that knocked out power lines and bridges in the exclusive Montego Bay enclave. In the resort city of Port Maria, Allen's enormous storm surge completely demolished two beachside hotels by undermining their concrete pilings. Witnesses claimed the five-story buildings simply toppled forward into the raging surf and disappeared.

After grazing the west coast of Cuba—where 110-MPH (177-km/h) winds forced the evacuation of 210,000 people from low-lying areas on August 7—Allen took aim at Mexico's vast Yucatán Peninsula. Luxury resorts on the shallow islands of Mujeres and Cozumel were hurriedly evacuated as Allen's eye swept inexorably up the Yucatán Channel. Although the picturesque islands did suffer some wind and water damage, they were largely spared the full fury of Allen's passage into the GULF OF MEXICO. On the peninsula's northern face, the hurricane's strengthening winds stripped foliage from trees and drove large waves onto the beachheads but caused no fatalities.

By the morning of August 8, Hurricane Allen's eye was centered 52 miles (84 km) due north of the Yucatán Peninsula and headed on a course that would take it across the Gulf of Mexico to a stormy landfall somewhere along the Texas lower coast. During the previous night, the storm's central pressure had seesawed between a vigorous 27.91 inches (944 mb) and an alarming 26.55 inches (899 mb), one of the lowest barometric readings ever recorded in an Atlantic hurricane. As NATIONAL HURRICANE CENTER (NHC) officials pored over computer-generated forecasts, trying to determine just where Allen would come ashore, civil defense authorities in Texas began to evacuate a quarter of a million people from the 450-mile (724-km) stretch of coastline between Brownsville and Corpus Christi. Over the agitated waters of the gulf, a squadron of chartered helicopters set to work airlifting hundreds of oil company personnel from deep-sea drilling rigs. In one tragic instance, Allen's extended gale-force winds caused a helicopter to crash into the sea off LOUISIANA, killing 13 workers.

Allen blasted ashore near Corpus Christi, Texas, on August 9, 1980, with a central barometric pressure of 27.91 inches (945 mb). Gusts of 160 MPH (258 km/h) drove before a 10-foot (3-m) storm surge, the highest seen on the coast in a half-century. Allen's whirring winds and pelting rains stripped buildings of their roofs and siding, crumpled billboards, toppled trees and telephone poles, and caused massive localized flooding. Mountainous waves drove a disabled Liberian tanker, loaded with 280,000 barrels of crude oil, onto a sandbar off Corpus Christi, raising fears of a dangerous oil spill. Coast Guard personnel bravely fought 40-foot (13-m) seas to rescue the tanker's crew of 37 and to secure the heavily laden ship against the relentless hammering of the ocean.

Steadily moving inland across the lower Rio Grande Valley, Hurricane Allen's downgraded winds spawned numerous tornadoes and caused nearly $1 billion damage to the region's citrus harvest. Conversely, the storm's 20-inch rains brought beneficial PRECIPITATION to countless farms, ending a prolonged drought that had threatened the valley's largely agrarian economy.

Allen's dramatic disintegration would herald the start of an unprecedented three-year hiatus in hurricane activity along the U.S. mainland coasts. Between August 10, 1980, when Allen blew itself out over Texas, and August 13, 1983, when Hurricane ALICIA surged ashore at Galveston, no hurricanes threatened U.S. coastal tranquility. The name ALLEN has been retired from the list of future HURRICANE NAMES.

Allison, Hurricane *Cuba–Southern United States, June 2–5, 1995*   A minimal Category 1 HURRICANE, Allison became the earliest mature-stage hurricane in recorded history to strike the mainland United States when it came ashore at Apalachicola, FLORIDA, on the morning of June 5, 1995. The 1995 hurricane season's first TROPICAL CYCLONE Allison originated as a TROPICAL DEPRESSION 150 miles (241 km) northeast of Honduras on June 2. Carried almost due north at between 10 and 14 MPH, the strengthening TROPICAL STORM slid across the western tip of CUBA on June 3, bringing steady rains and 50-MPH (81-km/h) winds to the island. One man was killed in Havana.

Clearly bound for the Florida Panhandle, Allison continued to intensify throughout that day and night, prompting that state's civil defense authorities to post HURRICANE WARNINGS from Pensacola to Clearwater and to commence the evacuation of some 5,000 particularly vulnerable coastal residents.

Shortly before 11 o'clock on Monday morning, June 5, Allison made landfall in Taylor County, approximately 45 miles (72 km) southeast of Tallahassee. With a fairly mild CENTRAL BAROMETRIC PRESSURE of 29.05 inches (984 mb), Allison's fitful 75-MPH (121-km/h) gusts buffeted trees and severed power lines, leaving an estimated 48,000 residents in seven counties with neither electricity or telephone service. Along a 150-mile (241-km) stretch of Florida's Big Bend, Allison's eight-foot (3-m) STORM SURGE flooded more than 65 surfside homes and caused extensive water damage to three hotels and a restaurant on the barrier island of St. George.

In Apalachicola, three fishing boats were swamped at their piers while five-inch (127-mm) rainfalls threatened the bay's fragile oyster beds. The soaring Bryant Patton Bridge, linking Apalachicola with St. George Island, was closed for much of the storm's duration as streaming seas washed debris onto its low-lying approaches. Further inland, isolated flooding damaged three houses and a grocery store and inundated a trailer park. Quickly moving into southwestern GEORGIA on the afternoon of June 5, Allison's downgraded winds spawned a TORNADO that touched down in the town of St. Marys, gutting an empty grade school at the Kings Bay Naval Submarine Base. No serious injuries or deaths were reported in either Florida or Georgia.

With some $785,000 in property damage to its credit, Allison was largely dismissed by survivors as one of the mildest hurricanes to have ever struck Florida. Nevertheless, it was in turn touted by scholars as the earliest storm of hurricane intensity to have made landfall in the United States since consistent record-keeping began in 1889. Although in the intervening 120 years no less than eight storms reached hurricane strength before June 10, only three—Hurricane ALMA, June 9, 1966, an unnamed tropical storm in late May of 1970, and Hurricane Allison—ever touched the nation, making them official, record-breaking landfalls. Bearing a central pressure of (1,001 mb) and 52-MPH (84-km/h) winds, Tropical Storm Allison made landfall in TEXAS on June 26, 1989. Formed on June 24, the system—the first to bear the name Allison—caused no deaths or major property losses in Texas.

The name Allison was retired from the rotating list of North Atlantic tropical cyclone names in 2001.

Allison, Tropical Storm *Southern United States, June 4–18, 2001*   The first named TROPICAL CYCLONE of the 2001 North Atlantic HURRICANE SEASON, Tropical Storm Allison delivered deluging rains to portions of southeastern TEXAS and LOUISIANA between June 5 and 11, 2001. Allison dropped between 10 inches (250 mm) and 30 inches (760 mm) of rain, claiming 24 lives and causing more than $5 billion in property losses.

The most costly TROPICAL STORM yet recorded in the United States, Tropical Storm Allison developed in the GULF OF MEXICO on June 4, 2001, from a subtropical low pressure system that had originated over Mexico's Bay of Campeche. On June 5, five days after the official start of the 2001 hurricane season and as the system meandered some 130 miles (200 km) south of Galveston, TEXAS, the NATIONAL HURRICANE CENTER (NHC) upgraded the system to tropical storm status, awarding it the name Allison. Carried westward, Allison's sustained 60-MPH (95-km/h) winds and heavy rains came ashore at Freeport, Texas, during the mid-afternoon hours of June 5. Bearing a central pressure of 29.52 inches (1,000 mb) and surging rains, Allison's poorly organized circulation system traveled slowly over northern Texas, then, on June 8, drifted southward, bringing enormous PRECIPITATION counts to the Houston metropolitan area. In the Port of Houston, one of the nation's busiest maritime trade nodes, some 37 inches (940 mm) of rain was recorded during Allison's erratic passage inland. By June 9, streams and drainage canals across eastern Texas were quickly overflowing their banks, and large sections of downtown Houston—including its city hall—were flooded when the Buffalo and White Oak bayous became saturated. Several medical facilities associated with the Texas Medical Center were evacuated as flash flooding caused by Allison's record-setting rainfalls disrupted utilities and destroyed medical equipment. In some instances, Allison's powerful downpours hampered the evacuations as they closed major arteries and freeways in the city. Some 70,000 residential and commercial structures in and around the Houston area experienced partial or total flooding, while in an underground parking garage in Houston, a woman drowned while trying to retrieve her automobile. All told, 21 people died in Houston, making Allison one of the deadliest tropical cyclones to have affected Texas in over a decade.

Shortly after noon on June 10, 2001, the remnants of a downgraded Tropical Depression Allison reentered the Gulf of Mexico and moved northeastward. Upgraded again, this time to classification as a subtropical storm, Allison finally ground ashore on June 10, at Morgan City, Louisiana. Still producing huge rains, Subtropical Storm Allison slowly swirled over MISSISSIPPI and ALABAMA before entering the NORTH ATLANTIC OCEAN over NORTH CAROLINA on June 17, 2001. Although the system had dissipated by June 19, its early-season trail of death and destruction lingered for many months thereafter. In Texas and Louisiana, some 55 people died and more than $5 billion in property losses were incurred. Two billion dollars worth of those losses were tallied at the Texas Medical Center, alone.

Replaced by the name Andrea, the identifier Allison was retired from the rotating list of North Atlantic tropical cyclone names in July 2001.

Alma, Hurricane *Honduras–Cuba–Southern United States, June 5–10, 1966* A large but fairly weak Category 2 HURRICANE, Alma lashed portions of north Honduras, CUBA and western FLORIDA between June 7 and 9, 1966. Brought on by a central pressure of 28.67 inches (967 mb), Alma's 100-MPH (161-km/h) winds and flooding rains claimed seventy-three lives in Honduras, seven in Cuba, and another five in Florida.

The first hurricane of the 1966 HURRICANE SEASON, Alma originated near the Rosario Bank off the east coast of MEXICO's Yucatán Peninsula on the night of June 5. Nursed by the Caribbean Sea's sun-drenched waters, the fledgling TROPICAL STORM wasted little time in strengthening: By the evening of June 6, as flanking gales lashed the Gulf of Honduras into nourishing spray, Alma's 72-MPH (116-km/h) core winds rapidly approached hurricane force. Weather reconnaissance flights indicated that the storm was now of considerable size—nearly 500 miles (805 km) across—and that it was moving in what appeared to be a north-northeasterly course at close to six MPH (10 km/h). ANALOGS based on the behavior of previous June hurricanes predicted that Alma would most likely continue to move toward west Cuba, but not before its water-laden fringes battered the Honduran coastline with nearly 30 inches (762 mm) of rain. Alma's incoming winds drove record high tides into San Rafael's harbor, smashing piers and flooding streets. In the surrounding hills, suffocating mudslides swept away several houses, burying more than 70 people.

After grazing the west coast of Cuba—where 117-MPH (188-km/h) winds decimated part of that year's sugarcane harvest, forced the evacuation of 125,000 residents from low-lying areas, and killed seven people on June 7—Alma took aim at Florida's southernmost tip. Luxury resorts on the Florida Keys shallow islands were hurriedly abandoned as Alma's EYE, now moving north-northeastward at close to 12 MPH (19 km/h), swept toward the GULF OF MEXICO. At Key West, units of the U.S. Army's Sixth Missile Battalion worked under the rain-driven cover of darkness to secure America's nuclear picket fence from the threat of possible flooding. On the peninsula's southern tip, Alma's gales stripped foliage from trees and drove large waves onto the beachheads but claimed no fatalities.

On the morning of June 8, 1966, as HURRICANE WARNINGS were dutifully issued for both coasts of Florida, Alma finally entered the Gulf of Mexico. Even though the hurricane's strongest winds—now clocked at 110 MPH (177 km/h)—were still miles away from any possible landfall in southern Florida, Miami was buffeted by 60-MPH (97-km/h) gusts and pelting rains. Road signs buckled and utility poles were laid over like windrows as more than seven inches (178 mm) of rain washed over the city. In the port of Miami, numerous pleasure craft, were wrenched from their moorings and driven ashore by Alma's vicious blasts; more than 40 were smashed to pieces by the pounding surf, littering several miles of the city's famed beaches with unsightly wreckage. One fatality—a teenage boy who was electrocuted by a downed power line—was reported in Miami, as were several injuries.

On Florida's normally tranquil west coast, Alma's predicted landfall just south of St. Petersburg prompted the mass evacuation of more than 100,000 people from a half dozen low-lying barrier islands. As 5-foot (2-m) STORM TIDES began to wash ashore at Fort Myers on the afternoon of June 9, fleeing vacationers clogged the area's bridges and highways. Local stores were emptied of canned foods, bottled water, flashlights, and candles. Homeowners raced to board up their windows, quickly depleting stocks of plywood at nearby lumberyards. Potentially lethal lawn furniture was securely stowed away, while small utility buildings were firmly anchored to the ground with stakes and chains. In the gulf itself, excursion boats laden with anxious sightseers quickly returned to the uncertain refuge of their homeports. One such vessel, the 50-foot (18-m) *Ranger III,* slammed into a bridge piling and sank; two people aboard the craft were lost.

On Wednesday evening, June 9, less than two hours before Alma's eye was to come ashore in the vicinity of Sarasota, the hurricane inexplicably veered away from the Florida coast and churned into the Gulf of Mexico. Despite an extensive examination of Alma's movements by meteorologists, no satisfactory explanation for the hurricane's sudden change of course has ever been determined. Indeed, Alma's erratic behavior blatantly defied the standard model for June hurricanes in the gulf.

Because most cyclonic storms of this type have a tendency to undergo RECURVATURE to the east, Alma by rights should have continued on its north-northeasterly track, passed over central Florida, and either continued up the eastern seaboard of the United States or else moved out in the Atlantic Ocean. As it was, Alma instead turned west away from Florida and its relieved residents. Passing in an almost linear course into the heart of the Gulf of Mexico, Alma quickly began to weaken; by week's end, it was completely gone.

Until Hurricane ALLISON made landfall on the shores of the Florida Panhandle on June 5, 1995, Alma held the unique distinction of being the earliest recorded hurricane to strike the mainland United States. Although preseason cyclonic activity was noted in May 1889, March 1908, and May 1951, none of these storms affected the U.S. coastline. Before the name Alma was retired following the 1974 hurricane season, there were (in addition to the 1966 system) four other North Atlantic Ocean tropical cyclones identified as Alma. The first existed between June 14 and 16, 1958, and was of moderate tropical storm intensity at its peak. Formed over Mexico's Bay of Campeche and bearing a central pressure of 29.44 inches (997 mb), Tropical Storm Alma tracked due northwestward before coming ashore in extreme northeastern Mexico on June 15. Sustained, 52-MPH (84-km/h) winds buffeted the coastline as Alma penetrated into southern Texas, dropping considerable amounts of rain. No lives were lost, however, and by June 10 the system had dissipated over the Texas flatlands.

The second tropical cyclone known as Alma grew into a Category 2 hurricane during the middle months of the 1962 North Atlantic hurricane season. Developed over the North Atlantic Ocean on August 26, Tropical Depression Alma grazed Florida's eastern coastline before transitioning to a hurricane while brushing across the Outer Banks of NORTH CAROLINA. With a central pressure of 29.11 inches (986 mb) and sustained winds of 98 MPH (158 km/h), Alma was poised to bring potentially devastating conditions to NEW YORK's Long Island and southern NEW ENGLAND before it recurved into the Atlantic and dissipated near Newfoundland, CANADA, on September 2.

The fourth Hurricane Alma was a preseason surprise, one that originated over the CARIBBEAN SEA on May 17, 1970. Quickly maturing into a Category 1 hurricane with sustained winds of 81 MPH (130 km/h) and a CENTRAL BAROMETRIC PRESSURE of 29.32 inches (993 mb), Alma recurved, weakened, and swept across the northwestern tip of Cuba on May 20 as a tropical storm. Five days later, on May 25, Hurricane Alma—now downgraded to a tropical depression—came ashore in northwestern Florida, bearing 30-MPH (48-km/h) winds.

The last tropical cyclone christened Alma inaugurated the ferocious 1974 North Atlantic hurricane season, one punctuated by Hurricane CARMEN's destructive passage in August and September. Of only tropical storm strength at its height, Alma is one of the few North Atlantic tropical systems to ever strike the South American continent. Born along latitude 10 degrees North on August 12, Alma followed a course due westward, veering only a few miles from its trace along 10 degrees North, before finally striking the islands of Trinidad and Tobago, and transiting into northern Venezuela. At the time of its landfall in Trinidad on August 13, Alma's central barometric pressure of 29.73 inches (1,007 mb) produced wind speeds of 63 MPH (101 km/h) across the islands, but no record of deaths or severe property damage.

Alpha, Tropical Storm *Hispaniola–Northern Caribbean Sea, October 22–24, 2005* The first North Atlantic TROPICAL CYCLONE in meteorological history identified with a name drawn from the Greek alphabet, a relatively weak Tropical Storm Alpha delivered 50-MPH (85-km/h) winds and torrential rains to HISPANIOLA between October 22 and 24, 2005. The 22nd named system of the exceptionally active 2005 North Atlantic HURRICANE SEASON, Alpha developed from a TROPICAL WAVE, some 180 miles (220 km) southwest of San Juan, PUERTO RICO, on October 22, before tracking to the northwest to strike the Dominican Republic early the following day. As with most tropical systems that interact with Hispaniola's towering mountain ranges, Tropical Storm Alpha quickly weakened, its circulation sheared by the island's ridges, as it trudged ashore near the coastal city of Barahona. Before entering the

Atlantic on October 23, Alpha deluged the Dominican Republic and neighboring Haiti with torrential rains, spawning flash floods and mudslides that killed more than 40 people in Haiti and the Dominican Republic. During the early morning hours of October 24, Alpha's brief but deadly debut came to an end as the system fell under the circulatory influence of the much larger, record-shattering Hurricane WILMA (which was then tracking eastward across the GULF OF MEXICO as a powerful Category 3 hurricane), and degenerated into a trough.

Although Tropical Storm Alpha was the first North Atlantic tropical cyclone to bear the name, it was not the first instance in which Alpha had been used as a tracking identifier for a storm system in the North Atlantic basin. Between May 23 and 29, 1972, FLORIDA and GEORGIA were directly affected by the 69-MPH (111-km/h) winds and rains associated with a subtropical system that had originated over the Gulf of Mexico and been dubbed Alpha by the National Weather Service (NWS); and in 1973, a slightly different permutation, Alfa, was used to

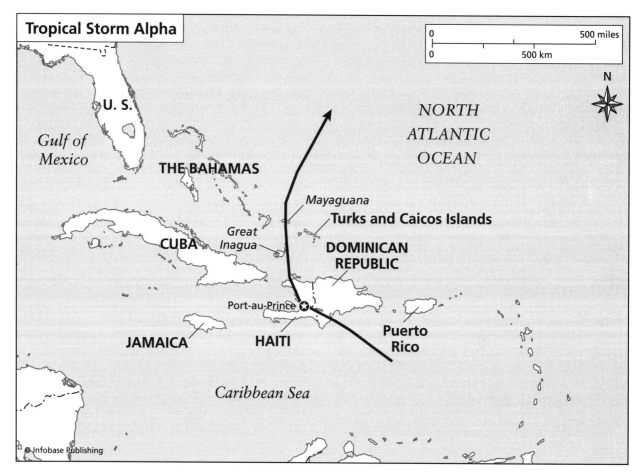

The first tropical cyclone to be identified under the standard naming system with a name taken from the Greek alphabet, Tropical Storm Alpha brought heavy rains to the island of Hispaniola in 2005.

identify a subtropical system that lingered off the NEW ENGLAND coastline.

Because the identifier Alpha is used during those very active seasons when the traditional male/female hurricane naming lists have been exhausted, it has been retained on the North Atlantic tropical cyclone naming lists.

Ana, Tropical Storm *North Atlantic Ocean, April 21–24, 2003*   A preseason surprise, Tropical Storm Ana formed to the south-southeast of Bermuda on April 21, 2003. It was the earliest TROPICAL CYCLONE formation in the NORTH ATLANTIC since Subtropical Storm One formed on January 18, 1978, and the first April tropical cyclone on record in the Atlantic basin. A more powerful system than the 1978 incarnation, Tropical Storm Ana developed from a subtropical storm that had originated on April 20 and possessed a central pressure of 29.41 inches (996 mb) and 52-MPH (84-km/h) winds. A short-lived system, Ana tracked almost due eastward into the mid-Atlantic, where it essentially merged into a frontal system and dissipated three days later. While Ana never made landfall, it did indirectly cause the deaths of two mariners in FLORIDA whose craft was overturned by its large swells on April 20.

The 2003 North Atlantic system was the fourth to bear the name Ana. The first appeared on June 19, 1979, in the form of a TROPICAL STORM that had originated in the western North Atlantic Ocean, drifted almost due westward across the Windward Islands, and entered the CARIBBEAN SEA before dissipating on the 24th of the month. At one time a relatively powerful tropical storm with a central pressure of 29.67 inches (1005 mb) and 58-MPH (93-km/h) winds, Tropical Storm Ana had weakened as it crossed the Windward Islands, its central pressure rising to 29.88 inches (1,012 mb) and its wind speeds dropping to 40 MPH (64 km/h).

While avoiding landfall for the duration of its existence, the second Tropical Storm Ana was a marginally more intense tropical cyclone than its 1979 predecessor. Born over the mid-Atlantic Ocean on July 15, 1985, Ana recurved to the northeast around Bermuda and sped into the extreme North Atlantic before fading off the Canadian Maritimes on July 19. A central pressure of 29.41 inches (996 mb) produced sustained winds of 69 MPH (111 km/h), but no deaths or damage assessments.

On June 30, 1991, a strengthening Tropical Depression Ana—the third such system to bear the name—sliced across the Florida peninsula, its central pressure of 29.88 inches (1,012 mb) producing 23-MPH (37-km/h) winds and pounding rains. Before dissipating over the Atlantic on July 5, 1991, Ana reached tropical storm intensity, with a central pressure of 29.52 inches (1,000 mb) and sustained winds of 52 MPH (84 km/h). No lives were lost due to Ana's early-season passage across Florida.

The name Ana has been retained on the rotating list of North Atlantic tropical cyclone identifiers, and is scheduled to reappear in 2009.

analogs   Derived from the word *analogous*, or any object that is similar to another in some way, analogs are computer-generated models that use historical, mathematical, and climatological information to forecast the track and intensity of TROPICAL CYCLONES. First developed in the early 1960s by the NATIONAL HURRICANE CENTER (NHC), analogs are essentially divided into two types of models—statistical and dynamical—and are based upon the nature of the data used to predict the TRACK of a tropical cyclone. In the case of *statistical* analogs, historical information from earlier HURRICANES, TYPHOONS, and CYCLONES—such as date and location of generation, course taken, and landfall made—is compared with that of a storm being presently tracked. *Dynamical* analogs, on the other hand, employ meteorological observations, such as barometric pressure, upper level winds, atmospheric configurations, and chaos theory, to simulate a hurricane's future movements mathematically. Computer programs that have been especially designed for this purpose include HURRAN, or *Hurr*icane *An*alog; CLIPER (*Cli*matology and *Per*sistence); NHC-67 (National Hurricane Center–1967); MFM (Movable Finemesh Model); and QLM, or the Quasi-Lagrangian Model. In addition, a number of these programs, such as NHC-73, operate by pooling the considerable resources of several different systems, including HURRAN, CLIPER, and NHC-67. Over the years, improvements in computer technology and modeling have allowed analogs to play an ever larger role in providing accurate, long-range tropical cyclone forecasts. On March 10, 2003, meteorologists with the National Oceanic and Atmospheric Administration (NOAA) announced that on May 15, the date on which the eastern North Pacific hurricane season traditionally begins, the agency would "extend the range of its hurricane forecasts . . . to five days from three." Enhancements in long-range forecasts not only allow emergency managers and other public safety officials sufficient time to evacuate vulnerable areas, but further reduce the indirect financial costs associated with tropical cyclone activity, such as premature and unnecessary evacuations, closed businesses and lost wages and revenue.

By late August 2005, as an embryonic Hurricane KATRINA was originating over the warm waters of the Gulf Stream between FLORIDA and the BAHAMAS, the various computer analogs used by the National Hurricane Center (NHC) to forecast the system's intensity patterns and eventual trajectory proved remarkably accurate in their five-and-three day windows. As early as August 23, 2005, the SHIPS model was forecasting that Tropical Depression 12 (TD 12) would strike Florida as a weak Hurricane Katrina, and this is precisely what happened two days later. During an initial run on August 23, the GFDL model erroneously predicted that TD 12 would remain a tropical depression as it reached Florida, while the SHIPS model called for sustained winds of 75 MPH (121 km/h) at landfall, making it a Category 1 hurricane. Forecasts issued during the early morning hours of August 24, 2005, also predicted that the system would weaken slightly while moving across Florida but reintensify once it had entered the Gulf of Mexico. This forecast for Katrina's strengthening pattern later proved accurate.

By the mid-morning hours of August 24, the GFDL and SHIPS models were confidently predicting that Tropical Storm Katrina would become a hurricane prior to landfall in southeastern FLORIDA; this represented a change by the GFDL model, which the day before had been limiting the system to tropical depression status at its time of landfall in the Sunshine State. As though overcorrecting for its earlier inaccurate intensity prediction, by the early evening hours of August 24, the GFDL model was forecasting that Katrina would make landfall in Florida with 128-MPH (206-km/h) winds—a dire prediction properly discounted by the NHC's skilled forecasters. When Katrina did eventually make landfall in southeastern Florida on August 25, 2005, it was a Category 1 hurricane with sustained winds of 80 MPH (129 km/h).

It was part of this same model run during the mid-afternoon hours of August 24, 2005, that the GFDN model initially predicted that a now upgraded Tropical Storm Katrina would eventually strike New Orleans, then several hundred miles away from the storm's center. "The GFDN is the westernmost model," the NHC wrote in its discussion, "and takes the cyclone to New Orleans . . . whereas the GFS and Canadian models are the easternmost models and take Katrina northeastward across the Northern Florida peninsula."

Five days from when Hurricane Katrina did actually make landfall in the New Orleans–Buras, LOUISIANA, area on August 29, 2005, the NHC's official track closely followed the model consensus, and placed the system less than 100 miles (161 km) to the east of where it eventually came ashore. In a probabilities assessment released simultaneously with the early evening discussion, the NHC essentially stated that there was a 4 percent chance that the system would make landfall in Buras and New Orleans between "2 pm on Friday [August 26] to 2 pm Saturday [August 27]," which did not occur. As though defying the most modern forecasting technology, Katrina instead spent August 26 and 27 rapidly intensifying over the GULF OF MEXICO.

On August 26—three days from Katrina's eventual landfall in southeastern Louisiana—almost all of the NHC's models were predicting that Katrina would grow into a major hurricane—in particular, the generally reliable GFDL and GFDN models. By the late-night model runs on August 26, the GFDL model was predicting that Katrina would produce 143-MPH (230-km/h) winds and a central pressure of 27.22 inches (922 mb). In its intensity predictions, the FSU Superensemble model was even more confident of Katrina's future strengthening configuration, bringing Katrina to 151 MPH (243 km/h). At the time that these runs were prepared, Katrina was a Category 2 hurricane on the SAFFIR-SIMPSON SCALE, with a central barometric pressure of 28.50 inches (965 mb) and sustained winds of 105 MPH (169 km/h).

It was also during this set of model runs on the night of August 26 that the variances between the different models markedly lessened, with the most recent dynamical model consensus sending the hurricane ashore on the border between Louisiana and Mississippi. Those models, which had been produced during the mid-afternoon hours of August 25, were very divergent, with a majority (GFDL, GFDN, and UKMET) sending Katrina ashore between Mobile, Alabama, and Grand Isle, Louisiana. Just under three days from when Katrina made landfall in southern Louisiana, the NHC's probabilities assessment indicated a 6 percent chance that Buras, Louisiana, and a 9 percent chance that New Orleans, would be directly hit by Katrina between 8:00 P.M. (EDT) on Sunday, August 28 and 8:00 P.M. (EDT) on Monday, August 29. Among the top three-highest rankings in the three-day probabilities assessment, these figures were essentially proven accurate when Hurricane Katrina clawed ashore in southern Louisiana on Monday, August 29, 2005, the most catastrophic natural disaster yet seen in American history.

Andrew, Hurricane *Bahamas–Southern United States, August 23–26, 1992* An exceedingly powerful Category 5 HURRICANE, Andrew remains one of the most destructive TROPICAL CYCLONES in U.S. history. With a CENTRAL BAROMETRIC PRES-

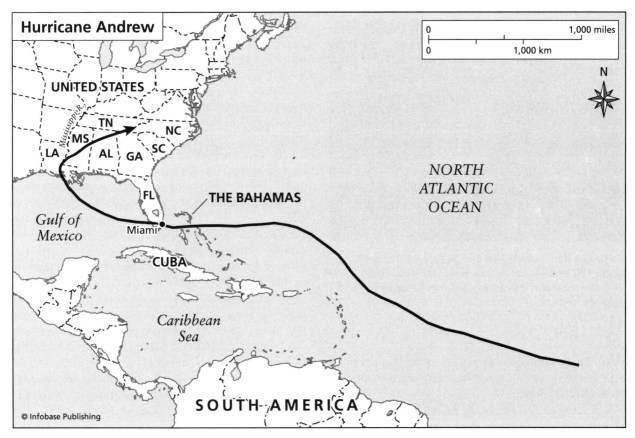

This track shows the passage taken by Hurricane Andrew as it visited record-setting destruction across Florida and the Gulf Coast of the United States in 1992.

SURE at landfall of 27.23 inches (922 mb) and sustained winds of 145 MPH (227 km/h), Andrew carved a 1,000-mile (1,609-km) swath of devastation through the BAHAMAS, SOUTH FLORIDA, and LOUISIANA between August 23 and August 26, 1992. In the northwest Bahamas, 120-MPH (190-km/h) winds drove a 23-foot (5-m) STORM SURGE ashore, drowning four people. In Florida's heavily populated Dade County, the burgeoning residential communities of Goulds, Homestead, Leisure City, Naranja, and Princeton were virtually demolished. Nearly 80,000 buildings were left in ruins, while more than 200,000 people were made homeless by Andrew's 175-MPH (282-km/h) gusts. The U.S. Air Force base at Homestead was severely damaged: collapsed hangers and wind-tossed aircraft littered the runways. Fifteen people in Florida were killed. In Louisiana, 140-MPH (225-km/h) winds crushed buildings, dropped power lines, uprooted thousands of trees and crops, and killed one person. Damage estimates were in excess of $20 billion.

Andrew's destructive legacy began on the afternoon of August 23, 1992, when it slammed into the low-lying Bahamas with winds of 120 MPH (190 km/h). Several hundred buildings, from quaint bun-

galows to high-rise luxury hotels, suffered smashed windows and significant structural damage. A 23-foot (5-m) storm surge, one of the largest on record, flooded ashore near the northwest tip of Eleuthera Island, washing away beachfront cottages and causing extensive coastal erosion.

From there, Andrew swirled across the narrow Florida Straits, steadily gaining strength from the warm waters and pushing before it a diminished, but no less dangerous, eight-foot (3-m) storm surge. Acting on the advice of NATIONAL HURRICANE CENTER (NHC) officials, civil defense authorities along Florida's east coast began the arduous task of mass-evacuating nearly a million people from Andrew's path. Fleeing vacationers clogged coastal bridges and highways, hastily abandoning vulnerable resort islands under the rainslick cover of night. Inland, thousands of Floridians labored to secure their property against Andrew's imminent onslaught and emptied stores of canned foods, bottled water, flashlights, and batteries. Stocks of plywood at local lumberyards were quickly depleted as homeowners raced to board up their windows. Potentially lethal lawn furniture was securely stowed, while small utility buildings were anchored to the ground

An intense Hurricane Andrew slices into southern Florida in this 1992 satellite photograph. Notice the darker band of colder cloud tops that defines Andrew's Category 5 eyewall. *(NOAA)*

with stakes and chains. Hundreds of residents in lightly constructed mobile homes were transported to nearby shelters.

Andrew came ashore 30 miles (48 km) south of downtown Miami during the early morning hours of August 24, 1992. Sustained hurricane-force winds accompanied by torrential tropical rains and TORNADO-like MINISWIRLS buffeted the entire southern half of the state. Major roadways were flooded, power and telephone lines were downed, and entire groves of palm trees were laid flat like windrows. Almost a third of the ecologically fragile Everglades were turned into a veritable wasteland of splintered trees and drowned wildlife. Scientists estimated that it would be at least a decade before the once-verdant oasis was restored to its pre-Andrew state.

At Cape Florida, the Bill Baggs State Park, noted for its Civil War-era lighthouse, sandy white beaches, and luxurious shade trees, was virtually denuded by Andrew's rigorous winds. Although the landmark stone lighthouse survived intact, more than 75,000 full-grown trees were later replaced in one of the most ambitious restoration projects ever undertaken in south Florida. Homes and businesses directly in the EYE's 30-mile (48 km)-wide path were completely stripped of their roofs, while a multistory condominium complex was damaged beyond repair when its outer walls were literally wrenched off and carried away by Andrew's fearsome gusts.

On the shores of Biscayne Bay, the picturesque Charles Deering estate, with its Renaissance palazzo, and scenic gardens, was flooded with 16.7 feet (8 m) of water. Overhead, Andrew's lashing gusts tore at the estate's famed colonnade of royal palm trees,

destroying 86 of them. At nearby Homestead Air Force Base, Andrew's wrath punched out control-tower windows, collapsed all the hangers, demolished the administration building and barracks, and overturned several F-18 fighter jets. In stormswept Miami, 3,300 National Guard troops were flown in to enforce a dusk-to-dawn curfew, and to discourage looting.

Its sustained winds barely diminished by its widespread assault on southern Florida, Andrew surged across the GULF OF MEXICO on August 25, roiling the warm waters of the gulf and kicking up tons of energy-laden spindrift. Forecasters at the National Hurricane Center predicted that the hurricane—with rapidly reintensifying winds and a cresting, 10-foot (3-m) storm surge—would make landfall somewhere in the vicinity of New Orleans, Louisiana. Fearing that New Orleans would sustain catastrophic flooding if the storm surge came ashore directly off the low-lying city, authorities began to mass-evacuate some 1.2 million people from the Mississippi Delta area. Several hundred workers stationed on oil-drilling rigs in the Gulf of Mexico were quickly airlifted to safety, while a large portion of Louisiana's shrimp fleet returned to the questionable safety of its crowded ports.

Andrew slammed into the Louisiana coast on the morning of August 26, bearing sustained winds of 140 MPH (225 km/h) and torrential rains. The hurricane's curving track took it ashore 90 miles (145 km/h) southwest of New Orleans, thereby sparing the historic city the destructive brunt of its seaborne fury. As it was, Andrew was responsible for extensive structural damage to buildings and piers in the sparsely inhabited region of Marsh Island and Lafayette, but caused only one fatality. Downgraded to a TROPICAL STORM later that evening, Andrew's lingering winds spawned an outbreak of tornadoes that uprooted trees and flattened crops. By the morning of August 27, Andrew had abruptly swung northeast, crossed into MISSISSIPPI, and almost completely dissipated.

Described as the single-most-expensive natural disaster to strike the United States up to that time, Andrew severely taxed emergency efforts to relieve the widespread suffering of hundreds of thousands of people left homeless in its wake. Units of the U.S. Army were deployed to set up tent cities, while insurance companies, staggered by the sheer breadth of Andrew's devastation, moved to create regional processing centers for the thousands of property-damage claims that poured in. Although then President George H. W. Bush quickly declared south Florida and parts of Louisiana disaster areas on August 24 and 26 respectively, his administration was roundly criticized for not responding vigorously enough to the election-year emergency. Interestingly enough, it was a charge that had been leveled at the first Bush administration several times before, most notably during the Hurricane HUGO disaster of September 1989.

Just in time for the 10th anniversary of its devastating landfall in southeastern Florida, 1992's Hurricane Andrew received a dubious promotion. On August 22, 2002, the National Hurricane Center's (NHC) Best Track Committee announced that based upon a number of surveys of Andrew's destructiveness, as well as the broad meteorological record, the system would be upgraded from a Category 4 to a Category 5 hurricane at landfall. According to the NHC's analysis, "When the storm was nearing Florida, an aircraft measured the wind at 10,000 feet above the ocean. The surface wind speed was then estimated at 145 MPH, about 80% of the high-level wind. New research on the wind structure of hurricanes has determined that surface winds average 90% of the 10,000-foot wind. The new estimated wind speed of 165 MPH makes Hurricane Andrew the third Category 5 hurricane to strike the United States."

An early-season tropical cyclone, the first tropical system in the North Atlantic named Andrew, formed between June 5 and 8, 1986. Originating off the northern Bahamian coast, Andrew deepened into a tropical storm with 52-MPH (84-km/h) winds and a central pressure of 29.50 inches (999 mb) as it brushed past NORTH CAROLINA's Outer Banks and dissipated off the coast of NEW ENGLAND. No deaths, injuries or major property losses were recorded.

Used in connection with only two North Atlantic tropical cyclones, the name Andrew has been retired from the list of North Atlantic hurricane identifiers.

anemometer An instrument used to measure WIND SPEED. Greek philosophers in the first century B.C. may have been the first scientists to devise an instrument for determining wind speed and direction. Spurred by Aristotle's philosophical investigations into the nature of weather, Athenian sky watchers constructed an octagonal "Temple of the Winds" in a rudimentary effort to link particular wind directions with certain atmospheric conditions. One end of a large rooftop weathervane faced into the prevailing winds, while the other simultaneously pointed to one of eight sculpted bas-reliefs on the temple's facade. Besides indicating the sort of weather to be expected from the wind in question, the reliefs—decorated with faces of different meteorological deities—gave a hint of wind speed by alternating their facial expressions from panel to panel. One god's tightly closed

mouth, for instance, might indicate no wind, while another's ballooning cheeks and pursed lips gave warning of an impending storm.

For all of its interpretive artistry, however, the Temple of the Winds ultimately failed to measure wind speed with any serious degree of quantifiable certainty. It was not until the British physicist Robert Hooke introduced his relatively simple anemometer in 1667 that anything close to an accurate measure of wind speed could be achieved. Hooke's rather ingenious anemometer, which operated on the same principle as a pendulum, featured a broad square plate, a swinging arm, and a calibrated arc-shaped scale. The flat plate was connected to one end of the pivot arm, which in turn was attached to the top of a vane that swiveled, thereby allowing the entire mechanism to remain turned into the wind at all times. As wind speed increased, the flat plate was lifted upward. The pivot arm rose with it to permit the user to gauge simply where the arm paused on the calibrated scale and the degree of its swing.

Hooke's highly effective anemometer remained in general use until the mid-1840s when a new prototype, designed by Irish astronomer John Robinson, was unveiled at Dublin. Robinson's anemometer, which recorded wind speed by measuring how many miles of wind moved past the device in a five-minute time period, featured the now-familiar quartet of four hemispherical cups fitted to the ends of two intersecting crossarms. These arms were in turn connected to a hub atop a spinning vertical shaft. The shaft itself was attached to a calibrated gear box that operated a small dial from which wind speed could be read with an accuracy never before possible.

While the Robinson-type anemometer remains standard equipment for weather bureaus across the world, numerous design modifications over the years have vastly improved its basic workings. There are now two-and three-cup models in use, and the geared dial has been replaced by digital readouts and sophisticated graphing systems like those used for measuring earthquakes. New, high-tensile strength mounting systems have also been devised to reduce the possibility that the fragile anemometer will be carried away (as has happened many times in the past) by a TROPICAL CYCLONE's relentless winds.

Because an increasing amount of information about HURRICANES, TYPHOONS, and CYCLONES is now being gathered by airborne reconnaissance missions, new anemometers were developed for aerial use. One such device, the *gust probe,* is specially designed to record a hurricane's dangerous low-level gusts (often hazardous to aircraft because of the presence of WIND SHEAR) by being trailed from an aircraft that is flying over, rather than through, the worst of the storm. Modern technology has also made it possible for meteorologists on the ground to employ hand-held anemometers, portable instruments that greatly aid hurricane study by permitting spot checks of wind speed in relation to particular geographical features.

aneroid barometer Aneroid barometers measure BAROMETRIC PRESSURE through mechanical rather than hydraulic means. First investigated in 1795 by Nicolas-Jacques Conté and patented in 1845 by French physicist Lucien Vidie, the basic aneroid barometer does not consist of a glass tube filled with mercury, but instead contains a metal chamber, a coiled spring, a shaft with bar arm, and a gauge calibrated in both inches and MILLIBARS. Any increase or decrease in atmospheric pressure causes the thin walls of the chamber—or sylphon, as it is correctly known—to contract or expand respectively. Tightly wound inside the sylphon, the spring in turn reacts by either opening or closing, while rotating the thin shaft and its attached arm to a given point on the dial. From there the current barometric pressure can be noted and its degree of change from its last position observed. Favored by meteorologists, scientists, and mariners because of its portability and ease-of-use, the basic aneroid model remains in widespread use around the world despite longstanding concerns regarding its precision and reliability in certain weather conditions.

More sophisticated aneroid barometers employed by the NATIONAL HURRICANE CENTER (NHC) and the JOINT TYPHOON WARNING CENTER (JTWC) are constructed from alloys (such as nickel-silver or silver-brass) to compensate for changes in external TEMPERATURE and are equipped with digital displays instead of the traditional gauge or dial configuration. Aneroid barometers have also been modified for use in MICROBAROGRAPHS and for deployment in HURRICANE-HUNTING weather-reconnaissance aircraft such as the *SUPERFORTRESS B-29* the *NEPTUNE P2V-3W*, and the *ORION WP-3*.

Angela, Typhoon *Philippines–Vietnam, October 25–November 6, 1995* At the time said by the Philippine Weather Bureau to have been the most severe SUPERTYPHOON to have passed over Luzon Island since 1987, Angela delivered sustained 141-MPH (227-km/h) winds, 10-inch (254-mm) PRECIPITATION counts, and an 11-foot (4-m) STORM SURGE to the central and northern provinces of the PHILIPPINES—and to the northeastern coast of Vietnam—between November 2 and 6, 1995. With a CENTRAL BAROMETRIC PRESSURE of 27.25 inches (922 mb) at landfall in the Philippines, Angela demolished more than 15,000 BUILDINGS, left some 300,000 people homeless and

claimed an estimated 740 lives in 25 of Manila's surrounding provinces. Before breaking apart over central Vietnam on the evening of November 6, Angela's 106-MPH, (171-km/h) winds and six-inch (152.4-mm) downpours caused an additional 20 deaths in that nation, making it one of the deadliest TROPICAL CYCLONES to have menaced southeastern Asia in recent memory.

The 27th tropical cyclone of a rigorously active TYPHOON SEASON (one which had seen the Philippines affected by no fewer than 13 TYPHOONS and TROPICAL STORMS since April), Angela originated over the subequatorial waters of the western NORTH PACIFIC OCEAN approximately 350 miles (563 km) southeast of Guam on the afternoon of October 25, 1995. Steadily sapping the ocean's surface of its heat and evaporative moisture, the budding TROPICAL DEPRESSION fitfully churned west-northwest at 15 MPH (24 km/h) before developing into a powerful tropical storm on the morning of October 28 and into a minimal Category 1 typhoon with a central pressure of 29.02 inches (984 mb) by the late-afternoon hours of October 30. Maximum sustained winds within Angela's EYEWALL now measured 80 MPH (129 km/h). After following a northwesterly trajectory until nightfall on October 31, Angela abruptly swung southward and then a few hours later to the west, its rain bands further deepening as it spiraled toward the low-lying cities and towns of the Bicol region on the southeast leg of Luzon Island.

Well acquainted with the archipelago's extensive history of tropical cyclone activity, Filipino authorities spent the morning of November 1 moving the residents of vulnerable shoreline communities to inland storm shelters in preparation for Angela's predicted arrival at noon the following day. As the typhoon inexorably strengthened, its central pressure continuing to drop and its sustained WIND SPEEDS exceeding 110 MPH (177 km/h), 135,000 people in the provinces of Catanduanes and Camarines were loaded onto buses and into cars and driven to government shelters further inland.

Whirring around a central pressure of 27.25 inches (922 mb) at landfall near the port town of Bicol shortly after noon on November 2, 1995, Angela's sustained, 130-MPH (209-km/h) winds, 168-MPH (270-km/h) gusts, and flooding rains devastated much of central Luzon Island. In Manila, Angela's 125-MPH (201-km/h) winds caused skyscrapers to sway and toppled a large billboard onto a nearby garage, crushing eight buses and two cars. On Catanduanes Island, where Angela's fury was felt the hardest, the governor stated to reporters that, "The winds are so powerful that people in tall buildings here feel they are being hit by an earthquake." In the town of Mabolo, 180 miles (290 km) southeast of Manila, Angela's 9-inch (229-mm) precipitation counts turned streets into waist-deep mires that kept several thousand people from their houses for nearly a week. In neighboring Sariayain, a four-lane highway was swept away by a landslide, and more than 5,000 buildings in surrounding villages collapsed.

animals and tropical cyclones Animals have long played a colorful if sometimes deadly role in the history of TROPICAL CYCLONES. Vultures, swans, seabirds, rabbits, snakes, alligators, cows, chickens, and even ants have at odd times and by various cultures been regarded as a HURRICANE's harbingers, scavengers, and weathervanes. Some animals warn of a coming TYPHOON, while others assist in cleaning up after it. Some, such as the sacred cows of India, are allowed to blow away like wind gauges, their divinity preserved by the CYCLONE whirring around them. On isolated occasions, fear-crazed animals have even been responsible for increasing a hurricane's tragic DEATH TOLL. When, for example, the now-legendary Hurricane CAMILLE thundered ashore in August 1969, hundreds of poisonous cottonmouth snakes poured out of MISSISSIPPI's flooded bayous and creeks, killing an estimated 40 people.

Seminole tribes who live along FLORIDA's lush Everglades have for years depended on the peculiar habits of indigenous wildlife to warn them of an impending hurricane. In an area where a complex ecosystem has of necessity become finely tuned to the periodic arrival of a hurricane's destructive winds and rains, the Seminole believe that shortly before a storm's landfall, small kites, wood storks, and other birds in the Everglades instinctively begin a mass migration, soon joined by rabbits and other small burrowing animals. Ants begin to reinforce their hills, adding as much as a half inch of sand to their height; huge alligators that slink through the fertile swamps move away from the shores and into the serenity of deeper water. Because the animals tend to flee at the same time and in the same direction, the Seminole assume they are escaping from the path of an oncoming CARIBBEAN SEA or GULF OF MEXICO hurricane and quickly follow to the safety of higher ground.

In the Caribbean basin itself, the sooty tern—or hurricane bird, as it has been nicknamed by the local population—is one of the first creatures to seek refuge from an approaching hurricane. At once, entire flocks of them lift off from their island nests and darken the azure sky as they grimly head inland to the subtropical forest's relative security. The tern's frantic migration, itself as densely packed and symbolic as a thunderhead, is generally viewed by Caribbean residents as a sure sign that a hurricane is due.

When evacuating in the face of a hurricane, it is important for pet owners to anticipate the needs of their pets. In this photograph, humankind's best friend is retrieved from a building damaged by 2005's Hurricane Katrina. Federal law now requires all jurisdictions to design and implement emergency animal control procedures. *(FEMA)*

The gentle swan is another bird that weather folklore claims can forecast a hurricane's arrival. It is said that when a swan flies directly into the wind, a hurricane is certain to come ashore the next day. The ancient Greeks had a similar, equally questionable method for predicting the onset of a tropical cyclone. In an area where tropical cyclonic systems do not usually make landfall, the ever-cautious Greeks nonetheless stated that if a wolf howls into the wind, a hurricane can be expected within three days.

In India and BANGLADESH, where the Asiatic monsoon's yearly advance and retreat signals the start of peak periods for cyclone generation in the Indian Ocean and BAY OF BENGAL, the indigenous people regard certain birds as aerial soothsayers, heralds of the rains and storms that so frequently bring enormous death tolls and significant destruction to the Indian subcontinent's upland shores. One such creature, the pied crested cuckoo—or monsoon bird as it is known to the local populace—arrives in India from its East African habitat by riding the monsoon winds, following their steady course across the Arabian Sea until it reaches the country's western slopes. Touching down just a few days before the first of the monsoon rains break, the "songbirds of the clouds" forewarn of the monsoon's imminent onslaught by gathering in trees and singing loudly. Another such bird, the hawk cuckoo or brain-fever bird, similarly chants its premonitions, calling out "the rains are coming, the rains are coming" in shrill notes that the village *Sabjantawallah,* or wise man, can interpret for the lifesaving benefit of the entire community.

Those seabirds that do, for one reason or another, find themselves overtaken by a tropical cyclone's encircling winds often land on nearby ships. During a powerful hurricane in August 1926, a steamship navigating the stormy waters of the Yucatán Channel was suddenly besieged by hundreds of seabirds. With their wings folded beneath them and their beaks turned out of the wind, they perched on the ship's rails and safely rode out the storm. A similar situation took place aboard a British steamship during the SANTA CRUZ DEL SUR HURRICANE of November 1932. The hurricane's almost-windless EYE swept over the ship, suddenly depositing hundreds of seabirds onto the vessel's decks. The ship's captain later stated that most of the birds were nearly dead from exhaustion, while several had had their wings broken by the hurricane's strenuous winds. During a 1919 typhoon off the coast of CHINA, the crew of a British destroyer traversing the Yellow Sea was astonished to see hundreds of kingfishers—seabirds who usually make their nest on barrier islands far to the east—descend on their ship.

There is also evidence to suggest that the hurricane's wind-driven action has actually assisted several species of birds in their migration to new habitats. It is believed that one or more hurricanes are responsible for bringing to the U.S. mainland the black anis from HISPANIOLA, the swallow and honeycreeper from the BAHAMAS, the cliff swallow from CUBA, and the plumed cattle egret from as far away as the French Antilles. These birds, caught up in the tropical cyclone's windless eye, travel with it until it makes landfall. There they either die with the dissipating system or survive to establish a new colony. During the GREAT SEPTEMBER GALE of 1815, hundreds of seagulls were carried into inland NEW ENGLAND by the strident winds of this particularly fierce hurricane. Contemporary newspaper accounts of the storm state that in the days following its passage, confused flocks of seagulls were seen wheeling above the fledgling mill city of Worcester, Massachusetts, 100 miles (161 km/h) from the nearest coastline.

Vultures, those wheeling, darting scavengers that most people regard with revulsion, have all the same played a largely beneficial role in the history of tropical cyclones. After a hurricane, typhoon, or cyclone has passed, vultures quickly move in and commence the gruesome but necessary process of disposing of the dead. In some cases, such as the

GREAT CALCUTTA CYCLONE of 1864 when nearly 100,000 head of cattle were killed, vultures saved hard-pressed authorities the Herculean task of collecting and burying so many decaying cows, goats, and sheep. Eyewitnesses reported that after a particularly destructive hurricane struck GEORGIA and SOUTH CAROLINA in August 1893, the only way the coroner's army of deputies could locate the hundreds of human corpses strewn about the isolated, storm-ravaged farms was by following the vultures that wheeled overhead. When a monstrous cyclone swept across India's Orissa District on October 12, 1967, thousands of decaying animal carcasses poisoned the air with their diseased stench, making it hazardous for rescue teams to enter the area and carry out their duties. The cyclone's deadly winds had killed every living creature in their path, including the vultures that would normally have stripped the carrion to the bone in a matter of days.

In places where sophisticated ANEMOMETERS and other measuring devices for determining wind speeds were not readily available, flying cows, bouncing pigs, and waterborne chickens were often held up in reports as solid evidence of a hurricane's ferocious strength. In the GREAT HURRICANE OF 1780, hundreds of cows that had been lifted up and carried inland by the high (possibly TORNADO-strength) winds were viewed with particular scientific interest—especially those few that had survived the ordeal. Indeed, several firsthand accounts of the BARBADOS storm are unanimous in stating their wonder and amazement at this colorful occurrence, and a great deal of attention was given in the local press to the cows plight. One plantation owner even offered to buy the surviving cows, so convinced was he of their hearty stock. Following the passage of a particularly powerful 1722 hurricane in LOUISIANA, surprised survivors were astonished to find several live deer, their limbs painfully splayed, cradled in the foliage of what few trees remained standing.

Hens, chickens, and other domestic fowl have also found their way into the descriptive accounts of past hurricanes. One young survivor of the remarkable SAVANNAH-CHARLESTON HURRICANE of 1893 that swept across South Carolina, recalled in a letter to a contemporary that as the hurricane's winds caused a nearby creek to overrun its banks, water "came through the house with many things bobbing up and down, including some white chickens." Here the tone of the letter not only indicates the degree of flood damage sustained by the family's homestead, but also a child's sense of fearful amusement at the incongruous sight of a harried flock of chickens being carried through the house on the crest of a wave. In addition, windborne chickens and roosters have been found,

This unique illustration from NOAA's archives shows a hail of rodents raining down on a hapless farmer and his grain crops. Although this illustration probably applies more closely to the special effects of a midwestern tornado, there are numerous accounts of birds and other creatures tumbling to Earth during tropical cyclone activity. *(NOAA)*

huddled but alive, in trees, in multistory belfries, and following the passage of a nasty hurricane in October 1896 skewered between the spokes of a wagon wheel.

The intensely low barometric pressures found in some tropical cyclones have also been evidenced by those animals unfortunate enough to be caught in the storm's path. When the extraordinarily powerful LABOR DAY HURRICANE washed ashore on the Florida Keys in September 1935, hundreds of beached fish literally exploded from the storm's record low pressure of 26.35 inches (892 mb). This same hurricane, with wind gusts in excess of 200 MPH (322 km/h), further stranded several pigs and two dogs in the tops of nearby palm trees, leaving them wet and howling but alive.

Sometimes, however, animals can significantly add to a tropical cyclone's human death toll. In Sep-

tember 1955, Hurricane JANET's torrential rainfall caused Mexico's snake-infested Panuco and Tamesi Rivers to overrun neighboring towns. Thousands of poisonous snakes attacked fleeing villagers, killing at least 30 of them. Likewise, the STORM SURGE from a July hurricane that struck ALABAMA's bayou country in 1819 caused numerous snapping turtles and alligators to crawl out of the swamps. Crazed by the sea salt, the animals went on a frenzied killing spree that left close to 100 people dead.

In an age when the issue of animal rights is gaining increasing cultural significance, concern for the well-being of farm animals and pets caught in a tropical cyclone's path can often assume unique manifestations. In West Palm Beach, Florida, animal guardians prepare their feline or canine companions for HURRICANE SEASON by conducting a series of "pet hurricane drills" at a local storm shelter. This "Hurricane Woof" program exposes crated and vaccinated animals to a host of hurricane simulations, including the loss of electricity, the roaring sounds of the wind, and the sudden bangs and booms brought on by flying debris. Many humane societies in south Florida further commemorate the start of storm season by publishing pamphlets and guides to assist owners in better preparing their pets for the potentially lethal effects of a major hurricane strike. Filled with information ranging from the size of different airline crates to those nearby shelters that accept pets, the animal hurricane guides provide a handy, lifesaving reference point at times of meteorological crises.

While in one regard such measures may be viewed as absurd or even useless, the fact remains that animals—particularly those that are on the loose both during and after a tropical cyclone's passage—are a matter of serious interest to civil defense and animal control authorities. Following Hurricane ANDREW's rampage across south Florida in 1992, thousands of animals from spooked horses to protective watchdogs to pet tarantulas freely roamed the shattered countryside. Many of these animals soon perished, either from wounds suffered during Andrew's Category 5 assault or in electrocution or drowning incidents shortly thereafter. Overwhelmed by the destruction's sheer human scale, rescue and relief teams were unable to collect animal corpses immediately, thereby allowing decomposition and a slew of communicable diseases to set in.

In an effort to prevent epidemics and to increase both public and animal safety, animal control organizations not only encourage owners to take whatever steps they deem viable to prepare their pets adequately beforehand, but also routinely deputize additional control officers in those areas most likely to be affected by the approaching storm. Dispatched with trucks, leashes, and nets, these individuals round up and bring to local shelters any live animals found wandering the streets. Recognizing that a majority of animals picked up after a hurricane's passage are lost pets, humane shelters afford the animals medical care, food, and housing until such time as their families reclaim them. Indeed, following the devastating passage of the GREAT NEW ENGLAND HURRICANE of September 21, 1938, an animal shelter in west Massachusetts collected 71 dogs, 45 cats, 31 hens, six turkeys, two guinea pigs, and a horse.

There is a prelude to this preparedness, however, one that begins with the posting of official hurricane warnings and ends with syringes and incinerators. Because many humane shelters and animal control pounds are of relatively small size, it sometimes becomes necessary to make room for lost pets by euthanizing those stray animals already being held for adoption. When Hurricane Erin, a fairly weak Category 1 system, threatened south Florida in late July 1995, 119 animals—48 cats and 71 dogs—being housed in an already overcrowded local shelter were put to death. Specified by the animal services disaster plan, the mass disposal of these animals took place 12 hours before Erin's 85-MPH (137-km/h) winds were due to strike the region, thereby allowing shelter personnel time to incinerate the bodies before possible power outages shut down their equipment. When Tropical Storm ALLISON struck eastern TEXAS in June 2001, nearly 3,000 animals used in laboratory studies at Baylor College of Medicine's Texas Medical Center were drowned.

The busy 2004 hurricane season in FLORIDA was even credited with reducing the number of shark attacks that occurred along the state's coasts during that year. In 2003, there were 30 shark attacks in Floridian waters; in 2004, only 12. Experts believed that the four hurricanes which struck the Sunshine State in 2004 reduced the number of attacks by limiting the number of people who were using the beaches and those areas where attacks usually occurred, and by keeping the sharks in deeper waters, away from the coasts where a tropical cyclone's destructiveness is most severe.

In Gulf Shores, Alabama, Hurricane IVAN's visit to the local zoo in September 2004 resulted in severe flooding that led to the escape of some 13 deer and six alligators from their enclosures; while one of the alligators was later shot and killed by animal control authorities, the other animals were never recovered. And Hurricane FLORENCE's grazing assault on BERMUDA in September 2006 claimed the lives of two flamingoes who were killed when a tree toppled onto their enclosure at the Bermuda Zoo.

The widespread destruction in LOUISIANA and MISSISSIPPI during Hurricane KATRINA's historic 2005 landfall resulted in enormous challenges for animal control and emergency management personnel. Thousands of animals in both states either perished or were injured, and many hundreds of others had to be left behind by their evacuated owners because local emergency shelters had neither the space nor the personnel available to accommodate them. As early as August 30, 2005, while a downgraded Tropical Storm Katrina still churned inland over Louisiana and Mississippi, humane societies as far away as San Antonio, TEXAS, were already receiving hundreds of dogs and cats that had been evacuated from already overcrowded shelters in Louisiana and Mississippi. In an effort to better coordinate animal control issues in the event of further catastrophic events, the federal government mandated in October 2005 that all jurisdictions would in future be required to develop and implement animal control plans and operations. An extreme solution to an extreme problem, the outcome of these animals' lives only underscores the tragedies, great and small, that too frequently accompany the rage and nature of tropical cyclones.

annular hurricane A relatively new term used by the U.S. NATIONAL HURRICANE CENTER (NHC), an annular (or annular-type) hurricane is a mature-stage TROPICAL CYCLONE that possesses meteorological characteristics that make them unique as compared to other tropical systems of similar size, intensity, or location. First identified in 2002, the annular hurricane generally contains a large EYE of between 25 miles and 45 miles (40 km–72 km) in diameter, with a symmetrical EYEWALL configuration, and convective activity that is evenly spread throughout the eyewall structure. An annular hurricane tends to survive and even intensify in environments that are not often considered favorable for tropical cyclone development, such as those with mild WIND SHEAR and below-average SEA-SURFACE TEMPERATURES (SSTs). Meteorologists believe that an annular hurricane is not as susceptible to eyewall rejuvenation cycles, and for this reason is able to maintain a certain level of intensity for much longer periods than statistical and other analogs would predict. In seeking to better define and understand annular hurricanes, meteorologists and climatologists have studied past tropical cyclones in the North Atlantic and eastern North Pacific Oceans, including 1996's Hurricane EDOUARD and 2003's Hurricane ISABEL in the North Atlantic; and Hurricane Daniel in the eastern North Pacific Ocean in July 2006. At the time that Daniel exhibited annular characteristics on July 20, the system was of Category 4 intensity, with a CENTRAL BAROMETRIC PRESSURE reading (estimated) of 27.99 inches (948 mb) and sustained winds of 125 MPH (205 km/h). Not all annular hurricanes are this powerful, however. In the example of Hurricane EPSILON in 2005, longevity rather than intensity, characterized the annular nature of this North Atlantic tropical system. It is possible that further research into the annular hurricane will indicate that Cyclone CATERINA, which surprisingly appeared in the SOUTH ATLANTIC OCEAN in March 2004 and made landfall in Brazil, was of the annular type.

Anita, Hurricane *Mexico, August 29–September 2, 1977* While Anita remains one of the most intense Category 4 HURRICANES on record, it is better remembered for its positive influence on the study of cyclonic weather systems. Before Anita crashed into the sparsely inhabited southwestern MEXICO coast on September 2, 1977, it was the subject of the most comprehensive scientific investigation into the mechanics of hurricanes yet attempted. For three days, Anita was tracked across the GULF OF MEXICO by a squadron of research organizations, including the National Oceanic and Atmospheric Administration (NOAA), the NATIONAL HURRICANE CENTER (NHC), and Texas A&M University. Dozens of experiments, ranging

from an examination of sea temperatures to a laser analysis of PRECIPITATION levels, were performed on the hurricane with significant success. Data collected from Anita conclusively answered a number of vital questions concerning hurricane-strengthening patterns and forever changed the fundamental model meteorologists have of hurricane structure and activity.

The first storm of an unusually quiet HURRICANE SEASON, Anita originated over the tepid waters of the eastern Gulf of Mexico, 250 miles (402 km) southwest of St. Petersburg, FLORIDA, on August 29, 1977. Moving in a westerly direction at nearly 10 MPH (16 km/h), Anita spent nearly two days as a TROPICAL STORM before being upgraded to Category 1 status on the morning of September 1. As the storm's CENTRAL BAROMETRIC PRESSURE continued to fall—from 29.94 inches (1,012 mb) on August 29 to 28.44 inches (963 mb) on September 1—evacuation warnings were issued for the southern Gulf Coast of TEXAS. A helicopter fleet hurriedly evacuated personnel from offshore oil rigs, while 100,000 people, some as far north as Corpus Christi, Texas, secured their shoreline property against the possible threat of high winds and torrential rains and then headed inland to safety.

A powerful Hurricane Anita bears down on the eastern coast of Mexico in 1977. *(NOAA)*

In the meantime, meteorologists from both the NOAA and the NHC quickly mobilized their resources to mount a full-scale scientific study of the burgeoning Anita. Specially modified hurricane hunting aircraft, such as the *NEPTUNE P2V-3W*, were dispatched to the storm's location. There, high technology ANEMOMETERS and BAROMETERS were systematically deployed to measure and record all aspects of Anita's complex construction. At sea, Texas A&M University's research vessel, the *Gyre*, bravely followed in Anita's tumultuous wake. Seeking to provide evidence of a direct correlation between water temperature and a hurricane's intensity, oceanographers aboard the *Gyre* used sensitive hydrometers to analyze sea temperature at various depths. They also made detailed observations regarding the state of the sea in a hurricane, recording the height and shape of the waves on the storm's outer flanks.

They were also in a perfect position to explain why, on the evening of September 1, Anita commenced a period of sudden and terrifying intensification. Now located less than 100 miles (161 km) from the Mexican coast and headed in a southwesterly direction that would take it ashore somewhere in that country's northern territory, Anita's central pressure rapidly dropped from 28.44 inches (963 mb) September 1 to 27.35 inches (926 mb) the next morning. This overnight intensification, one of the most dramatic ever observed, catapulted Anita from the status of a moderate Category 3 storm to that of a Category 4 hurricane of record intensity. Using data gleaned from their extensive temperature soundings, oceanographers aboard the *Gyre* determined that during the previous evening, Anita had crossed a large, isolated pool of warm water and been energized by its increased rate of evaporation. Preliminary studies indicated that area water temperatures were one to two degrees higher than the surrounding sea and that this differential might extend for up to 150 miles (300 km) in all directions. Because the Mexican coastline was now less than 100 miles (150 km) away, it was entirely possible that Anita would continue to strengthen and would come ashore as a rare Category 5 hurricane.

With a central pressure of 27.76 inches (940 mb), the small, tightly coiled EYE of Hurricane Anita surged ashore between the Mexican towns of San Fernando and La Pesca on the evening of September 2, 1977. Hardly weakened by a slight rise in barometric pressure from the night before, Anita's furious winds blasted the verdant palm groves of northern Mexico with 172-MPH (279-km/h) gusts and pelting rains. Entire forests were seemingly sheered off at the roots, and localized flooding swelled into shallow swamps. On the sparsely inhabited coast, Anita's

12-foot (4-m) STORM SURGE was felt as far away as Laguna Madre. Surprisingly, no lives were lost to such a powerful storm, although it would be years before the region's once-lush forests fully recovered their pristine beauty.

As Anita's tattered remnants finally dissipated over inland Mexico, meteorologists eagerly worked to assess the vast amount of data collected on the hurricane. In time, their discoveries would completely change the world's understanding of how cyclonic storm systems work. Aerial reconnaissance flights, for instance, confirmed that while Anita's most powerful winds—those of 150 MPH (130 km/h) or higher—were located to the north of the eye, those on the southern side did not even reach hurricane force. This understanding challenged the long-held belief that the asymmetrical distribution of a hurricane's wind speeds around the eye was due solely to the forward-moving influence of the storm's STEERING CURRENT. Even after Anita's 15-MPH (24-km/h) steering current was factored into the equation, it became apparent that other forces—perhaps a precipitous steepening of the storm's pressure differential—contributed to the unbalanced outlay of the hurricane's most dangerous winds.

Besides providing tangible evidence of a direct link between the presence of warm water eddies and a hurricane's abrupt intensification, oceanographers from the *Gyre* further indicated that in the case of Anita, a passing hurricane does indeed leave the ocean's surface several degrees cooler. Temperature experiments conducted before and after Anita's trek across the Gulf of Mexico revealed that in some places, passage of the storm had not only left the sea eight degrees (⁻13°C) colder, but had done so for close to 200 feet (70 m) down. Indeed, when the second hurricane of the 1977 season, Babe, tried to follow Anita's track two weeks later, it found the waters much too cold to sustain any significant degree of intensity and soon faded away. The name Anita has only been used once to identify a tropical cyclone in the North Atlantic. Following 1977's destructive Hurricane Anita, the name was retired from the rotating list of North Atlantic tropical cyclone names.

anticyclone An atmospheric disturbance in which a loosely defined center of high barometric pressure generates circular winds that flow outward, away from the center, and with diminishing degrees of intensity. Commonly referred to as a *high*, an anticyclone (as its name indicates) is the opposite of a CYCLONE, or a well-defined center of low pressure in which the winds roar inward. In an anticyclone, warm, sinking air does not allow sea-level moisture to rise high enough into the troposphere for rain clouds to form. For this reason, anticyclonic weather tends to be fair, with few clouds and warm temperatures, while cyclonic systems frequently produce the torrential PRECIPITATION and high winds associated with mature HURRICANES and TYPHOONS. Although meteorologists are not completely certain how anticyclones are formed, it is known that as isolated pockets of high pressure air expand, they begin to spin in a clockwise direction in the Northern Hemisphere and in a counter clockwise direction in the Southern Hemisphere. When combined with the relentless expansion of the high pressure air, this circular motion allows an anticyclone to fling its temperate influence over wide expanses of land and sea. This characteristic makes the anticyclone a much more powerful (though less intense) system than the smaller, more concentrated cyclone. Indeed, the intensity and TRACK of cyclonic storm systems are often determined by the size and proximity of high-level anticyclones. The BERMUDA HIGH, for instance, has a tendency to draw North Atlantic hurricanes westward along its southern flank and then up, over, and back toward the east again. Smaller, more localized anticyclones can serve to both vent a hurricane's critical OUTFLOW LAYER smoothly and distribute cool, dry air to the storm's outermost fringes. In both scenarios, the anticyclone aids in the hurricane's immediate intensification by either creating a space for the rapidly rising winds within the EYE to ascend to or by returning spent air to the storm's edges, where it sinks to sea level, becomes laden with evaporation, and is once again drawn into the developing cyclone.

Antigua *Eastern Caribbean Sea* This low-lying volcanic island, left virtually treeless by centuries of sugarcane production, has had a long, sometimes devastating history of TROPICAL CYCLONE strikes. Located in the eastern Caribbean Sea, Antigua (pronounced An-TEE-ga), is the largest of the 15 major islands that comprise the sprawling Leeward Islands group. The archipelago's position between latitudes 16 and 19 degrees North places it directly in the path of the mid-to-late-season hurricanes that spiral across the NORTH ATLANTIC OCEAN from Africa's Cape Verde Islands. Between 1664 and 2006, Antigua's sandy beaches and natural harbor at St. John's were directly affected by 80 recorded hurricanes. Of these, 23 were major hurricanes, with CENTRAL BAROMETRIC PRESSURES of 28.47 inches (964 mb) or lower and winds in excess of 111 MPH (179 km/h). At times this prosperous resort island has even suffered several hurricanes in quick succession, as happened in 1747 when two destructive storms came ashore within one month of each other and in 1837 when two hurricanes struck in less than

one week's time. Similar double strikes also occurred in 1792 and 1876.

Since 1664, Antigua has been struck by at least one June hurricane, although this particular storm was of minor intensity and caused little damage. In July, the probability of a hurricane landfall in Antigua increases slightly, with three reported strikes between 1664 and 1933.

But it is in August and September, when the first of the Cape Verde hurricanes arrives in the eastern reaches of the Caribbean, that the number of strikes jumps to a fearsome 19 and 29, respectively. Famous midseason hurricanes in Antigua include the August hurricane of 1681 that pulverized the sugarcane plantations and inundated the beachheads with a 13-foot (4-m) STORM SURGE. In September 1772, several British warships lying at anchor in St. John's Harbor were driven ashore when a powerful hurricane roared in from DOMINICA, while the legendary GREAT HURRICANE OF 1780 caused heavy rains that October that killed an estimated 6,000 people on the island. Other major hurricanes include an 1806 storm that leveled several hundred buildings, and an intense hurricane in early August 1835 that severely damaged the Royal Naval dockyards at English Harbor. In addition, Antigua had the unique distinction of being struck by a destructive out-of-season hurricane in late December 1954. Christened Alice, the storm started the New Year by crossing the Caribbean between December 30, 1954, and January 5, 1955. In Antigua, 75-MPH (121-km/h) winds lashed the countryside, knocking down road signs, scattering Christmas decorations, and killing three people.

On the night of September 16–17, 1989, Hurricane HUGO, considered the most damaging hurricane to have transited the Caribbean in nearly a half-century, pelted Antigua with sustained 130-MPH (200-km/h) winds and blistering, five–six-inch (127–152 mm) rains. Nearly 900 houses on the island were either destroyed or serious damaged, leaving thousands of people homeless. Downed trees and power lines blocked the streets, while intense rains spawned mudslides and flash floods. One man was killed, and another two reported missing.

During the remarkable 1995 HURRICANE SEASON, Hurricane Luis, a Category 4 storm of extreme violence, blasted Antigua with sustained, 145-MPH (233-km/h) winds, 12-inch (305-mm) rain squalls, and a six–nine-foot (2–3-m) storm surge on the night of September 5–6, 1995. An enormous cyclonic system that spanned nearly 700 miles (1,127 km) across, Luis's 60-mile-wide (97-km) EYE passed directly over the island, uprooting trees, severing power lines, buckling radio and television transmission towers, demolishing two beachfront hotels, and tearing the roof from the country's only hospital. More than 32,000 Antiguans were rendered homeless by the 12-hour tempest, while sporadic looting forced the government to cordon off shopping districts in downtown St. John's with armed soldiers. With more than $375 million in property damage and four deaths to its credit, Luis remains the single-most-destructive hurricane to have struck Antigua during the second half of the 20th century.

On September 5, 1995, Hurricane Luis brought 140-MPH (225-km/h) winds and a central pressure of 27.88 inches (944 mb) to Antigua. Four hotels were swept into the sea and two hospitals effectively destroyed.

Just under one year later, on July 9, 1996, Hurricane BERTHA delivered hurricane force winds and rains to Antigua as the system's EYEWALL passed directly over the island.

On September 21, 1998, Hurricane GEORGES passed close to the island, delivering 110-MPH (177-km/h) winds. Two people were killed and many BUILDINGS suffered roof damage. There was also severe coastal flooding along the southern coasts.

On October 20, 1999, Hurricane José passed directly over Antigua. Bearing a central pressure of 29.05 inches (983 mb), José delivered sustained winds of 80 MPH (129 km/h) and gusts topping 100 MPH (161 km/h). José was moving rather quickly to the northwest at 12 MPH (19 km/h) at the time of landfall in Antigua.

On August 22, 2000, Tropical Storm DEBBY delivered 70-MPH (90-km/h) winds to Antigua as it slipped to the north of the island.

Antje's Hurricane *United States–Mexico, September 8, 1842* This low-intensity Category 2 (estimated) HURRICANE crossed the CARIBBEAN SEA between August 30 and September 8, 1842. Named for the schooner *Antje*, which it dismasted in the east Caribbean on August 30, the hurricane followed an unusual straight-line TRACK west across the Caribbean before making landfall near the town of Victoria, MEXICO. Closely traced via ship's logs and weather reports by one of history's earliest hurricane watchers, WILLIAM REDFIELD, the storm did not follow the curving northward track normally taken by late-season hurricanes, but instead it sailed directly over the BAHAMAS before brushing past Havana, CUBA, and Key West, FLORIDA, on September 4. While Antje's Hurricane was of only minor intensity at Key West, in nearby Dove Key its swelling STORM SURGE carried away a small lighthouse and destroyed several outbuildings. Sources indicate that when the hurricane finally collided head-on with the Mexican coast on the afternoon of September 8, its hefty storm surge

flooded the shore as far north as the mouth of the Rio Grande. Its lowest barometric pressure, 28.93 inches (979 mb), was recorded by a weather station in Havana, while observers in Mexico noted that the calm passage of its EYE, which was apparently of small diameter, lasted no more than five minutes.

Because of a lack of weather maps that show synoptic conditions over North America for the year 1842, it is nearly impossible to say with any certainty why Antje's Hurricane followed the unusual course that it did. However, the better-tracked San Ciprian Hurricane of September 1932 followed a similar straight-line route west across PUERTO RICO and HISPANIOLA and into Mexico's Yucatán Peninsula. This-equally unusual course was caused by an intense high-pressure system that blanketed much of North America. With this scenario in mind, it is possible to see how Antje's Hurricane could have found itself near a similar high-pressure area in 1842, one that firmly kept it from swinging northward.

Arlene, Hurricane *Southern United States, June 8–13, 2005* The first named TROPICAL CYCLONE of the record-setting 2005 North Atlantic HURRICANE SEASON, and the ninth tropical system to bear the name, a downgraded Hurricane Arlene washed ashore in southern ALABAMA on June 11, 2005 as a moderately powerful TROPICAL STORM. Formed three days earlier from a tropical low over the GULF OF MEXICO, Arlene's central pressure of 29.26 inches (991 mb) brought 52-MPH (84-km/h) winds to Alabama, damaging trees and small buildings, but claiming no lives or serious injuries.

The name Arlene is one of the oldest tropical cyclone identifiers still in use in the North Atlantic basin. First appearing in 1959, with a preseason tropical storm that originated in the Gulf of Mexico before winding ashore in southern LOUISIANA on May 31, it has since been used no fewer than eight times between 1963 and 2005. In 1963, a powerful Category 2 Hurricane Arlene originated in the Atlantic before brushing past the islands of the northern CARIBBEAN SEA on August 4. At its peak bearing a central pressure of 28.61 inches (969 mb) and sustained, 104-MPH (167-km/h) winds, Arlene remained offshore for its entire existence, recurving northward off the U.S. east coast and causing widespread destruction in BERMUDA.

The 1967 North Atlantic hurricane season opened with another Hurricane Arlene, this one of less meteorological severity than its 1963 incarnation. Originating very close to latitude 15 degrees North on August 28, Arlene developed into a Category 1 system with sustained winds of 86 MPH (138 km/h), then recurved to the north before finally dissipating over the chill reaches of the NORTH ATLANTIC on September 4.

On July 4, 1971, Tropical Storm Arlene originated off the eastern seaboard of the United States, before trailing northward, remaining an offshore system until its dissipation four days later. At its height, Tropical Storm Arlene's central pressure of 29.52 inches (1,000 mb) generated sustained winds of 63 MPH (105 km/h) and heavy surf conditions along the coast.

A preseason North Atlantic tropical cyclone, the fifth tropical system christened Arlene developed on May 6, 1981, over the northwestern Caribbean Sea. Powered by a central pressure of 29.50 inches (999 mb) and producing 58-MPH (93-km/h) winds, Tropical Storm Arlene crossed CUBA and the BAHAMAS before penetrating deeply into the North Atlantic. While heavy rain and surf conditions were noted, no deaths or injuries ensued.

One of the longest-traveling North Atlantic tropical cyclones on record, Hurricane Arlene inaugurated the 1987 hurricane season with a sojourn across the Atlantic between August 9 and 17. At its peak a minimal Category 1 hurricane with sustained winds of 75 MPH (121 km/h) and a central pressure reading of 29.14 inches (987 mb), Arlene chiseled a bizarre path across the North Atlantic, one which saw it slide into the Iberian Peninsula as an extratropical depression on the night of August 16–17, 1987. On August 10, while still of tropical depression status, Arlene swept across the island of HISPANIOLA, its central pressure of 29.79 inches (1,009 mb) producing wind speeds near 23 MPH (37 km/h). While heavy rains fell across the island, no deaths or injuries were reported.

The seventh North Atlantic tropical cyclone dubbed Arlene—this one of tropical storm intensity—coursed through the Bay of Campeche and into southern TEXAS between June 18 and 21, 1993. Bearing a central pressure of 29.55 inches (1,001 mb) and sustained, 40-MPH (64-km/h) winds at landfall just above the Texas–Mexico border, Tropical Storm Arlene produced great quantities of rain, but claimed no lives or major damage tallies.

Between June 11 and 18, 1999, Tropical Storm Arlene quickly originated—and as quickly dissipated—to the east of Bermuda. With a minimum central pressure of 29.59 inches (1,000 mb) and sustained winds of 58 MPH (93 km/h), Arlene remained a moderately powerful tropical storm for its entire offshore existence, but caused no injuries or damages.

Due for its 10th incarnation during the 2011 North Atlantic hurricane season, the name Arlene has been retained on the rotating list of identifiers.

Arthur, Tropical Storm *Eastern United States, June 17–20, 1996* The first named TROPICAL CYCLONE of the 1996 North Atlantic HURRICANE SEASON, a weak and disorganized Tropical Storm Arthur brushed across NORTH CAROLINA's Outer Banks on June 20, 1996. A central pressure reading of 29.64 inches (1,004 mb) and 40-MPH (64-km/h) winds buffeted North Carolina's picturesque Cape Lookout, while two-to-four-inch (51–102-mm) rains and five-to-seven-foot (2.1-m) surf conditions lashed the coastal Carolinas before early-season Arthur drew to the northeast and dissipated over the mid-Atlantic. No deaths or major damage were recorded in Tropical Storm Arthur's wake.

The first North Atlantic tropical cyclone identified as Arthur appeared on August 28, 1984, to the east of the Caribbean's Leeward Islands. Of TROPICAL STORM intensity only, Arthur spent the next week tracking to the northwest along the Atlantic side of the Leeward Islands, but never made landfall. At its height, 1984's Arthur was a fairly powerful tropical storm, with a central pressure of 29.64 inches (1,004 mb) and 52 MPH (84 km/h).

The 2002 North Atlantic hurricane season began with the development of Tropical Storm Arthur off the coast of North Carolina on July 14. A moderately powerful tropical storm, Arthur possessed a minimum CENTRAL BAROMETRIC PRESSURE of 29.44 inches (997 mb) and sustained winds of 58 MPH (93 km/h). Moving to the northeast, away from the U.S. eastern seaboard, Tropical Storm Arthur quickly dissipated off the coast of Newfoundland.

The name Arthur has been retained, and is scheduled to next appear at the start of the 2008 North Atlantic hurricane season.

Atlantic Ocean   See NORTH ATLANTIC OCEAN and SOUTH ATLANTIC OCEAN.

Audrey, Hurricane *Southern United States, June 27–30, 1957* Classified as a Category 4 HURRICANE because of its extreme STORM TIDES, Audrey remains one of the deadliest TROPICAL CYCLONES in U.S. history. With a central pressure at landfall of 27.91 inches (945 mb) and sustained winds of 105 MPH (170 km/h), Audrey beat an early-season path of destruction up the GULF OF MEXICO, past Galveston and Port Arthur, TEXAS, and into the bayou lowlands of LOUISIANA between June 27 and 30, 1957. While damage in Texas was on the whole considered minimal, the languid Louisiana towns of Cameron, Creole, and Grand Chenier were virtually destroyed by Audrey's 97-MPH (156-km/h) winds, and pulverizing 20-foot (7-m) STORM SURGE. Despite telephone and radio technology that made it possible to warn coastal residents adequately of Audrey's imminent arrival, 518 people in Cameron Parish lost their lives to the first major hurricane to strike the area since 1918.

Audrey originated over the warm waters of the Gulf of Mexico, 350 miles (563 km) southeast of Brownsville, Texas, on June 25. Although it was the 1957 season's first hurricane, many gulf-coast residents received news of its sudden inception with an odd mixture of careful scrutiny and cautions inaction. After all, a majority of them had lived on the gulf for decades and had seen this sort of storm before. They believed that June hurricanes tend to be fairly small and of minor intensity and so were not overly concerned by the threat of torrential downpours and a few hurricane-force gusts. Only after it was reported that nine men were killed when a workboat being used to evacuate personnel from a deep-sea drilling rig was battered and sunk by Audrey's offshore winds did Galveston begin to fortify its famous seawall; the Louisiana shrimp fleet then hurriedly returned to its ports.

But in quiet delta towns like Cameron and Grand Chenier, Louisiana, where the population had over the years become seemingly inured to hurricane warnings, evacuation plans went largely unimplemented. Even as Audrey further intensified and as its barometric pressure slipped to 28.80 inches (973 mb) on the afternoon of June 26 and then to 27.91 inches (945 mb) in the early morning hours of June 27, hundreds of seaside residents continued to ignore the evacuation warnings issued by the weather forecasting office in New Orleans. Despite the fact that Audrey was a well-tracked hurricane, several survivors from Cameron and Grand Chenier would later state that there was some question in their minds as to the course the storm would finally take. According to the weather reports, Audrey was moving in a north-northeast direction at approximately 17 MPH (27 km/h). This meant it would either make landfall in northern Texas or spin out into the central gulf and threaten west CUBA and FLORIDA's Panhandle. For this reason, they did not believe that the hurricane—when and if it struck Louisiana—would be too severe.

By midnight of June 27, the EYE of Hurricane Audrey was centered 50 miles (80 km) southeast of Galveston and headed on a course that would take it ashore somewhere between High Island, Texas, and Morgan City, Louisiana. As Audrey's central pressure continued to drop, as the titanic mound of water beneath its eye continued to swell, alarmed civil defense authorities issued a final series of evacuation orders for the west Louisiana coastline. ANALOGS indicated that in some places Audrey's seething storm surge could rise as high as 20 feet (7 m) and that in

such a scenario its impact on the marshy lowlands, on the farms and single-story cottages that ringed the shoreline would be cataclysmic. In addition, the inundation of the bayous would begin long before the surge itself arrived with the hurricane's eye. The gulf would slowly but inexorably rise until it washed away the bridges and roadways that led inland. If a successful evacuation were to be completed, it would have to come at least six hours before Audrey's predicted landfall at daybreak. Any later, and trapped residents would find themselves standing alone against the malevolent fury of the onrushing storm.

Audrey barreled ashore at Cameron, Louisiana, on the morning of June 27, 1957. With a central barometric pressure of 27.91 inches (945 mb), Audrey's whirring winds and pelting rains stripped buildings of their roofs and siding, crumpled billboards, toppled trees and telephone polls, and caused massive flooding. Gusts of 130 MPH (209 km/h) drove before them a 20-foot (7-m) storm surge, the highest seen on the Louisiana coast in a most a century. The hurricane's cresting surge—so powerful that only hours before it had sheered nearly 60 feet (20 m) of shoreline off the northeast tip of Bolivar Island, Texas—thudded into Cameron. Laden with splintered tree trunks, steel beams, automobiles, and even a fishing boat, the surge rolled inland like an enormous ram, battering most of the town into oblivion. Hundreds of wooden cottages were swept off their foundations as the surge covered the sandy lowlands with 15 feet (5 m) of water. Some houses collapsed upon impact, spilling their families into the deadly froth, while others crazily bobbed through the center of town before breaking up. Frantic victims clambered onto floating wreckage as hundreds of poisonous snakes slithered from bayous. Deranged by the salt water, the snakes swarmed through the flooded streets of Cameron, killing five people.

By midafternoon on June 27, a weakened but still deadly. Audrey began to recurve northeast over inland Louisiana. With its fitful gales still strong enough to level patches of forest, the rain-laden storm underwent EXTRATROPICAL strengthening as it passed through eastern Ohio, Illinois, Indiana, NEW YORK, and Ontario, CANADA, on June 28. Before Audrey completely dissipated over the cold expanses of Canada on June 29, 30 more people had been killed.

For weeks thereafter, the death toll in storm-ravaged Cameron Parish, Louisiana, would grow much higher. By July 3, as federal relief agents sent by then President Dwight D. Eisenhower diligently combed through the wreckage of the communities, the DEATH TOLL stood at 296. By July 12, it had risen to 322, with 180 residents still missing. As a search of the waterlogged countryside continued, additional bodies were found in the bayous and marshes, bringing the final count to 518 dead on July 22. In addition, five other casualties were reported in northern Louisiana, and eleven from Texas. In Cameron Parish 1,900 buildings were utterly destroyed, while another 19,000 were badly damaged; 50,000 head of cattle were drowned.

At a time when the volatile civil rights issue occupied an increasingly larger share of U.S. consciousness, the fact that 90 percent of those lost to Hurricane Audrey were black Americans soon led to a heated barrage of criticism against local and state civil defense authorities. Virtually every organization tasked with alerting the public to the threat of incoming hurricanes from the Cameron Police Department to the National Weather Bureau was charged with bigotry, willful disregard for public safety, even mass murder. While it cannot be said for certain that in the process of warning coastal residents of Audrey's possible arrival some isolated instances of racial prejudice and neglect did not take place, a more likely cause for the disaster lies in both public apathy and communication shortfalls. Subsequent investigations into the exact sequence of evacuation procedures taken by authorities revealed that many local radio stations did not broadcast storm warnings in their entirety, but instead paraphrased them or left out whole segments of vital information. In turn, this practice left many coastal residents with the mistaken belief that there was still time for them to evacuate their bayou communities before Audrey's rising waters washed away the bridges and roadways that led to the elusive safety of higher ground. In view of the multifaceted significance of Hurricane Audrey, the name *Audrey* has been retired from the rotating list of North Atlantic tropical cyclone names.

Australia *Southwestern Pacific Ocean* Australia, the immense continent–nation that seemingly anchors the Pacific Ocean's southwest quadrant, has endured a long, continuous history of intense TROPICAL CYCLONE activity. With 12,210 combined miles of coastline bordering on three cyclone-prone bodies of water—the Coral Sea to the north, the South Pacific Ocean to the east, and the southern Indian Ocean to the west—Australia is particularly susceptible to mid- and late-season CYCLONE strikes on three of its shores. Between 1839 and 2006, approximately 462 mature cyclones and TROPICAL STORMS made landfall on the northeast and northwest coasts of Australia, delivering torrential rains and flailing winds to its marshy lagoons, arid tablelands, inland dunes, and salt bush deserts. On average, Australia's coral reefs, mangrove-lined beaches, and principal northern ports are likely to be inundated by at least one cyclone every

two years and by a major cyclone—storms characterized by CENTRAL BAROMETRIC PRESSURES of less than 28.50 inches (965 mb) and wind speeds in excess of 110 MPH (177 km/h)—every seven to ten years. In some instances, three or more cyclones have directly affected the expansive nation in a single year, as happened during the 1995 CYCLONE SEASON when three cyclones—Cyclone BOBBY, February 23; Cyclone Warren, March 7; and Cyclone Violet, March 9— claimed a total of eight lives and inflicted considerable property damage on both the continent's east and west coasts. On December 25, 1974, Australia suffered the worst peacetime disaster in its history when Cyclone TRACY utterly leveled the northwestern port city of Darwin. Bearing sustained 137-MPH (221-km/h) winds, 187-MPH (301-km/h) gusts, and four-inch (102 mm) rains, Tracy completely demolished 80 percent of the city's 12,000 lightly constructed BUILDINGS. Sixty-six people were killed, while an additional 160 victims—many of them fishing at sea when the cyclone struck—remained forever unaccounted for.

Northern Australia's position within the INTERTROPICAL CONVERGENCE ZONE (ITCZ), combined with the steady presence of subequatorial westerly winds over the continent's southern mantle, accords the country a brief but busy five-month cyclone season. Extending from December 1 to April 30, the Australian season includes two singularities, or periods in which cyclone activity is at its peak, these being the months of January and March. Between 1839 and 2006, for instance, 110 documented cyclones affected Australia's coasts in January, while 113 storms of mature intensity came ashore in March. February's averages do not lag far behind: 92 cyclones were observed during a 162-year span. Both December and April, or those months immediately preceding the advance and retreat of the Australian monsoon, witness a slight decrease in activity, as 37 and 51 cyclones, respectively, have made recorded landfalls during those two months.

Australia has also withstood its share of out-of-season cyclones, those Coral and Timor Sea storms that originate in October and November, and May and June, before tracking south. Strike counts for Australia's east and west coasts reveal that of the two minima, October and November are slightly more active months with a respective six and fourteen mature cyclones against May–June's nine and ten. Although a majority of pre- and post-season Australian cyclones are smaller, less intense systems than their mid- to late-season counterparts, a number of them have on occasion been responsible for inflicting substantial property damage and even loss of life. One such cyclone on May 28, 1988, drove the 115,000-ton bulk carrier *Korea Star* aground near Western Australia's rock-strewn Cape Cuvier. Pounded to pieces by the cyclone's 19-foot (6-m) waves and 95-MPH (153-km/h) winds, the *Korea Star*'s demise not only fouled nearby coves with hundreds of tons of spilled bunker fuel but also claimed the lives of three crew members.

There is no record of Australia having been struck by cyclones in July, August, or September or during the months when the southern winter is in progress.

Because most Australian cyclones form between latitudes 5 and 15 degrees south, Australia's broad position between latitudes 11 and 45 degrees south naturally makes for an uneven distribution of cyclone strikes on its various coasts. Early and late-season cyclones, those systems that usually develop over the 84-degree F (29°C) waters of the Arafura Sea during December and April, have a tendency to come ashore on Australia's two northern peninsulas, Cape York and Arnhem Land. Rarely penetrating inland as far south as the 23rd parallel, December and April cyclones generally do not travel in either an easterly or westerly direction but rather immediately undergo RECURVATURE, or a parabolic swing to the southeast or southwest. Sometimes achieving Category 3 status or better along the way, early and late-season cyclones in the western half of the Arafura Sea have been known to recurve to the southeast before coming ashore between Darwin and the Kimberley Plateau; recurve to the east and enter the Gulf of Carpentaria; and recurve to the southwest, where they make landfall between Cape Melville and Mackay.

Midseason Australian cyclones, or those storms that occur during the January and March singularities, essentially follow the same recurring trajectories as early and late-season cyclones but come ashore much farther south. Now originating over the sufficiently heated waters of the western Timor and eastern Coral Seas, January through March cyclones often swing away from their respective coasts, arc out into the Indian Ocean or the Coral Sea before progressing in a south-southeasterly or south-southwesterly direction. Bearing down on their respective coastlines, these cyclones either weaken and dissipate or rapidly intensify into storms of stunning intensity, bringing cyclone-force winds (74 MPH [119 km/h] or higher), flooding rains, and dangerous STORM SURGES to the voluminous regions of southern Australia.

While there have been recorded instances of Australian cyclones forming in the Indian Ocean's southeast quadrant and then moving east to pulverize the mangrove habitats of the North West Cape area, there has never been a documented cyclone strike on the continent's south coast or on that part of the country that extends from Flinders Bay in the west to Tasmania. Because Australian cyclones do not move northward, it is highly improbable that the region will ever sustain a direct strike.

Early accounts of Australian cyclones are largely culled from the country's north and east coasts, where the English were among the first European settlers to record the number of men, ships, and commerce looted from them by the sudden violence of the cyclone. During the latter half of the 19th century, a parade of cyclones struck in and around the fledgling seaport of Darwin, prompting the English to consider abandoning the seemingly inhospitable anchorage altogether. In January 1877, a powerful cyclone battered the town. More than 100 people were reportedly killed, while another 700 were injured. A March 1989 cyclone ravaged the city, claiming an estimated 139 lives and collapsing nearly every structure—from piers to warehouses—along the waterfront. Darwin would again be struck by major cyclones in December 1897 and January 1917. In each instance, dozens of lives were lost, while widespread property damage hindered rapid economic redevelopment of the area.

In 1881, and some 1,300 miles (2,092 km) to the southeast, the fishing haven of Cossack was virtually obliterated by an extraordinary midseason cyclone. Roaring across the Great Barrier Reef and into north-central Queensland on January 7, the cyclone produced an unconfirmed pressure reading of 27.00 inches (914 mb) at landfall, making it one of the nation's most intense. Estimated wind speeds of between 155 and 170 MPH (249–274 km/h) unroofed every last one of Cossack's 1,000 buildings, while an inescapable 19-foot (6-m) storm surge chased across the harbor. More than 230 people were listed as dead or missing, many of them pearl divers on small boats caught at sea by the cyclone's 30-foot (10-m) waves. A similar, slightly less intense cyclone again trounced Cossack on April 2, 1898. Coming ashore as a Category 4 system of extreme severity, this late-season cyclone quickly moved inland, its 137-MPH (221-km/h) winds wrenching entire trees from the ground before dropping them several yards away. Hundreds of dwellings were totally demolished, lifted from their foundations and then hurled to earth. Those vessels in the harbor not immediately overwhelmed by the cyclone's swamping onset were driven ashore with its storm surge and left impaled on nearby stone quays. Deluging 12- to 14-inch (305–356 mm) rains swept the Great Dividing Range's east slopes, producing fast-moving floods that transformed valleys into lowland lakes. Between 211 and 385 people perished during the 1898 cyclone, the majority of the victims again listed as missing either at sea or in crushing mudslides.

During the 1918 cyclone season, no less than two Category 4 cyclones laid waste to northeastern Queensland. The first, which came ashore near the port town of Mackay on January 20, killed 47 people and wrought more than $1 million in damage. The second, which blasted the neighborhoods of Innisfail on March 9, was of less intensity than the January storm, but nonetheless managed to claim another 100 lives. Between March 5 and 15, 1934, northern Queensland was again riven by twin cyclones, Category 3 systems that left 64 people dead and nearly $3 million in property and crop losses.

On January 6, 1897, the northern Australian city of Darwin was savaged by an exceptionally powerful cyclone that caused tremendous damage to the burgeoning settlement. Coming ashore during high tide, the cyclone spawned a storm tide condition that flattened buildings, uprooted trees, cast shipping aground, killed 15 people, and sank numerous smaller vessels, including several government-owned steam launches.

Australia's deadliest cyclone disaster occurred in March 1899, when more than 300 people were killed in the Bathurst Bay Hurricane, sometimes also known as Cyclone Mahina. According to records maintained by the Australian Bureau of Meteorology, the March 1899 cyclone brought with it an enormous storm surge that inundated low-lying coastal areas. Nearly 100 others died when a pearling fleet was lost offshore.

Just over one month after a powerful cyclone struck Port Douglass in February 1911, an even more severe tropical cyclone struck Port Douglas and virtually obliterated the struggling port town, as well as the neighboring settlement of Mossman. At least two people were killed. In the midst of the cyclone's churn, the steamship *Yongala* was lost on March 24. Constructed in 1903 and employed in service between Mackay, Melbourne, Townsville, and Cairns, the *Yongala* was overtaken by the cyclone as it neared the Whitsunday Islands' Cape Bowling Green. Although the exact circumstances of the *Yongala*'s demise will never be known, the discovery of the battered wreck on the seabed between Cape Bowling Green and Cape Upstart in 1947 indicates that the 324-foot (100-m) vessel was essentially overwhelmed by the cyclone's fury and sunk. All 120 aboard were killed.

Barely one year later, the Koombana Cyclone unleashed widespread destruction across the northwestern Australian reaches of Balla Balla and the Whim Creek district. Named for an Australian steamship, the *Koombana*, which was lost on March 26 with its entire complement of 140 passengers and crew, the cyclone was particularly hard on coastal

shipping, with more than 15 lives lost to grounded sailing ships and capsized steamships.

Two powerful cyclones struck Queensland in 1918. The first, which struck in January, was known as the Mackay Cyclone because of the tremendous damage it did in the coastal town of Mackay on the 20th of the month. At least 30 people lost their lives in Mackay and Rockhampton. The cyclone produced a central barometric pressure reading of 27.55 inches (933 mb), one of the lowest ever recorded in Australia. A storm surge of eight feet (2.7 m) inundated central Mackay, killing at least 19 residents. At sea, heavy winds and seas caused the loss of the schooner *Orete*, with several more lives lost.

An even more intense cyclone struck Australia on March 10, 1918, causing considerable destruction in Innisfail. More than 80 percent of the buildings in Innisfail were destroyed. The towns of Cairns, Babinda, and the Atherton Tableland were also affected. At Mission Beach, a storm surge measuring nine feet (3.6 m) rushed inland, claiming at least 37 lives in Innisfail and another 40 to 60 elsewhere in the region. In Mourilyan, the cyclone's barometric pressure produced a reading of 27.34 inches (926 mb).

Australia's second-deadliest cyclone on record struck the northwestern reaches of the continent-nation on the night of March 26–27, 1935. Ravaging the Lacepede Islands near Broome with ferocious winds, some 141 lives were lost.

In February 1954, the infamous Gold Coast Cyclone roared ashore in northeastern New South Wales. Wind gusts in excess of 62 MPH (100 km/h) struck the provincial capital of Brisbane on February 19, while the eye of the cyclone came ashore at Coolangatta, where the pressure dropped to 28.73 inches (973 mb). The town of Lismore was struck by heavy flooding caused by the 1954 system's 3.5 inches (900 mm) of PRECIPITATION. In all, 26 people were killed.

On December 25, 1951, a midseason Coral Sea cyclone progressed farther east than is usual before undergoing recurvature and crossing into the New Hebrides Islands. Located 1,122 miles (1,806 km) east of Cape Melville, the tiny spire of Epi Island was viciously ransacked by the Christmas Day tempest and was left with hardly a building or tree standing. One hundred men, women, and children perished in what was then judged to be the worst cyclone in the archipelago's history.

On April 4, 1989, Cyclone Aivu delivered sustained, 120-MPH (193-km/h) winds and 8-inch (203-mm) rains to the northeast coast of Queensland. Although no fatalities were reported, 10 percent of the territory's sugarcane harvest was destroyed. The following year, on December 24–26, 1990, portions of Queensland were again submerged by gushing rains, these associated with the Category 4 landfall of Cyclone Joy. Virtually defoliated by Joy's 145-MPH (233-km/h) winds and 9-inch (229-mm) precipitation counts, Queensland's losses exceeded $30 million.

Half a decade later, Cyclone Bobby snagged seven lives when it came ashore near the fishing village of Onslow, 850 miles (1,400 km) north of Perth, on February 23, 1995. Regarded as one of the most intense cyclones to have struck Australia since 1974, damage totals from Bobby soared into the tens of millions of dollars as breaching rains washed out major rail and road links in Western Australia, isolating entire tracts of the country. During the same season, Cyclone Chloe, a fairly weak Category 1 storm, rushed ashore over Western Australia's Kimberley Plateau on April 11, 1995. Born over the Timor Sea only two days before, Chloe's rain-shorn demise caused only minor property damage and no fatalities.

On the night of December 11, 1995, Cyclone Frank doused West Australia's sun-fired northern coast with 99-MPH (160-km/h) winds, 8-inch (6-m) precipitation counts, and 17-foot (203-mm) seas. In the fishing town of Exmouth, located some 700 miles (1,120 km) northwest of Perth, Frank's 120-MPH (193-km/h) gusts chopped thousands of mangrove trees into kindling, pumped tons of seawater into beachside swamps, and skimmed the wooden and tin roofs from more than 150 buildings. Traversing the Robinson mountain range at seven MPH (12 km/h), Frank's remaining rains sparked flash floods and landslides, mud-stained torrents that forced the nearby Gascoyne and Murchison rivers over their dusty banks. While an estimated $15 million in damage was tallied in the wake of Cyclone Frank, no lives were lost, and injuries were few.

The 1996 cyclone season was inaugurated on January 27 when Cyclone Celeste, a minimal cyclone with sustained winds of 85 MPH (137 km/h), brushed Queensland's northeast coast before veering eastward. A capricious system from the moment of its inception two days earlier, Tropical Storm Celeste startled meteorologists by rapidly deepening, growing into a mature cyclone with sustained winds of 76 MPH (122 km/h) in less than 12 hours. Drifting to the southwest at nearly 12 MPH (19 km/h), Celeste crossed the Great Barrier Reef during the early morning hours of January 27 and by midafternoon had penetrated as far inland as the town of Bowen, some 580 miles (933 km) northwest of Brisbane, before an advancing high-pressure front sent it back to sea for good. As it was, the cyclone's 100-MPH (161-km/h) gusts and driving rains severed power lines, toppled trees, defenestrated dozens of tropical houses, canceled passenger-train services, and transformed Police Camp Creek into a rapacious sluice-way. One

Bowen resident was drowned while attempting to cross a submerged bridge, and several local motorists were driven from the roads and injured by Celeste's overturning gusts and puncturing downpours. Preliminary damage estimates from Celeste ranged from $A 4–9 million.

Less than two weeks later, on February 5, 1996, the remote environs of northwestern Australia were again threatened by an even more powerful cyclone, Jacob. A Category 3 (on the Australian scale) system with sustained winds of 124 MPH (200 km/h), Jacob's leading gales and 32-foot (11-m) waves forced a large deep-sea drilling rig off the Pilbara Coast to halt its vital production of natural gas; otherwise, it spared the towns of Onslow and Exmouth a pulverizing direct strike. Recurving into the Indian Ocean at 15 MPH (25 km/h), Jacob blasted its way west, nearly reaching the island of Mauritius before dissipating on February 11. No fatalities or serious injuries were reported in its aftermath.

A coup within the hierarchy of intense Australian cyclones was successfully launched on the night of April 10, 1996, when Cyclone Olivia's central pressure of 27.75 inches (940 mb) released sustained 131-MPH (210-km/h) winds, 155-MPH (250-km/h) gusts, 4-inch (102-mm) rains, and 15-foot (5-m) breakers across the mouth of the Fortescue River, a quiet outlet to the Indian Ocean some 100 miles (161 km) northeast of Onslow. Judged by the Bureau of Meteorology to have been the second-most-powerful cyclone to have made a recorded landfall in Australia during the 20th century, Olivia's winds and rains ravaged much of Pilbara's southern coast, uprooting scores of trees, crumpling billboards, crushing automobiles, flattening entire houses, and running several dozen small pleasure boats ashore at Onslow and Karratha. Bested only by Cyclone Trixie, whose pressure reading of 27.16 inches (919 mb) brought 161-MPH (259-km/h) winds to the Pilbara fishing town of Mardie Station in March 1975, Olivia nevertheless wrought extensive damage on the ore-mining community of Pannawonica, a small company town located 60 miles (97 km), south of Dampier. Although Pannawonica's one-story, prairie-style houses were located well inland, Olivia's mowing winds and raking gusts had no difficulty in reducing dozens of them to rubble. As its fenced-in plots became littered with strips of tin sheeting, shards of window glass, and heaps of upended furniture, Pannawonica's police station, courthouse, and hospital were completely deroofed. Nearly every tree and telephone pole in the town was shorn off at ground level, while clutching rains and concomitant floods washed out roads, carried away bridges, and submerged nearby rail lines. Surprisingly, considering both Olivia's intensity and its destructive $50 million legacy, no human lives were reported lost in the tempest—a potent testament, perhaps, to the effectiveness of the warning and evacuation measures enacted in the hours preceding Olivia's near record landfall.

The 1996 tropical cyclone season in Australia concluded with two moderately powerful cyclones, Phil and Fergus, making landfall on the northern and eastern coasts of the continent in late December. During the late evening hours of December 27, Cyclone Phil swept across the Northern Territory's western coast, its sustained 94-MPH (150-km/h) winds uprooting trees and downing powerlines but causing no deaths or injuries in the sparsely inhabited region. Less than three days later, Cyclone Fergus, a Coral Sea upstart with sustained winds of 89 MPH (143 km/h), delivered a glancing blow to the port city of Cairns before abruptly veering to the northeast. Quickly recurving to the southeast, Fergus left Cairns wet but undamaged, as it sped ever deeper into the warm confines of the Coral Sea toward the neighboring nation of NEW ZEALAND. On December 30, several communities on New Zealand's North Island were placed on an emergency status as the downgraded remnants of Cyclone Fergus battered the picturesque island country. As thousands of vacationers packed their campsites and moved inland, Fergus's 50-MPH (81-km/h) gusts and three-inch (76-mm) downpours uprooted hundreds of palm trees, flooded highways, and knocked out electrical power to more than 20,000 people. Although no deaths or injuries were reported, Cyclone Fergus's dying passage over New Zealand tallied damage estimates of $5 million.

Australia's 1997 cyclone season opened with the appearance of another powerful Coral Sea cyclone, Justin. The largest cyclone observed off the Queensland coast since 1976's Cyclone David, Justin played an anxious, two-week waiting game with the residents of Australia's northeastern coast before finally pounding ashore in Queensland during the late-morning hours of March 24, 1997. Formed in the Coral Sea on March 7, Justin swiftly matured into a Category 2 (Australian scale) cyclone with sustained 81-MPH (130-km/h) winds, 104-MPH (167-km/h) gusts, and sweeping rains. Between March 8 and 12, Justin slowly menaced the eastern coast of North Queensland, its rain-drenched eyewall at times drifting to within 75 miles of landfall before an encroaching high-pressure dome to the northwest gently sent it spiraling back to sea. During its long offshore trek to the north, Justin's intensity slowly waned, its maximum sustained winds dropping below cyclone-strength on March 14. Briefly dubbed Tropical Storm Justin, the system glided to within 100 miles (161 km) of Papua New Guinea

before again changing its course and intensity. By the late-afternoon hours of March 16, as its forward trajectory shifted from nearly due north to the southwest, Tropical Storm Justin was upgraded to a Category 1 cyclone. Over the next two days, while it pursued a steady course for landfall in North Queensland, Cyclone Justin continued to strengthen over the Coral Sea's 86°F (30°C) waters. Graduated to the status of a Category 3 (Australian scale) cyclone on March 19, 1997, Justin's sustained 106-MPH (170-km/h) winds, threatened six-foot storm surge, and dousing rains compelled thousands of coastal residents to evacuate inland. And it was while the cyclone was at peak intensity that the two-masted wooden schooner *Queen Charlotte,* with two Australians and three New Zealanders aboard, was lost after being overwhelmed by the cyclone's towering seas. Chugging ashore between the port cities of Ingham and Cairns during the late-morning hours of March 24, 1997, Justin's weakened 78-MPH (125-km/h) winds destroyed banana, sugarcane, and pawpaw crops. The community of Townsville was struck by widespread flooding, while a large, $1.5 million marina at Cairns was splintered by Justin's 15-foot (5-m) waves. Near Townsville, a woman was killed when a landslide caused by Justin's torrential rains crushed her beneath a large boulder. In Innisfail, some 20 miles (32 km) north of Townsville, a young boy was electrocuted when a live wire dropped into the flood-waters in which he was struggling. While Justin did not succeed in postponing a joint military exercise between the United States and Australia, the expansive system did force 14 U.S. military aircraft from Guam to make a diversionary landing at Darwin, the rebuilt port city leveled during Cyclone Tracy's memorable assault in 1974, and 20 U.S. and Australian warships to quit their anchorage at Shoalwater Bay and seek the confineless safety of the open sea. All told, Cyclone Justin killed 11 people and caused damages estimates of $500 million.

In January 1970, Cyclone Ada caused severe damage in the Whitsunday Islands. Some 14 people were killed and damages were estimated at $390 million Australian dollars.

In 1971, Cyclone Althea struck Townsville and Magnetic Island in Queensland—the most powerful system to have struck the area since 1954. Cyclone Althea was closely followed by a cyclone that affected the Australian mainland for 17 days, traversing much of Australia's western coastline. The system first made landfall near Broome on February 18, where it sank small vessels and damaged banana plantations. Downgraded to a tropical depression, the 1971 cyclone caused heavy flooding in Kimberley and heavy stock losses. Fitzroy Crossing was isolated by record-setting floodwaters. On the 27th of the month, the cyclone emerged off the coast of Kimberley and regenerated into a tropical cyclone, causing coastal flooding and erosion in the Yampi Sound and Cape Leveque areas.

During Australia's 1974 Cyclone Season—the season which produced the legendary Cyclone Tracy—the continent was struck by Cyclone Wanda, which brought torrential rain and flooding to Brisbane. Some 5.1 inches (1,318 mm) of rain fell on Brisbane, causing flooding across one-third of the city's downtown center. Fifty-six houses were swept away and another 1,600 submerged. Nearly 10,000 people were left homeless, while 16 people died, and more than 300 were injured.

Despite its 160-year chronology of violent cyclone strikes, Australia has come to rely upon the depredations of cyclones for between 10 and 15 percent of its yearly rainfall needs. Because a moderately sized cyclone can deliver nearly 350 billion tons of precipitation in a single strike, Australia's verdant vineyards, sugarcane fields, and thriving deserts depend upon cyclones for their ongoing sustenance and renewal. Australia's two northern peninsulas, for instance, receive almost all of their annual rainfall during the five-month cyclone season, after which a seven-month drought characterized by searing sandstorms ensues. In 1995, Cyclone Bobby's 15-inch (381-mm) rains broke a prolonged drought over the Southwest Land Division, one which had seriously threatened the region's agricultural prosperity.

In an effort to quantify and record the size and strength of threatening cyclones, the Australian Meteorological Bureau has designed a five-point scale of cyclone severity. Similar in layout and purpose to the SAFFIR-SIMPSON SCALE (the system used to define the destructive potential of North Atlantic hurricanes), the Australian scale highlights the most salient aspects of cyclone formation, including wind speed, size of cloud cover, and respective barometric pressures.

In March 1999, Cyclone Vance produced the strongest wind gust yet recorded on the Australian mainland 166 MPH (267 km/h) at Learmonth, in Western Australia.

An intense Category 5 cyclone at landfall, Sam (also identified as TC 03S) caused heavy damage across northwestern Australia when it raced ashore near the town of Bidyadanga, approximately 25 miles south-southwest of La Grange, on December 8, 2000. Boasting sustained winds of 144 MPH (232 km/h) and gusts of 173 MPH (278 km/h), Sam caused severe damage to the main station homestead at Anna Plains, and uprooted hundreds of trees in and around Bidyadanga. Because Sam's landfall on

the Kimberley coastline was preceded by large EVAC-UATIONS, no deaths or serious injuries were reported. One of the most powerful cyclones to have affected Australia in several years, Sam dissipated over the Great Sandy Desert on December 10.

During the 2000–01 Australian cyclone season, two notable cyclones affected the continent-nation, Cyclone Terri on January 31, and Cyclone Alistair on April 24, 2001. Born from a tropical wave off the northern Kimberley coast on January 28, 2001, Cyclone Terri matured into a Category 2 cyclone before limping ashore, slightly weakened, on January 31, near the town of Pardoo. In Pardoo, wind speeds of 69 MPH (111 km/h) and gusts of 86 MPH (138 km/h) were observed, while rainfall in other parts of the region totaled less than four inches (100 mm). Late-season Tropical Cyclone Alistair stepped ashore near the town of Carnarvon on April 24, 2001. Formed over the Arafura Sea on April 17 and upgraded to a Category 2 cyclone on April 18, Alistair suffered considerable disruption of its circulation systems from WIND SHEAR before making landfall in Australia. Alistair's sustained winds, driven by a central pressure of 29.61 inches (1,003 mb), were clocked at 42 MPH (67 km/h), while other areas observed wind speeds of 68 MPH (110 km/h) and large rain fall counts, including 4.48 inches (114 mm) of precipitation in Ellavalla Station, and .94 inches (24 mm) in coastal Carnarvon. Although no deaths or serious injuries were reported, Cyclone Alistair did destroy between 30 percent and 40 percent of the crops on those plantations situated to the north of Carnarvon.

On March 9, 2005, an extremely large Cyclone Ingrid rolled ashore in Queensland, near the mouth of the Lockhart River. The most severe tropical system to have affected the province in more than 30 years, Ingrid's sustained 143-MPH (230-km/h) winds and the seven-foot (2-m) storm surge uprooted trees and caused beach erosion, but claimed no lives.

Just over one year later, on the morning of March 20, 2006, Cyclone Larry delivered a Category 5 punch to northeastern Queensland. A tightly coiled cyclone that had developed over the warm waters of the SOUTH PACIFIC OCEAN, due east of the coastal town of Innisfail, on March 18, Larry rapidly intensified as it swiftly churned due westward. Generating sustained winds of 120 MPH (190 km/h) and gusts up to 185 MPH (300 km/h) at landfall in Innisfail, Larry caused widespread damage to buildings and utility networks from Ingham in the south to Port Douglas in the north. More than 50 percent of the residences in Innisfail suffered roof or other structural damage, while some 80 percent of the buildings in Babinda, a hamlet located to the north of

Innisfail, were damaged or destroyed. Across the affected region, nearly 250,000 customers were left without electrical services. The nearby township of Doomadgee was flooded when the Nicholson River topped its banks, while nearly half a billion dollars (Australian) in losses were incurred by the region's banana and sugar producers. Because much of the vulnerable coastline population had been evacuated in advance of Larry's thundering landfall (two other cyclones had earlier struck northeastern Australia during the 2005–06 Australian cyclone season), no deaths were recorded in the wake of this remarkable tropical system. Some 30 people did, however, suffer minor injuries. As a downgraded Cyclone Larry swirled inland, thousands of people were left without shelter or basic services, prompting the Australian government, Emergency Management Australia, and the Queensland State Emergency Services Office to mount an enormous relief and response operation.

While still in the midst of recovery operations related to Cyclone Larry's destructive passage, northern Australia was affected by what many meteorologists described at the time as the most powerful cyclone yet observed in Australian waters. Born on April 17, 2006, over the heat-charged waters of the Coral Sea, Severe Tropical Cyclone Monica trudged across Cape York Peninsula on April 20, and while situated over the Gulf of Carpentaria, grew into a Category 5 meteorological bomb. On the afternoon of April 23, as Monica's eye passed less than 50 miles (31 km) off the Northern Territories, the system generated an exceedingly low central barometric pressure reading of 26.72 inches (905 mb), making it of equal intensity to Hurricane CAMILLE as it swept through the GULF OF MEXICO in August 1969. Wind speeds of 180 MPH (290 km/h) lashed the Wessel Islands, causing enormous seas that pummeled the coastline. The city of Cairns received nearly 22 inches (550 mm) of PRECIPITATION during Monica's offshore passage. Just as quickly as it had deepened, Severe Tropical Cyclone Monica weakened, becoming a Category 3 cyclone just before making landfall near Darwin on April 24. EVACUATIONS, coupled with the relatively sparse population density in the Northern Territory, mitigated Monica's human costs, and no lives were lost to the passage of this meteorologically significant system. In Australia, cyclones are tracked using a system of Tropical Cyclone Advices, including Cyclone Watches and Cyclone Warnings.

B

Backergunge Cyclone of 1876 *India, October 28–November 1, 1876* Viewed by a number of contemporary historians as the single-most-violent CYCLONE to have blasted across the BAY OF BENGAL during the 19th century, the Backergunge Cyclone delivered sustained 147-MPH (237-km/h) winds, flooding rains, and a remarkable STORM TIDE-STORM SURGE combination to the northeast coast of INDIA between October 31 and November 1, 1876. Some 100,000 people, many of them residents of the obliterated delta city of Backergunge (now Barisal, BANGLADESH), reportedly perished during the cyclone's initial landfall there, while an additional 100,000 people in surrounding regions soon succumbed to the dehydrating ravages of an ensuing cholera epidemic. A significant DEATH TOLL in a land of deadly cyclone strikes, the Backergunge Cyclone was by far the most disastrous TROPICAL CYCLONE to have affected India since Calcutta lost 300,000 residents to the colossal Hooghly River Cyclone of October 7, 1737.

The Backergunge Cyclone originated over the central Bay of Bengal, approximately 265 miles (426 km) northwest of the Andaman Islands, on October 28, 1876. Plotted in 1889 by cyclone scholar John Eliot, the storm's course took it steadily north-north-west, away from the islands and into the warm, open waters of the bay. There, fed by the strong south-westerly air currents that accompanied that summer's monsoon's retreat, the cyclone deepened very quickly. By the afternoon of October 29, a ship passing close to the cyclone's burgeoning EYE recorded a barometric pressure reading of 29.02 inches (982 mb), one that produced wind speeds of 74 MPH (119 km/h) or higher and considerable rains.

During the early morning hours of October 30, as 100-MPH (161-km/h) gusts lashed the middle of the Bay of Bengal, the Backergunge Cyclone slightly altered its course and began to head almost due north at approximately 13 MPH (21 km/h). Again encountering shipping later that afternoon, the cyclone chalked up a pressure reading of 28.00 inches (948 mb), elevating it to the status of a major system. As the storm's PRESSURE GRADIENT rapidly steepened and jacked its sustained winds to 128 MPH (206 km/h), its swirling storm surge—the dome of sea-water that it carried beneath its EYEWALL—began to loom ever larger. Though just a swell in deep water, the shoaling shores of the narrowing bay forced the cyclone's approaching 40-foot (12-m) surge to build on itself, cresting as friction with the seabed retarded its forward motion.

To the Backergunge storm surge's growth, however, was shortly added a further critical element, that of the high lunar tides normally experienced in the Bay of Bengal. On the morning of October 31, as the cyclone neared to within 150 miles (241 km) of India's northeastern coast, the first tide's ebb had just ended, and a 12-hour flow of water back into the bay began. Still moving due north at nearly 11 MPH (18 km/h), the cyclone's 40-mile-wide (48 km) surge inexorably swirled onward, compressing the rising tide against the walls of the Asian subcontinent. To the Burmese east and the Indian west, the tide that day rolled ashore as a series of enormous breakers, pounding waves that roared and hissed as they penetrated echoing coves and inlets.

But directly to the north, where the Megna River greeted the Bay of Bengal with a fragile delta fan, the situation was worsening by the hour. In the bayside anchorages of Backergunge, a city of jute and rice traders located 150 miles (241 km) due east of Calçutta, the tide came in as it should, but this time it did not go out. Rising throughout the afternoon of October 31, sometimes at a rate of two feet per hour, the tide eventually began to lap at the underside of piers, running as widening rivulets over the tops of the crude embankments that protected the prosperous city from the Megna's periodic floods. Hawser lines parted with a bang as heavily laden merchant ships suddenly rose on the tide, strained and listed at their piers until crew members cut them free with axes. Forced to beat to seaward against a stiffening southwest wind, many of the vessels were still clearing the mouth of the Megna River when the cyclone's surge came ashore some five hours later.

Shortly before dusk on October 31, as the leading gales of the oncoming cyclone began to raise whitecaps at the mouth of the Megna, British officials in Backergunge may have given rueful thought to the newly founded India Meteorological Department and the coastal telegraph network that the bureau's renowned director and cyclonic scholar Henry F. Blanford had recently constructed in Calcutta. Intended to one day link all of India's major ports, Blanford's highly publicized system would not only be used for the conveyance of meteorological data, but it would also serve as a warning device in the event of a tropical cyclone. Unfortunately, in 1876 the network did not yet extend as far east as Backergunge, leaving the doomed city dark and alone before one of the most cataclysmic meteorological events of the century.

Thundering ashore near the mouth of the river just before midnight on October 31, 1876, the immense cyclone quickly annihilated Backergunge, washing nearly all of its 1,700 buildings into the tide-swollen outlet. Adding nearly 40 feet (13 m) of water to already bloated anchorages, the cyclone's surge drove dozens of vessels—from swift tea clippers to jaunty paddle steamers—aground. Thousands of acres of low-lying plains were progressively submerged as the gaining surge held back the river, forced its watery thread to overflow its banks as far north as the town of Faridpur. More than 100 jute-farming communities along both sides of the Megna were swept away, plowed into blighted fields during the course of the night, while thousands of sacred cows were drowned, their decaying carcasses left to spread the cyclone's final contagion.

baguio A term used exclusively in the PHILIPPINES to describe the many TYPHOONS that strike the archipelago annually. Pronounced bag-YOU, the word is taken from the name of the Filipino town of Baguio, located some 80 miles (129 km) due north of Manila, where the United States built a major railway junction during its occupation of the country in the early 1900s. Although its exact derivation has since been lost, it is possible that the name's connection with the destructive effects of typhoons stems from an astounding instance in July of 1911, when no less than 88 inches (2,235 mm) of rain fell on the town of Baguio in four days. While in this particular case such an enormous PRECIPITATION count was not caused by the flooding strike of a TROPICAL CYCLONE, its tremendous rainfall did simulate those generally experienced in Filipino typhoons. It is also possible that adoption of the word *baguio*—or the name of the town constructed at the behest of the United States—as the name of a destructive being such as a typhoon indicates a cultural linkage between the two "leveling" entities of violent weather and the distant but seemingly omnipotent nation that during the past century had frequently embraced the Philippines in its sphere of influence.

Bahamas *North Atlantic Ocean* With some 2,700 limestone islands and shallow coral cays scattered across a 760-mile (1,223-km) range of the NORTH ATLANTIC OCEAN, the Bahamas has for centuries presented an expansive, low-lying landfall to incoming TROPICAL CYCLONES. Between 1528 and 2006, the group's six major islands—New Providence, Andros, Great Abaco, Grand Bahama, Long Island, and Inagua—have been directly affected by no less than 86 recorded HURRICANES and TROPICAL STORMS. In what is considered one of the most active counts in the entire Western Hemisphere, at least 51 of these were considered major hurricanes or storms that brought sustained winds in excess of 111 MPH (179 km/h) and BAROMETRIC PRESSURES below 28.47 inches (964 mb) to the archipelago's coral beaches, solitary hills, chic yacht marinas, and profitable surfside casinos. The Bahamas' position directly off the southeastern coast of FLORIDA makes it extremely susceptible to strikes from both westward-moving mid-Atlantic hurricanes that are undergoing RECURVATURE, or a turning to the north, and hurricanes that progress up from the CARIBBEAN to strike the gulf coast of the United States. As recently as 1992, the northwest coast of Andros Island was battered by the 120-MPH (190-km/h) winds and 18-foot (6-m) STORM SURGE associated with Hurricane ANDREW, then an intensely powerful Category 4 storm that later went on to devastate large portions of south Florida and LOUISIANA.

As with many of the other countries that ring the hurricane-prone Caribbean basin, the Bahamas has had its share of multiple strikes, as happened in 1837, when no less than three notable hurricanes and one tropical storm crossed the island chain between the months of July and October. The first of these hurricanes, which came ashore near Nassau on July 29, 1837, delivered scouring winds and driving rains to the port city for nearly 48 hours, causing significant structural damage to the Royal Naval dockyard located there. Just a few days later, on August 4–5, 1837, the second storm, known as the Los Angeles Hurricane, swept across Grand Bahama Island, bearing an enormous storm surge that washed away a beachhead and drowned several head of cattle. The third tropical system, which may have been of tropical storm strength, made a mid-September landfall in the central Bahamas, while the last of the quartet, the violent late-season hurricane of October 27, caused a great deal of damage to coastal shipping in the region of Long Island.

Less than a decade later, the fierce CUBAN HUR-RICANE OF 1844 claimed dozens—possibly hundreds—of lives as it passed over the Bahama Banks during the night of October 5. Completely taken by surprise by the fast-moving colossus, more than 100 small craft were caught offshore by its undulating seas and poor visibility. Collisions, dismastings, and swampings sank every one of them, resulting in one of Bahamian history's worst maritime tragedies. It would be another quarter-century at least before the GREAT BAHAMA HURRICANE OF 1866 exacted an equally severe toll when it slammed into Inagua Island on October 1. Touting a central pressure of 27.70 inches (938 mb) and 145-MPH (233-km/h) winds, the Great Bahama Hurricane launched a cresting 16-foot (6-m) storm onto the island's beaches, driving ships onto reefheads, smashing wharves, wracking buildings and trees to the ground, and flooding all major thoroughfares. Some 220 people were either pinned to death beneath the splintered wreckage of their collapsed houses or swept out to sea and never heard from again.

More recently, the Bahamas were struck by five hurricanes of varying intensity during the 1933 season. While the first four were considered to have been minor storms, those with barometric pressures of 28.47 inches (964 mb) and higher, the fifth achieved moderate intensity as it came ashore at Grand Bahama Island on November 3. Three people were killed, and numerous buildings were badly damaged. On September 6, 1965, Hurricane BETSY, one of the most vigorous Category 3 hurricanes on record, came ashore at Cat Island and then again at Nassau, bringing 146-MPH (235-km/h) winds and deluging rains on both sites. Millions of dollars in property damage to shipping and agriculture was

sustained. In August of 1988, Tropical Storm Chris took four lives and wrought a half-million dollars in damage as it crossed Andros Island on August 27. A moderate tropical storm, Chris's 50-MPH (81-km/h) winds and slight storm surge were but a hint of what would later accompany the gargantuan Hurricane Andrew as it rampaged over the island in August of 1992.

On the night of July 9–10, 1996, Hurricane BERTHA delivered hurricane force conditions to the eastern Bahamas. On it way to an eventual landfall in NORTH CAROLINA, Bertha's central pressure of 28.55 inches (967 mb) generated sustained 75-MPH (121-km/h) winds on San Salvador and Eleuthera, as well as torrential downpours and breaking surf.

The remarkable 2005 North Atlantic HURRICANE SEASON was an especially active one for the Bahamas. On July 21, 2005, Franklin delivered tropical storm conditions to the archipelago's northern islands. Dropping five inches (13 cm) of rain on Eleuthera and Great Abaco, Tropical Storm Franklin's sustained 45-MPH (72-km/h) winds lashed the popular tourist destination. At the time of its upgrade from tropical depression to tropical storm, Franklin was located about 100 miles (160 km) northeast of Nassau, the Bahamas, and was moving northward at 13 MPH (21 km/h).

One of the legendary hurricanes of history, KATRINA slowly strengthened as it passed over the Bahamas between August 23 and 25, 2005. A fairly disorganized tropical depression (TD 12) at the time it reached the northern Bahamas on August 23, northwestward-tracking Katrina matured into a tropical storm with wind speeds in excess of 40 MPH (64 km/h) on August 24, as it passed between the Bahamas islands. While WIND SHEAR hindered Tropical Storm Katrina's deepening cycle as its center passed beyond the Bahamas on August 25, the system continued to strengthen as it bore down on southeastern Florida, eventually coming ashore there as a Category 1 hurricane on August 25, 2005. Within days Katrina would grow into the most damaging hurricane yet observed in modern U.S. history.

On October 24, 2005, two months to the day after Tropical Storm Katrina romped through the Bahamas, an equally infamous Hurricane WILMA battered the islands with sustained winds of 96 MPH (154 km/h), heavy surf conditions and sweeping rains, as it soared northward at 29 MPH (46 km/h). A Category 3 hurricane with a central barometric pressure of 28.17 inches (954 mb), Grand Bahama Island bore the brunt of Wilma's speedy passage, with trees uprooted, utilities disrupted, and several small BUILDINGS damaged or destroyed. No deaths or injuries were reported, however.

Bangladesh *Southern Asia* Commonly referred to as "the nation born of a cyclone," Bangladesh has routinely sustained some of history's most catastrophic CYCLONE strikes. Formerly known as East Pakistan, this low-lying country of alluvial plains and expansive tea and jute plantations is on average affected by at least one TROPICAL CYCLONE per year. In some instances, such as the 1965 and 1970 cyclone seasons, two or more storms have come ashore with deadly results in a single year. Bangladesh's location on the northern mandible of the storm-ridden BAY OF BENGAL makes it dangerously susceptible to those early and late-summer cyclones that accompany the Asiatic monsoon's changing seasons. Swept out of the bay by the prevailing south-southeasterly winds, these cyclones, with CENTRAL BAROMETRIC PRESSURES of less than 28.47 inclics (964 mb), winds in excess of 111 MPH (179 km/h), and nine to 12 foot (3–4 m) STORM SURGES, frequently bring flooding rains and staggering DEATH TOLLS to the lowland stretch of the Ganges River delta. The narrowing shores of the Bay of Bengal—a body of water already beset by abnormally high lunar tides—funnel the storm surges of approaching cyclones into bulging rams of water, towering inundations at periodically sweep over the dozens of barrier islands that clog the country's southeast coast. Despite the fact that not one of these islands—known as *chars*—rises more than 12 feet (4 m) above sea level, Bangladesh's acute land shortage, coupled with its burgeoning population growth, makes it necessary for hundreds of thousands of people—principally fishers—to inhabit densely what would otherwise be regarded as hazardous sandbanks.

This unfortunate practice has long contributed to the astronomical death tolls often incurred by Bangladeshi cyclones. On October 16, 1942, a powerful cyclone raised a 15-foot (5-m) storm surge along the state's southwest coast, killing an estimated 40,000 people. During a similar cyclone on June 2, 1956, 199 people were lost when a ferry was driven ashore near the river town of Barisal, while another 20,000 were left homeless. Four years later, in October 1960, an even more intense cyclone drowned at least 5,000 people when its storm surge—estimated by survivors to have been some 20 feet (7 m) high—overwhelmed the shallow islands of Ramgati and Hatia. On May 28, 1963, a cyclone that came ashore near Comilla claimed the lives of another 22,000 people and left three times that number without homes. Just over two years later, another stunning series of three cyclones brought vast suffering to the country. The first struck on May 11–12, 1965, drowned 30,000 people in the Noakhali district. The second cyclone, which roared up from the Bay of Bengal less

than a month later, took another 10,000 lives in the neighboring district of Barisal. The third Bangladeshi cyclone of the 1965 season made landfall during the third week of December and added 10,000 more victims to the season's tally. Such record death tolls in Bangladesh would not be seen again until 1970, when two memorable cyclones, those of October 23 and November 12, added several hundred thousand more dead to the growing list. More recently, the cyclone of May 25, 1985, killed 10,000 people in the Noakhali, Sonagazi, and Chittagong districts.

On April 29, 1991, Bangladesh suffered its most catastrophic cyclone strike since the Great Cyclone of 1970, when Cyclone 02B tore into the Chittagong region as an extremely powerful Category 4 cyclone with sustained winds of 160 MPH (260 km/h). At one point a Category 5 cyclone with a record-setting central barometric pressure of 26.51 inches (898 mb), 02B was born on April 22, from a tropical depression over the Andaman Sea. The embryonic system steadily intensified as it first tracked westward, then northwestward, across the Bay of Bengal. As it neared the mouths of the Ganges River, Cyclone 02B grew in size until its entire outflow canopy covered the bay's northern reaches. Slashing ashore to the southeast of the Bangladeshi capital, Dhaka, Cyclone 02B raised a 20-foot (7-m) storm surge along the coast, while prying winds destroyed thousands of buildings, leaving more than 1 million people homeless. In the port city of Chittagong (which bore the brunt of 02B's winds), a towering dockside crane weighing tens of tons was sent hurtling into a nearby bridge, isolating large sections of the city. Dozens of aircraft belonging to the Bangladeshi Air Force were destroyed on the ground, while floodwaters inundated the country's principal naval facilities at Patenga. As damage estimates topped $1.5 billion, Cyclone 02B's death toll steadily escalated, finally reaching 138,000 people. It remains one of the top 10-highest tropical cyclone death tolls on record. On May 2, 1994, an intense cyclone boasting winds of up to 155 MPH (250 kph) came ashore near the seaside resort of Cox's Bazar. Although nearly 50,000 houses were torn apart by the storm's crushing winds, only 255 people lost their lives. On May 8, 1996, Tropical Storm 01B originated over the Bay of Bengal before recurving to the northeast and striking Chittagong with 46-MPH (74-km/h) winds. Just over five months later, on October 29, 1996, a more powerful tropical system—Tropical Storm 06B—moved north across the Bay of Bengal to trundle ashore at the mouths of the Ganges River. Central Bangladesh experienced sustained winds of 52 MPH (84 km/h) and heavy rains.

The 1997 Bay of Bengal cyclone season brought Bangladesh its most severe cyclone strike in years when Cyclone 01B slammed into central Bangladesh on May 19 as a Category 4 cyclone. Sustained winds of 132 MPH (212 km/h) and pounding surf conditions caused considerable damage along the mouths of the Ganges River, and scores of fatalities were reported. Later that season, Bangladesh was affected by a Category 1 cyclone, 02B, on September 26, 1997. This system, which crashed ashore near Chittagong with 75-MPH (121-km/h) winds, had earlier brushed the eastern coast of India as a tropical depression before recurving to the northeast and traveling along the Bay of Bengal coast before striking Bangladesh.

During the active 1998 Bay of Bengal cyclone season, Bangladesh was directly affected by two tropical cyclones. On May 19, 1998, Cyclone 01B delivered a dose of 81-MPH (130-km/h) winds and rains to the port city of Chittagong. Of minimal Category 1 strength at landfall, 01B had earlier (as a tropical depression) tapped the northern coast of Sri Lanka and the eastern coast of India, before recurving into the Bay of Bengal and slowly intensifying, becoming a mature cyclone directly before landfall in eastern Bangladesh. The second cyclone, 07B, originated over the Gulf of Thailand and as a tropical depression crossed the Malaysian peninsula and recurved into the Bay of Bengal on November 17, 1998. Intensifying into a tropical storm while situated over the central Bay of Bengal, 07B, tracked to the northeast, growing stronger along the way. Shortly before its forecasted landfall in Bangladesh, the system was of Category 1 intensity with sustained winds of 86 MPH (138 km/h), but as it neared the Asian subcontinent, the system rapidly weakened. When it flushed ashore at Chittagong on November 22, 1998, 07B was of tropical storm intensity with sustained winds of 63 MPH (101 km/h).

While Bangladesh enjoyed a reprieve in tropical cyclone activity during the 1999 Bay of Bengal cyclone season, it narrowly missed a direct strike by one of the most intense and deadly cyclones yet observed in the basin. Between October 25 (when it originated near the Malaysian peninsula) and October 29 (when it came ashore in India's Orissa district), Cyclone 05B churned to the northwest, maturing into a Category 5 over the nourishing waters of the Andaman Sea. While at any point the system could have recurved to the north and come ashore in Bangladesh as a dreadful reprise of the 1991 cyclone (02B), it closely followed its northwesterly track until making landfall near the city of Bhubaneswar. Bearing winds in excess of 161 MPH (259 km/h), Cyclone 05B killed nearly 10,000 people in India, but caused only minimal coastal flooding along the mouths of the Ganges and the Bangladeshi coastline.

During the 2000 and 2005 Bay of Bengal tropical cyclone seasons, Bangladesh was directly affected by three tropical cyclones. The first, on October 28, 2000, was of minimal tropical storm intensity as it rolled ashore near the Ganges, bearing sustained winds of 40 MPH (64 km/h). The second, Tropical Storm 03B, was a late-season system that deluged Bangladesh with rain on November 12, 2002. Bearing a central pressure of 29.05 inches (984 mb), and downgraded before landfall to a tropical depression, 03B's 63-MPH (101-km/h) winds nonetheless caused extensive damage as they crossed the mouths of the Ganges River. The third system, Tropical Storm 01B, brought a central pressure of 28.93 inches (980 mb) and 61-MPH (98-km/h) winds to Chittagong, on May 19, 2003. No deaths or injuries were reported.

Once part of what was then known as East Pakistan, Bangladesh gained its independence from that nation in December 1971 shortly after the cataclysmic GREAT CYCLONE of November 12–13, 1970, inundated several offshore islands, killing between 300,000 and 500,000 people. The slow pace of Pakistani-led relief efforts combined with subsequent allegations of rampant corruption and willful neglect within its emergency services soon served to catapult both nations into a vicious 10-month civil war, one that resulted in the formation of a sovereign but no less vulnerable Bangladesh a year later.

Barbados *Eastern Caribbean Sea* This flat, relatively dry island, left virtually deforested by more than three centuries of sugarcane production, has an extensive and frequently devastating history of TROPICAL CYCLONE activity. Situated in the CARIBBEAN SEA, Barbados is the easternmost member of the broad Windward Islands group. The archipelago—composed in part of MARTINIQUE, St. Lucia, and St. Vincent—is so positioned between latitudes 12 and 15 degrees North as to place it directly in the path of the mid-to late-season hurricanes that routinely whirl across the NORTH ATLANTIC OCEAN from Africa's CAPE VERDE Islands. Between 1666 and 2006, Barbados's white sand beaches, cheerfully painted villages, and spacious pastures were directly affected by 55 documented HURRICANES and TROPICAL STORMS. Of these, 23 were considered major hurricanes, storms with CENTRAL BAROMETRIC PRESSURES of 28.47 inches (964 mb) and lower and winds in excess of 111 MPH (179 km/h). On at least five occasions between 1786 and 1837, the fertile island weathered two or more hurricane strikes in quick succession: on October 10, 1780, when a considerably less-intense hurricane came ashore three weeks after the GREAT HURRICANE OF 1780 ravaged the country; in September 1818, when two very destructive storms made landfall within one week's time; in 1835, when two hurricanes came ashore in July and

September; and during the 1786 and 1837 seasons, when three hurricanes struck the island.

There is no record of Barbados having been affected by a June hurricane. Although it is possible that a rogue but now-forgotten June hurricane did once riddle the densely populated island, most early season storms develop in the distant reaches of the southwest Caribbean or lower GULF OF MEXICO and then progress rather weakly to the northeast, where the FLORIDA Panhandle awaits their lukewarm violence. In July, however, as the Sun-seeded waters of the Caribbean reach the critical 80–84°F (27–29°C) temperature threshold required for hurricane origination, the number of recorded strikes in Barbados portentously rises to six. The first of these July hurricanes occurred in "The Year of the Devil," 1666, and sank a heavily gunned British warship near Ragged Point on the eastern coast of the island. A heavy loss of life resulted. On July 29, 1757, an intense hurricane moved through the passage between Barbados and St. Vincent, bringing damaging winds and rains to the island's southeast shore. Contemporary reports state that at least eight ships were wrecked on the many coral reefs that spike the entrance to Carlisle Bay. During the twin 1837 hurricanes—the first coming ashore between July 9–10, the other on July 26—widespread destruction of the island's many houses, sugar mills, and granaries caused lingering food shortages and several hundred deaths.

But it is during the rainy months of August, September, and October, when the first of the Cape Verde hurricanes arrives at the Caribbean's eastern portal, that the number of strikes in Barbados leaps to a 15, 17, and 10, respectively. Notable midseason hurricanes in Barbados include the August 10, 1674, hurricane in which three hundred houses, eight ships, and more than two hundred people were destroyed. In late September 1694, an unwieldy convoy of 26 British merchant ships, laden with sugarcane, was overtaken by a furious hurricane at exposed Carlisle Bay. Despite a heroic effort by the ships' crews to save the vessels by jettisoning their valuable cargo and cannon, the entire assemblage was nonetheless sent to the bottom or driven onto the sandy shore by the hurricane's hounding winds. An estimated 1,000 sailors were lost. During a similar hurricane in October 1749, several British merchant ships were run ashore near the capital city of Bridgetown, claiming the lives of 230 men. A deadly hurricane in August 1765 inundated Barbados's northeastern beachheads with a 12-foot (4-m) STORM SURGE that swept away an entire village on North Point; an even greater surge during the height of the legendary Great Hurricane of October 10, 1780, demolished several hundred buildings that ringed Bridgetown harbor. During the

same storm, an army transport, two navy victuallers, and an ordnance vessel were driven from their anchorage and run ashore near the entrance to Carlisle Bay. Quickly dismasted by the Great Hurricane's shearing winds, the vessels hulls' wildly rocked in the thundering surf, opening up their seams, and eventually breaking their backs. More than 200 sailors were reportedly lost. Farther inland, both the hospital and the barracks for the island's military garrison were destroyed, forcing the frantic, rainstung soldiers to seek shelter wherever they could find it. On the ramparts of the imposing fortress that guarded the Bridgetown Harbor entrance, the Great Hurricane's heaving winds carried a 12-pound cannon from one side of the battery to the other, a distance of some 420 feet. In addition, the city's stone prison was first stripped of its sturdy roof, and then its walls collapsed, requiring the evacuation of more than 800 prisoners—some of them French and Spanish prisoners of war—into the debris-strewn fury of the storm. More than 700 people perished in the capital alone, while the estimated DEATH·TOLL across the entire island would eventually exceed 4,000. Barbados would later suffer an equally destructive hurricane strike on August 10–11, 1831, when more than 1,500 people perished and approximately $7 million in property damage was assessed.

More recently, Barbados was soundly pummeled by the 10th tropical cyclone of the very active 1955 season, Hurricane JANET. Between September 22 and 23, Janet brought 125-MPH (201-km/h) winds and nearly 12 inches (305 mm) of PRECIPITATION to the island's hillside communities. Suffocating mudslides and coursing flash floods claimed 24 lives and left another 22,000 people homeless. On September 14, 1995, the 80-MPH (129-km/h) winds and heavy rains associated with Hurricane Marilyn damaged two dozen homes, downed power lines, and uprooted scores of trees on the island but claimed no fatalities. On September 25, 2002, Tropical Storm Lili delivered 75-MPH (121-km/h) gusts to Barbados, but did not make a direct strike on the island.

Barbara, Hurricane *Caribbean Sea–Eastern United States, August 11–16, 1953* The first North Atlantic tropical system to be identified as Barbara under the CYCLONE naming system instituted by the U.S. Weather Bureau in 1953, Barbara spent several days in mid-August 1953, delivering Category 1 winds, rains, and seas to NORTH CAROLINA's Outer Banks. Formed over the northern Leeward Islands on August 11, Barbara deepened into a Category 2 HURRICANE with sustained winds of 109 MPH (175 km/h) before recurving to the north-northeast and grazing eastern North Carolina on August 14. Moving to the north-northeast at nearly 15 MPH (24 km/h), the system drew very close to the NEW ENGLAND coastline, causing heavy surf conditions along Cape Cod's beaches. No deaths or injuries were recorded as Hurricane Barbara transitioned into an extratropical system on August 15.

Barbara, Tropical Storm *Gulf of Mexico–Southern United States, July 27–30, 1954* A much weaker successor to its 1953 namesake, the second and final North Atlantic TROPICAL CYCLONE dubbed Barbara formed over the GULF OF MEXICO on July 27, 1954, but never deepened beyond a relatively weak TROPICAL STORM. Tracking to the northwest, Tropical Storm Barbara trundled ashore in LOUISIANA and TEXAS on July 29, bringing with it 40-MPH (64-km/h) winds and some rain.

While the name Barbara has not officially been retired, it is not in present use on the NATIONAL HURRICANE CENTER'S (NHC) North Atlantic tropical cyclone naming list. It was replaced on the 1955 list with Brenda.

bar of a tropical cyclone Also referred to as the bar of a HURRICANE, the bar of a TROPICAL CYCLONE is the gray-black wall of CUMULUS and CUMULONIMBUS CLOUDS that appears on the horizon in connection with the approach and passage of a mature-stage tropical system. Described in 1697 by the British adventurer WILLIAM DAMPIER and coined during the 19th century to describe the appearance from a distance of a tropical cyclone's EYEWALL over the ocean, the bar is preceded by the onset of high-level cirrus and CIRROSTRATUS CLOUDS, wispy collections of ice crystals that seem to the viewer to converge on the horizon. In turn, the advancing edge of the OUTFLOW LAYER slowly gives way to a thicker canopy of cumulus clouds, grayish-white tufts that fringe the lower reaches of the bar and closely resemble the shape and texture of an enormous mushroom cloud. Contingent upon the intensity of a respective system, the bar can range in color from light gray to indigo to black (a result of the vast quantities of moisture contained within the lower cloud towers of the eyewall) and may assume shades of yellow, red, green, and even purple depending upon the time of day it is observed. Numerous eyewitnesses have noted the vivid sunrises and sunsets that often precede a tropical cyclone's landfall, the prismatic result of the sun's orange rays being filtered through the system's thin outflow shield; and later, as the bar draws closer, through the successive layers of moisture that darken its core.

barometer An instrument used to measure BAROMETRIC PRESSURE. First described in 1643 by Florentine philosopher and mathematician Evangelista Torricelli, the barometer has in the past four centuries become established as an exceptionally valuable tool in the study and forecasting of TROPICAL CYCLONES. Initially intended as an investigation into the physical characteristics of a vacuum, Torricelli's primitive MERCURIAL BAROMETER not only served as a prototype for progressively more sophisticated types of barometers, but also revolutionized the science of meteorology by providing weather watchers with a quantifiable means of observing that most critical of climatological factors, barometric pressure. From the simple glass tubes that Torricelli filled with mercury one August morning, to the cumbersome wheel barometers designed by Robert Hooke in 1673, to the portable ANEROID BAROMETERS and MICROBAROGRAPHS in general use today, the barometer's versatile design has given meteorologists the keys with which to unlock many of the tropical cyclone's most destructive and deadly secrets. Now regularly deployed on board HURRICANE HUNTING aircraft and meteorological buoys, the barometer aids forecasters in predicting a tropical cyclone's track and probable intensity at landfall by producing data regarding its CENTRAL BAROMETRIC PRESSURE, its PRESSURE GRADIENTS, and its INVERSE BAROMETER.

barometric pressure Often referred to as atmospheric pressure, barometric pressure is the weight per unit area of the entire mass of air above a certain point of Earth's surface. Decreasing with altitude and measured in inches, MILLIBARS (mb) and kiloPascals (kPa) by an instrument called a BAROMETER, the gradations of barometric pressure are based on a bar, or the unit of pressure that at sea level equals the pressure of 29.53 inches (1,000 mb) of mercury. The result of differences in air TEMPERATURE and humidity levels within individual AIR MASSES, barometric pressure forms one of the major mechanical processes by which the world's weather systems are formed and sustained. For instance, an area of high atmospheric pressure (>29.60 inches or 1,002 mb)— a *high*—generally fosters those calm, dry conditions most frequently associated with ANTICYCLONES, while a *low*—an area of low barometric pressure, >29.53 inches (1,000 mb)—gives Earth its rain squalls, gales, blizzards, EXTRATROPICAL CYCLONES, mesocyclones, and TROPICAL CYCLONES.

In a tropical cyclone, where the mechanical principles of atmospheric pressure are further applied to the behavior of the system's PRESSURE GRADIENTS, INVERSE BAROMETER, and CENTRAL BAROMETRIC PRESSURE, the drop in barometric pressure begins slowly at first and then quickens as the storm's point of lowest pressure draws closer to the observing barometer. Concurrently, WIND SPEEDS caused by the system's steepening pressure gradient sharply increase, while WIND SHEAR zones within the EYEWALL may spawn TORNADO-like MINISWIRLS of considerable destructiveness. Survivors of some of history's most severe tropical cyclones have spoken of how abrupt changes in barometric pressure made their eardrums pop and their eyes involuntarily fill with tears.

The substantial changes in barometric pressure that accompany a tropical cyclone have been linked to recorded occurrences of EARTHQUAKE activity during landfall. Under normal atmospheric conditions, sea-level barometric pressure exerts approximately 2,160 pounds per square foot of pressure on Earth's surface, or 14.7 pounds per square inch. For every inch (33 mb) that the barometric pressure falls, some 70 pounds per square foot of weight is released. When multiplied across the 200-mile-wide (322-km) girth of an average HURRICANE or TYPHOON, the decrease in downward thrust on Earth's crust is measured in the millions of tons, thereby allowing for the possibility of shifting between nearby faults. This seismic phenomenon has been particularly pronounced in JAPAN and the PHILIPPINES, both of which lie in active earthquake zones that are subject to an extensive number of tropical cyclone strikes.

Barometric pressure in tropical cyclones is measured through a number of different technological systems, including dropsondes, barometers, satellite radiometers, and satellite estimates based on the Dvorak Scale (see DVORAK TECHNIQUE) as well as experienced intuition. In the United States, barometric pressure—in particular, the central barometric pressure of tropical cyclones—is measured in either millibars (mb; or mbar), inches (Hg), millimeters (mm) and hectopascals (hPa). Barometric pressures for Super Typhoon Tip, for instance, can be rendered as 25.69 inches or Hg; 870 mb or mbar; of 87.0 hectopascals, or hPa.

Barry, Hurricane *Southern United States–Gulf of Mexico–Mexico, August 23–29, 1983* Between August 23 and 29, 1983, Hurricane Barry tracked nearly due westward from its origination point off the eastern coast of the BAHAMAS, striking the central eastern coast of FLORIDA as a TROPICAL STORM, then passing into the GULF OF MEXICO, where it was downgraded to a TROPICAL DEPRESSION. As it moved westward across the warming waters of the Gulf, Tropical Depression Barry slowly reintensified, becoming a tropical storm again on August 28, and a HURRICANE with sustained winds of 81 MPH (130 km/h) just prior to making landfall on the MEXICO-TEXAS border. A strengthening tropical system at

landfall, Hurricane Barry produced a central pressure reading of 29.11 inches (986 mb) making it a relatively powerful Category 1 system. Tracking due west into northern Mexico, Barry dissipated by August 30.

Barry, Tropical Storm *North Atlantic Ocean, July 9–14, 1989* During a five-day period in July 1989, Tropical Storm Barry disrupted shipping traffic as it roiled the waters of the mid-NORTH ATLANTIC OCEAN. Formed on July 9 far to the east of the Leeward Islands, Barry spent over half of its existence as a weak and disorganized TROPICAL DEPRESSION as it steadily tracked to the northwest. Upgraded to TROPICAL STORM intensity, the system at its peak generated a minimum pressure reading of 29.67 inches (1,005 mb) and sustained winds of 52 MPH (84 km/h). Hampered by WIND SHEAR and poor circulation patterns, Barry dissipated to the northeast of BERMUDA on July 14, 1989.

Barry, Tropical Storm *Southern United States, August 2–8, 2001* The third North Atlantic tropical system named Barry originated in the eastern GULF OF MEXICO on August 2, 2001. A powerful TROPICAL STORM with sustained winds of 69 MPH (111 km/h) and a minimum pressure of 29.23 inches (990 mb) at eventual landfall, Barry tracked an erratic trajectory to the north-northwest, doubled-back on its course, moved almost due north, and finally limped ashore on the western tip of the FLORIDA Panhandle. Weakening into a TROPICAL DEPRESSION as it entered ALABAMA, Barry's remnants persisted as far north as Illinois before dissipating on August 8.

Barry, Tropical Storm *Southern United States, June 1–2, 2007* As though inspired by its 2001 predecessor, the fourth North Atlantic tropical system dubbed Barry originated in the eastern GULF OF MEXICO (near latitude 24 degrees North and longitude 85 degrees west) on June 1, 2007. A poorly organized early season system with maximum sustained winds of 52 MPH (83 km/h), Barry tracked to the northeast and made landfall near FLORIDA's Tampa–St. Petersburg during the daylight hours of June 2. Its minimum central barometric pressure of 29.44 inches (997 mb) at landfall produced precipitation counts in excess of six inches (152 mm) across much of northern Florida. After being downgraded to a TROPICAL DEPRESSION immediately upon landfall, Barry's low-pressure center moved across northern Florida and trailed northward along the eastern U.S. seaboard, dropping tremendous quantities of moisture in SOUTH and NORTH CAROLINA, VIRGINIA, and Pennsylvania. In New York City and across southern

NEW ENGLAND, a now-EXTRATROPICAL Barry delivered gale-force gusts and some three to four inches (76–102 mm) of precipitation, causing some localized flooding. Although more than seven inches (179 mm) of rainfall was recorded in CUBA, and another eight inches (203 mm) over Georgia, no direct deaths were reported in Barry's wake. Indeed, Barry's early season rains fortuitously broke a drought condition over northern Florida and helped firefighters battle an extensive wildfire in central Georgia.

The name *Barry* has been retained on the North Atlantic lists and is scheduled to reappear during the 2013 hurricane season.

Bay of Bengal Roughly situated between the equator and the tropic of Cancer (23.5 degrees North), the expansive, V-shaped Bay of Bengal stands as the northeasternmost extension of the Indian Ocean. Bordered to the southwest by the island of Sri Lanka, to the west and northwest by INDIA, to the north by BANGLADESH, and to the north-northeast by Myanmar (Burma), the Bay of Bengal's position within the INTERTROPICAL CONVERGENCE ZONE (ITCZ) has long made it one of the most active spots on Earth for CYCLONE generation. On average, eight percent of the world's TROPICAL CYCLONES originate over the warm waters of the bay in any given year. Between 1737 and 2006, some 742 recorded cyclones were formed; this represents an average of three storms annually. While cyclones have occurred in the Bay of Bengal during every month of the year, statistics indicate that peak periods for cyclone generation come in July and September, or during those periods when the Asiatic monsoon begins its advance and retreat. Furthermore, a number of these storms have been of significant intensity and duration, posing a grave hazard both to the shipping that regularly plies the bay and to those nations—in particular India and Bangladesh—that ring it.

Nearly three centuries of comprehensive study into the character of the Bay of Bengal's frequent cyclones have increasingly revealed a cyclical connection between such storms and the ever-changing seasons of the monsoon. Prompted by the sharp temperature and pressure differences between the landmass of the Indian subcontinent and the large bodies of water that surround it, the early summer arrival of the monsoon is characterized by both an abrupt reversal in the direction of the prevailing winds and the rapid formation of tropical cyclones. For this reason, cyclone generation in the Bay of Bengal tends to be most common during the change of seasons, or when the prevailing winds of May and June first shift to the north and flow up from the Indian Ocean and into the hot, humid regions of Bengal and northwest India.

Similarly, the retreat of the monsoon, which generally occurs between September and November, brings with it another peak in cyclone origination, one noted for its predilection toward violent cyclone strikes. Indeed, some of the most destructive cyclones on record have come ashore from the Bay of Bengal during September, October, and November, including the FALSE POINT CYCLONE of September 21, 1885, the HOOGHLY RIVER CYCLONE of October 7, 1737, the GREAT CALCUTTA CYCLONE of October 5, 1864, the BACKERGUNGE CYCLONE of October 31, 1876, and the GREAT CYCLONE OF 1970. Although DEATH-TOLL estimates from these storms vary from source to source, their cumulative losses can be accurately measured in terms of hundreds of thousands of people either left dead or missing.

A majority of early season Bay of Bengal cyclones tend to originate in the northern half of the bay during the months of June and July. These storms, which are usually of smaller size and intensity than those late-season cyclones that develop south of the 16th parallel, almost always move in a northwesterly direction, a course that takes them ashore on India's northeast coast or in the vicinity of the major port cities of Calcutta and Visakhapatnam. Although early season Bay of Bengal cyclones do not as a rule prove as destructive as their late-season cousins, they do remain better organized for a longer period of time, bringing huge PRECIPITATION counts and gusty winds to the inland regions of the Ganges River Valley. On the other hand, late-season cyclones in the Bay of Bengal, those storms that form between September and November, are generally of greater size and intensity at landfall because of the long distances over which they must travel. Such cyclones, which originate in either the south or southeast quadrants of the bay, are steadily carried to the north or northwest by the vast influx of moist air that sweeps up from the west-southwest as part of the retreating summer monsoon. Powered by both their energy-laden STEERING CURRENTS and the warm 86°F (30°C) waters over which they pass, these Bay of Bengal cyclones often grow into storms of enormous destructive potential. With CENTRAL BAROMETRIC PRESSURES below 27.91 inches (945 mb), winds in excess of 155 MPH (250 km/h), and record-breaking STORM SURGES, these cyclones frequently slam into India's southeast beaches or inundate the low-lying south coast of Bangladesh, with catastrophic results for the indigenous populations who make their lives along the shores of the Bay of Bengal.

Bay St. Louis Hurricane of 1819 *Southern United States, July 27–28, 1819* Touted as one of the most destructive TROPICAL CYCLONES to have affected the fledgling United States during the first half of the 19th century, the Bay St. Louis Hurricane unleashed pulverizing winds and a deadly five to six foot STORM SURGE on the coasts of ALABAMA, LOUISIANA, and MISSISSIPPI between July 27 and 28, 1819. In Mississippi, where the tightly coiled EYE of the 24-hour HURRICANE made landfall just before midnight on September 28, nearly every house, warehouse, and wharf along the banks of Bay St. Louis was destroyed. Eyewitness accounts state that the entire coastline from Pass Christian, Mississippi, to Mobile, Alabama, was littered with the remains of shattered buildings, uprooted fences, and snapped trees. Dozens of human bodies, along with the carcasses of several hundred head of cattle, festered on the beaches, in the swamped bayous, in the piles of wreckage that in some places formed an almost impenetrable wall. In Alabama, where the hurricane's broad surge was funneled inland across Mobile Bay's narrowing shores, deadly alligators, snakes, and snapping turtles were washed into the city's clogged streets, causing a number of gruesome fatalities. Several vessels—from small boats to 60-ton brigs—were driven ashore by the enormous six to ten foot (2–3 m) surge, one of them coming to rest with its bowsprit piercing the side of a quayside warehouse on Dauphin Street. At Cat Island, located 11 miles (18 km) to the southeast of Bay St. Louis, 39 sailors from United States sloop-of-war *Firebrand* were killed when the burgeoning storm first capsized their swift, 12-gun craft and then deposited it, overturned, on the shore. Farther inland, in the midst of the pine forests that blanketed Mississippi's midlands, the Bay St. Louis Hurricane overtook a contingent of U.S. soldiers who had encamped for the night in a shallow valley. First surprised, then terrified by the sudden tempest of wind and flooding rains that fell upon them during the evening hours of July 27, the soldiers fled to higher ground, into woods where trees were falling like nine-pins. At least one man was killed and twenty others seriously wounded as the hurricane shredded tents, scattered provisions, and showered the stricken party with arrowlike splinters. In Louisiana, where the hurricane's track first took it ashore just west of the lower Mississippi Delta, minimal but sustained winds buffeted New Orleans, running three ships aground at the mouth of the river and flooding many farms along the rims of lakes Borgne and Pontchartrain. Before dissipating over the cooling reaches of the continental United States on July 29, the Bay St. Louis Hurricane had claimed between 100 and 175 lives and caused untold thousands of dollars in property damage to the Mississippi Sound's budding trade and industry.

Belle, Hurricane *Eastern United States, August 8–10, 1976* A relatively mild Category 1 HURRI-

CANE whose 90-MPH (145-km/h) winds and torrential PRECIPITATION at landfall nonetheless claimed at least 12 lives and caused an estimated $23.5 million in property damage to communities in NORTH CAROLINA, New Jersey, NEW YORK, and NEW ENGLAND between August 9 and 10, 1976.

A midseason TROPICAL CYCLONE that had developed off the northeast coast of the BAHAMAS during the afternoon of August 8, 1976, Belle briskly intensified as it first underwent RECURVATURE and then sped up the United States's eastern seaboard on August 9. Aerial reconnaissance flights conducted by the NATIONAL HURRICANE CENTER (NHC) indicated that Belle, with a CENTRAL BAROMETRIC PRESSURE of 28.47 inches (964 mb), was packing sustained winds of 111 MPH (179 km/h) and gusts of 125 MPH (201 km/h) as it brushed past North Carolina's Outer Banks shortly after dawn on August 9. The state was fortuitously spared a direct strike by Belle's Category 3 winds and minimal four-foot (1.5-m) STORM SURGE; however, five people were drowned when the van they were traveling in was washed into a shallow gorge by one of the hurricane's rain-driven flash floods. Aside from a few isolated power outages and some minor beach erosion, material damage in North Carolina from Hurricane Belle's glancing passage was on the whole considered fairly moderate.

Quickly progressing up the coast in a wavering north-northeasterly direction, Belle's extensive spiral rain bands next grazed portions of New Jersey's eastern shoreline. In Atlantic City, the hurricane's showering winds and torrential rains brought with them a heavy surf, one that relentlessly pounded the resort community's famed wooden Boardwalk, clientless hotels, and clifflike stone jetties. More than 150,000 homes in New Jersey, some as far inland as Trenton, were left without electrical power. Three people were reportedly killed in storm-related automobile accidents, and damage estimates in Atlantic City alone totaled some $4.5 million.

Shortly after midnight on August 10, 1976, Belle's sprawling EYE finally came ashore on the south coast of New York's Long Island. Although the NORTH ATLANTIC OCEAN's cooling waters had by this time reduced the hurricane to Category 1 status, its sustained 90-MPH (145-km/h) winds managed to uproot hundreds of trees, knock out traffic lights and road signs, smash windows, and demolish several beachside cottages on the heavily settled island. Downed power lines sapped more than 275,000 homes of electricity, while isolated flood action washed out a small bridge near the Fire Island Inlet. One fatality—a man who was struck by a falling tree—was reported on the island. Damage estimates ranged from $8–9 million, making it the most costly hurricane to strike Long Island since Hurricane CAROL came ashore at Westhampton in August 1954.

With its wind and rain now sharply diminished, Belle spiraled inland, passing over southern and central New England during the evening hours of August 10. No casualties were reported in Connecticut, where the storm's gale-force winds scattered nearly 20 percent of that year's vital apple harvest like grapeshot, causing upward of $4 million in agricultural losses to the state. In Vermont's green hills, however, Belle's lingering rains spawned serious flash floods: In one tragic instance, a mother and her seven-year-old son drowned when the footbridge they were traversing collapsed, pitching them into the raging torrent below. A further $2.5 million in property damage was tallied in Vermont, where the dissipating storm's trampling effects on timber forests and dairy farms were judged to have been among the worst in living memory.

Belize *Central America* Thirty-one people perished on October 8, 2001, when a powerful but deadly Hurricane IRIS collided with the eastern coast of Belize. A compact system with a very small EYE-WALL, Iris blithely steamed across the north Caribbean basin between October 4 and 7, steadily gaining in intensity, and bound for a crushing landfall in Central America as a Category 4 system. Iris's CENTRAL BAROMETRIC PRESSURE of 27.99 inches (948 mb) at landfall generated sustained winds in excess of 140 MPH (225 km/h) and very high surf conditions; BUILDINGS were unroofed, trees uprooted or snapped, and many small craft run aground. Owing to the severity of Iris's assault on Belize, the name Iris was retired from the rotating list of North Atlantic TROPICAL CYCLONE identifiers.

Bermuda *North Atlantic Ocean* This hospitable collection of 150 coral islands and cays, renowned for its resorts, manicured hedgerows, pink houses, and shimmering horseshoe beaches, has endured an extensive history of notable TROPICAL CYCLONE strikes. Positioned in the NORTH ATLANTIC OCEAN some 1,000 miles (1,609 km) due east of Savannah, GEORGIA, Bermuda sits squarely in the path of those mid- to late-season HURRICANES and TROPICAL STORMS that either spiral across the North Atlantic from Africa's Cape Verde Islands or originate over the south and east CARIBBEAN SEA's warm waters before recurving to the northeast. Between 1551 and 2006, Bermuda's lush tropical forests, hillside roads, and turquoise bays were directly affected by 66 documented tropical storms and hurricanes. Of these, 24 were judged to be major hurricanes, or systems with

CENTRAL BAROMETRIC PRESSURES of 28.47 inches (964 mb) or lower and sustained winds in excess of 111 MPH (179 km/h). On at least four occasions between 1813 and 1894, Bermuda withstood several hurricane strikes in rapid succession. In 1837, two powerful storms came ashore between August 25 and October 3, while in 1845, three hurricanes made landfall in just over two weeks' time. Another double strike occurred during the 1894 HURRICANE SEASON when two late-season hurricanes battered the islands on October 19 and November 7.

Because most early-season hurricanes form in the southwestern Caribbean or lower GULF OF MEXICO before moving northeast, only one documented June hurricane has ever beset Bermuda. This particular hurricane, which occurred between June 3 and 6, 1832, was considered a fairly moderate storm by contemporary sources, one whose lukewarm winds and rains did little damage to the island's fledgling settlements.

In July, however, as the southeast Caribbean's waters begin to reach the critical 80–84°F (27–29°C) threshold required for hurricane origination, the likelihood of a hurricane landfall in Bermuda rises slightly, with three reported strikes between 1609 and 2006. The first recorded July hurricane in Bermudian annals, which occurred on July 25, 1609, was in fact just that; one of nine English ships en route to Jamestown, Virginia, the *Sea Venture,* was run aground at Bermuda and wrecked by the hurricane's tempestuous winds. Fifty-four survivors from the *Sea Venture,* including the squadron's commander, Admiral Sir George Somers, subsequently founded the island's first permanent settlement, thereby commencing Bermuda's long, illustrious habitation.

As though in deference to the 1609 hurricane's historic significance, two centuries passed before another two July hurricanes struck Bermuda, both during the 1813 season. Coming ashore within a week of each other, the first storm reportedly killed four people on July 20, while the second claimed six lives on July 26. In both instances, the capital city of Hamilton bore the brunt of the hurricanes' wrath, although a majority of the victims were on ships at sea.

But it is during the hot, humid months of August, September, and October that the first of the CAPE VERDE STORMS begins to recurve through the eastern Caribbean before traveling up the United States's eastern seaboard then the number of confirmed strikes climbs to a hefty 17, 11, and 19, respectively. The earliest midseason hurricanes in Bermudian history were experienced by the Spanish, whose treasure-carting convoys often found themselves cast ashore on one of Bermuda's many coral reefs by

CLIPPER SHIP "COMET" OF NEW YORK.

In this dramatic Currier and Ives print produced in 1855, the New York–built clipper ship *Comet* is seen partially dismasted, its sails tattered, as it attempts to navigate through "a hurricane off Bermuda on her voyage from New York to San Francisco in October 1852." According to existing records, the "hurricane" that the *Comet* encountered in October of 1852 was in reality a powerful tropical storm as it passed well north of the island on October 11. *(NOAA)*

their roaring sea-lane rivals. In August 1551, two vessels belonging to the Tierra Firme Armada were driven onto a reef near the island's southeastern tip. Although more than 200,000 gold ducats were later recovered by their crews, both ships became total wrecks. In September 1584, another Spanish convoy, this one comprised of 51 caravels, galleons, and naos, was overtaken by an intense hurricane while rounding Bermuda's southwest capes. Six treasure ships were dismasted, foundered, and then capsized by the storm's gargantuan seas, claiming the lives of nearly 500 sailors. Memorable 17th- and 18th-century hurricanes in Bermuda include the fanning fury of an August 16, 1669, hurricane that sank a large merchant ship at anchor in Castle Harbour and the colossal GREAT HURRICANE OF 1780, in which more than 50 vessels were thrown ashore all over the main island.

During the War of 1812, while Bermudian privateers boldly waged a futile campaign against U.S. shipping, a blustery hurricane came ashore at St. George's Harbour on August 1, 1814. More than 60 seized vessels lying at anchor were either sunk or run aground, beaten into driftwood by the hurricane's crushing seas. Some 45 sailors were

drowned, while the storm's toll in prize money was steep enough to raise a concerted call for the Royal Navy to develop better-sheltered anchorages if it expected Bermuda to serve as an effective military outpost.

Slightly more than four years later, on August 28, 1818, a violent, 110 MPH (177 km/h) hurricane drove the U.S. merchant ship *Hope* ashore on Bermuda's west coast near Wreck Hill, with the loss of seven crew members. Dozens of recently constructed BUILDINGS on the headland were blown down, while 15-foot (5 m) seas smashed people wharves and dockyards to splinters. Twenty people were killed.

During the second week of September 1836, an energetic hurricane dealt Great Britain's efforts to improve Bermuda's northeast harbors a considerable setback when it breached a large breakwater in three places. Tens of thousands of dollars worth of damage was inflicted on the ambitious project, while a number of small workboats and floating derricks foundered. So severe was the hurricane's destructive force that work on a nearby dockyard was not resumed for another 11 years while the British reconsidered Bermuda's hurricane threat.

Just more than three years later, on September 12, 1839, an even more intense hurricane again ravaged Bermuda's south coast. One of the first Cape Verde hurricanes of the season, the Great Bermuda Hurricane's central pressure reading of 28.30 inches (958 mb) brought 125-MPH (201-km/h) winds, monstrous rains, and an 11-foot storm surge to Port Royal and Castle Harbour. Several small vessels were wrecked, while hundreds of houses were unroofed. An estimated 100 people were killed.

Another August hurricane, this one in 1843, caused enormous suffering all over the island when its 102-MPH (164-km/h) winds came ashore near Hamilton on August 18. Hundreds of trees were uprooted, while two of Bermuda's distinctive catchments—the large, open tanks used to alleviate the island's notorious water shortages—collapsed during the hurricane's furious onslaught. An outbreak of yellow fever that claimed 114 lives soon followed blasting rains and treacherous flash floods.

During the latter half of the 19th century, a major hurricane on October 13, 1875, caused the 2,642-ton British warship *Irresistible* to nearly capsize at its mooring, while a devastating storm in late September 1878 badly damaged the Royal Navy's floating drydock at Grassy Bay. Christened the *Bermuda,* much of the celebrated drydock (touted at the time as one of the largest in the Western Hemisphere) was stove in, despite being flooded beforehand in order to prevent it from being blown ashore. Nine years later, the hurricane of August 19, 1887, sloshed past the northwestern pincer of Great Sound, its 11-foot (4-m) storm surge completely isolating Ireland Island from the rest of the country for several days. An estimated 22 people lost their lives to the offshore passage of this system, considered one of the most intense in Bermudian history.

Bermuda has also sustained strikes from at least seven November hurricanes, late-season furies that originate in the eastern Caribbean Sea before tracking to the northeast. While a majority of November hurricanes at Bermuda have been storms of minor intensity, one, on November 26, 1883, was sufficiently powerful to cast the 7,323-ton British warship *Northampton* ashore at Stagg Rocks. Although badly holed, the vessel was later refloated and returned to service. Its commanding officer, however, was court-martialed and subsequently beached.

During the second decade of the 20th century, two Category 1 hurricanes, storms with steady winds of between 74 and 95 MPH (119–153 km/h), struck Bermuda, inflicting heavy damage on many of the island's neighborhoods. On September 3, 1915, an 84-MPH (135-km/h) whirlwind sank a number of small boats and yachts at Grassy Bay, while another, on September

23, 1916, obliterated the Royal Navy's headquarters in Bermuda. Several lives were lost in each storm.

On September 21, 1922, a furious Category 3 hurricane with sustained 120-MPH (193-km/h) winds and 25-foot (8 m) seas slammed across the northeast portion of the island. Dozens of watercraft were wrecked as the storm, judged to have been one of the most violent in Bermuda's history, unroofed nearly 50 percent of the island's residences. Passing into the North Atlantic on the morning of September 22, the hurricane soon crossed the path of the 54,000-ton Cunard liner *Aquitania,* en route from Southampton, England, to New York. Bombarding the 990-foot (330-m) ship with pelting rains and 55-foot (30-m) waves, the hurricane shattered several portholes, twisted deckrails, and left five passengers with minor injuries.

On the afternoon of October 22, 1926, one of the most rapacious of all Bermudian hurricanes, the HAVANA-BERMUDA HURRICANE transited the western reaches of the island. Bearing sustained winds of 136 MPH (219 km/h), this memorable Category 4 system caused two British light cruisers, the *Calcutta* and the *Valerian,* to break away from their Ireland Island piers. Blown across the anchorage, the *Calcutta* noisily slammed against a breakwater and received some damage to its superstructure. The *Valerian,* on the other hand, drifted into the roiling confines of Staggs Channel and promptly capsized. Eighty-eight of the 109 officers and men aboard the cruiser were drowned, making the *Valerian*'s loss one of the worst disasters in the Royal Navy's history.

More recently, Bermuda was delivered a surprise blow by the fifth storm of the 1987 hurricane season, Emily. Abruptly changing direction as it rolled past the island on September 17, Hurricane Emily's 92-MPH (148-km/h) winds downed power lines, uprooted trees, flooded marinas, and planed at least 40 beachfront houses down to their foundations. Although some $35 million in property claims were filed, no casualties were reported. During the remarkable 1995 season, Hurricane Felix, one of the largest North Atlantic tropical cyclones on record, lashed southwest Bermuda with 85-MPH (137-km/h) winds and 2 inches of rain as its EYE passed 40 miles (64 km) south of the island on August 15. Postponing two days a planned referendum on independence from Great Britain, Felix's $2 million assault reminded the self-governing colony of one of the many costly hazards associated with complete independence, thus ensuring that the referendum ultimately failed by a margin of 3-1.

On September 20–21, 1999, the golden beaches of eastern Bermuda were pounded by 13-foot waves and gale-force winds generated by Hurricane Gert, as the fading Category 2 hurricane sped eastward past the island. At one time a mid-Atlantic behemoth with sustained winds of 150 MPH (241 km/h), Gert's diminished, 105-MPH (169-km/h) gusts and hammering seas demolished a shoreside restaurant, severely damaged a beachfront house, and destroyed a picturesque series of natural stone arches.

Bermuda's international airport and heavily visited cruise ship terminals were closed on September 5, 2003, as a very powerful Hurricane FABIAN passed to the west of Bermuda, its central pressure of 27.72 inches (939 mb), sustained 100-MPH (161-km/h) winds, and five to 10 inch (127–254 mm) rainfalls causing considerable damage along the island's northern coast. The most powerful tropical cyclone to have affected Bermuda since the Havana-Bermuda Hurricane of 1926, and the most destructive tropical system since 1987's Hurricane Emily, Fabian sped past Bermuda at 17 MPH (27 km/h), killing four people, and destroying a section of the Causeway, a long bridge that connected the island to its airport and an historic center, St. George's.

Less than one week later, the residual moisture from former Tropical Storm Henri delivered severe thunderstorms to western Bermuda, inundating the island's principal airport with some 2.44 inches (62 mm) of PRECIPITATION.

Bermuda High

The name given to the high-pressure ANTICYCLONE that dominates wind patterns over the NORTH ATLANTIC OCEAN. Influenced by the progression of the seasons, the clockwise-spinning Bermuda High serves to determine the tracks or trajectories of Atlantic HURRICANES as they move westward across the ocean's subtropical reaches. During the winter months when air and water temperatures in the Northern Hemisphere are low, the Bermuda High is relatively small and is positioned in the ocean's southeast quadrant. But during the summer months of July through September when water temperatures in the region are much warmer, the Bermuda High strengthens considerably, growing to encompass the entire center of the North Atlantic. In this position it influences a host of meteorological factors, including the STEERING CURRENTS of hurricanes. Because the Bermuda High is an area of settled high-pressure, low-pressure TROPICAL CYCLONES cannot encroach upon it. Instead, they must either progress along its southernmost flank and undergo RECURVATURE as they round the anticyclone's western edge, or else move up from the Caribbean Sea and continue with the northeasterly curl until it guides them into the North Atlantic's cooling spaces. If during a particular season the Bermuda High is of notably large size and is situated farther to the west than is normal,

that year's crop of hurricanes will be more likely to make landfall in the GULF OF MEXICO or the United States's eastern shores. Conversely, if the Bermuda High is less intense than usual, a number of Atlantic hurricanes can find their way to the west coast of EUROPE. Such an occurrence was seen during the 1966 HURRICANE SEASON, when Hurricane Faith brought 100-MPH (161-km/h) winds to Norway, and during the 1987 season, when TROPICAL STORM Arlene delivered flooding rains to Portugal.

Bertha, Tropical Storm *Southern United States, August 8–11, 1957*

Between 1957 and 2002, five North Atlantic TROPICAL CYCLONES were identified with the name Bertha. Three were of tropical cyclone intensity, and two were mature-stage HURRICANES, one a major hurricane.

The first North Atlantic tropical cyclone christened Bertha originated in the GULF OF MEXICO and tracked to the northwest before rumbling ashore on the LOUISIANA-TEXAS border on August 10, 1957. An intense tropical storm at landfall, Bertha's central pressure of 29.47 inches (998 mb) produced 69-MPH (111-km/h) winds and heavy rains. Penetrating inland, Bertha was downgraded to a TROPICAL DEPRESSION before dissipating on August 11, 1957, over the arid reaches of northern Texas.

Bertha, Tropical Storm *North Atlantic Ocean, August 30–September 4, 1984*

On August 30, 1984, Tropical Storm Bertha formed over the southeastern NORTH ATLANTIC OCEAN. Like most TROPICAL CYCLONES that originate south of 10 degrees North, the TROPICAL DEPRESSION that would eventually be upgraded and named Tropical Storm Bertha experienced some initial difficulty in organizing its complex circulation system. Meteorologists who have studied this phenomenon attribute it to the diminishing intensity of the CORIOLIS EFFECT, which diminishes the closer an object is to the equator. Bertha, which steadily moved to the northwest and away from the equator, did not intensify until it reached 15 degrees North, at which point it was upgraded to a TROPICAL STORM. At its peak, Bertha generated a central barometric pressure of 29.73 inches (1,007 mb) and sustained winds of 40 MPH (64 km/h), making it a very weak system. WIND SHEAR and cooler SEA-SURFACE TEMPERATURES again downgraded Tropical Storm Bertha to tropical depression intensity, and by September 4, 1984, Bertha had dissipated over the mid-North Atlantic Ocean.

Bertha, Hurricane *North Atlantic Ocean, July 24–August 2, 1990*

The 1990 North Atlantic HURRICANE SEASON featured yet another Hurricane

Bertha, a relatively weak Category 1 system that menaced the eastern North Atlantic basin between July 24 and August 2. Originating off the coast of NORTH CAROLINA on July 24, Bertha first tracked to the southeast, then to the southwest before executing a sharp loop and moving (almost over its previous track) to the north-northeast. At its height, Bertha generated a CENTRAL BAROMETRIC PRESSURE reading of 28.73 inches (973 mb) and sustained winds of 81 MPH (130 km/h) as it sped into the NORTH ATLANTIC, well away from the eastern U.S. seaboard.

Bertha, Hurricane *North Atlantic Ocean–Puerto Rico–Eastern United States, July 5–17, 1996* Of the five North Atlantic TROPICAL CYCLONES so far dubbed Bertha, the fourth was by far the most severe. The second named tropical system of the 1996 North Atlantic HURRICANE SEASON, Bertha began as a Cape Verde-bred TROPICAL DEPRESSION, south of 10 degrees North, on July 5. Slowly treading to the west-northwest, Bertha reached TROPICAL STORM intensity as it approached the northern Leeward Islands, and hurricane status as it neared PUERTO RICO. Deepening as it moved away from the equator, Bertha's central pressure dropped to 28.34 inches (960 mb), boosting its sustained wind speeds to 115 MPH (185 km/h). On the islands of ANTIGUA, Barbuda, Nevis, and St. Martin—as well as throughout the American and British VIRGIN ISLANDS—hurricane force winds generated by Bertha's EYEWALL unroofed small buildings, uprooted trees, destroyed power lines, and threw small watercraft ashore. One person was killed in St. Martin, another in Puerto Rico. Continuing its recurving track to the west-northwest, Bertha slowly grew larger in size, but weakened in intensity as it barreled past the BAHAMAS, some 75 miles (121 km) to the northeast, on the night of July 9–10. Sustained winds of 75 MPH (121 km/h) were observed on Eleuthera and San Salvador islands, and several small boats were cast ashore, but no deaths or serious injuries were reported. Its forward speed reduced from 23 MPH (37 km/h) to nine MPH (14 km/h), an enormous but disorganized Bertha slogged ashore in NORTH CAROLINA, between Wrightsville and Topsail Beaches, on July 12, 1996. A Category 2 system at landfall, Bertha's central pressure of 28.76 inches (974 mb) produced surface-level winds of 92 MPH (148 km/h), while the system's heavy surf and four to six foot (1–2 m) STORM SURGE caused widespread shoreline erosion, particularly along North Carolina's Outer Banks. Scores of beachfront structures in Topsail Beach were swept into kindling wood. Telephone poles and roadways in Wrightsville were essentially erased. As Bertha ground northward along the Carolina coast and entered VIRGINIA, as much as eight inches (203 mm) of PRECIPITATION fell across

the region, causing flash flood conditions in several highland communities. A total of seven TORNADOES were spawned by Bertha's constituent thunderstorms, including five in Virginia, one in North Carolina, and one in Maryland. Between landfall on July 12 and its final extratropical dissipation over northern NEW ENGLAND on July 14, a downgraded Tropical Storm Bertha sped to the northeast, along the northeastern U.S. coastline, dropping as much as five inches (127 mm) of rain and delivering 52-MPH (84-km/h) gusts to areas as far north as Maine. In New York City, Tropical Storm Bertha's low-lying cloud bands drifted northward over the city during much of the day on July 14, dropping more than three inches (76 mm) of rain, generating a central pressure reading of 29.38 inches (995 mb), and shrouding Manhattan's skyscraping peaks from view. A destructive early-season tropical cyclone, Bertha directly and indirectly claimed a total of 12 lives—eight of them in the United States—and caused some $250 million in property losses.

Bertha, Tropical Storm *Gulf of Mexico–Southern United States, August 4–10, 2002* The fifth and most recent incarnation for Bertha appeared on August 4, 2002, as a TROPICAL STORM that had originated from a strengthening TROPICAL DEPRESSION over the northern GULF OF MEXICO. Within one day of its upgrade, the system made landfall near New Orleans, LOUISIANA, bearing sustained, 35-MPH (56-km/h) winds and torrential, 10-inch (254-mm) rainfall counts. In the town of Slidell, Louisiana, some six inches (152.4 mm) of rain fell in less than 48 hours, causing severe flash flooding. Pushed to the southwest by an advancing high-pressure area, Tropical Depression Bertha returned to the Gulf of Mexico, where it reintensified to a tropical storm, eventually making a second landfall—this time in TEXAS—on August 9, 2002. Swirling inland over Texas, the system quickly dissipated.

The name Bertha has been retained on the North Atlantic basin naming lists and is scheduled to reappear during the 2008 HURRICANE SEASON.

Beryl, Tropical Storm *North Atlantic Ocean, August 28–September 6, 1982* At its peak a very powerful TROPICAL STORM, Tropical Storm Beryl remained over the waters of the mid-North Atlantic for its entire existence during the 1982 HURRICANE SEASON. Upgraded very quickly from TROPICAL DEPRESSION to tropical storm status on August 28, Beryl steadily intensified as it moved to the west-northwest over the mid-North Atlantic. It reached its maximum intensity on September 1, 1982, when its CENTRAL BAROMETRIC PRESSURE slipped to 29.20 inches (989 mb) and its sustained wind speeds topped 69 MPH (111 km/h). Pushed southward by the high-pressure BERMUDA HIGH, Tropical Storm Beryl's course shifted to due west, and the system was shortly downgraded to tropical depression status. Steadily weakening as it approached the northern islands of the CARIBBEAN SEA, Beryl dissipated to the northeast of PUERTO RICO on September 6, 1982.

Beryl, Tropical Storm *Gulf of Mexico–Southern United States, August 8–10, 1988* A short-lived TROPICAL CYCLONE that originated over the GULF OF MEXICO, Beryl achieved TROPICAL STORM intensity right before trundling ashore in extreme southeastern LOUISIANA on August 9, 1988. A CENTRAL BAROMETRIC PRESSURE of 29.55 inches (1,001 mb) and 52-MPH (84-km/h) winds swept across Louisiana as the system deeply recurved into the state, but no deaths or major damage was reported. By August 10, 1988, Tropical Storm Beryl had dissipated over north-central Louisiana.

Beryl, Tropical Storm *Mexico, August 13–15, 2000* Originating over the northern confines of MEXICO's Bay of Campeche (in approximately the same area that Hurricane BRET [1999] and Tropical Storm BARRY [2001] underwent cyclogenesis), Tropical Storm Beryl sprang into existence on August 13, 2000, from a dense collection of intense thunderstorms. Driven to the northwest by adjoining air masses, Beryl became a TROPICAL STORM on August 14, as its CENTRAL BAROMETRIC PRESSURE slipped to 29.73 inches (1,007 mb) and its sustained winds topped 52 MPH (84 km/h). It remained at tropical storm intensity until making landfall during the early morning hours of August 15, 2000, approximately 50 miles (80 km) south of the Mexico-TEXAS border. Moving inland over northern Mexico, Beryl dropped large amounts of rain before it dissipated on August 15, 2000.

Beryl, Tropical Storm *North Atlantic Ocean–Northeastern United States, July 18–20, 2006* The fourth North Atlantic TROPICAL CYCLONE identified as Beryl originated off the coast of NORTH CAROLINA on the morning of July 18, 2006. Initially dubbed Tropical Depression No. 2 (TD 2), the system slowly moved to the north-northeast, prompting the posting of TROPICAL STORM watches for the North Carolina coastline. While these watches were later dropped, the system continued to intensify, becoming Tropical Storm Beryl during the early morning hours of July 19, and steadily strengthening thereafter. As of 11:00 P.M. EDT on July 19, Beryl was a powerful tropical storm, with a CENTRAL BAROMETRIC PRESSURE of 29.58 inches (1,002 mb) and sustained wind speeds of 60 MPH (95 km/h). As Beryl drew away from

Carolina's Outer Banks and headed to the north-northeast at nine MPH (15 km/h), tropical storm watches were posted for southeastern Massachusetts, including Nantucket Island and Martha's Vineyard.

The name Beryl is scheduled to reappear during the 2012 North Atlantic HURRICANE SEASON.

Beta, Hurricane *Caribbean Sea–Central America, October 26–30, 2005* A record-setting TROPICAL CYCLONE of many degrees, Hurricane Beta was the first North Atlantic tropical system to be given a "B" identifier from the Greek alphabet. It was the 23rd named tropical system—and the 14th mature-stage HURRICANE—to develop during the historic 2005 North Atlantic HURRICANE SEASON. It also, on October 30, 2005, became one of a handful of tropical systems to make a direct landfall on the eastern coast of Nicaragua. A late-season tropical cyclone, Beta was born on October 27, 2005, from a TROPICAL DEPRESSION that had lingered in the southwestern CARIBBEAN SEA for several days before intensifying. Moving to the north, and then abruptly to the northeast, the system now a TROPICAL STORM with a central pressure of 29.20 inches (989 mb) deepened into a Category 1 hurricane on October 29, while still located less than 100 miles (161 km) off Central America's famed "Mosquito Coast." Between the mid-morning hours of August 29 and the early morning hours of August 30, Beta's CENTRAL BAROMETRIC PRESSURE slipped from 29.14 inches (987 mb) to 28.34 inches (960 mb), boasting its sustained wind speeds from 75 MPH (121 km/h) to 115 MPH (185 km/h) in just over a 24-hour period. On October 30, as Beta's gusts were clocked at nearly 140 MPH (225 km/h), it became the seventh major hurricane to develop during the 2005 hurricane season. Fortunately for those interests along the Costa Rican-Nicaraguan coastlines, Beta began to weaken as it turned due west, and then southwestward, and was of powerful Category 2 intensity as it came ashore in central Nicaragua, near the small coastal town of Sandy Bay, on October 30, 2005. A central pressure at landfall of 28.49 inches (965 mb) produced sustained wind speeds of 109 MPH (175 km/h), which uprooted trees, sank small watercraft, and caused extensive structural damage to small BUILDINGS and harbor facilities. A slow-moving hurricane, Beta's heavy rains caused several flash flood conditions in Nicaragua and Honduras. Dozens of people were injured on the offshore island of Providencia, while several injuries in Honduras and Nicaragua were also reported. Quickly downgraded to a tropical storm as it moved inland over Nicaragua, by October 30 Beta had dissipated. While Hurricane Beta was a powerful hurricane at landfall, local emergency management authorities attributed the lack of deaths to a combination of preparedness measures, most important among them the many evacuations that preceded the storm's landfall, and the fact that the system came ashore in a sparsely inhabited section of the coastline.

Because the identifier Beta is only used when the standard A–W naming list for a particular North Atlantic season is exhausted, it remains in use on the Greek alphabet tropical cyclone naming list.

Betsy, Hurricane *Bahamas–Southern United States, August 22–September 10, 1965* One of the most powerful Category 4 HURRICANES on record, Betsy blazed an unusual loop-the-loop track of death and destruction through the central BAHAMAS, south FLORIDA, and the gulf coasts of both LOUISIANA and MISSISSIPPI between August 27 and September 9, 1965. With CENTRAL BAROMETRIC PRESSURES at landfall ranging from 27.61 inches (935 mb) in the Bahamas, to 28.14 inches (953 mb) in Florida, to 27.99 inches (948 mb) in Louisiana, Betsy's 147-MPH (237-km/h) winds, flooding rains, and 6-foot (2-m) STORM SURGE wrought an estimated $1.4 billion in property damage, making it one of the most costly hurricanes to have affected the United States up to that time. Seventy-five people, a majority of them in Louisiana, lost their lives to the terrifying depredations of this unpredictable midseason storm.

The second TROPICAL CYCLONE of the rather tame 1965 HURRICANE SEASON, Betsy originated in the Windward Islands, 350 miles (563 km) east-southeast of BARBADOS, on August 27. Moving northwest at just over eight MPH (13 km/h), the embryonic hurricane slowly but inexorably gained strength from the 82 degree F (28°C) waters of the east CARIBBEAN SEA. On August 28, Betsy's TROPICAL STORM-force winds lashed the east coast of GUADELOUPE, driving a minimal one-foot storm surge into the harbor at Pointe-à-Pitre, but claiming no lives on the picturesque French dependency. Firmly committed to its northwest course, Betsy continued to intensify during the night of August 29, with a minimum pressure of 29.73 inches (1,007 mb) being recorded by aircraft reconnaissance just before dawn on August 30. However, during the early morning hours of August 31, while Betsy's EYE was located 275 miles (443 km). northeast of PUERTO RICO, the hurricane came under the steering influence of an enormous high-pressure ridge that had settled in over eastern United States. Now boasting a central pressure of 29.23 inches (990 mb) and sustained winds of 83 MPH (134 km/h), Betsy began to execute a sharp clockwise loop, first moving northeast, east, and southeast, before setting off for the southwest by

midafternoon on September 1. The hurricane's barometric pressure continued to fall during its course change, with an aerial reading of 28.93 inches (980 mb) being taken just before dusk on September 1.

Forced southeast by the encroaching high-pressure ridge, the swiftly intensifying Betsy continued to move in that direction throughout the afternoon and evening hours of September 1. With Betsy's central pressure rapidly approaching 28.10 inches (951 mb), civil defense officials in Haiti and JAMAICA prepared their respective organizations for the possibility of a destructive strike from the approaching hurricane. While there were indications that Betsy would eventually return to its initial northwest course once it had cleared the high-pressure ridge, both nations—each with its own history of deadly hurricane landfalls—stood waiting to evacuate their low-lying coastline communities, issue flash-flood warnings, and close down their airports. As it was, Betsy's expected course change came shortly before midnight on September 1 when the hurricane began to recurve to the northwest, returning it to its original course but nearly 500 miles to the west. Still intensifying, Betsy's central pressure dropped to 27.81 inches (942 mb) by noon on September 2, boosting its winds to a dangerous 110 MPH (177 km/h). Evacuation concerns now shifted to the Bahamas, the shallow basin of islands that lie off the east coast of Florida, where beachside resort hotels and sumptuous casinos were cleared of patrons and staff alike and then secured against Betsy's winds and rains with storm shutters and sandbag barricades.

With its forward speed reduced to a pedestrian four MPH (6 km/h), Betsy languidly moved northwest, skirting the east coast of Turks and Caicos Islands during the early morning hours of September 2. The squat islands were rigorously assailed by 130-MPH (209-km/h) winds and five inches (127 mm) of rain, uprooting coconut palms and telephone poles but causing no fatalities. On September 3, just as Betsy's roiling eye passed over Cat Island, the hurricane achieved its lowest recorded barometric pressure, 27.61 inches (946 mb). Cataracts of rain showered the sandy key, washing out bridges and roadways. Two small reef-diving boats, equipped with glass bottoms for undersea viewing, were driven ashore by the storm's bulging surge and quickly broken up. Now located 430 miles (692 km) south of Cape Hatteras, NORTH CAROLINA and bound for the Bahamian capital of Nassau, Betsy again stalled on September 4 as the western edge of the mounting high-pressure ridge continued to sag to the south. In an almost mirror-image of its earlier maneuver, the hurricane's eye performed another loop-the-loop course change. First swinging to the north, then to the north east, and then back toward the south again, Betsy pirouetted throughout the day of September 5, coming to rest shortly after midnight on a southwesterly tack that, if maintained, would take the hurricane's furious eye ashore somewhere in the vicinity of Havana, Cuba, on either September 7 or 8.

Now regarded as one of the most unpredictable hurricanes to have moved through the Caribbean in several years, Betsy maintained its southwesterly progression over the central Bahamas and toward Florida's southeast coast between September 6 and 7, 1965. During this time, the storm experienced the barometric vagaries generally associated with erratic hurricane behavior as its central pressure rose from 27.93 inches (946 mb) on September 4, to 28.11 inches (952 mb) on September 5, to 28.52 inches (966 mb) on September 6. By the morning of September 7, as the hurricane began to swing due west—in the process, threatening the southernmost tip of the Floridian peninsula—its barometric pressure again began to drop, reaching 28.26 inches (957 mb) by midafternoon. This ominous development, along with Betsy's strident course to the west, prompted civil defense authorities in Florida to order a costly evacuation of all low-lying coastal communities from Miami to Key West. Tens of thousands of tourists, many of them in attendance for the state's Labor Day festivities, were hurriedly cleared from palatial hotels and shorefront bungalows, while native Floridians rushed to board up their windows, stockpile food and medical supplies, and secure their boats from Betsy's anticipated seven-foot storm surge.

With a central pressure of 28.14 inches (953 mb), Betsy blasted ashore near Biscayne Bay, just south of Miami, Florida, on the morning of September 8, 1965. Gusts of 135 MPH (217 km/h) drove before them five to nine–foot (2–3 m) storm surge, the highest seen on Florida's southeast coast in nearly a quarter of a century. Betsy's brutal winds and pounding rains stripped trees of their foliage and buildings of their roofs and siding, crumpled billboards, toppled telephone poles, and caused considerable localized flooding. The Orange Bowl Stadium, home of the University of Miami's Hurricanes football team, was heavily damaged when Betsy's blocking winds overturned bleachers, buckled tower lights, and blew down the marquee. On Florida's west coast, a transiting Betsy not only devastated large portions of the Big Cypress National Preserve but also spawned a five-foot (2.5-m) storm surge in Everglades City that caused several hundred thousand dollars worth of water damage to posh gulfside villas and yacht-straddled marinas. Three small TORNADOES, whistling black funnels that trailed Betsy's well-organized cloud bands, further augmented the hurricane's destructive

legacy. Numerous mobile homes in Dade County were violently wrenched from their foundations and then peeled apart panel by panel while a number of vehicles—among them a refrigerated citrus truck—were catapulted through the stormy skies by the tornadoes' whirring intensity. A total of thirteen people were killed in Florida, nine of them listed as "missing at sea" after Betsy's tumultuous waves overtook a squadron of fleeing excursion craft near Barnes Key.

Its sustained winds barely weakened by its stunning assault on south Florida, Betsy then spiraled across the Gulf of Mexico during the evening hours of September 8. Forecasters at the National Hurricane Center (NHC) predicted that the hurricane—with rapidly reintensifying winds and a cresting, nine-foot (3-m) storm surge—would make landfall somewhere in the vicinity of New Orleans, Louisiana. Fearing that New Orleans would suffer catastrophic flooding if the raging, fast-moving storm came ashore directly over the low-lying city, authorities began the mass evacuation of some 250,000 people from the Mississippi Delta region. Louisiana's famed shrimp fleet hauled in its nets, secured its gear, and returned to the crowded safety of its ports, while dozens of oil company personnel stationed on exploratory drilling rigs in the Gulf of Mexico were promptly airlifted to safety.

Bearing a central pressure of 27.99 inches (948 mb), sustained winds of 131 MPH (210 km/h), and drenching rains, Betsy crashed into the coast of Louisiana shortly before midnight on September 9, 1965. The hurricane's curving track took it ashore 90 miles (145 km) southeast of New Orleans in the vicinity of the Grand Terre Islands, thereby sparing the historic port city the eradicating brunt of its seaborne fury. As it was, Betsy's fringe winds, measured at 124 MPH (200 km/h), lashed neighborhoods in the western quadrant of New Orleans, causing some damage to chimneys, porches, fences, trees, and telephone poles. Closer to the shore, Betsy's assault was far more severe: In the Golden Meadows region, more than 27,000 houses were completely destroyed, leaving twice that number of people homeless, more than 2,000 businesses, from farms to factories to shopping centers, were likewise ruined, hindering efforts to provide hurricane victims with vital relief supplies. Sixty-one people in Louisiana were killed. In nearby Mississippi, Betsy's 74-MPH (120-km/h) winds drove an eight to 10-foot (2–3 m) storm surge into Bay St. Louis, washing away entire beaches as it flooded most of the bay's hurricane-prone north shore; one man was drowned. Quickly moving inland, Betsy brought 99-MPH (160-km/h) winds and two inches (51 mm) of rain to Louisiana's capital, Baton Rouge, causing isolated power outages and a small tornado. Downgraded to a tropical storm later that evening, Betsy's lingering winds spawned a further outbreak of tornadoes over Arkansas, killing another four people. By the morning of September 10, Betsy had peeled off to the northeast, crossed into Tennessee, and almost completely dissipated. Remembered as the first "billion dollar" hurricane in U.S. history, Betsy's rampage through the Bahamas, Florida, and Louisiana claimed 75 lives and left an estimated $1.4 billion dollars in property damage.

Hurricane Betsy of 1965 was not the first TROPICAL CYCLONE by that name to terrorize the North Atlantic basin. During the first week of August 1956, the first Hurricane Betsy originated over the eastern NORTH ATLANTIC as a tropical wave before tracking westward to develop into a dreaded Cape Verde system. On August 11, Betsy—now a Category 3 HURRICANE with sustained wind speeds of 121 MPH (195 km/h)—slid across the northern Leeward Islands, its CENTRAL BAROMETRIC PRESSURE of 28.17 inches (954 mb) producing torrential rains and a large STORM SURGE. Weakening as it sliced across the highlands of PUERTO RICO and HISPANIOLA on August 12, Betsy's 92-MPH (148-km/h) winds downed trees and power lines, while its heavy rains caused many instances of flash flooding. Maintaining its intensity as it recurved to the east of the BAHAMAS, Betsy followed the circulation patterns of the BERMUDA HIGH, and was drawn to the northeast, remaining well off the U.S. eastern seaboard. By August 20, the system had dissipated over the extreme North Atlantic.

It was fortunate that the second Hurricane Betsy remained an offshore tropical system for its entire existence because for a time it was one of the most powerful tropical cyclones yet observed in the North Atlantic basin. Originating south of 15 degrees North on September 2, 1961, Betsy chugged northwestward, steadily growing in severity until September 5, when it achieved Category 4 status. With a central pressure of 27.90 inches (945 mb) and sustained winds of 138 MPH (222 km/h), Betsy tore the surface of the Atlantic into windrows of spray and chased weary shipping from the sea lanes, but caused no damage or loss of life before weakening to Category 3 intensity on September 6. Passing well to the east of BERMUDA on September 9, 1961, Betsy rapidly recurved into the North Atlantic and underwent extratropical transitioning by September 12.

Following the devastating passage of the third Hurricane Betsy in 1965, the name was struck from the cyclical list of North Atlantic hurricane identifiers.

Beulah, Tropical Storm *Caribbean Sea–Mexico, June 15–18, 1959* Between 1959 and 1967, three

North Atlantic TROPICAL CYCLONES were identified with the name Beulah. The first, a relatively powerful TROPICAL STORM, originated over MEXICO's Bay of Campeche on June 15, 1959. Slowly intensifying to tropical storm intensity, the system recurved to the northwest and came ashore along the mid-Gulf coast of Mexico on June 17 with a central pressure of 29.14 inches (987 mb) and 69-MPH (111-km/h) winds.

Beulah, Hurricane *North Atlantic Ocean, August 20–28, 1963* The second system christened Beulah eventually matured into a major HURRICANE during the course of the 1963 North Atlantic HURRICANE SEASON. Originating on August 20 while over the mid-Atlantic, Beulah soon deepened into a Category 3 hurricane with a central pressure of 28.28 inches (958 mb) and sustained, 121-MPH (195-km/h) winds. A deepwater hurricane for its entire existence, Beulah recurved to the north-northeast and dissipated over the northern reaches of the eastern North Atlantic on August 28. Hurricane Beulah (1963) had the unique distinction of being one of a handful of North Atlantic TROPICAL CYCLONES to undergo experimentation as part of PROJECT STORMFURY.

Beulah, Hurricane *Gulf of Mexico–Southern United States, September 5–22, 1967* During the 1967 North Atlantic HURRICANE SEASON, the third and final system identified as Beulah became one of the most powerful tropical systems yet observed in the Atlantic basin, and proved so destructive in northern MEXICO and TEXAS that its name was retired. Of Cape Verde origin, Beulah formed to the east of the CARIBBEAN SEA on September 5, 1967. Traveling to the west-northwest, Beulah passed over the Windward Islands as a TROPICAL STORM (29.70 inches [1,006 mb], 58 MPH [93 km/h] winds) entered the Caribbean Sea, where it intensified into a Category 1 hurricane (29.20 inches [989 mb], 86-MPH (138-km/h) winds), and swept along the southern shores of PUERTO RICO and HISPANIOLA, growing into a very powerful Category 4 hurricane (27.96 inches [947 mb], 150-MPH (241-km/h) winds) along the way. On September 12, 1967, as it drew to within 200 miles of JAMAICA's eastern coast, Beulah fell under the steering influence of a high-pressure ridge to the north, and abruptly recurved to the southwest, in the process steeply weakening to tropical storm intensity (29.52 inches [1,000 mb], 69 MPH [111 km/h] winds). Moving well to the south of Jamaica, Beulah began another recurvature on September 14, 1967, and shifted its trajectory to the northwest, setting it on an eventual collision course with the eastern tip of Mexico's Yucatán Peninsula. Beulah also began

to reintensify, deepening into a Category 3 hurricane (28.46 inches [964 mb], 115 MPH [185 km/h] winds) just prior to its landfall in the Yucatán during the late-evening hours of September 16. Weakening slightly as it traversed the peninsula's dense tropical foliage, Beulah emerged into the Bay of Campeche on September 17 as a Category 2 system (28.55 inches [967 mb], 98 MPH [158 km/h]). Once in the Bay of Campeche, Beulah's trajectory carried the system to the northwest, and then more northerly, as it bore down on the Mexico–Texas border, steadily gaining in intensity. By the late evening hours of September 18, Beulah had again deepened to Category 3 status, with a central pressure of 28.55 inches (967 mb) and sustained winds of 115 MPH (185 km/h). By the mid-afternoon hours of September 19, Beulah had deepened to a Category 4 system, with a central pressure of 28.02 inches (949 mb) and 144-MPH (232-km/h) winds. Now within 100 miles (161 km) of landfall along the mouths of the Rio Grande, Beulah escalated into a Category 5 hurricane—with a central barometric pressure reading of 27.26 inches (937 mb) and sustained wind speeds exceeding 160 MPH (257 km/h). The most intense tropical cyclone to have been observed in the area since 1961's Hurricane CARLA, Beulah steadily loped to the northwest, weakening to Category 3 status just prior to blasting ashore on the Mexico–Texas border on the morning of September 20, 1967. A large and hazard-filled system, Beulah generated more than 100 TORNADOES across southern Texas, causing widespread property damage and several fatalities. On Texas's South Padre Island, sustained winds of 136 MPH (219 km/h) were recorded, while rainfall counts exceeded 25 inches (635 mm) in some places. A central pressure of 28.07 inches (951 mb) and a 12-foot (4 m) storm tide were observed in Brownsville, Texas, well to the north of Beulah's landfall location. Driven to the southwest and over the mountains of northern Mexico, Beulah was rapidly downgraded to a tropical storm, and finally a tropical depression before dissipating on September 22, 1967. A low-pressure remnant of Beulah eventually recurved into the GULF OF MEXICO on September 25, but did not regenerate. All told, some 59 people lost their lives to Hurricane Beulah, and more than $1 billion in property losses were incurred.

Following the destructive passage of 1967's Hurricane Beulah, the name Beulah was retired from the cyclical naming lists.

Bill, Hurricane *North Atlantic Ocean, July 11–13, 1997* A deepwater TROPICAL CYCLONE for its entire existence, Hurricane Bill churned across the mid-North Atlantic Ocean between July 11 and 13, 1997.

Originating to the southeast of BERMUDA as a TROPICAL DEPRESSION, the system tracked to the northeast, intensifying into a TROPICAL STORM during the early morning hours of July 11, and a Category 1 hurricane on July 12. At its peak, Hurricane Bill generated a CENTRAL BAROMETRIC PRESSURE reading of 29.14 inches (987 mb) and sustained winds of 75 MPH (121 km/h). Nearing 45 degrees North and its chilling waters on July 13, Bill faded into a tropical storm, and within a day had dissipated.

Bill, Tropical Storm *Mexico–Gulf of Mexico–Southern United States, June 28–July 3, 2003* The second named TROPICAL CYCLONE of the 2003 North Atlantic HURRICANE SEASON, Tropical Storm Bill delivered high winds and heavy rains to sections of MEXICO's Yucatán Peninsula, and to southeastern LOUISIANA, between June 28 and July 3, 2003. Of TROPICAL DEPRESSION intensity as it crossed to the north-northwest over the Yucatán Peninsula on June 28, Bill entered the southern GULF OF MEXICO and intensified to TROPICAL STORM status on June 29. Recurving to the northeast a day later, Bill came ashore in Louisiana, near the mouths of the Mississippi River, on June 30, bearing a CENTRAL BAROMETRIC PRESSURE of 29.44 inches (997 mb) and sustained winds of 58 MPH (93 km/h). On July 1, a TORNADO generated by Bill's constituent thunderstorms touched down near New Orleans, while another weak tornadic outbreak was observed in Plaquemines Parish. The system remained at tropical storm intensity as it tracked across the southeastern communities of Louisiana, then entered MISSISSIPPI, where it was downgraded to a tropical depression. Tropical Depression Bill remained fairly intact as it moved across ALABAMA, the northwestern tip of GEORGIA, and into Tennessee and VIRGINIA, where it finally dissipated on July 3, 2003.

The identifier Bill has been retained on the North Atlantic tropical cyclone naming lists. It is scheduled to be reused during the 2009 hurricane season.

Black River Hurricane *South–Northern Caribbean, November 11–19, 1912* One of the most intense TROPICAL CYCLONES in Jamaican history, the Black River HURRICANE pummeled the southwest coast of JAMAICA and the southeast coast of CUBA with 120-MPH (193-km/h) winds, 35-foot (11.6-m) seas, and a 12 to 14-foot (4–5 m) STORM SURGE. In Jamaica, where the Black River Hurricane's Category 3 assault was felt the hardest, some 5,000 BUILDINGS were either damaged or destroyed, forcing the displacement of an estimated 35,000 people. Listed as dead were 124 men, women, and children on the island, while another 2,000 were injured, making the

Black River Hurricane the deadliest storm to have struck Jamaica since an unnamed hurricane claimed 43 lives there in late September 1896. In Cuba, where the Black River Hurricane's diminished 76-MPH (122-km/h) gusts and whipping rains caused huge flood emergencies in and around Guantanamo Bay, five people were reportedly drowned or slain by airborne debris. At least $50 million in property damage was assessed in both Jamaica and Cuba, characterizing the Black River Hurricane as one of notable violence and duration.

The seventh and final storm of the 1912 HURRICANE SEASON, the Black River system developed over the subequatorial waters of the southwestern CARIBBEAN SEA, some 200 miles (322 km) north of the unfinished Panama Canal, on the afternoon of November 11, 1912. Initially just another of the 10 or more TROPICAL WAVES that seasonably drift northward with the Caribbean trade winds, the Black River's young vortex quickly deepened into a TROPICAL DISTURBANCE, or a heavy massing of thunderstorms, by midnight on November 12 and into a TROPICAL DEPRESSION, a now-revolving collection of thunderstorms, by the midmorning hours of the following day.

Slowly spiraling around an invisible center of low barometric pressure, the Black River's CUMULONIMBUS CONVECTION cells extracted vast strength from the Caribbean's evaporating waters before expending part of that heat energy as wind, gales that swiftly transformed the sea's surface into a rich field of spindrift. Putting in gear the almost mechanical cycle of rising and condensing air that would tragically buoy the system for the next week, the Black River storm steadily intensified, saw its CENTRAL BAROMETRIC PRESSURE drop from 29.58 inches (1,001 mb) on the afternoon of November 13 to 29.10 inches (985 mb) by midnight of the same day, to 28.96 inches (980 mb) by dawn of November 14.

Still classified as a TROPICAL STORM, or a cyclonic system with wind speeds of between 39 and 73 MPH (63–118 km/h), the Black River storm slowly coursed to the northwest at nearly four MPH (6 km/h), riding the nourishing winds ever closer to the limestone uplands of unsuspecting Jamaica. On the evening of November 15, as its central pressure sank to an estimated 28.79 inches (974 mb), the Black River Hurricane's sustained 104-MPH (167-km/h) winds, interspersed with 111-MPH (179-km/h) gusts, furiously fanned the Caribbean's surface. Tons of heat-laden spray were hoisted into its towering, 40,000-foot (13,333 m) cells and then dispersed as rain—as the light drizzle that in a matter of days would begin to settle ominously across Jamaica's south shores.

Meanwhile, the Black River Hurricane inexorably churned across the sea, its strengthening 115-MPH

(185-km/h) winds forcing dozens of crippled vessels to limp painfully into nearby ports. On the morning of November 16, 1912, as RECURVATURE to the northeast brought the hurricane within 100 miles (161 km) of Jamaica, heavy breakers began to roll ashore at Savanna-La-Mar and Kingston, providing stormwise islanders with their first undeniable evidence that a powerful hurricane was indeed on the approach. In Jamaica, a country that has withstood an extensive history of destructive hurricane strikes, an endless round of evacuation warnings and meteorological alerts did little to save people who were killed when the hurricane's blistering winds finally came ashore near the mouth of the Black River, 70 miles (113 km) west of Kingston, just before dusk on November 18, 1912.

The most intense tropical cyclone to have affected the island in nearly a quarter of a century, the Black River Hurricane's EYE continued curling to the northeast shortly after landfall, bringing increasing quantities of destruction to inland communities. Raking the coastal towns of Lover's Leap and Treasure Beach with 120-MPH (193-km/h) winds and torrential rains, damage was particularly severe in Savanna-La-Mar, where the hurricane-prone harbor was clogged with shipping run aground by the hurricane's 35-foot (11 m) breakers. Bayside warehouses swayed and then collapsed, casting hundreds of casks of Jamaican rum into the frothing surf. Thousands of palm trees toppled into the streets, linked together to form minidams that trapped the hurricane's seven inches (178 mm) of rain behind them. Nearly 5,000 buildings were either damaged or destroyed by the ensuring flash floods, and 124 people lost their lives to drowning or mudslide incidents.

Slowly pulling away from Jamaica on the morning of November 19, 1912, the Black River Hurricane trod to the northeast, bound for a second landfall in either southeast Cuba or perhaps the northwestern crag of HISPANIOLA. Severely weakened by its spirited bombardment of Jamaica, the hurricane rapidly lost intensity, seeing its sustained winds diminish to 69 MPH (111 km/h) by mid-morning of November 19. Unable to reintensify over the confined waters of the Windward Passage, the gusty remains of the Black River Hurricane nonetheless managed to cause extensive flooding in and around the entrance to Guantanamo Bay. Four fishermen were drowned, while another man died of injuries received in a building collapse. The dissipating traces of the Black River Hurricane were not tracked beyond the western Caribbean Sea.

Blanche, Hurricane *Eastern United States–Canada, August 11–13, 1969* As though following the same set of tracks some six years apart, Hurricanes Blanche (1969 and 1975) directly affected the extreme eastern

Canadian provinces of Nova Scotia and Newfoundland. The first Hurricane Blanche originated to the north of the BAHAMAS on August 11, 1969. Trailing along the eastern U.S. seaboard, Blanche rapidly spun northward, deepening into a Category 1 hurricane (with a central pressure of 29.44 inches [997 mb]) and delivering glancing, 86-MPH (138-km/h) winds to the Canadian Maritimes on August 13.

Blanche, Hurricane *North Atlantic Ocean–Canada, July 24–28, 1975* The second Hurricane Blanche dealt the forests of Nova Scotia and Newfoundland an even more direct strike in late July 1975. Formed deep into the Atlantic, off the northern Leeward Islands on July 24, Blanche first recurved to the northeast, then to the north, where it clanged ashore in Nova Scotia on July 28. At the time of its landfall, Blanche's central pressure was observed at (980 mb), and sustained winds of 86 MPH (km/h).

Although never officially retired, the name Blanche is not in current use on the North Atlantic TROPICAL CYCLONE lists.

Bob, Hurricane *Southern United States, July 9–12, 1979* The first Atlantic HURRICANE to be given a male name, Bob killed one person, forced the evacuation of 80,000 others, and caused $2.3 million in property damage as it thudded into southeast LOUISIANA on July 11, 1979. Bob's CENTRAL BAROMETRIC PRESSURE of 28.98 inches (981 mb) at landfall made it a weak Category 1 hurricane, bringing sustained 76-MPH (122-km/h) winds and steady rains to the entire Timbalier Island region. On the southeast coast, the storm's minimal three-foot (1-m) STORM SURGE swamped two small pleasure craft, smashed a fueling pier, and flooded a number of low-lying roads but caused no fatalities. In New Orleans, 64-MPH (103-km/h) gusts closed businesses, uprooted dozens of trees, crumpled carports, and severed power lines; more than 53,000 people in and around the city were left without electricity. During the height of the storm, a man lost his life to a heart attack while attempting to prevent his garden shed from being carried away by the hurricane's strenuous winds. Before dissipating into a range of thunderstorms over central MISSISSIPPI on July 12, Bob further wrought extensive damage on Louisiana's rice and sweet potato crops, costing farmers nearly $1 million in market losses.

When the early season hurricane first originated over the lukewarm waters of southwestern CARIBBEAN SEA on July 9, it had already been decided by officials at the World Meteorological Organization (WMO) that it would be christened *Bob*. Responding to both societal pressures and a dwindling list of available female names, the WMO decided to begin to incorporate male names into their alternating lists of HURRICANE NAMES with the 1978 North Pacific HURRICANE SEASON. When this experiment proved successful with both meteorologists and the general public, a similar practice was quickly adopted for the 1979 North Atlantic hurricane season. Thus, 26 years of hurricane tradition came to an end on the morning of July 10, 1979, when the burgeoning TROPICAL STORM then located 100 miles off the northeast tip of MEXICO's Yucatán Peninsula was officially dubbed *Bob*. Despite the historic significance attending the naming event, it was later decided that Bob's relatively minor tantrum in Louisiana was not destructive enough to warrant retiring the name to the so-called Hurricane Hall of Fame at that time.

Bob, Hurricane *Gulf of Mexico–Southern and Eastern United States, July 21–26, 1985* Between July 21 and 26, 1985, the second North Atlantic TROPICAL CYCLONE to be identified as Bob carved a swerving path of wind and rain through FLORIDA, SOUTH CAROLINA, NORTH CAROLINA, and into VIRGINIA. Formed over the eastern GULF OF MEXICO on July 21, Bob was of TROPICAL STORM intensity as it skipped ashore in southwestern FLORIDA on July 22, and crossed the peninsula to enter the NORTH ATLANTIC OCEAN. Moving due north and fueled by the warm SEA-SURFACE TEMPERATURES of the Gulf Stream, the system quickly intensified into a Category 1 HURRICANE and maintained its intensity as it came ashore in southern South Carolina on July 25. Bearing a central pressure of 29.58 inches (1,002 mb) and wind speeds of 75 MPH (121 km/h), Bob was quickly downgraded to a tropical storm as it continued to move northward over central South Carolina and the western highlands of North Carolina. Bob further weakened to a TROPICAL DEPRESSION as it crossed the border into Virginia, and within less than 24 hours had essentially dissipated.

Bob, Hurricane *Eastern United States–Canada, August 15–20, 1991* The second North Atlantic HURRICANE to be christened Bob, this moderately powerful Category 2 system raked the eastern seaboard of the United States and Canada between August 18 and 19, 1991. In NORTH and SOUTH CAROLINA, Bob's 70-MPH (113-km/h) winds grazed barrier islands, raising a six-foot (2 m) STORM SURGE that drowned two people. In NEW YORK State, the central and eastern reaches of Long Island were bruised by sustained gale-force winds, 70-MPH (113-km/h) gusts, and two minor TORNADOES; three people in New York were killed. In NEW ENGLAND, Bob's CENTRAL BAROMETRIC PRESSURE of 28.14 inches (953 mb) at landfall brought deadly 125-MPH (201-km/h) winds and nine inches of rain to portions of Connecticut, Rhode Island, Massachusetts, New Hampshire, and Maine. Ten people in the region—a number of them summer vacationers—lost their lives to Bob's seaborne fury. Progressing into eastern CANADA on August 19, Bob's downgraded 81-MPH (130-km/h) winds and two to four-inch (50–100 mm) rains assaulted southwestern Nova Scotia and New Brunswick, claiming the lives of two storm watchers who were washed from a pier and drowned. Damage estimates from this, the most intense hurricane to strike northeastern United States since 1985, totaled more than $780 million.

The second storm of an unusually peaceful HURRICANE SEASON, Bob developed as a TROPICAL DISTURBANCE over the midsummer waters of the western NORTH ATLANTIC OCEAN, 175 miles (282 km) east of Grand Bahama Island, on the morning of August 15, 1991. Moving northwest at nearly nine MPH, Bob rapidly underwent intensification, becoming a full-fledged TROPICAL DEPRESSION on the evening of Friday, August 16, and a 40 MPH (64 km/h) TROPICAL STORM by late morning of the next day. On August 18, as its central barometric pressure slipped to 28.76 inches (974 mb), the storm was upgraded to hurricane status. Its clearly defined EYE, then located 50 miles east of South Carolina, was moving in a north-northwesterly direction at 11 MPH (18 km), bringing it ever closer to the mid-Atlantic coast of the United States. By late evening on August 18, as HURRICANE WARNING flags wildly snapped at their clubhouse mastheads, the storm's 70 MPH (113 km/h) EYEWALL winds passed just by the North Carolina capes, bringing four inches (102 mm) of rain to both Cape Hatteras and its treacherous brood of barrier islands. During the course of the seven-hour tempest, trees were uprooted, power lines were downed, and flash floods in both the Carolinas claimed two lives through road accidents.

Swiftly continuing up the eastern seaboard, Hurricane Bob further intensified as it raged past VIRGINIA, Delaware, and New Jersey during the night of August 18. Reconnaissance data from the storm now indicated a central pressure of 28.14 inches (953 mb), upgrading it to a Category 2 hurricane, or one capable of delivering a six to eight foot (2–2.6 m) storm surge and heavy rains at landfall. Bob's offshore course, some 100 miles (161 km) to the east of Delaware and New Jersey, spared both states serious damage, although heavy surf action and rains were recorded in Atlantic City. Steadily curling to the northeast, Bob's eye passed over Rhode Island's Block Island before making landfall near Narragansett Bay, Rhode Island, at shortly before ten o'clock on the morning of August 19, 1991. To the west of Bob's eye in central and eastern Long Island, sustained 54-MPH (87-km/h) winds gusted to more than 70 MPH (113 km/h), whipping the foliage from trees and buffeting the airport at Farmingdale. Two small tornadoes touched down in eastern Long Island, causing isolated fence and roof damage to the towns of Noyak and Wading River. In New York City, 200 miles (322 km) to the west of Bob's track, torrential rains and intermittent gusts shivered the stately hardwoods of Central Park but otherwise caused no damage. In the days following the hurricane's transit, insured damage assessments in New York State steadily rose to $75 million.

In Rhode Island, 125-MPH (187-km/h) winds drove Bob's six-foot (2-m) storm surge into the tapering narrows of Narragansett Bay. Dozens of yachts and lobster boats were torn from their anchorages and carried ashore. Widespread flooding transformed neighboring streets into small rivers, restricting efforts to rescue those individuals trapped in their houses by the rising waters. In nearby Newport, Bob's sheering passage quickly marred the manicured elegance of the summer resort's cliffwalk estates. Fragrant gardens were blighted, gaily striped window awnings were shredded, and Breakers—the famed Vanderbilt residence—lost several of its terra-cotta roof tiles to the hurricane's ferocity. All told, insured losses in Rhode Island were in excess of $100 million.

Maintaining its northeasterly course, Bob then passed into southern Massachusetts, bringing record low barometric pressures and PRECIPITATION counts to Boston. Logan International Airport suspended all flight operations as hurricane-force gusts drove blinding thunderstorms into the region, spawning numerous instances of severe tree and structural damage within the confines of the city. Farther to the east, Bob's 115-MPH (185-km/h) winds battered evacuated seaside villages all along the sandy crook of the Cape Cod peninsula. Streaming winds and following seas toppled expensive summer cottages into the surf, scrubbed the wooden-shingled roofs from quaint boutiques and restaurants, and caused extensive dune erosion. In the town of Ipswich, located on Massachusetts Bay's west coast, several houseboats were tossed ashore at Plum Island, while nearby Crane's Beach suffered huge washouts. Downed power lines cast more than 150,000 residents into total darkness. Insured damage estimates in Massachusetts would eventually top $525 million.

Now moving almost due north at 30 MPH (48 km/h), Bob's eye passed into east Maine during the early afternoon hours of August 19. At Portland, the hurricane's central pressure of 29.09 inches (985 mb) brought with it 39–54-MPH (63–87-km/h) gusts and six to 10-inch (152–254 mm) rainfalls, causing extensive injury to the state's maturing potato harvest. Entire pine groves were leveled, numerous small buildings collapsed, and two people were killed as the hurricane inflicted $21 million in property damage. Downgraded to tropical storm status by late afternoon on August 19, Bob moved into New Brunswick, Canada, with 80-MPH (130-km/h) winds and two to four-inch precipitation counts. In Nova Scotia, two additional people were killed when they were plunged into the sea by Bob's onrushing surge and drowned. Bob's destructive legacy was 14 dead and $1.3 billion in insured and uninsured losses. The name Bob has been retired from the rotating list of North Atlantic tropical cyclone identifiers.

Bobby, Cyclone *Western Australia, February 19–26, 1995* An intensely powerful Category 4 (Australian scale) CYCLONE, Bobby battered the Australian towns of Onslow and Gascoyne Junction with 175-MPH (280-km/h) winds and deluging rains between February 24 and 26, 1995. With a minimum BAROMETRIC PRESSURE of 27.13 inches (919 mb) at landfall, Bobby was the most intense cyclone to strike northwestern AUSTRALIA since Cyclone TRACY came ashore at Darwin on Christmas Day, 1974. Although Bobby's fury would eventually claim the lives of seven people, its torrential rains—estimated to be in excess of 15 inches (38 cm) in some places—broke a prolonged drought over the Southwest Land Division.

The second cyclone of the 1995 season, Bobby was spawned over the Timor Sea, approximately 200 miles (322 km) northwest of Cape Londonderry, Australia, on February 19. Carried steadily south-southwest at 10 MPH (16 km/h), the cyclone entered the energized 86 degree F (30 degrees C) waters of the east Indian Ocean and rapidly intensified. Aerial reconnaissance flights conducted by the Australian Bureau of Meteorology on the afternoon of February 21 indicated that Bobby's central pressure had dropped to an alarming 28.47 inches (964 mb) and that its cloud mass with accompanying gales extended all the way to Indonesia.

As Bobby continued to strengthen through the afternoon of February 22, 125-MPH (201-km/h) winds swept the surface of the Indian Ocean, causing five-foot (1.5-m) swells to come ashore at Port Hedland, Australia. High wind and flood warnings were posted for much of the remote, mangrove-lined Pilbara coast as Bobby's EYE, now less than 400 miles (644 km) from the Dampier Archipelago, began to move in a tight south-southwesterly direction. A large area of high pressure had begun to curl up and around the cyclone's southwestern flank, blocking its parallel track along the coast and redirecting it inland. ANALOGS based on the behavior of previous cyclones predicted that the system would make land-

fall somewhere in the vicinity of Barrow Island and that it would most likely continue to intensify as it came ashore.

In the small fishing town of Onslow, located approximately 850 miles (1,400 km) north of Perth, residents began the anxious task of securing their warehouses, storefronts, and holiday cottages against Bobby's imminent onslaught. Hundreds of sandbags were deployed in low-lying areas around the village as a precaution against flash flooding. In the harbor itself, numerous prawn trawlers—steel-hulled 60-foot (20-m) vessels—were warped alongside their piers, secured against the cyclone's anticipated 14-foot (4.6-m) STORM SURGE. According to newspaper accounts, local stores were emptied of canned foods, bottled water, flashlights, and kerosene lanterns. Stocks of plywood a nearby lumberyards were quickly depleted as homeowners worked through the night to construct storm shutters for their windows. Potentially lethal lawn furniture was stowed away, while small utility buildings were firmly anchored to the ground with stakes and chains. In the stormy Indian Ocean, where 15-foot (5-m) seas had already begun to crest, several stray prawn trawlers quickly altered their courses and headed for the uncertain refuge of their homeports.

With a central barometric pressure of 27.13 inches (919 mb), Cyclone Bobby blasted ashore at Onslow, Western Australia, just after midnight on February 23, 1995. Sustained winds of 132 MPH (212 km/h) brought with them a 13-foot (4.3-m) storm surge, the highest seen on the Pilbara coast in a half-century. For 12 hours, Bobby's 175-MPH (282-km/h) gusts and pelting rains stripped buildings of their roofs and siding, splintered storm shutters, crumpled billboards, toppled trees and telephone polls, sank several moored boats, and caused massive localized flooding. Mountainous wind-driven waves pummeled more than 100 miles (161 km) of shoreline, causing isolated mudslides and extensive beach erosion. Just outside the relative safety of Onslow's harbor, two prawn trawlers were quickly overwhelmed by Bobby's swamping seas. One vessel, the *Harmony,* was later found abandoned and capsized 10 miles (16 km) off the coast, while the sunken wreck of the other, the 50-ton *Lady Pamela,* was discovered by divers less than five miles (8 km) north of the harbor's entrance. Of their crews—three men and one woman aboard the *Harmony* and three men from the *Lady Pamela*—no trace was ever found.

Steadily moving across the mining fields of Western Australia, Cyclone Bobby's slightly diminished winds and intense PRECIPITATION spawned raging floods that downed power lines, demolished mobile homes, closed airstrips, crippled rail links, and hampered gold-mining operations. On the morning of February 25, the small town of Gascoyne Junction, located 400 miles (644 km) southeast of Onslow, was buffeted by 100-MPH (161-km/h) winds and searing rains. Thirty people trapped on the roof of a roadhouse had to be airlifted to safety after the swollen Ashburton River overran its banks. Several of the region's major thoroughfares, including the Pilbara access road, the North West Coastal Highway, and the Eyre Highway—Western Australia's principal road link with the country's eastern states—were washed out by Bobby's torrential rains. A week-long halt in traffic created massive shortages of food, medical supplies, and building materials on the west coast and severely restricted the speed and effectiveness of rescue operations. As the cyclone continued to penetrate inland, further floods in the north and northwestern reaches of the state isolated several Aboriginal communities, forcing the government to commence a program of aerial food drops to the affected areas.

On the morning of February 26, 1995, as Bobby was first downgraded to a Category 1 cyclone and then completely dissipated into the warm embrace of the spreading high-pressure area, Western Australia's shaken residents began to tally their losses—and count their blessings. Although virtually every building in Onslow had received some degree of injury from Cyclone Bobby, the devastation could have been much worse. Cyclone Tracy, for instance, had completely leveled 7,200 buildings as it crossed over Darwin, Western Australia, on December 25, 1974. Aside from the seven unfortunate fishers who lost their lives to the opening stages of Bobby's assault, no other casualties were reported. In addition, while Bobby's deluging floods were responsible for close to $50 million in property damages to Western Australia, they conversely brought beneficial precipitation to countless farms and mining communities, thus ending a prolonged summer drought that had threatened the region's economic and agricultural prosperity.

Bonnie, Hurricane *North Atlantic Ocean, August 14–19, 1980* In mid-August 1980, Hurricane Bonnie tracked an unusual course from south to north across the middle reaches of the NORTH ATLANTIC OCEAN. Born just north of 10 degrees North on August 14, 1980, Bonnie became a TROPICAL STORM as it crossed 15 degrees North on August 15, and a hurricane as it passed over 25 degrees North during the early morning hours of August 17. At its peak, Hurricane Bonnie generated a central pressure reading of 28.79 inches (975 mb) and boasted sustained winds of 98 MPH (158 km/h), making it a powerful Category 2 system on the SAFFIR-SIMPSON SCALE.

Doggedly maintaining its almost due northward trajectory, Bonnie penetrated as far north as 55 degrees North before undergoing extratropical transitioning on August 19, 1980.

Bonnie, Hurricane *Gulf of Mexico–Southern United States, June 23–28, 1986* The second named TROPICAL CYCLONE of the 1986 North Atlantic HURRICANE SEASON, Bonnie grew into hurricane intensity before bouncing ashore in northeastern TEXAS on June 25, 1986. Developed over the central GULF OF MEXICO from a TROPICAL DEPRESSION on June 23, an early-season Bonnie first grew into TROPICAL STORM intensity while swirling to the northwest. Several hours later, as it drew to within 200 miles of southern LOUISIANA, Bonnie matured into a Category 1 hurricane with a minimum central pressure of 29.29 inches (992 mb) and sustained, 86-MPH (138-km/h) winds. Bonnie maintained hurricane intensity after making landfall in Texas, and recurving due north. It then spent several hours as a tropical storm before being downgraded to a tropical depression and recurving across central Arkansas. The system reached as far north as southeastern Missouri before dissipating on June 28.

Bonnie, Hurricane *North Atlantic Ocean, September 17–October 2, 1992* On September 17, 1992, Hurricane Bonnie developed in the mid-North Atlantic Ocean, well to the northeast of the island of BERMUDA. The first named tropical system in the Atlantic following the legendary Hurricane ANDREW (August 16–28, 1992), and trailing an erratic course across the North Atlantic, Bonnie reached peak intensity when its CENTRAL BAROMETRIC PRESSURE slipped to 28.49 inches (965 mb) and its sustained wind speeds topped 109 MPH (175 km/h), making it a powerful Category 2 system on the SAFFIR-SIMPSON SCALE. A deepwater tropical system for its lifetime, Bonnie continued to track to the east-northeast, and finally dissipated on October 2, 1992.

Bonnie, Hurricane *North Atlantic Ocean–Eastern United States, August 19–31, 1998* The most powerful of the five North Atlantic tropical systems so far named Bonnie, Hurricane Bonnie delivered Category 3 winds, rain, and surf conditions to the eastern shores of NORTH CAROLINA on August 26, 1998. Essentially a CAPE VERDE STORM, Bonnie originated well to the east of the Leeward and Windward Islands on August 19, 1998. It passed to the north of the outer Caribbean islands as a TROPICAL STORM between August 20 and 21, and then deepened into a Category 1 hurricane while nearing the southeastern BAHAMAS during the early morning hours of

August 22. Nourished by the warm SEA-SURFACE TEMPERATURES of the Gulf Stream, Bonnie continued to intensify, becoming a Category 2 system later the same day, and a Category 3 system during the morning hours of August 23. Moving to the northwest, the system was expected to make a direct landfall in the Wilmington, NORTH CAROLINA, area when it suddenly recurved to the northeast, and moved into the Atlantic on August 26, 1998. Both North Carolina and VIRGINIA bore the brunt of Bonnie's oblique strike, experiencing wind gusts of between 90 and 100 MPH (145–161 km/h). ANEMOMETERS at the enormous U.S. naval base in Norfolk, Virginia, observed sustained wind speeds of 100 MPH (161 km/h) during Bonnie's passage. At its peak intensity on August 24, Bonnie generated a central pressure reading of 28.17 inches (954 mb) and sustained winds of 115 MPH (185 km/h), making it a powerful and potentially dangerous tropical system. Not surprisingly, tremendous rainfall accompanied Bonnie's passage along the eastern seaboard; after weakening into a tropical storm, Bonnie dropped 10 inches (254 mm) of rain on eastern North Carolina, causing flash flooding and extensive property damage. Two people in the Carolinas lost their lives to Bonnie's fury, while damage estimates neared $2 billion. Bonnie's tropical remnants became extratropical on August 30, and the system largely dissipated the following day.

Bonnie, Tropical Storm *Caribbean Sea–Gulf of Mexico–Southern and Eastern United States, August 3–14, 2004* On August 9, 2004, the fifth North Atlantic tropical system dubbed Bonnie—this one a TROPICAL STORM—developed over the waters of the southern GULF OF MEXICO. Spawned by the remnants of a TROPICAL DEPRESSION that had transitioned into a tropical wave while crossing the southern CARIBBEAN SEA between August 3 and 9, 2004, Bonnie reintensified to tropical storm intensity while located some 175 miles (282 km) west-southwest of Apalachicola, FLORIDA. Slowly regaining strength, Bonnie tracked to the northwest, intensifying into a powerful tropical storm as it did so. By August 11, 2004, as the U.S. Navy transferred several dozen training aircraft to inland airfields, Bonnie was generating sustained winds of 65 MPH (105 km/h) and heavy rains, and was bound for landfall somewhere on the Florida Panhandle. Weakening slightly as it approached the Florida coast, Bonnie delivered sustained, 50-MPH (80-km/h) winds at its landfall in Florida on August 12, 2004. Again downgraded, this time to a tropical depression, the system recurved to the northeast, over GEORGIA, the Carolinas, and emerged into the NORTH ATLANTIC just to the east of Maryland and New Jersey on August 14, 2004. The system dissipated shortly

thereafter. No deaths or injuries were reported in the wake of Tropical Storm Bonnie's mid-season passage.

There have also been tropical systems in the western North Pacific Ocean that have carried the moniker Bonnie. Between August 9 and 12, 1978, Tropical Storm Bonnie caused severe disruption in Hong Kong, including the destruction of several small watercraft.

The name Bonnie has been retained on the North Atlantic tropical cyclone naming lists, and is scheduled to reappear in its sixth incarnation in 2010.

breakpoints Used by the NATIONAL HURRICANE CENTER (NHC), the U.S. National Weather Service (NWS), and emergency managers, breakpoints are predetermined geographical positions along a coastline within which hurricane and TROPICAL STORM watches and warnings could, if necessary, be issued. Maintained by the NHC for the Atlantic and Gulf coasts, CUBA, MEXICO, and the BAHAMAS, the use of breakpoints essentially allows the NHC and those organizations responsible for tropical cyclone-related relief operations, to identify a finite section of coastline along which an existing hurricane or tropical storm might come ashore. Since there remains considerable variance in forecast models and tropical cyclone activity, it is not always possible to pinpoint where a tropical cyclone will make landfall; the use of breakpoints establishes the boundaries within which this could occur, and thereby permits the activation of more effective evacuations and other emergency preparations. For instance, Cape Lookout, NORTH CAROLINA, is a breakpoint, while the next to the north is Ocracoke Inlet; after that, Cape Hatteras. Depending on the nature of a representative tropical system and the variances in the guidance envelope, the NHC might choose to set HURRICANE WATCHES for locations between the Cape Lookout and Ocracoke Inlet breakpoints, and a HURRICANE WARNING for the coastline between the Ocracoke Inlet and Cape Hatteras breakpoints. As there are hundreds of breakpoints along the U.S. coastlines, tremendous flexibility in pinpointing breakpoint zones can be achieved.

Brenda, Hurricane *Gulf of Mexico–Southern United States, July 31–August 3, 1955* Because Hurricane Alice (Alice 2) endured from December 26, 1954, until January 6, 1955, the first TROPICAL CYCLONE of the official 1955 North Atlantic HURRICANE SEASON was identified as Brenda. A moderately powerful TROPICAL STORM that originated over the GULF OF MEXICO on July 31, Brenda delivered 69-MPH (111-km/h) winds and heavy rains to extreme southeastern LOUISIANA on August 1. No deaths or injuries were reported.

The second named tropical cyclone of the 1960 North Atlantic HURRICANE season, Tropical Storm Brenda brought TROPICAL DEPRESSION force winds and rains to several eastern states between July 28 and August 1. Formed over the Gulf of Mexico, Brenda leaped ashore in northwestern FLORIDA on July 29 with sustained winds of 35 MPH (56 km/h) before recurving to the northeast. For the next several days, Tropical Storm Brenda swirled northward, passing over GEORGIA, the Carolinas, VIRGINIA, New Jersey, NEW YORK and into central NEW ENGLAND, producing 52-MPH (84-km/h) winds and driving rains. No deaths or injuries were reported before Tropical Storm Brenda transitioned into an extratropical system over eastern Canada on August 1.

The 1964 North Atlantic hurricane season produced yet another Tropical Storm Brenda, this a mid-Atlantic system that moved almost due east, producing a central pressure estimate of 29.70 inches (1,006 mb) and 52-MPH (84-km/h) winds, between August 7 and 10.

On June 17, 1968, Hurricane Brenda originated in the Florida Straits. Rolling ashore in southeastern Florida as a tropical depression, Brenda tracked northward along the state's centerline before recurving into the Atlantic near the Florida–Georgia border on June 20. Once over the heat-nourished waters of the Gulf Stream, the as-yet unnamed tropical depression steadily strengthened, becoming Tropical Storm Brenda on June 21, and a Category 1 hurricane with a central pressure of 29.23 inches (990 mb) on June 23. Penetrating deep into the mid-North Atlantic, Brenda eventually weakened and dissipated by June 26, 1968.

The fifth and last North Atlantic tropical cyclone christened Brenda—this one a hurricane—followed an unusual trajectory through the northwestern CARIBBEAN SEA and into MEXICO between August 18 and 22, 1973. Formed to the east of Mexico's Yucatán Peninsula, Brenda reached tropical storm intensity before gliding ashore on the peninsula on August 19. First recurving to the west, then to the southwest, Brenda steadily weakened until emerging into the Bay of Campeche on August 20, where it promptly deepened into a Category 1 hurricane. Making its second landfall in Mexico, Brenda—with a central pressure of 28.85 inches (977 mb)—delivered 92-MPH (148-km/h) winds to Mexico's eastern coastline on August 21. Several small BUILDINGS were damaged or destroyed, and dozens of small boats run ashore.

While the name Brenda has not been officially retired from the rotating list of North Atlantic tropical cyclone identifiers, it is not in current use on the region's naming lists.

Bret, Tropical Storm *North Atlantic Ocean–Eastern United States, June 29–July 1, 1981* Between 1981 and 2005, five North Atlantic tropical systems—four tropical storms and one very intense hurricane—have carried the identifier Bret.

A powerful TROPICAL STORM when it came ashore in the Chesapeake Bay area between VIRGINIA and Maryland on July 1, 1981, the first Tropical Storm Bret originated over the warm, nourishing waters of the Gulf Stream, some 300 miles north of BERMUDA, on June 29. At peak intensity at the time of landfall, Tropical Storm Bret produced a CENTRAL BAROMETRIC PRESSURE reading of 29.41 inches (996 mb) and sustained winds of 69 MPH (111 km/h), but no lives were lost or major property damage incurred. Tropical Storm Bret of 1981 is the first North Atlantic tropical system identified with the name Bret.

Bret, Tropical Storm *North Atlantic Ocean, August 18–24, 1987* Between August 18 (when it originated off the northwestern coast of Africa) and August 24 (when it dissipated over the mid-NORTH ATLANTIC OCEAN), the second Tropical Storm Bret curled a long, disruptive passage across one of the world's most heavily traveled oceans. The second named tropical system of the 1987 North Atlantic HURRICANE SEASON, Bret was a classic example of a CAPE VERDE STORM that never quite managed to reach full maturity. Of TROPICAL STORM intensity for most of its existence, Bret generated a minimum pressure reading of 29.52 inches (1,000 mb) and 52-MPH (84-km/h) winds, but no deaths or injuries.

Bret, Tropical Storm *North Atlantic Ocean–South America–Central America, August 4–11, 1993* Following a several-hundred-mile course due west, directly along 10 degrees North, between August 4–11, 1993, Tropical Storm Bret became one of the deadliest TROPICAL CYCLONES to affect the South American continent in the 20th century. Originating in the NORTH ATLANTIC OCEAN some 300 miles (500 km) east of Aruba on August 4, the third tropical system identified as Bret steadily intensified as it spun westward, delivering tropical storm-force rains and breaking seas to the islands of Trinidad and Tobago, and to Curaçao, on August 7. Driven by a central pressure reading of 29.58 inches (1,002 mb) and 58-MPH (93-km/h) winds, Tropical Storm Bret treaded ashore in eastern Venezuela on August 8, 1993. Some six inches (152.4 mm) of precipitation fell on the capital city of Caracas as the system steadfastly maintained its westward trajectory across the northern coast of South America. Widespread flash flooding destroyed thousands of houses in Venezuela, and numerous bridges and roadways collapsed or were undermined. Steadily weakening as its eyeless center passed through Venezuela and into Colombia, Bret emerged into the extreme southwestern CARIBBEAN SEA on August 9, having dropped enormous quantities of rain on Colombia, and killing one person. Despite its close proximity to land, the system made an inconsistent effort to reintensify, slipping to TROPICAL DEPRESSION status during the late evening hours of August 9, but deepening to its pre-South American landfall intensity on August 10. Now powered by a central pressure of 29.58 inches (1,002 mb) and bearing 46-MPH (74-km/h) winds, Bret gingerly began a slight recurvature to the northwest as it came ashore in Bahia Punta Gorda, Nicaragua, on August 10, 1993. Ten people in Nicaragua were killed during Bret's rain-laden passage over the Central American isthmus, most of them in flash flood incidents. Maintaining its low-level circulation patterns, Tropical Depression Bret entered the eastern North Pacific Ocean and in keeping with identification practices was renamed Tropical Depression No. 8-E. It effectively dissipated on August 11.

Bret, Hurricane *Gulf of Mexico–Southern United States, August 18–25, 1999* The fourth North Atlantic tropical system dubbed Bret remains the most powerful HURRICANE yet to carry the identifier. Pounding ashore on TEXAS's Padre Island during the mid-afternoon hours of August 22, 1999, Hurricane Bret delivered Category 4 conditions to a wide swath of coastline between the port cities of Brownsville and Corpus Christi, Texas. Born over the extreme southern waters of the Bay of Campeche on August 18, 1999, Bret rapidly intensified into a TROPICAL STORM as it tracked due north. Within one day, Bret's central barometric pressure had deepened to hurricane status, and the system began to track to the northwest, finally recurving into southern Texas on August 22. During its trek across the Bay of Campeche and into the GULF OF MEXICO, Bret continued to intensify, finally reaching Category 4 status just prior to landfall. Along with a central barometric pressure of 27.87 inches (944 mb) and sustained winds of 144 MPH (232 km/h), Bret delivered enormous amounts of PRECIPITATION to the region. DOPPLER RADAR installations in southern Texas observed some 20 inches (508 mm) of rain that fell over Kennedy County alone. Bret further generated several TORNADO outbreaks, and caused extensive wind damage to trees and other small structures. While Hurricane Bret was the first hurricane to strike the Texas coast since Hurricane JERRY in October 1989, it made landfall in a sparsely inhabited portion of the

state, one primarily employed in cattle grazing. Four deaths in Texas were indirectly attributed to automobile accidents caused by Bret's torrential downpours. Recurving to the southwest and into northern MEXICO on August 24, Bret was downgraded to a TROPICAL DEPRESSION, and by the following day had dissipated.

Bret, Tropical Storm *Mexico, June 28–30, 2005* The fifth North Atlantic system named Bret appeared as the second named tropical system of the furiously active 2005 HURRICANE SEASON. Originating over the Bay of Campeche (and in roughly the same area as 1999's Hurricane BRET) on June 28, Bret tracked to the northwest and came ashore near Tuxpan, MEXICO, on June 29 as a weak TROPICAL STORM. While large rainfalls across southern Mexico caused severe flooding in the state of Veracruz, Bret's CENTRAL BAROMETRIC PRESSURE reading of 29.58 inches (1,002 mb) generated sustained winds of 40 MPH (64 km/h) at landfall. Bret quickly weakened after landfall and dissipated by June 30.

The name Bret has been retained on the rotating list of North Atlantic hurricane identifiers. In its sixth incarnation, it is scheduled to reappear in the year 2011.

Bridget, Hurricane *Western Mexico, June 17, 1971* See MEXICO.

buildings and tropical cyclones In the hierarchy of human priorities and interests, the often-destructive effect of a TROPICAL CYCLONE on buildings and other manufactured structures ranks second only to human DEATH TOLLS in its universal ability to provoke a profound variety of cultural responses. Founded on humankind's dark regard for the terrors of homelessness and sustained by a spate of eyewitness accounts, graphic photographs, and dramatic aerial news footage, the glaring losses in property and shelter incurred by many HURRICANES, TYPHOONS, and CYCLONES, have led civilizations in virtually every storm-prone region around the globe to devise laws and other social mechanisms through which the physical and cultural impact of destructive storms on their collective experience can be lessened—or at least better understood.

From JAPAN to the United States, from the islands of the CARIBBEAN SEA to AUSTRALIA and INDIA, the shattered houses, stricken fortresses, toppled steeples, and submerged lighthouses have, at various times and by various people, been viewed as an integral part of history, of meteorology, of physics, economics, government, and sheer human triumph in the midst of ferocious adversity.

A rooftop message from a homeowner attests to the damage and destruction that can be inflicted on buildings by a tropical cyclone's prying winds and pounding rains. Hurricane Andrew swept into southern Florida in 1992 as a Category 5 hurricane. *(NOAA)*

For the historian, the 60,826 houses reportedly destroyed during the Makurazaki, or Occupation, Typhoon of September 17, 1945, became an important dimension in any analysis of those MILITARY OPERATIONS conducted in south Japan during the first few months of the U.S. post–World War II occupation.

For the meteorologist, detailed narrative passages from personal diaries, letters, weather journals, and newspapers that describe the devastation wrought on entire neighborhoods by previous tropical cyclones provide a means by which the unmeasured intensity of long-extinct storms can be in some way observed and recorded for future study as a sort of historical ANALOG.

To those individuals whose interests lie with aerodynamics, with the manner in which a cyclonic system's powerful winds act on different types of structures, the depredations of hurricanes, typhoons, and cyclones present an endless variety of alternating landscapes, roof layouts, and street configurations through which to gather tangible intelligence about the exact physics of wind and about those misengineered structures that wind tunnel tests claimed would never collapse in reality.

Along with the continuing mission of reducing human DEATH TOLLS from hurricanes, typhoons, and cyclones, the desire to minimize damage to property by the high winds, heavy PRECIPITATION counts, and pounding STORM SURGES associated with these furious storms glares at the top of the list of concerns faced by developers, local officials, civil defense authorities, insurance company executives, and building owners alike. In a century that has paid for its share of "billion-dollar" tropical cyclone strikes—storms that have damaged or destroyed tens of thousands of buildings in a single day—the need for ongoing achievements in building design and construction continues to be of critical importance as coastal communities around the world further blossom in the shade of their respective storm seasons.

Because even the weakest of tropical cyclones generally comes ashore with some degree of structural damage—be it as minor as a few lost roof shingles or knocked out windowpanes to the more spectacular failures connected with coastal flooding—the value of stringently enforced building codes directly influences the levels by which a particular structure will survive a storm. This, in turn, decides how safe and effective a shelter a certain building will prove to be to those individuals huddled inside, an issue of paramount concern not only to the disadvantaged homeowner but to harried civil defense authorities as well.

Even some of the strongest structures—stone fortresses, brick lighthouses, and towering churches—have suffered severe damage in past hurricanes. The Great Carolina Hurricane of 1752, for instance, severely mauled Charleston, SOUTH CAROLINA's Fort Johnson, collapsing the stone barracks and dismounting most of the guns. During the GREAT HURRICANE OF 1780, the governor of BARBADOS recorded that his "family took to the center of the building [the Governor's residence], imagining from the prodigious strength of the walls, they being three feet thick, and from its circular form, it would have withstood the wind's utmost rage; however, by half after eleven o'clock, they were obliged to retreat to the cellar, the wind having forced its way into every part, and torn off most of the roof." A much lesser hurricane that came ashore near the Mississippi River Delta in 1839 caused the 87-foot (29-m) brick lighthouse that gave the LOUISIANA town of Balize (*beacon* in Spanish) its name, to tumble into the sea, sparing the life of its keeper but posing a temporary navigational hazard for the waters adjacent to the town's busy south point. In the midst of the NORFOLK–LONG ISLAND HURRICANE of September 2, 1821, the steeple of the Presbyterian meeting house was blown down, while in nearby Wethersfield, Connecticut, a brick schoolhouse was similarly destroyed. The Louisiana hurricane of August 19, 1812, in New Orleans proved that brick buildings provide no certain protection against the ferocity of a hurricane's winds. Eyewitness accounts state that the Market House in the center of the city "was blown down and its 24-inch (60-cm) diameter columns sent tumbling. The roof of the Convent Church was rolled up and blown away . . . many brick houses were demolished, or severely damaged."

In Japan, where centuries of pounding typhoons have resulted in some of the strictest building codes in the modern world, particularly damaging typhoons include the First Muroto Typhoon of September 21, 1934, which utterly razed some 43,000 houses; the Makurazaki Typhoon, in which 60,826 houses were reportedly ruined; and the infamous Typhoon VERA, September 26, 1959, in which 36,109 houses were destroyed. Here, where on any given year typhoons are responsible for 80 percent of the nation's construction costs, government regulated building codes stipulate that houses must be able to withstand sustained wind speeds of 134 MPH (216 km/h). During Typhoon Nancy, September 17, 1961, 137,000 people were left homeless in Japan, prompting the gov-

ernment to institute more stringent building codes, In 1966, however, the 175-MPH (282-km/h) winds associated with Typhoons IDA and Helen left a total of 174 people dead and more than 17,000 houses leveled.

Buys-Ballot's Law Also known as the Law of Storms, this rule allows those individuals caught in the midst of an approaching TROPICAL CYCLONE to determine the approximate position and direction of both the storm's calm EYE and its dangerous EYEWALL winds. First posited in 1860 by Christopher Buys-Ballot, director of the Netherlands Meteorological Institute, the law states that in a tropical cyclone, the deflection of the wind increases as the velocity increases, thereby causing the system's upper level winds to blow parallel to the ISOBARS, or the circular lines that link points of equal barometric pressure. This in turn permits those observers of HURRICANES and TYPHOONS to face into the counterclockwise winds, observe the speed and direction of the lower clouds, and know that the storm's eye and eyewall are located to the right; conversely, travelers in the Southern Hemisphere can face into the *clockwise* winds to discover the center of the cyclone on the left. In both hemispheres, a tropical cyclone's point of lowest barometric pressure rides toward the rear of the system, some 90 to 120 degrees to the right for a hurricane or typhoon and an equivalent eight to 10 points to the left in the case of an Indian Ocean or Australian cyclone. Application of Buys-Ballot's Law to the relationship between wind velocity and the rate of fall of barometric pressure can further be used to determine roughly the distance of the eye relative to the observer's position, thus indicating how much time one has in which to devise a route for avoiding the very heart of a storm's fury.

Buys-Ballot's Law has saved the lives of countless sailors by plotting the system in terms of DANGEROUS and NAVIGABLE SEMICIRCLES, quadrants in which the decision between life and death is very clearly delineated by significant changes in the severity of wind and sea activity. Coming upon an approaching hurricane, typhoon, or cyclone, a ship's master only has to apply Buys-Ballot's Law to the meteorological intelligence gained from observation to know in which direction to navigate safely away from the center of the oncoming storm. While any number of factors can subsequently aid or impede a vessel's progress, the fact that a ship is not blindly steered into the very worst of a tropical cyclone's conditions can easily increase the odds of its survival.

In December 1944, a series of inaccurate observations of wind speed and direction taken by a junior weather officer resulted in a series of course changes that eventually saw Admiral William Halsey's powerful Third Fleet steam into the clutches of Typhoon COBRA. Believing that they were on a course to safety, the fleet's 48 vessels were completely surprised when the worst of the typhoon began to rage on the morning of December 17. By dawn of the following day, three destroyers had been lost—simply overwhelmed by the typhoon's 70-foot (23-m) seas—taking nearly 800 men to the bottom. In a letter dated 13 February 1945, Fleet Admiral Chester W. Nimitz eloquently upbraided commanding officers on the dangers of ignoring Buys-Ballot's Law when he stated, "Seamen of the present day should be better at forecasting weather at sea, independently of the radio, than were their predecessors. The general laws of storm and the weather expectancy for all months of year . . . are now more understood, more completely catalogued, and more readily available in various publications."

C

California *Western United States* Although California's 790-mile (1,271-km) coastline borders on the storm-prone waters of the eastern NORTH PACIFIC OCEAN, its heavily developed canyonside communities, sprawling cities, and major tourist and film attractions have not suffered a particularly active history of HURRICANE strikes. Between 1906 and 2006, California's citrus groves and mountain mining towns were directly affected by six recorded TROPICAL CYCLONES of varying degrees of size and intensity. Representing an average of one TROPICAL DEPRESSION, TROPICAL STORM, or hurricane every 20 years, only one system was actually of hurricane-strength (winds of 74 MPH [119 km/h] or higher) when it came ashore, in San Diego county on August 18, 1977. Dubbed Doreen, the hurricane dropped seven inches (178 mm) of rain on the state's southwest portion, spawning flash floods that swept away more than 300 hillside houses and claimed five lives. Prior to 1977, four storms of tropical storm intensity (winds of between 39 and 73 MPH [63–117 km/h]) struck the state, while in September 1978, Tropical Depression Norman delivered torrential rains and 25-MPH (40-km/h) winds to the metropolitan Los Angeles area. Ninety-five percent of that year's valuable raisin crop was ruined, costing growers nearly $1 billion.

In one regard, California's dearth of hurricane activity is rather surprising. Positioned between latitudes 32 and 46 degrees North, the curvaceous spread of the state roughly shares the same parallels as those other storm-ridden regions of the planet, including the southeastern states of GEORGIA and SOUTH CAROLINA, and the TYPHOON-laden shores of central CHINA. Bordered by warm, 79-degree F (26°C) Pacific waters that spawn on average seven tropical cyclones per June-to-November season, it would stand to reason that California would have as extensive a history of hurricane activity as do those other areas.

But the state does not, partly because of its location on the western edge of the North American continent. During the months (August through November) of peak hurricane generation in the eastern North Pacific, large high-pressure ANTICYCLONES develop near latitude 30 degrees North, the waters adjacent to the Hawaiian Islands. Because anticyclones like the BERMUDA HIGH are known to influence the tracks of respective hurricanes, the dominating presence of similar anticyclones north and west of California has a tendency to deflect incoming North Pacific hurricanes to the southwest, thus sparing the densely populated coastline the fury of storms that have historically been very intense. Furthermore, most North Pacific hurricanes originate southwest of California, off the coast of northern MEXICO. Because the clockwise-spinning Hawaiian anticyclones tend to influence wind patterns between 10 and 20 degrees North by directing them in a westward direction, a majority of North Pacific hurricanes follow a westerly course that takes them several hundred miles to the south of HAWAII before dissipating over the central Pacific's cooling eddies.

No meteorological buffer is entirely constant, however, and it is during those early summer and out-of-season months when the Hawaiian anticyclones are at their weakest, or gone altogether, that California has suffered some of its most memorable cyclonic strikes. On August 18, 1906, a rain-laden tropical storm dropped 5.66 inches (127 mm) of rain on Needles, California, causing severe flash floods and some tree damage but no fatalities. During the last week of

September 1939, a Pacific-bred tropical storm—picturesquely identified as "El Cordonazo" or the "Lash of St. Francis"—delivered 50-MPH (81-km/h) winds and 11 inches (279 mm) of PRECIPITATION to the hills and mountains of southern and central California. At least 45 lives were lost (many of them at sea), while mud and rock slides destroyed several bridges and road networks. As it moved northward on September 26, El Cordonazo swept the dusty streets of Los Angeles with some 5.66 inches (127 mm) of precipitation. According to contemporary media reports, more than $2 million (in 1939 currency) in property losses were incurred. More recently, Tropical Storm Hyacinth came ashore near Los Angeles County on September 6, 1972. Downgraded to a tropical depression, Hyacinth's 30-MPH (48-km/h) winds and heavy rains were remnants of the first tropical cyclone to strike the state since the unnamed tropical storm of September 25, 1939. On September 10, 1975, Tropical Storm KATHLEEN pounded San Diego and its extensive patchwork of naval and marine installations with 50-MPH (81-km/h) winds and extensive floods. Two people were killed; damages were estimated at $333 million. In October 1982, a year that also saw Hawaii

struck by the rare late-season Hurricane Iwa, Tropical Depression Olivia sent heavy rains into the valleys of San Diego County.

On September 22–23, 1999, moisture from a dissipating Tropical Storm Hilary sparked a band of severe thunderstorms across southern California, causing flooding and minor mudslides. Intensifying to a Category 2 hurricane, Hilary caused heavy rains and localized flash flooding in the southern Rocky Mountains but no deaths or injuries.

Camille, Hurricane *Southern–Northeastern United States, August 17–22, 1969*   A North Atlantic HURRICANE of extraordinary intensity and rare meteorological violence, Camille stands as one of the most memorable TROPICAL CYCLONES in U.S. history. The second of only three Category 5 hurricanes to have yet hit the Continental U.S., Camille's record-low CENTRAL BAROMETRIC PRESSURE at landfall of 26.84 inches (909 mb), wailing 175-MPH (km/h) winds, 30-inch (762-mm) PRECIPITATION counts, and thunderous 22-foot (7-m) STORM SURGE visited unprecedented devastation on the LOUISIANA and MISSISSIPPI coasts and across the highlands of western VIRGINIA between

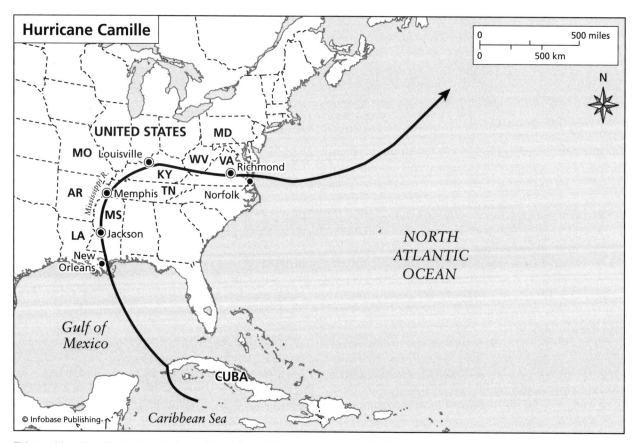

This tracking chart illustrates Hurricane Camille's whirring passage across the Gulf of Mexico and into Mississippi in August 1969.

August 17 and 20, 1969. Small, tightly coiled, and particularly deadly, Camille's 50-mile (80 km)-wide EYEWALL drove the frothing waters of the GULF OF MEXICO nearly 25 feet (8 m) above sea level in Pass Christian, Mississippi. Hundreds of BUILDINGS were utterly demolished, their shattered remains left to litter the Mississippi River Delta bayous. Although downgraded to a TROPICAL STORM as it sheered through inland Mississippi on the morning of August 18, Camille's fury would be felt across western Tennessee, central Kentucky, and western Virginia, where dying rains spawned deluging rains and sweeping landslides on August 19 and 20. One hundred forty-three people perished along the Louisiana-Mississippi coast, while another 113 lives were lost in the northeast. Nearly $1 billion in property losses were assessed, at the time making Camille—in the words of Dr. Robert H. Simpson, former director of the NATIONAL HURRICANE CENTER (NHC) and co-creator of the SAFFIR-SIMPSON SCALE—"the greatest storm of any kind ever to have affected the mainland of the United States."

HURRICANE CAMILLE CHRONOLOGY

August 5: Immediately upon exiting northwestern Africa, one of several low-pressure TROPICAL WAVES begins to organize over the NORTH ATLANTIC OCEAN's warm waters approximately 100 miles (161 km) due east of the Cape Verde Islands. Slowly spinning to the west at nearly 14 MPH (23 km/h), the deepening ring of thunderstorms generates 20-MPH (32-km/h) gusts and intermittent rain showers. If the seedling system should reach maturity, it will become the 1969 HURRICANE SEASON's second hurricane, one that had previously witnessed the formation of a healthy Tropical Storm Anna on July 23 and a fairly mild Hurricane Blanche on August 6. Both Anna and Blanche had remained over the Atlantic, thereby sparing nearby landmasses the full brunt of their fury.

August 10: Having spent the past four days moving due west across the North Atlantic, the as-yet-unnamed and unnumbered tropical disturbance skates across the northern Leeward island of GUADELOUPE to enter the eastern CARIBBEAN SEA. While sustained 12-MPH (19-km/h) winds and gusty rains cause Guadeloupe's palm trees to sway and its residents to seek shelter, the island remains happily unscathed by the disturbance's inclement passage. A relatively disorganized tropical system, storm watchers decide that it is unlikely the disturbance will ever intensify into a serious meteorological crisis.

August 12: The decision is made by officials at the United States Weather Service that the tropical disturbance, currently located less than 100 miles (161 km) south of Cabo Beata, HISPANIOLA, is not a suitable candidate for that year's PROJECT STORMFURY seeding operations. It is judged that the system is too close to land and its circulation too weak to benefit from seeding with silver iodide crystals.

August 13: Virtually unchanged after its two-day jaunt across the Caribbean Sea's, warm, humid feeding grounds, the seemingly frail tropical disturbance brushes along the south coast of JAMAICA. Gale-force gusts, two-foot (1-m) seas, and occasional downpours lash Jamaica's lush vegetation and rocky coves for almost 12 hours but cause no injuries or fatalities. As is commonly seen with tropical systems reaching this latitude (19 degrees North), the disturbance begins to recurve, or alter its course, to the northwest.

Some 1,000 miles (1,609 km) to the north, an intense upper-level AIR MASS begins to spread its cooling influence across the mouth of the Mississippi River. Itself having developed a cyclonic circulation while elongating across the southern United States during the past three days, the frigid air mass—complete with closed ISOBARS—will soon deeply penetrate the confines of the Gulf of Mexico, reaching as far south as MEXICO's Yucatán Peninsula.

August 14: Now centered some 320 miles (515 km) south of Havana, CUBA, and maintaining its meandering course west-northwest, the tropical disturbance suddenly begins to intensify shortly after noon. Indeed, it will deepen so quickly (from a central pressure of 29.50 inches [999 mb] on the morning of August 14 to 29.14 inches [988 mb] at midnight of the same day) that astonished storm trackers at the NHC in Miami, Florida, will barely have enough time to observe its progression through the intermediate stages of tropical depression and tropical storm. Just before midnight on the August 14, as its sustained winds top 65-MPH (105-km/h), the still-intensifying tempest is officially christened Tropical Storm Camille.

Although its cyclonic circulation has begun to show some preliminary signs of subsidence, the upper level air mass centered over Mississippi and ALABAMA continues its migration southward throughout the day and night of August 14. As its eastern leading edge nears to within 50 miles (80 km) of northwestern Cuba, meteorologists speculate on what effect its upper level winds and TEMPERATURE will have on Camille's future intensification, course, and forward rate of speed.

August 15: As Tropical Storm Camille's barometric pressure continues to plummet, falling to 28.67 inches (971 mb) by noon on August 15, the system is upgraded to hurricane status. Whirring some 480 miles (772 km) southwest of Miami, the hurricane's 82-MPH (132-km/h) winds, thickening rains, and six-foot (2-m) seas are felt for several miles along the shores of extreme western Cuba. Firmly engaged by recurvature rules, Camille pursues its swing to the north-northwest—a trajectory that, if maintained, will send the Category 1 storm ashore on western Florida's hurricane-prone Panhandle sometime in the next 36 to 48 hours. Accordingly, late-afternoon HURRICANE WATCHES are posted for the 110-mile (177 km) stretch of coastline between Apalachicola and Fort Walton Beach.

In the meantime, the ever-encroaching dome of the upper level air mass to the north has inexplicably begun to disintegrate over the Gulf of Mexico's subtropical waters. As the barometric pressure within its cold core continues to rise, as its elliptical isobars grow progressively wider before collapsing into individual cold pockets, the upper level air mass gradually sinks across north Cuba. In later months, long after Camille has blown itself out over the chilling expanses of eastern CANADA, tropical meteorologists will closely study the young hurricane's remarkable interaction with the subsiding cold air mass. After poring over weather charts, satellite photographs, and other meteorological data from August 15, they

Still the second-most intense hurricane to have made landfall in the United States, Camille blasted the United States Gulf coast in August 1969, with Category 5 winds and storm surge conditions. Here a fishing boat sits firmly aground amid the wreckage of Camille's fury. *(NOAA)*

will conclude that in the process of sinking, the air mass fortuitously aided in the hurricane's subsequent hyperintensification by releasing its potential energy directly into the tropical cyclone. It is also probable that in the act of subsiding, the air mass left behind a low-pressure zone in the upper atmosphere, a vast "hole" that Camille's ever-increasing OUTFLOW swiftly expanded to fill. Having developed a more efficient excretory system, the hurricane was free to draw escalating quantities of warm, moist air into its ascending EYE, thereby increasing the height and density of the towering CUMULONIMBUS CLOUDS in its dangerous EYEWALL.

Whatever the cause of its sudden deepening, Camille does not pass up an opportunity to shave another 23 millibars promptly off its central pressure within the space of 12 hours. Between noon and midnight of August 15, as Camille whirls northwestward through the Yucatán Channel, its barometric pressure falls from 28.67 inches (971 mb) to 27.99 inches (948 mb), making the system of firm Category 3 status on the Saffir-Simpson Scale. Hiking its sustained winds to 115 MPH (185 km/h), Camille lashes northwestern Cuba with 130-MPH (209-km/h) gusts, four inches (102 mm) of rain, and pounding surf.

August 16:   Still bound for a possible landfall on the Florida Panhandle, Hurricane Camille's shrinking eye and tightening eyewall slip into the Gulf of Mexico during the early morning hours of August 16. Now centered less than 400 miles (644 km) southwest of Pensacola, Florida, Camille's intense circulation patterns are subjected to an equally severe round of aerial and satellite surveys by an increasingly anxious NHC. A barometric pressure reading taken by reconnaissance aircraft just before midday indicates that Camille has further strengthened during the night and seen its pressure tumble from 27.99 inches (948 mb) at midnight on August 15 to 27.13 inches (919 mb) at noon the following day. Photographs taken of Camille by the *NIMBUS III* weather satellite harrowingly illustrate the hurricane's compact, 158-MPH (254-km/h) eyewall, tiny 11-mile-wide (18-km) eye, and wispy splay of CIRROSTRATUS CLOUDS that comprise its fully operational OUTFLOW LAYER.

In an updated bulletin issued shortly thereafter, the NHC warns that Camille is "small but dangerous" and due for a possible landfall in northwest Florida at dawn the following day. Thousands of Floridians, mindful of past hurricanes that had ravaged their white-crescent beaches and delicate oyster beds, heed posted HURRICANE WARNINGS and leave their oceanfront homes by car, boat, and plane. As though reinforcing the abiding senses of urgency and caution that overhang dinner tables along the entire gulf coast on this particular evening, further weather bulletins issued during the night stress the "extremely dangerous" nature of this "very intense and dangerous storm." A capricious hurricane from the start, Camille's continued intensification, coupled with a late-afternoon course change to the northwest, keeps storm trackers fearful as well as fascinated.

Neatly sliding into the trough created by the sinking of the cold air mass only the day before, Camille steadily swings northward across the Gulf of Mexico and bears down on the island-lined bays and harbors of Alabama and Mississippi. Just after dusk on August 16, hurricane warnings are summarily dropped in northern Florida but rehoisted between Mobile, Alabama, and Grand Isle, Louisiana. For the second time in 12 hours, tens of thousands of gulf-coast residents are advised to board up their cottages, their store fronts, their restaurants, and their motels and evacuate to higher ground. Thousands of them do and are driven deep inland by the still, humid air—heavily scented with sea salt—that during the day has gradually settled over their low-lying bayou communities like a portent and only now has finally convinced them.

Some 310 miles southeast of the languid delta town of Pass Christian, Mississippi, Camille concludes its final day at sea by spawning a brood of 170-MPH (274-km/h) gusts, torrential rain showers, and 40-foot deepwater swells. Droning beneath the turbulent darkness, a solitary WP-3 Orion aircraft first jolts its way through Camille's violent eyewall into the storm's calm eye and then straight through the roiling eyewall on the opposite side. Inside the plane, an array of sophisticated instruments—from ANEROID BAROMETERS, to digitized ANEMOMETERS, to the sweeping arcs of radar screens in search of rain bands—gather as many of Camille's vital statistics as the safety of the mission will allow. Of them, one in particular—the hurricane's central barometric pressure—is of most interest to Camille's trackers because it will not only reveal how sharply the system has further intensified but will evidence just how severe its winds, rains, and tides are likely to be at imminent landfall. Promptly radioed to the NHC in Miami for analysis, the data does not look very promising; in fact, it appears positively grim at a catastrophically low 26.73 inches (905 mb)—one of the top 20 lowest barometric pressures yet recorded in a North Atlantic hurricane.

August 17: Seemingly insatiable in its desire to shatter all existing low-pressure records, Hurricane Camille spends the morning of August 17 further intensifying. A late-afternoon Air Force reconnaissance flight observes a barometric pressure reading of 26.61 inches (901 mb), while high-altitude winds within Camille's eyewall now exceed 210 MPH (338

Two of the most famous photographs in the study of tropical cyclones, these before-and-after photographs of the Richelieu Manor apartment building in Pass Christian, Mississippi, are often held as an example of the unmitigated destructive potential of intense hurricanes. The Richelieu's Gulfside location made it an unhappy target for Camille's 25-foot (8-m) storm surge. *(NOAA)*

km/h). As Camille stalks to within 250 miles (402 km) of Mobile, Alabama, the early morning preparations for its forecasted arrival on the gulf coast assume a more diligent—even quickened—pace. Schools, churches, and county courthouses are hurriedly converted into crowded shelters for the lines of men, women, and children being chased from their homes by the very real threat of Camille's crushing, 20-to-25-foot (7–8 m) storm surge. Emergency water wells are uncapped. Small tin-roofed outbuildings, mobile homes, and radio antennas are anchored to the ground with cable or rope. Cattle are herded into barns. Thousands of tons of shipping—cargo vessels, shrimp boats, excursion cruisers, private yachts—either make for the open sea or dangerously crowd the exposed anchorages of Lake Pontchartrain and Mobile Bay. The clanging steel works of nearby Biloxi, Mississippi, close down early, posting just a skeleton crew to oversee the mighty cranes and towering gantries that are already creaking and swaying before Camille's scouting gusts. Presiding over a midmorning press conference at the National Hurricane Center, then-director Dr. Robert Simpson solemnly concludes that "Never before has a populated area been threatened by a storm as extremely dangerous as Camille."

Its crew earlier airlifted to safety, an abandoned deep-sea oil drilling rig anchored off Louisiana's east coast begins to rock beneath Camille's mounting, 45-foot (15-m) swells and rising winds shortly before dusk. An automatic anemometer on board the rig records an extreme wind speed of 172 MPH (277 km/h), one of the fastest ever observed in the Gulf of Mexico.

In the Pass Christian, Mississippi, some 60 miles (97 km) northeast of New Orleans, 11 devout parishioners take refuge in a large stone church. Terrified by the reports of the meteorological armageddon now bearing down on them, they decide to huddle between the shadowy pews and pray.

On the adjacent side of Pass Christian, overlooking the coastal highway that winds its way from Biloxi to Bay St. Louis, 23 hearty residents of the Richelieu Apartments—a three-story brick horseshoe with a large in-ground swimming pool in the forecourt—gather in an upper-story residence for an impromptu HURRICANE PARTY. Amid music, beer, potato chips, and candlelight, the almost two-dozen revelers dance before the apartment's picture windows, oblivious to the slow but perceptible rise of the Mississippi Sound just across the street.

By midevening on August 17, Camille's rampaging eyewall is rapidly spreading its blistering, 190-MPH (306-km/h) winds and thunderous 23-foot (7-m) storm surge across east Louisiana's Boudreau Islands. In the Louisiana delta town of Boothville, Camille's 15-foot (5-m) tides inundate the U.S. WEATHER BUREAU's radar station with 4 feet of water as screeching winds carry away its anemometer. Four meteorologists at work tracking Camille's progress are trapped by the rising seas, while downed power lines cast the station's radar screens into ominous darkness. Shoaling against the steep approaches to Lake Borgne, Camille's surge soon forces its way into the shallow lake. Ten-to-twelve-foot waves undermine seawalls and sweep scores of cottages from their pilings, drowning nine people. Several spans of the Bay St. Louis bridge tumble into the frenzied surf. It will later be determined that Camille's shrieking winds tore across the mouth of the Mississippi River with the sound intensity of 120 decibels, a relentless whine equivalent to that of a jet engine.

Slightly diminished by its passage across Louisiana, Camille wails ashore near Gulfport, Mississippi, a few minutes before midnight on August 17. The most astonishing hurricane to have then affected to the southern United States in living memory, Camille's central pressure of 26.84 inches (909 mb), conjoined with its 183-MPH (295-km/h) winds and bulleting rains, quickly carve a 32-mile-wide (51-km) trail of calamitous devastation through west Mississippi's Hancock and Harrison counties. To the west in the bayside towns of Waveland and Bay St. Louis, the onset of Camille's NAVIGABLE SEMICIRCLE uproots hundreds of trees, utility poles, fence posts, and road signs and then sends them shooting into nearby buildings at impaling speeds. After Camille's 23-foot (8-m) storm surge has blasted ashore to the east, much of southern Hancock County will be submerged beneath 18 feet (6 m) of debris-swollen water. A litany of statistics later released by the state of Mississippi confirms Camille's grim toll in the west: 12 people dead, another 2,095 injured, 602 houses completely destroyed, 1,775 houses seriously damaged, some 4,375 families left homeless.

August 18: To the east, however, in those Harrison County settlements subjected to the full fury of Camille's DANGEROUS SEMICIRCLE, the human and material cost proves much higher. During the early morning hours of August 18, almost all of Pass Christian is destroyed, first by Camille's 200-MPH (322-km/h) winds and then by its cresting 25-foot (8-m) surge. The 11 panicked parishioners gathered beneath their quivering church roof are instantaneously killed when the entire structure collapses beneath Camille's 210-MPH (338-km/h) gusts, burying them beneath tons of rain-soaked stone and timber. Gatecrashing the hurricane party at the Richelieu Apartments, Camille's winds shower the almost two dozen startled guests with broken glass and palm fronds before systematically tearing the large building to pieces. Feeling for bathrooms, closets, and corridors, 20 frantic celebrants are not able to reach safety before Camille's gathering surge overwhelms the apartment complex, completely demolishing it and drowning them. Three lucky survivors miraculously escape by *swimming* out a third-floor window. Aerial photographs later taken of the scene stunningly attest to the thoroughness of Camille's destruction—of the Richelieu Apartments, only the waterlogged foundation and in-ground pool remain in place.

Some 15 miles (24 km) east of Pass Christian, three large freighters on the 3,800-ton *Alamo Victory*—are beached at Gulfport. Scores of mobile homes in neighboring Biloxi are swept from their foundations before coming together as one gigantic barricade. All the windows in Long Beach, Mississippi's, City Hall are punched out by Camille's 200-MPH (322-km/h) gusts, while the building's steps are piled high with splintered planks, porch columns, and asphalt roof shingles. Mile-long sections of the busy coastal highway are left buckled behind crumbled seawalls. One hundred fourteen people in Harrison County die, and more than 2,000 others are injured, 234 of them seriously; 3,000 dwellings are demolished, as are 116 mobile homes; some 45,000 families are rendered homeless.

Moving across the timbered inlands of Mississippi, Camille is downgraded to a tropical storm with maximum wind speeds of less than 74 MPH (118 km/h) by midafternoon of August 18. Amid growing reports of looting, Mississippi's governor declares martial law. Appalled by the apparent extent of Camille's destruction, President Richard M. Nixon declares Mississippi a federal disaster area and allocates $1 million for immediate aid. Equally awed by Camille's double-eyewall, the National Hurricane Center states that Camille was the strongest hurricane to strike the United States since 1935. Tons of food, clothing, and medicine pour in from sympathetic communities across the United States.

August 19: Again downgraded, this time to a tropical depression, the disintegrating core of former-Hurricane Camille progresses across west Tennessee and the rolling hills of central Kentucky and into the highlands of south Virginia. Slowly moving east over Virginia's Blue Ridge Mountains, Camille spawns 30-inch (762-mm) downpours that force the gushing James River to overrun its banks. In Virginia and West Virginia 155 people die, a majority of them drowned in flash floods or buried beneath murky landslides. Damage estimates top $1 billion, making Camille one of the single most-expense hurricanes in U.S. history.

September 8: Returning from a state visit to Mexico, then President Richard Nixon tours those areas of Louisiana and Mississippi ravaged by Camille. Fourteen federal agencies, in addition to the United States military and the American Red Cross, are already at work erecting tent cities, repairing roads and bridges, restoring water, electrical and communications networks, and burying the dead. Landing by helicopter at Gulfport, the president meets with Mississippi's governor and two senators, promising "a continuation of interest by the federal government" in reconstruction efforts. In late December 1969, Congress appropriates more than $180 million in aid to victims of Hurricane Camille.

Canada *North America* An immense nation of rocky mountains, fertile lowlands, and deepwater bays, Canada has over the course of the past two centuries witnessed a procession of notable TROPICAL CYCLONE activity. With more than 36,000 (57,936 km) combined miles of coastline bordering on two storm-prone bodies of water—the NORTH ATLANTIC OCEAN to the east and the eastern NORTH PACIFIC OCEAN to the west—Canada is particularly vulnerable to mid- and late-season HURRICANE strikes on its Atlantic shores. Between 1752 and 2006, an estimated 83 mature hurricanes, TROPICAL STORMS, and dissipating TROPICAL DEPRESSIONS made landfall on the northeast provinces (Newfoundland, Nova Scotia, Quebec, and Ontario), delivering torrential rains and flailing winds to its frigid tundra, spidery rivers, and hearty cities.

On average, Canada's cool beaches, squat islands, and principal eastern ports are likely to be inundated by at least one tropical cyclone of varying size and intensity every two years and by a major hurricane—storms characterized by CENTRAL BAROMETRIC PRESSURES of less than 28.50 inches (965 mb) and wind speeds greater than 110 MPH (177 km/h)—every 50 years. In some instances, two hurricanes have directly affected the expansive nation in a single year, as seen during the 1869 HURRICANE SEASON when two strong storms ravaged the Maritimes on September 8 and October 4; during the 1904 season when two unnamed hurricanes in their decaying stages—September 16 and November 14—claimed six lives and inflicted significant property damage on Newfoundland and Nova Scotia; and during the remarkable 1954 season when two violent hurricanes—EDNA of September 11, and HAZEL of October 15—left almost 90 people dead.

On September 9, 1775, Canada suffered what is generally considered its deadliest confirmed hurricane landfall when the Independence Hurricane blasted Halifax, Nova Scotia, and the eastern ports of Newfoundland with seering winds and heavy rains. As hundreds of buildings were demolished, their steep colonial roofs and leadpaned windows violently torn asunder, more than 200 vessels, struggling to return to port against the furious northeasterly winds, were sunk. Some 4,000 mariners, from Grand Banks fishers to Royal Navy press members, were drowned, while tens of thousands of pounds of damage was tallied.

On September 29, 2003, Canada suffered its most destructive and deadly tropical cyclone strike in nearly a century as Hurricane JUAN delivered Category 2 winds, rains, and surge conditions to the Atlantic Maritime provinces of Nova Scotia and Prince Edward Island. In total, eight people lost their lives, while nearly a quarter of a billion dollars in property losses were assessed in Juan's paralytic wake.

Canada's position north of the 43rd parallel (where sea-surface temperatures rarely reach the minimum 79-degree threshold needed for tropical cyclone generation) affords the country a short, relatively mild hurricane season. Ostensibly ranging from June 1 to November 30, there is no record of Canada having been affected by a mature June or July hurricane on either of its coasts. On June 26, 1972, the soggy remnants of Hurricane AGNES delivered nearly an inch of rain to east Ontario, while Tropical Storm Barry lashed Nova Scotia with 3-inch (76-mm) rains and 50-MPH (80-km/h) gusts on July 9, 1995. Of the two storms, Barry proved the more severe as it overflowed storm drains, flooded a church basement, and inundated acres of crop land in Digby County. No casualties were reported in either event.

It is during the late-summer months of August, September, October, and November when the first of the CAPE VERDE STORMS begins to transit the North Atlantic that the number of documented hurricane or tropical storm strikes in Canada begins to rise sharply. On October 1, 1752, more than 50 vessels, from supine merchant ships to spitfire warships, were run ashore and stranded when a tropical cyclone of undetermined intensity made landfall near Nova Scotia's Louisburg Harbor. As with a number of early North Atlantic hurricanes, no confirmed DEATH TOLL exists for this particular storm, although it is safe to assume that some loss of life did ensue.

During the following season, yet another hurricane or tropical storm swept the eastern reaches of Canada, this one on October 7, 1753. Contemporary accounts state that some 40 vessels, mostly merchant ships, were wrecked at Cape Breton Island, Nova Scotia, and that an estimated 100 lives were lost in the tempest. Less than four years later, on September 24, 1757, Louisburg Harbor was affected by a tropical cyclone that sank two British warships, drowning scores of sailors and leaving much of the anchorage clogged with wreckage.

Toward the end of the 18th century, another memorable Canadian hurricane came ashore near Halifax Harbor on September 25, 1798. More than a dozen watercraft were disabled or sunk, including an American brig and a British frigate.

Nova Scotia's thriving seaports were again pummeled by a powerful tropical cyclone on August 18, 1830. Roaring northward over NEW ENGLAND's coastal waters, the storm caused substantial damage on land as well as at sea. Dozens of fishing schooners and coastal traders were lost in the heavy surf, while fierce rains and scudding winds unroofed settlements from Cape Sable to Sydney. Loss of life from this

cyclonic system was considered severe, with more than 200 people listed as unaccounted for.

More than a quarter of a century later, the gale of September 8, 1869, inundated the mouth of the St. Lawrence River with a destructive STORM SURGE, sinking two ships and retarding water transport between Quebec and the rest of the world for nearly a week. On October 4 of the same year, Saxby's notorious hurricane forced Category 2 tides in the Bay of Fundy to rise six feet above recorded high water, inflicting heavy damage on shipping and coastal villages. Not long after Saxby's Hurricane claimed an estimated 40 lives in Canada, an even deadlier unnamed hurricane caused more than 1,200 vessels to founder as it blasted ashore at Newfoundland's Avalon Peninsula on the night of August 25, 1873. An estimated 598 seafarers were caught at sea by the hurricane's unexpected arrival and drowned, making the 1873 hurricane one of the deadliest of all Canadian furies.

The first decade of the 20th century proved equally stormy for the Canadian maritimes as, between October 1903 and October 1909, nothing short of eight tropical cyclones of differing sizes and intensities directly affected the region. These cyclonic systems all had earlier made landfall in the CARIBBEAN SEA or along the eastern seaboard of the United States. Now, in their decaying stages, they underwent RECURVATURE and crossed Canada's borders, meaning that they delivered large PRECIPITATION counts and some high winds to their respective areas but little else. Again, no casualty figures are available for these storms, most likely because few lives, if any, were lost.

More recently, an unnamed Category 1 hurricane struck London, Ontario, on September 25, 1941, with 81-MPH (130-km/h) winds and two inches of rain. First blasting ashore near Cape Hatteras, NORTH CAROLINA, on the morning of September 24, the system initially lost its hurricane-force winds as it bounced over the Appalachian Mountain chain and penetrated west Pennsylvania, but it again strengthened as it crossed the placid waters of Lake Erie. Seven people in London were killed, while hundreds of thousands of dollars of property losses were recorded.

In 1954, Canada suffered serious consequences from two of that year's major Atlantic hurricanes, Edna and Hazel. Edna, which swung into Nova Scotia from Maine on September 11, ruined nearly all of Nova Scotia's ripening apple crop, while Hazel claimed 80 lives and wrought more than $100 million in damages as it offloaded what meteorologists mathematically calculated to be some 300 million tons of rain onto the city of Ontario on October 15. Now regarded as Canada's most renowned hurricane, Hazel remains the most destructive hurricane to have struck Canada during the latter half of the 20th century.

On August 16, 1971, the second hurricane of that season, Beth, delivered more than 11.5 inches (279 mm) of rain to Nova Scotia, causing widespread flash flooding, power outages, and mudslides. One woman was killed and another 30 people injured in traffic accidents and bridge washouts.

During the 2000 North Atlantic hurricane season, the southeastern coast of Newfoundland was directly affected by one tropical and one post-tropical system; Hurricane FLORENCE (which delivered tropical storm force winds on September 17), and the extratropical remnants of Hurricane Michael, which brought late-season winds and rains to the province on October 19–20. Of moderate tropical storm intensity at landfall, Tropical Storm FLORENCE's 58-MPH (93-km/h) winds shivered trees and downed power lines across the southern reaches of Newfoundland, but claimed no lives or serious injuries. Bearing a central pressure of 28.55 inches (967 mb), the waterlogged remnants of Hurricane Michael, downgraded to an extratropical system on October 18, burnished Newfoundland's rocky southern coast with sustained 86-MPH (138-km/h) winds and 107-MPH (172-km/h) gusts. No major damage or loss of life was reported.

The 2001 North Atlantic hurricane season was an active one for eastern Canada, with no less than three tropical systems affecting the region between September and October. Between September 14 and 15, 2001, Hurricane Erin delivered 81-MPH (130-km/h) winds, 30-feet (9.3-m) seas, and five-inch (131-mm) rains to Newfoundland's Avalon Peninsula; within one week, the extratropical remains of Hurricane Gabrielle brushed past Canada's eastern coast, bringing up to 6.8 inches (175 mm) of PRE-CIPITATION to the Avalon Peninsula. An extreme gust of 81 MPH (130 km/h) was observed at Cape Race. Gabrielle's flooding rains caused widespread disruption in Newfoundland's largest city, St. John's, where many residents claimed it was the worst storm in living memory. One month later, on October 15, Tropical Storm Karen became the second tropical cyclone to strike eastern Canada in as many years. Powered by a central pressure at landfall in Liverpool, Nova Scotia, of 29.44 inches (997 mb) and sustained winds of 46 MPH (83 km/h), the former Category 1 hurricane did not claim any lives or cause major property losses. An extreme gust of 65 MPH (104 km/h) was observed near Halifax, while some 1.73 inches (46 mm) of precipitation was recorded in Yarmouth.

The 2003 North Atlantic hurricane season was an historic one for Canada, with two tropical systems—one of them the most destructive tropical cyclone

witnessed in Canada in many decades—affecting the eastern Maritime Provinces and southern Ontario during the month of September. Following a similar trajectory to deadly Hurricane HAZEL of 1954, the former Hurricane ISABEL—now downgraded to a tropical storm—traversed Lake Erie and swept into southern Ontario, bringing sustained 34-MPH (55-km/h) winds, 45-MPH (73-km/h) gusts, and 1.9 inch (50 mm) precipitation counts on September 19, 2003. As its powerful winds generated 16-foot (4-m) wave heights across Lake Ontario, Tropical Storm Isabel destroyed power lines, felled hundreds of trees, and spawned numerous localized flood emergencies. No lives were lost to Isabel's windy passage. Less than two weeks later, Hurricane JUAN twirled ashore to the west of Halifax, Nova Scotia, on September 29, 2003. Born nearly 300 miles (500 km) to the southeast of Bermuda on September 25, Juan quickly deepened as it spun to the north-north-west, becoming a hurricane during the afternoon of September 26, and reaching its maximum intensity (28.64 inches [970 mb]) on September 27. Following a more northerly trajectory, Juan subsequently weakened as it neared the Canadian Maritimes, but was of minimal Category 2 intensity as it rushed ashore in Nova Scotia during the early morning hours of September 29. Powered by a central barometric pressure of 28.76 inches (974 mb), Juan's sustained, 80-MPH (129-km/h) winds caused considerable damage in and around Halifax, and on neighboring Prince Edward Island. Meteorological equipment in Halifax Harbour recorded a sustained wind speed of 98 MPH (158 km/h), and a maximum gust of 144 MPH (232 km/h). Some 1.49 inches (38 mm) of precipitation fell at Halifax International Airport, while a six-foot (2-m) storm surge established a new record for Halifax Harbour. Juan's pounding seas flooded piers and other harborside installations, while 105-MPH (169-km/h) gusts uprooted or destroyed some 100 million trees across the island province, downing utility lines (leaving over a quarter of a million people without electricity), unroofing small BUILDINGS, and decimating marinas. The most destructive and deadly tropical cyclone strike in Halifax's contemporary history, Juan directly and indirectly claimed a total of eight lives and caused nearly $200 million (in 2003 dollars) in property losses.

During the record-shattering 2005 North Atlantic hurricane season, central and eastern Canada were affected by three post-tropical systems, including Hurricanes KATRINA and RITA. On August 31, the watery remnants of historic and deadly Hurricane Katrina passed over the Niagara Peninsula, dropping enormous quantities of precipitation across Lake Ontario and the St. Lawrence River

Valley. Downgraded to an extratropical cyclone as it churned northward, Katrina's fading RAIN BANDS dropped some 3.93 inches (100 mm) of precipitation across southern Canada, creating numerous flash floods. Less than one month later, residual moisture from former Hurricane Rita merged with an extratropical system then traveling eastward. Heavy rains inundated portions of Quebec and the Maritimes; in Stephenville, Newfoundland, some 5.9 inches (150 mm) of precipitation fell between September 26 and 27, causing widespread flood emergencies. While transitioning to an extratropical cyclone, the former Hurricane OPHELIA cast herself ashore in eastern Nova Scotia during the early morning hours of September 18, 2005. Although forecast to make landfall as a tropical storm, Ophelia steadily weakened as it approached the Canadian Maritimes; it nevertheless produced gusts of 50 MPH (80 km/h) and drowning, 3.14 inch (80 mm) rainfalls across much of eastern Nova Scotia. Continuing its extratropical deepening, Ophelia's remains crossed the cooling waters of the Cabot Strait and coursed through Newfoundland, leaving one person dead in Halifax, Nova Scotia.

On September 13, 2006, the extratropical remnants of Hurricane Florence delivered high winds and heavy rains to the Maritime provinces of Newfoundland and Nova Scotia. Under a tropical storm warning for most of the 13th, portions of eastern Canada received wind gusts of 100 MPH (161 km/h), three inches (76.2 mm) of precipitation and high surf conditions. No significant property losses were recorded.

Canada's west coast, on the other hand, has not endured such a violent history of hurricane activity. Although more than 800 linear miles (1,287 km) of Canada's west coast adjoins the storm-active expanses of the eastern North Pacific Ocean, the steady presence of the westward-moving tradewinds has ensured that only one tropical storm has ever managed to make a recorded landfall in the region. On October 12, 1962, a dying TYPHOON Freda brought 90-MPH (145-km/h) gusts and blistering rains to the island city of Victoria, killing seven people and leaving $10 million in property damage as a testament to the course of this wholly contrary storm.

During the remarkable 1996 hurricane season the east Canadian provinces of Nova Scotia and Newfoundland took a rare direct strike from a mature North Atlantic hurricane on September 14, 1996. During the night, Hortense, named for one of the wicked stepsisters from *Cinderella*, slipped her skirt of 100-MPH (161-km/h) winds and five-inch (127 mm) rains ashore near the Cabot Strait, some 20 miles (32 km) east of Halifax. The most powerful hurricane to have made landfall in Canada since 1975's Blanche, Hortense's 75-MPH (121-km/h) winds, added to the forward momentum of its 35-MPH (56-km/h) STEERING CURRENT, gave it windspeeds of 110 MPH (177 km/h) and higher, qualifying it as a powerful Category 2 system. Downing power lines, uprooting acres of pine forests, lifting roofs, crushing cars, and stranding boats as its EYEWALL skipped across Nova Scotia and into the rocky coves of Newfoundland, Hortense's 20-foot (7-m) breakers shaved dunes, underminded seawalls, and destroyed half of a house in East Lawrencetown, Nova Scotia. Earlier responsible for claiming 22 lives in Puerto Rico and the Dominican Republic, Hortense caused no deaths in Canada but tallied some $20 million in insured property losses, making it one of Canada's most expensive tropical cyclone strikes.

Headquartered in Dartmouth, Nova Scotia, as a unit of the Atlantic Storm Prediction Centre (ASPC), the Canadian Hurricane Centre (CHC) is responsible for tracking all tropical and posttropical systems that enter Canada's Atlantic (and Pacific) maritime areas. For the purposes of issuing advisories to emergency management personnel, the media, and the general public, the CHC has established offshore "Response Zones" (RZ) that are activated when a tropical storm, hurricane, or posttropical extratropical system enters the RZ's respective parameters. Staffed by experienced meteorologists and emergency response personnel, the CHC closely coordinates with the U.S. NATIONAL HURRICANE CENTER (NHC), as well as Canadian emergency response agencies, to ensure that the most up-to-date information is provided during tropical cyclone activity in the North Atlantic and North Pacific Oceans.

Cape Verde Storm This term is given to those North Atlantic HURRICANES that originate in the vicinity of the Cape Verde Islands, some 400 miles (644 km) west of Senegal, Northern Africa. The archipelago's position between the 15th and 19th parallels places it directly in the path of the mid- to late-summer trade winds, powerful air currents that relentlessly flow from east to west over the baking sands of the Sahara and into the moist reaches of the NORTH ATLANTIC OCEAN. Fueled by the enormous evaporation rates found over the heated waters of the open Atlantic, the roiling pockets of warm, highly unstable air contained within the equatorial trough frequently develop into tropical disturbances, rain-laden clusters of thunderstorms that either dissipate or mature into low-pressure TROPICAL DEPRESSIONS. Steadily carried west toward both the vulnerable spread of the Caribbean Islands and the coastal lowlands of North and South America, these depressions in turn either dissolve or further band together as embryonic TROPICAL CYCLONES. Those Cape Verde disturbances that eventually do grow into cyclonic systems are often noted for their large size and ferocious intensity. More than 1,000 miles (1,609 km) of uninterrupted sea before landfall provides Cape Verde hurricanes with both the space and distance needed to develop into storms with minimum CENTRAL BAROMETRIC PRESSURES of less than 28.00 inches (948 mb) and sustained winds of 125 MPH (201 km/h) or higher. While not every hurricane that develops between 10 and 15 degrees North and 40 to 50 degrees West ultimately achieves such destructive potential, those that do often inflict stunning damage and tragic DEATH TOLLS on the nations that ring the western edge of the North Atlantic. Notable Cape Verde storms that have affected the United States include the GREAT HURRICANE OF 1780, the SAVANNAH-CHARLESTON HURRICANE of 1893, the GREAT GALVESTON HURRICANE OF 1900, the SAN FELIPE HURRICANE of 1928, the GREAT NEW ENGLAND HURRICANE of 1938, and Hurricane GILBERT of September 1988.

Although instances are rare, the Cape Verde Islands have themselves been directly affected by tropical cyclone activity. Because the TROPICAL WAVES that provide the seed moisture and convective instability for many North Atlantic tropical cyclones become seaborne to the south of the islands, the possibility always exists that these systems can become quickly organized, and when blocked from moving westward across the Atlantic by the semi-permanent high pressure ridge of the BERMUDA HIGH, turned northward toward the Cape Verde Islands.

Between August 28 and August 30, 1982, Tropical Storm BERYL delivered torrential rains and gale-force winds to the southern Cape Verde Islands while moving to the north-northwest across the Atlantic. Some 115 Cape Verdeans were killed, primarily by flash flood conditions on the island of Brava. Beryl's toll remains the deadliest natural disaster in the islands' history.

Some two years later, an intensifying Tropical Storm FRAN swept past the islands, visiting upon them large rainfall counts and gale-force winds, on September 17, 1984. According to media sources, 31 people were killed in flash flooding and landslide incidents.

On August 22, 2006, the Cape Verde Islands were directly affected by the RAIN BANDS associated with Tropical Depression No. 4 (TD 4), later Hurricane DEBBY. As the government of the Cape Verde Islands issued TROPICAL STORM WARNINGS, the incipient tropical cyclone's CENTRAL BAROMETRIC PRESSURE of 29.74 inches (1,007 mb) delivered gale-force winds and as much as three inches (76.2 mm) of rain to the island of Fogo. Unlike past tropical cyclone activity in the islands, no deaths, injuries, or property losses were reported.

Caribbean Sea   Situated between latitudes 10 and 30 degrees North, the island-rimmed basin of the Caribbean Sea stands as the southwesternmost extension of the NORTH ATLANTIC OCEAN. Bounded to the south by the continent of South America, to the west by Central America and MEXICO, to the north by the islands of CUBA, HISPANIOLA, PUERTO RICO, and the VIRGIN ISLANDS, and to the east by the Lesser Antilles (ANTIGUA, GUADELOUPE, and BARBADOS), the Caribbean Sea's position within the INTERTROPICAL CONVERGENCE ZONE (ITCZ) has long made it one of the most active places on Earth for TROPICAL CYCLONE generation. On average, 4 percent of the world's HURRICANES and TROPICAL STORMS develop over the heated waters of the Caribbean Sea in any given year. Between 1494 and 2006, approximately 138 recorded tropical cyclones were formed, for an average of four storms per year. While hurricanes and tropical storms have occurred in the Caribbean Sea during December and May, statistics reveal that peak periods for the region's tropical cyclone formation come in June, July, and November or when the subequatorial trough over the southwestern Caribbean is at its strongest. Early season systems, which tend to be of smaller size and intensity than their mid- to late-summer counterparts, almost always move in a north-northeasterly direction, a trajectory that carries them ashore in extreme western Cuba, Mexico's Yucatán Peninsula, or into the states of FLORIDA, ALABAMA, MISSISSIPPI, LOUISIANA, and TEXAS. During the 1969 and 1996 North Atlantic tropical cyclone seasons, no fewer than five systems each originated over the Caribbean Sea. In 1969, Hurricane CAMILLE, Hurricane Francelia, Tropical Storm Jenny, Hurricane Laurie, and Hurricane Martha all became TROPICAL DEPRESSIONS while over the basin. The same occurred during the 1996 season, and included Hurricane CESAR, Hurricane DOLLY, Hurricane Kyle, Hurricane Lili, and Hurricane Marco.

During the very active 2003 North Atlantic HURRICANE SEASON, Tropical Storm Odette became the first tropical storm on record to form over the Caribbean Sea during the month of December. Originating on December 4, Odette churned to the northeast and deepened before blustering ashore in eastern Hispaniola (over the Dominican Republic) with 63-MPH (101-km/h) winds on December 6. Like most tropical systems that interact with Hispaniola's mountainous terrain, Odette quickly dissolved into a rain-laden low pressure system, dropping some seven inches (177.8 mm) of PRECIPITATION across the island. While June, July, and November tropical cyclones in the Caribbean do not typically prove as destructive as similar systems in August, September, and October, a number of them have been responsible for inflicting significant DEATH TOLLS on those low-lying land masses that ring the area. Notable tropical storms and hurricanes that have received their start in the Caribbean include ALLISON (1995), ALMA (1966), AUDREY (1957), BAY ST. LOUIS (1819), BLACK RIVER (1912), GORDON (1994), OPAL (1995), and MITCH (1998).

Carla, Hurricane *Southern United States, September 5–14, 1961*   A powerful Category 5 HURRICANE, Carla's 150-MPH (241-km/h) winds, 12-inch (305-mm) rains, 10-foot (3-m) STORM SURGE, and 26 confirmed TORNADOES battered large portions of TEXAS, LOUISIANA, Kansas, and Missouri between September 5 and 14, 1961. With a CENTRAL BAROMETRIC PRESSURE of 27.49 inches (931 mb) at landfall, Carla was the most intense hurricane to have struck the east coast of Texas since the GREAT GALVESTON HURRICANE OF 1900 (September 8) that killed nearly 12,000 people in the island city that gave the hurricane its name. Mercifully much lower than that of the earlier tempest, Carla's DEATH TOLL eventually mounted to 46 people, a tribute to the wisdom of the mass evacuations—involving some 500,000 people—that immediately preceded Carla's brutal landfall. More than $400 million in 1961 property damages were assessed in Hurricane Carla's twisted path, making it one of the most expensive storms to have afflicted the region in almost a half-century.

A waning-season hurricane that received its start over the southwestern Caribbean Sea's warm waters approximately 200 miles north of the Panama Canal on the afternoon of September 5, 1961, Carla rapidly intensified before striking the Texas cattle town of Port Lavaca, 120 miles (193 km) southwest of Galveston,

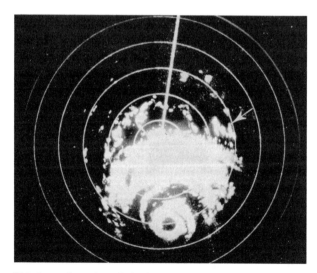

This image from the relatively early days of weather radar shows Hurricane Carla's storm bands rolling ashore near Galveston, Texas. This 1961 tropical fury claimed some 43 lives. *(NOAA)*

with staggering fury during the early afternoon hours of September 11, 1961. Whirling around a central barometric pressure of 27.49 inches (931 mb) at landfall, Carla's 61-mile-wide (98-km) EYEWALL cradled beneath it a lurking 10-foot (3-m) storm surge that shoaled into a 12 to 18 foot (4–6 m) water ram as it neared the gently sloping approaches to Matagorda and Lavaca bays. Gushing ashore just after one o'clock in the afternoon, the hurricane's 11-foot (4-m) surge was eventually felt as far north as Galveston, where it submerged nearby Follets Island and seriously damaged the west side of Galveston's legendary seawall.

Damage in the hurricane-prone towns of Port Lavaca and Port O'Connor, on the other hand, was cataclysmic. Besides sheering nearly 400 feet (133 m) off the southern tip of Matagorda Peninsula, Carla's sustained 150-MPH (241-km/h) winds gutted nearly every building in Port O'Connor, an industrious fishing village 30 miles (48 km) southeast of Port Lavaca. All of its churches, its town hall, its schools, stores, cattle pens, and houses were unroofed or collapsed while Carla's onrushing surge slammed three 72-foot (24-m) shrimp boats several hundred feet inland. Utility poles, mowed down in rows, crisscrossed flooded streets as humming transformers popped and sparked. Hundreds of trees, their foliage and branches whittled away by Carla's 175-MPH (282-km/h) gusts, swayed against the hurricane's chopping rains, threatened at any moment to crash down on those few dwellings miraculously left intact. Cars and trucks were overturned while road signs were twisted by Carla's racing passage. In Port Lavaca, a patrol boat assigned to the nearby Matagorda Island Air Force base and used to keep curious boaters from straying too close to the base's bombing range was trundled ashore with the same ease as a rowboat and was left firmly aground some feet from the bay. Two stone-and-concrete causeways linking the town with its resort beaches were undermined by Carla's 19-foot (6-m) STORM TIDES. Entire spans dropped into the roiling waters of Lavaca Bay, where the heavy seas quickly broke them apart. Twenty-three people in the Port Lavaca area lost their lives to Carla's Category 4 rampage, principally in drowning accidents and building failures.

Rapidly weakening as the EYE spiraled inland, Carla's winds swiftly diminished to 27-MPH (44-km/h) but delivered mounting rain squalls and lightning strikes to Austin and its environs just before midnight on September 11, 1961. Merging into a stationary front then positioned over north Texas, the disintegrating hurricane spent the next two days deluging more than a million acres of panhandle farmland with sudden gusts, nine to 12 inches (227–305 mm) of rain, and screaming TORNADOES. In Galveston,

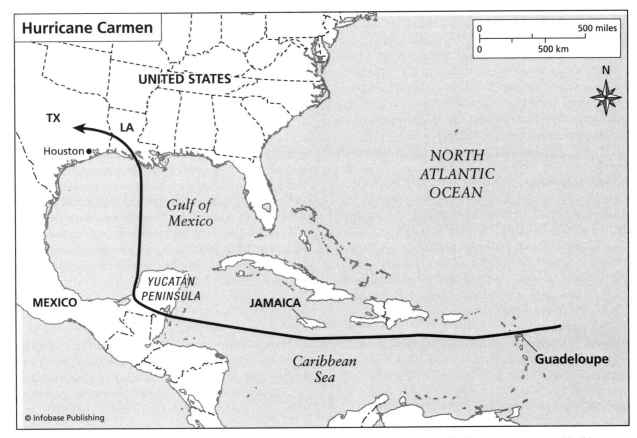

Hurricane Carmen

Featured in the movie *Forrest Gump,* Hurricane Carmen's stormy passage into the American Gulf Coast is represented in this tracking chart.

where some 60,000 people had taken refuge behind the seawall, two vicious tornadoes touched down during the early morning hours of September 12. The first, which originated in the GULF OF MEXICO, sprang over the seawall before whirring across the barrier island, destroying a café, several hotels, a school, the county courthouse, and more than 200 residences. The second, spawned over the gulf three hours later, demolished several more buildings in the city. Eight people in Galveston were killed, while $15 million in 1961 losses were incurred.

Its low-pressure core elongating over the United States's central plains on September 13, Carla's dissolving remains spawned yet another tornado outbreak, this one in west Louisiana. Six people perished when a particularly fierce twister set down in a residential community near Leesville just after dawn, while eight inches (203 mm) of rain created isolated power outages and some flash flooding as far east as Louisiana's capital, Baton Rouge. Indeed, terrific quantities of rain would become synonymous with Carla's EXTRATROPICAL demise over the heartland of the United States as more than 12 inches (305 mm) of rain were dropped on Kansas and Missouri between September 13 and 14. Flash floods in Kansas killed

five and injured sixty; in Missouri, one man was killed when his truck was swept into a rain-gorged river. More than $50 million in insurance claims were filed, making Carla one of the worst natural disasters in the region's history.

On September 17, then President John F. Kennedy declared large parts of Texas, Louisiana, Kansas, and Missouri disaster areas, making available several hundred million dollars in federal emergency relief. The American Insurance Association later set the total insured property damage from Carla at $408 million, ensuring that the name *Carla* was retired from the revolving list of HURRICANE NAMES the following year.

Carla, Typhoon *Taiwan, October 11–19, 1967*
See TAIWAN.

Carmen, Hurricane *Northern Caribbean Sea–Mexico–Southern United States, August 28–September 10, 1974* Said by the NATIONAL HURRICANE CENTER (NHC) to have been the most intense TROPICAL CYCLONE to have stalked the CARIBBEAN SEA and GULF OF MEXICO since 1969's Hurricane CAMILLE, Carmen delivered damaging winds and heavy rains to the VIRGIN ISLANDS, PUERTO RICO,

JAMAICA, MEXICO, and central LOUISIANA between August 29 and September 9, 1974. At several points in its existence a Category 4 hurricane of considerable severity, Carmen progressively weakened before coming ashore in Louisiana, a development that helped keep the DEATH TOLL in that state to a minimum. The most powerful tropical system of the 1974 HURRICANE SEASON, Carmen killed one person in Louisiana and tallied total damage estimates in excess of $175 million. The name Carmen has been retired from the revolving list of HURRICANE NAMES.

HURRICANE CARMEN CHRONOLOGY

August 23: A low-pressure TROPICAL WAVE enters the NORTH ATLANTIC OCEAN near North Africa's Cape Verde Islands during the late-afternoon hours. Moving west at nearly 16 MPH (26 km/h), the wave spends the next four days transiting the 79 degrees F (25.5 degrees C) waters of the mid-Atlantic, slowly deepening as dark clusters of CUMULUS and CUMULONIMBUS CLOUDS begin to crowd its northeastern flank.

August 28: TROPICAL DEPRESSION No. 4 begins to assume shape over the western North Atlantic Ocean, approximately 250 miles (402 km) northeast of the island of DOMINICA, during the late-evening hours. Pursuing its course due west, the thickening band of thunderstorms produces sustained winds of 27 MPH (44 km/h) and light rain showers. At the NHC in Miami, FLORIDA, a series of aerial reconnaissance missions to the depression by the HURRICANE HUNTERS service are scheduled for the following day.

August 29: Shortly before dusk the as-yet-unnamed tropical depression sweeps across the northern Leeward islands of Dominica and GUADELOUPE and enters the Caribbean Sea. While sustained 29-MPH (47-km/h) winds and gusty rains cause palm trees to sway and residents to seek refuge in government shelters, both islands remain unscathed by the tropical depression's inclement passage. A fairly large system, the deepening RAIN BANDS of its DANGEROUS SEMICIRCLE extend as far north as the Virgin Islands, where six-inch (152-mm) PRECIPITATION counts cause outbreaks of flooding but claim no lives. Property losses in the Virgin Islands are considered minimal. A round of late-afternoon reconnaissance flights indicates that top winds now measure 37 MPH (60 km/h), placing the tropical depression on the cusp of becoming a TROPICAL STORM. Civil defense authorities in nearby Puerto Rico and Jamaica prepare their islands for possible rain squalls from the grazing system.

August 30: Bearing 52-MPH (84-km/h) winds and torrential tropical downpours, a newly christened Tropical Storm Carmen passes some 80 miles (129 km) south of Puerto Rico during the early morning hours. Although Carmen's strengthening EYEWALL remains well offshore, the system's cumulonimbus rain bands deliver large quantities of precipitation to the island's southern cities and towns. Counts of five inches and higher are observed in and around Mayaguez, while inundated streams in the central highlands flood dozens of low-lying roads with several inches of water. During the midmorning hours, a TORNADO touches down near the southern port city of Ponce, uprooting trees and damaging scores of buildings. Even though property losses across Puerto Rico exceed $2 million, no casualties are reported in Carmen's wake.

August 31: Just before dawn, as its CENTRAL BAROMETRIC PRESSURE slides to 29.11 inches (985 mb) and its sustained winds approach 77 MPH (124 km/h), Tropical Storm Carmen is upgraded to hurricane status. Centered 90 miles (145 km) south of Kingston, Jamaica, the well-organized system unleashes gale-force gusts, three-foot seas and steady rains across the island for more than nine hours but causes no injuries or deaths.

September 1: Now centered over the northwestern Caribbean Sea some 500 miles (805 km) southeast of Chetumal, Mexico, Hurricane Carmen spends much of the morning delivering sustained 34-MPH (55-km/h) winds and 41-MPH (66-km/h) gusts to Swan Island (Islas del Cisne), a small coral atoll located off the eastern tip of Honduras. During the night the hurricane has undergone a slight course change to the northwest and is presently poised to make landfall in central BELIZE, perhaps on the border between Belize and Mexico, sometime in the next 24–36 hours.

September 2: A reconnaissance flight conducted shortly after midnight into Carmen's EYE—now drawn to within 200 miles (322 km) of Chetumal—indicates that the hurricane has commenced a period of rapid intensification. Subsequent missions conducted before dawn confirm the initial readings, with central pressures tumbling to 27.40 inches (928 mb) in a matter of hours. As sustained 130-MPH (209-km/h) winds whip the Caribbean's surface into long swells, a haphazard attempt is made to evacuate those coastal villages most vulnerable to Carmen's predicted 11-foot (4-m) STORM SURGE. Lashed by intermittent rain squalls, hundreds of people along the east shores the Yucatán Peninsula clamber into automobiles and trucks, while nearby schools and government buildings are hastily converted into storm shelters. Evacuation measures are further hampered by uncertainty as to Carmen's point of landfall; sharply sheering to the northwest, Carmen appears to forecasters to be about to skirt Belize entirely and instead come ashore in Mexico.

Bearing a greatly risen central pressure of 28.23 inches (956 mb), Hurricane Carmen barrels ashore in the Mexican province of Quintana Roo, five miles (8 km) north of Chetumal, during the midmorning hours. A powerful Category 3 system, Carmen's sustained 130-MPH (209-km/h) winds cause widespread damage to small boats and piers as well as the uprooting of hundreds of trees. Positioned within Carmen's NAVIGABLE SEMICIRCLE, the cities of Chetumal and Corozal lose much of their electrical and telephone services to Carmen's 100-MPH (161-km/h) winds but are otherwise spared the full brunt of the hurricane's fury. Slowly penetrating the sparsely populated heart of Yucatán, Carmen produces 7.5-inch (178-mm) downpours in places, triggering flash floods and bridge washouts. Although property damages reach well into the tens of millions of dollars, Carmen's initial assault on Mexico causes few injuries and claims no lives.

September 3: Virtually stalled over the dense jungles of the Yucatán Peninsula, Hurricane Carmen continues to deintensify through the daylight hours. Downgraded to a tropical storm shortly before noon, the system maintains its slow drift northwest, spawning 65-MPH (105-km/h) gusts that snap groves of small trees and collapse dozens of rain-weakened structures in the western province of Campeche. After sharply changing direction to the north during the midafternoon hours, Tropical Storm Carmen enters the Bay of Campeche after dusk. Barely making headway north, Carmen settles in over the bay's warm coastal waters and slowly begins to regain strength.

September 4: Seemingly committed to its grinding course to the north Tropical Storm Carmen spends the entire day languishing 50 miles (80 km) off the western shores of the Yucatán Peninsula. Though slowly reintensifying—its sustained WIND SPEEDS now clocked at 60 MPH (97 km/h)—the tropical storm's forward speed has fallen to less than five MPH (eight km/h). After drenching much of the upper Yucatán with a third day of tropical rains, Carmen fitfully approaches the peninsula's northwestern capes shortly before dusk. A series of late-night reconnaissance flights into the eyewall indicates that Carmen is continuing to intensify and may reattain hurricane status by the following morning.

September 5: Returned to hurricane status during the midmorning hours, Carmen slowly tracks northward, away from the waterlogged, gale-battered environs of the Yucatán Peninsula, at speeds of six to eight MPH (10–13 km/h). Moving into the open feeding grounds north of the Tropic of Cancer just before nightfall, Hurricane Carmen's central pressure resumes its tumble and reaches 28.50 inches (965 mb) by midnight. Sustained winds within Carmen's eyewall now measure 110 MPH (177 km/h).

September 6: Centered approximately 300 miles (483 km) of Mérida, Yucatán, Mexico, Carmen's roiling eyewall spends its last full day at sea maintaining both its cycle of deepening and its plodding northward trajectory. As computer ANALOGS are run at NHC in an effort to determine where and when Carmen might make a landfall on the gulf coast of the United States, a late-evening reconnaissance mission finds that the hurricane's central pressure has dropped to 27.99 inches (947 mb), in the process boosting the system's maximum wind speeds to 120 MPH (193 km/h). HURRICANE WATCHES are posted from Grand Isle, Louisiana, to Mobile, ALABAMA.

September 7: While civil defense organizations in Louisiana, MISSISSIPPI, and Alabama scramble to activate their coastal evacuation procedures, Carmen undergoes an abrupt course change to the northwest and then draws to within 300 miles (483 km) of Morgan City, Louisiana, during the predawn hours. Now matured into a Category 4 hurricane of extreme destructive potential, Carmen's central pressure of 27.61 inches (935 mb), coupled with its sustained 149-MPH (240-km/h) winds, makes it of equal intensity to the deadly Miami Hurricane of 1926. At the NHC, midmorning analogs all clearly point to Carmen's probable landfall in southwestern Louisiana sometime in the next 24 hours—a prediction that results in the immediate hoisting of HURRICANE WARNINGS for Louisiana's entire gulf coast. With the legacy of Hurricane Camille's violent passage over eastern Louisiana still fresh in their memories, 75,000 people spend the afternoon and early evening hours fleeing the state's low-lying resort islands.

Wheeling around a central pressure of 28.93 inches (980 mb), Hurricane Carmen slips ashore near Atchafalaya Bay, about 10 miles (16 km/h) north of Morgan City, shortly before midnight. Seriously weakened upon contact with a cool, dry AIR MASS moving southward over the Mississippi River, Carmen's sustained 86-MPH (138-km/h) winds cause widespread damage to thousands of acres of sugarcane fields. Numerous shrimp boats, wrenched from their moorings by Carmen's 6-foot (2-m) STORM TIDES, are cast aground near Patterson, while piers as far west as Marsh Island are splintered beneath 15-foot (6-m) breakers. Situated within Carmen's navigable semicircle, Morgan City loses its electrical and telephone services to the hurricane's 95-MPH (153-km/h) gusts but is otherwise spared the worst of the system's wrath. Arcing northwest over central Louisiana, Carmen produces rainfall counts of four to six inches (102–152 mm), sparking isolated incidents of flash flooding.

September 8: Downgraded to a tropical storm just before daybreak, Tropical Storm Carmen's spiral rain bands continue to weaken as they drift west toward the TEXAS border during the daylight hours. In Louisiana, reports list Carmen's toll as one person dead, with $90 million in agricultural losses and another $60 million in structural damages assessed. By midnight, Carmen has again been demoted, this time to a tropical depression. Its remaining rain bands completely dissipate over north-central Texas during the afternoon hours of September 9.

Carmen, Typhoon *Korea, August 26, 1960   See* KOREA.

Carol, Hurricane *Northeastern United States, August 25–31, 1954*   The first of three major TROPICAL CYCLONES to have affected NEW YORK and NEW ENGLAND during the notably active 1954 HURRICANE SEASON, Carol's 105-MPH (169-km/h) winds, 130-MPH (209-km/h) gusts, five-inch PRECIPITATION counts, and 14-foot (5-m) STORM TIDE raked large portions of Long Island, Connecticut, Rhode Island, Massachusetts, and central New Hampshire during the day and night of August 31, 1954. Followed ashore by Hurricanes EDNA (September 11) and HAZEL (October 15), Carol's CENTRAL BAROMETRIC PRESSURE of 28.77 inches (974 mb) at landfall near Saybrook, Connecticut, brought on meteorological conditions that eventually killed 60 people, injured another 1,000, and caused some $500 million in property losses across the region.

A quintessential midseason HURRICANE, Carol developed over the subtropical waters of the NORTH ATLANTIC OCEAN, approximately 289 miles (465 km) east of the Leeward Islands, on the morning of August 25, 1954. Slowly spiraling to the northwest at three MPH (5 km/h), the young TROPICAL DEPRESSION rapidly deepened and was upgraded to tropical-storm status while brushing past the BAHAMAS on the afternoon of August 26. Now officially dubbed Carol, the still-strengthening system at first showed no intention of hurrying up the eastern seaboard of the United States, spending all of August 27, 28, and 29 tramping to the west-northwest at no more than

One of the tropical cyclone's deadliest hazards—flooding—is dramatically illustrated in this photograph taken during Hurricane Carol's waterlogged passage up the eastern seaboard of the United States in August 1954. *(NOAA)*

four MPH (6 km/h)—but in the meantime growing into a mature, potentially destructive hurricane.

On the evening of August 30, as a wallowing oil tanker off NORTH CAROLINA's Cape Hatteras radioed that Carol's sustained winds had surpassed 100 MPH (161 km/h), the hurricane began to gather forward momentum, bearing down on the northeast United States at nearly 41 MPH (66 km/h). An intense Category 3 hurricane at first landfall on central Long Island during the midmorning hours of August 31, Carol's 125-MPH (201-km/h) gusts, scouring rains, and 12-foot (4-m) seas toppled huge elm trees in East Hampton, washed out roads on the island's west end, and stranded 4,500 people in the town of Montauk. An extreme gust of 135 MPH (217 km/h) was noted on Block Island (Rhode Island), some 20 miles (32 km) east of Long Island. Two people perished, both of them in New York City. After traversing Long Island, Carol quickly crossed Long Island Sound and screamed ashore in southern Connecticut, the most powerful hurricane to strike the area since 1938.

As is fairly common in New England hurricanes of Carol's type, the shallow waters of Narragansett Bay were pushed over their embankments and into downtown Providence, Rhode Island, flooding hundreds of stores with eight feet of water. The casino at Newport, Rhode Island, was demolished by 30-foot (10-m) breakers, while several of the resort town's famous mansions lost roof tiles, windows, and awnings. Three hundred cottages were swept into oblivion at Misquamicut Beach. In nearby Point Judith Bay, 30 tuna boats gathered for an annual fishing tournament were chased ashore by Carol's giant surge and wrecked.

In Boston, east of Carol's path, the upper sections of a 667-foot (222-m) television tower were pitched to the ground, crushing a transmitting station and a parked automobile below. Buffeted for hours by Carol's 90-MPH (145-km/h) winds, the wooden spire of historic Old North Church finally toppled into the street, its spectacular 100-foot (33-m) drop fortunately sparing the treasured belfry from which hung the lanterns that spurred Paul Revere on his midnight ride. Such was the power of Carol's seas at Boston that a 100-foot (33-m) boat was thrown onto a riverbank.

Before dying out over inland Canada on September 1, Carol uprooted more than 63,000 pine trees across central New Hampshire, providing storm-weary property owners with a windfall supply of lumber for new sheds, garages, and porches.

Caroline, Hurricane *North Atlantic Ocean–Cuba–Mexico, August 24–September 1, 1975* The first and as yet only North Atlantic TROPICAL CYCLONE to be identified as Caroline, Hurricane Caroline spent nearly two weeks of the 1975 North Atlantic HURRICANE SEASON bringing deluging rains and high winds to several Caribbean islands and the eastern coast of MEXICO between August 24 and September

Marooned in a swirl of raging seas, a yacht club in Rhode Island is inundated by Hurricane Carol's hammering storm surge. Note the tattered hurricane warning flags in the upper right hand corner of the photograph. *(NOAA)*

1. Originating as a TROPICAL DEPRESSION some 200 miles (450 km) north of HISPANIOLA on August 24, Caroline tracked to the southwest, haphazardly strengthening and weakening as it crossed extreme southeastern CUBA and penetrated as far south as the Cayman Islands. On August 26, the system recurved to the west-northwest and weakened slightly. On August 27, as its central pressure rose to 29.94 inches (1,014 mb), the system brushed past the northeastern tip of Mexico's Yucatán Peninsula. Because the bulk of Caroline's nascent EYEWALL remained offshore, the system did not experience any significant weakening during this time period. Indeed, as it hiked to the northwest and entered the GULF OF MEXICO on August 28, Caroline steadily began intensifying. It achieved TROPICAL STORM status—and its unique name—during the late afternoon hours of August 29, as its central pressure dipped to 29.61 inches (1,003 mb) and its sustained winds rose to 40 MPH (64 km/h). Within 24 hours, Caroline had matured to Category 1 intensity, with a central pressure of 29.23 inches (990 mb) and sustained winds of 75 MPH (121 km/h). Forecast to make landfall in northern Mexico, Caroline rapidly intensified as it approached the Mexican Gulf coast, deepening to a Category 3 hurricane (with a central pressure of 28.73 inches [973 mb]) during the early morning hours of August 31. Now bearing sustained winds of 115 MPH (185 km/h) and situated less than 200 miles from landfall in Mexico, Caroline continued to strengthen. When it burst ashore in Mexico, approximately 125 miles south of Brownsville, TEXAS, on August 31, 1975, its central pressure of 28.43 inches (963 mb) produced 120-MPH (193-km/h) winds and heavy seas. A slow-moving system, Caroline drifted inland over Mexico, being downgraded to tropical storm status just before midnight on August 31. By the following morning, September 1, 1975, Caroline was of tropical depression intensity (29.58 inches [1,002 mb]) and rapidly dissipating over northeastern Mexico.

Although it is not in current use, the identifier Caroline has not officially been retired from the list of North Atlantic hurricane names.

Caterina, Cyclone *South Atlantic Ocean–Brazil, March 26–28, 2004* Perhaps the most unusual TROPICAL CYCLONE yet observed in the world's storm zones, Cyclone Caterina (some sources refer to the system as Hurricane Caterina) enjoys the rare distinction of being the first mature-stage tropical cyclone yet observed in the South Atlantic Ocean. Bearing sustained winds of 80 MPH (129 km/h) and 95-MPH (153-km/h) gusts, Caterina howled ashore on the eastern coast of Brazil on March 28, 2004. The isolated town of Santa Caterina, where the system made landfall, was severely

Un-named hurricane in South Atlantic comes ashore in southeast Brazil 1745 UTC on 27 March 2004 GOES-12 visible channel
GOES Project NASA-GSFC

Although very rare, tropical cyclones do form in the South Atlantic Ocean. In 2004, a very controversial Cyclone Caterina made landfall in Brazil. *(NOAA)*

damaged, and owing to the absence of a codified naming list for South Atlantic tropical cyclones, in turn lent its name to its destructive visitor. At least three people were killed and another 38 injured as Caterina howled inland, causing widespread flooding and structural damage across a section of eastern Brazil. Although geostationary SATELLITE photographs and other data clearly support Caterina's tropical antecedents, some controversy exists as to whether or not the system was really a meteorological hurricane. Studies conducted by the NATIONAL HURRICANE CENTER (NHC) indicate that Caterina was, indeed, a rare South Atlantic tropical cyclone, while the Brazilian meteorological services maintain that Caterina was a severe subtropical or extratropical system.

Celia, Hurricane *Cuba–Southern United States, July 23–August 5, 1970* An exceptionally severe Category 3 HURRICANE, Celia delivered sustained 130-MPH (209-km/h) winds, 6-inch (152-mm) PRECIPITATION counts, and 9-foot (3-m) STORM TIDES to the eastern gulf coast of TEXAS on the afternoon of August 3, 1970.

One of the most powerful midseason TROPICAL CYCLONES ever observed in the GULF OF MEXICO,

Celia originated as a TROPICAL WAVE some 300 miles (483 km) off the west coast of Africa on the morning of July 23. Unable to undergo deepening over the expanses of the subequatorial NORTH ATLANTIC OCEAN, Celia remained a TROPICAL DISTURBANCE as it rapidly crossed the Leeward Islands and entered the northern CARIBBEAN SEA during the late-night hours of July 27.

Gradually recurving northwest at speeds of up to 13 MPH (21 km/h) and subject to increasing scrutiny by the HURRICANE HUNTERS service, Celia continued to show few signs of deepening into a TROPICAL DEPRESSION until the evening of July 30, at which time it abruptly changed course due north and swiftly began to intensify.

Having spent most of July 30 sideswiping southern JAMAICA, and now within the influence of an encroaching low-pressure trough over the southern United States, Celia trained its sights on western Cuba. Steadily strengthening as it slid toward the extreme southwest tip of the island, Celia was a powerful tropical depression—with a minimum BAROMETRIC PRESSURE of 29.73 inches (1,007 mb) and sustained winds of 36 MPH (58 km/h)—by the time it came ashore near the city of Pinar del Río on the evening of July 31. While the tropical depression's winds were of little destructive consequence, its inundating rainfalls killed five people in Cuba and caused isolated damage to roads and crops.

Shortly after dawn on August 1, while positioned some 850 miles southeast of Corpus Christi, Texas, Celia was upgraded to a TROPICAL STORM. Officially awarded the name *Celia,* the system promptly underwent another period of deepening, its CENTRAL BAROMETRIC PRESSURE dropping from 29.23 inches (990 mb) on the morning of August 1 to 28.49 inches (965 mb) by seven o'clock the same evening. Now upgraded to a hurricane and positioned 500 miles (805 km) southeast of Corpus Christi, Celia continued to swing northwest, a course that if maintained would take the system ashore somewhere in northern MEXICO or southern Texas sometime in the next two days.

On the morning of August 3, with its EYE now centered less than 100 miles (161 km) from Corpus Christi and its advancing RAIN BANDS showering hundreds of thousands of fleeing Texans with sporadic downpours and gale-force gusts, Celia's central pressure again began to tumble. Sliding from 28.67 inches (971 mb) to 28.14 inches (953 mb) just before landfall, the fall in Celia's pressure hiked its maximum windspeeds to well over 130 MPH (209 km/h).

Attended by a central pressure of 28.03 inches (949 mb), recorded wind gusts of 160–180 MPH (258–290 km/h), and destructive MINISWIRLS, Celia crashed ashore at Port Aransas, Texas, during the

A veritable fleet of luxury yachts and other small craft lie grounded in the wake of Hurricane Celia's unwelcome visit to Corpus Christi, Texas, in 1970. *(NOAA)*

late afternoon hours of August 3. According to figures provided by the American Red Cross, 8,950 houses and 310 farm structures were destroyed, while 55,850 others suffered varying degrees of damage. Some 252 businesses were demolished, along with the loss of 331 shrimp and pleasure boats cast ashore at Rockport, Texas. Several highrise buildings in downtown Corpus Christi, which *The New York Times* reported "looked as if it had been bombed," lost their glass curtainwalls, while hundreds of National Guard troops were dispatched to maintain order. All told, Celia's rampage through Texas left 12 people dead and damage estimates of at least $500 million. The name Celia has been retired from the list of HURRICANE NAMES.

Central America Hurricane *North Atlantic, September 10–13, 1857* Although the grazing fringes of this large offshore HURRICANE would eventually bring 24-MPH (39-km/h) winds and light rains to Charleston, SOUTH CAROLINA, this storm is better remembered for the tragic role it played in one of the 19th-century's worst maritime disasters, the foundering of the steamship *Central America* and the drowning of more than 400 of its passengers, on September 12, 1857.

Now lost to the decay of time, the Central America Hurricane's exact origination point is unknown. A mid-September storm, its large size and considerable intensity upon entering South Carolina's coastal waters on September 10 would seem to indicate that it was of the CAPE VERDE variety, a hurricane that had developed over Northern Africa's Cape Verde Islands some days before. It is also known from surviving members of the *Central America*'s crew that the hurricane overtook the steamship from the south on midnight of September 10, coming up on them two days after the vessel departed Havana, CUBA, bound for New York. This would seem to indicate that the hurricane had already undergone RECURVATURE and further intensified as it rounded the western edge of the BERMUDA HIGH and heading northeast. A minimum BAROMETRIC PRESSURE reading of 29.61 inches (1,003 mb) taken near Charleston on the afternoon of September 12 indicates that while the hurricane was of minor intensity along the shore, the bulk of its fury firmly remained over the NORTH ATLANTIC OCEAN's seething waters.

There, the 1,500-ton sidewheeler *Central America,* deeply laden with 578 passengers and crew and almost 13 tons of gold bars and coins, labored against the rising seas and foaming gusts that assailed it. Although the *Central America* was a fairly large ship for her day—280 feet (93 m) long, powered by two hulking, steam-shrouded reciprocating engines—her sleek black hull, graceful clipper bow, triple-crown of masts, and towering red capped funnel were pitched from wave to wave, corkscrewed into 30-foot (10-m) troughs, and subjected to stresses that even the strongest of vessels would be hard pressed to withstand. On Friday morning, September 11, as the hurricane's EYE drew closer to where the *Central America* writhed beneath the escalating winds, the ship sprung a serious leak somewhere along her keel.

The steamer's bilges, cargo holds, and coal bunkers, buried deep within the wooden hull, instantly began to fill with water. Desperate attempts by both crew and passengers to find the source of the leak soon proved futile. By early evening on September 11, as the gaining sea threatened to douse the engine-room fires leaving the ship adrift without either power or steerageway, a bailing brigade was formed. While this heroic effort on the part of the vessel's complement ultimately provided the *Central America* with a few additional hours of life, it did nothing to forestall the relentless onslaught of the hurricane—or the encroaching sea. As the *Central America*'s list to starboard continued to worsen and as the hurricane's most-strident winds steadily bore down on the paralyzed craft, her 72-foot (24-m) wooden foremast was cut away to improve stability.

The hurricane's eye passed directly over the stricken *Central America* shortly after dawn on Saturday, September 12. As the brief but welcome lull set in, weary passengers and crew rested from their frantic bailing, ate a meager breakfast of waterlogged biscuits, and attempted to signal their distress to other ships by flying the steamer's national ensign upside down. They had no sooner accomplished these tasks than the hurricane's eye, judged by survivors to have been some 20 to 30 minutes in duration, finally passed. With this came a return of the storm, an increased tempo in the winds and rains that eyewitnesses claimed was of "great fury."

The terror-worn passengers onboard had apparently put the hurricane's respite to wise use, for shortly after midday on September 12, a passing brig spotted the sidewheeler's inverted flag and ventured to her rescue. Registered as the *Marine*, the equally battered vessel railed for almost an hour against the hurricane's winds before maneuvering itself close enough to allow the *Central America* to launch its boats into the midst of storm-pitched seas. In what would become one of the most spectacular deep-sea rescues of the 19th century, more than 153 passengers—a majority of them women and children—were safely transported by open boat to the buffeted yet seaworthy decks of the *Marine* before the hurricane's unrelenting convulsions made it impossible for the rescue effort to continue. With more than 425 men still on board, the *Central America* went down stern-first shortly before dusk on September 12, leaving no further survivors. As the overloaded *Marine* painfully struggled to NEW YORK with the *Central America*'s shaken survivors housed within her holds, the hurricane that now bears the unfortunate ship's name continued to move northeast, bringing steady breezes to the eastern shores of NEW ENGLAND before dissipating over the cold northern springs of the Atlantic Ocean.

central barometric pressure　The term given to those readings and gradations, achieved through the use of a BAROMETER, of the minimum BAROMETRIC PRESSURES found within the EYE and EYEWALL of a TROPICAL CYCLONE. Now understood to be linked proportionally to the speed of a cyclonic system's winds, the central barometric pressure has become a reliable means by which meteorologists and civil defense authorities can measure a particular tropical system's intensity. Defined in part by the PRESSURE GRADIENT, or the decrease in atmospheric pressure over the distance of the tropical cyclone's mass, the rise and fall of a storm's central barometric pressure can indicate whether the system is strengthening, weakening, or remaining stationary. By studying how rapidly the central pressure moves in a HURRICANE, observers can take into account a host of unseen factors—such as changes in SEA-SURFACE TEMPERATURE or the presence of an unknown air pocket—and include them in their ANALOGS, their predictions as to the storm's future movements. The central barometric pressure has also become a way for scholars to convey to history something of a storm's true fury, an easily quantifiable indication of the screaming violence that accompanied a long-passed strike by one of nature's most awesome events.

The physical relationship between the central barometric pressure of a tropical cyclone and the speed of its winds is not difficult to follow. Because the eye of a tropical cyclone is a low-pressure area around which the atmosphere—defined in this case by spiral bands of thunderstorms—has gathered to fill, any change in central pressure—be it a weakening rise or a deepening fall—is going to cause the atmosphere to either rush in at a faster rate or slow its windy progress. To this model is added the storm's pressure gradient, the steepening drop of barometric pressures within the entire system's cloud cover. Beginning with the 29.5-inch (1,000-mb) ISOBAR range, the barometric pressures in tropical cyclones are measured inward, toward the eye, where the lowest pressures are to be found. Recorded on charts as isobars, or lines that connect points of equal barometric pressure, the pressure gradient of a mature hurricane, TYPHOON, or CYCLONE presents a series of concentric rings, broadly spaced at the storm's outer edges but moving closer together as they near the center. The distance between the isobars, which in some intense storms had been only a few miles wide, indicates both the scope of the pressure drop and the commensurate speed at which the tropical cyclone's winds are moving to equalize it.

For these reasons, barometric pressures within a tropical cyclone can tumble very quickly. During the Snow Hurricane of October 9, 1804, barometric pressure readings taken near Boston, Massachusetts, indicated a pressure drop of 0.44 inches (1.49 mb) in a matter of hours, while a Royal Navy barometer on the Caribbean island of ANTIGUA recorded a 1-inch (33-mb) fall in pressure over the space of 90 minutes during a hurricane on August 18, 1835. In 1869, two powerful hurricanes produced record-setting pressure gradients over northeastern United States. The first, on September 8, saw barometric pressures stand at 29.94 inches (1,013 mb) at sunrise, fall to 29.14 inches (986 mb) by dusk, and then decrease still further, to 29.02 inches (982 mb) by midnight. Just over a month later, on October 4, the 12-hour onset of Saxby's Gale caused a barometer in Augusta, Maine, to fall from 29.70 inches (1,005 mb) at dawn to 28.99 inches (981 mb) by early evening. Barometric pressures in the LAST ISLAND HURRICANE of August 10–11, 1856, recorded by a schooner passing through the hurricane's eye, fell some 1.7 inches (35 mb) in less than 24 hours to 28.20 inches (954 mb).

The first known barometer was not invented until 1643; there are, of course, no central barometric pressure readings in existence from before that year. While it is entirely possible that ancient cultures did in fact possess some sort of manufactured instrument for measuring atmospheric pressure, it is more likely that they depended on the behavior of ANIMALS to alert them to unseen changes in the air. Instead of a series of numbers couched in terms of inches of mercury observed, early accounts of tropical cyclones often feature the erratic flight patterns of birds, the restlessness of horses and cows and the deepwater retreat of snakes and alligators as evidence of an approaching storm. To this day in Barbados, at a time when the most sophisticated of barometers are readily available to record a passing hurricane's intensity, the local population can still be

overheard describing the blast as one which made "the pigs dance."

Indeed, it was not until 1743 that the first barometric pressure readings for hurricanes began to appear in contemporary U.S. accounts. During that year, John Winthrop, a professor of Natural Philosophy at Harvard College, described the passage of an October 22 hurricane over east New England as having a "lowest" barometric pressure of 29.35 inches (993 mb) at two o'clock in the afternoon. Like Professor Winthrop and his 1743 pressure measurement experiment, meteorologists, physicists, historians, and emergency management personnel around the world's tropical cyclone regions have had the often harrowing opportunity to observe some of the lowest central barometric pressure readings ever recorded. During the remarkably active 2005 North Atlantic HURRICANE SEASON, no fewer than three pressure records were broken—including the record for the lowest central barometric pressure yet observed in the North Atlantic basin.

On the morning of August 28, 2005, as it drew to within 500 miles (805 km) of New Orleans, LOUISIANA, Hurricane KATRINA generated a central barometric pressure reading of 26.63 inches (902 mb), which was at the time the fourth-lowest on record behind Hurricane GILBERT (1988), the 1935 LABOR DAY HURRICANE, and Hurricane ALLEN (1980). Driving sustained surface wind speeds within the eyewall of 167 MPH (269 km/h), Katrina's central pressure was tied with that observed in 1969's Hurricane CAMILLE, and made the system of firm Category 5 status.

Less than three weeks later, an even more intense hurricane, RITA, charged across the GULF OF MEXICO, following a trajectory similar to that of Katrina. On the morning of September 22, 2005, Hurricane Rita generated a central barometric pressure reading of 26.48 inches (897 mb), displacing Katrina and Camille one slot. Attributed primarily to the hurricane's interaction with the heat-charged waters of the Loop Current, Rita's record-setting deepening configuration became the most rapid yet observed in a North Atlantic tropical cyclone. At noon on September 21, Rita was a Category 4 hurricane with a central pressure of 27.99 inches (948 mb); by 9:00 P.M., its pressure had plummeted to 26.99 inches (914 mb), and by midnight had slipped further still, to 26.51 inches (898 mb).

Rita, Katrina, and Camille would all slide down another notch in the central barometric pressure records list when on October 19, 2005, Hurricane WILMA broke Hurricane Gilbert's 17-year record for producing the lowest barometric pressure reading yet observed in the North Atlantic basin. During the early morning hours of October 19, as Wilma neared the Yucatán Straits and entered the Gulf of Mexico, it underwent one of the most rapid deepening cycles yet observed in a North Atlantic tropical cyclone. Between the afternoon of October 18, when it reached Category 1 status (28.85 inches [977 mb]), and the early morning hours of October 19, when it achieved Category 5 intensity, Wilma's central barometric pressure plummeted 2.80 inches (95 mb) in less than 12 hours. Driven by exceptionally warm SEA-SURFACE TEMPERATURES in the Yucatán Straits, as well as a favorable environment with no WIND SHEAR, Wilma produced a central barometric pressure reading of 26.04 inches (882 mb), a record for the North Atlantic basin. Simultaneously, Wilma's sustained wind speeds eclipsed an astonishing 170 MPH (274 km/h).

The following three lists accompanying this entry present interesting facts on central barometric pressure.

Five Lowest Central Pressures In North Atlantic Ocean Tropical Cyclones

Hurricane Wilma, Caribbean Sea, 2005: 26.04 inches (882 mb)

Hurricane Gilbert, Caribbean Sea, 1988: 26.22 inches (888 mb)

Labor Day Hurricane, North Atlantic, 1935: 26.34 inches (892 mb)

Hurricane Rita, Gulf of Mexico, 2005: 26.48 inches (897 mb)

Hurricane Allen, Gulf of Mexico, 1980: 26.54 inches (899 mb)

Five Lowest Central Pressures In Western North Pacific Ocean Tropical Cyclones

Super Typhoon Tip, Western Pacific, 1979: 25.69 inches (870 mb)

Super Typhoon Zeb, Western Pacific, 1998: 25.75 inches (872 mb)

Super Typhoon Gay, Western Pacific, 1992: 25.75 inches (872 mb)

Super Typhoon Keith, Western Pacific, 1997: 25.75 inches (872 mb)

Super Typhoon Joan, Western Pacific, 1997: 25.75 inches (872 mb)

Five Lowest Central Pressures In Indian Ocean And South Pacific Ocean Tropical Cyclones

Cyclone Zoe, South Pacific, 2002: 25.95 inches (879 mb)

Cyclone Danielle-Agnielle, South Indian, 1995: 26.13 inches (885 mb)

Cyclone Gafilo, South Indian, 2004: 26.42 inches (895 mb)

Cyclone Ron, South Pacific, 1998: 26.57 inches (900 mb)

Cyclone Susan, South Pacific, 1998: 26.57 inches (900 mb)

Cesar, Tropical Storm *North Atlantic Ocean, August 31–September 2, 1984* Three North Atlantic TROPICAL CYCLONES were identified as Cesar between 1984 and 1996. Of these, two were TROPICAL STORMS and one was a minimal Category 1 hurricane.

A deepwater tropical system, Tropical Storm Cesar roiled the waters of the NORTH ATLANTIC OCEAN between August 31 and September 2, 1984. Originating to the north of BERMUDA and well away from the U.S. eastern seaboard from a subtropical depression, Cesar was upgraded to a tropical storm on August 31, when its central pressure was observed as 29.55 inches (1,001 mb). Passing to the northeast and penetrating deeply into the waters of the extreme North Atlantic, Cesar reached its maximum observed intensity on September 2, 1984, when its central pressure dipped to 29.35 inches (994 mb) and its sustained winds peaked at 58 MPH (93 km/h). With SEA-SURFACE TEMPERATURES cooling by the degree every few hours, Cesar was unable to sustain its maturation cycle and within 24 hours had transitioned into an extratropical storm. Interestingly, its central pressure as an extratropical system—29.20 inches (989 mb)—was more intense that at any recorded point along its history as a tropical system.

Cesar, Tropical Storm *North Atlantic Ocean, July 31–August 7, 1990* A weak and disorganized system, Tropical Storm Cesar tracked northeastward across the NORTH ATLANTIC OCEAN between July 31 and August 7, 1990. Formed very close to the western coast of North Africa, and near 10 degrees North, Cesar moved to the northwest before becoming a named TROPICAL STORM (with a central pressure of 29.67 inches [1,005 mb]) on August 2. Wobbling slightly as it scudded northeastward, Cesar reached its peak intensity of 29.52

inches (1,000 mb) late on the night of August 2, and remained at that intensity for the next two days. Gradually weakening as it moved across colder sea-surface temperatures (SST), Cesar was downgraded to a TROPICAL DEPRESSION (with a central pressure of 29.79 inches [1,009 mb]) on August 6, and dissipated the following day.

Cesar, Hurricane *Caribbean Sea–Central America, July 24–28, 1996* On July 24, the third TROPICAL CYCLONE of the 1996 North Atlantic HURRICANE SEASON, Cesar, originated in the extreme southeastern CARIBBEAN SEA, some 75 miles (100 km) west of the island of Curaçao. Destined to become one of the deadliest tropical cyclones in southern Caribbean history, Cesar scudded nearly due west, intensifying to TROPICAL STORM status and grazing the northern coast of South America July 26–27. In Colombia, 11 people lost their lives to Cesar's rain-generated flash floods. Reaching Category 1 status on July 26, while situated over the southwestern Caribbean Sea, Cesar gently recurved to the northwest and came ashore in Bluefields, Nicaragua, on July 28. Armed with a central pressure of 29.23 inches (990 mb) and 81-MPH (130-km/h) winds, Cesar caused considerable damage and killed nine people. Elsewhere in Central America, Cesar's torrential downpours claimed 34 lives in Costa Rica and 13 in El Salvador. Retaining hurricane intensity until it neared the Pacific coast of Central America, Cesar emerged into the western NORTH PACIFIC OCEAN and in keeping with identification tradition, assumed the next available name on the list of North Pacific tropical cyclones. Renamed Tropical Storm Douglas, the former Hurricane Cesar moved westward into the Pacific, eventually deepening into a Category 4 hurricane with a central pressure of 27.93 inches (946 mb) and 140-MPH (220-km/h) winds on August 1, 1996. By August 6, the system had dissipated over the reaches of the eastern North Pacific Ocean. All told, Cesar claimed 67 lives and caused more than $40 million (1996 dollars) in damage.

The name Cesar has been retired from the revolving list of North Atlantic tropical cyclone identifiers and replaced with Cristobal.

Chantal, Hurricane *North Atlantic Ocean–Eastern United States, September 10–15, 1983* A deepwater tropical system for its entire existence, Chantal originated over the NORTH ATLANTIC OCEAN, approximately 200 miles south of BERMUDA, on September 10, 1983. A TROPICAL DEPRESSION, the system deepened to a TROPICAL STORM on September 11, and achieved HURRICANE status later

the same day. Chantal quickly reached its peak intensity of 29.35 inches (994 mb), and remained a Category 1 hurricane with sustained winds of 75 MPH (km/h) for the next two-and-a half days. On September 15, 1984, as it crossed latitude 40 degrees North, Chantal was downgraded to a tropical depression. Within hours it had begun its final dissolution.

Chantal, Hurricane *Gulf of Mexico–Southern United States, July 30–August 3, 1989* Born off the northern shores of MEXICO's Yucatán Peninsula as a TROPICAL DEPRESSION on July 30, 1989, Chantal rapidly swept to the northwest across the GULF OF MEXICO, deepening as it fed on the high SEA-SURFACE TEMPERATURES found in that area. On July 31, as its CENTRAL BAROMETRIC PRESSURE dropped to 29.64 inches (1,004 mb), Tropical Depression No. 3 became Tropical Storm Chantal. Now situated less than 500 miles (689 km) due south of New Orleans, LOUISIANA, Tropical Storm Chantal was forecast to continue intensifying as it maintained its north-northwesterly trajectory toward the northern rim of the Gulf of Mexico. By the early morning hours of August 1, 1989, Chantal had deepened to a Category 1 hurricane with a central pressure of 29.26 inches (991 mb) and sustained winds of 75 MPH (121 km/h). With its outer RAIN BANDS washing ashore in northern TEXAS and southeastern Louisiana, Chantal steadily if slowly intensified as it drew closer to landfall in northeastern Texas. At landfall on High Island, Texas, on August 1, the system's central barometric pressure reading of 29.05 inches (984 mb) produced 81-MPH (130-km/h) winds and heavy rains. Three people in Texas were killed, while another 10 perished in Louisiana when an oil services vessel capsized in Chantal's heavy seas near Morgan City. Some $100 million in property losses were incurred.

Chantal, Tropical Storm *North Atlantic Ocean, July 12–22, 1995* An offshore tropical system that developed from a TROPICAL DEPRESSION located over the extreme southern NORTH ATLANTIC OCEAN on July 12, Tropical Storm Chantal spent the next 10 days of the remarkably active 1995 North Atlantic hurricane season recurring to the north-northeast, and reaching tropical storm intensity while passing to the north of PUERTO RICO and HISPANIOLA on July 14. With a minimum pressure of 29.26 inches (991 mb) and wind speeds of 69 MPH (111 km/h), Chantal reached its peak intensity—well to the east of the BAHAMAS—on July 16, 1995, but never posed any serious threat to land. Steered to the northeast, then eastward, by the

This satellite photograph captures the sprawling yet immature stage rain bands of Tropical Storm Chantal. _(NOAA)_

BERMUDA HIGH, a downgraded Tropical Depression Chantal dissipated over the chilly expanses of the North Atlantic on July 22, 1995.

Chantal, Tropical Storm *North Atlantic Ocean–Caribbean Sea–Mexico, August 14–22, 2001* A CAPE VERDE STORM that never reached its possible potential, Tropical Storm Chantal originated over the reaches of the NORTH ATLANTIC OCEAN, well to the east of the Windward Islands, on August 14, 2001. A TROPICAL DEPRESSION until it reached Martinique, Chantal slowly but haphazardly intensified as it tracked westward across the Caribbean Sea. With a central pressure at its peak of 29.44 inches (997 mb), Chantal delivered 69-MPH (111-km/h) winds to MEXICO's Yucatán Peninsula at landfall on August 22, 2001. Retaining TROPICAL STORM intensity as it moved inland over the peninsula, Chantal sharply recurved to the southwest, and was downgraded to a tropical depression just before the system dissipated in a hail of thunderstorms on August 22.

The identifier Chantal has been retained on the North Atlantic TROPICAL CYCLONE identification lists, and is scheduled to reappear in its fifth incarnation in 2007.

Charley, Hurricane *North Atlantic Ocean–British Isles–Northern Europe, August 13–30, 1986* One of the most unusual North Atlantic TROPICAL CYCLONES yet observed, Hurricane Charley charted a unique course eastward across the NORTH ATLANTIC OCEAN between August 13 and 30, 1986. Born as a subtropical depression in the GULF OF MEXICO, Charley would eventually make landfall thousands of miles distant, in the British Isles, as a powerful extratropical storm, in the process becoming one of the most destructive meteorological events in British history. Formed as a subtropical system near the Florida Panhandle on August 13, 1986, Charley (then still unnamed) maintained its subtropical characteristics until the early morning hours of August 15 when its interaction with a dissipating frontal system over northern FLORIDA and GEORGIA provided it with sufficient energy to transition to a TROPICAL DEPRESSION. Sliding to the north-northeast, the system emerged over the heat-laden waters of the Gulf Stream and rapidly began intensifying.

Upgraded to tropical storm status on August 15, a newly minted Charley matured to Category 1 hurricane status on August 17. Situated just off NORTH CAROLINA's Outer Banks, Charley reached peak intensity on August 17, 1986, as its central barometric pressure slid to 29.14 inches (987 mb) and its sustained winds eclipsed 80 MPH (130 km/h). High winds, heavy seas and drenching rains pummeled eastern North Carolina and VIRGINIA, claiming five lives, three of them in an aviation crash. Before drawing into the Atlantic, Hurricane Charley delivered a rare strike to Maryland and Delaware, crossing over the Delmarva Peninsula on August 18 with a central pressure of 29.23 inches (990 mb) and 75-MPH (121-km/h) winds. Some $60 million in damages were incurred in the United States from Charley's erratic passage. Undergoing extratropical deepening on August 21, while situated several hundred miles east of Canada, Charley became a powerful extratropical storm, one that grew in intensity as it spun to the northeast, nearing the British Isles. By the time Extratropical Storm Charley passed the southern coast of Ireland and entered England on August 26, its central pressure had deepened to 28.96 inches (981 mb) and wind speeds of 52 MPH (84 km/h) were recorded across much of the region. Nearly five inches (127 mm) of rain fell across Ireland, while high winds downed power lines and uprooted trees in England. Several million pounds worth of damage was incurred, although no deaths or injuries were reported. Downgraded to an extratropical depression on August 30 while over the North Sea, Charley soon dissipated.

Charley, Hurricane *North Atlantic Ocean, September 21–29, 1992* At peak intensity a Category 2 hurricane on the SAFFIR-SIMPSON SCALE, Hurricane Charley spent September 21–29, 1992, wandering across the shipping lanes of the northeastern NORTH ATLANTIC OCEAN. Formed near the Azores on September 21 from a weak tropical wave, Charley was upgraded to TROPICAL STORM status when its central pressure dipped to 29.67 inches (1,005 mb) on the afternoon of September 22. The following day, as the BERMUDA HIGH shifted Charley's trajectory sharply east-northeastward and pushed it into the eastern North Atlantic, the system began to mature, finally achieving hurricane status on September 23 (28.96 inches [981 mb]). Continuing to track to the northeast, Charley further strengthened, becoming a Category 2 hurricane on September 4, as its central pressure slipped to 28.49 inches (965 mb) and its sustained winds topped 109 MPH (175 km/h). Shortly after reaching peak intensity, Charley steadily began weakening, finally transitioning to an extratropical

cyclone on September 27, 1992, while it rapidly neared the southwestern coast of Ireland.

Charley, Tropical Storm *Gulf of Mexico–Southern United States, August 21–24, 1998* Born over the GULF OF MEXICO on August 21, 1998, Tropical Storm Charley recurved to the northwest, intensifying as it haphazardly moved toward the southeastern TEXAS coastline. Still deepening, Charley rolled ashore near Port Aransas, Texas, on August 22, bearing a central pressure of 29.52 inches (1,000 mb), 58-MPH (93-km/h) winds, and a three-to-five-foot (1–2 m) STORM TIDE. Like most TROPICAL STORMS, Charley was a ferocious rainmaker. On August 23, some 16.83 inches (406 mm) of precipitation fell on the town of Del Rio, Texas, while widespread flash flooding occurred elsewhere in the Lone Star State. Before dissipating on August 24, Charley claimed 13 lives in Texas, another seven in MEXICO, and left six people missing. Almost all of the fatalities were due to inland flooding. Some $50 million in property losses were tallied.

Charley, Hurricane *Gulf of Mexico–Southern United States, August 9–15, 2004* Judged at the time to have been the most intense and costliest hurricane to make landfall in the United States since 1992's Hurricane ANDREW, the fourth Hurricane Charley ravaged FLORIDA with Category 4 winds, rains, and surge conditions on August 13, 2004. Spawned as a TROPICAL DEPRESSION to the north of the islands of Trinidad and Tobago on August 9, Charley slowly strengthened into a TROPICAL STORM as it tracked west-northwestward across the CARIBBEAN SEA. On August 11, while positioned 115 miles (275 km) to the south of JAMAICA, Charley was upgraded to a Category 1 hurricane, and gently began a slow recurvature to the northwest. In preparation for a possible grazing landfall by the deepening hurricane, Jamaican authorities closed the island's airports and diverted cruise ship traffic, but Charley, moving at 24 MPH (39 km/h), passed Jamaica with considerable damage to crops and one fatality. Within two days (having crossed western CUBA as a Category 3 hurricane with a CENTRAL BAROMETRIC PRESSURE of 28.93 inches [980 mb] on August 12), and while moving to the northnorthwest over the southern GULF OF MEXICO, Charley began a harrowing deepening cycle. During the midafternoon hours of August 13, Hurricane Charley's central barometric pressure stood at 28.49 inches (965 mb). By the early evening hours of the same day, the system's lowest observed pressure had dropped to 28.17 inches (954 mb), and by the late evening hours, had plummeted to 27.78 inches (941 mb); a drop of .70 inches (24 mb) in just over six hours. During the same period, Hurricane Charley's sustained winds increased from 109 MPH (175 km/h) to 144 MPH (232 km/h). At its strongest, Hurricane Charley barreled into southwestern Florida, in the town of Cayo Costa (just north of Fort Myers), during the afternoon hours of August 13; it shortly thereafter made a second, slightly weaker landfall in Port Charlotte, Florida. A powerful system at each landfall, Charley produced sustained, 115-MPH (185-km/h) winds; an extreme gust of 111 MPH (179 km/h) was recorded at the Punta Gorda airport. At least 20 people were killed in Florida and damages estimates ranged as high as $15.4 billion. Charley retained hurricane intensity as it crossed the Florida peninsula and emerged into the Atlantic to strike SOUTH CAROLINA—between the coastal cities of Charleston and Georgetown—as a Category 1 hurricane. When the system entered the Atlantic its sustained winds were approximately 86 MPH (137 km/h). Charley quickly weakened after landfall, but maintained tropical storm intensity while moving northward across southern and eastern NEW ENGLAND. By August 15, Charley rapidly dissipated as it moved across Massachusetts and toward the Canadian Maritimes. Precipitation totals along the New England coast totaled one to three inches (25–75 mm), with one death being reported in Rhode Island.

Charley's record as the second-costliest hurricane to have struck the United States did not last long. Less than one year later, Hurricane KATRINA seemingly obliterated the Gulf Coast, its $75 billion dollar price tag becoming the most expensive natural disaster yet seen in American history, and displacing Andrew (and 1989's Hurricane HUGO) one place on the list of most expensive landfalling U.S. tropical cyclones on record. In the end, Charley would also prove one of the deadliest, with 35 direct and indirect deaths.

The name Charley has been retired from the rotating list of North Atlantic HURRICANE NAMES. It has been replaced for the 2010 season with the identifier Colin.

Charlie, Hurricane *Eastern Caribbean–Jamaica–Mexico, August 13–23, 1951* One of the last major North Atlantic HURRICANES to be identified with a name taken from the military phonetic alphabet, Charlie delivered 130-MPH (209-km/h) winds, 8.5-inch (203-mm) rains and 30-foot (10-m) seas to the eastern Caribbean islands of GUADELOUPE and Montserrat, to the entire south coast of JAMAICA, and to Tampico, MEXICO, between August 16 and 23, 1951. In Jamaica, where Charlie's Category 3 punch was felt the hardest, some 40,241 BUILDINGS were either damaged or destroyed, forcing the displacement of an estimated 50,000 people. One hundred sixty-two

men, women, and children on the island were killed, while another 2,000 were injured, making Charlie the deadliest storm to have struck Jamaica since the BLACK RIVER HURRICANE claimed 164 lives in mid-November 1912. In Mexico, where Charlie's unmitigated 160-MPH (258-km/h) gusts and whipping rains caused massive flood emergencies in and around the gulf-coast city of Tampico, 115 people were drowned or struck down by airborne debris. At least $150 million in property damage was assessed in both Jamaica and Mexico, characterizing Charlie as a hurricane of remarkable violence and duration.

The first CAPE VERDE STORM of the 1951 HURRICANE SEASON, Charlie originated over the subequatorial waters of the mid-Atlantic, 2,194 miles (3,531 km) east of the Lesser Antilles, on the afternoon August 13, 1951. Initially just another of the 50 or more TROPICAL WAVES that seasonally drift westward with the North African trade winds, Charlie's embryonic vortex quickly developed into a TROPICAL DISTURBANCE, or a heavy massing of thunderstorms, by midnight on August 13 and into a TROPICAL DEPRESSION, or a now-revolving collection of thunderstorms, by the mid-morning hours of August 14. Slowly spiraling around an invisible center of low BAROMETRIC PRESSURE, Charlie's CONVECTION cells drew enormous strength from the evaporating waters of the Atlantic before expending part of that heat energy as wind, near-gales that swiftly transformed the sea's surface into a rich field of spindrift. Setting in motion the almost mechanical cycle of rising and condensing air that would tragically endure for the next nine days, Charlie steadily intensified and its CENTRAL BAROMETRIC PRESSURE dropped from 29.58 inches (1,001 mb) on the evening of August 14, to 29.10 inches (985 mb) by midnight that same night, to 28.96 inches (980 mb) by dawn the next day.

Still classified as a TROPICAL STORM, or a cyclonic system with wind speeds of between 39 and 73 MPH (63–118 km/h), Charlie coursed due west at nearly 22 MPH (35 km/h), riding the constant trade winds ever closer to the volcanic beaches of the Lesser Antilles, the arc of islands that enclose the CARIBBEAN SEA's east end. On the evening of August 15, when it at last became apparent to those authorities tracking the now-powerful hurricane that Charlie would most certainly cross the island chain, storm warnings were posted from Anguilla in the north to MARTINIQUE in the south. As thousands of people hastened to board up their cottages, their hotels, and their sugar plantation drying houses and to secure their boats against the inevitable surf, Charlie rushed onward, its central pressure sinking to a considerable 28.81 inches (976 mb) during the early morning hours of August 16. Now located less than 150 miles (241 km) from the east peaks of Guadeloupe, Charlie's sustained 104-MPH (167-km/h) winds interspersed with 111-MPH (179-km/h) gusts to fan furiously the surface of the sea, scooping the heat-laden spray into its towering 45,000-foot (15,000-m) canopy and then dispersing it as rain, as the light drizzle that settled over the island just before an overcast daybreak.

On the nearby island of Montserrat, Charlie's portentous showers carved deep rivulets in the 3,000-foot (1,000-m) slopes of Chances Peak, the volcano that dominates the "Emerald Isle's" principal city, Plymouth. As last-minute precautions were being taken all over the island, the wind began to rise in gusts, driving the rain nearly horizontal against walls, windows, and doors. Quickly clearing the streets, the tiny dependency's entire population of 12,500 people—from the British governor to the fishers of Rendezvous Bay—retreated indoors. There, tucked within a seemingly safe cocoon of hurricane lanterns, food and water stocks, and radios, the people of Montserrat anxiously awaited Charlie's arrival. The eldest among them certainly would have recalled the last hurricane to strike the island directly, the infamous SAN FELIPE HURRICANE of September 12, 1928, in which 28 people succumbed to the storm's 150-MPH (241-km/h) winds and 18-foot (6-m) STORM SURGE.

But Hurricane Charlie was much kinder to Montserrat than the San Felipe storm had been, partly because of an abrupt course change that took the hurricane's roiling EYEWALL well south of the island. Although both Montserrat and Guadeloupe were visited by three to five inches (76–127 mm) of rain and gale-force winds, the brunt of Charlie's destruction remained at sea, gaining even greater strength from the warm, semienclosed waters of the Guadeloupe Passage. No casualties or serious injuries were reported in either Montserrat or Guadeloupe.

In the meantime, Charlie inexorably churned across the southern Caribbean, chasing shipping out of the area as its 130-MPH (209-km/h) winds spiraled toward Jamaica's east coast. In that country, which had sustained a long history of destructive hurricane strikes, an endless round of evacuation warnings and meteorological bulletins did little to save the lives of some 162 people, killed when Charlie's rampaging winds came ashore near Kingston on the evening of August 17, 1951. The most powerful hurricane to have affected the island in nearly a quarter of a century, Charlie's EYE curved southwest shortly after landfall, delivering further destruction to the entire south coast. In the posh bayside hamlet of Port Morant, dozens of prized coconut palms that had been painstakingly trained to form a picturesque promenade along the white-sand beaches were so completely swept away that not even their stumps

remained. In nearby Port Royal, where an earlier barrage of 17th- and 18th-century hurricanes had severely damaged the notorious pirate communes once homeported there, Charlie's 30-foot (10-m) waves and cresting 18-foot (6-m) storm surge shattered wharves, shifted navigation buoys, and drove six vessels—among them a 10,000-ton freighter—onto the narrow spit of land separating Port Royal from the rest of the island. Across the harbor in Kingston itself, hundreds of tin-roofed bungalows immediately collapsed, knocked to the ground by falling trees, splintered telephone poles and the buffeting intensity of Charlie's 125-MPH (201-km/h) gusts. Six inches of rain turned flat roads into ponds and entire neighborhoods into islands, stranding thousands of people on rooftops. When three men in the town of Ferry, some 15 miles (24 km) west of Kingston, attempted to climb trees in an attempt to escape Charlie's rising rains, the hurricane's sheering winds cut the trees down and cast the screaming men into the seething torrent below, where they died of drowning or electrocution. Similar horrors were witnessed in Spanish Town, where Charlie's Category 3 passage destroyed nearly every one of the town's 9,000 structures, including the brick-and-stucco town hall and all of the principal churches.

Swinging back into the Caribbean, Charlie headed northwest, raking the coastal communities of Lover's Leap, Treasure Beach, White House, and Negril with 125-MPH (201-km/h) winds and heavy rains. Damage was particularly severe in Savanna-La-Mar, where the hurricane-prone harbor was choked with shipping run aground by Charlie's 20-foot (6-m) breakers. Warehouses buckled and radio towers swayed and then toppled, crushing the transmitting stations below. Hundreds of trees tumbled into the streets, forming minidams as Charlie's five inches of rain piled up behind them. Cars and trucks, unwisely parked near beachheads, were washed into the sea. Dented and crushed, their windows smashed, their fuel tanks leaking gas slicks, they slowly settled to the bottom, in time becoming life-giving artificial reefs.

Quickly pulling away from Jamaica on the morning of August 18, 1951, Charlie continued moving northwest, possibly threatening the Cayman Islands atoll. Relief officials from Great Britain and the United States, horrified at the devastation wrought by Charlie in Jamaica, issued urgent warnings to Caribbean governments, advising them of Charlie's severity. Although stringent precautions were taken in both the Cayman Islands and along CUBA's southwest coast, Charlie eventually spared them its burgeoning fury by maintaining its trek northwest. Bound for either the Yucatán Channel or the northeast tip of the Yucatán Peninsula itself, Charlie's approach forced mass evacuations along the lower Mexican coastline. Shallow Cozumel was cleared, as were several seaside fishing towns between Cancun and Tulum. When Charlie crossed the Yucatán Peninsula's northeast tip on the evening of August 21, 1951, its 160-MPH (256-km/h) winds and rejuvenated eight-inch (203-mm) rainfalls claimed only three lives at Cabo Catoche, but caused tremendous property damage in and around greater Cancun.

Arcing into the GULF OF MEXICO, Charlie altered its course for the last time, returned to its northwesterly heading and a concerted assault on the gulf-coast city of Tampico. Noted for its oil refineries and tanker piers, Tampico creaked and hissed as Charlie's sustained 140-MPH (225-km/h) winds whistled ashore on the afternoon of August 22, 1951. Thousands of buildings, from plywood shanties to oil company offices, were either damaged or destroyed by the high winds and flash floods. A stricken refinery, its web of piping twisted and shorn by the hurricane, caused a minor pollution incident when one of its holding tanks was punctured by flying debris. In Cardenas, a valley town 64 miles (103 km) west of Tampico, 42 people were drowned when flood waters breached a dam; in nearby La Palona, another 50 fatalities were reported when the Guayalijo River overran its banks. Power outages in the region persisted for almost three weeks following the disaster, hampering efforts to reconstruct the city. With a total of 115 people dead in Mexico, Charlie was the deadliest hurricane to have struck that country in just over a decade.

Although the name Charlie was subsequently retired when the Weather Service abandoned the military phonetic alphabet system of HURRICANE NAMES at the conclusion of the 1952 season, it was returned to the post-1979 lists of male and female names in a slightly different incarnation, *Charley.*

Charlotte, Typhoon *North Pacific Ocean–Japan, October 11–17, 1959* An exceptionally powerful late-season TYPHOON, Charlotte delivered sustained 115-MPH (185-km/h) winds and heavy rains to the south Japanese island of Okinawa on October 16, 1959. Proving very destructive to U.S. military installations on the island, Charlotte claimed 28 lives and caused damage estimates of $15 million.

Charter Oak Hurricane *Northeastern United States, August 21, 1856* A relatively mild TROPICAL CYCLONE (its CENTRAL BAROMETRIC PRESSURE at landfall was 29.13 inches (986 mb), the Charter Oak Hurricane released sawing 77-MPH (124-km/h) gusts and three-inch (76-mm) downpours over Connecticut, Rhode Island, eastern Massachusetts, and southern Maine. In Connecticut, far to the west of where

the HURRICANE's nascent EYE dragged ashore near Providence, blustery outer gales caused Hartford's historic Charter Oak, an ancient oak tree within which the Connecticut Charter had been hidden during a period of political instability in the 1680s, to break off some six feet (2 m) from the ground. Several ships were run ashore near Boston, while heavy rains akin to those found in TROPICAL STORMS caused destructive flood emergencies over Massachusetts's and Maine's inland farms and forests.

China *East Asia* The world's most populous nation, China's 2,300-mile (3,701-km) coastline, coupled with its position on TYPHOON ALLEY's west end, has for centuries made it dangerously vulnerable to those early and late-summer TROPICAL CYCLONES that stampede westward across the NORTH PACIFIC OCEAN every TYPHOON SEASON. Records maintained by the Chinese meteorological services indicate that between 1915 and 2006 the tidal plains, misty river valleys, timbered hills, and marshy rice paddies of southern and eastern China were struck by no less than 348 typhoons and TROPICAL STORMS, for an average of four tropical systems per year. In some instances, such as the 1956, 1964, 1995 and 2005 seasons, five or more tropical cyclones have rushed ashore between June and September. On numerous occasions, China's principal seaports of Shanghai, Wenzhou, and the former British crown colony of Hong Kong have suffered catastrophic strikes by typhoons of severe intensity, whirring killers armed with 135-MPH (217-km/h) winds, 10 to 20-inch (254–508 mm) PRECIPITATION counts, and devastating STORM SURGES and STORM TIDES. While China has endured its ration of early and late-season typhoons, tropical storms, and TROPICAL DEPRESSIONS, its peak periods for tropical cyclone activity occur in June and July and in August and September, when the first of the North Pacific's SUPERTYPHOONS begin to recurve north-northwest into its low-lying east flank. Although not as frequent as the rainmaking *BAGUIOS* of the PHILIPPINES or as consistent as the wind typhoons that yearly scale the volcanic shores of JAPAN, several Chinese typhoons have in the past been responsible for inflicting enormous DEATH TOLLS and immense property damage on the country's densely populated east provinces. Indeed, between 1915 and 2006, an astounding 83,000 people in China were killed by tropical cyclones, with some 60,000 of them perishing in a single storm—the monstrous typhoon of August 2–3, 1922 that all but extinguished the northern fishing village of Swatlow.

On the night of July 21–22, 1841, a slow-moving midseason wind typhoon accompanied by a 12-foot (4-m) storm surge and confused seas inundated Hong Kong. Wheeling around a CENTRAL BAROMETRIC

PRESSURE of 28.50 inches (965 mb), the typhoon's 110-MPH (177-km/h) winds wrecked nearly every ship moored in the harbor. Dramatic eyewitness accounts later published in Great Britain described how scores of terrified Chinese fell to their knees in frantic supplication upon the typhoon's arrival, imploring "their gods in vain for help" while the storm's merciless onslaught steadily gained on them. Either swept into the roiling waters of the harbor or buried beneath the collapsed rubble of their tiny bungalows, some 1,000 Chinese eventually perished in this, the first major typhoon experienced by the British in Hong Kong.

More than 37,000 people were reportedly killed when an unnamed typhoon lashed the southeastern port city of Canton (Guangzhou) with 107-MPH (172-km/h) winds and seven inches of rain during the afternoon hours of July 27, 1862. Canton's thriving waterfront was virtually obliterated, its sturdy wooden piers, stone quays, and spacious warehouses reduced to rubble. Caught at anchor by the typhoon's surprise arrival, six coal-fired steamships were promptly cast ashore; two of them were later declared total wrecks. Dozens of smaller vessels, from sleek tea clippers to delicate sampans, were dismasted and sunk, overwhelmed by the typhoon's raging breakers. Hundreds of those lost were fishers, drowned at sea during the tempest's opening stages.

Less than a quarter of a century later, another particularly deadly typhoon sliced into southern China just north of the mouth of the Red (Hong) River on the morning of October 5, 1881. Generating sustained winds of 115 MPH (185 km/h) over the Gulf of Tonkin, the typhoon's whirling EYEWALL cut a broad swath of destruction across the city of Haiphong (located in present-day Vietnam), uprooting thousands of trees and triggering horrific flash floods. Preceded ashore by a thundering 20-foot (7-m) storm surge, the 1881 typhoon wrought such devastation that an accurate accounting of the dead could not be compiled. Some sources place the death toll at 300,000 people, while others are more conservative in their estimates, listing the loss at 30,000. In either instance, the 1881 Haiphong typhoon ranks as one of the deadliest North Pacific tropical cyclones on record.

On July 24, 1896, a blustery midseason typhoon bisected southern TAIWAN before slamming with terrifying violence into the central port city of Shantung (Shantou), China. Torrential rains, clutching seas, and wind-borne debris claimed 490 lives in China and left thousands more homeless, almost destitute.

On the rain-stung night of September 18, 1906, the celebrated Second Hong Kong Typhoon wailed ashore in southwestern China, blighting thousands of acres of rice fields, sinking scores of ships and razing to the ground hundreds of houses in and around the neighboring outpost of Kowloon. Said by survivors to have been one of the most violent typhoons to have impacted the area in living memory, the Second Hong Kong Typhoon threw no less than 33 vessels—from 2,000-ton tramp steamers to a French torpedo boat—aground on the jagged reefs that guard the entrance to Hong Kong harbor; some 2,000 lightly built sampans and junks were also wrecked. At least 10,000 people perished, while damage estimates quickly topped $20 million.

At least 600 people died on the evening of July 25, 1915, when a wind typhoon jolted ashore at Hangchou Bay, an important trading post located on China's south coast. Joined in the carnage by a 17-foot (6-m) storm tide, the 1915 typhoon collapsed hundreds of houses, trapping their unfortunate inhabitants beneath the wreckage. Every vessel at anchor in the bay was either run aground or sunk, with many suffering the loss of their entire crews. Gust-sharpened debris knifed through the air, striking down scores of fleeing villagers, while tremulous rockslides carried away roads and bridges, only increasing Hangchou's submerged isolation.

Just one of many rain-laden tropical storms to have lashed Hong Kong during the past 200 years, the expansive tropical system of July 17, 1925, delivered nearly 21 inches (533 mm) of rain to the crowded foothills of southern China. More than 100 people perished as the tropical storm's spiral bands inundated Hong Kong for 11 straight hours, a precipitation record that remains among the highest in the region.

Not more than a year later, on September 30, 1926, a vastly more intense tropical cyclone raced ashore near Shanghai. Containing slight rains, the typhoon's 112-MPH (180-km/h) winds and cresting storm surge nevertheless managed to submerge several hundred acres of coastal plains. Shanghai's lighthouse was lost to the invading surf, as were some 2,000 men, women, and children.

Another 5,000 people died the following year when a gigantic midseason typhoon laid waste to the central port city of Kwangtuang. Slung beneath its tightly wound EYE, the typhoon's pulverizing 22-foot (7-m) storm surge swept hundreds of houses from their foundations, carved deep furrows through pine groves and sank nearly every ship in the harbor. The disruption of water supplies, combined with mounting shortages of food, medicine, and clothing, fueled widespread riots across the province. Similar outbreaks of looting and violence later followed the murderous passage of typhoons in 1935 (August 5,

Fukien Province, 900 dead) and 1936 (August 16, Hong Kong, 100 dead).

On October 7, 1947, a gargantuan late-season typhoon ashore at Hong Kong. Sustained, 110-MPH (177-km/h) winds stripped houses of their clay-tile roofs, crumpled signs, and uprooted thousands of bamboo trees. Deluging rains turned Hong Kong's narrow streets into temporary rivers, rushing outlets that soon emptied tons of wreckage into the harbor's frothing confines. Hong Kong's famed fleet of fishing junks, their folding sails torn to tatters by the typhoon's 133-MPH (214-km/h) gusts, was nearly annihilated while still in the South China Sea. In total, some 2,411 lives were lost to this storm, many of them drowned fishers.

During the course of the 1948 season, China was struck by three immense typhoons between June and October. On June 17 an unnamed typhoon splashed ashore near Shantou. Wading as far inland as Dongting Lake, some 450 miles (724 km) north of Kowloon, the June typhoon triggered crushing mountain mudslides that killed 190 people. Less than a month later, a slightly more intense tropical system drifted ashore at Shanghai on July 4. Not as durable a rainmaker as the June 17 storm, the July typhoon's strident winds killed 34 people and injured another 100. A third, that blasted Hainan Island on October 2, killed an additional 30 people and left more than $5 million in property losses.

At least 29 people were killed and much of Shanghai left in flooded ruins after an unnamed typhoon whisked over the city on the night of July 24–25, 1949. A strapping system that only the day before had demolished 41,000 buildings and killed 38 people on the island of Okinawa, the 1949 typhoon left a quarter of a million people in Shanghai without shelter and caused enormous losses to that year's rice harvest.

The summer of 1956 saw another two typhoons strike China with deadly consequences—Virginia on the evening of July 31 and Wanda on the morning of August 3. The weaker of the two storms, Category 2 Virginia killed 57 people and injured another 102 as it spun through the south-central provinces of Chekiang, Honan, and Hopeh. Virginia's heavy rains caused rivers to overrun their banks, destroying 38,000 homes and almost a third of the region's staple rice stocks. Less than one week later, Typhoon Wanda, a Category 3 system with extreme winds of 113 MPH (182 km/h), washed ashore at Hangchou. A voluminous rainmaker, Wanda's flash floods soon submerged large portions of Anhwei, Chakiang, and Kiangsu provinces, killing 83 people, injuring 300 others, and rendering another 20,000 homeless.

On July 18, 1957, Typhoon Wendy, a relatively mild tropical cyclone that had formed east of the PHILIPPINES nearly a week earlier, swept into Hong Kong harbor in a corona of spindrift and 109-MPH (175-km/h) gusts. Bearing a mantle of 12-foot (4-m) seas and copious rains, Wendy killed 16 people and injured 200.

One of the most violent of all 20th-century Chinese typhoons, Iris visited catastrophic property damage and huge death tolls on the southeast province of Fukien on August 20, 1959. Earlier responsible for shortening six lives on the nearby island of TAIWAN, Iris's tragic legacy proved much higher on the Chinese mainland: 2,334 people dead, twice that number injured, and more than 30,000 buildings either damaged or completely destroyed. Some 10 months later, much of Fukien Province was again ravaged by wind and water as Typhoon Mary made a severe midmorning landfall near Hong Kong on June 9, 1960. Although only 90 people in the city perished beneath Mary's 112-MPH (180-km/h) winds and drumming downpours, the typhoon's lingering rains and concomitant landslides claimed an additional 1,600 lives elsewhere in the province. Nearly 19,000 people across the region were left homeless, with starvation and disease soon gaining the upper hand on thousands of weary survivors.

In 1964, Hong Kong was forced to take cover from two very different North Pacific tropical cyclones. The first, a minimal Category 1 rain typhoon code-named Winnie, shot across the city on the morning of June 30. Its sustained winds hovering just above typhoon intensity (75 MPH [120 km/h] at landfall), Winnie's seven-inch (178-mm) precipitation count fostered dozens of gushing floods in the hillside communities surrounding the city. Ten people were killed and another 41 injured. Hong Kong had no sooner completed repairs from Winnie when a much greater maelstrom, Typhoon Ruby, stomped ashore near Kowloon during the early morning hours of September 5. A Category 3 typhoon of exceptional severity, Ruby's 118-MPH (190-km/h) winds and 12-foot (4-m) surge killed an estimated 700 people and wrought more than $30 million in 1964 property losses.

Boasting top winds of 130 MPH (209 km/h) and 12-inch (305-mm) rainfalls, Super Typhoon Rose plowed into southwest China on the morning of August 17, 1971. Passing some 60 miles west of Hong Kong (thus placing the city within the typhoon's DANGEROUS SEMICIRCLE), Rose's clipping winds and rocking seas capsized 40 fishing boats, overturned a ferry in Hong Kong harbor, and drove 26 large cargo ships aground on nearby beachheads. Ninety people died—88 of them drowned in the foundering of the ferry—while another 200 lay injured, several of them critically.

After raking the south Korean peninsula with 110-MPH (177-km/h) winds and six-inch (152-mm) rains, Typhoon Judy crossed the Yellow Sea to make a devastating landfall near the Chinese port city of Qingdao on the afternoon of August 26, 1979. Scudding rains spawned capricious floods and landslides that buried villages, swept away bridges, and inundated thousands of acres of rice paddies. At least 600 people were killed, with another 289 listed as missing.

Bound for an eventual landfall near the Shanghai's industrial center, the blossoming EYEWALL of Typhoon Orchid overwhelmed the 169,000-ton British ore-carrier *Derbyshire* on August 13, 1980. Snapped in two by Orchid's massive 52-foot swells, the 1,000-foot (333-m) *Derbyshire,* then positioned 800 miles south of Japan, quickly plunged to the seabed, taking its entire 44-member crew with it. Considerably weakened by time of landfall, Orchid released 90-MPH (145-km/h) gusts and ribbons of rain over Shanghai but claimed only seven deaths on the Chinese mainland.

Pirouetting about a record-low central pressure of 27.28 inches (924 mb), Super Typhoon Clara bounced ashore near Kowloon during the early morning hours of September 21, 1981. Before disintegrating over Guangdong Province's wooded hillsides two days later, Clara's 138-MPH (222-km/h) winds and seeping rains demolished 2,734 buildings, toppled hundreds of utility poles and signs, and claimed an estimated 120 lives. Of firm super typhoon status at landfall, Clara was among the most-intense recorded typhoons to have battled China during the latter half of the 20th century.

In a season that had witnessed deadly strikes on China's east shores from six separate tropical cyclones, Typhoon Ken darted into the central province of Zhejiang just before dawn on July 30, 1985. A typical wind typhoon in any number of meteorological respects, Ken's 120 MPH (193 km/h) ADVECTION currents characteristically scattered 20,000 lightly constructed buildings, decimated 74,000 acres of rice paddies, winnowed 30,000 acres of bamboo and mahogany forests down to mere stumps, and killed 177 people and injured 400 others.

On August 21, 1994, the winds and rains associated with Typhoon FRED struck down more than 1,000 people in Zhejiang province.

During the notably active 1995 typhoon season, China's hill-guarded harbors and forest-lined beaches endured the deleterious effects of three mature typhoons and two tropical storms. On August 15, Tropical Storm Helen dropped 20 inches (508 mm) of rain on Hong Kong. Even though floods and landslides eventually killed 23 people, the death toll was kept low by the advanced typhoon warning system then in operation in the former British colony. On August 26, a westward-moving Typhoon Irving, complete with 110-MPH (177-km/h) winds and 15-foot seas, also fell upon Hong Kong. A substantial tropical system, Irving's rains penetrated as far north as the city of Chiai. No deaths or injuries were reported. Two days later, a deepening Tropical Storm Janice delivered blustery gales and localized flooding to Shanghai on August 28. Closely following Janice ashore, Super Typhoon Kent sent rolling breakers crashing against the southeast coast of Guangdong province during the day and night of September 1–2. Causing extensive surge damage to the coastal communities of Hainan Island, Kent later claimed the lives of six Chinese fishers. Just barely a typhoon as it skidded into south China on October 3, 1995, Typhoon Sybil's 74-MPH (120-km/h) winds and five-inch (127-mm) precipitation counts wrought widespread tree, billboard, and scaffold damage on Hong Kong. One of the milder tropical cyclones in the city's stormy history, Sybil injured 12 people—two of them seriously—but claimed no fatalities.

The 1996 typhoon season was no gentler to China than the 1995 season had been, with five powerful tropical cyclones raiding the nation's Pacific coastline between July 23 and September 20. On July 23, Typhoon Frankie struck the island province of Hainan with 90-MPH (145-km/h) winds and torrential rains, disrupting electrical and water supplies, crumpling billboards, pushing small boats ashore, and killing one person. Less than two weeks later, on August 1, Typhoon Herb raced into the southeast province of Fujian. Targeting the port city of Pingtan, Herb's 87-MPH (140-km/h) winds swept one victim, into the sea, while eight-inch (203-mm) rains forced already flooded rivers and streams to top their banks. On the evening of August 8, Tropical Storm Lisa added an additional five inches (127 mm) of rain to the still-bloated reservoirs of Fujian province when its 60-MPH (97-km/h) eyewall winds came ashore east of Hong Kong. Flash flooding claimed four lives.

After a brief one-month lull in tropical cyclone activity, China suffered its deadliest typhoon strike of the 1996 season on the afternoon of September 9 when Typhoon Sally (the third Chinese typhoon to bear that name), twirled ashore in Guangdong province. Sustained winds of 110 MPH (177 km/h), combined with 4-inch (101-mm) precipitation counts, devastated the industrial cities of Zhanjiang and Maoming. Newspaper reports indicated that nearly all the trees in Zhanjiang were uprooted, while warehouses and other substantial manufacturing facilities were utterly demolished. Some 700 fishing boats were run ashore at Zhanjiang. Another 60 small watercraft foundered at the entrance to Maoming harbor. Nearly 200,000 houses were either damaged or destroyed, leaving several hundred thousand people without shelter. Typhoon Sally's visit to southeast China left 114 dead, 110 missing, and $1 billion in property damages.

Said to have been the strongest typhoon to have struck Hainan Island in a decade, Typhoon Willie delivered 100-MPH (177-km/h) winds and lashing seas to the provincial capital of Haikou on the night of September 20, 1996. Five people died and another 28 were injured.

On August 23, 2000, Super Typhoon Bilis ("Swift" in Philippine Tagalog) pounded Taiwan and eastern China with 125-MPH (205-km/h) winds, heavy surf conditions, and considerable PRECIPITATION. At least 12 people in China were killed, primarily in flood-related accidents.

During the 2001 eastern North Pacific typhoon season, three tropical systems of consequence affected China, including Typhoon Chebi on June 23, Typhoon Durian (Thai for a type of southeast Asian fruit) on July 1, and less than one week later, Typhoon Utor on July 6. Of the three, Typhoon Utor proved the most damaging to China. Chebi's early-season downpours, in some places reportedly measuring four inches (100 mm), fell across Guangdong and Fujian provinces, producing landslides, destroying crops, and leveling thousands of buildings. Chinese authorities stated that 73 people lost their lives to Typhoon Chebi, including 22 who were killed in the collapse of a wall in Hangzhou, while damage estimates were in the hundreds of millions of yuan. While Typhoon Durian produced sustained wind speeds of 85 MPH (137 km/h) across southeastern China on July 1, 2001 (reportedly claiming 78 lives), Typhoon Utor was preparing to make landfall over Hong Kong on July 6. Utor's streaming caused several hundred million yuan in property losses, and killing some 144 people. Widespread flooding in the Guangdong province and Guangxi Autonomous Region destroyed 3,700 buildings and damaged another 8,390.

On September 5, 2002, Typhoon Sinlaku (a Micronesian goddess) rushed ashore in eastern China, leaving 23 people dead and another five missing. The system made landfall near Wenzhou, Zhejiang Province, with sustained winds of 81 MPH (130 km/h) and three and one-half inches (76.2 mm) of rain. Shortly thereafter, Typhoon Sinlaku moved southward to Fujian Province, then inland to Jiangxi Province. On August 6, 2002, Tropical Storm Kammuri swirled into the southern Chinese province of Guangdong, bringing widespread damage to the coastal cities of Shanwei, Shantou, and Lufeng. Chinese state media indicated that 13 people were killed, 14 others were injured, and another 20 listed as miss-

ing. Ten of those lost were farmers reportedly killed in a massive landslide in Wuhua County.

The 2003 western North Pacific typhoon season was an active one for China. The nation was struck by Typhoon Parma, Koni (0308; July 16–23, 2003), Imbudo (0307; July 17–25, 2003), Krovanh (0312; August 17–26, 2003), and Dujuan (0313; August 29–September 3, 2003). The most severe of the 2003 typhoons in China was Typhoon Dujuan.

For China, the 2005 western North Pacific typhoon season was the most active in decades. More than 20 tropical cyclones directly affected the country, with 19 of them being mature-stage tropical systems. On July 19, 2005, a downgraded Tropical Storm Haitang struck the southeastern coast of mainland China. At one time a powerful typhoon, Haitang caused considerable damage in Taiwan before spiraling into the Chinese coastal town of Huangqi, in storm-swept Fujian province. Nearly 1 million people were evacuated (*see* EVACUATION) from their homes in Fujian and Zhejiang provinces in advance of the storm's landfall, and some 5,000 police officers were mobilized to oversee preparation and relief operations. In Shanghai, rain-swollen rivers threatened to top their embankments and flood low-lying neighborhoods. Haitang's sustained, 74-MPH (119-km/h) winds made landfall during the early evening hours on July 19. During Haitang's landfall, the Sai River in Fujian Province surged six feet above the official flood level. Nearly 50 people were killed in traffic accidents, flash floods, mudslides, and electrocutions. During the month of September alone, no fewer than three typhoons—Talim, Khanun, and Damrey—struck the country's eastern coast. More than 130 people perished in flash floods, BUILDING collapses, and mudslides. On September 12, 2005, Typhoon Khanun made landfall just to the south of the port city of Shanghai. At least seven people were killed, more than 1 million others were evacuated, and significant property damage occurred in the coastal city of Taizhou. While the city's airports were closed, Shanghai authorities evacuated some 160,000 people. Downgraded to a tropical storm, Khanun delivered heavy rains to the port city of Qingdao, dropping nearly three feet (1 m) of rain in some mountainous locations. Typhoon Longwang (Dragon King) slid into China's eastern coast on October 1, 2005, causing enormous damage across Fujian province. Despite the evacuation of some 537,000 people from vulnerable coastal cities, Longwang's sustained 74-MPH (119-km/h) winds and torrential claimed dozens of lives as it drilled inland after coming ashore near the port city of Longyan during the early evening hours of October 1. Some 5,400 houses were destroyed and several thousand acres of crops ruined. In the coastal city of Fuzhou, a flash flood caused by Longwang's pounding rains rolled down a hillside and destroyed a training school barracks, killing 80 paramilitary security officers. Elsewhere in the province, a landslide claimed three lives. All told, 85 people died during Longwang's passage over China.

On July 14, 2006, the second western North Pacific Ocean typhoon identified as Bilis—this one a Severe Tropical Storm—drifted into eastern China and collided with that season's southerly monsoon. In a seeming flash, large sections of eastern China were inundated with torrential precipitation; some 3 billion yuan in property losses were recorded. In Hong Kong alone, some four and one-half inches (101.6 mm) of precipitation fell during a one-hour period, establishing a new rainfall record. A slow-moving system of considerable size, Tropical Storm Bilis produced an extreme gust of 96 MPH (155 km/h) in the city of Zhejiang, as well as rainfall measurements totaling 14.1 inches (355.6 mm), at Guangdong. The system wrought enormous destruction in Hunan, and the largest fraction (346) of the 637 deaths recorded in the storm occurred among the province's mountain villages. Spawned by Bilis's relentless downpours, raging landslides swept away some 30,000 buildings and destroyed railway systems and bridges. Across southeastern China, nearly half a million people were left homeless, proving an enormous logistical challenge for the Red Cross Society of China and other relief organizations.

Just over two weeks later, on July 25, 2006, the southern Chinese coastline was battered by Typhoon Kaemi, one of the most powerful of recent western North Pacific Ocean tropical cyclones to affect Fujian province, and the fifth major typhoon to strike China during the 2006 season. In preparation for Kaemi's arrival, Chinese authorities evacuated some 643,000 people from vulnerable locations, while a newly implemented early warning system (using text messages sent to cellular telephones) alerted 6 million people in the region of the typhoon's impending landfall.

Chris, Tropical Storm *Gulf of Mexico–Southern United States, September 9–12, 1982* Born from a subtropical depression that had lingered over the central GULF OF MEXICO, Tropical Storm Chris delivered heavy rains and 63-MPH (101-km/h) winds to southern LOUISIANA and northeastern TEXAS on September 11, 1982. Originating on September 9, some 300 miles south of New Orleans, Chris was upgraded to a TROPICAL DEPRESSION on the morning of September 10, and reached TROPICAL STORM intensity during the late evening hours. First tracking to the west, Chris sharply recurved to the north as it intensified. The system was at its peak intensity as it came ashore on the Texas-Louisiana border, near

Sabine Pass, Texas, with a central pressure reading of 29.35 inches (994 mb) and 63-MPH (101-km/h) winds. Passing over Texas and into central Louisiana, the system was downgraded to a tropical depression by the time it entered Arkansas, and dissipated. Although heavy rains accompanied Chris's passage, no deaths or injuries were reported.

Chris, Tropical Storm *Caribbean Islands–Hispaniola–Southern and Eastern United States, August 21–30, 1988* Identified as Tropical Depression No. 3 for much of its early existence, Tropical Storm Chris was of Cape Verde origin when it emerged from the North Atlantic (as a TROPICAL DEPRESSION) on August 21, 1988. Gently recurving to the northwest, then to the northeast, over the Leeward and Windward Islands, PUERTO RICO, and HISPANIOLA, between August 23 and August 27, Tropical Depression No. 3 dropped nearly five inches (127 mm) of PRECIPITATION on Puerto Rico, killing three people on the island. Upgraded to Tropical Storm Chris on August 28 while grazing the eastern coast of FLORIDA, the system reached its peak intensity of 29.67 inches (1,005 mb) just before making landfall on the GEORGIA-SOUTH CAROLINA border—near Savannah, Georgia—with 52-MPH (84-km/h) winds later the same day. As it was returning to TROPICAL DEPRESSION status over South Carolina, Chris generated several TORNADOES, claiming one life in that state. Under the steering influence of the BERMUDA HIGH and broadly recurving over NORTH CAROLINA, VIRGINIA, Pennsylvania, NEW YORK, and southern NEW ENGLAND, a downgraded Tropical Depression Chris retained its tropical characteristics well into eastern CANADA, where Newfoundland was lashed with heavy rains and 30-MPH (48-km/h) winds. A rainmaking tropical system, Tropical Storm Chris caused nearly $500,000 in property losses.

Chris, Hurricane *North Atlantic Ocean, August 16–23, 1994* The first hurricane of the 1994 North Atlantic season, Hurricane Chris remained an offshore system for its entire existence. Formed on August 16 from a TROPICAL DEPRESSION, Chris passed to the east of the Windward and Leeward Islands as a TROPICAL STORM, then a hurricane, between August 19 and 20. A Category 1 hurricane at peak intensity, Chris generated a central barometric pressure reading of 28.90 inches (979 mb) on August 19. Sustained winds were estimated at 81 MPH (130 km/h). Undercut by WIND SHEAR, Chris returned to tropical storm status on August 20, and spent the remainder of its lifespan trailing to the north-northeast, and slowly weakening. Nearing the eastern coast of Newfoundland on August 22, Chris underwent extratropical deepening and dissipated the following day.

Chris, Tropical Storm *North Atlantic Ocean, August 17–19, 2000* A short-lived TROPICAL STORM, Chris originated and dissipated to the east of the Windward and Leeward Islands between August 17 and 19, 2000. Of tropical storm intensity for less than one day, Chris reached its maximum intensity on August 18, when its central pressure slipped to 29.76 inches (1,008 mb) and its sustained winds reached 40 MPH (64 km/h). Within hours of these observations, Chris weakened into a TROPICAL DEPRESSION, and dissipated shortly thereafter.

Cindy, Hurricane *Gulf of Mexico–Southern United States, September 16–20, 1963* The second North Atlantic TROPICAL CYCLONE known as Cindy formed over the northern GULF OF MEXICO on September 16, 1963. Fed by warm SEA-SURFACE TEMPERATURES (SST) and favorable upper-altitude conditions, the system strengthened into a TROPICAL STORM, then a Category 1 hurricane, in less than 24 hours. According to observations taken by the NATIONAL HURRICANE CENTER (NHC), Cindy achieved peak intensity just before midnight on September 17, as its CENTRAL BAROMETRIC PRESSURE deepened to 29.41 inches (996 mb). Slightly recurving to the north-northwest as it moved across the GULF OF MEXICO, Hurricane Cindy came ashore at High Island, TEXAS, during the early evening hours of September 18, 1963. Sustained winds were clocked at 75 MPH (121 km/h). After making landfall and being downgraded to a tropical storm, Cindy sharply recurved to the southwest and rolled deep into southern Texas. Again downgraded, this time to a TROPICAL DEPRESSION, Cindy spent September 19 and part of September 20, delivering heavy rains and gusty winds to central Texas. Three people in the Lone Star State were killed, and damage estimates were near $13 million (1963 dollars).

Cindy, Tropical Storm *North Atlantic Ocean, August 2–5, 1981* Originating from a subtropical depression situated over the mid NORTH ATLANTIC OCEAN, approximately 500 miles (700 km) north of BERMUDA, on August 2, 1981, Tropical Storm Cindy first matured into a TROPICAL DEPRESSION while tracking to the north-northeast on August 3, and into a TROPICAL STORM (with a central pressure of 29.61 inches [1,003 mb]) just before midnight. It reached its maximum intensity of 29.58 inches (1,002 mb) during the morning hours of August 4, producing sustained winds of 58 MPH (93 km/h). On August 5, as the system passed well off the coast of Newfoundland, CANADA, Tropical Storm Cindy lost its tropical characteristics, and was reclassified as an extratropical depression. It completely dissipated by August 6.

Cindy, Tropical Storm *North Atlantic Ocean, September 5–10, 1987* Formed near latitude 15 degrees North on September 5, 1987, Tropical Storm Cindy remained a mid-North Atlantic system for its entire existence. Tracking to the northwest, the system was upgraded to a TROPICAL STORM on September 7, 1987, when its CENTRAL BAROMETRIC PRESSURE dropped to 29.73 inches (1,007 mb). The system reached peak intensity on September 8, with a central pressure reading of 29.52 inches (1,000 mb) and sustained winds of 52 MPH (84 km/h). Recurving to the north-northeast, Tropical Storm Cindy quickly weakened and underwent extratropical transitioning as it harmlessly neared the Canary Islands on September 10, 1987.

Cindy, Tropical Storm *Caribbean Sea–Hispaniola, August 14–17, 1993* Born from a TROPICAL WAVE that had originated off the western coast of Africa on August 6, Tropical Storm Cindy became the third named tropical system of the 1993 North Atlantic HURRICANE SEASON when it achieved TROPICAL STORM intensity over MARTINIQUE on August 14. Intense tropical downpours inundated sections of the island, claiming two lives. Churning to the west-northwest, Cindy slowly intensified, finally reaching peak intensity on August 16, with a central pressure reading of 29.73 inches (1,007 mb) and 46-MPH (74-km/h) winds. Within a day, Cindy lurched ashore in the southern Dominican Republic, near the coastal town of Barahona, showering the highlands of HISPANIOLA with treacherous flash floods. Two people in the Dominican Republic drowned. Unable to survive passage over Hispaniola's mountainous terrain, Tropical Storm Cindy first returned to TROPICAL DEPRESSION status and dissipated by August 17.

Cindy, Hurricane *North Atlantic Ocean, August 19–31, 1999* The fifth North Atlantic tropical system to be dubbed Cindy also became one of the most powerful of the 1999 HURRICANE SEASON. Originating on August 19, very near the western coast of Africa as a Cape Verde system, Cindy rode the trade winds due westward, intensifying into a TROPICAL STORM before recurving to the north-northeast on August 21. On August 22, Cindy briefly reached HURRICANE intensity (29.14 inches [987 mb]) before weakening to tropical storm intensity later the same day. Trailing along the lower edge of the BERMUDA HIGH, Cindy moved well to the northeast of the Leeward Islands before intensifying into a Category 1 hurricane on August 26. Fueled by a low WIND SHEAR environment and warm SEA-SURFACE TEMPERATURES associated with the Gulf Stream, Cindy spent the following two days rapidly intensifying. In a 24-hour period, Cindy's central barometric pressure plummeted from 28.85 inches (977 mb) to 27.87 inches (944 mb), hiking its sustained winds from 104 MPH (167 km/h) to 138 MPH (222 km/h). A large system, enormous waves rolled away from Cindy's powerful EYEWALL, eventually washing ashore in Bermuda and the northern Caribbean islands. Never a serious threat to land, Cindy steadily began to weaken as it crossed 35 degrees North, and by August 31 had dissipated.

Cindy, Hurricane *Mexico–Gulf of Mexico–Southern United States, July 3–7, 2005* During an early morning landfall on July 6, 2005, Hurricane Cindy delivered 74-MPH (119-km/h) winds and a 4–6 foot STORM SURGE to the southern coasts of LOUISIANA and MISSISSIPPI. Bearing a CENTRAL BAROMETRIC PRESSURE of 29.47 inches (998 mb) at landfall, Cindy dropped between four and six inches (102–152 mm) of rain and prompted TORNADO watches to be posted across four U.S. states. Cindy originated off the eastern coast of MEXICO's Yucatán Peninsula before rumbling across it as a TROPICAL DEPRESSION. The system remained of tropical depression intensity as it moved northward across the GULF OF MEXICO, finally reaching TROPICAL STORM classification on July 5, 2005, and hurricane intensity directly before landfall in Louisiana. Initially judged to have been of tropical storm intensity, Cindy's classification at landfall was upgraded to that of a Category 1 hurricane by the NATIONAL HURRICANE CENTER (NHC) in January 2006. As in the example of Hurricane ANDREW, which was upgraded to Category 5 status in 1999, Cindy's promotion was due to a lengthy review and analysis of meteorological data, such as barometric pressure readings and other observable factors. Recurving to the northeast, Tropical Depression Cindy dissipated over the northwestern corner of GEORGIA on July 7. All told, Cindy directly and indirectly claimed three lives in Georgia and ALABAMA, and caused $320 million (2005 dollars) in property losses.

cirrostratus clouds Found in the upper levels of the atmosphere, cirrus-type clouds are thin, wispy collections of frozen water vapor. Formed when moist ascending air reaches the cold topside of the troposphere and turns to ice crystals, cirrostratus clouds form a fibrous, high-level canopy over TROPICAL CYCLONES, sometimes completely obscuring a hurricane's, typhoon's, or cyclone's towering EYE. A by-product of the storm's exhaust system, cirrostratus clouds make up a large part of the OUTFLOW, or venting layer, where spent rising air that leaves the eye crests over and spreads out, forming a tangible "cloud-roof" over the hurricane's CUMULONIMBUS spiral RAIN BANDS. Cirrostratus clouds also tend to gather on the approaching side of a tropical cyclone and to fan out before it in ominous parallel lines. It is now understood

that many of the spectacular sunsets that precede the arrival of intense hurricanes are due in part to the reflection of light by the crystalline cirrus canopy.

Claudette, Tropical Storm *North Atlantic Ocean–Hispaniola–Cuba–Southern United States, July 15–29, 1979* Between 1979 and 2003, five tropical systems in the NORTH ATLANTIC OCEAN have carried the identifier Claudette. Of these five, two were TROPICAL STORMS and the other three, mature-stage hurricanes. Two of the systems named Claudette directly affected TEXAS, in 1979 and 2003.

The first TROPICAL CYCLONE dubbed Claudette originated in the North Atlantic Ocean, some 500 miles east of the Leeward Islands, on July 15, 1979. Formed from a tropical wave with Cape Verde antecedents, the system remained at TROPICAL DEPRESSION intensity until July 17, when it intensified into the third tropical system of the 1979 North Atlantic season as it passed into the CARIBBEAN SEA to the north of the island of GUADELOUPE. Like all tropical systems a tremendous rainmaker, Claudette produced nearly nine inches of PRECIPITATION over Guadeloupe's green hills, causing scores of flash floods, but no deaths or serious injuries. Unfavorable winds aloft hampered Claudette's intensification as the system pushed due westward, and by the time the system reached PUERTO RICO on July 18, it had been downgraded to a tropical depression with a central pressure of 29.85 inches (1,011 mb) and 35-MPH (56-km/h) winds. Nine inches (228.6 mm) of precipitation fell across the length of Puerto Rico, killing one person and causing nearly $1 million in damage. Claudette's passage over the Puerto Rican highlands aggravated its already unstable circulation systems, and when the system limped ashore on the eastern coast of HISPANIOLA during the late-night hours of July 18, 1979, it had essentially weakened to tropical wave intensity. Between July 18 and July 21, as it scudded westward across Hispaniola's mountainous spine and recurved along the deep foothills of southern CUBA, Claudette's central pressure rose from 29.85 inches (1,011 mb) to 29.91 inches (1,013 mb). Heavy rains fell across Hispaniola and Cuba, but no lives were lost and property damage was minimal. Claudette again began to strengthen as it entered the GULF OF MEXICO during the early evening hours of July 22. Now tracking to the northwest on a trajectory that if maintained would take the system ashore on the MEXICO-Texas border, Claudette slowly reorganized, its central pressure dropping from 29.82 inches (1,010 mb) on July 21 to 29.64 inches (1,004 mb) on July 23. Shortly before noon on July 23, as the system drew to within 300 miles (500 km) of the northeastern Texas coastline, Claudette again recurved—this time to the north-northwest—and again intensified into a tropical storm. By the time Claudette made landfall on the Texas-Louisiana border during the early morning hours of July 25, its central pressure had deepened to 29.44 inches (997 mb), the most intense pressure reading in the system's entire existence. Winds of 46 MPH (74 km/h) lashed Texas and Louisiana, but it was Claudette's heavy precipitation that caused the most destruction. As it rolled inland over Texas, Claudette's swollen RAIN BANDS produced torrential downpours; the town of Alvin, Texas, recorded some 42 inches (1,066 mm) of rain—this remains a record for the most rainfall recorded in a 24-hour period in the continental United States. One person in Texas was drowned, as property losses in the state soared to nearly $300 million (1979 dollars). Perhaps not surprisingly for a system that had spent a majority of its lifespan as a tropical depression, Claudette doggedly maintained its tropical depression status for the next four days. Recurving to the north, then to the east-northeast, Tropical Depression Claudette delivered heavy rains to Oklahoma, Arkansas, Missouri, Illinois, Indiana, Ohio, Kentucky and VIRGINIA before dissipating on July 29.

Claudette, Hurricane *North Atlantic Ocean–Eastern United States, August 9–17, 1985* Originating off the coast of GEORGIA as a subtropical depression on August 9, 1985, the second system identified as Claudette slowly strengthened as it passed east-northeast across the Gulf Stream. Upgraded to a TROPICAL STORM on August 11 (with a central pressure of 29.67 inches [1,005 mb]), Claudette meandered eastward across the Atlantic, finally maturing to HURRICANE status on August 14. The system achieved its peak intensity on August 15, 1985, when its central pressure slipped to 28.93 inches (980 mb) and its sustained winds topped 86 MPH (138 km/h). Recurving to the northeast on August 16, a downgraded Tropical Storm Claudette passed less than 200 miles north of the Canary Islands. On August 17, Claudette transitioned into an extratropical system, and by the following day had dissipated.

Claudette, Hurricane *North Atlantic Ocean, September 4–14, 1991* Between September 4 and 14, 1991, Hurricane Claudette carved a rather unusual track across the mid-Atlantic reaches. Originating to the southeast of BERMUDA on September 4, Claudette first tracked to the south-southwest before gradually looping to the west-northwest, then northeastward. Along the way, Claudette rapidly intensified, becoming a Category 1 hurricane (with a central pressure of 28.90 inches [979 mb]) during the early evening hours of September 6, and a Category 4 hurricane (with a central pressure of 27.93 inches [946 mb]) by the following morning. Steered due eastward by the block-

ing BERMUDA HIGH, Claudette's RAIN BANDS steadily weakened as the powerful anticyclone caused disruptive WIND SHEAR along the system's northern edge. Downgraded to a TROPICAL STORM on September 9 (29.26 inches [991 mb]), and a TROPICAL DEPRESSION on September 11 (29.79 inches [1,009 mb]), Claudette continued to track nearly due east across the Atlantic before finally dissipating near the Canary Islands on September 14, 1991.

Claudette, Tropical Storm *North Atlantic Ocean, July 13–16, 1997* A mid-Atlantic tropical system for its entire existence, Tropical Storm Claudette originated several hundred miles off the eastern seaboard of the United States on July 13, 1997. Quickly intensifying into a TROPICAL STORM, the system first tracked northward before sharply recurving due east and churning deeper into the NORTH ATLANTIC OCEAN. Tropical Storm Claudette achieved peak intensity on July 13–14, when its central pressure dropped to 29.61 inches (1,003 mb) and its sustained winds topped 46 MPH (74 km/h). Downgraded to a TROPICAL DEPRESSION July 15, 1997, Claudette dissipated the following day.

Claudette, Hurricane *North Atlantic Ocean–Caribbean Sea–Mexico–Southern United States, July 8–18, 2003* The first hurricane of the 2003 North Atlantic season, Hurricane Claudette brought Category 1 winds and torrential rains to south-central TEXAS on July 16–17, 2003. Born to the east of the Windward Islands on July 7, Claudette remained a TROPICAL DEPRESSION until reaching the relatively enclosed waters of the CARIBBEAN SEA, where it was upgraded to TROPICAL STORM intensity on July 9. Slowly tracking to the northwest across the Caribbean basin, Claudette followed a fairly standard course past JAMAICA and into the Yucatán Straits, where it continued to slowly intensify. Its circulation patterns hampered by Central America to the west and by CUBA to the east, Claudette slumped across the northeastern tip of the Yucatán on July 11 as a weak tropical storm, with a central pressure of 29.79 inches (1,009 mb) and sustained winds of 52 MPH (84 km/h). Heavy rains inundated portions of Cancun and Cozumel. Upon entering the GULF OF MEXICO, Claudette established a north-northwesterly trajectory, one that if followed would take the system ashore on the TEXAS-LOUISIANA border. On July 15, while located less than 200 miles from the mouths of the Mississippi River, a large but disorganized Tropical Storm Claudette abruptly recurved due west, and intensified to hurricane status. In less than 24 hours, the system deepened from 29.17 inches (988 mb) to 28.96 inches (981 mb), boosting its sustained wind speeds from 69 MPH (111 km/h) to

81 MPH (130 km/h). A relatively mild hurricane at landfall, Claudette came ashore near Port Lavaca, Texas, during the late evening hours of July 15. Bearing a central pressure of 28.99 inches (982 mb), sustained winds of 81 MPH (130 km/h), and gusts topping 104 MPH (167 km/h), Claudette sank a shrimp boat near Sabine Pass, Texas, produced eight-foot (3-m) waves in Galveston, and toppled trees and power lines. As a downgraded Tropical Storm Claudette ground inland at 14 MPH (23 km/h), between five and eight inches (127–203 mm) of rain fell across two dozen Texas counties, leaving some 74,000 businesses and residences without electrical service. Before dissipating over extreme western MEXICO on July 18, Tropical Depression Claudette's powerful thunderstorms prompted TORNADO watches to be issued in several Texas counties. Two people in Texas lost their lives to Hurricane Claudette's passage, while damage estimates topped $25 million.

The name Claudette has been retained on the revolving list of North Atlantic tropical cyclone identifiers. It is scheduled for its sixth incarnation during the 2009 season.

Cleo, Hurricane *Guadeloupe–Hispaniola–Southern United States, August 21–27, 1964* One of the deadliest Category 3 HURRICANES on record, Cleo spent five tremulous days during the 1964 HURRICANE SEASON hacking a path of destruction through the east Caribbean island of GUADELOUPE, its northern islands of HISPANIOLA and CUBA, and the southern half of FLORIDA, between August 23 and 27.

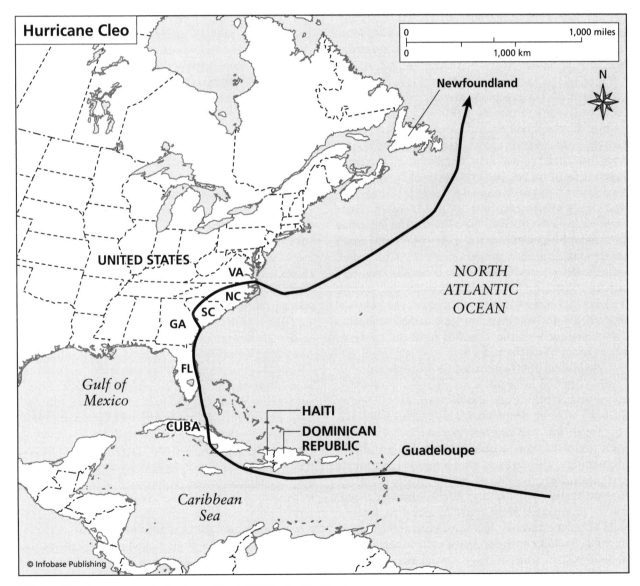

This tracking chart represents Hurricane Cleo's 1964 landfall in Florida.

In Guadeloupe, where Cleo's sustained 110-MPH (177-km/h) winds and 8-inch (203-mm) rainfalls demolished hundreds of houses and spawned terrible mudslides, an estimated 14 people lost their lives and another 100 were reported injured. In Haiti, which Cleo besieged with 135-MPH (217-km/h) gusts and 20-foot (7-m) seas on August 24, more than 120 people were killed, many of them drowned when the southwest fishing village of Chardonniere was partly submerged by the hurricane's eight-foot (3-m) STORM SURGE.

Recurving northwest during the night of August 24, Cleo maintained its strength as it bore down on Cuba, blasting ashore near the southern city of Condado on August 25, 1964. Although the Cuban government never released any official casualty or damage figures, Cleo's DEATH TOLL in that storm-prone country was believed to have exceeded 35. Trailing downed power lines, uprooted trees, and sailing roofs behind it, Cleo rampaged across central Cuba and emerged into the rejuvenating waters of the Florida Straits on the morning of August 26, 1964.

Just after midnight on August 27, Cleo made landfall near Miami Shores, Florida. Barreling ashore as a Category 2 hurricane, Cleo's CENTRAL BAROMETRIC PRESSURE of 28.58 inches (968 mb) produced sustained 106-MPH (171-km/h) winds, heavy rains, and pounding surf. Dozens of Gold Coast hotels were seriously damaged, their scenic windows sprung from their frames, their turquoise-blue swimming pools muddied by oil slicks and castaway wreckage. Elegant lobby furnishings sat half-buried on the beach, while trampled palm fronds carpeted staircases. More than 1,200 BUILDINGS, from multimillion-dollar seafront mansions to towering office blocks, were either destroyed or damaged. Dozens of sailboats and motoryachts were driven ashore by Cleo's roaring surf and then left stranded on front lawns and city streets.

Thirteen people in Florida lost their lives to Hurricane Cleo. More than $200 million in property losses—much of it to the state's citrus harvest—were likewise suffered. Responding to the crises, President Lyndon B. Johnson declared portions of south and central Florida federal disaster areas on September 10, making them eligible for emergency assistance. The identifier Cleo had been used before, for an extremely powerful hurricane that transited the central NORTH ATLANTIC OCEAN between August 11 and 22, 1958. Shortly before midnight on August 16, while moving to the northwest across latitude 20 degrees North, Cleo briefly intensified to Category 5 status, with a central barometric pressure reading of 27.99 inches (948 mb) and estimated wind speeds in excess of 161 MPH (259 km/h). A potent menace only to shipping, the first Hurricane Cleo caused no deaths or injuries.

But it was the second Hurricane Cleo, with 147 confirmed deaths to its credit and damage estimates as high as $290 million (in 1964 U.S. dollars), that relegated the name itself to history; the identifier Cleo was retired from the cyclical list of HURRICANE NAMES at the conclusion of the 1964 season. It was replaced with Candy for the 1968 season.

Cobra, Typhoon *Western North Pacific Ocean, December 17–18, 1944* See BUYS-BALLOTS LAW and MILITARY OPERATIONS.

condensation   In the broadcast sense of the definition, condensation is the physical process by which a vapor becomes a liquid. More specifically, the meteorological function by which warm water vapor, collected from an ocean's surface through EVAPORATION, cools to the point of saturation as it rises through the low-pressure core of a TROPICAL CYCLONE. It is through the process of condensation that a tropical cyclone's CUMULONIMBUS CLOUD bands are formed and that LATENT HEAT is simultaneously released into the system's CONVECTION currents, thereby adding to the thickness and strength of the EYEWALL cell. Secondary stages of condensation at altitudes above 25,000 feet (8,333 m) further yield the CIRROSTRATUS CLOUD canopy associated with a tropical cyclone's OUTFLOW LAYER. Conversely, the increased condensation rates found in dissipating tropical cyclones generally result in the formation of large quantities of PRECIPITATION.

condensation level   This is the altitude at which warm water vapor rising through the EYE of a TROPICAL CYCLONE undergoes CONDENSATION into CUMULUS and CUMULONIMBUS CLOUDS. In the EYEWALL of a mature-stage tropical cyclone, where its lowest BAROMETRIC PRESSURES are most often located, the condensation level can range from 1,200 feet (400 m) above the ocean's surface in systems with central pressures of between 29.53 and 28.50 inches (1,000–965 mb), to 800 feet (267 m) or less in those with minimum pressures of 28.05 inches (950 mb) or below. Beyond the eyewall, where a tropical cyclone's spiral RAIN BANDS grow thinner and are spaced further apart, condensation levels tend to increase in altitude to between 1,500 and 2,000 feet (500–667 m).

Connie, Hurricane *Northeastern United States, August 4–14, 1955*   The first of three major HURRICANES to strike the eastern seaboard of the United States during the 1955 HURRICANE SEASON, Connie killed 41 people, forced the evacuation of some

75,000 others, and tallied more than $50 million in property damages as it careened across NORTH and SOUTH CAROLINA, VIRGINIA, and Maryland between August 12 and 13, 1955. Preceding by less than a week Hurricane DIANE's multibillion-dollar landfall in South Carolina on August 18 and by more than a month Hurricane Ione's September 19 barrage on North Carolina, Connie's 125-MPH (201-km/h) winds, 10 foot (3 m) swells, and 12-inch (305 mm) rainfalls damaged or destroyed hundreds of BUILD-INGS, capsized a large schooner in Chesapeake Bay, and dropped such enormous quantities of PRECIPITA-TION on NEW YORK's La Guardia Airport that the facility was forced to suspend operations.

The third-named storm of the very active 1955 hurricane season, Connie was first marked as a TROP-ICAL DISTURBANCE approximately 659 miles (1,061 km) east of PUERTO RICO on the morning of August 4, 1955. Scudding northwest at nearly 11 MPH (18 km/h), the disturbance speedily strengthened, was upgraded to a TROPICAL DEPRESSION on the morning of August 5 and a 53-MPH (85-km/h) TROPICAL STORM by midnight on the same day.

On August 6, as its CENTRAL BAROMETRIC PRES-SURE fell to 29.05 inches (983 mb), the storm was upgraded to hurricane status. Connie's rotating EYE, then located 1,400 miles (2,253 km) southeast of Wilmington, North Carolina, continued spiraling in a north-northwesterly direction at 11 MPH (18 km/h), a course that kept its 92 MPH (148 km/h) EYE-WALL winds well off the southeast shores of Florida but posed a grave threat to the 600-mile (966-km) stretch of coastline between Savannah, GEORGIA, and Norfolk, Virginia.

Reconnaissance flights conducted by the newly organized NATIONAL HURRICANE CENTER (NHC) in Miami, FLORIDA, indicated that by the morning of August 8, Connie's central pressure had further dropped to 28.87 inches (977 mb), a strengthening pattern that promised further intensification as well as sustained 100-MPH (161-km/h) winds and a potential six to eight foot (2–3 m) STORM SURGE at landfall.

Hurricane Connie, accompanied by an entourage of 125-MPH (201-km/h) gusts, seven-inch (179-mm) rains, and 17-foot (6-m) waves, first arrived in the eastern United States during the early morning hours of August 12, 1955. Disembarking in North Carolina near the mouth of Cape Fear River, Connie swept the sandy Outer Banks as a Category 3 system of considerable intensity. Curving to the northwest, the hurricane's track unleashed pounding seas on the hundreds of inlets and rivers that frill the state's scenic coastline. The industrial city of Wilmington, located near Cape Fear River's mouth, had several of

its low-lying neighborhoods submerged by Connie's upriver surge, while the nearby resort haven of Myrtle Beach, South Carolina, suffered the splintered loss of more than 40 oceanfront cottages, restaurants, and gaming arcades. Twenty-seven people in North Carolina were killed, victims of traffic accidents, drownings, building failures, and electrocutions.

Maintaining its swing to the northwest, Connie briefly reemerged over the coastal shallows of Virginia before coming ashore again in Maryland, near the tip of Cape Charles, on the evening of August 12. There, spreading across the mouth of Chesapeake Bay, the hurricane's 110-MPH (177-km/h) gusts and diminishing seas convulsed the normally placid waters of the bay, pushing the undulating effects of its surge the entire length of the 200-mile (322-km) inlet.

As five-foot (2-m) rollers began to pile up against the piers of Baltimore, the 125-foot wooden schooner *Levin J. Marvel,* having failed a Coast Guard safety inspection the year before, now found itself struggling to stay afloat in sawing 12-foot (4-m) waves. Originally built in 1891 for the fishing trade, the picturesque two-masted vessel had since found work as an excursion boat, carrying tourists on day cruises along Maryland's dune-fileted North Beach.

Indeed, the 150-ton vessel was in the process of returning to North Beach from just such a happy occasion when Connie's dismasting winds and blinding rains violently overtook it. Within seconds its canvas sails burst into ribbons, while its top hamper, undermined by the dry rot of age, crashed to the deck. Badly buffeted by the rising waves, the schooner twisted in the seaway, sending its 16 terrified passengers and crew scurrying for the vessel's one lifeboat. They were frantically trying to release it from its chocks when the entire ship was suddenly thrown on her beam-ends, capsized so quickly that everyone on board was tossed into the wailing sea. The *Marvel,* its ancient sides stove in, quickly sank, leaving 16 people, many of them without life jackets, tossing in the midst of a hurricane. Gagging on seawater, their eyes stung shut by the salty spindrift, their exhausted limbs now as heavy as rocks, 14 of the *Marvel*'s complement slowly drowned in what would become one of the worst maritime accidents in the annals of Chesapeake Bay.

Laden with enormous quantities of precipitation, Connie surfed through Maryland, central Pennsylvania, west NEW YORK, and into southern CANADA on a tide of torrential rainstorms between August 13 and 14, 1955. Its sustained winds now returned to tropical-storm status, the dying system delivered a veritable deluge of rain with 7–13-inch (178–330 mm) precipitation counts across the heavily wooded northern landscape. Even though no casualties were reported, Connie's drenching demise did inflict $20 million in 1955 property losses, rating it as a hurricane of memorable intensity. The name Connie has been retired from the list of HURRICANE NAMES.

convection   In the strictest sense of the definition, convection refers to the transfer of any atmospheric property, be it heat or moisture, through upward or downward movement. In a more general sense, convection pertains to one of two meteorological processes (the other being ADVECTION) that creates the huge, often deadly CUMULONIMBUS CLOUD systems found in a developing or mature-stage TROPICAL CYCLONES.

In a TROPICAL STORM, HURRICANE, TYPHOON, or CYCLONE, convection currents are the vertical currents of air that either rise from, or descend toward, Earth's surface. Created when warm surface air rises high into the troposphere and then cools and falls, convection directly results in the formation of *convective cells,* or those self-sustaining cloud systems most often associated with thunderstorms and cyclonic activity.

This cyclical motion, which ensues from the uneven heating of the air by unequal surface temperatures, in turn serves to power the convective cell and will remain in continuous motion as long as the two contrasting surface temperatures persist. In tandem with advection, convection forms one of the primary mechanisms by which all atmospheric circulations, from EXTRATROPICAL CYCLONES to tropical cyclones to localized thunderstorms, are born and sustained.

Coriolis effect   Sometimes referred to as the *Coriolis force,* the Coriolis effect states that when a moving object (such as wind) is observed from a point of reference that is itself turning, the object will appear to deflect, curving in accordance with Earth's rotation. First posited in 1835 by the French physicist Gustave Coriolis (1792–1843), the effect has since been adapted by scientists to explain a number of physical and meteorological processes, including why TROPICAL CYCLONES possess a distinctive counterclockwise or clockwise spin and why they do not as a rule originate within 5 degrees north or south of the equator.

Acting as a counterbalance to a tropical cyclone's PRESSURE GRADIENT, or the rate at which its barometric pressure decreases over a certain distance, the Coriolis effect serves to start a developing TROPICAL WAVE or TROPICAL DEPRESSION spinning by relentlessly deflecting those air currents drawn into the system by its low-pressure EYE. Like a ballerina born to a union between Earth's rotation in one direction and the respective spin of the low-pressure core itself in the other, the pirouetting tropical cyclone

begins when warm, moist air that has been drawn in by the pressure center does not reach its revolving target but is instead curved to the right (or east) in the Northern Hemisphere and to the left (or west) in the Southern Hemisphere by the Coriolis effect. Concentrating the force of the air currents in one of these two directions, the Coriolis effect literally organizes the winds around a low-pressure center, establishing from the very beginning the circular pattern that will eventually dominate the mature tropical cyclone.

Because the Coriolis effect is proportionally linked to increases in wind velocity, the faster the air moves into the tropical cyclone, the more deflection the effect will bring to bear on its direction, thereby aiding the nascent system's cohesiveness. It is also known that the Coriolis effect grows more pronounced in higher latitudes, meaning that those tropical cyclones that undergo RECURVATURE, or a swing to the north-northeast or south-southwest, increasingly experience its influence the farther away from the equator they progress. In such instances, tropical cyclones spin 10 to 20 MPH (16–32 km/h) faster and travel very quickly toward Earth's polar killing fields.

Meteorological models based on the observed characteristics of the Coriolis effect have further illustrated why it is that no known tropical cyclones have originated between 5 degrees North and 5 degrees South or in that region of the globe immediately adjoining either side of the equator. At the equator, over the long, windless stretches of sea and land called the doldrums, the Coriolis effect is minutely counteracted by high, stable BAROMETRIC PRESSURES. Because wind is a result of air moving from areas of high pressure to low pressure, the sustained high pressure belt that girdles the first 10 degrees of the planet prohibits all semblance of wind. Without wind, there is nothing for the Coriolis effect to act upon, leaving it to slumber in serene, unrealized potential.

Cristobal, Tropical Storm *North Atlantic Ocean, August 5–8, 2002*   The first tropical system in the North Atlantic basin to bear the name, Tropical Storm Cristobal formed to the north of the BAHAMAS, over the reaches of the eastern North Atlantic, on August 5, 2002. A fairly mild TROPICAL STORM for most of its offshore existence, Cristobal was noted for its unusual southeasterly track, and its abrupt RECURVATURE to the northeast on August 7, which took it into the NORTH ATLANTIC and its dissipation. At its peak, Tropical Storm Cristobal generated a central pressure of 29.50 inches (999 mb) and wind speeds of 52 MPH (84 km/h).

The name Cristobal has been retained on the North Atlantic tropical cyclone naming lists.

Cuba *North Caribbean Sea* A sprawling island of towering mountain ranges, timbered valleys, slow-winding rivers, and hospitable natural harbors, Cuba has during the past five centuries defended itself against a relentless, often deadly onslaught of NORTH ATLANTIC OCEAN, CARIBBEAN SEA, and GULF OF MEXICO TROPICAL CYCLONES. From 1494 to 2006, Cuba's sweeping beaches, prolific copper mines, generous sugar fields, and fading colonial cities were directly affected by 161 documented HURRICANES and TROPICAL STORMS, some 54 of these being considered major hurricanes. Serving as a veritable checkpoint at the western end of storm-ridden HURRICANE ALLEY, the huge, 44,662 square-mile (71,876 square km) collection of Cuban islands have on at least 33 occasions between 1768 and 2006 even endured strikes from two or more tropical cyclones during the course of a single HURRICANE SEASON, giving it one of the most consistent strike records in the Caribbean. In the years 1796, 1810, and 1886, three tropical systems lashed Cuba from June to November, with the final tempest of 1810—a late-season wonder that raged from October 23 to 26—sinking 32 vessels in Havana harbor, shattering piers, uprooting hundreds of trees, and demolishing more than 60 BUILDINGS in and around the capital city.

Cuba has also suffered strikes from at least three recorded Category 5 hurricanes, meteorological marvels that possessed CENTRAL BAROMETRIC PRESSURES of 27.17 inches (920 mb) and below, sustained winds of 155 MPH (249 km/h) and higher, five–11-inch (127–279-mm) PRECIPITATION counts, and pulverizing STORM SURGES and STORM TIDES. The first Category 5 hurricane, known to history as the GREAT HAVANA HURRICANE, decimated the city of the same name on the morning of October 11, 1846. Less than a century later, on the morning of November 9, 1932, Cuba's second Category 5 hurricane slammed ashore near the port city of Santa Cruz del Sur, some 325 miles (523 km) southeast of Havana. Eventually named for the city it virtually obliterated, the Santa Cruz del Sur Hurricanes's central pressure of 27.01 inches (914 mb) brought with it a cresting 21-foot (7-m) storm surge that left the waterfront a littered wasteland of collapsed buildings, stranded steamships, and festering corpses. A more powerful hurricane than the 1846 storm, Santa Cruz del Sur's 210-MPH (338-km/h) winds and drowning surge claimed some 2,500 lives, placing it among the deadliest of all Cuban hurricanes.

Reposing on the northwestern apex of the Caribbean Sea, Cuba's position between latitudes 16 and 23 degrees North sets it squarely in the path of both the early season tropical cyclones that originate over the contained waters of the Caribbean Sea and west-ern Gulf of Mexico before recurving to the north-northeast and the mid- to late-season furies that frequently boil westward across the Atlantic from Africa's Cape Verde Islands. Thus endowed with a full six-month hurricane season (June 1–November 30), Cuba has, from 1494 to 2006, withstood at least seven tropical cyclones in June, 6 in July, eighteen in August, thirteen in September, forty-six in October, and six in November. The remarkable spike in hurricane and tropical storm activity in October has long been known to residents of the island, with the famed 19th-century observer of Cuban hurricanes, Father Benito Viñes, recording in his *Cyclonic Circulation and Translatory Movement of West Indian Hurricanes,* that "ecclesiastical authority knew by experience that the cyclones of October were very much to be feared in Cuba, but not those of August. . . ."

Most often of tropical-storm intensity as they pass over the western extremities of the island, May, June, and July tropical cyclones in Cuba usually contain vast quantities of precipitation, torrential rainfalls that have in numerous instances transformed the rugged mountain terrain into a seething mire of death and destruction. On June 21, 1791, a powerful tropical storm struck the inland town of Puentes Grandes, its heavy rains and resulting floods carrying a large house belonging to a local landowner into the Almendareo River. According to eyewitness accounts, the house was at the time of the tragedy filled with mourners, faithful retainers who had earlier come to pay their respects to the recently deceased nobleman. As though hinting at a greater symbolic significance for the storm, survivors further stated that the wooden casket—swept downriver by the apocalyptic flood waters—was never recovered. On June 8, 1966, Hurricane ALMA passed over west Cuba as a minimal Category 1 storm, its central barometric pressure of 28.93 inches (980 mb) at landfall producing prodigious 6-inch downpours. On June 3, 1968, the 62-MPH (100-km/h) winds and blinding rains associated with Tropical Storm Abby lashed extreme west Cuba, triggering flash floods and landslides that killed five people. Finally, on July 1, 1994, Tropical Storm ALBERTO dropped nearly nine inches (229 mm) of rain on Isle de la Juventud (the Isle of Youth) as it trekked north to FLORIDA, claiming no lives but causing extensive damage to roads and crops across the satellite island.

Cuba was in 1494 and 1970 struck by two May tropical systems of memorable consequence. Classification of the first storm, which 19th-century hurricane chronicler Andreas Poey listed as a hurricane that raged off the southeast coast of Cuba between May 19 and 21, 1494, has long been debated by cyclonic scholars. Poey, who resided in Havana for most of his life, never stated his source for the 1494 storm, although it is now believed to have been taken from the writings of Bartolomé de Las Casas, a perceptive Spanish bishop whose 1563 *Historia de las Indias* (History of the Indies) recorded the presence of thunderstorms along Cuba's south coast during the third week of May 1494. Las Casas, in turn, had based his studies on the logbooks of Christopher Columbus, who had reportedly encountered the system while surveying Cuba's south shores during his second voyage to the New World (1493–1496). Whether it was indeed a hurricane, tropical storm, or thundersquall, the uncertain passage of the 1494 system was perhaps mirrored by a much later May system, the second Hurricane Alma. Swinging ashore near the southwest town of Cienfuegos on May 22, 1970, then Tropical Storm Alma delivered 63-MPH (101-km/h) winds and eight inches of rain to the central and western highlands of the island, spawning flash floods and killing seven people.

Conversely, Cuban hurricanes in August, September, and particularly October tend to generate faster winds than heavy rains, scouring currents of air that leave the land's surface withered, seemingly burned. Rampaging across the North Atlantic Ocean from the Cape Verde Islands, these mid-to-late-season furies often achieve great intensity before making landfall in south or east Cuba. A chronology of Cuban hurricane history includes a number of such memorable storms:

CUBAN HURRICANE HISTORY

1525: A severe late-season hurricane quickly overtakes an unidentified Spanish treasure ship near the northwest trading post of Marianao on the afternoon of October 12. En route from Havana to Veracruz, MEXICO, with a valuable cargo of gold, spices, and mercury, the ungainly vessel is swiftly dismasted and then sunk, drowning more than 70 people. Numerous buildings in recently founded (1519) Havana are unroofed, while enormous storm-driven waves cause widespread damage to the newly constructed piers and quays of Havana harbor.

1527: Another particularly violent hurricane—this one most likely of Caribbean origin—trounces the southcentral port town of Trinidad de Cuba, some 190 miles southeast of Havana, on the morning of October 18. Two heavily laden Spanish merchant ships are lost, claiming the lives of more than 100 seamen.

1537: On the evening of September 9, a hurricane of considerable size and intensity inundates Havana harbor with a thundering storm surge and heavy seas. Several vessels are either dismasted or sunk, including two large Spanish merchant ships lying at anchor. Dozens of lives are lost.

1636: An exceptionally severe hurricane strikes Havana just after midnight on October 14. Touted by contemporary sources as one of the worst tropical cyclones to have pounded the city in living memory, the 1636 tempest washed scores of houses into the sea. Havana's Morro Castle is seriously damaged, many of its cannon and stone ramparts being swept into the harbor. At least 200 people perish, a majority of them Spanish sailors.

1692: After delivering devastating winds, blistering rains, and deadly seas to the island of PUERTO RICO on October 21, the San Rafael Hurricane transits southeast Cuba on the morning of October 24. Hundreds of buildings in the former capital city of Santiago de Cuba are utterly obliterated, and thousands of palm and mahogany trees are either stripped of their foliage or uprooted. Acres upon acres of sugarcane fields are ruined, causing severe financial hardship across much of the island. An estimated 235 people are killed, making San Rafael one of the deadliest hurricanes in early Cuban history.

1772: Long considered one of the most destructive of all 18th-century Cuban tropical cyclones, the early season hurricane of July 20 sends more than 150 ships ashore on the island's rock-strewn north coast. Some 349 sailors are drowned. Less than two months later, another ferocious hurricane (that had only days earlier ravaged the Leeward islands of DOMINICA and St. Kitts), stampedes ashore near Santa Cruz del Sur on the morning of September 5. Trailed inland by a storm surge of remarkable height—one unconfirmed report states that the surge topped out at 72 feet (22 m)—the September 1772 hurricane strands nine ships and produces numerous smothering rockslides in the mountains of central Cuba. An estimated 400 people perish, many of them carried to their deaths by the hurricane's flooding rains.

1810: A large, slow-moving hurricane destroys 32 ships in Havana harbor between October 23 and 26. Accompanied by 15-foot (4.6-m) seas and scudding downpours, the 1810 hurricane shatters piers, levels churches, and flattens some 60 buildings in the Cuban capital. Many of these failed structures are two- and three-story buildings, their design not introduced to Havana until 1779—for reasons that Cuba's active storm history make obvious. Scores of lives are lost in the 1810 hurricane, while the bodies of dozens of victims go missing, never to be found.

1856: Considered the most severe hurricane to affect Cuba since the Great Havana Hurricane of 1846, the tempest of August 27 and 28 causes widespread damage in and around Havana. Triggered by a minimum pressure reading of 28.62 inches (969 millibars) over Havana, the hurricane's 116-mph (187-mph) winds and torrential rains undo much of the rebuilding commenced in the wake of the 1846 catastrophe. Another 200 lives are reportedly lost, a majority of them sailors at sea.

1873: A powerful hurricane in western Cuba on October 6 leaves much of Pinar del Río, a mountain community situated 100 miles southwest of Havana, in total ruins. Blinding rains transform ravines into treacherous sluiceways that quickly become littered with fallen trees, building debris, and human corpses. Provided with free passage by the Cuban railroads, Father Viñes travels to Pinar del Río to survey the damage.

1875: Father Viñes, having spent every day for the last five years astutely observing the vagaries of Cuban weather, issues his first HURRICANE WARNING on the morning of September 11 for the island's south shores. On the evening of September 12, a fairly severe hurricane grazes southeastern Cabo Cruz (Cape Cruz) before swinging to the northwest and moving inland the following day. Sailing directly over Havana on September 14, the hurricane's 105-MPH (169-km/h) winds and five-inch (127-mm) rains collapse scores of houses and blight sugarcane stocks but otherwise claim few fatalities on the island—a testament, it is said, to the wisdom and accuracy of Viñes's forecast. This particular hurricane will, after an abrupt course change west, later ravage the town of Indianola, Texas.

1888: A monster tropical cyclone pummels much of western Cuba with 115-MPH (185-km/h) winds and inundating seas on the night of September 4 and 5. Considered a truly "great" hurricane by any number of contemporary sources, the 1888 tempest causes widespread destruction in Havana, with many of the capital's principal buildings left waterlogged and uninhabitable. Despite the Belén observatory's best efforts to alert the populace to the storm's approach, the toll is particularly steep; more than 600 people are left dead and another 2,500 are injured. Damage estimates quickly top several million pesos, with more than 10,000 valuable mahogany and pine trees destroyed. The island is soon propelled into a stiff but temporary economic depression.

1910: A slow-moving, late-season hurricane carves a loop-the-loop course of destruction over western Cuba between October 12 and 21. One-hundred-mile-per-hour winds, followed by nearly a week of submerging rains, leaves much of Pinar del Río and Havana in

sodden ruins. Some 100 people are reportedly killed, a majority of them buried beneath roaring mudslides.

1944: A hurricane of even greater intensity than the 1926 system, the GREAT ATLANTIC HURRICANE unravels its firm Category 3 winds and rains over Havana on the morning of October 17. Brought on by a central pressure of 28.02 inches (949 mb), the Great Atlantic Hurricane's 125-MPH (201-km/h) winds, 163-MPH (262-km/h) gusts, and seven-inch (178-m) rains collapse 269 buildings in Havana, down hundreds of electrical and telephone lines, sheer thousands of palm and laurel trees at ground level, and poison city water supplies for more than a week. Marching ashore during World War II, the hurricane's widespread destructiveness places a nearly insurmountable burden on the material resources of Cuba, and it is several months before all the damage is repaired. Clearly one of Cuba's most notable tropical cyclones, the Great Atlantic Hurricane kills 330 people and injures another 1,200.

1963: In what is probably the deadliest hurricane disaster in Cuban history, Hurricane FLORA gushes ashore near Santiago de Cuba on the morning of October 4. Stalled for the next three days over Oriente and Camagüey provinces, Flora's torrential rains and resulting landslides kill some 1,300 people and leave thousands of others homeless.

1964: Hurricane CLEO crosses central Cuba on the gust-riven night of August 25 and 26. More than $70,000 worth of damage is tallied in the mountain towns of Taguasco and Chambas, while deaths across the island exceed 50. On October 13, Tropical Storm Isbell traverses western Cuba. On the cusp of becoming a hurricane, Isbell's 70-MPH (113-km/h) winds kill three people and down sevral hundred trees and utility poles.

1968: A strengthening Hurricane Gladys passes over Havana during the night of October 16 and 17. A relatively mild Category 1 system (a pressure reading of 29.26 inches (990 mb) is noted in Havana), Gladys claims six lives in the city and causes extensive structural damage to piers and embankments.

1970: Only of tropical depression-strength as it lumbers ashore near Pinar del Río on the night of July 31, Hurricane CELIA spawns copious rains and at least two tornadoes over the western enclaves of the island. Originating in Cape Verde Islands, Celia kills five people in Cuba and injures several hundred others. Rain and wind damage to crops and buildings soon exceeds $25 million. Much like the hurricane of September 1875, Celia will eventually slide northwestward into

Texas, bringing Category 4 winds to the greater Corpus Christi area on August 3.

1973: Trudging to the northeast at 8 MPH (13 km/h), Tropical Storm Gilda delivers 57-MPH (92-km/h) winds and some rain to central Camagüey Province on the afternoon of October 18. A fairly severe tropical storm (a pressure reading of 29.38 inches is noted at landfall), Gilda causes no significant damage and claims no lives—one of Cuba's more benign October cyclones.

1990: Tropical Depression Marco drops flooding rains over north Cuba between October 8 and 10. Three people are drowned.

1994: Slicing across east Cuba on the night of November 13 and 14, Tropical Storm GORDON generates an extreme gust of 120 MPH (193 km/h) at Guantánamo Bay. Two people are killed.

1996: Hurricane Lili, the first mature-stage tropical cyclone to affect Cuba directly since 1985's Hurricane KATE, delivers sustained 90-MPH (145-km/h) winds, 110-MPH (177-km/h) gusts, and 22-foot (7-m) seas to west and central Cuba on the morning of October 18. Earlier responsible for claiming 14 lives in Nicaragua, Costa Rica, and Honduras, Lili's late-season flowering over Cuba forces the closure of Havana's José Martí International Airport, prompts the EVACUATION of some 247,000 islanders from vulnerable seaside villages, unsound inland buildings, and flood-prone valleys and results in profound distress for an economy already buffeted by the ever-shifting winds of global politics. A powerful Category 1 hurricane at landfall near the southwest port city of Cienfuegos, Lili's rainladen storm bands pluck some 16,000 tons of grapefruit and oranges from trees on Isla de la Juventud as they transit the island toward the northeast at nine MPH (15 km/h). Accompanying seas, said to have been in excess of 20 feet (7 m), sweep cinder-block beach cabanas into rubble at a posh resort on Cayo Largo del Sur, while all 28 sugar refineries in the central-northern province of Villa Clara lose their roofs and windows and have their cisterns filled and then overturned by nearly 19 inches (482 mm) of rain within a 48-hour period. A dozen neglected buildings in Old Havana collapse as Lili raids the fading capital city's colonial heritage, toppling trees and flattening houses while leaving nearby lampposts and fences unscathed. One of Cuba's milder October hurricanes, Lili destroys 2,300 buildings, damages another 47,000 structures, undermines a bridge in the town of San Nicolas, and causes widespread ruin to sugar, coffee, banana, and citrus harvests. No deaths and only one serious injury is reported—a miraculously low

casualty figure for a tropical system of such costly destructiveness.

2000: During the 2000 North Atlantic hurricane season, Cuba is directly affected by a downgraded Hurricane DEBBY, which delivers mild tropical storm conditions to the extreme southeastern regions of the island on August 24. Weakening as it passes along the northern highlands of Hispaniola, Debby's central pressure of 29.85 inches (1,011 mb) produces heavy PRECIPITATION counts and large surf, but claims no lives and causes little property damage. Just under a month later, on September 20, Tropical Storm Helene brings heavy rainfalls to the southwestern tip of Cuba while still of tropical depression intensity. Recurving to the north-northeast and entering the Gulf of Mexico, the tropical depression's central pressure of 29.82 inches (1,010 mb) and 29-MPH (47-km/h) winds cause minimal damage on the island.

2001: On November 4, 2001, Cuba suffers a direct strike from late-season Hurricane MICHELLE, a moderately powerful Category 4 tropical system that had originated in the extreme southwestern Caribbean Sea on October 29 before recurving to the north-northeast and rapidly intensifying. Driven by a central barometric pressure of 28.02 inches (949 mb) at landfall near the Isle of Youth, Michelle's sustained, 132-MPH (212-km/h) winds cause considerable damage to trees, crops, and buildings along the south-central coast of Cuba. In Punte del Este, Michelle produces up to 11.83 inches (279 mm) of rain, while the neighboring island of Cayo Largo del Sur is pounded by a nine-foot (3-m) storm surge. Nearly 10,000 residences and businesses are destroyed in Matanzas and Cienfuegos provinces, and structural damage is reported in Havana. Downgraded to a Category 3 system (with a central pressure of 28.20 inches [955 mb]) as it traverses the island, Michelle emerges near the Straits of Florida as a Category 2 system with sustained, 109-MPH (176-km/h) winds the following day. Although Michelle will prove the most intense tropical cyclone to strike Cuba since 1952, the preemptive evacuation of some 740,000 people from vulnerable coastal and mountain areas (where landslides and flash floods can occur), keeps the death toll in Cuba to five people. Nearly $2 billion in property losses are tallied.

2002: During the early morning hours of September 20, Cuba endures a direct strike from the ninth-named tropical cyclone of the 2002 North Atlantic hurricane season, Hurricane ISIDORE. Formed to the south of the Windward and Leeward Islands chains, Isidor slowly recurves to the west-northwest before crossing northwestern JAMAICA on September 18 and subsequently deepening into a mature-stage tropical cyclone with a minimum pressure of 28.90 inches (979 mb) at landfall near Cabo Frances. A ferocious rainmaker, Isidor produces some 21.7 inches (533 mm) of precipitation over Pinar del Río province, causing severe damage to farming structures and tobacco crops. No lives are lost to Hurricane Isidor's waterlogged passage over Cuba. Less than one month later, on September 30, the island is directly hit by another Category 1 tropical cyclone, Hurricane Lili. Of Cape Verde origin, and the second tropical cyclone identified as Lili to strike Cuba in its history (*see* 1996), Lili (with a central pressure of 29.11 inches [986 mb]) slithers ashore on the southwestern tip of Cuba, bringing 75-MPH (121-km/h) winds and heavy surf conditions. After causing minimal damage in Cuba, Lili spins into the Gulf of Mexico, and swiftly intensifies, becoming a Category 4 hurricane (with a central pressure of 27.69 inches [938 mb]) on October 2.

2004: Although Cuba escapes direct tropical cyclone activity during the 2003 North Atlantic hurricane season, it is not so fortunate during the 2004 theater. On August 13, Hurricane Charley makes a Category 4 landfall near the southern coastal town of Playa del Cajio, situated in the western half of the island. The most intense tropical cyclone to have affected Cuba during the month of August in many years, Charley's central pressure of 28.46 inches (964 mb) generates 110-MPH (179-km/h) winds and heavy surf conditions. The system weakens slightly as it recurves to the north over the island and emerges into the Gulf of Mexico (near the Florida Keys) on August 14 as a Category 1 hurricane. One person in Cuba is killed.

2005: The unprecedented activity witnessed during the 2005 North Atlantic hurricane season results in Cuba being directly and indirectly affected by no fewer than five tropical cyclones, including Hurricane DENNIS, the most intense and destructive tropical cyclone to strike the island since 1963's Hurricane FLORA. On the morning of June 10, 2005, the first of these systems—a disorganized Tropical Storm ARLENE—delivers its central pressure of 29.52 inches (1,000 mb) and sustained, 46-MPH (74-km/h) winds to Pinar del Río province, dropping enormous quantities of rain, but causes no deaths or injuries. Not one month later, on July 7, Hurricane Dennis blasts itself into Cuban meteorological history when it makes two separate landfalls along the island's southern coastline as a powerful Category 3 and 4 tropical system. Originating in the southeastern Caribbean Sea on July 5, Dennis recurves to the west-northwest, brushing the northern coast of Jamaica on July 6 as a Category 1 hurricane before rapidly intensifying. When it comes ashore near Punta del Ingles, Cuba, during the late evening hours of July 7, Dennis is of Category 3 intensity, with a central barometric pressure of 28.26 inches (957 mb) and sustained wind speeds topping 115 MPH (185 km/h). A relatively fast-moving system, Dennis slightly weakens as it passes over the peninsula and enters the heat-fueled waters of the Gulf of Guacanayabo during the early morning hours of July 8. By mid-afternoon of the same day, Dennis swiftly strengthens into an exceptionally-powerful Category 4 system with a central pressure of 27.69 inches (938 mb) and sustained winds of 150 MPH (241 km/h). Weakening as it draws closer to its landfall near Punta Mangles Altos, Dennis is still of firm Category 4 status as it roars ashore in central Cuba, its central pressure of 28.02 inches (949 mb) generating 132-MPH (212-km/h) winds and an extreme gust of 149 MPH (240 km/h) at Cienfuegos. As it crosses south-central Cuba, bound for an eventual landfall in Florida, Dennis again weakens, dropping enormous quantities of precipitation across much of the island's western quadrant. Nearly all of the nation's electrical utilities are disrupted, and some 15,000 buildings are demolished. All told, Hurricane Dennis kills 16 people in Cuba and causes $1.4 billion (in 2005 dollars) in damages. On August 29, while on course to eventually become the most expensive natural disaster yet witnessed in U.S. history, an offshore Hurricane KATRINA brings tropical storm conditions (55 MPH [89 km/h] winds) to the northwestern coast of Cuba. At the time Katrina's eyewall brushes past the island, the system is of severe Category 4 intensity, with a harrowingly low central barometric pressure of inches 26.87 inches (910 mb) and sustained 155-MPH (250-km/h) winds. Some 7.87 inches (200 mm) of precipitation falls across the western province of Pinar del Río, forcing thousands of people to evacuate. While Katrina uproots trees and disrupts utility systems, no deaths or major injuries are reported in Cuba. In just over a month's time, the northern coast of Cuba is again menaced—and affected—by a significant North Atlantic hurricane, this one known as RITA. Between September 20 and 21, Rita brings tropical storm force conditions to the northwestern coast as its eyewall passes into the Gulf of Mexico, some 150 miles (km) to the north. While at the time of its passage Rita is a much weaker hurricane (with a central pressure of 28.99 inches [982 mb]) than Katrina had been, Cuban authorities waste little time in ordering preemptive evacuations of low-lying coastal areas. Heavy surf conditions damage several waterfront piers and facilities, but no deaths or injuries are reported. Between October 23 and 24, 2005, Hurricane WILMA nears to within 200 miles of landfall in western Cuba before recurving to the north-northeast and entering the Gulf of Mexico. At the time a Category 3 hurricane (with a central pressure of 28.28 inches [958 mb]) of growing intensity, Wilma had, while churning through the Yucatán Straits only four days earlier, established a new record as the most intense tropical cyclone yet observed in the North Atlantic basin. Fearing that Wilma would make landfall on the storm-swept western tip of the island, Cuban authorities evacuate nearly 1 million people; as it is, sections of Havana are flooded with six feet (2 m) of water by Wilma's radiating seas, and several buildings—as well as coastal transportation networks—are damaged.

2006:   The first named tropical system of the 2006 North Atlantic hurricane season, ALBERTO forms off the western coast of Cuba on June 10. A disorganized system with a weak convective center beset by WIND SHEAR, Alberto remains a tropical depression (with a central pressure of 29.61 inches [1,003 mb]) for its entire existence in Cuban waters before recurving to the northeast and eventually making landfall in Florida. In anticipation that the tropical depression will come ashore in western Cuba, authorities evacuate thousands of civilians. Like most tropical depressions and tropical storms, Alberto is a powerful rainmaker, dropping significant amounts of precipitation on western Cuba. Some communities in Pinar del Río province report receiving up to 16 inches (406 mm) of rain, while torrential downpours cause two buildings in Havana to collapse. No deaths or injuries are reported in Cuba.

cumulonimbus clouds   Also known as thunderheads, cumulonimbus clouds comprise the largest part of a TROPICAL CYCLONE's complex cloud system. Towering some 40,000 to 50,000 feet (13,333–16,666 m) above Earth, cumulonimbus clouds form both the dense gray wall of thunderstorms that surround the cyclone's EYE and those spiral RAIN BANDS that radiate out from its low-pressure center. Created when highly unstable, moisture-laden cumulus clouds are pushed into the upper atmosphere by ascending air currents, cumulonimbus clouds frequently develop into vast convective cells that can produce huge rains, lightning, hail, gale-force winds, and TORNADOES. Over the open stretches of the world's tropical waters, where squadrons of cumulus clouds are steadily carried west by the trade winds each day, fluffy cumulonimbus systems can develop over a matter of hours into enormous, anvil-headed storms of cyclonic potential. Provided that the sea is warm enough and that the upper- and lower-level winds are moving in the same direction at the same speed, cumulonimbus clouds will begin to cluster around a point of low pressure. Spun into a circular course by Earth's rotation, the clouds continue to thicken and move closer together, rising to the point where the upper-atmosphere winds sheer downward. Now capped by the familiar Pileus anvil,

Towering like castles in the sky, cumulonimbus towers reach several thousand feet into the atmosphere. Created by very unstable conditions, cumulonimbus clouds produce thunderstorms, hail, tornadoes, waterspouts, and volumes of rain. They also, when spun into synchronicity by the Coriolis effect, form a tropical cyclone's cloud bands. *(NOAA)*

the cumulonimbus clouds further spiral inward as the CENTRAL BAROMETRIC PRESSURE of the burgeoning TROPICAL STORM continues to drop. In time, the convective cumulonimbus cells within the mature hurricane, typhoon, or cyclone will spread outward for several miles, forming rain bands 30 miles wide and 100–300 miles long containing pelting winds and rains of significant intensity.

cyclone   A *cyclone* is the regional name given to those mature TROPICAL CYCLONES that develop over the Arabian Sea, the BAY OF BENGAL, and the Indian Ocean and within the coastal waters of AUSTRALIA. Like its meteorological rivals the North Pacific TYPHOON and the North Atlantic HURRICANE, the cyclone is a tightly organized low-pressure system surrounded by warm winds that rotate in a clockwise direction in the Southern Hemisphere and counterclockwise in the Northern. Characterized by sustained surface winds in excess of 74 MPH (119 km/h) and barometric pressures ranging from a fairly mild 29.25 inches (991 mb) to a catastrophic 27.17 inches (920 mb) or less, a cyclone usually contains tremendous amounts of PRECIPITATION and may at times spawn TORNADOES. The cyclone also totes beneath its EYE a huge dome of seawater known as a STORM SURGE, which in the event of a landfall can be responsible for considerable property damage and loss of life.

Cyclones form over the warm, 84–86°F (25–27°C) waters of the Indian Ocean (including the Arabian Sea and Bay of Bengal) and the western South Pacific Ocean (south of the equator and west of the International Date Line), every year. Although the official cyclone season in the north Indian Ocean extends from May to November, there are recorded instances of mature cyclones striking both coasts of INDIA, the east coast of Madagascar, and the assorted northeastern islands of the region in April and December. Australia's cyclone season ranges from December to March, with January and February being the peak months for cyclonic generation.

During the early part of the season, cyclones principally form between latitudes 5 and 20 degrees South over the sun-swamped waters that wash the Australian continent's shores. While a percentage of these early storms tend to have less intense winds than late-season cyclones, they do carry greater amounts of precipitation. Relentlessly moving in a south-southeasterly direction at speeds of between five and 15 MPH (3–24 km/h), a number of them have over time brought major flood emergencies to Australia's northern states. Conversely, late-season cyclones—those storms that primarily develop north of 5 degrees in the Arabian Sea and Bay of Bengal—have much higher death tolls due in part to their enormous size and more organized nature. For some of them, it is a several-hundred-mile trek across energized waters before a stormy landfall is made in India, BANGLADESH, Myanmar (Burma), and, periodically, the island of Sri Lanka. This gives late-season cyclones plenty of time to grow into pressure-driven machines of enormous destructive potential.

When the Australian cyclone season is merged with that of Indian Ocean storms, cyclone generation becomes far more frequent than either North Atlantic hurricanes or North Pacific typhoons. On average, cyclones account for 43 percent of the world's annual tropical cyclone activity, whereas typhoon generation accounts for another 30 percent, and hurricane formation an additional 27 percent.

In terms of their trajectories, early-season cyclones that develop north of Australia in the Timor and Coral Sea areas generally move in a south-southwesterly direction, while those few that originate over the Bay of Bengal move west in a linear direction and lash India with their lighter winds but gargantuan rainfalls. Late-season cyclones, on the other hand, will often travel great distances across the south Indian Ocean before swinging southeast, curving in a parabolic fashion around the high pressure ANTICYCLONES that dominate the weather patterns of the Southern Hemisphere during the summer months. Like a majority of hurricanes and typhoons, cyclones do not travel in perfectly straight lines but instead wobble back and forth over an imaginary track. It is not unusual for a cyclone's eye to drift 30 or more miles in one direction and then float back to its original position a few hours later.

Some of the lowest barometric pressures ever recorded have been observed in cyclones. In 1833 an extraordinarily powerful cyclone near Kedgeree, India, produced a minimum pressure reading of 26.30 inches (891 mb) on board the British merchant ship *Duke of York*. The Australian cyclone that dismasted the clipper ship *Houqua* as it transited the Timor Sea in February 1848 rendered a reading of 27.60 inches (935 mb). The infamous False Point Cyclone of September 22, 1885, blasted ashore on India's northeast coast with a minimum pressure reading of 27.08 inches (917 mb). A southern Indian Ocean cyclone that struck Madagascar in February of 1899 produced a reading of 28.69 inches (972 mb) as it came ashore near the town of Vohemare. More recently, Cyclone TRACY, which pulverized the port city of Darwin, Australia, on Christmas Day, 1974, did so with a CENTRAL BAROMETRIC PRESSURE of 27.00 inches (914 mb), while the GREAT CYCLONE of November 12–13, 1970, that killed in excess of 300,000 people in Bangladesh, registered a pressure reading of 27.13 inches (919 mb). On the afternoon of April 23, 2006, Severe Tropical Cyclone Monica generated a central barometric pressure reading of 26.72 inches (905 mb) as it moved off the northern coast of Australia's Northern Territory.

Such remarkably low barometric pressures are responsible for the considerable size and intensity of late-season cyclones. Although a minimal or moderate cyclone rarely exceeds 200 miles across, a routine number of major cyclones—those storms with barometric pressures lower than 28.47 inches (964 mb)—have widths of between 300 and 400 miles (483–644 km). Although a large cyclone's maximum. winds are found directly around its tightly coiled (10–30 miles (16–48 km) across) eye, its extensive cloud mass allows for a greater distribution of the PRESSURE GRADIENT, or the drop in barometric pressure over a certain distance from the eye. This brings gale-force winds, many with 74-MPH (119-km/h) gusts, to a wider ocean area, thereby increasing the rate of evaporation though wind-driven spray.

Those smaller countries that ring the Indian Ocean—Bahrain, Myanmar, Mozambique, Pakistan, and Madagascar—all have sporadic histories of cyclone strikes. A number of these storms have been of exceptional severity, making them of meteorological or historical interest. One Arabian Sea cyclone, born near the Gulf of Oman on December 27, 1957, traveled directly west, where it showered Bahrain with 110-MPH (177-km/h) winds. Twenty employees of Shell Oil lost their lives when the storm capsized an offshore drilling rig. A similar Arabian Sea cyclone in February 1959 left an additional 500 people in Bahrain dead.

Situated on the Bay of Bengal's eastern edge, the nation of Myanmar, formerly known as Burma, lost 36 of its citizens and 72,000 of its houses to a moderate, earlyseason cyclone on May 1, 1936. A similar storm, on May 11, 1975, took an additional 187 lives. Although in this instance the cyclone's recorded wind speeds did not exceed 90 MPH (145 km/h), its 16-foot (5-m) storm surge completely inundated the delta lowlands that brace the Irrawaddy River's mouth. Hundreds of buildings and thousand of acres of rice paddies were utterly destroyed.

On the opposite side of the bay, the island country of Sri Lanka, which seemingly hangs pendantlike from the lobe of the Indian subcontinent, was struck by a cyclone on December 23, 1964. In a nation that is unaccustomed to cyclone strikes (its position in the Bay of Bengal places it below the area most favored for cyclone generation), Sri Lanka suffered 842 deaths and another 100 missing. A similar cyclone on November 24, 1978, killed 150 people and left almost a million people on the island homeless.

South of the equator and aligned with Africa's east coast, the islands of Madagascar and Mauritius have suffered their share of deadly cyclone strikes. In March 5, 1927, the eastern port city of Toamasina was virtually obliterated as an impressive cyclone crossed Madagascar reportedly killing 500 people. On February 8, 1932, a powerful cyclone careened across Reunion Island before coming ashore in central Madagascar; 45 people were killed, and another 38,000 were rendered homeless. During the first week of January 1951, a southward-moving cyclone laid waste to the Comoro Islands as it rocked its way into northern Madagascar, more than 500 people were reportedly killed, and two coastal trading vessels were sunk. On April 24, 1959, more than 83,000 people were left without shelter when a killer cyclone struck Madagascar directly; 300 to 350 people perished. For nearly a week in mid-February of 1896, Mauritius was raked by a particularly wet cyclone, that dropped more than 20 inches (508 mm) of rain on the island in just one night; less than a century later, on March 1, 1960, a cyclone crossed Mauritius at midnight, killing 42 people and destroying 19,000 buildings and residences. As recently as the 1994 season, Madagascar was directly affected by no less than five cyclones, the worst of them—Cyclone Geralda—claiming 70 lives as its 125-MPH (201-km/h) winds lashed the port city of Toamasina on February 2, 1994. More than 90 percent of the city was reportedly destroyed, and the entire island suffered severe damage to its crucial rice stocks.

Periodically, those Indian Ocean cyclones that make landfall in Madagascar and Mauritius continue across the Mozambique Channel and into Africa's southeast coast. In Mozambique, the former Portuguese protectorate that is somewhat shielded from heavy wind and rain by the offshore mountain ranges of Madagascar, a resilient cyclone came ashore on April 14, 1956, killing some 107 people in the Nampula and Niassa districts. A majority of the victims were residents of the town of Memba, a small fishing port that the cyclone's tearing winds left in complete ruins. More than three decades later, Cyclone Nadia's 105-MPH (169-km/h) gusts killed 240 people and left approximately a million others homeless as it spiraled into northeastern Mozambique on March 24, 1994.

Those satellite island groups that orbit Australia's tropical waters are also subject to sporadic cyclonic activity. On March 2, 1982, cyclone Isaac swept out of the Cook Islands and into Tonga, killing nearly 100 people and leaving tens of thousands of others homeless for weeks thereafter. Judged to have been the worst cyclone to have struck the archipelago in its history, nearly 80 percent of the buildings in the capital of Nukualofa sustained major damage as Isaac's 170-MPH (274-km/h) winds furiously enveloped the 45-island group, downing power lines and furrowing croplands.

The intensely low barometric pressures found in many cyclones have also been responsible for creating enormous, often deadly, storm surges. Caused when low central pressures literally allow the sea to form a dome of water beneath the cyclone's eye, these storm surges have frequently crashed ashore with catastrophic results. During the BACKERGUNGE CYCLONE of October 31, 1876, a towering storm surge, estimated by witnesses to have been 40 feet (13 m) high, swept across the mouth of India's Megna River. Rolling unimpeded across the low-lying fan of silt and debris, the storm surge washed through the city of Backergunge, drowning 100,000 people. Other notable storm surges include that of the cyclone that came ashore at Coringa, India, in December 1789. Described by survivors as a succession of three waves, the initial wave drove ashore all the shipping in the harbor, while the second and third simply flowed in, inundating the delta for miles. Some 20,000 people are said to have been lost, buried beneath the huge mounds of sand that the retreating waters left behind.

Although the word *cyclone* appears to be a comparatively recent addition to the language, there are some questions surrounding its etymological history. For years it was believed that HENRY PIDDINGTON, curator at the Calcutta Museum of Oceonomic Geology, coined the term, meaning "coil of a snake," with the 1839 publication of his first treatise on storms in the *Journal of the Asiatic Society of Bengal*. Ivan TANNEHILL, in his 1939 study of hurricanes, wrote that the word was derived by Piddington "about the middle of the last century," with subsequent authors closely following suit. However, in his 1963 masterpiece, *Early American Hurricanes*, DAVID M. LUDLUM quotes an 1821 source that describes the eye of the September 2 hurricane that passed with great violence over Cape May, New Jersey, as: "The vortex or center of this cyclone as laid down in Blunt's Coast Pilot. . . ." This indicates that the word was in use in North America as early as 1821—or 18 years before Piddington's initial publication.

In order to quantify and record the size and strength of approaching cyclones more easily, the region's meteorological and civil defense agencies have adopted a five-point scale of cyclone intensity. Similar to both the SAFFIR-SIMPSON SCALE and the North Pacific Typhoon Scale, it details the most salient highlights of cyclone generation and organization.

Cyclone 01A *India, June 9, 1998*   On June 9, 1998, a rare cyclone transited the Arabian Sea, delivering 115-MPH (185-km/h) winds and pounding rains to the coastal state of Gujarat. One of only 11 TROPICAL CYCLONES to strike the Indian subcontinent from the west, the June cyclone claimed some 1,126 people, mostly salt workers, and left several thousand others destitute. As is frequent in catastrophes of this kind, relief efforts were hampered by flooded roads and washed-out bridges, thereby preventing emergency crews from speedily reaching the battered districts. Fifteen thousand people were evacuated. The cyclone also killed nine people in the neighboring Pakistani port city of Karachi.

Cyclone 04B *India, November 6, 1996*   The deadliest cyclone of the 1996 BAY OF BENGAL TROPICAL CYCLONE season, Cyclone 04B delivered 112-MPH (180-km/h) winds, torrential rains, and 10-foot waves to INDIA's storm-riven Andhra Pradesh state. Demolishing more than 10,000 houses spread among 300 villages, the cyclone's nine-inch rainfalls similarly inundated many of the most fertile rice paddies in the Godavari districts, destroying several hundred million dollars worth of food stocks. At sea, hundreds of fishers were lost, as their small, open boats were overwhelmed by the cyclone's 17-foot (6-m) swells. All told, Cyclone 04B's rampage through southeastern INDIA cost that nation 1,320 lives and $1.5 billion in property losses.

cyclone season   In AUSTRALIA, the cyclone season runs from November to May of each year. On average, Australia is affected by 10 tropical cyclones per season, with a majority crossing the northern and western coasts of the continent. In 1963, some 16 cyclones affected the continent-nation.

On January 13–14, 2003, Cyclone Ami brought considerable destruction to Fiji's second largest island, Vanua Levu. Ami's gusts were clocked at 106 MPH (170 km/h) and intensifying. Cyclone Ami formed south of Tuvalu during the previous weekend. As the system approached Fiji, its sustained winds were clocked at 68 MPH (110 km/h).

At least 15 people on Vanua Levu were killed by flash floods caused by Cyclone Ami. Just over two weeks later, Cyclone Cilla passed over the affected areas while still a TROPICAL DEPRESSION, with winds of 35 knots.

D

Dampier, William (1652–1715) A talented author, naturalist, navigator, and sometimes adventurer, William Dampier was among the first of European observers to describe the circulatory nature of a TROPICAL CYCLONE correctly. Born the son of a wealthy landowner in Somerset, England, and educated at the village school, Dampier first went to sea in 1668. After making commercial voyages to Indonesia (1669), JAMAICA (1674), and MEXICO (1675)—all of them ending without significant financial reward—Dampier decided to change his ill-fortune by engaging in piracy. Between 1676 and 1690, Dampier raided logging camps in Mexico, sacked the town of Portobelo in Panama (1680), endured poverty in VIRGINIA (1682), sailed the NORTH PACIFIC OCEAN under the command of buccaneer captain John Cook (1684), and was eventually marooned on the Nicobar Islands (1688). In May of that year, while on the second-to-last leg of his historic 12-year circumnavigation of the world (1679–1691), Dampier encountered a powerful, early season TYPHOON in the South China Sea and lived to tell about it. First published in 1697 in his *A New Voyage Round the World,* and embroidered with riveting scientific insights into a tropical cyclone's behavior. Dampier's account of the typhoon remains one of the most important narratives in the early canon of storm literature. Stating that "typhoons are a sort of violent whirlwinds" preceded by weather that is clear and calm and generates "light winds," Dampier defined the 1688 tempest as having two halves: one in which winds came on with great violence from the northeast and then after a serene calm that lasted for just under one hour, resumed with equal fury from the southwest. Dampier's narrative further describes the BAR OF A TROPICAL CYCLONE, the appearance of the dark line of clouds that comprised the system's EYEWALL: "Before these whirlwinds come on, there appears a heavy cloud to the northeast which is very black near the horizon, but toward the upper part is a dull reddish color." After safely arriving in Sumatra, Dampier resumed his travels, living in INDIA and sailing the BAY OF BENGAL before returning to England in 1691. A disciplined diarist who had spent more than a decade compiling observations on dozens of subjects ranging from geography to oceanography to meteorology and sociology, Dampier later authored *Voyages and Descriptions* (1698), "Discourse on the Trade Winds," (1699) and, after a voyage to AUSTRALIA (1699–1701), *A Voyage to New Holland* (1703).

dangerous semicircle This term, now somewhat archaic, is given to that portion of a TROPICAL CYCLONE in which the winds and rain are most intense. Primarily used by mariners, the term is derived from the practice of dividing a HURRICANE, TYPHOON, or CYCLONE into dangerous and NAVIGABLE SEMICIRCLES, or right and left halves, based upon the system's forward motion. In the Northern Hemisphere, where tropical cyclones spin in a counterclockwise direction, an observer facing into the winds of an approaching hurricane will find the dangerous semicircle on the right, or eastern side, and the navigable semicircle on the left, or western side. The reverse is true, of course, in the Southern Hemisphere, where the clockwise spinning of a cyclone will yield a dangerous semicircle on the left, or western side, and a navigable semicircle on the right, or eastern half. While in actuality both halves of a tropical cyclone are dangerous, the half that finds itself strengthened by both the forward speed of the system's STEERING

CURRENT and the storm's own forward velocity will possess significantly faster winds and higher seas. For this reason, those mariners who have divided an oncoming hurricane, typhoon, or cyclone into dangerous and navigable semicircles have been better able to guide their vessels away from the storm's most furious aspects, thus greatly improving their odds of surviving it.

Danielle, Tropical Storm *Gulf of Mexico– Southern United States, September 4–7, 1980* Between 1980 and 2004, five tropical cyclones in the NORTH ATLANTIC OCEAN were identified with the name Danielle. Of these, three were TROPICAL STORMS, and two were hurricanes—both reaching Category 2 status during their respective lifespans. The identifier Danielle has been retained on the rotating list of North Atlantic hurricane names, and is scheduled to next appear during the 2010 season.

The first tropical cyclone in the North Atlantic basin to carry the name Danielle originated over the warming waters of the northern GULF OF MEX- ICO during the late afternoon hours of September 4, 1980. Upgraded from a TROPICAL DEPRESSION (with a CENTRAL BAROMETRIC PRESSURE of 29.76 inches [1,008 mb]) during the evening of September 5, Tropical Storm Danielle continued to slowly intensify as it tracked due west across the Gulf of Mexico, eventually reaching its peak central barometric pressure of 29.64 inches (1,004 mb) shortly before making landfall during the late-evening hours on September 6, 1980. Drifting ashore near the major port city of Galveston, Danielle's 58-MPH (93-km/h) winds rustled trees and crumpled billboards, but caused no major damage or loss of life. As the system penetrated into TEXAS, torrential precipitation caused significant flooding between Beaumont and Port Arthur on September 7, with some 17.16 inches (432 mm) of precipitation establishing a new rainfall record for the region.

Danielle, Tropical Storm *North Atlantic Ocean– Caribbean Sea, September 7–10, 1986* The second North Atlantic tropical system identified as Danielle originated over the eastern NORTH ATLANTIC, some 200 miles east of the Windward Islands, on September 7, 1986. Formed from an African-born tropical wave, Danielle's strengthening cycles were hampered by weak CORIOLIS vorticity and unfavorable wind shear conditions. With a CENTRAL BAROMETRIC PRESSURE of 29.61 inches (1,003 mb), Danielle was upgraded to TROPICAL STORM intensity during the early evening hours of September 7, and shortly thereafter drifted across the Windward Islands and entered the CARIB-

BEAN SEA. At its peak, Danielle generated a central barometric pressure of 29.52 inches (1,000 mb) and sustained, 58-MPH (93-km/h) winds on September 8. By noon on September 10, however, the system had weakened (with a central pressure of 29.88 inches [1,012 mb]) to the point of dissipation. Although a relatively weak tropical system, Danielle did cause considerable damage on the islands of Barbados, St. Vincent and the Grenadines, and Trinidad and Tobago, where several BUILDINGS were destroyed. In Trinidad and Tobago, Danielle's heavy PRECIPITATION spawned flash flood and landslide conditions that washed out bridges and disrupted electrical and water utilities. Although no deaths or injuries were reported, Tropical Storm Danielle's passage through the eastern Caribbean caused more than $9 million (in 1986 U.S. dollars) in crop and other property losses.

Danielle, Tropical Storm *North Atlantic Ocean– Eastern United States, September 22–26, 1992* A short-lived TROPICAL CYCLONE, the third North Atlantic tropical system identified as Danielle originated some 200 miles (325 km) off the eastern seaboard of the United States on September 22, 1992. Moving to the north-northeast, Danielle was upgraded from a TROPICAL DEPRESSION (with a central pressure of 29.79 inches [1,009 mb]) to a tropical storm (with a central pressure of 29.70 inches [1,006 mb]) during the early evening hours of September 22. On September 24, however, the system fell under the influence of a high-pressure steering current to the north, which sent the TROPICAL STORM into a westward loop before setting it on a new trajectory to the north-northwest. Now poised to make an eventual landfall in the Chesapeake Bay area, Danielle slowly intensified, reaching its peak intensity of 29.55 inches (1,001 mb) during the daylight hours of September 25. By nightfall on September 25, the system was trundling ashore on the Delmarva Peninsula, its sustained, 63 MPH (102 km/ h) causing considerable damage to trees, small coastal BUILDINGS, and watercraft. An ardent rainmaker, Danielle produced average PRECIPITATION counts of 4.0 inches (102 mm) over portions of VIRGINIA, Delaware, and Maryland, while the system's minor storm surge flooded low-lying areas in Norfolk, Virginia. Downgraded to a tropical depression during the mid-morning hours of September 26, Danielle's remnants delivered several inches of rain, gusty winds, and high surf to southern Pennsylvania and New Jersey. While damage losses were light, Danielle did claim two lives in a boating accident in New Jersey.

Danielle, Hurricane *North Atlantic Ocean– Europe, August 24–September 8, 1998* The longest-lived and farthest-traveled of the five North Atlantic

TROPICAL CYCLONES thus far identified as Danielle, this system originated well to the east of the Leeward Islands, south of latitude 15 degrees North, on the morning of August 24, 1998. Steadily plodding to the northwest, the system was upgraded to TROPICAL STORM intensity (and given the identifier, Danielle) later the same day. Within 24 hours, as its CENTRAL BAROMETRIC PRESSURE dropped to 29.23 inches (990 mb), Danielle was promoted to a Category 1 hurricane with sustained surface winds of 75 MPH (120 km/h). Positioned well into the NORTH ATLANTIC OCEAN, but maintaining its northwesterly trajectory toward the eastern United States, Danielle continued to intensify, its growth spurts determined in large part by high SEA-SURFACE TEMPERATURES (SSTs), reduced wind shear, and the increased vorticity (due to the CORIOLIS EFFECT) undergone by tropical cyclones as they move away from the equator. By dawn on August 26, as its central pressure dipped to 28.79 inches (975 mb), Danielle was upgraded to a Category 2 hurricane on the SAFFIR-SIMPSON SCALE. Sustained surface winds of 92 MPH (148 km/h) roiled the Atlantic waters to the east of the Windward Islands, sending large rollers ashore as far south as Barbados. As is frequently seen in long-lived tropical cyclones, Hurricane Danielle's intensity patterns continually fluctuated throughout its existence, and within 12 hours of reaching Category 2 intensity, the system had returned to Category 1 status, with a downgraded central pressure observation of 29.08 inches (985 mb). Between August 27 and September 1, an increasingly disorganized Danielle's central pressure seesawed from 29.35 inches (994 mb) on August 27, to 28.49 inches (965 mb) on September 1. Hurricane Danielle reached its recorded peak intensity shortly before dawn on September 3, 1998, when its central pressure slipped to 28.34 inches (960 mb). By that time, the system (steered by the BERMUDA HIGH and a developing trough along the eastern United States) had skirted to the north of the BAHAMAS, recurved sharply away the FLORIDA coast, brushed past Bermuda, and sped to the northeast, well into the North Atlantic. At midnight on September 3, while located to the east of Newfoundland, CANADA, the system began its extratropical transition, and by the same time the following day, was a mature-stage extratropical cyclone with a central pressure of 28.79 inches (975 mb) and surface winds of 69 MPH (111 km/h). As an extratropical system, Danielle scudded along the northern periphery of the Bermuda High, and accelerated to the northeast, into those waters adjacent to the British Isles. By September 6, it had deepened to 28.46 inches (964 mb), and was delivering heavy seas and gusty winds to the western British coastline. Several small watercraft were damaged or sunk, and the media reported a rogue wave along the southwestern Cornish coastline. By the morning of

September 8, the extratropical storm that had been Hurricane Danielle was downgraded to an extratropical depression, and was soon absorbed by a more robust extratropical system over western EUROPE. A long-distance tropical cyclone of the first order, Danielle remained an offshore system for most of its existence and caused no deaths or major property losses.

Danielle, Hurricane *North Atlantic Ocean, August 13–21, 2004* The fourth-named tropical cyclone of the very active 2004 North Atlantic HURRICANE SEASON (and the fifth system to bear the name Danielle), Hurricane Danielle originated near the Cape Verde Islands during the midnight hours of August 14, 2004. By midnight of the same day, the system was upgraded from a TROPICAL STORM to a Category 1 HURRICANE with a CENTRAL BAROMETRIC PRESSURE of 29.14 inches (987 mb). Traveling away from the western coast of Africa on a northwesterly trajectory, Danielle continued to intensify. It reached Category 2 status during the early evening hours of August 15, and by the same time the following day, was generating a central barometric pressure of 28.46 inches (964 mb) and sustained winds of 109 MPH (176 km/h). Unable to round the southern periphery of the subtropical ridge known as the BERMUDA HIGH, Hurricane Danielle continued to track to the northwest where, on August 18, it underwent a half-loop track change to the northeast before coursing due westward to resume its northwesterly direction. During this abrupt track shift, Danielle first weakened to Category 1 intensity, followed by a downgrade to tropical storm; this was no doubt due to the strong westerly shear caused by the same high-pressure ridge that had altered the system's initial northwesterly track. By August 20, 2004, Danielle was classified as a TROPICAL DEPRESSION with a central pressure of 29.88 inches (1,012 mb), and finally dissipated, some 855 miles west-southwest of the Azores. An offshore system for its entire existence, Hurricane Danielle caused no deaths or property losses.

Danny, Hurricane *Gulf of Mexico–Southern United States, August 12–20, 1985* Between 1985 and 2003, four tropical cyclones in the North Atlantic basin have been identified as Danny. Three were of hurricane intensity, while the fourth was a TROPICAL STORM. The 1997 incarnation—Hurricane Danny—was the most destructive, with 10 deaths reported. The identifier Danny has been retained on the revolving list of North Atlantic hurricane names, and is scheduled to reappear in 2009.

The first North Atlantic tropical cyclone to bear the identifier Danny originated as a TROPICAL DEPRESSION over the heat-charged waters of the northern CARIBBEAN SEA on August 12, 1985. Moving in a northwesterly direction, the system remained a rain-pumping tropical depression as it grazed the western tip of CUBA on August 13, entered the Yucatán Straits, and streamed into the GULF OF MEXICO; by the early morning hours of August 14, while still situated several hundred miles to the south of LOUISIANA, the system was upgraded to a tropical storm. Firmly maintaining its northwesterly trajectory (one that if followed, would eventually take the system ashore in Galveston, TEXAS), Tropical Storm Danny gradually strengthened, becoming a Category 1 hurricane (with a CENTRAL BAROMETRIC PRESSURE of 29.44 inches [997 mb]) shortly before midnight on August 14. The system continued to deepen as it began a more northerly RECURVATURE, eventually reaching its peak intensity of 29.17 inches (988 mb) just before making landfall in southern Louisiana—near Lake Charles—on the evening of August 15, 1985. Rapidly downgraded to a tropical depression, Danny chugged through central Louisiana before sharply recurving to the east-northeast and deluging MISSISSIPPI, ALABAMA, GEORGIA, and the Carolinas with tropical depression force precipitation. Not surprisingly, Danny's destruction was the most severe in Louisiana, where 80-MPH (129-km/h) winds unroofed BUILDINGS, uprooted trees, destroyed utility systems, and caused extensive coastal wave flooding. Two people in Louisiana were killed, and property losses topped $275 million (in 1985 U.S. dollars).

Danny, Tropical Storm *North Atlantic Ocean, September 7–11, 1991* The second North Atlantic TROPICAL CYCLONE known as Danny—this one of TROPICAL STORM intensity—originated well into the Atlantic on September 7, 1991. Born from a tropical wave of Cape Verde origin, the system remained a TROPICAL DEPRESSION (with a minimum central pressure of 29.70 inches [1,006 mb]) until the morning of September 9, when it was upgraded to Tropical Storm Danny. Hindered by strong vertical WIND SHEAR and low CORIOLIS EFFECT vorticity, Danny only marginally deepened during the next three days, reaching its peak intensity of 29.50 inches (999 mb) on September 10. During the daylight hours of September 11, 1991, while situated just to the east of the Leeward Islands, Danny was downgraded to a tropical depression and soon dissipated. An offshore system, Danny caused no deaths or injuries.

Danny, Hurricane *Gulf of Mexico–Southern United States, July 16–27, 1997* On July 16, 1997, Hurricane Danny—the third North Atlantic tropical system to carry the name—originated over the northwestern GULF OF MEXICO from a weak and disorganized TROPICAL DEPRESSION. An early season upstart, Danny slowly tracked to the northeast, growing into a TROPI-CAL STORM (with a central pressure of 29.61 inches [1,003 mb]) shortly before noon on July 17. Within 12 hours, it had deepened to a Category 1 hurricane, generating a pressure reading of 29.29 inches (992 mb) and sustained winds of 75 MPH (120 km/h). Crossing directly over Lake Charles, LOUISIANA, Danny continued to strengthen as it maintained its northeasterly track, intensifying to a minimum pressure of 29.05 inches (984 mb) just before landfall on the ALABAMA-FLORIDA border, near Mullet Point, on July 19, 1997. While sustained winds of 81 MPH (130 km/h) caused tree and powerline damage in an around the Mobile Bay area, it was Danny's driving rains that wreaked the most havoc. On Dauphin Island, Louisiana, Danny produced a DOPPLER-RADAR verified reading of 43 inches (1,092 mm) of precipitation, while torrential downpours caused flash flooding in parts of Alabama and Georgia. Two TORNADOES were also reported to have touched down in Alabama. One person in Louisiana was drowned. Sandwiched between two high pressure zones and spun steadily to the northeast, Tropical Depression Danny also caused massive flooding in NORTH and SOUTH CAROLINA, and in VIRGINIA. Numerous tornadoes were reportedly spawned by Danny's constituent thunderstorms, including a particularly powerful funnel that destroyed several businesses near Norfolk, Virginia. Three people lost their lives in the Carolinas. Just before midnight on July 20, Tropical Depression Danny entered the NORTH ATLANTIC OCEAN near the Delmarva Peninsula. Energized by the warm waters of the Gulf Stream, the system wasted little time in reintensifying, again becoming a tropical storm (with a central pressure of 29.35 inches [994 mb]) by July 25. Although Danny's northeasterly trajectory took it well offshore, its sprawling rain bands delivered enormous quantities of precipitation to southern and eastern NEW ENGLAND; the picturesque Massachusetts island of Martha's Vineyard received nearly three inches (76.2 mm) of PRECIPITATION, while gusty winds rustled trees and traffic signals. Undergoing extratropical transitioning on July 27, Danny dissipated shortly thereafter. In the end, Hurricane Danny tallied property losses of $100 million (1997 U.S. dollars) and 10 direct and indirect fatalities.

Danny, Hurricane *North Atlantic Ocean, July 16–27, 2003* Although the fourth North Atlantic TROPICAL CYCLONE dubbed Danny briefly reached hurricane intensity on July 18, 2003, this early-season, offshore system spent much of its existence as a TROPICAL DEPRESSION. Born near latitude 30 degrees North on July 16, Danny followed an unorthodox course, scribing an enormous semicircle across the mid-Atlantic. First tracking to the northwest, it underwent a recurvature to the north-northeast on July 17, and was upgraded to a TROPICAL STORM. Slowly intensifying, the system reached hurricane status just before crossing the 40th parallel on July 18, and achieved its peak intensity of 29.52 inches (1,000 mb) during the early morning hours of July 19. Pushed southward by its interaction with the semi-permanent high pressure ridge of the BERMUDA HIGH, Danny weakened to a tropical storm, and by August 20, a tropical depression. The system performed a tight northeasterly loop between July 24 and July 26, steadily weakening, and finally dissipated (with a CENTRAL BAROMETRIC PRESSURE of 29.94 inches [1,014 mb]) during the early morning hours of July 27. An offshore tropical system, Danny caused no death, injuries, or property losses.

David, Hurricane *Eastern and Northern Caribbean Southeastern United States, August 28–September 7, 1979* Considered one of the most powerful Category 5 HURRICANES to have swept through the CARIBBEAN SEA during the latter half of the 20th century, David spent nine harrowing days, from August 30 to September 7, 1979, forging a violent legacy of death and destruction through the eastern islands of DOMINICA, GUADELOUPE, and MARTINIQUE and the northern islands of HISPANIOLA and the BAHAMAS before dealing southeastern United States, not one but two damaging strikes. The first, in eastern FLORIDA, killed five people, while the second, on the border between GEORGIA and SOUTH CAROLINA, brought 90-MPH (140-km/h) winds and eight-foot (3-m) STORM TIDES to Savannah. In Dominica, where David's sustained 135-MPH (217-km/h) winds completely destroyed that year's bumper grapefruit and banana harvest on August 30, an estimated 37 people lost their lives. In the Dominican Republic, which David besieged with 150-MPH (240-km/h) winds and scudding rains on August 31, an estimated 2,000 people died, placing the hurricane among the top 25 of the deadliest NORTH ATLANTIC hurricanes on record. Of the $1.2 billion property damage assessed in the wake of Hurricane David, $487 million of it was sustained by the United States.

A quintessential CAPE VERDE STORM, David originated over eastern North Atlantic waters, some 475 miles (746 km) west of Senegal, Northern Africa, on the morning of August 28, 1979. Carried steadily westward by the equatorial trade winds, the embryonic TROPICAL DEPRESSION quickly sapped the mid-ocean surfaces of their heat and moisture, growing into a powerful TROPICAL STORM on the afternoon of August 29 and a mature Category 1 hurricane with a CENTRAL BAROMETRIC PRESSURE of 28.89 inches (978 mb) and 85-MPH (137-km/h) winds by midnight of the same day.

By the early morning hours of August 30, as David relentlessly bore down on the palm-trimmed eastern Caribbean islands of Dominica, Guadeloupe, and Martinique, weather reconnaissance flights through its EYE indicated barometric pressures as low as 27.91 inches (945 mb) and intermittent wind gusts of 138 MPH (222 km/h). Heralded as one of the most potentially dangerous hurricanes in recent times, David roared across Dominica on the morning of August 30, 1979. One-hundred-thirty-mile-per-hour (209-km/h) winds tore into Dominica's ripe banana plantations, uprooting trees, leveling warehouses, and killing 37 people. In the capital city of Roseau, nearly every BUILDING—from the stately mansion that held the island's legislature, to the Botanic Garden's splendid arboretum and orchid house, to the rows of storefronts that lined the Roseau River—were either destroyed or heavily damaged. Sixty thousand people were displaced as David's 12-foot (4-m) STORM SURGE pounded low-lying towns, sweeping away entire blocks of houses. All water, electrical, and telephone services across the island were interrupted, while dozens of small boats were wrenched from their moorings and cast ashore, choking the beaches with wreckage. Several hundred million dollars in damages were incurred, making David the most costly hurricane to have struck Dominica in nearly a quarter of a century. On flanking Guadeloupe and Martinique, David's 300-mile (483-km) cloud canopy brought hurricane-force gusts and several hours of heavy rain to both islands but caused no fatalities.

Emerging into the southeastern Caribbean Sea's sun-charged waters on the evening of August 30, David underwent a rapid, rather ominous period of intensification. Round-the-clock reconnaissance flights indicated that its central barometric pressure had dropped to an alarming 27.55 inches (933 mb), boosting its sustained wind speeds to a hefty 150 MPH (240 km/h). Now classified as a moderate Category 4 hurricane, David inexorably whirred northwest, bound on a course that, if maintained, would take it ashore somewhere in eastern Hispaniola, probably into the Dominican Republic. United States civil defense authorities, well aware of how devastating a hurricane strike of David's intensity could prove on the crowded shores and slopes of the Dominican Republic, immediately notified Santo Domingo of the potential threat. As pertinent weather intelligence was passed to the Dominican government, evacuation efforts in that country proceeded at a clumsy, haphazard rate. While several vulnerable oceanside towns surrounding the capital of Santo Domingo were safely evacuated, the refugees were housed in hillside shelters that later proved to be death traps as rain-swollen rivers triggered huge mudslides, undermining their foundations and sending them crashing into the narrow valleys.

As a blanket of broken palm fronds and lethal roof shingles scattered itself across the countryside, downed power lines popped and sparked in the flooded streets of Santo Domingo, presenting a fatal misstep to a number of survivors who unwisely ventured out during the storm's height. In the mountain town of Padre Las Casas, 75 miles (121 km) west of to Domingo, a church and neighboring school that was being used as a storm shelter were swiftly overrun by the nearby Yaque River.

First driven onto chairs and then tabletops by the everrising waters, and estimated 438 people—a majority of them women and children—decided to abandon the coffinlike confines of the church and school before the rain-gorged river seated them inside forever. Forming a human chain, they stepped into the waist-high mire and began to wade outside. Still bound together by knotted belts, shirts, and curtain ties appropriated from the school's office, they had almost reached the center of town when a roaring flash flood carried nearly every one of them to death.

Spawned by the failure of one of the river's dikes, the flood further damaged a quarter of the dwellings in Padre Las Casas, making the town one of the hardest hit by David's fury. In other parts of the Dominican Republic, the hurricane's flattening winds devastated much of the country's staple coffee crop, while its torrential rains created other flash floods and building collapses that killed an additional 380 people. In Santo Domingo and its attendant neighborhoods, an estimated 152,000 people were left homeless, prompting the U.S. government to begin a massive aerial relief campaign.

Considerably weakened by its inland passage over mountainous Hispaniola, David's maximum 85-MPH (137-km/h) winds nevertheless delivered seven-inch (178-mm) rains and 15-foot (5-m) seas to the low-lying coral cays of the eastern Bahamas during the daylight hours of September 1, 1979. While above-average tides did cause some minor flooding and beach erosion on Inagua and Cat Islands, no fatalities or serious injuries were reported. Seemingly committed to its present north-northwesterly course, David fermented over the Bahamas, bringing hurricane-force gusts and rocking seas to Nassau before gradually turning west late on September 1. As the outer fringes of its 290-mile mass dropped a light but steady rain on southeastern Florida, civil defense workers in that state began the mass evacuation of some 300,000 residents and summer tourists from shoreline communities on the morning of September 2.

A majority of them had been safely removed to higher ground when David finally did come ashore in southeastern Florida, 70 miles (113 km) north of Miami, on September 3, 1979. Downgraded to a moderate Category 2 hurricane, David's 89-MPH (143-km/h) winds wrought fairly extensive property damage in the vicinity of Riviera Beach. Penetrating

only slightly into the Florida Peninsula, the hurricane's 35-mile (56-km) eye sheered up the east coast, gutting a multistory condominium complex in Melbourne Beach and causing widespread power outages from Vero Beach to Daytona before arcing back into the Atlantic near the city of Saint Augustine. A survey of insurance claims covering squashed mobile homes to flooded restaurants indicates that David's DANGEROUS SEMICIRCLE, or that part of the storm in which its winds and rain were most intense, traveled right along the densely developed east coast, precipitating an estimated $60 million in property damage.

Now located 49 miles (79 km) off Georgia's coast, David was positioned to become one of those rare Atlantic hurricanes whose eyes have come ashore more than once on the U.S. mainland. Bound for the border between Georgia and South Carolina, David's central barometric pressure slightly deepened to 28.91 inches (979 mb) as it moved across the temperate shallows that wash the anchorages of Savannah and Charleston. Surging past Savannah with 90-MPH (140-km/h) winds and six-to-eight-foot (2–3-m) storm tides, David's eye came ashore for the second, and final, time on September 4, 1979, between Hilton Head and Charleston, South Carolina. At Folly Island, a popular resort town 12 miles (19 km) southeast of Charleston, David's pounding seas destroyed a number of beachside buildings, including several dance clubs and a portion of the famed turn-of-the-century Atlantic House restaurant. As far north as Myrtle Beach, David's ripping currents and digging waves overwhelmed makeshift seawalls, causing extensive shoreline erosion. While some $280 million damage was assessed in South Carolina, mass evacuations kept the hurricane's DEATH TOLL to six.

Rapidly disintegrating over the inland reaches of VIRGINIA and Pennsylvania on September 5–6, David's lingering rains would further bring major power failures and widespread flash floods to New Jersey and NEW YORK. All told, 16 people in the United States lost their lives to David, while another 2,052 deaths were tallied in the Caribbean. In view of Hurricane David's fearsome record of death and destruction, the name David has been permanently retired from the alternating list of HURRICANE NAMES. It was replaced on the 1985 list with Danny.

Dean, Tropical Storm *North Atlantic Ocean–Eastern United States, September 26–September 30, 1983* Between 1983 and 2007, five North Atlantic TROPICAL CYCLONES have been identified as Dean. Of the five, three were tropical storms, while the other two were strong hurricanes.

On September 26, 1983, Tropical Storm Dean became the first North Atlantic tropical cyclone to carry this particular identifier. Originating to the northeast of the BAHAMAS during the late afternoon hours, Dean meandered to the north-northeast, its CENTRAL BAROMETRIC PRESSURE of 29.82 inches (1,010 mb) slowly sinking until reaching its minimum pressure (29.50 inches [999 mb]) on the morning of September 28. Recurving to the northwest while off of Cape Hatteras on September 29, Tropical Storm Dean began to weaken as it drew closer to its eventual landfall on the southern tip of the Delmarva Peninsula. When the system came ashore near Salisbury, Maryland, during the early morning hours of September 30, its central pressure of 29.79 inches (1,009 mb) produced heavy rains and 46-MPH (74-km/h) winds, but little property damage. Quickly downgraded to a TROPICAL DEPRESSION, Dean had dissipated over the central northeastern United States by the following day.

Dean, Hurricane *North Atlantic Ocean–Bermuda–Canada, July 31–August 9, 1989* The second most powerful of the five Atlantic tropical cyclones thus far named Dean, this mid-season, Category 2 hurricane underwent cyclogenesis over the warm waters of the western NORTH ATLANTIC on July 31, 1989. Of Cape Verde origin, the system passed from TROPICAL DEPRESSION intensity (29.70 inches [1,006 mb]) on August 1, to a powerful TROPICAL STORM intensity of 29.41 inches (996 mb) on August 2, and achieved Category 1 hurricane status (29.35 inches [994 mb]) shortly thereafter. Recurving to the northwest while north of PUERTO RICO and the VIRGIN ISLANDS, Hurricane Dean first assumed a more northerly trajectory on August 3—one which set it on a direct course for BERMUDA. During this time, Dean continued to strengthen. On the evening of August 6, the system achieved Category 2 status, with a central pressure of 28.67 inches (971 mb) and sustained 98-MPH (158-km/h) winds. By midnight of August 6, Hurricane Dean had generated its peak intensity reading of 28.58 inches (968 mb), hiking its sustained wind speeds to 104 MPH (167 km/h). It had also passed directly over Bermuda, causing considerable damage to BUILDINGS, trees, and support infrastructure. Although no deaths were reported in Bermuda, Dean's fury injured scores of people and tallied nearly $10 million (1989 U.S. dollars) in property losses. Weakening as it ground to the northnortheast, Dean was of tropical storm intensity as it crossed the southeastern tip of Newfoundland, CANADA, on August 8, 1989, with a central pressure of 29.26 inches (991 mb) and 52-MPH (83-km/h) winds. While heavy rains were recorded in eastern Canada, no deaths, injuries, or significant damage estimates were reported.

Dean, Tropical Storm *Gulf of Mexico–Southern United States, July 28–August 2, 1995* Of TROPICAL

STORM intensity for less than 24 hours, Tropical Storm Dean—the third North Atlantic tropical cyclone to carry the name—brought heavy rains to Freeport, TEXAS, on July 31, 1995. Formed over the central GULF OF MEXICO from a TROPICAL DEPRESSION on July 28, and steadily sent to the northwest by its steering currents, the system remained unnamed until just prior to its landfall in Texas on July 31, when its central pressure dipped to 29.61 inches (1,003 mb) and it was promoted to tropical storm status. Tropical Storm Dean made landfall while at its peak intensity of 29.50 inches (999 mb) and produced 46-MPH (74-km/h) winds across the state's eastern coastline. As much as 17 inches (432 mm) of precipitation fell on Chambers County alone, causing several localized evacuations, but no deaths or injuries.

Dean, Tropical Storm *North Atlantic Ocean–Puerto Rico–Virgin Islands, August 22–29, 2001* The fourth North Atlantic tropical system identified as Dean originated just off the coast of PUERTO RICO on August 22, 2001. Beset by unfavorable WIND SHEAR conditions, Dean struggled to deepen, achieving a CENTRAL BAROMETRIC PRESSURE of 29.79 inches (1,009 mb) during the early evening hours of August 22 before commencing a weakening trend. By noon on August 23, Tropical Storm Dean's central pressure had risen to 29.88 inches (1,012 mb) and the system was downgraded to Tropical Depression Dean. Some six hours later, while Tropical Depression Dean was located to the east of the BAHAMAS, it lost much of its central convective activity and was downgraded further to a Tropical Wave. Undergoing a prolonged RECURVATURE to the north and northeast, and away from the Bahamas, the tropical wave strengthened, regained a closed ISOBAR, and was upgraded to Tropical Depression Dean during the late night hours of August 24. For the next two days, Dean intermittently intensified, and returned to tropical storm classification on August 27, 2001, while situated well to the north of BERMUDA. Later the same day, the system reached its minimum observed central pressure of 29.35 inches (994 mb), which produced sustained 69-MPH (111-km/h) winds. Off the coast of the Canadian Maritimes by August 28, Dean quickly spun to the northeast and underwent extratropical transitioning by the early evening hours of August 28. It remained an extratropical system until the following day, when it dissipated.

Dean, Hurricane *North Atlantic Ocean–Caribbean Sea–Mexico, August 13–25, 2007* The most intense and destructive of the five Atlantic tropical cyclones identified as Dean, this midseason, Category 5 hurricane brought considerable death and destruction to several islands in the CARIBBEAN SEA and to the eastern coast of MEXICO between August 13 and August 23, 2007. Dean was the first tropical system to make landfall in the North Atlantic basin since Hurricane Andrew in 1992.

The first mature-stage tropical cyclone of the 2007 North Atlantic hurricane season, Dean matured from a TROPICAL WAVE over the extreme eastern North Atlantic Ocean on August 13, 2007. Initially identified as TROPICAL DEPRESSION No. 4, the system (with a CENTRAL BAROMETRIC PRESSURE of 29.7 inches [1,005 mb]) continued to intensify as it scudded westward at 21 MPH (34 km/h). On August 14, as its central pressure dropped to 29.65 inches (1,004 mb) and its sustained winds topped 40 MPH (65 km/h), the system was upgraded to TROPICAL STORM status and given the name Dean. On the morning of August 16, Dean was upgraded to hurricane intensity as it continued to slide to the west-northwest at 24 MPH (39 km/h). Located some 590 miles (950 km) east of MARTINIQUE, the system's intensification (it was now producing a central pressure of 29.1 inches [985 mb]) prompted a flurry of TROPICAL STORM WARNINGS, HURRICANE WARNINGS, and WATCHES across the islands of the Lesser Antilles, including BARBADOS, Guadeloupe, Martinique, DOMINICA, and St. Lucia.

August 17, 2007, would prove a pivotal day for Hurricane Dean. During the early morning hours the system was classified as a Category 2 hurricane with sustained winds of 100 MPH (160 km/h) and a central pressure of 28.8 inches (976 mb). By midmorning, as it schooled the islands of Martinique and St. Lucia with tropical storm–force winds, 87 MPH (143 km/h) gusts, and five-inch (127-mm) rains, Dean's central pressure dropped to 28.6 inches (970 mb), portending continued intensification. Shortly after noon, the system reached Category 3 status, with a central pressure of 28.4 inches (961 mb) and maximum sustained winds of 125 MPH (205 km/h). Now located within the Caribbean basin, Hurricane Dean thundered westward at speeds approaching 22 MPH (35 km/h). Classified as a major hurricane, Dean's progress across the Caribbean was met with concern and preparation across JAMAICA, the Cayman Islands, and along the southern coast of CUBA. Preliminary DEATH TOLLS indicated that at least seven people had perished during Dean's passage across the Windward Islands; two people in Dominica, two in Martinique, and two in HAITI. In St. Lucia, a man was drowned while attempting to rescue a cow from a rain-swollen river.

Hurricane Dean graduated to Category 4 status on the SAFFIR-SIMPSON SCALE during the evening hours of August 17, 2007. The system's maximum sustained winds of 145 MPH (233 km/h) paddled the Caribbean's surface, drawing enormous amounts of energy from the heat and moisture contained within the 81°F (27°C) water.

Generating comparisons to 1988's Hurricane GILBERT, Dean spiraled toward Jamaica on a trajectory that threatened to take the extremely dangerous system westward across the island. As the nation's electrical grid was shut down to prevent electrocutions and to protect its infrastructure, Dean slipped to the south of the island. Pushed southward by a high-pressure ridge positioned over the southeastern United States, Dean passed some 50 miles (80 km) south of Kingston, Jamaica, between August 17 and 18, 2007. As Dean's heavy surf tore into Jamaica's coastal resorts, the system's northern eyewall produced wind speeds of 114 MPH (183 km/h) at Kingston International Airport.

After leaving a sodden trail of destruction across Jamaica, Dean intensified to Category 5 status during the evening hours of August 20, 2007. With sustained winds of 160 MPH (257 km/h) and a central pressure of 27 inches (915 mb), Dean became one of the most intense tropical cyclones yet observed in the Caribbean basin.

As witnessed in earlier tropical cyclones, petroleum and natural gas production in the Gulf of Mexico and the Bay of Campeche was largely disrupted by Dean's energetic passage. Some 14,000 workers were evacuated by ship and helicopter from hundreds of offshore drilling platforms, reducing output by nearly 3 million barrels per day. The Mexican government deployed 4,000 troops to assist with evacuation and sheltering operations across the Yucatán Peninsula; the coastal resort cities of Cancun and Cozumel were virtually emptied as airlines increased their flights to evacuate tourists.

With a central pressure of 27 inches (914 mb), Hurricane Dean tore ashore on the Yucatán Peninsula, near Chetumal, during the early morning hours of August 21, 2007. Extrapolated from measurements gathered by an Air Force Reserve reconnaissance aircraft, Dean's sustained surface wind speed of 160 MPH (257 km/h) caused significant damage to trees, crops, and small buildings across the peninsula. Preliminary media reports indicate that Hurricane Dean claimed at least eight lives across the Caribbean and Mexico.

death tolls and tropical cyclones No other subject in the cultural hierarchy of TROPICAL CYCLONES elicits greater fascination and resolve on the public's part than the matter of death tolls. Rooted in humanity's painful regard for the dark terrors of death by violence and reinforced by an endless play of eyewitness accounts, poignant photographs, and weepy newsreel interviews, the graphic loss of life incurred by many HURRICANES, TYPHOONS, and CYCLONES has become their most identifiable cultural feature. Interpreted by historians, quantified by meteorologists, and happily postponed by emergency managers, death tolls in tropical cyclones are the emotional touchstone by which the human element inherent in most of these storms can finally be witnessed, seemingly experienced, as the numbing shock of distant death from the skies in some way reaches uneasy consciences all over the world.

That death tolls, indeed, *gargantuan* death tolls, are powerful cultural triggers is evidenced by the fact that many of the world's major newspapers do not publish articles detailing the landfall of tropical cyclones in other regions of the planet unless such a landfall is accompanied by a relatively large death toll, usually in excess of 100 people. There is also evidence to suggest that many of the death tolls reported in past CARIBBEAN hurricanes and BAY OF BENGAL cyclones may have been deliberately inflated in an effort to garner sympathy, and hence more financial assistance, from foreign governments. When, two weeks after an April 1991 cyclone slammed Bangladesh with a 20-foot STORM SURGE the Bangladeshi government announced that 138,000 of its citizens had perished and that it would require at least $1.41 billion for relief and reconstruction efforts, the world's governments wasted little time in opening their purses. Moving about as quickly as the cyclone's storm surge had, $150 million in aid poured into low-lying Bangladesh. Just as quickly, stories began to circulate in the Western press that flatly stated that the initial casualty figure of 138,000–139,000 was deliberately exaggerated by Bangladeshi officials as a way of increasing financial empathy for its flooded plight. Claiming that the actual death toll was closer to 60,000–70,000, the articles soon precipitated both a Bangladeshi admission and a worldwide drought of further compassion—monetary and otherwise.

But in some respects, the Bangladeshi authorities can be forgiven their exaggerations, for accurate death tolls in hurricanes, typhoons, and cyclones have often proved difficult to compile. In the GREAT HURRICANE OF 1780, believed by many scholars to be the most deadly NORTH ATLANTIC hurricane on record, death toll estimates widely range from 11,000 to 22,000 people. Among the individual Caribbean islands—BARBADOS, MARTINIQUE, and St. Vincent—that the Great Hurricane ravaged, death counts become even murkier, with some contemporary sources quoting three or four different tallies in the course of the same account. While rightly attributing such discrepancies to uncertainties regarding the exact number of soldiers and sailors who perished at sea, many of these same accounts take great care in recording the particular race of the deceased. John Fowler, in his 1781 treatise, *A General Account of the Calami-*

ties, states that during the earlier SAVANNA-LA-MAR HURRICANE of 1780, the death toll in Montego Bay, JAMAICA, included, "63 whites, 50 mulattoes, and 250 negro slaves."

Firmly arranged along the stratified lines of 18th-century Caribbean society, Fowler's death toll reveals some details concerning class and survivability but little in the way of how the victims actually perished. Of the 63 whites who died, how many were crushed beneath their supposedly safe "hurricane cellars"? Of the 250 slaves, how many died of head injuries caused by the hurricane's 135–145 MPH (217–233 km/h), debris-filled winds? How many were drowned by its 12–18-foot (4–6 m) storm surge? Modern meteorologists, tasked with programming the particulars of past tropical cyclones into their ANALOG computer systems, find such information useful in determining the most fatal aspects of a storm's landfall. This allows for greater improvements in seawalls and other storm-resistant structures, thereby reducing future death tolls.

More recently, similar difficulties in accurately accounting for the lost arose in the case of the 1991 Bangladeshi cyclone. Some 100,000 fishers at sea on 2,000 small boats when the cyclone came ashore were promptly reported missing, hampering attempts to achieve a reasonably correct count of the dead while offshore search-and-rescue operations were still under way. Likewise, many Caribbean hurricanes and Chinese typhoons of the past have only approximate death-toll counts on record, a fact due in part to the difficulties inherent in reconstructing the lives and actions of an entire village of people that had literally been swept away by rising seas and ravenous winds. For years, rescue teams in both Haiti and the PHILIPPINES have been dogged by the dilemma of accounting for those bodies buried beneath their respective mudslides, while the Chinese government has traditionally kept very quiet about all such natural disasters, giving no clues as to what exactly happened, let alone how many people it exactly happened to. There is, for instance, evidence that suggests that a 26-hour rainfall from a typhoon on August 7, 1975, caused the huge Banqiao and Shimantan dams in China's Henan Province to burst. Although the Chinese government has never officially confirmed the death toll, various sources have placed it between 83,000 and 230,000 people. If the latter estimate were confirmed, it would catapult the August 1975 typhoon's toll onto the list of the top five largest death counts ever incurred by a tropical cyclone. It would also, many believe, embarrass the Chinese government.

Even in the United States, where local and federal emergency agencies have for some time kept fairly accurate track of death rates, some variance in hurricane death toll counts does exist. Several published accounts of Hurricane DAVID, for example, state that the 1979 storm claimed one life when it came ashore in south FLORIDA on September 3, while a number of others place the toll at four. Similarly, some histories place Hurricane ANDREW's toll at 61 people, while others lower it to 23. This confusion is due in part to the modern practice of dividing death tolls into direct and indirect columns, or those deaths directly attributable to the dangerous meteorological conditions of the hurricane and those that occurred under concurrent, but different, circumstances. Of the four documented deaths in Florida during Hurricane David, only one was actually due to the fury of the storm itself. The other three deaths—two heart attacks and one electrocution—occurred during the frantic period of evacuation and preparedness that immediately preceded David's moderate landfall. In Hurricane Andrew, emergency relief agencies stated that fifteen people in Florida and another eight in LOUISIANA died as direct victims of the hurricane's 16-foot (5-m) seas and 140-MPH (225-km/h) winds. The additional 38 fatalities between the respective states were considered indirect deaths, casualties brought on by heart attacks, automobile accidents, isolated drowning incidents, disease, and sheer fright.

Although for these reasons any list of significant death tolls, arranged according to size, is a tenuous effort at best, there are a number of recorded hurricanes, typhoons, and cyclones that do clearly contend to head the list of history's deadliest. These include the Hooghly River Cyclone, in which an estimated 375,000 people perished near Calcutta, India; the GREAT CYCLONE OF 1970, in which an estimated 300,000 people perished in Bangladesh; the notorious April 30, 1991, cyclone in Bangladesh that killed upward of 70,000; the typhoon of August 2–3, 1922, that drowned 60,000 people in the northern Chinese city of Swatlow; the October 15, 1942, Indian cyclone that claimed nearly 42,000 souls; and the twin Bangladeshi cyclones of May and June 1965 during which 17,000 and 30,000 people died, respectively.

On a more regional basis, it is generally conceded that the Great Hurricane of 1780 was the deadliest North Atlantic or Caribbean hurricane on record, with between 13,000 and 22,000 deaths attributed to its remarkable passage. Between October 27 and 30, 1998, the flooding rains associated with Category 5 Hurricane MITCH claimed some 12,000 lives in the Central American nations of Nicaragua and Honduras. On September 14–19, 1974, Hurricane FIFI claimed as many as 10,000 lives when it blasted ashore in east Honduras, while another 7,186–8,000 died as a result of Hurricane FLORA's rampage across Haiti on October 3, 1963. During the first week of September 1930, an unnamed hurricane spawned floods and mudslides that killed nearly 8,000 people in the Dominican Republic. Between September 12 and 16, 1928, 1,530 men, women, and children in GUADELOUPE and PUERTO RICO were slain when the brutal SAN FELIPE HURRICANE knifed through the east Caribbean basin, while an additional 3,369 people perished in the San Ciriaco Hurricane that pulverized the Puerto Rican capital of San Juan on August 8, 1899.

In the Far East, where centuries of intense typhoons have mercilessly lashed the volcanic shores of the Philippines, JAPAN, and TAIWAN, death tolls from tropical cyclones have been of equal size and severity. Between 1934 and 1961, 11 Japanese typhoons killed an estimated 19,591 people, one of the highest quarter-century totals in cyclonic history. In the Philippines, where the archipelago's reputation for being the planet's most storm-ridden country is somewhat ameliorated by the fact that a majority of Philippine typhoons are *bean,* or small rain typhoons, 184 typhoons between 1948 and 1994 claimed the lives of 9,104 people. In November of 1991, Tropical Storm Thelma swept across the Philippines, claiming as many as 6,000 lives, and making it one of the deadliest tropical systems in Philippine history. On September 17, 1828, the most deadly recorded typhoon in Japanese history, the Nagasaki Typhoon, claimed approximately 14,429 lives along the shores of the Ariake Sea. In 1281, the legendary Hakata Bay typhoon reportedly drowned between 45,000 and 65,000 Mongolian and Korean soldiers as they gathered their war junks to launch Kublai Khan's amphibious invasion of central Japan. Although few contemporary sources survive to corroborate the toll, Japanese legend holds that the typhoon's timely winds—now honored in the language as *kamikaze,* or the "divine winds"—completely destroyed some 2,200 ships, thereby saving Japan from a long, bloody period of repulse. In the case of the Hakata Bay typhoon, however, the divergent number of available "accurate" death tolls has really become secondary to the coincidental timing of the storm as its principal cultural identifier. While the Japanese do recognize the Hakata Bay typhoon's considerable death toll as one of the storm's more salient cultural aspects, its essential allure remains with the massacre's serendipitous nature, with the way it came ashore so completely and so quickly as to leave not one junk afloat or one invader alive.

As the victims of the Hakata Bay typhoon so tragically discovered, there are countless ways to die in a tropical cyclone. While in the past, coastal drownings have accounted for some 80 percent of all deaths due to tropical cyclones, improvements in evacuation procedures have in recent years generally reduced this number. They have not, however, been able to eradicate completely the fires, electrocutions, building collapses, crazed ANIMALS, mudslides, flash floods, decapitations, and impalements that in a tropical cyclone routinely claim their gruesome share of victims. Statistics released in the wake of Hurricane Andrew reveal that of the 23 deaths directly attributable to the storm, two were due to the severe head trauma associated with flying objects, while another two victims were washed off anchored boats and drowned. One 47-year-old man was killed when a pine tree squashed his camper near the Everglades National Park, while a truck trailer that was being used as a shelter by 13 warehouse workers in Homestead was overturned by Andrew's 130-MPH (209-km/h) winds. Two men were killed; another five were seriously injured.

On September 2, 1935, during the height of the extraordinary LABOR DAY HURRICANE, a government road worker who had sought shelter behind a pine tree was impaled through the lower abdomen by the wind-thrown spike of a two-by-four. Lingering until dazed rescuers found him the next morning, he is reported to have requested that two beers be brought to him before the stake was pulled out. Gratefully drinking the beers, he nodded once to the expectant rescuers. Without further ceremony, they gave the stake one sharp tug and the man slumped down, instantly dead.

Such a poignant litany of death has led sympathetic cultures in nearly every storm-prone region around the world to devise laws and other social mechanisms through which the sacrifice of human lives to these destructive phenomena can be lessened—or at least better understood. In Japan, where death rates from typhoons have historically been quite severe, a stringent series of building codes has served to make the island nation's millions of houses more storm resistant, better able to substitute as storm shelters for the families huddled within them. In the United States, where nearly 200 years of constant hurricane activity has left a conservative estimate of 42,198 people dead, public awareness programs and elaborate civil defense procedures developed during the 20th century have thus far significantly curtailed any major escalation in this figure. In 1900, for instance, between 8,000 and 12,000 people perished in the GREAT GALVESTON HURRICANE of September 8. Ninety-four years later, Hurricane Andrew, a Categery 4 hurricane of even greater intensity than the Galveston storm, directly claimed only 23 lives as it pummeled south Florida and Louisiana on August 24 and 26, 1994.

While for decades the storm surge existed as the primary killing mechanism in landfalling tropical cyclones, the recent history of tropical storm and hurricane activity in the United States has evidenced a change in that paradigm. When Tropical Storm FLOYD scudded up the eastern seaboard between September 16 and 19, 1999, its vast PRECIPITATION releases caused widespread inland flooding that killed 35 people in NORTH CAROLINA, three in VIRGINIA, two in Delaware, six in Pennsylvania, six in New Jersey, two in NEW YORK State, and one person each in Connecticut and Vermont. At the time, it was the deadliest tropical cyclone to have affected the United States since Hurricane AGNES in June 1972—itself a system whose inland floods killed a total of 72 people in Pennsylvania and New York State alone.

The catastrophic death toll incurred during Hurricane KATRINA's clawing assault on the central Gulf coast of the United States in August of 2005 graphically illustrates the dangers that ancillary and tertiary hazards can present during periods of hurricane activity. Although Katrina was at one point an exceptionally powerful hurricane, it was not at landfall as intense a system as either the LABOR DAY HURRICANE of 1935, or the legendary Hurricane CAMILLE in 1969—both of which, (while more intense at landfall than Katrina), claimed far fewer lives. In Louisiana's example, Katrina's preliminary toll of 1,577 people included many indirect deaths attributed to inadequate medical care, incomplete EVACUATIONS, various traffic and electrocution accidents, and drowning. At least 50 of the deaths tallied in Louisiana's accounting actually died in places far removed from Katrina's landfall location, such as the Texas cities of Houston and San Antonio to which they had been evacuated. Additionally, hundreds of people perished when much of the City of New Orleans was flooded following the failure of several dykes along the shores of Lake Pontchartrain and the surging Mississippi River. In neighboring MISSISSIPPI, another 238 lives were lost, a majority of them in the Gulf-side counties of Harrison and Hancock, where Katrina's 27-foot (8-m) surge caused widespread destruction. All told, Hurricane Katrina claimed 1,836 lives in seven states, making it the deadliest U.S. hurricane strike since the SAN FELIPE II (Lake Okeechobee) Hurricane in August of 1928. With more than 700 people still listed as missing more than two years after the disaster, it is possible that Katrina will eventually become the deadliest American hurricane since the 1900 Galveston strike.

Likewise, in the cyclone-susceptible countries of INDIA and BANGLADESH, improvements in communication and road systems are for the first time making it possible to conduct effective mass evacuations of the densely populated villages and barrier islands that have traditionally borne the brunt of both nation's most deadly storms. Considered among the highest in the world, the death tolls from any number of BAY OF BENGAL cyclones range from between 10,000 and

375,000 dead. While such counts may seem enormous by Western standards, they are often deemed "inconsequential" by those governments faced with the difficult problem of population control.

Despite the sometimes appalling death tolls that accompany tropical cyclone activity, emergency managers and other public safety officials in the United States and overseas continue to make substantive progress in mitigating the potential death tolls from severe tropical weather. Improved evacuation plans and procedures are continually compiled and implemented, and enhanced communications and logistics placement speeds response and recovery operations. As evidenced by Katrina, however, increased commercial and residential development along the world's coastlines and inland areas, along with public apathy and ignorance about the many dangers associated with tropical cyclone activity, significantly increases the risk that large death tolls may result from future tropical cyclone strikes.

Debbie, Hurricane *North Atlantic Ocean, August 11–21, 1969* See PROJECT STORMFURY.

Debby, Tropical Storm *North Atlantic Ocean, September 13–September 20, 1982* Between 1982 and 2006, five North Atlantic tropical cyclones were identified with the name Debby. Of these, three were classified as hurricanes, and two as TROPICAL STORMS. The name Debby has been retained on the list of North Atlantic TROPICAL CYCLONE identifiers and is scheduled to reappear in 2012.

The first of these systems, which existed from September 13 to September 20, 1982, eventually reached Category 4 status while recurving to the northeast, well into the NORTH ATLANTIC, on September 18, 1982. Originating directly off the coast of the Leeward Islands and PUERTO RICO on September 13, and delivering tropical storm force winds to the island of BERMUDA on September 16, Hurricane Debby's peak intensity of 28.05 inches (950 mb) produced 132-MPH (213-km/h) winds over the waters of the northeastern Atlantic Ocean, but no deaths or property damage. Hurricane Debby, after gradually weakening, transitioned to an extratropical system shortly before its dissipation over western EUROPE on September 20, 1982.

Debby, Hurricane *Gulf of Mexico–Mexico, August 31–September 8, 1988* The second North Atlantic tropical system identified as Debby existed between August 31 and September 8, 1988. Formed in the extreme eastern Bay of Campeche from a MESOSCALE CONVECTIVE COMPLEX (MCC) on the evening of August 31, the system slowly intensified as it drifted to the west-northwest, and toward the eastern coast of MEXICO. On the morning of September 2, the system was upgraded to TROPICAL STORM classification, and less than 12 hours later, to Category 1 intensity. Now less than six hours from landfall near Tuxpan, Mexico, Hurricane Debby reached its peak (observed) intensity of 29.14 inches (987 mb). Wind speeds of 75 MPH (120 km/h) uprooted trees, downed utility lines, and drove small watercraft ashore, while heavy rains caused localized flooding. First downgraded to a tropical storm, then to TROPICAL DEPRESSION intensity, Debby churned across Mexico, emerging into the eastern NORTH PACIFIC OCEAN during the early morning hours of September 5, 1988. Maintaining its tropical identity but not its name, Tropical Depression Debby was renamed TD 17-E; the system sharply recurved to the north-northwest on September 6, and delivered heavy rains as far north as the Baja Peninsula before dissipating. All told, at least 10 people were killed.

Debby, Tropical Storm *North Atlantic Ocean–Martinique, September 9–11, 1994* On September 9, 1994, Tropical Storm Debby—the third North Atlantic system to bear the name, and the shortest-lived of the five—originated from a tropical wave to the east of the island of BARBADOS. Beset by extremely unfavorable WIND SHEAR conditions, the system slowly intensified, reaching TROPICAL STORM status (with a CENTRAL BAROMETRIC PRESSURE of (29.82 inches [1,010 mb]) shortly before crossing the lower Leeward Islands during the late evening hours of September 10. A ferocious rainmaker, Tropical Storm Debby delivered more than seven inches (178 mm) of rain to MARTINIQUE and St. Lucia, damaging or destroying hundreds of BUILDINGS and decimating crops and foliage. Media reports state that 11 people were either killed or listed as missing, most of them on St. Lucia. Unable to improve its convective organization over the southeastern CARIBBEAN SEA, Tropical Storm Debby rapidly began to weaken, and by midnight on September 11, 1994, had dissipated. Some $230 million (1994 U.S. dollars) in property losses were tallied.

Debby, Tropical Storm *North Atlantic Ocean–Puerto Rico–Hispaniola, August 19–24, 2000* The fourth North Atlantic system dubbed Debby originated from a Cape Verde system on August 19, 2000. Upgraded to a TROPICAL DEPRESSION while located well to the east of the Windward and Leeward Islands, the system grew into a TROPICAL STORM during the early morning hours of August 20. Carried to the west-northwest by the trade winds, Tropical Storm Debby achieved Category 1 hurricane status on August 21, at the same time as it began a shallow RECURVATURE to the northwest. Skirting the Windward Islands, the VIRGIN ISLANDS, and PUERTO RICO (but dropping up to 12 inches [305 mm] in some places as it did so), Hurricane Debby's EYEWALL first continued to deepen (setting a minimum barometric pressure of (29.32 inches [993 mb] on August 22), then weakened as it approached the mountainous ranges of HISPANIOLA. On the afternoon of August 23, 2000, Hurricane Debby slid across the northern coastline of Hispaniola and promptly weakened, its convective strength sheared by Hispaniola's deeply furrowed terrain. Its convective cloud towers seemingly crumbling as it moved almost due west across northern HISPANIOLA, Tropical Storm Debby continued to weaken, and by the time it reached the Jamaica Channel and the eastern tip of CUBA, it had been downgraded to a tropical depression, with a central pressure of 29.85 inches (1,011 mb), and winds of 35 MPH (56 km/h). The system dissipated shortly thereafter. A relatively mild tropical cyclone, Debby caused some damage in the U.S. Virgin Islands, on Puerto Rico, in the Dominican Republic, and along the coast of eastern Cuba. In Puerto Rico, landslides destroyed bridges and roadway systems, while several small boats were swept ashore elsewhere in the Greater Antilles. Media reports attributed one indirect death to Debby in Puerto Rico and damage estimates totaling nearly $1 million (in 2000 U.S. dollars) in property losses.

Debby, Hurricane and Tropical Storm *North Atlantic Ocean, August 21–27, 2006* Tropical Storm Debby—the fifth Atlantic system to be so identified—originated off the southern CAPE VERDE ISLANDS on August 21, 2006. Moving to the west-northwest at nearly 16 MPH (26 km/h), Debby, while still a TROPICAL DEPRESSION (TD 4), brought rains squalls to the southern Cape Verde Islands and prompted the issuance of TROPICAL STORM WARNINGS for the first time since deadly Tropical Storm FRAN affected the islands in 1984. A large, but incomplete tropical system, Debby's convective organization improved during the late-afternoon hours of August 22, and the system was upgraded to a TROPICAL STORM. It achieved its minimum central pressure on August 24, with a central pressure of 29.52 inches (1,000 mb). Subsequent interaction with a warm, dry AIR MASS, along with a weakening convective infrastructure due to southerly WIND SHEAR, hindered Tropical Storm Debby's continued development, and by August 26, the system had been downgraded to a tropical depression. Now positioned well to the west of the Cape Verde Islands, Debby was absorbed into an approaching front on August 27, 2006. In its wake, no deaths or property losses were reported.

Debra, Hurricane *Gulf of Mexico–Southern United States, July 21–25, 1959* A weak Category 1 HURRICANE, Debra brought sustained 78-MPH (126-km/h) winds, 10-inch (254-mm) PRECIPITATION counts, and a minimal 2-foot (3-m) STORM SURGE to the central TEXAS gulf coast on July 25, 1959. Originating over southwestern GULF OF MEXICO on the afternoon of July 21, Debra haphazardly deepened while undergoing RECURVATURE to the north-northwest. Bearing a CENTRAL BAROMETRIC PRESSURE of 29.05 inches (984 mb) at landfall just north of Galveston, Debra downed power lines, bent traffic signs in U-turns, and damaged BUILDINGS in Texas City and Freeport but caused no fatalities. An early season rainmaker of the first order, Debra's localized flooding caused property losses of more than $2 million. The name Debra has been retired from the revolving list of North Atlantic HURRICANE NAMES.

Delia, Tropical Storm *Southern United States, August 29–September 4, 1973* See TEXAS.

Della, Typhoon *North Pacific Ocean–Japan, August 23–30, 1960* An extremely powerful midseason TYPHOON, Della brought sustained, 113-MPH (182-km/h) winds and eight-inch PRECIPITATION counts to the volcanic islands of southern JAPAN on the morning of August 29, 1960. Proving very destructive to hundreds of lightly constructed BUILDINGS across the Ryukyus Islands, Della claimed 41 lives and caused damage estimates of $19 million.

Delta, Tropical Storm *North Atlantic Ocean, November 27–29, 2005* The first North Atlantic TROPICAL CYCLONE to be identified as Delta, and the 25th named tropical system of the exceptionally active 2005 HURRICANE SEASON, Tropical Storm Delta originated from an extratropical low, some 1,000 miles (1,500 km) southwest of the Azores Islands, on November 19, 2005. A late-season system, Delta's extratropical core tepidly deepened until the evening of November 20, when it was upgraded to an extratropical storm with a central pressure of 29.61 inches (1,003 mb). By the late-afternoon hours of November 22, the extratropical storm had acquired tropical characteristics, was reclassified as a TROPICAL STORM, and was given the name Delta. Following an erratic course across the eastern NORTH ATLANTIC OCEAN that included a counterclockwise loop, a leg almost due southward, followed by a sharp recurvature that took Delta *eastward,* toward Africa, the system reached its minimum (observed) barometric pressure of 28.93 inches (980 mb) during the afternoon hours of November 24, 2005. At this time, Tropical Storm Delta was producing sustained wind speeds of 69 MPH (111 km/h). It thereafter weakened as it drew closer to northwestern Africa, losing its tropical characteristics on November 28, and transitioning back to a severe extratropical storm for the remainder of its existence. Between November 27 and November 28, Extratropical Storm Delta pummeled the Canary Islands with 95-MPH (153-km/h) winds, heavy rains, and large surf conditions. Although the system passed 100 miles (162 km) to the north of the Canaries, its effects were severe, and resulted in landslides and flash flooding on the islands of Tenerife and La Palma. According to media reports, Delta's grazing passage left seven people dead, and another 12 missing, in the Canary Islands. Property losses in excess of $360 million (in 2005 U.S. dollars) were recorded, primarily to crops and fishing stocks. Steadily weakening but nevertheless managing to maintain its extratropical circulation, Extratropical Storm Delta rolled ashore in northern Africa, in Morocco, on November 29, with a central pressure of 29.29 inches (992 mb) and sustained winds of 58 MPH (93 km/h). Before dissipating over inland Africa, the system delivered much needed PRECIPITATION to the area.

During the 1972 North Atlantic hurricane season, the NATIONAL HURRICANE CENTER (NHC) tracked Subtropical Storm Delta, a late-season subtropical system that had originated from an extratropical low situated over the mid-Atlantic. In existence November 1–7, Subtropical Storm Delta scribed a fairly unique track, passing through nearly 300 degrees of the compass before dissipating short of its origination point. One of the first times that the NHC had given names to North Atlantic subtropical cyclones, the practice has rarely been used since.

Dennis, Hurricane *North Atlantic–Caribbean Sea–Southern and Eastern United States, August 7–23, 1981* Between 1981 and 2005, five North Atlantic TROPICAL CYCLONES were identified as Dennis. Of these, two were TROPICAL STORMS, and three were HURRICANES. The fifth Dennis, which originated during the remarkable 2005 HURRICANE SEASON, became the most powerful tropical cyclone yet observed in the North Atlantic basin during the month of July. Its 14-day tantrum through the northern CARIBBEAN SEA islands and the southern United States was so severe that the identifier Dennis has been retired from the revolving list of North Atlantic tropical cyclone names. It has been replaced for 2011 by Don.

The first North Atlantic tropical cyclone dubbed Dennis formed as a TROPICAL DEPRESSION during the early morning hours of August 7, 1981. Then located near longitude 25 degrees West, this CAPE VERDE STORM tracked to the west-northwest, slowly gaining intensity as it did so. By the late-night hours of August 8, the system had matured to tropical storm intensity and within a day had generated its first minimum central pressure of 29.55 inches (1,001 mb) before commencing a prolonged weakening cycle. By midnight on August 11, while situated some 140 miles

(210 km) east of the Leeward Islands, Tropical Storm Dennis had been downgraded to Tropical Depression Dennis and was showing signs of further weakening. Later that day, the system crossed the Leeward Islands and entered the northeastern Caribbean Sea, where it rapidly lost its closed circulation and was downgraded to a TROPICAL WAVE. Maintaining its steady curl to the northwest, Tropical Wave Dennis brought heavy rains to JAMAICA on August 14 before sharply recurving to the northwest and traveling ashore along CUBA's southern coastline. Just prior to landfall in Cuba, Tropical Wave Dennis initiated a new strengthening configuration, one which saw it upgraded to Tropical Depression Dennis on August 15, and to Tropical Storm Dennis during the night of August 16, 1981. Interestingly, Tropical Storm Dennis continued to intensify as it moved across central Cuba and approached the Florida Keys, most likely because of warmer SEA-SURFACE TEMPERATURES (SSTs) in the Florida Straits. Bearing a central pressure of 29.47 inches (998 mb) and sustained winds of 40 MPH (65 km/h), Tropical Storm Dennis came ashore on the southern tip of FLORIDA—near Dade County—during the early morning hours of August 17, 1981. Gorging forth enormous quantities of precipitation, the system crawled due north, and had traversed nearly the entire length of the Florida peninsula before reemerging into the Atlantic Ocean on August 19. Held firmly against the eastern seaboard of the United States by a strong high-pressure ridge to the east, Tropical Storm Dennis recurved along the SOUTH CAROLINA–NORTH CAROLINA coastlines, quickly passing directly over the Outer Banks before recurving to the northeast and away from the Carolina coastline on August 20. Almost immediately upon reaching the warmer waters of the Gulf Stream, Tropical Storm Dennis intensified into a Category 1 hurricane, with a central pressure of 29.38 inches (995 mb), and sustained winds of 81 MPH (130 km/h). Within six hours, however, increasing vertical WIND SHEAR, coupled with cooler SSTS, reduced Hurricane Dennis to tropical storm status; by midnight on August 22, while positioned several hundred miles to the north of BERMUDA, the system underwent extratropical transitioning and shortly thereafter, dissipated. Like all tropical storms a robust rainmaker, Hurricane Dennis (while still of tropical storm intensity) delivered some 25.26 inches (635 mm) of PRECIPITATION to southern Florida, and another 10 inches (254 mm) to the Carolinas. Several million dollars in property losses were incurred, primarily in Florida.

Dennis, Tropical Storm *North Atlantic Ocean, September 8–20, 1987* The second tropical system identified as Dennis originated on September 8, 1987, just a few hundred miles off the western coast of Africa. Forming as a TROPICAL DEPRESSION very close to 10 degrees North, the system first tracked to the west, then to the northwest, before being upgraded to a TROPICAL STORM during the mid-morning hours of September 10. Maintaining its northwesterly trajectory, Tropical Storm Dennis spent the next eight days traversing the mid-NORTH ATLANTIC, reaching its peak intensity of 29.52 inches (1,000 mb) during the early evening hours of September 11. On September 18, after completing a sharp RECURVATURE to the north-northwest while situated well to the east to the Greater Antilles, Tropical Storm Dennis encountered hampering WIND SHEAR from the semi-permanent ridge of the BERMUDA HIGH, and was downgraded to Tropical Depression Dennis. On September 20, having recurved eastward into the Atlantic, Tropical Depression Dennis underwent extratropical transitioning, and within 24 hours, had been absorbed into an approaching frontal system. A long-distance traveler among North Atlantic tropical cyclones, Tropical Storm Dennis remained an offshore system and caused no deaths or property losses.

Dennis, Tropical Storm *North Atlantic Ocean, August 23–28, 1993* Tropical Storm Dennis—the third North Atlantic system to bear the name—originated on August 23, 1993, from a TROPICAL WAVE in southeastern NORTH ATLANTIC OCEAN. An offshore system for its entire existence, it reached TROPICAL STORM intensity during the mid-morning hours of August 24, and its minimum CENTRAL BAROMETRIC PRESSURE of 29.52 inches (1,000 mb) the following day. Traveling along a north-northwesterly trajectory, Tropical Storm Dennis nearly reached latitude 25 degrees North before sharply recurving to the west, and weakening to a TROPICAL DEPRESSION. By August 28, the system had dissipated, causing no deaths or property losses.

Dennis, Hurricane *North Atlantic Ocean–Bahamas–Eastern United States, August 24–September 8, 1999* On August 24, 1999, the fourth Dennis—this one destined to become a HURRICANE—formed from a TROPICAL WAVE just to the northeast of the Leeward Islands. Slowly and gracefully recurving to the northnorthwest and trailing along the coastlines of the BAHAMAS, the TROPICAL DEPRESSION was upgraded to Tropical Storm Dennis during the late morning hours of August 24, and a Category 1 hurricane, with a central pressure of 29.38 inches (995 mb) on August 26. By the afternoon of August 28, as the system approached the northern Bahamas islands of Eleuthera and Abaco, it was upgraded to Category 2 status, with a central pressure of 28.73 inches (973 mb) and sustained winds of 98 MPH (158 km/h). Some

4.4 inches (102 mm) of PRECIPITATION was recorded on Abaco Island. Abruptly pulled away from the eastern seaboard by an encroaching low pressure trough between August 29 and August 30, Hurricane Dennis slowed its forward motion and drew vast quantities of moisture from the warm waters of the Gulf Stream; on the morning of August 30, the system reached its peak intensity of 28.40 inches (962 mb). Sustained winds of 104 MPH (167 km/h) lashed the Atlantic waters, feeding additional energy into the system. Passing off the Outer Banks of NORTH CAROLINA on August 30, Hurricane Dennis began to weaken as it interacted with the BERMUDA HIGH—an interaction that sent the system on a triangular trajectory due north, then across its previous track to due south, then back to the west-northwest, where it finally came ashore in North Carolina as a powerful TROPICAL STORM on the night of September 5, 1999. A central pressure of 29.08 inches (985 mb) caused 58-MPH (93-km/h) winds across sections of coastal North Carolina. Wrightsville Beach, North Carolina, observed an extreme gust of 100 MPH (161 km/h), while precipitation counts of between 10 inches (254 mm) and 19 inches (483 mm) were observed across much of eastern North Carolina. Dennis was noted for its extensive shoreline erosion, and the destruction of Highway 12 hindered recovery operations for residents on Hatteras and Ocracoke Islands. Later that day, Tropical Storm Dennis became Tropical Depression Dennis while positioned over central North Carolina. After a jog in its trajectory to the west, the tropical depression sharply turned northward and passed through VIRGINIA (where an F-2 TORNADO caused structural damage in Hampton, and 13.82 inches [330 mm] of precipitation fell in Allisonia); Maryland, and Pennsylvania, then transitioned into an extratropical depression as it crossed NEW YORK State on September 8. Skirting the shores of Lake Ontario, Extratropical Depression Dennis drifted into CANADA and dissipated. A once potentially dangerous tropical system, Dennis's tropical storm force landfall indirectly killed three people (in FLORIDA and North Carolina), and caused some $150 million (in 1999 U.S. dollars) in property losses.

Dennis, Hurricane *North Atlantic Ocean–Jamaica–Hispaniola–Cuba–Southern United States, July 4–18, 2005* The fifth and final North Atlantic TROPICAL CYCLONE identified as Dennis was not only the most severe of the five, but was also one of the most notable storms yet observed. The fourth named tropical system of the extraordinary 2005 North Atlantic HURRICANE SEASON, Dennis was an earlyseason contender, forming from a TROPICAL WAVE during the afternoon hours of July 4 in the extreme southeastern CARIBBEAN SEA. Scudding to the westnorthwest as a TROPICAL DEPRESSION, the as yet unnamed Dennis dropped heavy rains on the Dutch protectorates of Aruba and Curaçao on July 5, before strengthening to TROPICAL STORM intensity during the afternoon hours of July 6. Gradually recurving to the north-northwest, Tropical Storm Dennis bore down on the islands of JAMAICA, HISPANIOLA, and CUBA at 14 MPH (23 km/h)—and swiftly intensified in seeming anticipation of its eventual landfall passage through them. At midnight on July 7, Tropical Storm Dennis was upgraded to a Category 1 hurricane, with a central pressure of 28.99 inches (982 mb) and estimated wind speeds of 81 MPH (130 km/h); within 12 hours it was upgraded to a Category 2 system, with a central pressure of 28.55 inches (967 mb) and 104-MPH (167-km/h) winds. By the time the NATIONAL HURRICANE CENTER (NHC) issued its 1800Z bulletins on September 7, Hurricane Dennis had deepened to a Category 3 hurricane, with a central pressure of 28.14 inches (953 mb) and estimated wind speeds of 115 MPH (185 km/h). Shaving along Jamaica's northern coast in a shower of rain on July 7, Hurricane Dennis indirectly killed one person on the island, and 56 others in Haiti, where strident winds destroyed dozens of residences, and demolished a large bridge with numerous fatalities.

Maintaining its RECURVATURE to the north-northwest, Hurricane Dennis entered the sheltered waters of the Jamaica Channel and continued its harrowing strengthening pattern. At midnight on July 8, Dennis first reached Category 4 intensity, with a central pressure of 28.08 inches (951 mb) and sustained winds in the EYEWALL of 150 MPH (241 km/h). At this time, the system was crossing the southernmost tip of Cuba, near Punta del Inglés, with 140-MPH (225-km/h) winds. This interaction weakened Dennis's convective continuity, and the system was downgraded to Category 3 status (central pressure: 28.34 inches [960 mb]) during the early morning hours of July 8. Within 12 hours, however, Dennis's eyewall had cleared the Cuban coastline and promptly deepened at an astonishing rate over the semi-enclosed waters of the Gulf of Guacanayabo. By noon on the same day, Hurricane Dennis's CENTRAL BAROMETRIC PRESSURE had dropped to 27.69 inches (938 mb), hiking its sustained wind speeds to 150 MPH (241 km/h). Just five MPH (8 km/h) short of Category 5 status, Hurricane Dennis relentlessly churned to the northwest, finally coming ashore in south-central Cuba, near the town of Punta Mangles Altos, during the evening hours of July 6, 2005. Sustained winds of 138 MPH (222 km/h), and gusts as high as 149 MPH (240 km/h), caused widespread damage in northern and central Cuba; more than 10,000 houses were destroyed and another 120,000 damaged, and much of the island's utilities systems were inoperable. Sixteen people in Cuba were killed, and damage esti-

mates topped $4 billion (in 2005 U.S. dollars). By all accounts, Hurricane Dennis was the most destructive tropical cyclone to have struck Cuba since Hurricane FLORA in 1963.

Weakened to Category 1 status by its passage over Cuba, Hurricane Dennis entered the GULF OF MEXICO during the early morning hours of July 9. Armed with a central pressure of 28.73 inches (973 mb) and 86-MPH (139-km/h) winds, the system arced into the Gulf of Mexico, its parabolic recurvature (which had barely shifted since it began more than 1,000 miles [1,200 km] to the south), setting it on course for a landfall on upper Gulf coast. A tenacious tropical cyclone, Dennis again began to intensify, again becoming a Category 4 hurricane during the daylight hours of September 10, with a central pressure reading of 27.46 inches (930 mb), its lowest pressure yet. At this time, the EYE of Hurricane Dennis was less than 300 miles (500 km) south of Pensacola, FLORIDA, and still moving to the northwest. Huge, 40-foot (13-m) swells radiated from Dennis's eyewall, causing the world's then largest deep-sea oil drilling platform, the *Thunder Horse*, to list more than 20 degrees. Costing more than $1 billion (in 2005 U.S. dollars) to construct, the Thunder Horse, although badly damaged, was later salvaged. Weakening slightly as it neared the Florida Panhandle, a menacing Hurricane Dennis charged ashore near Navarre Beach, Florida, during the early morning hours of July 10, 2005. A Category 3 system at landfall, Dennis's central pressure of 27.81 inches (942 mb) produced eye wall wind speeds of 127 MPH (204 km/h), and an 11-foot (3.5-m) STORM SURGE in Wakulla County, Florida. Some six inches (152 mm) of precipitation fell across portions of ALABAMA, Florida, and GEORGIA, as a now downgraded Tropical Storm Dennis swirled across MISSISSIPPI and western Tennessee. More than 1 million customers were without power in Florida, Alabama and Mississippi, and five people in Florida and Georgia were indirectly killed by Dennis. According to media reports, the American Red Cross was providing shelter to 50,000 people in 158 temporary locations. Downgraded to a tropical depression on July 12, Dennis's remnant low executed a clockwise loop over the Ohio River Valley between July 13 and July 15, dropping several inches of rain and causing some flash flooding. Maintaining its recurvature to the northeast, Tropical Depression Dennis crossed the Great Lakes on September 17, and within a day had dissipated over central CANADA. One of the most destructive tropical cyclones in recent memory, Dennis killed a total of 89 people—15 of them in the United States—and caused more than $2.5 billion (2005 U.S. dollars) in American property losses.

Diana, Hurricane *Eastern United States, September 7–13, 1984* The first major HURRICANE to have drifted ashore in the United States during the unusually active 1984 HURRICANE SEASON, Diana unfurled its erratic, 135-MPH (215-km/h) winds, 6-inch (152-mm) downpours, and breaking, 15-foot seas across the sandy Outer Banks of NORTH and SOUTH CAROLINA between September 11 and 13, 1984. In what would later prove to be a banner year for TROPICAL CYCLONE generation–thirteen tropical systems were formed, four of them achieving hurricane intensity—Diana would, however, ultimately shine as a powerful yet unrealized meteorological star.

At first touted by the NATIONAL HURRICANE CENTER (NHC) as one of the most dangerous hurricanes to have threatened the United States during the second half of the 20th century, Diana's alarmingly low CENTRAL BAROMETRIC PRESSURE of 28.02 inches (949 mb) sharply rose as the hurricane edged to within 30 miles (48 km) of North Carolina's Cape Fear on the evening of September 11, thereby sparing coastal communities the brunt of its former fury. Alerted to the hurricane's potentially destructive approach by an anxious round of HURRICANE WATCHES and HURRICANE WARNINGS, hundreds of thousands of coastal residents had earlier boarded up their summer cottages, securely moored their watercraft, and then headed their packed vehicles inland for the safety of hotels and shelters. For nearly three days they were compelled to await patiently Diana's whimsical pleasure, helplessly standing by in frustration as the hurricane first neared the central seaboard on September 11 and then abruptly peeled to the northeast on the morning of September 12.

By early evening of the same day, Diana—now downgraded from Category 3 to Category 2 on the SAFFIR-SIMPSON SCALE—completed a tight clockwise loop east and then headed due west, its 106-MPH (171-km/h) winds and flooding rains bound for a midnight landfall on the border between North and South Carolina. Pressing into the highlands of central North Carolina, Diana swiftly weakened and was demoted to a TROPICAL STORM by dawn on September 13. Two people in North Carolina died during the hurricane's passage, while damage estimates to houses, roads, boats, and trees topped $36 million. The name Diana has been temporarily retired from the cyclical list of HURRICANE NAMES.

Diane, Hurricane *Eastern United States, August 15–19, 1955* The first "billion-dollar hurricane" in U.S. history, Diane killed 191 people, displaced some 35,000 families, and wrought a record-breaking $3.25 billion in property damage as it arced across the United States's eastern seaboard between August 18 and 19, 1955. The second North Atlantic HURRICANE to strike the northeastern United States in less than a week, Diane's sustained 82-MPH (132-km/h) winds were of firm Category 1 status as the hurricane accelerated toward the SOUTH CAROLINA coastline but immediately lost their ominous intensity upon landfall. Suddenly encountering waters left three to five degrees cooler by Hurricane CONNIE's August 13 passage into NORTH CAROLINA, Diane's CENTRAL BAROMETRIC PRESSURE rapidly began to rise, and the system's winds subsided to a gusty but inconsequential 40 MPH (64 km/h) as it first trundled ashore near Myrtle Beach, South Carolina. The hurricane's phenomenal 12.5–19.76 inch (304 mm–483 mm) rainfalls did, however, cause deadly flood emergencies in ten states, including VIRGINIA, Pennsylvania, NEW YORK, New Jersey, and the southern and central NEW ENGLAND states of Rhode Island, Connecticut, and Massachusetts.

The fourth named TROPICAL CYCLONE of the busy 1955 HURRICANE SEASON, Diane originated as a TROPICAL DISTURBANCE 262 miles (422 km) southeast of BERMUDA on the morning of August 15. Moving on a northwesterly course at 11 MPH (18 km/h), the disturbance swiftly intensified, becoming a full-fledged TROPICAL DEPRESSION on the morning of August 16 and a 40 MPH (64 km/h) TROPICAL STORM by late afternoon on the same day. On August 17, as its central barometric pressure skidded to 29.14 inches (987 mb), the storm was upgraded to hurricane status. Diane's elliptical EYE, then located 450 miles (724 km) east of Charleston, South Carolina, continued churning in a north-northwesterly direction at nearly 16 MPH (26 km/h), a course that kept its 88-MPH (142-km/h) EYEWALL winds well off the southeast shores of Florida but posed a grave threat to the 600-mile (966-km) stretch of coastline between Savannah, GEORGIA, and Norfolk, Virginia. Reconnaissance flights, conducted every hour now that Diane was approaching the mainland, indicated that the hurricane's central pressure had further dropped to 28.98 inches (981 mb), a development that hinted at further intensification while simultaneously promising 95-MPH (153-km/h) gusts and a four-to-five-foot (1–2 m) STORM SURGE at landfall. On the evening of August 17, at a time when most families were sitting down to supper, the first of Diane's evacuation warnings were issued for those low-lying coastal communities that flanked either side of the North and South Carolina border. Early computer ANALOGS had predicted that Diane would most likely come ashore in South, rather than North, Carolina, thereby sparing the Hatteras Capes a repeat of Connie's devastating visit just five days before.

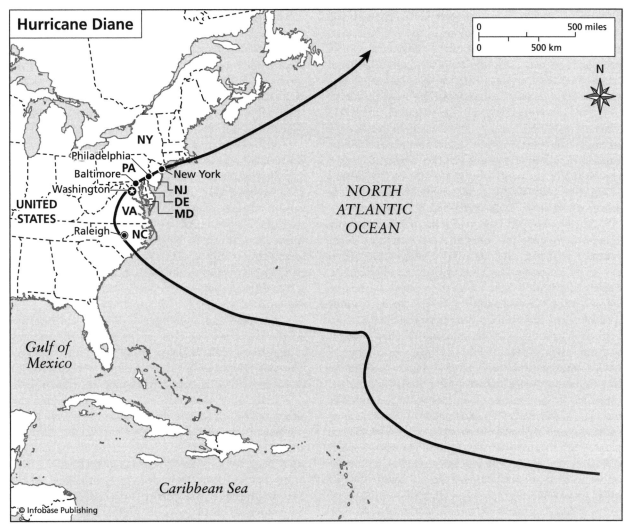

Hurricane Diane

NORTH
ATLANTIC
OCEAN

Gulf of
Mexico

Caribbean Sea

© Infobase Publishing

Weakened by its passage across the track taken by earlier Hurricane Connie, Hurricane Diane brought tremendous rains to the central section of the U.S. eastern seaboard in 1955.

On the afternoon of August 18, 1955, Hurricane Diane embarked on a parabolic trajectory through the inland heart of the northeast United States when it came ashore 12 miles (19 km) southeast of Myrtle Beach, South Carolina. Severely weakened by its last-minute passage over cooler waters, Diane made landfall as a minimum tropical storm, one with wind gusts of 55 MPH (89 km/h). Deeply laden with enormous quantities of PRECIPITATION, Diane huffed through North Carolina, central Virginia, eastern Pennsylvania, and New Jersey before reemerging into the North Atlantic at the western tip of New York's Long Island. In South Carolina, where Diane made the first of its two landfalls, damage was on the whole considered minimal. Aside from a half-dozen blown electrical transformers and a few downed trees, Diane's assault on the Myrtle Beach area was not as severe as it would

have been had the storm not held back its precipitation. Quickly carrying these into central Virginia, Diane unleashed inches of rain that spawned flash floods and mudslides. Three people in Virginia were killed, while an estimated $80 million in damage to property and agriculture was assessed.

In Pennsylvania, which Diane transited shortly before midnight on August 18, the picturesque valleys of the Pocono Mountains were transformed into deadly spillways as Diane's 13.79 inches (330 mm) of rain caused the widespread flooding of rivers and streams. In the mountain resort town of Stroudsburg, 47 vacationers at Camp Davis huddled in their 14 wooden cabins, wondering when the rain's incessant drumming on the roof, on the windows, on the screen doors that opened onto quaint porches, was going to end. Anxiously hovering over the glow of their kerosene hurricane lamps, the 47 men, women, and children told stories, quoted the Scriptures, and complained about how the rain had ruined their vacation, never suspecting for a moment that in the distance along the fern-carpeted banks of Brodhead Creek, a 30-foot (10-m) piston of water surged toward them along the riverbed. The wave cascaded into Stroudsburg, uprooting ancient pine trees and scarring boulders before obliterating Camp Davis. Sweeping the matchbox cottages right off their stone foundations, the seething torrent carried them down the mountainside, smashing them into pulpwood along the way. Thirty-eight people were killed in what would prove Diane's most tragic incident.

After leaving 113 people dead in Pennsylvania, Diane completed its hemispherical track by turning due east and passing over New Jersey during the early morning hours of August 18. Delivering eight inches (203 mm) of rain to Jersey City and its neighboring communities, Diane would eventually kill six people and destroy or damage some $5 million worth of property in the state before exiting into New York. Steadily curling to the northeast, Diane's disintegrating eyewall passed over the length of Long Island, whipping the foliage from trees while gouging the beaches with huge washouts. In New York City, torrential rains and intermittent gusts shivered the hardwood colonnades of Central Park but otherwise caused no damage. In the days following Diane's passage, insured damage estimates in New York State steadily climbed to $15 million.

Tracing along New England's south shores, Diane's gale-force winds dropped 12.5 inches (305 mm) of rain on Hartford, Connecticut, causing the Naugatuck, Still, and Mad rivers to spill over their banks. In Rhode Island, lingering floods not only caused historic cemeteries in Woonsocket to burp up their coffins but also left 6,000 workers idle after completely swamping their wire and rubber manufacturing plants. In the nearby town of Putnam, Connecticut, Diane's seemingly unfathomable depths first set fire to, then collapsed, a magnesium processing facility belonging to the Metal Sellers Corporation. As the first light of dawn flared across the darkened, submerged environs of the town, hundreds of barrels of burning magnesium drifted through the streets, staging a perverse pyrotechnic celebration every time one of them exploded. Fifty people in Connecticut and Rhode Island were killed, with some $1.1 billion in damage sustained by thousands of residences and businesses.

In Massachusetts, where Diane's record, 20-inch (508-mm) rainfall counts caused extensive structural and crop damage from Worcester to Cape Cod, the famed Revolutionary-era wooden bridge that spanned the Concord River was sadly swept away, reduced to splinters by the raging waters. In the central Massachusetts mill city of Worcester, an already depressed local economy was further weakened by the idling of some 10,000 garment workers as Diane's rising flood waters turned blocks of shoe mills into abandoned brick islands. Along the outer reaches of Cape Cod, Diane's fading eye brought 35-MPH (56-km/h) winds, five-inch precipitation counts, and spectacular surf to Martha's Vineyard and Nantucket Island on the afternoon of August 19, causing power failures and some injury to trees but no deaths. Elsewhere in the state, 13 people were killed, a majority of them drowning victims. More than $1 billion in estimated damages were incurred, prompting the Massachusetts Legislature to float a $55 million bond issue to assist in paying for the reconstruction of public water and highway systems.

Appalled at the severity of Diane's toll, dozens of world organizations from the federal government to the Russian Red Cross contributed to a number of relief funds being established for survivors. One day after, then President Dwight D. Eisenhower declared all of Connecticut and Pennsylvania and portions of New England federal disaster areas, the Australian government donated $50,000 to the fund. In an attempt to make it easier for Britons to contribute, Great Britain announced that it was temporarily lifting its currency restrictions, while the former president of the Dominican Republic, Rafael Trujillo, contributed $100,000 to the pool. Interrupting a family vacation in Colorado to inspect reconstruction efforts in the afflicted areas personally, President Eisenhower decided to award an additional $71 million in defense contracts to the northeast region. They were, he stated in his request for the appropriation, to be used as an added relief measure.

Diane's destructive legacy of 191 dead and $3.25 billion in insured and uninsured losses was sufficiently important enough to justify the striking of the name Diane from the rotating list of HURRICANE NAMES.

Dinah, Typhoon *Taiwan, June 9–18, 1965* *See* TAIWAN.

doldrums *See* INTERTROPICAL CONVERGENCE ZONE.

Dolly, Hurricane *North Atlantic Ocean–Europe, September 8–17, 1953* Between 1953 and 2002, six North Atlantic TROPICAL CYCLONES have been identified as Dolly. Besides the names FLORENCE and IRENE, Dolly is the oldest name in continued usage in the North Atlantic tropical theater, appearing in 1953 as the first "D" name on the list. Of the six systems, four were HURRICANES and two were TROPICAL STORMS. As yet, the deadliest and most destructive of the North Atlantic Dollys occurred during the 1996 season. Seemingly unwilling to let the tropical cyclone parade pass it by, the name Dolly is scheduled to reappear in its seventh incarnation during the 2008 North Atlantic season.

The first North Atlantic tropical cyclone given a "D" identifier under the newly implemented tropical cyclone naming system, Dolly—eventually to become a hurricane—originated on September 8, 1953, off the northern coast of PUERTO RICO. First traveling to the southwest, the system (upgraded to a tropical storm), sharply recurved just north of HISPANIOLA, and swung to the north-northwest. During the midday hours of September 9, while passing along the eastern coastline of the BAHAMAS, Tropical Storm Dolly was upgraded to hurricane status. Maintaining its arcing dance to the north-northwest, Hurricane Dolly achieved Category 2 status during the early evening hours of September 9, and within another 12 hours, its peak, Category 3 intensity of 29.38 inches (995 mb) on September 10. Pushed to the northeast by an encroaching high pressure AIR MASS over the southern United States, Hurricane Dolly curled into the NORTH ATLANTIC OCEAN, weakening along the way. By the early evening hours of September 11, as Dolly crossed latitude 35 degrees North, it weakened to a Category 1 hurricane with estimated wind speeds of 86 MPH (139 km/h). Now tracking to the east-northeast, its forward speed quickly increasing, Hurricane Dolly faded to tropical storm intensity during the midnight hours on September 12, and later that day underwent extratropical transitioning. Moving almost due east along the "corridor" between the 40th and 45th parallels, Extratropical Storm Dolly spent the next four days gradually weakening, eventually becoming an extratropical depression on September 17, 1953, shortly before dissipating just off the Iberian peninsula's Atlantic coast.

Dolly, Hurricane *North Atlantic Ocean, August 31–September 4, 1954* Eventually following a north-northwesterly/north-northeasterly trajectory very similar to 1953's Hurricane Dolly, the second North Atlantic system so identified proved mildly intense, but short-lived. Originating on August 31 and dissipating by September 4, 1954, Hurricane Dolly reached its peak intensity during the midday hours of September 1. The system was then located off the northern BAHAMAS, and was producing 98-MPH (158-km/h) winds. Recurving into the colder waters off CANADA's Maritime Provinces and gaining

forward speed, Hurricane Dolly underwent extratropical deepening during the early evening on September 2, but maintained its hurricane force winds for another six hours. Progressively weakened by its passage over colder SEA-SURFACE TEMPERATURES (SSTs), Extratropical Storm Dolly had dissipated by noon on September 4, 1954.

Dolly, Hurricane *Southern United States, August 10–17, 1968* The third North Atlantic system dubbed Dolly formed over the southwestern BAHAMAS August 10, 1968. As a strengthening TROPICAL DEPRESSION, the system recurved over southeastern FLORIDA, bringing 29 MPH (46 km/h) to the Sunshine State before entering the NORTH ATLANTIC OCEAN just to the south of Jacksonville. Over the next two days, the tropical depression gradually intensified under the shearing hindrance of an approaching cold front, finally reaching TROPICAL STORM intensity during the midday hours of August 12. By the following morning, while Tropical Storm Dolly was marching far into the northeastern Atlantic, it was upgraded to a Category 1 hurricane, with an estimated CENTRAL BAROMETRIC PRESSURE of 29.35 inches (994 mb) and sustained 75-MPH (120-km/h) winds. Thereafter weakening, Hurricane Dolly was demoted to a tropical storm on August 14, but by the 15th had reintensified to hurricane strength. Dolly remained a Category 1 hurricane until the noon hours of August 16, 1968, when its circulation crumbled and it was downgraded to a tropical depression. By August 17, it had dissipated.

Dolly, Tropical Storm *North Atlantic Ocean, September 2–5, 1974* An offshore tropical cyclone for its brief lifespan, the fourth North Atlantic tropical cyclone identified as Dolly originated several hundred miles southwest of BERMUDA on September 2, 1974. Recurving to the north-northeast, it was, like its 1954 predecessor, a short-lived system that reached TROPICAL STORM intensity on September 3, and by September 5 had undergone extratropical transitioning.

Dolly, Hurricane *Caribbean Sea–Mexico, August 19–25, 1996* One of five tropical systems to form over the CARIBBEAN SEA basin during the 1996 North Atlantic HURRICANE SEASON, Hurricane Dolly—the fifth incarnation—was also to prove the deadliest Dolly yet. Formed on August 19, 1996, off the southwestern coast of JAMAICA, the system was upgraded to Tropical Storm Dolly later the same day. Firmly following a west-northwesterly track toward MEXICO, Dolly was upgraded to Category 1 intensity just as it made its first landfall in Mexico—near Chetumal on the Yucatán Peninsula—during the early evening hours of August

20. A central pressure at landfall of 29.50 inches (999 mb), 75-MPH (120-km/h) winds, and more than five inches (127 mm) of precipitation caused widespread destruction to small BUILDINGS and foliage in Quintana Roo province. Weakening as it twirled across the Yucatán, Dolly had been downgraded to a TROPICAL DEPRESSION by the time it reached the Bay of Campeche on August 21. Carried to the northwest, Dolly rapidly intensified, becoming a TROPICAL STORM on August 22, and a Category 1 hurricane (with a central pressure of 29.20 inches [989 mb]) during the midday hours of August 23. It was during this second peak period that Hurricane Dolly made its second landfall, along Mexico's eastern coastline. The port cities of Tampico and Veracruz were deluged with as much as six inches (142.4 mm) of PRECIPITATION, and 81-MPH (130-km/h) winds caused considerable damage to buildings, signs, and electrical services. Now moving due west, Dolly was downgraded to a tropical storm during the early evening hours of August 23, and to a tropical depression by midnight of the 24th. A tremendous rainmaker, Dolly dropped up to 13 inches (330 mm) of precipitation over central Mexico, causing widespread flash flooding. Before emerging into the eastern NORTH ATLANTIC as a dissipating tropical depression on August 25, 1996, Dolly levied a hefty toll on Mexico—14 lives lost and tens of millions of dollars in property losses. A large tropical cyclone, Dolly delivered up to six inches (142.4 mm) of rain to the southeastern TEXAS cities of Corpus Christi and Brownsville.

Dolly, Tropical Storm *North Atlantic Ocean, August 29–September 4, 2002* Like its 1974 forerunner, the sixth TROPICAL CYCLONE named Dolly—this one a TROPICAL STORM—remained over the waters of the open Atlantic for its entire existence. Formed on August 29, 2002, near 10 degrees North, 30 degrees West, the system hastened to the west-northwest, becoming a tropical storm during the early evening hours of the 29th, and achieving its peak intensity of 29.44 inches (997 mb) the following day. A fairly robust tropical storm for much of its life, Tropical Storm Dolly finally dissipated on September 4, 2002, some 750 miles (1,000 km) east of the Windward Islands.

Dominica *Eastern Caribbean Sea* This small, mountainous island—famous for the mists that frequently shroud its rugged inland peaks—has during the past two centuries withstood a long, often ravaging history of TROPICAL CYCLONE activity. Situated in the eastern CARIBBEAN SEA, Dominica (pronounced Dom-in-*ee*-ca), is the largest of the five major islands that comprise the extensive Windward Islands group.

The archipelago's location between latitudes 12 and 15 degrees North positions it directly in the path of the mid- to late-season TROPICAL STORMS and HURRICANES that swirl across the sun-charged expanses of the NORTH ATLANTIC OCEAN from Africa's Cape Verde Islands. Between 1766 and 2005, Dominica's elfin forests, scenic sulfur springs, steaming freshwater lakes, and hospitable capital city of Roseau were directly affected by 46 recorded tropical cyclones. Of these, 22 were considered major hurricanes, systems with CENTRAL BAROMETRIC PRESSURES of 28.47 inches (964 mb) or lower, sustained winds in excess of 111 MPH (179 km/h), and torrential, six–12-inch (132-305 mm) PRECIPITATION counts. On at least seven different occasions between 1787 and 1968, the island endured several hurricane strikes in rapid succession, as witnessed in 1787 when three destructive storms trampled ashore from August 3 to August 29; in 1806 when two hurricanes bounced across the island on September 9 and September 24; and in 1816 when another two tempests made landfall between September 15 and October 16. Among the most violent of Dominican tropical cyclones, the three August hurricanes of 1787 cumulatively sank three British slave traders, capsized five schooners, and wrecked two rum-laden brigs at Roseau.

Because a majority of early season tropical systems form over the west Caribbean Sea, there is no record of Dominica being struck by a tropical storm or hurricane during the month of June. In July, however, the probability of a hurricane landfall in Dominica sharply increases, with six reported strikes between July 26, 1769, and July 20, 1987. Judged to have been the worst July hurricane in Dominica's history, the 1769 tempest uprooted most of the island's flowering trees, collapsed scores of single-story houses, and drove 15 merchant ships ashore at Roseau and Prince Rupert Bay. During the 1813 HURRICANE SEASON, Dominica was lashed by two July hurricanes, both fairly large and powerful and both racing ashore within one week's time.

It is in August, when the first of the CAPE VERDE STORMS begins their RECURVATURE into the eastern Caribbean, that the number of tropical cyclone strikes in Dominica escalates—to 14 documented storms between 1772 and 2006. Notable August hurricanes in Dominica include the fantastic whirlwind of August 30, 1772, that wrecked 18 British merchant ships at Roseau, undermined several of the island's stone fortresses (casting their heavy iron cannon into the raging surf), and claimed dozens of lives. Not two decades later, on August 1, 1792, an equally strident tropical system completely enveloped the 305-square-mile (491-square-km) island in its EYEWALL, mowing down acres of palm groves, demolishing houses, and destroying 14 British trading vessels that had sought safe anchorage on Dominica's west coast. On the same date in 1809, another fearsome hurricane scudded across the southern half of the island, causing severe structural damage to most of the principal BUILDINGS in Roseau, as well as sinking two vessels in the harbor. In what would become this particular hurricane's single worst incident, the 18-gun British sloop *Lark*, was blown ashore near Point Palenqua and wrecked. Three members of its crew were thrown into the sea during rescue operations and tragically drowned. Another hurricane, this one on August 11, swept over the island in 1830, snapping trees at ground level, peeling the tiled roofs from houses, and throwing 15-foot seas ashore from Roseau in the south to Cape Melville in the north. The latter half of the 20th century in Dominica was marked by two lethal August hurricanes, DAVID and ALLEN. On the morning of August 30, 1979, Hurricane David's 135-MPH (217-km/h) winds prostrated much of the island's lush rain forests, while blinding downpours ruined vital agricultural stocks. Some 37 lives were lost. Less than a year later, on August 4, 1980, an even more intense Hurricane Allen delivered 135-MPH (217-km/h) winds and an 11-foot (4-m) STORM SURGE to central Dominica, uprooting thousands of trees, collapsing hundreds of lightly constructed houses, shattering windows, severing power lines, and destroying much of Dominica's major airport, including its air-traffic-control tower. Twelve islanders perished, while millions of dollars of property losses were tallied. Between August 25 and 26, 1995, Tropical Storm IRIS blossomed near the island, producing a pressure reading of 29.47 inches (998 mb) and wind speeds in excess of 58 MPH (93 km/h). While Iris went on to become a powerful hurricane elsewhere in the CARIBBEAN, no deaths, injuries, or major property losses were reported in Dominica.

September is the peak month for hurricane activity in Dominica, with at least 22 of them having made landfall on the island between 1776 and 2005. Often systems of substantial size and memorable destructive consequence, these midseason furies generally follow the trade winds west across the Atlantic before blazing into Dominica with roaring winds, pelting rains, and eroding seas. On the evening of September 6, 1776, a powerful hurricane trailed its considerable rains north along Dominica's west coast, triggering terrifying mudslides that obliterated numerous hillside villages. Two heavily laden British merchant ships were beached near Prince Rupert Bay, although their crew and cargo were eventually saved. Slightly more than 30 years later, an even greater September hurricane, this one occurring on September 9, 1806, whipped across south Dominica. A true rainmaker, the 1806 hurricane's DANGEROUS SEMICIRCLE released enormous precipitation counts over Roseau, forcing the Roseau River to top its crude earthen embankments swiftly. Much of the capital was subsequently submerged, its narrow streets choked with mud, splintered trees, and debris from fallen houses. Said to have been the worst hurricane to have affected the island since 1772, the 1806 strike left at least 131 people dead and dozens of others missing. In the sober words of a contemporary newspaper account of the tragedy, "the whole island offers a scene of devastation and ruin."

In the years 1833 and 1834, Dominica was pummeled by two horrific hurricanes, both of them occurring on September 20. The first, which spent almost two full days terrorizing the northern half of the island with blistering winds and rocking seas, claimed some 40 victims, most of them sailors on three merchant ships that went down off Cape Melville. The second hurricane, which stricken survivors unanimously stated was the worst tropical cyclone to have riddled the island up to that time, flattened nearly every tree; chiseled wide furrows through fields and crops; demolished 600 houses; collapsed churches, hospitals, and warehouses; sank 21 ships; and killed a total of 239 people. A durable system of extreme meteorological violence, the 1834 hurricane would later go on to decimate large portions of Hispaniola, where it killed several hundred people alone in the thriving trading hub of Santo Domingo. Dominica would suffer further strikes from two other September hurricanes, these coming ashore on September 6, 1865, and September 11, 1887. Of the two, only the 1865 hurricane was judged to have been severe on the island, with numerous houses gutted and several ships either dismasted, wrecked, or run aground. Although it did not make a direct landfall on Dominica, Hurricane Marilyn caused some damage as it passed close to the island on September 15, 1995. At the time a Category 1 hurricane, Marilyn's central pressure of 29.06 inches (984 mb) generated 92-MPH (148-km/h) winds at its EYEWALL, but these were largely of tropical storm intensity on Dominica, where small buildings were unroofed, power lines downed, and trees uprooted. No deaths were reported, however. On September 7, 1996, a strengthening Tropical Storm Hortense gusted to the north of the island, producing a pressure reading of 29.64 inches (1,004 mb) and 60-MPH (97-km/h) winds.

Statistically, the late-season months of October and November are not a particularly dangerous time for hurricanes in Dominica. Since 1766, only two hurricanes have come ashore on the island in October (October 7, 1766, and October 11, 1780), while none have made landfall in November. Both October hurricanes, however, were systems of considerable severity, with the second, known to posterity as the GREAT HURRICANE OF 1780, causing widescale damage to property and significant loss of life.

Dominican Republic *See* HISPANIOLA.

Donna, Hurricane *Eastern and Northern Caribbean–Southern and Eastern United States, September 1–13, 1960* Regarded by the United States Weather Service as one of the 15 most intense recorded HURRICANES to have passed through the CARIBBEAN SEA and the eastern NORTH ATLANTIC OCEAN during the 20th century, Donna delivered sustained 140–150-MPH (225–241-km/h) winds, five to 11-inch (127–279 mm) rainfalls, and four to 11-foot (1–4 m) STORM TIDES to the islands of ANTIGUA, PUERTO RICO, HISPANIOLA, and CUBA between September 4 and 9, 1960. In the United States, where Donna initially came ashore on the FLORIDA Keys on September 10, no less than 12 states, including GEORGIA, NORTH and SOUTH CAROLINA, VIRGINIA, NEW YORK, Connecticut, Rhode Island, and Maine, were directly affected by the hurricane's powerful, Category 2–3 winds and torrential rains. Some 118 people in the Caribbean and another 50 in the United States lost their lives to Donna, while an estimated $1.8 billion in property damages were assessed.

One of the greatest CAPE VERDE STORMS on record, Donna originated over the mid-Atlantic's temperate waters 581 miles east of the Leeward Islands during the early morning hours of September 1, 1960. Carried west at speeds of up to 10 MPH (16 km/h), Donna, as with most hurricanes of the Cape Verde variety, deepened very quickly. An aerial reconnaissance flight conducted at noon on September 2 found that during the night Donna had intensified to TROPICAL STORM status, boosting wind speeds within its EYEWALL to a strident 57 MPH (92 km/h). Another, conducted at 7 P.M. the same evening, discovered that Donna was rapidly approaching Category 1 status, or a hurricane with wind speeds of between 74 and 95 MPH (119–153 km/h).

As Tropical Storm Donna's CENTRAL BAROMETRIC PRESSURE continued to drop, slipping from 28.97 inches (981 mb) at 3 P.M. to 28.87 inches (978 mb) just after midnight on September 3, the system climbed the SAFFIR-SIMPSON SCALE of destructive potential, alarming weary meteorologists, civil defense personnel, and property owners along the eastern rim of the Caribbean basin. Although many of them understood that it was not unusual for TROPICAL CYCLONES to experience rapid fluctuations in barometric pressure, Donna's voracious promotions—from Category 2 on

the morning of September 3 to Category 3 by midafternoon of the same day—had police and evacuation officials in the vulnerable Leeward Island nations of Antigua, Barbuda, St. Kitts, Nevis, and the VIRGIN ISLANDS, scurrying through the streets, issuing urgent warnings for citizens to evacuate their coastal homes and businesses.

Then in their infancy, statistical ANALOGS based on the behavior of documented storms had predicted that Donna's steady RECURVATURE northwest would most likely take the hurricane over the Leeward Islands sometime during the day of September 4. HURRICANE WARNINGS advised that Donna, now touting a harrowing central pressure of 27.9 inches (947 mb), had sustained winds of between 125 and 130 MPH (201–209 km/h) and gusts of between 145 and 150 MPH (233–241 km/h), making it one of the most dangerous hurricanes to have threatened the Leeward Islands in nearly a quarter of a century.

Familiar with the archipelago's extensive history of violent hurricane strikes, thousands of inhabitants rushed to secure houses, boats, mainstay vacation resorts, and sugarcane plantations against Donna's promised fury. Wooden shutters closed over tall colonial windows as shaded verandas were cleared of their airy wicker furniture. Laundry was taken in, while frail outbuildings were anchored to the ground with ropes and stakes. Emergency service crews, tasked with repairing damaged power lines and water mains, were placed on standby. In the harbors and inlets of Antigua, northern GUADELOUPE, Anguilla, and St. Kitts, all small watercraft from sport diving vessels to fishing boats were tightly warped alongside their piers. Those vessels with individual, deepwater moorings prepared to ride out the storm by deploying additional bow chains and rigging life lines. Sea anchors, designed to keep a boat's head into the waves, splashed over the side. On board the larger vessels, watch schedules were set, generally through the drawing of straws. By midnight of September 3, as Donna's advancing seas raised a heavy surf along the Leeward's rocky eastern coves, the stalwart islands stood ready.

On the morning of September 4, 1960, Hurricane Donna, with a slightly weakened central pressure of 28.11 inches (952 mb), roared ashore in Antigua. Crossing the island at nearly 12 MPH (19 km/h), Donna's 115 MPH (169 km/h) sustained winds and torrential five-inch (127-mm) rains mowed down palm trees, uprooted telephone poles, wrenched storm shutters from their hinges, stripped shingles and tiles from roofs, and blew down chimneys. On the island's southeast coast, where Donna's nine-foot (3-m) STORM SURGE inundated the black sand beaches of Half Moon and Willoughby bays, 17 small boats were rendered total losses after they broke away from their anchorages and were driven onto rocks. Three mariners drowned. Passing directly over the Antiguan capital of St. Johns as it exited the island, Donna dropped eight-inch (203-mm) rainfalls on the city's narrow streets, spawning serious flash floods. Almost 40 percent of St. John's commercial property suffered some degree of structural damage from broken windows to collapsed roofs, later necessitating a massive reconstruction effort. On the neighboring islands of Monsterrat, Barbuda, and Guadeloupe, Donna's fringe rains and gale-force gusts forced the closure of all airports and other transportation facilities but otherwise caused little damage. One woman in Barbuda was killed, however, while investigating for damage, when she fell off the roof of her house and broke her neck.

Firmly maintaining its swing to the northwest, Donna blasted into the Caribbean Sea, stirring up tons of spray as it spiraled toward the nearby volcanic sentinels of St. Kitts and Nevis. In St. Kitts, where a steadfast drizzle of rain had for hours announced Donna's imminent arrival, evacuation and other safety concerns had resulted in a dark, motionless landscape of shuttered towns and abandoned seaside villages. Earlier crowded with hundreds of refugees, the island's mountain roads were now virtually empty, ready to become deadly spillways for Donna's downhill rains and inevitable floods. Sequestered in their houses, storm cellars, and mass shelters, the people of St. Kitts closely monitored their radios, tracking Donna's progress as though it was the legendary hurricane of September 1, 1667, and another 700 lives were hanging in the balance.

Under citrine-colored skies, Hurricane Donna brushed across the northeast tip of St. Kitts on the afternoon of September 3, 1960. One-hundred-ten-mile-per-hour (177-km/h) winds plowed deep furrows in the island's expansive sugarcane fields and heavily forested hillsides. Seven-inch (178-mm) PRECIPITATION counts turned dozens of roads into riverbeds, while raging floods carried away a stone bridge near Rawlins Plantation. Although no lives were lost to Donna's passage over St. Kitts, several million dollars worth of agricultural losses were reported.

After lashing the northern Virgin Islands with 116-MPH (187-km/h) winds and heavy rains just before midnight of September 4, Donna spent the next five days pummeling the north coasts of Puerto Rico, Hispaniola, and Cuba. Still committed to its north-northwesterly trajectory, Donna sailed past Puerto Rico during the early morning hours of September 5. Although the island was spared a direct strike by Donna's still rising (28.29 inches) central pressure and commensurate 105-MPH (169-

In recent years, inland flooding has become one of the leading causes of death in tropical cyclones. In this photograph taken on the flooded streets Queens, New York, a team of rescuers performs lifesaving operations on a casualty of Hurricane Donna's destructive passage up the eastern seaboard in September 1960. *(NOAA)*

km/h) winds, the hurricane's 12-foot (4-m) storm tides and nine-inch downpours initiated flood and mudslide emergencies that claimed the lives of 102 people. Six hundred others were injured, many of them seriously. In neighboring Hispaniola, both the Dominican Republic and Haiti were besieged by hurricane-force winds and rains during the late-afternoon and evening hours of September 5. A mudslide just outside the Haitian capital of Port-au-Prince claimed three lives as it swept seven hillside houses into the ravine below and then buried them. Lingering rains caused by Donna's abrupt course change to the west on September 6 hampered rescue and repair operations in the Haitian capital for nearly 24 hours, leaving many parts of the nation without electricity.

Curving slightly southwest, Donna spent the better part of September 6 and 7 reintensifying. As the hurricane drew ever closer to Cuba's northeast coast, its central barometric pressure teetered between 27.7 inches (940 mb) on the 6th and 27.8 inches (944 mb) by noontime on the 7th, elevating it to the vaunted status of a Category 4 hurricane. Now firmly recognized as one of the most dangerous storms of recent times, Donna's snaking trail of destruction over the southern BAHAMAS on September 8 and 9 prompted hurricane contingency plans to be hurriedly activated in both Cuba and southern Florida. As word filtered out that Donna's central pressure was still continuing to drop (it would eventually reach 27.6 inches [936 mb] by the afternoon of September 9), hundreds of thousands of people, many of them summer tourists on the Florida Keys, began their slow, anxious retreat to higher ground.

In communist-controlled Cuba, where it was initially believed that the hurricane would come ashore on September 9, evacuation instructions were haphazardly broadcast over the state-owned radio network, creating waves of confusion and panic in oceanside cities and towns across the island. Even though the U.S. embargo against Cuba did not extend to the trading of weather intelligence, the Cuban government distrusted U.S. weather advisories, considering them vehicles of economic conspiracy. Because in any nation the cost of mass evacuating several hundred thousand people is often quite considerable, the use of false hurricane alerts by an opposing government would be an effective way to disrupt local economies—and, in Cuba's case, to embarrass the Castro regime financially. In response, U.S. hurricane bulletins were duly received by Havana, patiently scrutinized, compared with information gleaned by Cuba's own meteorological services, and then heavily edited before being broadcast to the population as a revolutionary call to arms. On the whole it was an unwise policy, one that would later significantly contribute to the even more costly devastation wrought by 1963's Hurricane FLORA.

Much to the well-publicized delight of the Cuban government, Hurricane Donna did not come ashore in that country on September 9, but instead spent the day roaring north toward the delicate coral trail of the Florida Keys. Now located 279 miles (449 km) southeast of Key West, Donna's furious EYE contained barometric pressures of 27.6 inches (936 mb), classifying the hurricane as one of moderate Category 4 status. Truly alarmed by the meteorological spectacle now bearing down on them, many Floridians packed their cars, trailered their boats, and abandoned the shallow keys before Donna's predicted 15-foot (5-m) storm surge washed out the Overseas Highway. Others remained behind, seeking shelter in whatever brick or concrete building they could find.

They would need it: When Donna finally crossed the central Florida Keys shortly after three A.M. on September 10, 1960, it was a mature hurricane of exceptional intensity. Drawing upon a central barometric pressure at landfall of 27.9 inches (948 mb), Donna's churning 150-MPH (241-km/h) winds and explosive 180-MPH (290-km/h) gusts ravaged Vaca, Long, Matecumbe, and Plantation keys, utterly destroying almost every structure in the towns of Marathon and Tavernier. Entire spans of the old Florida Keys Highway 3 were swept away as the sea, whipped into a blinding lather, undermined their concrete piers.

Inexorably arcing to the northeast, Donna brought 180-MPH (290-km/h) gusts to Key Largo and 12-inch (305-mm) rains to Miami as it thundered onto the peninsula shortly before dawn on September 10. Coming ashore just south of Naples, Donna's four to 11 foot (1–4 m) storm tides severely battered a number of western Florida's resort islands, entirely submerging many of them. Sarasota and Collier counties sustained more than $26 million in damages as Donna's 115-MPH (185-km/h) winds buffeted posh marinas, bayside restaurants, and hundreds of residential communities. Over the lush green enclave of the Everglades National Park, Donna's slowly abating vortex nevertheless managed to snap off, or uproot, half of the preserve's treasured mangrove trees. Sadly surveying the damage, park officials later discovered that nearly 40 percent of the state's white heron population had also perished, making Donna one of the deadliest ecological disasters in Florida's history.

Sharply continuing its recurvature northeast, Donna passed directly through the peninsula's center, bringing 46-MPH (74-km/h) winds with hurricane-force gusts to the citrus groves of Orlando just before midnight on September 10. Reconnaissance missions had earlier indicated that Donna's central pressure had quickly begun to rise shortly after its landfall, increasing from 27.75 inches (940 mb) on the morning of September 10 to 28.58 inches (968 mb) on the following morning before settling back to 28.29 inches (958 mb) on reentering the Atlantic later that afternoon near Daytona Beach. In its wake, millions of dollars in damages to the state's grapefruit, orange, and tangerine harvests were recorded, portending a meager winter for citrus growers and consumers alike.

Not surprisingly, Donna briefly reintensified after returning to the temperate Atlantic, seeing the rapid drop in its central pressure once again boot it from a Category 2 to a Category 3 hurricane on the evening of September 11. With its eye positioned well offshore, Donna shot to the northeast, traveling in a straight line up the U.S. eastern seaboard at speeds of up to 25 MPH (40 km/h) before coming ashore between Wilmington and Morehead City, North Carolina. Coastal sections of Georgia and South Carolina received heavy rains and gales throughout most of the day and night of September 11, killing three people in each state. Another six were killed in North Carolina when Donna's 97-MPH (156-km/h) winds and five-foot (2-m) storm surge ground across the Outer Banks, destroying dozens of cottages and causing significant beach erosion along Pamlico Sound.

Again positioned over the Atlantic, a weakening Donna's still-treacherous eyewall continued to skirt north-eastward, bringing three to four-inch (76–102-mm) rains and gusty winds to Virginia, Delaware, New Jersey, and New York's Long Island (where an

11-foot (4-m) storm tide in New York Harbor caused extensive pier damage); then Donna made its ninth, and final, landfall near New London, Connecticut, on the morning of September 12, 1960. With a central pressure of 28.55 inches (967 mb) and 95-MPH (153-km/h) winds, Donna's Category 2 trek through Connecticut, Rhode Island, eastern Massachusetts, southeastern New Hampshire, and central Maine left seven people dead, thousands of trees uprooted, and tens of millions of dollars damage to valuable apple and potato crops. Crossing into eastern CANADA on the morning of September 13, Donna died in a hail of gusty rains over the St. Lawrence River.

Truly a marathon hurricane by any standard, Donna's 5,000-mile (8,047-km) excursion through eight countries and 13 states left 168 people—50 of them Americans—dead, some 37,000 others homeless, and an estimated $1.8 billion in agricultural and property losses behind. The name Donna has been retired from the rotating list of HURRICANE NAMES.

Doppler radar   Developed in 1971 by the National Severe Storms Laboratory in Norman, Oklahoma, a division of the U.S. National Weather Service (NWS), Doppler radar observes meteorological activity in TROPICAL CYCLONES by bouncing radio waves off a system's spiral RAIN BANDS. A highly sophisticated technological achievement, Doppler radar allows meteorologists and scientists to study HURRICANES and TORNADOES with unprecedented scientific detail.

Developed by the University of Oklahoma, the National Severe Storms Laboratory and the National Center for Atmospheric Research in 1995, Doppler on Wheels (DOW) was used to study tornadoes in TEXAS and to survey Hurricane Fran as it trundled ashore in NORTH CAROLINA in September 1996. Anchored to the ground by hydraulic pins, the prototype DOW truck and its six-foot (2-m) dish was retired in late 1996 and replaced by DOW 2 and DOW 3, two 1/2-ton trucks fitted with upgraded transmitters and eight-foot (3-m) dishes.

Dora, Hurricane *Southeastern United States, August 28–September 16, 1964*   This moderately powerful Category 2 HURRICANE's sustained 125-MPH (201-km/h) winds, eight-inch (203-mm) PRECIPITATION counts, and five to eight-foot (2–3-m) STORM TIDES walloped large sections of northeastern FLORIDA and southern GEORGIA between September 9 and 13, 1964. Bearing a CENTRAL BAROMETRIC PRESSURE of 28.52 inches (957 mb) at landfall in St. Augustine, Florida, just after midnight on September 10, Dora damaged or destroyed more than 1,000 BUILDINGS, uprooted hundreds of oak and palm trees, mowed down telephone poles, smashed piers,

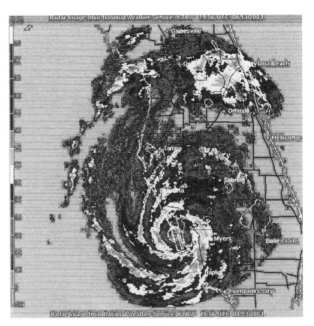

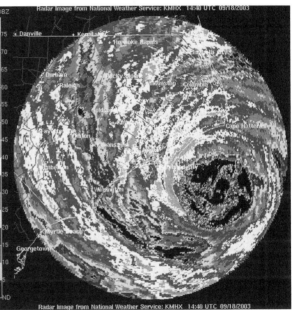

Before the development of meteorological satellites, radar was used as a primary tool in monitoring and studying the mechanics of tropical cyclones. In these two views, modern Doppler radar stations in Florida and North Carolina capture the rain bands associated with two landfalling tropical cyclones in 2001 and 2003, respectively. *(NOAA)*

and caused widespread coastal erosion as far north as Jacksonville. Said by contemporary newspaper accounts to have been the first mature-stage TROPICAL CYCLONE to have come ashore at St. Augustine during the 20th century, Dora claimed five lives in Florida and tallied some $200 million (in 1964 dollars) in property damage across the northern third of the state.

The third of five tropical cyclones to have affected Florida directly between June 7 and October 14 of the 1964 HURRICANE SEASON (an unnamed TROPICAL STORM and Hurricanes CLEO, HILDA, and Isbell were the others), Dora originated southeast of the Cape Verde Islands some 300 miles off the North African coast on the afternoon of August 28, 1964. At first Dora seemed a classic CAPE VERDE STORM in the making; however, its first three days over the evaporative feeding grounds of the NORTH ATLANTIC OCEAN were instead hampered by an unfavorable passage beneath that season's BERMUDA HIGH, a southwesterly TRACK that took the weak TROPICAL DEPRESSION to within 50 miles (80 km) of the 10th parallel on the morning of August 31. Pinned between the impenetrable high pressure of the ANTICYCLONE to the north and the dissipating latitudes below 10 degrees North (where the CORIOLIS EFFECT is not strong enough to set tropical cyclones spinning), the bands of thunderstorms that formed the tropical depression struggled to maintain their cohesiveness as they scudded west-southwest at speeds of between 11 and 16 MPH (18–26 km/h).

During the early morning hours of September 1, while positioned some 800 miles (1,287 km) east of BARBADOS Island, the nearly exhausted tropical depression suddenly began to undergo RECURVATURE, or a parabolic course change, to the north-northwest. Spiraling away from the 10th parallel at nearly 15 MPH (24 km/h), the tropical depression's splintering CUMULONIMBUS CLOUD banks quickly regenerated their convective cells. Shortly after dawn on September 1, as its sustained WIND SPEEDS topped 43 MPH (69 km/h), the tropical depression was upgraded to a tropical storm.

Given the name *Dora* by meteorologists at the NATIONAL HURRICANE CENTER (NHC), the system slowly continued to strengthen during the afternoon and evening hours of September 1. By the morning of September 2, as it crossed the 15th parallel some 400 miles (644 km) east of MARTINIQUE, Dora's maximum winds were clocked at 60 MPH (97 km/h), classifying it as a powerful tropical storm. Aerial reconnaissance flights conducted into its embryonic EYEWALL by the famed HURRICANE HUNTERS observed that Dora's central pressure had slipped to 29.20 inches (988 mb) and was showing signs of further deepening. From its orbit several hundred miles above Earth, the RADIOMETER aboard the TIROS VII weather satellite recorded above-normal SEA-SURFACE TEMPERATURES in Dora's path, while photographic images taken by the satellite's powerful camera and then transmitted to a receiving station in New Jersey were already indicating a perceptible tightening of the cloud ring at the tropical storm's center. Carefully

scrutinized by NHC meteorologists, the data conclusively pointed to Dora's eventual maturation into a full-fledged hurricane.

On the evening of September 2, as hurricane hunting aircraft noted sustained wind speeds of 76 MPH (122 km/h) within its thickening eyewall, Tropical Storm Dora was upgraded to a Category 1 hurricane. Now positioned 600 miles (966 km) east of PUERTO RICO, Dora's north-northwest progression was once again beginning to sputter, presenting the system's forecasters with yet another degree of uncertainty. Between midnight on September 2 and midnight on September 6, Dora's curl to the northwest took it across only 600 miles (966 km) of ocean for an average speed of six MPH (10 km/h). During the same period the hurricane had steadily continued to intensify, its central pressure dropping to 28.50 inches (965 mb) on the morning of September 4, to 28.02 inches (948 mb) just after dawn on September 5, and 27.90 inches (917 mb) on the morning of September 6. Virtually stalled over the subequatorial waters east of the BAHAMAS, Dora's spiral RAIN BANDS furiously fed on the high EVAPORATION rates generated there and were eventually able to develop ADVECTION winds that whipped the surface of the sea at speeds of 129 MPH (208 km/h).

Shortly after daybreak on September 7, Hurricane Dora–now producing gusts of 140 MPH (225 km/h)—fell under the influence of an encroaching high-pressure AIR MASS to the north. Unable to pursue its languid recurvature to the northwest, the exceedingly powerful Category 3 hurricane altered course west, initially set itself on a track that if maintained would take it ashore near the border between Florida and Georgia sometime in the next 72 to 96 hours. As HURRICANE WATCHES were hurriedly posted along the coasts of both states, Dora crept westward at speeds of between five and 12 MPH (8–18 km/h). A steady round of reconnaissance flights into the hurricane's narrowing EYE indicated that the system's central pressure was continuing its slide, eventually falling to its lowest recorded point—27.82 inches (942 mb)—just before dusk on September 7.

A Category 4 hurricane of significant intensity, Dora spent much of September 8 steadily advancing on Florida's Daytona Beach at nearly 11 MPH (18 km/h). As though unwilling to commit itself to a potentially fatal landfall on the peninsula, the hurricane once more slowed its forward speed during the evening hours of September 8. Slowly deteriorating as it pulled to within 300 miles (483 km) of Daytona Beach, Dora's eyewall abruptly commenced a course shift to the northwest, sparing Daytona a direct strike by its Category 3 fury, but in turn threatening the extensive waterfront installations of Jacksonville with their first hurricane landfall in nearly a century.

Crawling toward northeastern Florida at just under 10 MPH (16 km/h), Hurricane Dora progressively weakened throughout the daylight hours of September 9. Hourly reconnaissance missions revealed a gradual rise in Dora's central pressure from 27.99 inches (947 mb) during the evening of September 8 to 28.30 inches (958 mb) just after midday on September 9. Wind speeds, once sustained at 135 MPH (217 km/h), had decreased to 129 MPH (208 km/h). By the time Dora's Category 2 eyewall passed over the mouth of the St. Augustine Inlet shortly after midnight on September 10, its maximum wind speeds had fallen to between 110 and 125 MPH (177–201 km/h) with 115–120-MPH (185–193-km/h) gusts. While it is not uncommon for tropical cyclones to weaken upon approaching within 50 miles (80 km) of a coastline's various surface currents, Dora's stand-down was in large part hastened by the presence of upper-level WIND SHEAR, a condition brought on by the passage of the high-pressure ridge to the north.

St. Augustine, a city that had earlier in the season suffered localized flooding in the wake of Tropical Storm Cleo's 40-MPH (64-km/h) winds and heavy rains, endured the equally voracious brunt of Dora's historic incursion into northeast Florida. Pounded by eastward-moving hurricanes in October 1944 (GREAT ATLANTIC HURRICANE) and October 1950 (King), St. Augustine was subjected to more than 15 continuous hours of Dora's hurricane-force winds, rains, and seas. The beaches of Anastasia Island, a seven-mile-long (11 km) barrier island north of St. Augustine Inlet, shuddered beneath the crashing of 20–30-foot (10-m) waves, while 12-foot (4-m) storm tides carved more than 100 feet off the shoreline of St. Augustine Beach. Extensive flooding topped seawalls along the banks of the Halifax and San Sebastian rivers, submerging the streets and lower stories of entire neighborhoods. Parts of U.S. Highway A1A between Salt Run and St. Augustine Beach were completely washed out. A half-dozen wooden fishing piers were splintered as Dora's raging breakers collapsed a section of St. Augustine's famous boardwalk; a bayfront restaurant, its wooden pilings knocked down like ninepins, was nearly toppled into the maelstrom; entire groves of cedar trees were turned into kindling wood; miles of telephone and electrical wires were downed. While few of St. Augustine's colonial buildings sustained damage from Dora's assault, hundreds of newer structures crumpled beneath the hurricane's surf or lost their roofs, windows, and outbuildings to its 129-MPH (208-km/h) gusts. Hundreds of families in the greater St. Augustine area were driven from their houses.

On the morning of September 12, while positioned over extreme southwest Georgia, Dora sharply changed course and began to move east-northeast at almost 17 MPH (27 km/h). Downgraded to a tropical storm, Dora chugged northeast, bringing 50-MPH (81-km/h) winds and heavy rains to central Georgia, NORTH and SOUTH CAROLINA, and VIRGINIA before returning to North Atlantic waters on the evening of September 13. Developing into an EXTRATROPICAL CYCLONE, Dora's offshore remains swiftly sped northeast, bypassing NEW YORK's Long Island and the rocky coves of eastern NEW ENGLAND before sailing across Newfoundland, CANADA, during the afternoon of September 15. After delivering almost three inches (76 mm) of rain to the Canadian Maritimes, Dora's fading rain bands dissipated over the Arctic Ocean's frigid waters on September 16. All told, damage estimates from Dora's passage through Florida and Georgia and up the eastern seaboard totaled some $250 million in 1964 currency values.

The name Dora was added to the revolving list of North Atlantic HURRICANE NAMES in 1956, following the retirement of the name *DIANE* the year before. Although the name *Dora* was in turn replaced on the list by the name *DEBBIE* in 1965, it was later transferred to the eastern NORTH PACIFIC list where it was given to a tropical storm that existed between July 15 and 20, 1987.

dropsonde An updated technological cousin of the RADIOSONDE, a dropsonde is an instrument that simultaneously observes and transmits meteorological data concerning BAROMETRIC PRESSURE, air TEMPERATURE, and humidity levels as it is dropped

A crew member aboard an aircraft belonging to the Hurricane Hunters prepares for a dropsonde launch. As they fall through a tropical cyclone's rain bands, dropsondes collect detailed information about the system, thereby providing more data for forecasting and emergency management purposes. *(NOAA)*

from an aircraft into the spiral RAIN BANDS of a TROPICAL CYCLONE. Encased and fitted with a parachute and a radio transmitter, a dropsonde can slowly drift through a tropical cyclone's EYEWALL and gather information regarding a system's vital signs with pinpoint accuracy. On an average aerial reconnaissance mission, high-altitude HURRICANE HUNTING aircraft such as the GULFSTREAM IV-SP may deploy up to 100 dropsondes, while specially modified ORION WP-3 and HERCULES WC-130 aircraft flying at lower altitudes receive the dropsondes' transmitted data before relaying it to the NATIONAL HURRICANE CENTER (NHC) in Miami, FLORIDA.

Dvorak Technique Originally known as the "Dvorak Scale," and developed in 1974 by Vernon F. Dvorak, a meteorologist then assigned to the National Environmental Satellite, Data, and Information Service, the Dvorak Technique is a method of using imagery from meteorological SATELLITES (primarily polar-orbiting satellites with microwave technology) to systematically estimate the intensity of a TROPICAL CYCLONE. Employed by many of the world's primary meteorological forecasting operations (including the NATIONAL HURRICANE CENTER), the Dvorak Technique uses infrared and visible satellite imagery to determine cloud patterns in deepening and mature-stage tropical cyclones. When visible satellite images are used, their cloud patterns are compared to a set of predetermined templates, each of which has been extrapolated from many decades of meteorological observation. When infrared satellite imagery (which is based on heat pattern identification) is added, it is possible for meteorologists to use the Dvorak Technique to determine the difference between the temperature of the warm EYE, and the colder cloud tops of the surrounding EYEWALL. Once this has been completed, a T-number is issued and a Corresponding Intensity (CI) value determined.

The Dvorak Technique divides tropical cyclone organization into five base patterns, including the curved band pattern; the shear pattern; the central dense overcast pattern; the banding eye pattern; and the eye pattern. Because each of these base patterns reflect cloud distribution at different points during a tropical cyclone's life cycle, it is possible to assign quantitative values—ranging from 1.0 to 8.0—to each. For the curved band pattern, the T-numbers

range from 1.0 to 4.5; for the shear pattern, the T-numbers range from 1.5 to 3.5; the central dense overcast pattern carries T-numbers of between 2.5 and 5.0; the banding eye pattern has T-numbers of 4.0 to 4.5; while the eye pattern ranges from 4.5 to 8.0.

Since its development, the Dvorak Technique has been sufficiently refined and enhanced to allow meteorologists to accurately compare the T-numbers to a chart listing wind speed and CENTRAL BAROMETRIC PRESSURE (as based on sea level readings). For example, a tropical system with a T-number of 3.5 will generally have 63-MPH (102-km/h) winds, and a central pressure of 29.35 inches (994 mb). A system with a T-number of 5.5 will have 104-MPH (167-km/h) winds, and a central barometric pressure of 28.34 inches (960 mb); and a system with a T-number of 8.0 will have wind speeds of a whopping 196 MPH (315 km/h) and a central pressure of 26.28 inches (890 mb).

The Dvorak Technique is particularly appealing because it allows meteorologists to compile more accurate intensity forecasts for distant tropical cyclones without having to dispatch an expensive HURRICANE HUNTER aircraft on a time-consuming (and potentially dangerous) reconnaissance mission. The National Hurricane Center draws on a number of resources to provide it with Dvorak Technique intensity estimates, including its Tropical Analysis and Forecast Branch (TAFB), its Satellite Analysis Branch (SAB), and the U.S. Air Force Weather Agency (AFWA). In recent years, the Dvorak Technique's human interpretive element has been augmented by the use of computer ANALOGS; one such system, developed by the University of Wisconsin's Cooperative Institute for Meteorological Satellite Studies (CIMSS), is known as the Objective Dvorak Technique, and is primarily used to determine the CI, or the value that results when the T-number is compared to wind speed and barometric pressure.

For all of its usefulness, the Dvorak Technique does have, at present, some relatively minor limitations. Because it relies on cloud patterns to determine the T-numbers and CI, embryonic tropical features such as TROPICAL WAVES, TROPICAL DEPRESSIONS, and even some TROPICAL STORMS can be too indistinct for accurate forecasting. If in stronger systems the eye is obscured by cloud cover, the use of IR satellite technology is the only method for obtaining a T-number and the subsequent CI.

Earl, Hurricane *North Atlantic Ocean, September 4–11, 1980* Between 1980 and 2004, five TROPICAL CYCLONES in the North Atlantic basin have been identified with the name Earl. Of these, three eventually reached HURRICANE status while the remaining two were ranked as TROPICAL STORMS. Only one of the five made landfall—in FLORIDA in 1998—while the last observed system named Earl (in 2004) lingered long enough to reach the eastern NORTH PACIFIC OCEAN where it was given a new identifier, Frank. It was the first time since the 1996 North Atlantic hurricane season that a North Atlantic system (in that instance, Cesar) had retained sufficient tropical characteristics to allow regeneration over the world's largest ocean. The name Earl has been retained on the North Atlantic tropical cyclone naming lists and is scheduled to make its next appearance in 2010.

The first North Atlantic tropical cyclone identified as Earl originated in the far eastern Atlantic, relatively close to the eastern coast of Africa, on September 4, 1980. Of TROPICAL DEPRESSION intensity, the system began a recurvature to the north-northwest, and during the early morning hours of September 5, was upgraded to a tropical storm. Recurving well into the NORTH ATLANTIC, the system spent the next three days slowly intensifying before achieving Category 1 hurricane status on the morning of September 8. Sheared by its interaction with the semi-permanent anticyclone positioned over the mid-North Atlantic, Hurricane Earl reached its minimum intensity of 29.08 inches (985 mb) within six hours; thereafter, the system began a weakening trend. Now recurving to the north-northeast and with increasing forward speed, Hurricane Earl dropped to tropical storm intensity early on September 10, and by night-fall of the same day, had undergone extratropical deepening. The system dissipated on September 11, 1980.

Earl, Hurricane *North Atlantic Ocean, September 10–20, 1986* Like its predecessor, the second North Atlantic TROPICAL CYCLONE to be identified as Earl remained an off-shore system for its entire existence. It remains most notable for the unusual path it traced across the mid-North Atlantic—one that when viewed from above, looks not unlike an incomplete *m*. Originating just north of 20 degrees North on September 10, 1986, the system quickly intensified as it drew to the north-northwest, becoming a TROPICAL STORM during the early morning hours of September 11, and a Category 1 HURRICANE (with a central pressure of 29.50 inches [999 mb]) during the early evening hours of the same day. By noon on September 12, Hurricane Earl had deepened to a Category 2 hurricane, with a central pressure reading of 29.17 inches (988 mb), and had commenced a change in forward direction to north-northeast. Over the following two days, Hurricane Earl continued to strengthen, while simultaneously recurving to the south-southeast, as though about to trace a loop-the-loop trajectory. On September 14, Hurricane Earl reached its peak intensity of 28.90 inches (979 mb), and began a sharp recurvature to the north-northeast; its new trajectory nearly paralleled its course for the previous three days. Sheared by the same high-pressure ridge that had altered its course, Hurricane Earl weakened, its central barometric pressure rising to 29.02 inches (983 mb), on September 16. Once again set on a northeasterly track, Earl churned across the mid-North Atlantic, finally undergoing extratropical deepening on September 19,

1986. It remained a fairly intense extratropical storm (with a minimum central pressure of 29.29 inches [992 mb]) until September 20, when it dissipated over the extreme NORTH ATLANTIC OCEAN.

Earl, Tropical Storm *North Atlantic Ocean, September 26–October 3, 1992* Born on September 26, 1992, off the northeastern coast of the BAHAMAS, Tropical Storm Earl—the third North Atlantic tropical system to bear the name—eventually came very close to making landfall along the eastern seaboard of the United States as a TROPICAL DEPRESSION before sharply recurving into the NORTH ATLANTIC and passing into meteorological history well to the south of BERMUDA on October 3, 1992. Of tropical depression intensity for the first two days of its existence, the system first tracked to the southwest (and past the northern Bahamian islands) before recurving to the west-northwest and away from central FLORIDA on September 29. Upgraded to a TROPICAL STORM—and dubbed Earl in the process—on September 29, the system fell under the blocking influence of the BERMUDA HIGH and began a RECURVATURE to the east-southeast. Tropical Storm Earl achieved its lowest (observed) pressure reading during the early morning hours of October 1, with a central pressure of 29.41 inches (996 mb) and sustained winds of 58 MPH (93 km/h). Its development hampered by WIND SHEAR aloft and a diminishing CORIOLIS EFFECT, Tropical Storm Earl was downgraded to Tropical Depression Earl on October 3, 1992, and an extratropical depression later the same day. As with its two predecessors, Earl remained an offshore system and caused no death, injuries, or property losses.

Earl, Hurricane *Gulf of Mexico–Southern United States–North Atlantic Ocean, August 31–September 8, 1998* The fourth North Atlantic TROPICAL CYCLONE identified as Earl was also the first of them to make landfall—in this instance, along the northern coast of the GULF OF MEXICO. It is also notable for spending a majority of its existence as an extratropical storm that was considerably stronger than that observed while the system was a mature stage tropical cyclone. Born over the warm waters of the Bay of Campeche on August 31, 1998, the system quickly deepened into a TROPICAL STORM while recurving to the north-northeast. It fairly rapidly developed over the following two days, achieving Category 1 hurricane status during the daylight hours of September 2, with a CENTRAL BAROMETRIC PRESSURE reading of 29.35 inches (994 mb). Traveling to the northeast and bearing down on the FLORIDA Panhandle, Hurricane Earl continued to strengthen, reaching Category 2 intensity (with a central pressure of 29.17 inches [988 mb]) during the afternoon hours of September 2. Now

situated less than 100 miles (161 km) from Panama City, Florida, Earl's sustained wind speeds suddenly weakened while its central pressure continued to drop. At landfall near Panama City during the early morning hours of September 3, Earl's central pressure was observed at 29.08 inches (985 mb), while its sustained winds topped 81 MPH (130 km/h). This made it of Category 1 intensity at landfall. Before weakening first to a tropical storm, then to an extratropical storm, Hurricane Earl caused several million dollars in damage to communities in the Florida Panhandle. Panama City, Florida, received 12 inches (305 mm) of precipitation over a 48-hour period, while TORNADOES were reported in parts of northern Florida that destroyed and damaged scores of residential and commercial structures. Two people in Florida were drowned in offshore boating accidents attributed to Earl's landfall. Within hours of its landfall in Florida, Earl traveled across central GEORGIA, where it transitioned into an EXTRATROPICAL CYCLONE before trailing across the eastern Carolinas and southeastern VIRGINIA, and entering the NORTH ATLANTIC OCEAN. The system spawned tornadoes in Georgia, and several others in SOUTH CAROLINA, where one person was killed. Moving into the North Atlantic in a northeasterly direction, Extratropical Storm Earl's central pressure continued to drop, reaching its minimum intensity of 28.46 inches (964 mb) on September 6, 1998, just as the system buffeted the eastern coastline of Newfoundland, Canada, with heavy rains and 58-MPH (93-km/h) winds. All told, Earl claimed three lives in the United States, and caused damage tallies of nearly $80 million (in 1998 U.S. dollars).

Earl, Tropical Storm *North Atlantic Ocean–Windward Islands, August 13–15, 2004* A veritable pauper among the North Atlantic TROPICAL CYCLONES dubbed Earl, the fifth system to bear the name was also the shortest-lived, existing only from August 13 to August 15, 2004. Born in the southwestern Atlantic, to the south of 10 degrees North, the system was, owing to its location and unfavorable environment, never able to solidly organize. Bearing down on the Windward and Leeward Islands, a weak Tropical Depression No. 5 (TD 5) tamely intensified, becoming a minimal TROPICAL STORM (with a central pressure reading of 29.82 inches [1,010 mb]) on August 14. It achieved its lowest CENTRAL BAROMETRIC PRESSURE (29.79 inches [1,009 mb]) on the morning of August 15, while passing over the Leeward Islands of Grenada and St. Vincent and the Grenadines, where several lightly constructed buildings were damaged by Tropical Storm Earl's 52-MPH (83-km/h) winds. The system entered the eastern CARIBBEAN SEA, where, beset by WIND SHEAR, it degenerated into a TROPICAL WAVE during the early evening hours of August 15, 2004—at which point the NATIONAL HURRICANE CENTER (NHC) declared it a dissipated tropical system. Considering its tepid existence in the NORTH ATLANTIC, it is perhaps ironic that Earl's remnant tropical wave, carried westward by the trade winds, ghosted across Central America, and entered the eastern NORTH PACIFIC OCEAN on August 18, 2004, to eventually become a Category 1 hurricane. Reinvigorated by its emergence over warm waters and in an environment with minimal WIND SHEAR, the tropical wave grew into Tropical Storm Frank on August 23, and a full-fledged hurricane later the same day. Frank remained a North Pacific tropical system until August 27, when it dissipated near the northwestern Mexican coast.

earthquakes and tropical cyclones Although quite rare, there are recorded instances in the annals of cyclonic history in which significant seismic activity has reportedly coincided with the landfall of a TROPICAL CYCLONE. During the monstrous SAVANNA-LA-MAR HURRICANE of October 3, 1780, numerous eyewitnesses claimed that at the storm's height, twin earthquakes jolted that part of the Jamaican coast where the hurricane had come ashore. One of these quakes reportedly lifted a capsized British man-of-war, righted it, and then firmly set it down upright on a sandbank. At the time, this unique occurrence was regarded with some relief by survivors, many of whom found shelter from the storm inside the abandoned hulk. Less than two weeks later, victims of the GREAT HURRICANE OF 1780 in BARBADOS spoke of feeling the earth "quake" beneath the onslaught of the ferocious hurricane. British admiral George Rodney—who inspected the ravaged island shortly thereafter—likened the destruction to that caused by an earthquake. Survivors of a powerful 1783 hurricane strike on Charleston, SOUTH CAROLINA, claimed that the earth moved beneath them during the storm's course, while a similar hurricane in 1813 provoked one Charlestonian to irritably snap that his house's foundations were shaking so badly in the wind that he could not get to sleep until morning. During the first week of August 1837, the Los Angeles Hurricane claimed more than 500 lives in the VIRGIN ISLANDS as its 112-MPH (180-km/h) winds were conjoined with an early evening earthquake. Nearly every building on the island of St. Thomas collapsed, while a raging tsunami (tidal wave), due perhaps to the hurricane's storm surge, wrecked every vessel in the harbor. Despite the tons of rain that fell on the island during the earlier, six-hour tempest, massive fires quickly erupted. Burning out of control, they significantly added to the DEATH TOLL as they consumed those victims still trapped beneath fallen residences. More than a century later and half a

world away, another dual CYCLONE and earthquake killed 22 people on Manua Island, Samoa, on February 12, 1915. On December 26, 1947, as an intense TYPHOON shot into the PHILIPPINES, a minor earthquake and tsunami rocked eastern Luzon, directly in the path of the raging storm. Forty-nine people were killed, including 34 of 63 passengers and crew aboard the Danish motorship *Kina*, wrecked when the tsunami rushed ashore off northern Samar. In September 1966, Typhoon Trix reportedly precipitated no less than six small earthquakes as it came ashore in JAPAN, killing more than 50 people.

On August 14, 2007, as Hurricane Flossie slipped to within 400 miles (644 km) of the Hawaiian Islands, the big island of Hawaii was struck by a magnitude 5.4 earthquake. While no damage from the earthquake was reported, the connection between the seismic activity and Hurricane Flossie's approach was studied by researchers.

Despite many years of extensive research into the question, scientists have not yet fully defined the relationship between earthquakes and tropical cyclones. It is entirely possible that no such relationship in fact exists, that the supposed earthquakes were in truth nothing more than rhetorical devices, a way to illustrate graphically the destructive effects of the hurricane's winds and rains. One effusive writer, describing the Galveston Hurricane of October 3, 1867, stated that as the hurricane approached the barrier island, ". . . the Heavens [were] filled with portent; the earth seems to tremble. . . ."

On the other hand, there is the research of C. F. Brooks, former president of the American Meteorological Society, to consider. During the first half of the 20th century his study into earthquakes and hurricanes indicated that a two-inch (51-mm) drop in CENTRAL BAROMETRIC PRESSURE relieves 2 million tons of air pressure from each square mile of Earth's surface. To this he added the weight of a 10-foot (3-m) STORM SURGE that would add nine million tons of surface pressure per square mile. Putting the equation together, Professor Brooks believed that the loss of air pressure coupled with the increased weight of the encroaching sea was enough to cause seismic activity. A contemporary account of the Savanna-La-Mar Hurricane in Jamaica supports this conclusion by stating that the first earthquake occurred just minutes before the storm surge finally swept ashore, while the second occurred as the waters were abating.

Easy, Hurricane *Cuba–Southern United States, August 29–September 6, 1950* Named for the fifth letter in the military phonetic alphabet, a rather loopy HURRICANE Easy deluged western CUBA and northwestern FLORIDA with 120-MPH (193-km/h) winds and astounding 39-inch (991-mm) rainfalls between

September 3 and 6, 1950. In the fishing villages of Cedar Key and Yankeetown, some 90 miles (145 km) north of Tampa, Easy's tearing winds and record-breaking PRECIPITATION counts destroyed hundreds of BUILDINGS, sank numerous small boats, and initiated one of the most prolonged flood emergencies in Floridian history. Lingering for nearly three weeks after the storm's passage, Easy's watery remnants fortunately claimed no lives but committed staggering amounts of property damage to the citrus groves of central and northern Florida.

One of the first major NORTH ATLANTIC hurricanes to be christened with a name from the phonetic alphabet, Easy formed over the northwest CARIBBEAN SEA, approximately 800 miles (1,287 km) southwest of Cedar Key, Florida, on the morning of August 29, 1950. Swiftly intensifying as it tracked to the northeast, Easy grew into a TROPICAL STORM while traversing the Yucatán Channel on August 30. By midmorning of the following day, as the system's CENTRAL BAROMETRIC PRESSURE slumped to 29.39 inches (995 mb), Easy was upgraded to mature hurricane status (one with sustained winds of 74 MPH [119 km/h] or higher). It was at this time that the western Cuban towns of Mantua and Mendoza, grazed by the hurricane's DANGEROUS SEMICIRCLE, were assailed by 110-MPH (177-km/h) gusts and driving rains. Surprisingly, no lives were reported lost in Cuba, although considerable damage was sustained by those houses, piers, and roads over which Easy passed.

Easy underwent RECURVATURE, or a parabolic swing to the northeast, as it entered the GULF OF MEXICO during the early morning hours of September 1, 1950. Its central pressure still deepening, the hurricane's sustained wind speeds climbed to 105 MPH (169 km/h) as it seemingly laid a course for the tempest-prone Florida Panhandle. Weather reports obtained from shipping in the gulf indicated to storm watchers that the slow-moving TROPICAL CYCLONE was producing barometric pressure readings as low as 28.57 inches (967 mb) and maximum gusts eclipsing 125 MPH (201 km/h). Because Easy was still a vigorous young hurricane, no hint of the remarkable rainfalls that would later accompany its death was apparent to observers, making the hurricane's flooding final act all the more surprising.

A show-stopping hurricane in a number of respects, Easy then proceeded to spend September 3–5 not only intensifying but also performing a pair of loops over the warm, energized waters of the Gulf of Mexico. The first came on September 3, when Easy's 115-MPH (185-km/h) EYE, now positioned 150 miles (241 km) west of Cedar Key, sharply swung west, then south, and then east, where it stayed. The second loop came on September 5 during the hurricane's

landfall in northwestern Florida. Having spent September 4 traveling east-northeast, Easy continued to strengthen, its central pressure subsiding to 28.29 inches (958 mb) by midnight, September 4th. With sustained winds of 128 MPH (206 km/h), Easy first came ashore near the Withlacoochee River's mouth in the vicinity of Yankeetown on September 5. Penetrating as far inland as the river town of Inglis, the hurricane immediately embarked on another loop, this one taking it as far north as Cedar Key and then back out into the gulf (where it may have briefly intensified) for an eastward return to Yankeetown.

Struck twice in the same day by the same hurricane both Cedar Key and Yankeetown were left as veritable ghost towns. As the wreckage of sheared-off roofs, snapped trees, and snaking power lines littered streets, 90 percent of Cedar Key's wooden-hulled fishing fleet found itself reduced to kindling. In 72 hours more than two feet (1 m) of rain fell on Cedar Key, while precipitation counts as high as 38.6 inches (990 mm) were measured in Yankeetown. The Withlacoochee River, swollen to record heights by Easy's unprecedented rains, quickly overran its meager embankments, inundating more than 5,000 acres of farmland. Unable to handle the runoff, the river created a small lake dotted with the island remains of houses, trees and utility poles. More than three weeks would pass before the water finally subsided, during which time those buildings and vehicles submerged by Easy's rains were utterly ruined.

On September 6, a dying Easy, downgraded to TROPICAL DEPRESSION status, crossed northern Florida and entered the North Atlantic Ocean. Irrevocably exhausted by its strenuous assault on the Leisure State, Easy continued to dissipate and was nothing more than a toss of thunderstorms and a waterlogged memory by the morning of September 7. The name *Easy* was retired from the list of HURRICANE NAMES in 1953 when the United States Weather Service adopted the practice of naming tropical systems after female first names.

Eclipse Hurricane *Northeastern United States, November 2–3, 1743* Sometimes referred to as Franklin's Hurricane because Benjamin Franklin reportedly used the respective appearances of a lunar eclipse to time and track the northeast movement of this particular TROPICAL CYCLONE from Philadelphia, Pennsylvania, to Boston, Massachusetts, the Eclipse Hurricane relayed "excessive high" winds, heavy rains, and a flooding STORM SURGE to the saltmarsh coasts of east NEW ENGLAND. The first of only three recorded hurricanes to ever pass over New England during the month of November, the Eclipse Hurricane proved notably severe in Boston, where several heavily laden merchant ships were pushed ashore by the surge and then left stranded on waterlogged streets. Small boats were tossed onto piers as once-grand elm and walnut trees first creaked and groaned in the wind and then toppled over. In Portsmouth, New Hampshire, the flow of the Piscataqua River was turned back on itself by the hurricane's passing surge and rose to such a height that several piers and bankside houses were completely swept away. Hay fields and apple orchards in western Massachusetts were riddled by gusts and turned into veritable wastelands by the distant hurricane's far-reaching gales.

A notable hurricane for any number of reasons (the first barometric pressure reading taken in a New England hurricane—29.35 inches [990 mb]—was observed at Cambridge, Massachusetts, on the afternoon of November 2), the Eclipse Hurricane is best known for the role Benjamin Franklin played in plotting its course up the U.S. east coast. In Philadelphia on November 2, Franklin the polymath was denied the opportunity of studying a predicted lunar eclipse by the early evening arrival of the hurricane's attending gales, lightning strikes, and drenching rains. Later informed by letter that the eclipse had indeed been seen at Boston, some 290 miles (467 km) northeast, between nine P.M. and one A.M. (or just before the onset of the hurricane in Philadelphia), Franklin deduced that the storm must have been moving in a northeasterly direction when it transited Philadelphia; thus, it had to have originated to the south-southwest or over the CARIBBEAN SEA's tropical expanses. Posited when the workings of tropical cyclones had yet to be fully or accurately defined (numerous 18th-century narratives, for instance, mistake the passing of a hurricane's calm EYE for a break between *two* separate but equally powerful storms), Franklin's theory was quickly subjected to the most critical of debates before posthumous developments in the Law of Storms proved his findings correct.

Edna, Hurricane *Northeastern United States, September 6–11, 1954* The second in a trio of major HURRICANES to strike the northeastern United States during the memorable 1954 HURRICANE SEASON, Edna battered NORTH CAROLINA, southeastern NEW ENGLAND, and the maritime provinces of CANADA with sustained 115-MPH (185-km/h) winds, six-inch (152-mm) rains, and heavy seas between September 10 and 11, 1954. Twenty-two people, a number of them Maine fishers, were killed, while more than $50 million in property damage to BUILDINGS, boats, piers, and Nova Scotia's ripening apple harvest were assessed.

Preceded by Hurricane CAROL, which had soared into New England on August 31, and followed by Hurricane HAZEL, which would trounce the U.S. eastern seaboard on October 15, Edna originated off the north coast of PUERTO RICO some 1,700 (2,736 km) southeast of Cape Cod, Massachusetts, on September 6, 1954. Recurving northeast, Edna intensified erratically, growing into a TROPICAL STORM on the morning of September 7 and a Category 1 hurricane, with a CENTRAL BAROMETRIC PRESSURE of 29.26 inches (991 mb) by nightfall on September 9.

Maintaining its curl to the northeast, Edna passed 230 miles east of GEORGIA on September 10, delivering pounding swells to the marshy lowlands around Charleston. Gaining speed as it coasted north, Edna further intensified as it brushed past NORTH CAROLINA's Outer Banks during the night of September 10, slipping to 28.64 inches (970 mb) as it primed the sand shoals with 10-foot (3-m) seas and intermittent rain squalls.

Its forward speed suddenly accelerated by a 30-MPH (48-km/h) STEERING CURRENT, Edna's central pressure further deepened, skidding toward 28.17 inches (954 mb) as it bore down on southeast New England during the early morning hours of September 11. A midrange Category 3 hurricane in its own right, Edna's scouring 115-MPH (185-km/h) winds were further augmented by the steering current's speed, thereby boosting its sustained strength to a harrowing 145 MPH (233 km/h).

Edna thudded ashore near Martha's Vineyard, Massachusetts, just after dawn on September 11, 1954. Its DANGEROUS SEMICIRCLE, complete with 125-MPH (201-km/h) winds and seven-foot (2-m) STORM TIDES, passed over entire length of Cape Cod, causing severe damage to piers, small boats, and seaside villages. Six people in Massachusetts were killed, four of them fishers lost when their trawler went down in Edna's choppy seas.

Emerging into the chilling confines of the Gulf of Maine on the afternoon of September 11, Edna maintained its strength as it curved sharply northwest, following this course until it barreled ashore near Maine's Arcadia National Park on the same evening as did a Category 2 hurricane of significant intensity. Accustomed to only periodic hurricane activity, Maine found Edna's central pressure of 28.64 inches (970 mb) and strident 105-MPH (169-km/h) winds most vicious as they gutted portions of the popular, timber-lined park, hammered several dozen ancient cottages into the surf, and collapsed a portion of the sweeping veranda on the celebrated Somerset Hotel. Nine people in Maine perished, while $17 million damage in 1954 currency value was tallied, making Edna the most destructive—and deadly—hurricane to have struck Maine since the Great Atlantic Hurricane of 1944.

During the early morning hours of September 12, Edna exited Maine and moved into Canada. Still classified as a Category 2 hurricane, durable Edna lashed Newfoundland and Nova Scotia for most of the day, raising a heavy surf along the Prince Edward Island's east coast. Farther inland, three-quarters of Nova Scotia's ancient apple orchards were riven by Edna's 115-MPH (185-km/h) gusts, uprooted and destroyed in a shower of wind-blown fruit and overturned farm implements. Millions of Canadian dollars in damage was reported, and it would take decades for Nova Scotia's orchards to again take root.

On September 13, 1954, long after Edna had dissipated over central Canada, President Dwight D. Eisenhower declared five counties in Maine disaster areas, making them eligible for federal disaster aid. The name *Edna* has been retired from the list of HURRICANE NAMES.

Edouard, Tropical Storm *Gulf of Mexico, September 14–15, 1984* Between 1984 and 2002, four TROPICAL CYCLONES in the NORTH ATLANTIC OCEAN have been identified with the French male name Edouard. Of the four, only one reached cyclonic maturity—Category 4 hurricane Edouard in 1996—and only one (Tropical Storm Edouard of 2002) made landfall. The name Edouard has been retained on the naming lists, and is next to appear during the 2008 season.

The first North Atlantic tropical cyclone to be identified as Edouard could not have had a more inauspicious existence. Lasting just over 48 hours before dissipating, the system originated over the deep Bay of Campeche during the early morning hours of September 14, 1984. As a TROPICAL DEPRESSION the system tracked to the north, but was quickly turned back on itself by an entrenched high-pressure ridge over the southern GULF OF MEXICO. Simultaneously, the system strengthened to TROPICAL STORM intensity, and reached its minimum observed pressure of 29.47 inches (998 mb) during the early morning hours of September 15. Within six hours, unfavorable WIND SHEAR conditions returned the system to tropical depression status and by midday on the 15th, the first Edouard was a disorganized, open-low (meaning there were no closed ISOBARS within the low) entry into meteorological history.

Edouard, Tropical Storm *North Atlantic Ocean– Azores, August 2–13, 1990* The second North Atlantic TROPICAL CYCLONE known as Edouard remained an offshore system for its entire existence, yet nonetheless displayed an unusual meteorological biography. Formed from a subtropical depression on August 2, 1990, while situated near 37

degrees North and 23 degrees East, the subtropical depression first traveled northward before sharply recurving to the west-northwest and strengthening to TROPICAL STORM status during the afternoon of August 3. As a tropical storm, it achieved a minimum pressure of 29.67 inches (1,005 mb) on August 4, but by the early evening hours of the same day (and after executing a tight, clockwise loop), had regained its subtropical characteristics. Its westward progress hampered by the semi-permanent BERMUDA HIGH, the system remained a subtropical depression until the early evening hours of August 6, 1990, when it was reclassified as a tropical depression. It remained at tropical depression intensity until the evening of August 8, when it regained tropical storm status. Although Edouard produced its lowest observed barometric pressure reading during this time (29.61 inches [1,003 mb]), it was to be a brief regenerative interlude for this system. Within two days, Edouard was again downgraded to a tropical depression, and it continued to weaken until August 11, when it underwent extratropical deepening. Now traveling nearly due east, the fading system brought rains and heavy surf to the islands of the Azores, an area generally unaffected by tropical cyclone activity. By August 13, 1990, the system had dissipated.

Edouard, Hurricane *North Atlantic Ocean– Northeastern United States, August 19–September 6, 1996* Unlike its two tepid predecessors, the third North Atlantic TROPICAL CYCLONE identified as Edouard became one of the most intense HURRICANES observed in the Atlantic basin. In existence between August 19 and September 6, 1996, Hurricane Edouard twice strengthened to Category 4 status during its marathon trek across the NORTH ATLANTIC OCEAN. Born as a TROPICAL DEPRESSION less than 200 miles (322 km) off the northwestern coast of Africa, Edouard—as a classic example of a CAPE VERDE STORM—steadily churned to the west-northwest, strengthening as it did so. By the early morning hours of August 22, it had deepened into a TROPICAL STORM with a central pressure of 29.70 inches (1,006 mb) and by the same time the following day, was on the verge of reaching hurricane intensity. At noon on August 23, as it neared longitude 40 degrees West, Tropical Storm Edouard was upgraded to a Category 1 hurricane with a central pressure of 29.17 inches (988 mb). By noon of the following day, Edouard had achieved Category 2 classification, with a central pressure of 28.64 inches (970 mb) and sustained winds of 104 MPH (167 km/h). In what would become one of the swiftest deepening cycles yet observed in the North Atlantic, Hurricane Edouard spent the next 24 hours shaving 1.06 inches (36 mb) from its CEN-

TRAL BAROMETRIC PRESSURE. During the early morning hours of August 25, 1996, and benefiting from an upper level anticyclonic flow, Hurricane Edouard produced a pressure reading of 27.55 inches (933 mb), and sustained winds of 144 MPH (232 km/h). The system was at this time situated some 350 miles (563 km) north of PUERTO RICO, and was firmly moving to the northwest. If maintained, this trajectory would have taken Edouard ashore near the FLORIDA-GEORGIA border. On August 28, however, a downgraded Hurricane Edouard (of Category 3 intensity) began a broad RECURVATURE to the north and began moving northward and nearly parallel to the eastern seaboard of the United States. By the midday hours of August 29, 1996, Edouard's location over the fueling waters of the Gulf Stream, coupled with a low WIND SHEAR environment, allowed the system to again intensify to Category 4 status. At shortly before dusk on August 30, Edouard generated a pressure reading of 27.58 inches (934 mb), the third-lowest observation in its history. Maintaining its parallel course to the northeastern United States, Edouard—again downgraded to a Category 3 hurricane—recurved to the north and northeast on August 31 and September 1, and rapidly began to weaken. By the morning of September 2 it was a Category 1 hurricane (with a central pressure of 28.46 inches [964 mb], and by midnight on September 3, of tropical storm intensity, with a minimum pressure of 28.88 inches (978 mb) and sustained winds in the range of 69 MPH (111 km/h). At this time the system was some 95 miles (153 km) southeast of Nantucket, Massachusetts, and it dropped heavy rains on the island, downing power lines, and generating large surf conditions. Later on the 3rd, Tropical Storm Edouard underwent extratropical transitioning, and grew into a fairly intense extratropical storm that passed well to the east of the Canadian Maritime provinces. As an extratropical storm, Edouard's lowest observed pressure was 29.08 inches (985 mb), which was generated on September 3, during the system's transition from tropical to extratropical storm. An intense North Atlantic tropical cyclone that for much of its existence possessed the potential to wreak catastrophic damage had it come ashore, Edouard was harmlessly absorbed into a larger, stronger extratropical system over the northern extremes of the North Atlantic on September 6, 1996. All told, Edouard claimed two lives through drowning incidents in New Jersey, and some $4.5 million in 1996 property losses.

Edouard, Tropical Storm *North Atlantic Ocean–Southeastern United States, September 1–6, 2002* The fourth North Atlantic TROPICAL CYCLONE identified as Edouard, September 1 to September 6, 2002, never topped TROPICAL STORM intensity,

but delivered rains and gusty winds to FLORIDA on September 5. Generated over the extreme western NORTH ATLANTIC OCEAN, near 29 degrees North, 79 degrees West, during the evening hours of September 1, Tropical Depression No. 5 first tracked northward before executing an almost perfect clockwise loop-the-loop maneuver that left it traveling to the west-southwest. Growing into a tropical storm during the early morning hours of September 2, Tropical Storm Edouard reached its minimum pressure of 29.58 inches (1,002 mb) on September 4. Never able to organize into a significant tropical cyclone, Tropical Storm Edouard slowly weakened, and generated a central pressure reading of 29.76 inches (1,008 mb) at the time of its landfall along Florida's central eastern coast, near Ormond Beach. Quickly downgraded to a TROPICAL DEPRESSION, Edouard dropped large quantities of PRECIPITATION on central Florida, producing winds in excess of 20 MPH (37 km/h) as it swung to the southwest across the peninsula. Emerging into the GULF OF MEXICO on September 5, 2002, Tropical Depression Edouard's organized circulation lasted until September 6, when it was absorbed by a strengthening Tropical Storm Fay, then active in the northwestern Gulf of Mexico. In Florida, no lives were lost to Tropical Edouard's passage, and property damage was light.

Elena, Hurricane *Southern United States, August 27–September 4, 1985* The first of three "billion-dollar" HURRICANES to have come ashore on mainland United States during the tumultuous course of the 1985 HURRICANE SEASON, Elena's loop-the-loop, four-day track along the northern GULF OF MEXICO brought 125-MPH (200-km/h) winds, five inches (127 mm) of rain, and a steady round of mass evacuations to the coasts of FLORIDA, ALABAMA, MISSISSIPPI, and LOUISIANA between August 30 and September 3, 1985. In Florida, where Elena's offshore waiting game forced a million tourists and residents to not once, but twice, flee their vulnerable beachfront communities, three people were killed in BUILDING failures, automobile accidents, and drownings. In Alabama, where 175,000 people were hurriedly vacated from gulfside beachheads on August 30, Elena's torrential rains and 110-MPH (177-km/h) winds battered seawalls, wrenched roofs from houses, downed power lines and trees, and sent a five-foot (2-m) STORM SURGE spinning into Mobile Bay. On the border between Mississippi and Louisiana, where Elena's burgeoning Category 3 fury finally came ashore on the afternoon of September 2, 1985, nearly 300,000 people—a majority of them inhabitants of east New Orleans—were sent to inland storm shelters. Of the two states, Mississippi was the hardest hit by Elena's 138-MPH (222-km/h) gusts: More than 13,000 homes in Harrison, Hancock, and Jackson counties either damaged or destroyed. Some 60,000 people were left homeless, and widespread power outages hampered relief and reconstruction efforts for nearly a week. As insurance claims quickly topped $1 billion, Elena became one of the more costly hurricanes to have struck the United States up to that time.

A midseason TROPICAL CYCLONE, Elena developed over the southeast Gulf of Mexico's temperate waters, roughly 320 miles (515 km) south of Tampa, Florida, on the morning of August 27, 1985. The fifth storm of an active season—one that had seen Hurricane Danny strike Louisiana and TEXAS on August 15 and would less than two weeks later witness Hurricane GLORIA's blast up the eastern seaboard and into NEW ENGLAND on September 26–27—Elena strengthened leisurely as it drifted north-northwest at nearly 11 MPH (18 km/h). Reconnaissance flights conducted by the NATIONAL HURRICANE CENTER (NHC) indicated that Elena's CENTRAL BAROMETRIC PRESSURE had deepened to 29.31 inches (992 mb) by the afternoon of August 28, upgrading it to a TROPICAL STORM, with sustained winds of 53 MPH (85 km/h). By the same time on August 29th, Elena had blossomed into a Category 1 hurricane, with sustained 79-MPH (127-km/h) winds that hinted at still further intensification.

On the morning of August 30, as emergency management authorities in New Orleans consulted with the National Hurricane Center on Elena's continued swing northwest, thousands of Labor Day vacationers, year-round residents, fishers, and oil rig personnel began to evacuate the entire south coast from Morgan City, Louisiana, to Tampa, Florida. While disappointed families hastily overpacked their station wagons, minivans, and campers, the major petroleum companies that own drilling platforms in the Gulf of Mexico—and routinely monitor weather broadcasts—dispatched a squadron of supply boats and helicopters to airlift oil rig crews to the relative safety of land.

By the afternoon of August 30th, as Elena further strengthened into a Category 2 hurricane, the eastern reaches of New Orleans were ordered cleared. From the dank shores of the bayous to the fashionable neighborhoods of Gretna and Algiers, thousands more people locked their doors, closed their businesses, and went a day without pay as they headed inland, away from the fertile delta that in the past two centuries had been inundated no less than 14 times.

In Tampa, some 430 miles (692 km) southeast of New Orleans, Elena's powerful fringe gales and steady rains were already being felt as the hurricane, abruptly coming under the spreading influence of an eastward-moving warm front, began to move due east. Hurricane-force winds and a rising tide spilled into surge-prone Tampa Bay, washing over seawalls and inundating coastal roadways. Already occupied by those evacuees Elena had forced out of the Panhandle, local storm shelters throughout north Florida were taxed to capacity as thousands more refugees from Tampa northward wearily shuffled in. Indeed, the shelter shortage was such that many evacuees from the Panhandle were told to return home, the belief being that Elena was going to continue across central Florida, sparing the Panhandle's valuable oyster beds, wooden cottages, and tree-lined neighborhoods from yet another gulf tempest.

But no sooner had the Tampa people been moved into the shelters and the Panhandle residents allowed to use the highways than Elena, now confronted by the westerly flow of the trade winds, began to move away from Florida. Spinning back into the gulf on the morning of September 1, the intensifying Category 2 storm returned to its former northeasterly course, once again steadily churning toward the Panhandle. Completely exasperated by Elena's teasing track, nearly a million Florida residents played a sort of musical chairs as the Panhandlers returned to the shelters and Tampa residents went home to pay for the $32 million in swamped streets, downed trees and road signs, and extensive beach erosion left by Elena's passage.

In a move that would mirror Hurricane JUAN's equally haphazard stroll across the Gulf of Mexico just seven weeks later, Elena crossed the Florida coast and blasted as far inland as the Panhandle town of Lynn Haven before stalling. Slightly weakened, Elena slowly withdrew, returning to the more hospitable gulf waters during the night of September 1. Held off the coast by a high-pressure ridge, a still intensifying Elena, now upgraded to a Category 3 hurricane, careened along the U.S. gulf's northern coast. Unable to penetrate the high-pressure AIR MASS to the north, Elena spent the next 12 hours rolling along the gulf coasts of Alabama and Mississippi, delivering 135-MPH (217-km/h) gusts to Biloxi, Gulfport, and Mobile Bay before coming ashore for the last time on the Mississippi-Louisiana border on the morning of September 2, 1985.

In Mississippi, punishing rains uprooted trees, overturned cars, sank mobile homes, and rattled against Biloxi's hulking steelworks. Sustained winds of 125 MPH (201 km/h) sheared across the landscape, splintering telephone poles, toppling church steeples, splitting fences, and driving several small sailboats ashore at Long Beach. Some 13,000 houses—from historic plantation mansions to barrier island cottages—in three Mississippi counties—were either damaged or utterly destroyed. Five-inch downpours precipitated flash floods, turning numerous townships into extensions of the local bayou; more than 100,000 households across the state lost electricity and telephone services. No lives in Alabama, Mississippi, or Louisiana were lost, however, a statistic that civil defense officials later attributed to the effectiveness with which the harrowing series of evacuations were carried out.

Coursing north across the timbered highlands of Mississippi, Elena quickly slackened, its sustained winds dropping to 38 MPH (61 km/h), or TROPICAL DEPRESSION stage, as it entered east Arkansas on September 3, 1985. Splitting into a line of powerful thunderstorms on September 5, the day then President Ronald Reagan declared southern Mississippi eligible for federal disaster relief, Hurricane Elena was by dawn of September 6 just a billion-dollar memory. The name Elena has been retired from the cyclical list of HURRICANE NAMES.

El Niño A broad current of warm water that periodically flows from north to south in the southeast Pacific Ocean off the west coast of Peru. First identified in 1891 by Dr. Luis Caranza, president of the Lima Geographical Society, the oscillating existence of the current had long been known to mariners plying the coastal trade between the Peruvian port towns of Paita and Pacasmayo and had been christened by them *El Niño* (the Child Jesus) because it most often appeared a few days after Christmas every three to five years. Intense El Niño periods have been noted in 1941, 1957, 1958, 1972, 1978, 1982, 1986, 1997, 1998 and 2002. A temperate influence on the cooler waters of the southeastern Pacific Ocean, the onset of El Niño's cycle usually produces intense thunderstorms along the Peruvian coastline, tropical downpours that have on occasion spawned mudslides and flash floods. In those years when the El Niño cycle is weak or nonexistent (known as the La Niña cycle), SEA-SURFACE TEMPERATURES in the southeastern Pacific remain below 75°F, BAROMETRIC PRESSURE readings remain high, and PRECIPITATION levels over eastern South America are minimal. Subjected to intense scientific research in recent decades and now systematically classified in intensity as strong, moderate, weak, and very weak, El Niño's influence on global weather patterns has been determined by William Gray and other meteorologists to play a significant role in TROPICAL CYCLONE generation in the NORTH ATLANTIC OCEAN. Called teleconnections, the relationship between El Niño and North Atlantic HURRICANES lies in the suppression of the westerlies, the powerful

midlevel winds that sweep eastward across the North American continent and keep tropical systems well out to sea. In years when El Niño is at its weakest, high barometric pressure dominates the eastern half of the NORTH PACIFIC OCEAN, thereby deflecting the westerlies and allowing westward-moving tropical systems to come ashore along the U.S. eastern seaboard. Such a scenario was witnessed during the 1996 North Atlantic HURRICANE SEASON when a strong La Niña condition existed over the eastern North Pacific Ocean and two mature-stage hurricanes, BERTHA and FRAN, made landfall in NORTH CAROLINA, between July and September.

The full scope and influence of the El Niño/La Niña cycle (also known as the El Niño-Southern Oscillation or ENSO) on tropical cyclone activity in the North Atlantic and North Pacific basins continues to provoke debate among meteorologists. Studies undertaken by the NATIONAL HURRICANE CENTER (NHC) in 2003 indicate that while there is some connection between El Niño/La Niña cycles and tropical cyclone activity in the North Atlantic and Pacific Oceans, ". . . no significant long-term trend in hurricane strength or frequency has been observed in the Atlantic Basin." Other studies provide evidence that while El Niño's influence in 1997 and early 1998, resulted in diminished tropical cyclone activity in the North Atlantic basin, but the onset of the La Niña during the 1998 North Atlantic hurricane season resulted in the generation of 14 named tropical systems, seven of which caused nearly $7 billion in 1998 dollar losses to the United States. Meteorologists have also attributed the ability of late-season Hurricane MITCH—the deadliest North Atlantic tropical cyclone since 1974—to generate from a MESOSCALE CONVECTIVE COMPLEX in the southwestern CARIBBEAN SEA to the waning influence of El Niño and the strengthening of La Niña's. During the 2002 North Atlantic hurricane season, a time when El Niño's teleconnections were at their most influential, there were relatively few tropical cyclones (12), with a majority of them (eight) being of tropical storm intensity. In their studies of El Niño's and La Niña's effects on tropical cyclone activity in the North Atlantic and North Pacific Oceans, meteorologists at the National Hurricane Center (NHC) have concluded that the above-active 2003 North Atlantic hurricane season (16 named storms, seven of which were hurricanes) was in part due to the onset of a La Niña cycle.

Scientists have further observed that a strong La Niña influence was in place during the precedent-setting 2005 North Atlantic hurricane season, when 27 named tropical cyclones developed, 15 of them of hurricane intensity. The 2005 season also produced no fewer than four Category 5 tropical systems, including Hurricane Wilma, which in October of that year established a new central barometric pressure record for the North Atlantic basin. Conversely, the 2006 North Atlantic hurricane season was one of the most inactive on record. Due to the onset of an El Niño pattern in early 2006, tropical cyclone activity in the North Atlantic that year was moderate. As of October, nine named storms had appeared in the North Atlantic, five of them hurricanes, but only two that reached Category 3 intensity.

Australian meteorologists have produced additional studies that provide evidence that El Niño and La Niña cycles can affect tropical cyclone activity along Australia's shores. During summer seasons when El Niño is at its strongest in terms of sea-surface temperatures and upper altitude wind patterns, fewer tropical cyclones are generated in the Coral and Timor Seas, and in the southwestern Pacific Ocean.

Eloise, Hurricane *Northern Caribbean–Southern Northeastern United States, September 17–27, 1975* One of the most destructive Category 3 hurricanes on record, Eloise spent over a week of the 1975 hurricane season beating a ghastly course of devastation through the northern Caribbean islands of Puerto Rico and Hispaniola and into the southern and eastern United States between September 17 and 27. In Puerto Rico, where Eloise's sustained, 60-MPH (97-km/h) winds and two-inch (51-m) rains were only of tropical storm strength at landfall, roaring mudslides nevertheless managed to bulldoze dozens of buildings and claim the lives of 34 people on September 17. In Haiti and the Dominican Republic, both of which Eloise decimated with 75-MPH (121-km/h) gusts and torrential rains on September 19, more than 26 people were killed, many of them drowned in seething flash floods along Hispaniola's south coast.

Maintaining its westward advance for the next two days, Eloise intensified as it traversed the late-season Caribbean Sea waters. Portentously gaining on the resort beaches of eastern Mexico's Yucatán Peninsula, Eloise deepened slightly on the evening of September 20, its central barometric pressure dropping to 29.38 inches (994 mb). This boosted its sustained winds, causing advection currents in the system's eyewall to exceed 71 MPH (114 km/h). Heavy rains scudded across the wind-washed surface of the sea, sending large breakers rolling ashore as far north as Kingston, Jamaica.

This spectacular photograph taken from the archives of the National Weather Service (NWS) shows Hurricane Eloise's storm tides approaching Fort Walton Beach, Florida, in 1975. While storm-generated waves are a tempting target for surfers and surf watchers, their dangerous undertows and sheer inundating power have claimed many lives in the past. (NOAA)

Booming seas were also recorded to the west, where civil defense organizations in Cancun, Mexico, had already implemented coastal evacuation warnings. As azure lagoons gradually darkened to blue, tour boats, their masters' attention firmly fixed on the ink-black clouds now gathering on the horizon, hastily returned to their piers. According to newspaper accounts, towering surf-side hotels, aquatic playgrounds, and dance halls were grudgingly abandoned by ferry, leaving behind bands of workers who would now secure them against Eloise's potential stampede. Steel storm shutters were unrolled, and brightly colored awnings were taken down and folded, stowed in concrete sheds along with beach furniture and pool umbrellas. By midnight September 20, 1975, as a still-intensifying Hurricane Eloise neared to within 75 miles (121 km) of the Yucatán Peninsula, grumbling hotel operators remained defiant in their determination to end the season with a profit.

On the morning of September 21, 1975, Eloise trundled ashore at Puerto Morelos, Mexico, as a moderate Category 1 hurricane. In the hierarchy of Mexican tempests, Eloise was not a particularly violent contender for first place. Bearing a central pressure of 29.4 inches (990 mb), the hurricane's winds hovered at 80 MPH (129 km/h); strong enough to lift small roofs, down power lines, and damage trees but relatively mild compared to those found in systems of Category 3 status or higher. Eloise's rains, later to play such havoc over the mountain ranges of northeast United States, were also mild, measuring just under an inch at Cozumel. No lives were reportedly lost in Mexico, a matter of great relief to local hoteliers and U.S. travel agents alike.

Pursuing its recurring course northwest, Eloise first entered the GULF OF MEXICO on the morning of September 21. Not at all weakened by its passage over Yucatán, the hurricane's central pressure further intensified as it chugged toward central LOUISIANA, drooping to 29.3 inches (993 mb) by the morning of September 22. With eyewall gusts now exceeding 110 MPH (177 km/h), it was clear to storm watchers that Eloise was poised to grow very quickly and would within the next 24 to 48 hours graduate to Category 2, or perhaps even Category 3, status. In view of the hurricane's fairly steady progress northwest, it was decided that HURRICANE WATCHES should be announced for Louisiana's southeast shores.

By the afternoon of September 22, however, a stream of aerial and satellite reconnaissance missions revealed that Eloise's course had relented somewhat, now the maturing storm was directed toward MISSISSIPPI or perhaps even west Florida. As dusk set in Eloise's central pressure produced a reading of 28.6 inches (968 mb), elevating the hurricane to that of a powerful Category 2 TROPICAL CYCLONE. Sustained winds in the eyewall blasted away at 110 MPH (177 km/h), while a domed STORM SURGE slowly formed right and rear of the EYE, swelling to the point that tides along the Florida Panhandle began to rise very quickly just before nightfall. Responding to on-scene reports from Fort Walton Beach, Pensacola, and Panama City, the NATIONAL HURRICANE CENTER (NHC) determined that Eloise would indeed come ashore somewhere on the Florida Panhandle, most likely within the following 24 hours. This portion of the Florida Panhandle had not been struck by a powerful hurricane, whose winds were in excess of 111 MPH (179 km/h), in more than 75 years, leading many residents to grow complacent about the inherent dangers of such systems.

All along the Florida coast, postseason guests hurriedly packed their bags, loaded their cars, campers, and vans, and then took to the road in a long, blaring caravan of disappointment, fear, and exhaustion. Conducted by the glare of headlights and under the watchful flash of red-and-blue emergency lights, the mass evacuation of some 50,000 people from a 60-mile (97-km) stretch of coastline was not without its difficulties. Rain-slick roads made driving hazardous, while lifelong residents, unable to recall the last major hurricane strike to affect the area, hung back in their doorways, boldly stating to civil defense personnel that their beachfront cottages were their own and that they were going to ride this one out.

Just before dawn on September 23, 1975, Eloise made landfall near Fort Walton Beach, Florida. Crashing ashore as a Category 3 hurricane, Eloise's central pressure of 28.2 inches (955 mb) produced 130-MPH (209-km/h) winds, 5.5-inch (127-mm) rains, and a pounding 11–18-foot (4–6-m) storm surge. Fourteen-foot STORM TIDES swept the Panhandle coast from Pensacola to Panama City, undermining hotels, toppling restaurants, crushing houses, overwhelming seawalls, and washing out numerous bridges and causeways. At the evacuated Roundtowner Motel, Eloise's 16-foot (5-m) storm surge rose as high as the complex's second story, punching out doors and windows, tearing away roofs, and collapsing two of its spreading wings. Elegant lobby furnishings sat half-buried in the sand, while trampled palm fronds carpeted skewed staircases. Almost left a total wreck, most of the Roundtowner Motel would subsequently be torn down and never rebuilt.

In Pensacola, the four-story Holiday Inn lost 150 feet (50 m) of its duned beach as Eloise's battering seas swamped several hundred acres of nearby waterfront. Although the blocklike, balconied structure did not collapse, subsequent investigations revealed that Eloise's clawing approach had removed much of the sand from beneath the building, uncovering its concrete pilings and leaving them without support. Dozens of other establishments on the "Riviera of the South," from expensive gulfside winter houses to squat office blocks, were likewise damaged or destroyed. Some 18 sailboats and motoryachts were driven ashore by Eloise's raging surf and then left incongruously stranded on front lawns and city streets.

Four people in Florida lost their lives to Hurricane Eloise. More than $150 million in property losses were inflicted, prompting then President Gerald R. Ford to declare Florida a major disaster area on September 26. In fact, on the very day that President Ford signed the emergency bill into law, Eloise's precipitation-laden remnants were bombarding Washington, D.C. and its neighborhoods with eight inches (203 mm) rain, swollen rivers, and gusty winds. In Pennsylvania, the Susquehanna and Juniata rivers overran their banks, forcing more than 19,000 people to seek the safety of higher ground. Roads, bridges, and reservoirs in VIRGINIA, Maryland, and NEW YORK were also submerged by Eloise's five-day rains, making the dying hurricane one of the most expensive natural disasters to strike the region in almost a half-century. Ten people in the northeast were killed; an additional $50 million in property losses were assessed. With 76 confirmed deaths to its credit and damage estimates as high as $200 million, the name Eloise was removed from the recycled list of HURRICANE NAMES upon the conclusion of the 1975 season.

Emily, Hurricane *North Atlantic Ocean–Bermuda, August 31–September 12, 1981* Between 1981 and 2005, five tropical cyclones in the NORTH ATLANTIC OCEAN have been identified as Emily. All but one reached hurricane intensity, while 2005's Hurricane Emily briefly matured into a Category 5 hurricane—the earliest (July) tropical cyclone of such intensity on record in the Atlantic basin. The identifier Emily has also been used for five tropical systems in the eastern NORTH PACIFIC OCEAN, including a hurricane in 1963; a hurricane in 1965; a tropical storm in 1969; a hurricane in 1973; and a tropical storm in 1977. The identifier Emily has been retained on the North Atlantic tropical cyclone naming lists, and is scheduled to reappear in 2011.

The first North Atlantic tropical cyclone to be identified as Emily originated from a subtropical depression, some 350 miles (563 km) northeast of the Windward Islands on August 31, 1981. First tracking almost due east, the system commenced its recurvature to the northeast during the early morning hours of September 1, and was upgraded to a tropi-

cal storm (with a central pressure of 29.64 inches [1,004 mb]) shortly thereafter. On September 2, 1981, Tropical Storm Emily passed directly over the island of BERMUDA, dropping large quantities of rain but otherwise sparing the island a destructive interlude. According to meteorological records, sustained wind speeds across Bermuda were 58 MPH (93 km/h). Slowly intensifying as it spun to the northeast, Tropical Storm Emily performed a counterclockwise loop-the-loop between September 3 and September 4, gaining intensity as it did so. Shortly after midnight on September 4, the system was upgraded to Category 1 status, with a central pressure of 28.82 inches (976 mb) and wind speeds of 75 MPH (120 km/h). Trailing along the western edge of the BERMUDA HIGH, Hurricane Emily steadily intensified, achieving its lowest (observed) barometric pressure of 28.52 inches (966 mb) during the early evening hours of September 5. Emily maintained its hurricane status until September 7, when it was downgraded to a tropical storm while located several hundred miles to the south of Newfoundland, CANADA. Carried eastward by the clockwise-spinning Bermuda High, Tropical Storm Emily continued to weaken over the cooling waters of the extreme North Atlantic Ocean, and transitioned to an extratropical depression on September 11, 1981. It soon dissipated.

Emily, Hurricane *North Atlantic Ocean–Windward Islands–Hispaniola–Bermuda, September 20–26, 1987* The second North Atlantic tropical system identified as Emily formed as a TROPICAL DEPRESSION near latitude 9 degrees North and longitude 51 degrees West during the midnight hours of September 20, 1987. A classic example of a CAPE VERDE STORM, the system steadily intensified as it drew to the northwest, becoming a TROPICAL STORM (with a central pressure of 29.67 inches [1,005 mb]) late on September 20. It crossed the Leeward Islands and entered the CARIBBEAN SEA during the midday hours of September 21, bringing sustained, 52-MPH (83-km/h) winds and torrential rains to the volcanic uplands of St. Vincent. Doggedly maintaining its northwesterly trajectory, Tropical Storm Emily bore down on the island of HISPANIOLA, and during the early morning hours of September 22, as its anticyclonic outflow improved, began a rapid deepening cycle. By daybreak on the 22nd, Emily had reached Category 1 status, with a central pressure of 28.88 inches (978 mb); some six hours later, Hurricane Emily was promoted to a Category 2 system with a central pressure of 28.67 inches (971 mb); and within six hours of that milestone, Emily reached Category 3 status, with a central pressure of 28.28 inches (958 mb) and sustained wind speeds of 127 MPH (204 km/h). WIND SHEAR from a strong high-pressure ridge to the east of Emily's center weakened the system's circulation, and Emily had weakened to Category 1 intensity as it crashed ashore in the southeastern Dominican Republic during the early morning hours of September 23, 1987. A minimum central pressure of 29.05 inches (984 mb) generated sustained winds of 81 MPH (130 km/h); three people were killed and some 6,200 others left homeless. Damage estimates to crops and other property topped $30 million (in 1987 U.S. dollars). Downgraded to a tropical storm as it tripped across Hispaniola's mountain ranges, Emily passed through the southern BAHAMAS and began a broad recurvature to the north-northeast on September 24. Now carried over warmer waters, Tropical Storm Emily reintensified, again becoming a Category 1 hurricane during the early morning hours of September 25, 1987—just as its EYEWALL passed within 25 miles (40 km) of BERMUDA. As it lashed the idyllic British colony with sustained winds of 92 MPH (148 km/h), causing more than $35 million (in 1987 U.S. dollars) in property losses, Emily continued to intensify, reaching a minimum pressure of 28.76 inches (974 mb) during the midday hours of the 25th. Now moving to the northeast at forward speeds in excess of 30 MPH (48 km/h), the system quickly entered the extreme NORTH ATLANTIC OCEAN and very quickly underwent extratropical transitioning on September 26, 1987.

Emily, Hurricane *North Atlantic Ocean–Eastern United States, August 22–September 6, 1993* At one time in its existence a moderately powerful Category 3 hurricane, the third North Atlantic system to bear the name, Emily, delivered a glancing blow to the low-lying inlets and sandy Outer Banks of NORTH CAROLINA between August 31 and September 1, 1993. A Cape Verde system that had scribed an unusual course across the North Atlantic Ocean since August 22, Emily first reached TROPICAL STORM intensity on August 25—and Category 1 status (with a central pressure of 29.64 inches [1,004 mb])—during the early evening hours of August 26. Although its central pressure continued to drop (reaching 29.29 inches [992 mb] on the morning of August 27), the system's wind speeds remained below the 74 MPH (119 km/h) threshold for hurricane classification. By the early evening hours of August 27, however, as its central pressure dropped to 28.99 inches (982 mb) and its wind speeds hiked to 75 MPH (120 km/h), Emily was again classified as a hurricane. Blocked from turning northward by the BERMUDA HIGH, Hurricane Emily spun to the northwest, pursuing a trajectory that if maintained, would take it ashore on the SOUTH CAROLINA-North Carolina border. Of Category 1 intensity from

August 27 to just after midnight on August 31, Emily began a deepening cycle during the early morning hours of August 31, reaching Category 2 status (with a central pressure of 28.64 inches [970 mb]) during the 31st, and Category 3 status (with a central pressure of 28.40 inches [962 mb]) by the close of the day. Hurricane Emily generated its lowest pressure reading of 28.34 inches (960 mb) during the early morning hours of September 1, while located some 25 miles (40 km) east of Cape Hatteras, North Carolina. Carried back into the deep Atlantic by an encroaching high pressure ridge over the central U.S. seaboard, Emily gradually weakened, and was downgraded to a Category 1 hurricane (with a central pressure of 28.79 inches [975 mb]) just after midnight on September 3, 1993. At this point, Emily was located far into the Atlantic, and was headed almost due south when it sharply recurved to the northeast and was downgraded first to a tropical storm, then to a TROPICAL DEPRESSION on September 4. Undergoing extratropical deepening on September 6, the system soon dissipated. While Hurricane Emily's 115-MPH (185-km/h) winds and scudding, five-inch (127-mm) rains either damaged or destroyed a number of beachfront properties in North Carolina, an extensive round of EVACUATIONS preempted any casualties or serious injuries. More than $10 million in financial losses were reported, a majority suffered by North Carolina's tourist industry.

Emily, Tropical Storm *North Atlantic Ocean, August 24–28, 1999* A short-lived tropical system, the fourth North Atlantic TROPICAL CYCLONE to carry the identifier Emily originated to the east of the Windward Islands, near latitude 11 degrees North, during the early morning hours of August 24, 1999. Moving almost due north, the system intensified, reaching TROPICAL STORM status (with a central pressure of 29.70 inches [1,006 mb]) during the midday hours of the 24th. By the evening hours of the same day, Tropical Storm Emily produced its minimum pressure reading of 29.64 inches (1,004 mb). Largely inhibited from further intensification by the outflow from nearby Hurricane CINDY (which was at the time quickly approaching Category 4 intensity), Tropical Storm Emily continued churning northward, quickly weakening to a TROPICAL DEPRESSION by the midday hours of August 28, 1999. Emily's tattered remnants were subsequently drawn into Hurricane Cindy's circulation, providing additional moisture and heat energy to that system. An offshore system for its entire existence, Tropical Storm Emily caused no deaths or injuries.

Emily, Hurricane *North Atlantic Ocean–Windward Islands–Mexico, July 11–21, 2005* The fifth named storm of the wholly remarkable 2005 North Atlantic

HURRICANE SEASON, Hurricane Emily was also the most intense—and destructive—of the five North Atlantic systems to thus far carry the name. First identified as Tropical Depression No. 5 (TD 5), Emily originated on July 11, 2005, near latitude 11 degrees North, longitude 42 degrees West, or some 900 miles (1,448 km) west of the Windward Islands. Tracking to the west-northwest at 15 MPH (24 km/h), it reached TROPICAL STORM intensity during the early morning hours of July 12. Benefiting from above-average SEA-SURFACE TEMPERATURES (SSTs) and low vertical WIND SHEAR, Tropical Storm Emily continued to intensify, and was upgraded to Category 1 status (with a central pressure of 29.50 inches [999 mb]) shortly after midnight on July 14. The still strengthening system sailed over the Windward Islands during the early morning hours of July 14, its 81-MPH (130-km/h) winds causing considerable damage on the island of Grenada, where one person was killed. Immediately upon entering the confined reaches of the CARIBBEAN SEA, Hurricane Emily embarked on a rapid deepening cycle, growing to Category 2 status (with a central pressure of 28.93 inches [980 mb]) at noon on July 14; by the early evening hours of the same day, Hurricane Emily achieved Category 3 status as its central pressure dipped to 28.67 inches (971 mb) and its sustained winds jumped to 115 MPH (185 km/h). By the early morning hours of the following day, July 15, Hurricane Emily had graduated to Category 4 status, with a central pressure of 28.11 inches (952 mb) and wind speeds in excess of 130 MPH (209 km/h). Pursuing a shallow recurvature to the west-northwest, Hurricane Emily commenced an EYEWALL rejuvenation cycle during the early evening hours of July 15, which saw its central pressure rise to 28.61 inches (969 mb) and its sustained wind speeds drop to 109 MPH (176 km/h). The system was now located some 300 miles (483 km) south of HISPANIOLA, and firmly maintaining its north-westerly trajectory. By the following morning, as Hurricane Emily completed its eyewall rejuvenation cycle, the system again began intensifying, becoming a Category 3 system (with a pressure of 28.28 inches [958 mb]) shortly after midnight on July 16, and a Category 4 hurricane later the same day. During this time, Emily, with its eyewall extending some 45 miles (75 km) from the center, grazed the southern coastline of JAMAICA, inundating the island with between five and 10 inches (127–254 mm) of precipitation that spawned landslides across its eastern reaches and claimed five lives. Grinding to the northwest at nearly 18 MPH (29 km/h), Emily continued to deepen, finally becoming a Category 5 hurricane—with a central pressure reading of 27.43 inches (929 mb)—shortly after midnight on July 17, 2005. With sustained wind speeds of 161 MPH (259 km/h), Emily was now the most intense

tropical cyclone yet observed in the North Atlantic basin during the month of July. Fortunately for the storm-prone nations that ring the Caribbean Sea, Hurricane Emily began to slightly weaken; by the time it crashed ashore in MEXICO's Yucatán Peninsula on the night of July 18–19, 2005, the system had weakened to Category 4 intensity. Driven by a central pressure of 28.20 inches (955 mb), Emily's sustained, 135-MPH (217-km/h) winds battered the resort cities of Playa del Carmen, Cancun, and Cozumel, destroying small BUILDINGS, smashing windows, flooding streets, and causing residents and tourists to huddle in darkened shelters. Across the Quintana Roo province, Emily crumpled billboards, and toppled communications towers, but surprisingly, caused no direct deaths or major injuries. Two people were killed when their helicopter, which was being used to evacuate offshore oil drilling platforms, crashed, and in another instance, a resident of Playa del Carmen was electrocuted. Analysts attributed the nonexistent direct death toll to a comprehensive series of EVACUATIONS that had occurred over the preceding two days. Following a similar trajectory to 1988's Hurricane GILBERT, and weakening further as it ground across the Yucatén's jungle spaces, Emily entered the Bay of Campeche during the early morning hours of July 19. Recurving more to the west, Emily again intensified over the warm waters of the Bay of Campeche, reaching Category 2 status during the early evening hours of July 19, and Category 3 status within another six hours. With a minimum pressure observation of 27.87 inches (944 mb), an intense Category 3 hurricane Emily's eyewall plowed into the eastern coastline of MEXICO, some 75 miles (121 km) south of the United States/Mexican border, during the mid-morning hours of July 20, 2005. In the town of San Fernando, sustained winds of 127 MPH (204 km/h) uprooted trees, destroyed communications and transportation networks, and threw small craft ashore, but claimed no lives or major injuries. While damage estimates of $400 million (in 2005 U.S. dollars) were tallied in Mexico, a downgraded Tropical Storm Emily lurched inland over northern Mexico, dropping large quantities of rain on the Sierra Madre mountain ranges between July 20 and 21. A fairly large tropical system, Emily further dropped between one and three inches (25.4–76.2 mm) across southeastern Texas. A tropical cyclone noted for its precocious early season intensity, Hurricane Emily directly and indirectly claimed a reported total of 15 lives, and caused several hundred million in property losses in the Caribbean, Mexico, and the United States.

Emma, Typhoon *Japan–Korea, September 6–11, 1956* A particularly severe late-season TYPHOON, Emma spent September 8–10, 1956, bringing sustained 140-MPH (225-km/h) winds and 22-inch (559-mm) PRECIPITATION counts to the island of Okinawa, mainland Japan, and the south coast of Korea. Millions of dollars in property losses were incurred between the two countries, a majority of them by U.S. military forces stationed on Okinawa. Seventy-seven people were killed, 27 of them U.S. service members, in separate drowning and air-crash incidents, making Emma one of the more costly typhoons to have addled the NORTH PACIFIC OCEAN up to that time.

The fifth TROPICAL CYCLONE of the 1956 TYPHOON SEASON, Emma originated as a tropical disturbance near the Mariana Islands, approximately 1,300 miles (2,092 km) east of Luzon Island, the PHILIPPINES, on September 3. By the afternoon of September 5, as the typhoon's CENTRAL BAROMETRIC PRESSURE continued to drop and its associated winds and seas further increased in intensity, meteorologists on Guam began to transmit a comprehensive series of warning messages to subsidiary forecasting offices positioned around the North Pacific rim. Designed to alert shipping and shore stations to the possible threat of a developing typhoon, these messages were in turn received by U.S. Air Force and Marine personnel on the occupied Japanese island of Okinawa, some 420 miles southwest of mainland Japan. Now well aware of the fact that Typhoon Emma's barometric pressures had not only dipped below 28.3 inches (959 mb) but also that its sustained wind speeds had escalated to 98 MPH (158 km/h), authorities quickly moved to prepare the island's military installations for a probable strike by the still-strengthening typhoon. According to newspaper accounts, aircraft were either moved to hangars or securely tied down in wind-breaking rows by the edge of the runways. Storm shutters were bolted across windows, while small vehicles and other light equipment were stowed under cover. In front of respective headquarters, the private pennants of Okinawa's resident flag officers were taken in, while the national ensign remained symbolically aloft, a steadfast morale booster before the coming storm.

On Okinawa's eastern beaches, where Emma's groundswells had already begun to roll ashore, typhoon-warning flags—two bright red standards with ominous black squares in their centers—suddenly appeared at local mastheads. Several dozen Marines, until then unaware of Emma's dangerous advance, had come to the beaches, eager to surf in the amazing breakers. Hazy sunlight illuminated the soldiers as they began to leave the sea, gathering up their gear for a return to their stations. As the typhoon warning flags lightly beckoned, 11 Marines who had paddled out beyond the breakers turned around and strenuously began to swim back to the

beach until a deep undercurrent caught them and began to drag them to sea. With frantic cries for help ringing in their ears, fellow Marines on shore hurriedly telephoned for assistance, for boats and rescue planes to be sent out, but it was too late. All 11 Marines drowned, their bodies lost to the storm-churning waters that now swirled around the island.

Less than 12 hours later, the EYE of Typhoon Emma passed along the Okinawa's west coast. Wind gusts of 143 MPH (230 km/h) tore across runways, flipping aircraft and knocking out landing lights. Entire walls were stripped from Air Force barracks, forcing personnel to seek shelter in quivering hangars. The four massive legs that supported a nearby watertank began to buckle and then finally sheer away, plunging the entire structure to the ground. As the steel watertank exploded in spray, its contents merged with the torrents of rain spilled by Emma. Within a 12-hour period, no less than nine inches (229 mm) of rain fell, gorging streets with flash floods and sudden mudslides. Part of a hangar wall collapsed when raging waters undermined one of its corners. Hundreds of houses were flooded with several inches of brackish water as early septic systems backed up, leaving a thin veil of mud over everything it touched. By the time Emma sped away from Okinawa on the evening of September 7, it had inflicted more than $8 million in property damage, including $4.1 million in aircraft losses to the Air Force, $2.5 million to the United States Marines, and another $1.8 million to the Army.

After battering the Japanese island of Kyushu with 135-MPH (217-km/h) winds and astounding 22-inch (559-mm) rainfalls on the night of September 8 and all through the next day, Emma was again positioned to claim the lives of military personnel. Although the typhoon's assault on mainland Japan had been formidable—34 people dead, hundreds of houses destroyed, millions of yen in damage assessed—Emma's intensity had barely been checked as it slipped into the Yellow Sea on September 10. Now bound for the Korean Peninsula's south coast, the typhoon began to strengthen, prompting the U.S. Air Force to dispatch a Boeing RB-50 to study wind velocities in the storm's dangerous northern quadrants. Laboriously rising through the spiral cloud bands, jolted by riveting rains and sporadic bolts of lightning, the RB-50 and its crew of 16 eventually disappeared.

Exactly what happened has never been determined, partly because no trace of the RB-50 or its crew was ever found. A subsequent inquiry into the disaster decided that either wind sheer or faulty altimeters were responsible for the plane's loss, but no unanimous verdict was ever reached. In the first scenario, winds moving at sharply different speeds on different levels could have plunged the aircraft into a fatal tailspin, while in the second, the intensely low barometric pressures found in Emma could have caused the RB-50's altimeters (which are, in essence, barometers) to give a false reading. The plane's crew, believing that they were in fact much higher than they really were, would have continued flying at their assumed altitude until they simply slammed into the sea. Such an occurrence was believed to have been responsible for the earlier loss of a U.S. Navy NEP-TUNE P2V-3W during Hurricane JANET on September 27, 1955.

The eye of Typhoon Emma, bearing 140-MPH (225-km/h) winds and pounding seas, came ashore near Pusan, Korea, on the evening of September 10, 1956. In the typhoon-prone port city, cargo vessels were slammed against their piers or else broken away and driven aground. Dozens of houses were blown down, causing isolated fires that were quickly extinguished by Emma's 16-inch (406-mm) rains. In the end, 42 people were killed, and another 35 were listed as missing. Many of those missing were fishers, caught at sea in their small boats.

After threading its way up the Korean Peninsula's east coast on September 11, a now-weakened Emma dissipated over the cooling waters of the northern Sea of Japan. The name Emma has been retired from the rotating list of TYPHOON NAMES.

Enigma Hurricane *Northeastern United States, August 19–20, 1788*   A meteorological riddle of uncertain classification, this small, fast-moving disturbance visited staggering destruction on southern and central NEW YORK State. Called a "most violent hurricane" by one survivor, the Enigma Hurricane's documented characteristics are not always in keeping with known cyclonic behavior. In New York City, for instance, the Enigma's winds blew with "incredible fury" for some 23 minutes and then suddenly abated, while the waters of the harbor, risen to a "very great height," submerged much of Manhattan's Front and Water streets. The Battery's west ramparts were knocked into the harbor; dozens of vessels, blown into a tangled dam at the Hudson River's outlet, formed a net for the ominous tons of debris—tree trunks, sections of houses, casks, crates of apples and pumpkins, even the carcasses of dead ANIMALS—now beginning to float downriver.

In Poughkeepsie, a Hudson River valley town located some 60 miles (97 km) north of New York City, the Enigma's surprising onset lasted for almost an hour; its fearsome winds and flooding rains uprooted the largest hardwood trees, swept away bridges, and mowed down acres upon acres of orchards and corn-

fields. In the nearby settlement of Hillsdale a number of sheep and oxen were killed, and fences, mill dams, potash works, and other sites suffered severe damage. According to an account of the storm in August 26, 1788's *Hudson Weekly Gazette*: "The day had passed . . . attended by showers, the air remarkably thick and sultry, the wind at southeast, when it chopped around, as it were in an instant, and blew a most violent hurricane from the northwest, attended by an amazing deluge of rain; it could hardly, however, with propriety be called rain—it rather wore the appearance of large rivers precipitated from huge mountains, and driven through the atmosphere by the irresistible force of the wind."

Soon crossing into western NEW ENGLAND, the Enigma Hurricane left the populous banks of the Hudson River in gloomy, rain-washed ruin. Although contemporary newspaper accounts did not report any known fatalities, enormous amounts of tree and structural damage did occur along a 75-mile-wide (121-km) swath of central and eastern New York. A meteorological mystery of enduring fascination, the question of whether the Enigma Storm was a tightly coiled hurricane or a TORNADO of extraordinary size and duration may remain unsolvable until a similar disturbance retraces its remarkable path.

Epsilon, Hurricane *North Atlantic Ocean, November 29–December 9, 2005*   The 26th named tropical system of the 2005 North Atlantic HURRICANE SEASON, and the first to be identified as Epsilon, late-season Hurricane Epsilon originated in the central North Atlantic, some 820 miles (1,290 km) east of BERMUDA, on November 29, 2005. Like a majority of late and postseason tropical systems in the North Atlantic, Epsilon both formed from an extratropical storm, and subsequently carved an erratic course as it slowly intensified to Category 1 status in the central Atlantic. Beginning at latitude 32 degrees North on the 29th, Epsilon tracked to the southwest as a TROPICAL STORM, passing very close to Bermuda on the night of November 29–30, 2005. Epsilon's offshore pressure of 29.32 inches (993 mb) delivered sustained 50-MPH (85-km/h) winds to Bermuda, but caused no damage or deaths. The system then performed a counterclockwise loop-the-loop maneuver, one that again took it past Bermuda on December 2 as a moderately severe tropical storm, with a central pressure of 29.26 inches (991 mb) and 58-MPH (93-km/h) winds. Well to the northeast of a sodden Bermuda by the following day, Epsilon became a Category 1 hurricane and shortly thereafter executed a course shift almost due eastward. Hurricane Epsilon achieved its lowest pressure reading—29 inches (981 mb) during the night of December 5–6—and despite the presence of colder SEA-SURFACE TEMPERATURES (SSTs) and unfavorable winds at higher altitudes, was able to maintain a well-defined EYE. Described by forecasters at the NATIONAL HURRICANE CENTER (NHC) as an ANNULAR HURRICANE, the system nevertheless steadily began to weaken. On December 7, 2005, located well to the southeast of where it had originated nine days earlier, the former Hurricane Epsilon was downgraded to a tropical storm, then to a TROPICAL DEPRESSION; by the early evening hours of December 9, 2005, it had dissipated.

Europe Although the southernmost reaches of the European continent lie on the same parallel as hurricane-prone Cape Hatteras, NORTH CAROLINA, the major coastal nations of Europe—Great Britain, France, Norway, Germany, Holland, and those of the Iberian Peninsula—have suffered only periodic incidents of TROPICAL CYCLONE activity. While a percentage of late-season North Atlantic HURRICANES and TROPICAL STORMS do develop east of 30 degrees or in those southwestern waters directly adjacent to Spain and Portugal, the trade winds' westward flow generally carries them to the New World where they devastate former European colonies in both the Caribbean basin and North America. On occasion, one of these storms, after having progressed to the United States's east coast and undergone RECURVATURE, will slide up the northwestern edge of the, BERMUDA HIGH and back into the Atlantic. There, the hurricane usually rapidly weakens, allowing the eastern course of the jet stream to carry its rain-laden gales into northern Europe.

Such a scenario took place in September 1966 when Hurricane Faith traveled nearly 8,000 miles (12,875 km) across the Atlantic, up the coast of North America, and back across the Atlantic again to lash Trondheim, Norway, with hurricane-force gusts and moderate rains. A central pressure estimated at 28.3 inches (960 mb) produced 104-MPH (167-km/h) winds over the North Sea during the early morning hours of September 6; within hours, Faith had undergone extratropical transitioning, but continued to deliver 46-MPH (74-km/h) winds to northern Europe. One person was drowned when a ferry boat foundered off the coast of neighboring Denmark. Those late- or out-of-season hurricanes that do manage to move to the north-northeast can only do so after the Bermuda High, or the ANTICYCLONE that dominates Atlantic wind patterns during the summer months, has subsided to its winter position south of latitude 40 degrees. So situated, it often blocks the westward passage of CAPE VERDE STORMS, forcing them to move north or into those waters that crest along Western Europe's rocky shores.

For this reason, a number of European hurricanes—those storms of true *tropical* origin, not simply EXTRATROPICAL CYCLONES with hurricane-force winds—have been offshore storms. In this capacity they have had the ability to bring both high winds and heavy seas to the coast's ancient industries, from dockyard workers to fishers. On December 1, 1947, at least 165 fishers were drowned when an out-of-season hurricane passed Oporto, Portugal, sinking dozens of steam-and-sail-driven fishing smacks. Less than a decade later, a similar hurricane swept the European coast from Spain to Scandinavia between December 29 and 30, 1951. Numerous vessels, from oil-laden coasters to holiday-bedecked ferry boats, were caught unaware by the surprise storm. Several were disabled, while others were run ashore and broken up. Despite the huge rescue operation launched by Great Britain, Holland, and Norway, more than 30 people lost their lives to the frigid, wind-stung waters of the North and Norwegian seas.

Of course, there have been those infrequent examples of hurricanes that have directly come ashore in Europe, some with catastrophic results. Dutch history holds that on October 1, 1574, a powerful hurricane drowned 20,000 Spanish soldiers as they laid siege to Leyden, Holland. Coming ashore shortly after dusk, the hurricane's ramlike STORM SURGE battered the lowland dikes, breaching them in places and flooding the countryside. Entire battalions of Spanish soldiers, along with their horses, tents, provisions, and weapons, were catapulted into a twisted tide of wreckage that eventually ebbed miles from where it had originally been encamped. The blow was so devastating, in fact, that the Spanish soon lifted the siege of Leyden, abandoning the recovering city to William the Silent.

One hundred twenty-nine years later on November 27, 1703, Great Britain would in turn be ravaged by what many eyewitnesses mistook for a hurricane. Known in England as the Great Storm of 1703, this intense extratropical system reportedly drowned more than 30,000 sailors and leveled some 5,000 houses in Gloucestershire, Sussex, Kent, and Suffolk. Recorded in lurid detail by Daniel Defoe, the two-day tempest produced winds of such intensity that they embedded stray roofing tiles and bricks 8 inches into the earth. Hundreds of head of cattle were killed, and an enormous wave, attributed by Defoe to the storm's passage, rose at the mouth of the River Thames. Spreading itself across the Medway, the resulting flood was hallmarked by great destruction all the way to London. At least 2,500 people perished.

More recently, in September 1961, Hurricane DEBBIE, a CAPE VERDE STORM that originated off

Africa's north coast, broke away from the trade winds's influence and managed to make it all the way to Ireland's west coast. There, Debbie's 106-MPH (171-km/h) winds claimed 11 lives as hundreds of houses in Ireland and Great Britain collapsed. Damage estimates ran into the millions of pounds, representing one of the costliest natural disasters in the region's history.

As recently as August 1987, Hurricane ARLENE seemingly changed all the rules when it originated off the SOUTH CAROLINA coast, traveled due east against the trade winds as a tropical storm, blossomed into a hurricane while leap-frogging across the mid-Atlantic, and then subsided into an extratropical storm as it neared the Iberian Peninsula. Deemed of extratropical depression intensity, Faith came ashore in central Portugal on August 26, bringing with it 39-MPH (63-km/h) winds and five inches (127 mm) of PRECIPITATION to the port city of Oporto. Arlene's influence would eventually be felt as far east as Madrid.

Although it was popularly believed in Great Britain to have been a hurricane, the Great October Storm of October 15, 1987, was not a cyclonic system of tropical origin. An exceptionally powerful extratropical cyclone that rapidly deepened as it crossed England from the southwest, the Great October Storm generated an extreme gust of 85 MPH (137 km/h) at Dover and severe, widespread flooding in the coastal counties of Sussex and Essex. Adequately forecast by the British Meteorological Office the previous day, the storm's relentless gales nevertheless downed hundreds of ancient oak, maple, and chestnut trees throughout central England and caused injury to numerous picturesque park settings by 18th-century landscape architect Capability Brown. Entire groves of rare trees in the Royal Botanical Gardens at Kew, some 20 miles (32 km) west of London, were uprooted or defoliated, prompting one curator to declare the Great October Storm's visit "the worst day in the entire history of Kew." Several chimneys and traffic signs in London itself were blown down, crushing automobiles and snarling transportation patterns for days. Well over 2 million people in southern England lost electrical and telephone service, and flooded highways and debris-strewn railroad tracks hindered the pace of relief operations. The most violent extratropical weather system to have affected the British Isles in modern times, the Great October Storm claimed 19 lives and caused 1.4 billion pounds, or $2.8 billion, worth of property losses.

On October 28–29, 1996, the meteorological tranquility of southern and western Great Britain was severely disrupted by what the London *Times* referred to as the "arrival from America of Hurricane Lili." A 15-day-old interloper that had first taken shape over the southwestern CARRIBEAN SEA on October 14, Lili carried its 90-MPH (145-km/h) winds and heavy seas across some 4,400 miles (7,081 km) of the NORTH ATLANTIC OCEAN before casting them against the stony shores of Wales, Cornwall, Dorset, the Isle of Wight, and Kent. The most powerful storm of tropical origin to have affected the island nation since 1961, Lili produced an extreme gust of 92 MPH (148 km/h) at Swansea, South Wales, and raging 15-foot breakers in the Bristol Channel. A four-foot (1-m) STORM TIDE inundated the Thames River estuary's muddy confines flooding scenic walkways and forcing the famous Thames Barrier to be raised. Five hundred holiday cottages in Somerset were seriously damaged by Lili's pounding waves. A U.S. oil drilling platform, under tow in the North Sea, snapped free in Lili's 40-foot (13-m) swells, nearly carrying its crew of 69 aground at Peterhead before a badly jostled tugboat could reestablish the line. On the cliff-shorn southwest coast of the Isle of Wight, a 75-foot (25-m) sailboat was beached at Chale Bay, necessitating the dramatic rescue of five passengers by a local lifeboat service. Electrical service to thousands of homes and businesses in Wales and the West Country was interrupted for several days. The most expensive meteorological disaster to have struck Great Britain since the Great October Storm of 1987, Hurricane Lili killed five people in England and Wales and tallied damage estimates of 150 million pounds, or about $300 million. Its three-inch (mm) rainfalls also, fortuitously enough, broke an expanding four-month drought over southwest England.

On September 13, 1993, the extratropical remnants of Hurricane FLOYD delivered hurricane-force winds to France's Atlantic coastline, buffeting buildings, boats, and causing high surf conditions. At one point in its existence a Category 1 hurricane (with a central pressure of 29.23 inches [990 mb]), Floyd transitioned to extratropical characteristics on September 10, and continued to deepen as it moved due eastward. By the time the system had reached the French coast, its central pressure of 28.52 inches (966 mb) produced 81 MPH (130 km/h) across the province of Brittany.

On September 27, 2005, an extratropical storm formed from the remnants of Hurricane Karl struck in and around the North Sea, bringing high winds, rains, and surf conditions to Denmark's Faroe Islands and southern Norway. Powered by a central pressure of 28.9 inches (980 mb), the extratropical system's 46-MPH (74-km/h) to 70-MPH (110-km/h) winds caused little property damage and no loss of life.

On October 11, 2005, Tropical Depression Vince—formerly Hurricane VINCE, and the 21st named storm of the 2005 North Atlantic season—

trundled ashore on the western coast of Spain, near the town of Huelva. The system's central pressure of 29.5 inches (1,000 mb) unfolded 48-MPH (77-km/h) winds across southwestern Spain, delivering some 3.3 inches (76.2 mm) of precipitation to the Cordoba Plains alone.

evacuation Perhaps the most effective preventive measure to the destructive and deadly effects of TROPICAL CYCLONE activity, evacuation is the process by which people, animals, and property are physically removed from those areas—along a coastline, riverbank, tidal strait, or inland valley—that are most at risk from the intense winds, coastal flooding, TORNADOES, and heavy precipitation associated with a tropical cyclone's impending landfall. Often costly, emotionally charged, and logistically challenging, evacuations have nonetheless measurably reduced the DEATH TOLLS from tropical cyclone activity around the world. They are the subject of intense planning and logistical operations on the part of emergency management agencies in the United States and overseas, and are often a highlighted topic in media coverage of a tropical cyclone strike. Several films, documentaries, and literary works have also focused on the concept and drama of tropical cyclone evacuations, and these often inform the public's perception of the need for, and the effectiveness of, preventive evacuations.

Although its scope and operational manifestations vary from state to state and country to country, the evacuation process for tropical cyclones has over decades become more refined, more ritualized through a jurisdictional desire for enhanced preparedness, lower death tolls, and reduced property and insurance losses. At the same time, it has been informed by improvements in meteorological technology and forecasting, and enhancements to emergency management operations; it has also suffered those unfortunate instances in which tropical cyclone evacuations have been incomplete, or never undertaken, or never anticipated. Although people have been moving out of the paths of tropical cyclones for centuries, large death tolls still occurred in the late 20th and early 21st centuries, as Hurricanes MITCH (1998) and KATRINA (2005) so graphically illustrated. In the first example, a majority of those lost in Honduras were killed in mountainous areas where Mitch's heavy rains and their concomitant landslides and flash floods were unexpected or underestimated, and no evacuations were undertaken. In the second example, several hundred people perished in LOUISIANA and MISSISSIPPI despite the largest and most complex mass evacuation yet seen in American history. These losses were attributable to a slew of emer-

gency management challenges and gaps, including the evacuation of nursing homes and the economically disenfranchised.

In one respect, it is very easy to understand the magnitude of difficulties that confront the tropical cyclone evacuation process. There are so many planning and operational vagaries, in fact, that emergency managers are often forced to triage priorities according to the ready availability of financial and logistical resources. There are, for example, the sheer numbers of people requiring evacuation. In preparation for Hurricane FRANCES's potential landfall as a Category 4 system, the state of FLORIDA undertook the evacuation of some 2.8 million people, the largest yet seen in a state well accustomed to tropical cyclone evacuations. There are also financial considerations. In the United States, it generally costs $2 to 3 million to evacuate one mile of coastline. These direct and indirect figures, which are generally a compilation of lost wages, increased overtime for public safety personnel, increased equipment deployment costs, increased sheltering and transportation costs, insured and uninsured losses, and lost revenue (especially tourist and energy production revenue), are often overwhelming considerations when deciding when and where to conduct a tropical cyclone evacuation. Tropical cyclones are all too frequently unpredictable phenomena, and their cumulative history has shown that just as hurricanes, TYPHOONS, and cyclones do come ashore in unprepared areas, they are just as likely to go north, or south, or out to sea—and all while thousands of people are waiting impatiently to safely return to their houses, their businesses, and their lives. It is not surprising, therefore, to find emergency management and public decision makers sometimes hesitant to commit expensive and often scare resources to a preventive measure that may not be needed.

While somewhat tempered by public awareness programs and improved evacuation practices, the cultural phenomenon of those individuals who wish to remain behind to secure and protect their property during a tropical cyclone strike likewise remains a challenge to evacuation planning organizations. A reminder of a time when the most effective property protection was provided by the owner, the proliferation of well-trained and efficient public safety agencies around the United States has provided an alternative approach to this dilemma. In recent years, the United States has adopted a policy of mandatory evacuations, a practice that essentially prohibits any personnel from remaining within the boundaries of an evacuated zone. When Hurricane OPHELIA threatened NORTH CAROLINA in September 2005, for instance, state officials ordered the mandatory evacuation of six counties, and voluntary evacuations for parts of eight others. Some 30 shelters were opened and 350 National Guard troops were activated to assist with response operations. Still recovering from the catastrophic effects of Hurricane Katrina less than two weeks earlier, the Federal Emergency Management Agency (FEMA) dispatched a further 250 personnel to North Carolina to assist with evacuation and sheltering operations. Mandatory evacuations are not a guarantor of safety, however; when Hurricane WILMA set its Category 3 sights on southwestern Florida in October 2005 (less than two months after Hurricane Katrina), a mandatory evacuation order was issued for Monroe County that some 80 percent of the population failed to heed.

In addition to sheltering and mass care concerns, transportation is a primary logistical challenge in tropical cyclone evacuation operations. History is fraught with examples of evacuations that have gone horribly awry, including the destruction of an evacuation train on the Florida Keys during the LABOR DAY Hurricane of 1935, and the fiery deaths of 23 evacuees on a bus being used to evacuate nursing home patients in Texas in advance of 2005's Hurricane RITA. As there are limits to the effectiveness of mass transit evacuation in suburban and coastal environments, the use of automobiles, trucks, and other vehicles is paramount. Several states that are subject to tropical cyclone activity have adopted special transportation and road regulations during these evacuations—including the use of designated lanes for gasoline purchases, roadway closures, and high-occupancy traffic—in order to reduce the jams and bottlenecks that characterize mass evacuations.

Military organizations also employ tropical cyclone evacuations as a preventive measure. The United States, which maintains dozens of installations in storm-vulnerable areas around the globe, routinely evacuates ships, aircraft, and personnel in preparation for tropical cyclone activity. In September 1998, for example, the U.S. Air Force evacuated nearly three dozen fighter aircraft to an airbase in Oklahoma in advance of Hurricane EARL's forecasted passage near Florida's Eglin Air Force Base. Ships homeported at the U.S. Navy's principal base at Norfolk, VIRGINIA, are regularly sent to sea in the event of tropical cyclone activity along the eastern seaboard, as seen when Hurricane EMILY drew to within 200 miles (322 km) of Norfolk in early September 1993. In preparation for Hurricane FELIX's potential landfall in Norfolk in 1995, the U.S. Navy sent more than 60 vessels to sea. As the events of the Great Samoan Cyclone of 1883 amply taught the navy, vessels caught within the confines of a harbor are in greater danger of being run aground or driven into collision with piers or other vessels than they are

of being overwhelmed by a tropical cyclone's fury while on the open sea.

Tropical cyclone evacuations also occur on those deep sea oil drilling platforms that dot the world's oceanic storm zones, particularly in the GULF OF MEXICO. A standard oil rig can house up to several hundred personnel, and in preparation for a tropical cyclone passage over or near these rigs, mass evacuations are undertaken by the respective petroleum companies that own and operate the rigs. Expensive, time-consuming, and often dangerous to undertake, there are strong economic ramifications that accompany the evacuation of deep-sea petroleum-producing installations. There is also an overwhelming need to protect rig personnel from severe tropical cyclones; when Hurricane Katrina blasted across the Gulf of Mexico in August 2005, 30 oil drilling platforms were damaged or destroyed, including one that was wedged beneath a bridge in Mississippi. The shutdown of these rigs, coupled with the closure of coastline refineries, reduced energy production operations over a six-month period by more than 75 percent, causing petroleum, gasoline, and natural gas supplies to dwindle, and prices to soar.

Tropical cyclone evacuations can also occur after a particularly severe landfall. In the days following Cyclone Tracy's devastating Christmas visit to Darwin, AUSTRALIA, in 1974, government authorities encouraged the evacuation of the storm-ravaged city in order to reduce the threat of an epidemic and to permit wide-scale recovery efforts to begin. Government-issued financial incentives, such as relocation and evacuation costs for evacuees, were covered, and access to the city was strictly limited by physical security measures.

A survey of tropical cyclone evacuation practices around the world reveals a number of different theoretical and operational concepts for pre-, peri-, and poststrike evacuations. These are, of course, tailored to the unique traditions and rituals of a particular jurisdiction, as well as to the availability of resources and the solidity of the political and administrative system. CHINA, for example, routinely uses its military to evacuate citizens from those coastal cities and towns that are vulnerable to TYPHOON activity. In mid-September 2005, the Chinese army evacuated more than 1 million people to railway stations, hotels, and other shelters as Category 4 Typhoon Kanun bore down on the port city of Taizhou. Although Kanun weakened before making landfall on September 11, its 104-MPH (167-km/h) winds, heavy surf, and flooding rains caused considerable damage, but no loss of life. Had such a large evacuation not been undertaken, the consequences could have been appalling. In September 2006, as Hurricane FLORENCE bore down on

BERMUDA, hotels issued written evacuation and other emergency preparedness information to their guests, thereby proactively ensuring that some degree of situational awareness was preserved.

CUBA is another nation that uses its military resources to affect coastal evacuations. According to media sources, the Cuban tropical cyclone evacuation protocols—known as the Cuba Emergency Response System—are unique in that they not only stipulate the mandatory evacuation of all people and animals, but much of the personal property of the evacuees, as well. Employing a version of the Citizens Emergency Response Team (CERT) model adopted in the United States after the terrorist attacks of September 11, 2001, Cuba's hurricane evacuation protocols call for the posting of wardens in each neighborhood to monitor the safety of evacuees and to provide sheltering information. Particular care is also given to special needs communities, such as the elderly and those with mobility limitations. There are controversial and perhaps unwise aspects of the Cuban Emergency Response System; part of the program calls for preventing electrocutions by terminating electrical services in evacuated areas and predeploying military resources. While a great deal of political posturing influences Cuba's record of effective hurricane evacuations, there is little doubt that the island nation—which has suffered a long history of deadly hurricane strikes—is seriously addressing its tropical cyclone preparedness.

There is a second preventive response to tropical cyclone activity aside from evacuation, and that is to shelter in place. In the Solomon Islands, for example, the archipelago's remote location makes an effective mass evacuation for an impending cyclone almost impossible. Consequently forced to shelter in place, the Solomon islanders have developed stronger building codes and practices, such as constructing low-lying buildings that pose less of a profile to a cyclone's winds, several cyclone shelters, and buried food and water supplies. On the other hand, sheltering in place has not always proven as effective as emergency management standards require. During Hurricane Katrina's landfall in New Orleans, thousands of nonevacuated people took refuge in what were called "shelters of last resort," such as the Louisiana Superdome, which while previously used as a hurricane shelter, was not properly equipped or stocked for an event of Katrina's magnitude or duration. The failure of several levees allowed much of the city—including the area immediately around the Superdome—to flood, thereby hindering rescuers access to the structure. Media reports state that three people perished in the Superdome, and dozens of others were injured or made ill.

The historically active 2005 North Atlantic HURRICANE SEASON (and Hurricane Katrina's catastrophic assault on the City of New Orleans) rekindled debate on the respective preparedness levels of those states and communities that lie within the nation's hurricane zones. Large oceanfront cities like NEW YORK and Miami were again confronted with the reality of their potential vulnerability to powerful tropical cyclone activity, as well as the innumerable challenges that would have to be addressed in order for an effective preventive evacuation program for their populations to be designed and implemented. Because of its sky-high population density, New York City's hurricane evacuation plan called for zoned evacuations, meaning that those communities that lie closest to the Atlantic would be evacuated first, and so on inland, based upon the forecasted category of a threatening hurricane. In addition to posting evacuation route signs in vulnerable coastal areas, the City of New York also adopted a solar system sheltering model, whereby evacuees are first sent to a reception center, and from there to a shelter. This controversial evacuation program was reinforced through a public relations campaign that did as much to highlight the challenges in effectively evacuating such large numbers of people as it did to educate the public on how to successfully accomplish this objective. In one telling instance, New York City residents evacuating one oceanfront neighborhood by automobile would have three options; two of them involving crossing crowded bridges, and the third by driving several miles into a neighboring county before circling back to reenter the city's limits.

Since Hurricane Katrina, the United States has made significant strides toward improving the effectiveness of its overall tropical cyclone evacuation procedures. One improvement requires that local evacuation systems now accommodate animals—principally family pets like dogs and cats—in reception centers and shelters. Although cumbersome and potentially difficult to implement, the adoption of this and other such measures reduces some of the emotional personal trauma inherent in mass evacuations, and consequently makes them easier to implement and control.

To better improve tropical cyclone evacuation readiness for individuals, families, and businesses, emergency management specialists recommend the following:

• Possess an awareness of the environment and its vulnerability to tropical cyclone activity. Consult local emergency management authority in order to determine whether an area is susceptible to flooding or other hazards associated with a tropical cyclone landfall;

- If residing or vacationing in an area susceptible to hurricane activity, preplan evacuation routes and project timetables for accomplishing those routes. If possible, make previous arrangements to shelter with family or friends outside of the evacuation zone;

- Learn local shelter and reception center locations, as well as primary and secondary transportation routes to those locations;

- During an evacuation, be aware of the secondary hazards associated with tropical cyclone activity, such as landslides, flash floods, roadway ponding, downed power lines (electrocution), collapsed roadbeds and bridges, damaged or falling trees, flying debris, loose animals, tainted water, and chemical spills;

- Prepare a Go-Bag or other portable resource that contains items essential in an emergency, including medicines, spare eyeglasses, flashlight, light sticks, a first aid kit, food, water, insurance and other important documents, and personal contact information. If evacuating with children, special needs individuals, or pets, include relevant survival items for each;

- In the event a tropical cyclone threatens to make landfall, evacuate as early as possible;

- When a hurricane evacuation order is issued, obey it in an orderly and efficient manner.

evaporation In the broadest sense of the definition, evaporation is the physical process by which a liquid becomes a vapor, more specifically, the meteorological function by which warm ocean water changes into water vapor before rising into Earth's TROPOSPHERE to undergo the opposite processes of CONDENSATION and the resulting production of clouds. Varying from season to season and between hemispheres, evaporation rates over the NORTH ATLANTIC, NORTH PACIFIC, and Indian Oceans have long been recognized by meteorologists and scientists as one of the climatological seeds from which TROPICAL CYCLONES spring and by which they are in turn sustained. Determined in large part by SEA-SURFACE TEMPERATURES evaporation has in a very real sense the power of life and death of a tropical cyclone because it is through the process of evaporation that the enormous quantities of water vapor required to feed the condensing RAIN BANDS are generated.

explosive deepening *Explosive deepening* is a term employed by the NATIONAL HURRICANE CENTER

(NHC) to describe the swift intensification of a TROPICAL CYCLONE in the NORTH ATLANTIC and eastern NORTH PACIFIC OCEAN basins. By definition, explosive deepening occurs when the minimum central pressure of a tropical cyclone drops at least 0.073 inches (2.5 mb) per hour for a 12-hour period, or 0.14 inches (5 mb) per hour for at least six hours. Explosive deepening is used in conjunction with the term *rapid deepening,* which constitutes a drop in a tropical cyclone's central pressure of 0.05 inches (1.75 mb) per hour for a 24-hour period, or 1.2 inches (42 mb) for a 24-hour period. In recent history, Hurricanes KATRINA, RITA, and WILMA all underwent periods of explosive deepening during the 2005 North Atlantic season. As illustrated by Hurricane Wilma, explosive deepening can be an awesome but terrifying phenomenon to observe. During the early evening hours of October 18, Hurricane Wilma's central pressure stood at 28.8 inches (975 mb), making it a powerful Category 1 hurricane. Within six hours, Wilma's central pressure had plummeted to 27.9 inches (946 mb), pegging it at Category 4 intensity. And within six hours of that observation, the system's minimum central pressure had skidded to 26.3 inches (892 mb), and within another six hours, to the record-setting intensity of 26.04 inches (882 mb)—the most intense Category 5 hurricane yet observed in the North Atlantic basin. In one 18-hour period, Hurricane Wilma shaved 2.7 inches (93 mb) from its central pressure; and within that period, lost 0.85 inches (29 mb) of pressure during a six-hour window.

extratropical cyclone This specifically defined term is used to describe those macroscale cyclonic systems that originate outside the world's tropical weather zones. One of three types of cyclonic storms—TROPICAL CYCLONES, and mesocyclones are the others—extratropical cyclones differ from tropical cyclones in any number of important ways. First, whereas the low-pressure center of a HURRICANE, TYPHOON, or CYCLONE possesses a core that is warmer than the surrounding air, the low-pressure center of an extratropical cyclone is colder than its adjoining atmosphere. Second, the strongest winds in a counterclockwise-spinning (in the Northern Hemisphere) extratropical cyclone are found in the upper atmosphere (above 20,000 feet [6,667 m]), while those in a counterclockwise tropical cyclone occur close to Earth's surface. Although a large tropical cyclone rarely exceeds 500 miles (805 km) across, the average cell of an extratropical cyclone can reach widths of 800–1,000 miles (1,287–1,609 km), thereby allowing it to affect a much greater area for a much longer (several days to a week) period of time. Also, unlike tropical cyclones, extratropical systems contain *fronts,* lines where air masses of different TEMPERATURES and varying humidities come together as localized storms. For these reasons, extra tropical cyclones are often responsible for creating some of the world's most severe weather, including those notorious blizzards known as Nor'easters that occur over the U.S. eastern seaboard during the winter months.

Following landfall, it is not unusual for a tropical cyclone to either merge into, or evolve into, an extratropical cyclone. This is known as extratropical deepening. In the former scenario, a hurricane, typhoon, or cyclone interacts with the cold, low-pressure core of an advancing extratropical cyclone and swiftly intensifies, as dramatically evidenced by both Hurricane CAMILLE (1969) and the GREAT NEW ENGLAND HURRICANE OF 1938. In the latter scenario, the warm center of a tropical cyclone in the decaying stage gradually cools, sinks, and elongates as it tracks away from the equator, thereby transforming it into a macroscale cyclonic system with extratropical characteristics. It is also not unusual for tropical cyclones to form from or within large extratropical cyclones positioned over the world's oceans. According to meteorologists, large extratropical cyclones generally possess uniform baroclinic zones, shallow temperature gradients and low WIND SHEAR conditions that when added to warm SEA-SURFACE TEMPERATURES (SSTs), can permit the formation of a tropical cyclone within an extratropical system. On November 25, 1980, for instance, Tropical Storm Karl formed at the center of a deep layer extratropical cyclone located over the central Atlantic, and retained hurricane intensity until November 27, 1980, at which point it regained its extratropical characteristics. At its most intense, Hurricane Karl produced a pressure reading of 29.1 inches (985 mb). Shortly after midnight on October 10, 2004, Sub-Tropical Storm Nicole originated from an extratropical storm (with a central pressure of 29.5 inches [1,000 mb]) situated near Bermuda. By the mid-morning hours, Nicole was classified as a tropical storm, and by midnight on the 11th, had produced a pressure reading of 29.4 inches (994 mb). At its most intense, Tropical Storm Nicole produced a pressure observation of 29.1 inches (986 mb) during the afternoon of October 11, and just hours before it was absorbed into another large extratropical low pressure system swirling off the Canadian Maritimes.

A majority of the tropical cyclones that dissipate over the open oceans first undergo extratropical deepening—or the system's adoption of extratropical characteristics. Some become extratropical storms, while others fade to extratropical depression status. While only periodically affected by direct tropical cyclone activity, Europe—particularly the nations of western Europe—are frequently struck by extratropical cyclones that originated from North Atlantic hurricanes and tropical storms. In late August 1986, for example, the British Isles were battered by Extratropical Storm Charley, a former Category 1 hurricane that had undergone extratropical deepening while churning eastward across the extreme NORTH ATLANTIC. On the morning of August 26, 1986, and while passing over Ireland, Extratropical Storm Charley generated a pressure reading of 28.9 inches (980 mb); as a Category 1 hurricane it had produced only a minimum pressure reading of 29.1 inches (987 mb) on August 18. Severe damage to trees and small BUILDINGS resulted, and parts of Ireland received some five inches (127 mm) of precipitation. Considering Charley's extratropical fury, it is perhaps noteworthy that Hurricane Charley had itself originated from a subtropical depression that had formed over the Florida Panhandle on August 13.

As practiced by the NATIONAL HURRICANE CENTER (NHC), once a tropical cyclone in the North Atlantic is awarded an identifier from the naming list, it maintains that name until dissipation—even if the system undergoes extratropical deepening in the process. For instance, 1992's Tropical Storm Earl was identified as Extratropical Depression Earl after it had lost its tropical characteristics and closed circulation.

In some cases, tropical cyclones can undergo extratropical deepening before reverting to tropical cyclone characteristics. During the 2004 North Atlantic hurricane season, a tenacious Hurricane IVAN first reached Category 5 status twice before making landfall in northwestern FLORIDA as a Category 3 system. It then recurved in a clockwise direction across the eastern United States, being downgraded to a tropical depression along the way. Upon reemerging over the North Atlantic near the Virginia/Delaware border on September 18, the tropical depression assumed extratropical characteristics and was reclassified as an extratropical depression. Carried almost due southward, by the early morning hours, Ivan was upgraded to an extratropical storm, a classification that it maintained until the midday hours of September 20, at which point it was downgraded to an extratropical depression. On September 21, as it neared landfall in southeastern Florida, Ivan was again classified as a tropical depression, and was later upgraded to a tropical storm while moving across the GULF OF MEXICO to its final landfall, as a tropical storm, on the TEXAS-LOUISIANA border. According to records maintained by the NHC, the decision to rename the system Ivan after it had undergone extratropical deepening was not an easy one, and considerable time was spent reviewing the system's characteristics and history before it was determined that sufficient energy remained from Hurricane Ivan's remnants to allow the regenerated system to carry the name.

eye　Sometimes referred to as the vortex or calm center of a TROPICAL CYCLONE, the eye stands as the most distinguishable physical feature of a mature HURRICANE, TYPHOON, or CYCLONE. Generally visible to weather satellites, aerial reconnaissance aircraft, and those Earth-bound observers over which it passes, the circular or elliptical eye not only serves as the rotation point for a tropical cyclone's clockwise or counterclockwise circulation but also contains its meteorological core, its lowest barometric pressures. Ascending 40,000–50,000 feet (13,333–16,667 m)—from Earth's surface to the top of the OUTFLOW LAYER—and ranging in width from three to 65 miles (5–105 km) the eye is encased by a broad band of CUMULONIMBUS CLOUDS, billowing black CONVECTION cells that constitute the violent parameters of a tropical cyclone's EYEWALL. Within the eyewall, wind speeds are at their fastest, and heavy rains sweep the sea or land into frothy destruction. But in the eye itself, a veritable oasis endures. The tropical cyclone's winds, once so turbulent, quickly decrease to a breezy 15 MPH (24 km/h) or less. Its rains, formerly so deluging as to reduce visibility to zero, abruptly cease, leaving the air still but humid. During the day, sunlight penetrates the open eye, while on moonlit nights pale white shadows paint the tops of the convulsive seas below. The eye's temperature often rises and falls and can escalate to 85°F (29°C) or higher before dropping as much as seven to 10 degrees, and the system's CENTRAL BAROMETRIC PRESSURE, its point of lowest barometric pressure, stabilizes and remains unchanged until the tropical cyclone undergoes another deepening or weakening cycle.

The genesis of the eye occurs in the early stages of a tropical cyclone's development when a TROPICAL WAVE, or an isolated pocket of low barometric pressure, finds itself positioned over the moist, heated expanses of the world's INTERTROPICAL CONVERGENCE ZONE (ITCZ). Once a thunderstorm (or, in some instances, the rejuvenating remains of a dissipated hurricane, typhoon, or cyclone), the highly unstable tropical wave hesitantly migrates west with the trade winds. Surrounded by a posse of climatological enemies, the wave will either undergo deepening as a favorable alliance of evaporation and conduction feeds heat-laden water vapor into the troposphere's lower levels or else succumb to upper-level WIND SHEAR, low evaporation rates, cool seawater temperatures, and rising internal barometric pressures.

Should the tropical wave endure, witnessing a continued drop in its barometric pressure, it will become the void on which a series of meteorological forces will soon converge. Nearby cumulus and nimbus cloud systems will move to fill the low-pressure center, gather around it in ever-tightening coils that begin to yield thunder, lightning, rain, and 10–15-MPH (16–24-km/h) winds. Now rising some 5,000–10,000 feet (1,667–3,333 m) above the sea, these light-gray convection cells steadily blossom into 25,000-foot (8,333-m) cloud towers as the low-pressure pump draws in humid, sea-level air and then sends it spiraling upward. Bound for the top of the troposphere, or that point at which the stratosphere begins, this warm, moist air steadily cools as it rises, condensing into more PRECIPITATION-filled cumulonimbus clouds. These, in turn, fold their way into those cells already present, their added rains progressively turning the burgeoning TROPICAL DEPRESSION battleship gray in color.

With sustained wind speeds of less than 39 MPH (63 km/h), the embryonic tropical depression is not yet sufficiently organized for the circular eye associated with mature cyclonic systems to become clearly defined. Satellite photographs of developing depressions generally reveal a crescent-shaped splay of cumulonimbus and nimbus clouds grouped around 90–180 degrees of the low-pressure center. These strengthening rain bands, organized in part by the CORIOLIS EFFECT, follow the eye's rotation, spinning counterclockwise in the Northern Hemisphere and clockwise in the Southern, outlining in a sense its approximate diameter and shape. Because even the eye of a mature tropical cyclone can experience wide variances in its configuration, swelling and contracting as much as 10 miles (16 km) in response to minimal changes in its central barometric pressure, it is not surprising to find the same behavior in the eye of a formative storm. Elongating to the west, bulging to the north, spreading to the south, a tropical depression's airborne eye wobbles as it spins, forcing its thickening mantle of convection cells to do the same. At times this curious effect may even make it appear as though the tropical depression is suddenly breaking apart, returning to its base thunderstorms by splitting its very core in two. Although the eyes of those tropical depressions that either make landfall or lose their intensity while still over open waters do indeed manifest this type of collapse, those eyes slated to continue strengthening simply pull back, wobbling another few degrees before slipping into transition again.

Beneath such seeming chaos, however, the tropical depression's eye is swiftly organizing itself into an efficient airshaft, an umbilical cord through which the storm will grow from the inside out. As the eye's central pressure further decreases, sinking to 29.5 inches (999 mb) and below, increasing quantities of warm, wet, nourishing air are taken in, dispersed in rising currents to those cloud bands nearest its outer edges. The increased velocity of the ascending currents, now clocked at between 35 and 38 MPH (56–61 km/h), drives them even higher into the troposphere, allowing them to deliver more residual moisture to upper elevations, thereby increasing through condensation the height of the cumulonimbus towers. Not only forming a lengthened, 29,000-foot (9,667-m) flue for the eye's increased venting, these cumulonimbus bands will also shortly crowd the troposphere's underside, curling over and downward to begin to connect the lower cloud bands's convective circulations into one enormous, 45,000–50,000-foot-high (15,000–16,667-m) convective cell.

With its sustained eyewall winds now exceeding 39 MPH (63 km/h), the tropical depression and its ill-defined eye is upgraded to TROPICAL STORM status, or to that classification in a tropical cyclone's life when its eye begins to assume definitively the circular shape so familiar to storm watchers. Although the eye remains unclosed, its adjoining eyewall now encircles up to 280 degrees of its gale-force perimeter. Within the eye itself, the system's point of lowest barometric pressure unevenly floats from side to side, allowing the strengthening eyewall to protrude momentarily into the tightening ring. Vagrant wisps of cloud drift into the center, trailing along the inner edge of the eyewall before rising air currents shear them apart, carrying their remnants into the upper atmosphere. Robbed of most of their moisture along the way, they join the eye's emerging outflow layer as cirrus and CIRROSTRATUS CLOUDS.

Created by the increased venting of the eye, the outflow layer's icy cloud canopy can sometimes obscure the eye, making it invisible to satellite photography. The cirrus cover's presence, however, is a significant indicator of a tropical storm's strengthening configuration, a sign that the eye is now digesting such large quantities of tropical air that it has had to develop a complex excretory system to deal with the overflow. Should the eye of a tropical storm—indeed, any class of tropical cyclone—not vent itself efficiently, rising air currents will eventually inundate the eye, force up the barometric pressure, and break the system to pieces.

Conversely, if an eye is venting at peak performance, if its ascending currents are moving at speeds greater than 110 MPH (177 km/h), its outflow layer further aids in its survival by carrying spent air much farther away from the eye, delivers it to the storm's fringe bands where it condenses into rain, falls to the sea, and is drawn back into the eye.

In this way, the eye's circulatory preeminence adds yet another story to a tropical cyclone's self-sustaining construction. Constantly augmenting its mass from the inside, the eye builds every time its barometric pressure further deepens. As though using one crane to hoist another, it again increases the height of its towering flue by tunneling elevated amounts of humid sea air into the upper atmosphere. Above 25,000 feet (8,333 m), where the air turns much colder, cumulonimbus blocks continue linking the shorter, thinner rain bands beneath them into a unified circulation. Those very same bands, in the meantime, are proceeding with their own intensification, drawing moisture and latent heat from the eye's whirling currents and applying it to the organization of their own convection cells. Growing more and more saturated by the minute, the storm's lower 3,000 feet (1,000 m) will eventually burst into roiling rain showers, wind-driven torrents that will add their own moisture to the eye's relentless intake, and the outflow layer, soaring above like an enormous round roof, will further envelop the entire assembly, sheltering it within its symbiotic embrace.

In time and if aided by favorable climatological factors, the tropical cyclone will grow into a well-organized meteorological machine. Controlling from the hub a hurricane's various circulation patterns, the eye simultaneously develops as it feeds its surroundings and can on rare occasions even achieve the same cyclonic vorticity as a TORNADO. Category 1 tropical cyclones, or those with sustained wind speeds of between 74 and 95 MPH (119–153 km/h), are certainly not of this class, although at this point in a tropical cyclone's development the eye is most certainly organized enough for satellites to record its location and basic shape. Sometimes still unclosed by the eyewall, the eye of a Category 1 tropical cyclone is most often obscured by the outflow canopy and may in fact be filled with stray clouds that are sinking back to Earth from higher elevations at a rate of 30 feet (10 m) per minute. Because most strengthening tropical cyclones spend less than two days as a Category 1 system, any intrusion of clouds into the eye itself does not generally prove fatal to its organization; further decreases in central barometric pressure, in tandem with increased air intake, soon drive such cloud masses into the thickening eyewall, nourishing the process that will eventually upgrade the system to that of Category 2 status.

Category 2 tropical cyclones, or those with barometric pressures of between 28.9 inches and 28.5 inches (979–965 mb), tend to have better organized circular eyes, although in some noted cases the eyewall has remained unclosed for most of the second stage, giving the storm in question a curiously elliptical appearance. Fewer clouds penetrate the eye of a Category 2 system, while increased venting aloft further strengthens the outflow layer's efficiency, thereby making the top of the eye visible from space. The average Category 2 hurricane, typhoon, or cyclone possesses an eye between 15 and 30 miles (24–48 km) across: this of course increases or decreases in size depending on the system's deepening or weakening configuration. During the Category 2 stage, a tropical cyclone's eye significantly increases the amount of warm, moist air it pumps into the troposphere, thereby adding to its cumulonimbus eyewall towers and hence its convective, or heat-exchanging, power.

At the point when a tropical cyclone reaches Category 3 and 4 stages, the tight, circular eye becomes clearly delineated. Like an enormous well-head at the storm's heart, the eye of a Category 3 or 4 system indicates at once, and with ominous certainty, how efficiently the meteorological pump below is beating. Ranging from 13 to 60 miles (21–97 km) across, the eye of a Category 3 or 4 tropical cyclone is always completely enclosed by the eyewall, although shifts in barometric pressure and course changes can on occasion cause the wall of larger storms to be pierced. Quickly filled by the cyclone's rotating cloud bands, the break in the eyewall is sealed even tighter as the system's central pressure continues to drop, proportionally increasing the amount of nourishing

Hurricane Isabel illustrates its pinwheel eye on September 13, 2003. At the time this photograph was taken, Isabel had just been downgraded from a Category 5 to a Category 4 system. Note the spokelike cloud structure in the storm's eye; with its four points it resembles a pinwheel. *(NOAA)*

fuel being drawn into the bottom of the eye. Ascending air currents, some moving as fast as 50 MPH (81 km/h), spiral around the eye, rising up to spin away into the dispensing outflow layer. For this reason, the eye of a Category 3 or 4 tropical cyclone is usually free of stray clouds, permitting satellites to peer into the very essence of the storm and photograph the glint of moonlight on the writhing night seas below.

Category 5 tropical cyclones, those uncommon but recorded furies that redefine cyclonic theory every time they appear in the world's storm zones, contain eyes of consummate sophistication, organization, and efficiency. Powered by central pressures of 27.17 inches (920 mb) or *less,* the eyes of most documented Category 5 systems are often on the small side, extending between eight and 12 miles (13–19 km) across. In some respects more akin to tornadoes than mature cyclones, these eyes are surrounded by tightly bound eyewall rings, dense cumulonimbus cloud cells that produce sustained wind speeds of 155 MPH (249 km/h) or higher but fairly mild rains.

In some extremely rare instances, such as 1969's Hurricane CAMILLE, two eyes can form within the same system. Reflecting an extraordinarily low central barometric pressure of 26.8 inches (909 mb), Camille's pressure gradient eventually created a hurricane within a hurricane by dividing the system's 100-mile-wide (161-km) eyewall into two halves; the first, which surrounded the hurricane's initial eye, extended outward for 50 miles (80 km) until a clear band, or a second eye, carried the eyewall's remainder to the point where the outflow layer began. In other words, for a time the outer eye vented the outflow layer while the inner eye, tightly enclosed by a second even more-violent eyewall, spiraled inside. A similar scenario was witnessed in the hours immediately preceeding Cyclone TRACY's cataclysmic landfall at Darwin, AUSTRALIA, on Christmas Day, 1974. At the time of its landfall in Darwin, Tracy's eye measured some seven miles (12 km) in diameter.

Enhanced observational and monitoring technology has allowed meteorologists unprecedented access to the eye of the tropical cyclone. Along the way, the mechanics of the eye have led to new terms and classifications for the relationship between the eye and eyewall, such as a *moat* (a clear, concentric band of cool, sinking air that surrounds the eyewall or eyewalls of intense tropical cyclones; the *stadium effect,* where the eye opens upward, with its narrowest point above the Earth's surface and striations above; and the *pinhole eye,* or the very tightly coiled eyes observed in the most intense tropical cyclones. Examples of pinhole eyes can be found in the LABOR DAY HURRICANE (1935; 10 miles [16 km] in diameter), Cyclone

Tracy (1974; seven miles [12 km] in diameter), Super Typhoon Maemi (15W, 2003, six miles [10 km] in diameter) and Hurricane WILMA (2005; four miles [6 km] in diameter). In Wilma's example, the system's central barometric pressure at the time it generated a pinhole eye (October 19) was 26.04 inches (882 mb), while the Labor Day Hurricane's central pressure was estimated at 26.34 inches (892 mb). Super Typhoon Maemi's central pressure at the time it produced a pinhole eye was 26.87 inches (910 mb), while Super Typhoon Lupit's (November 27, 2003) was an astonishingly low 26.13 inches (885 mb).

The eye of a tropical cyclone is at its healthiest while over the open ocean's humid expanses. Its forward motion hindered by only minimal amounts of surface friction, the eye maintains its cylindrical form until it either passes over cooling waters or makes landfall. In the first scenario, cold water pools and eddies restrict the flow of warm, humid air into the eye, thereby denying it its convective nourishment. As a consequence, the tropical cyclone's central pressure begins to rise, ascending currents weaken, descending currents strengthen, and the eyewall is starved into collapse as the eye itself rapidly starts to expand. In time it will either disintegrate altogether or undergo EXTRATROPICAL deepening, becoming one of the many high-pressure centers that daily swirl above Earth.

Upon making landfall, however, a similar destructive chronology besets the eye, except that friction between the bottom of the eye and the surface of the land tends to retard the eye's forward progress, causing it to elongate and weaken. Drawn into an ellipse, the eye's dissipation is increased by the fact that overland evaporation rates are hardly large enough to sustain a tropical cyclone's voracious need for tons of moisture. Denied this, the tropical cyclone's cumulonimbus pylons begin to fold in on one another, losing their convective elasticity in a shower of rain squalls. Again, the eye's central pressure begins to rise, forcing the outflow canopy to melt into isolated cirrustratus wisps. Before long it will have expanded to perhaps 200 miles (322 km) across, a complete inversion of the precise point of low barometric pressure that once existed at the heart of the mightiest meteorological phenomenon known to humankind.

However, for many centuries the exact nature of the tropical cyclone's eye was not fully understood, leading to a spate of misconceived stories concerning the circular shape of tropical systems. In their journals, letters, and diaries, many survivors of 17th- and early 18th-century hurricanes describe the passing of the eye as though the storm had doubled around and come back. Others actually mistook the eye's calm for the end of one storm and its returning fury as the start of a second storm. While they were certainly aware of the eye's tranquil attributes and of the great number of birds and other ANIMALS found trapped within its windless confines, they remained seemingly oblivious to its purpose or even its approximate shape.

It was not until the late 1830s, when pioneering hurricane scholar WILLIAM C. REDFIELD published the first of his famous treatises on the law of cyclonic storms, that the eye was seen as an integral part of a single storm, recognized for the towering meteorological core it is. Quickly accepted by the public, Redfield's model for cyclonic storms was by 1850 well enough understood for a newspaper in Apalachicola FLORIDA, to liken the passage of a hurricane's eye over the city on August 23, 1850, as being on the "verge of the 'annulus,'" or the orifice through which the hurricane vented its powerful winds. The term *annulus* was eventually abandoned in favor of *vortex,* a term that remained in widespread use until it was adopted into the lexicon of tornadoes, and used to describe the writhing prairie pythons that routinely terrorize the flat-lands of the U.S. Midwest. Subsequently, *vortex* was superseded by *eye,* a term that remains in widespread use at the present time.

eyewall   This name is given to the thick wall (or walls) of CUMULONIMBUS CLOUDS that immediately surrounds the EYE of a TROPICAL CYCLONE. Towering some 40,000–50,000 feet (13,333–16,667 m) above Earth, the eyewall of a HURRICANE, TYPHOON, or CYCLONE contains the system's most violent winds, and its heaviest rains. Sometimes referred to as BAR OF A TROPICAL CYCLONE the eyewall appears on the horizon as a gray-black barricade, or bar, of roiling clouds, one punctuated by periodic flashes of lightning. Generally ranging in width from 30 to 150 miles (48–241 km), it contains the vast CONVECTION cells which power the circulation patterns of a tropical cyclone. Seeking at all times to reach the center of low barometric pressure contained within a cyclone's eye, the eyewall's burgeoning cloud banks frequently penetrate the confines of the eye, trace its wobbling shape before being repulsed.

In maturing North Atlantic hurricanes, Western North Pacific typhoons, and Australian cyclones, the eyewall tends to grow more concentric as the respective system's CENTRAL BAROMETRIC PRESSURE continues to drop. Spun into either a clockwise or counterclockwise rotation by the deflecting force of the CORIOLIS EFFECT, a typical tropical cyclone's eyewall follows the ISOBARS, the closed rings that denote points of equal barometric pressure, thereby

The size of an amphitheater built for giants, the cloud rings that form the inner eyewall of a mature-stage tropical cyclone ascend into the upper atmosphere. Although peaceful in appearance, they can contain wind speeds in excess of 200 miles per hour. *(NOAA)*

adding to the circular nature of a tropical cyclone's construction.

The eyewall further contains a tropical cyclone's heaviest rains, the result of rising and cooling air drawn in by the low-pressure eye. In the Northern Hemisphere, where western North Pacific typhoons, BAY OF BENGAL cyclones, and Atlantic hurricanes spin in a counterclockwise direction, these early RAIN BANDS collect on the eye's north and west edges, while in the Southern Hemisphere, where Indian Ocean and Australian cyclones turn clockwise, the concentration of convective activity will be along the south and east quadrants.

Not surprisingly, the dimensions and characteristics of a tropical cyclone's eyewall varies with size, environment, and intensity. As is the case with the relationship between the size of a tropical cyclone's eye and the system's intensity, the relative size of an eyewall is indicative of its organizational strength and severity. In those tropical cyclones of Category 4 and higher, the eyewall tends to be smaller in diameter. At the time that Cyclone TRACY delivered its historic landfall to northern AUSTRALIA in 1974, its eyewall measured only 31 miles (50 km) in diameter. At the time that Hurricane WILMA was at its most intense (and setting records as the most powerful Atlantic hurricane on record), its eyewall was only 15 miles (30 km) in diameter. Tropical storm force winds extended 160 miles (260 km) from the system's center while within the eyewall itself, sustained winds topped 175 MPH (280 km/h). In larger tropical cyclones, the eyewall can often be much larger. When it came ashore in NORTH CAROLINA in September 2003, Hurricane ISABEL's hurricane-force winds extended up to 115 miles

(185 km) from the center, with tropical storm force winds another 205 miles (330 km) outward. At one point in its existence, Isabel was a Category 5 hurricane, but its weakening resulted in diminished convective circulation and a slow disintegration of the eyewall core.

During the last decade of the 20th century, meteorologists using SATELLITE technology and aerial reconnaissance aircraft identified several unique phenomena related to eyewall generation, regeneration, and structure. As early as 1975, meteorologists studying typhoons in the western NORTH PACIFIC OCEAN noted that many of the region's most intense tropical cyclones developed a double eyewall configuration, where one eyewall surrounded the eye, and a second eyewall—separated by a moat—surrounded the first. On October 12, 1975, Super Typhoon Elsie (17W) generated wind speeds of 155 MPH (250 km/h) while passing between TAIWAN and the PHILIPPINES. Meteorologists noted a double eyewall configuration and attributed the super typhoon's intensity to this characteristic. While a similar double eyewall configuration had been noted in 1969's Hurricane CAMILLE, it was not until the 1990s and early 2000s, that advanced satellite and reconnaissance data confirmed the presence of multiple eyewalls in intense tropical cyclones. In the western North Pacific basin, the double eyewall configuration was noted in Super Typhoon Winnie (Ibiang or 14W) in 1997, in Typhoons Kirogi and Kai-Tak (2000), and in Typhoon Dujuan (14W), which on September 2, 2003, produced double eyewalls of 12 miles (20 km) and 62 miles (100 km), in respective diameter. At the time that Super Typhoon Winnie produced its double eyewall configuration on August 12, wind speeds were estimated at 161 MPH (259 km/h). Supertyphoon Dujuan's double eyewalls appeared as the system was beginning to weaken, with a central pressure of 27.2 inches (922 mb). Just a few weeks later, on September 10, 2003, Super Typhoon 15W produced a double eyewall configuration with a central pressure of 26.1 inches (885 mb). Double eyewall configurations are not confined to western North Pacific tropical cyclones. On January 5, 2004, Cyclone Heta generated a double eyewall configuration as it passed to the west of the Samoan Islands in the South Pacific Ocean with sustained winds nearing 161 MPH (259 km/h). On August 29, 1996, as its central pressure dipped to 27.6 inches (934 mb), Hurricane EDOUARD produced three concentric eyewalls, a feature meteorologists attributed to its rapid intensification.

Meteorologists have also suggested that double eyewall configurations occur during periods of eyewall rejuvenation or replacement cycles. In technical terms kindly provided by the Hong Kong Observatory: "When strong typhoons show a double eye-walled structure, they are often in the process of undergoing an eye wall replacement cycle where a new eye wall develops and replaces an existing one. The cycle begins with a concentrated ring of convection that develops outside the eye wall. The ring of convection then propagates inward leading to a double-eye. The inner eye wall eventually dissipates while the outer intensifies and moves inward. The double-eye-walled structure usually marks the end of an episode of intensification and may last for a day or two."

In a discussion issued on July 15, 2005, U.S. Air Force reconnaissance aircraft observed concentric eyewalls of nine miles (14 km) and 28 miles (45 km), respectively, in a strengthening Hurricane EMILY, and suggested that, ". . . Emily may be going through an eyewall replacement cycle . . . which could result in fluctuations in intensity." During the 24-hour period covered by this discussion, Emily's central pressure dropped from 28.3 inches (959 mb) to 28.1 inches (952 mb) before rising again to 28.6 inches (969 mb) by early evening. Just prior to midnight, Emily's minimum pressure had again dropped to 28.3 inches (958 mb), returning the system to Category 3 status.

Meteorologists additionally believe these eyewall rejuvenation cycles are not simply confined to only the most intense tropical cyclones, but are often responsible for slowing a system's forward motion. On July 5, 2005, Tropical Storm (later Hurricane) DENNIS, while the system's core convection was growing more organized and intensification into a tropical storm was occurring, underwent an eyewall rejuvenation cycle. Prior to the cycle's commence-

The inner wall of Hurricane Katrina's eyewall. In time, this pristine-white collection of clouds would be responsible for the most expensive natural disaster yet suffered in American history. *(NOAA)*

ment, Dennis was moving to the west-northwest at nearly 20 MPH (32 km/h); by the late evening hours of July 5, as the rejuvenation cycle was completed, the storm's forward speed had dropped to 15 MPH (24 km/h).

F

Fabian, Hurricane *North Atlantic Ocean–Bermuda, August 27–September 6, 2003* The sixth-named tropical cyclone of the 2003 North Atlantic HURRICANE SEASON, Fabian was one of the most powerful tropical systems observed in the North Atlantic basin. At one point in its history a Category 4 hurricane on the SAFFIR-SIMPSON SCALE, Fabian became the most severe hurricane to directly affect BERMUDA in half a century when it passed very close to the island on September 5, 2003.

A perfect example of a classic CAPE VERDE STORM, Fabian originated from a TROPICAL DEPRESSION on the afternoon of August 27, 2003, while situated some 370 miles (596 km) west of the Cape Verde Islands. Within 24 hours, the system had deepened into a TROPICAL STORM (with a central pressure of 29.7 inches [1,006 mb]), and had commenced its long, stormy trek westward across the central NORTH ATLANTIC. At midnight on August 30, with a central pressure of 29.1 inches (987 mb), and sustained winds of 75 MPH (120 kp/h), Fabian was upgraded to a Category 1 hurricane. Less than six hours later it was upgraded to a Category 2 system (with a central pressure reading of 28.7 inches [973 mb]), and within another six hours, achieved Category 3 intensity, with a central pressure of 28.34 inches (960 mb). By the early evening hours of August 31, 2003, while positioned several hundred miles east of the Windward Islands, Fabian reached Category 4 intensity, with a central pressure of 28 inches (948 mb). At this time, Fabian's sustained wind speeds were estimated at 132 MPH (213 kp/h).

A healthy tropical system, Fabian continued to intensify as it churned to the west-northwest at a forward speed of 10 MPH (17 kp/h). Shortly after midnight on September 2, Fabian produced a CENTRAL BAROMETRIC PRESSURE reading of 27.8 inches (943 mb), its lowest so far; sustained wind speeds were clocked at a searing 144 MPH (232 kp/h). Now located some 200 miles (320 km) northeast of the Caribbean island of Barbuda and moving firmly to the northwest, Fabian underwent an EYEWALL rejuvenation cycle, one which dropped it to Category 3 intensity. By the following night, however, the system had boldly reintensified, generating a peak barometric pressure reading of 27.7 inches (939 mb) just after midnight on September 4, making it of firm Category 4 status. While situated about 155 miles (215 km) south of Bermuda on the morning of September 5, Fabian began a long recurvature to the north-northeast, a trajectory that brought its well-defined EYE within 10 miles (30 km) of the British colony. Of similar characteristics to 1995's Hurricane FELIX, Fabian's sustained winds of 121 MPH (195 kp/h) and heavy surf conditions pounded Bermuda's pink sand beaches and coastal environs. Although preparations on the island had been comprehensive and complete, four lives were lost to Fabian's glancing fury. Some $300 million (in 2003 U.S. dollars) in property losses were incurred on the island, including extensive damage to all of its world-class golf courses. According to media reports, Fabian's radiating seas were directly responsible for the loss of the vessel *Pacific Attitude* off the coast of Newfoundland, CANADA, on September 7, killing three crew members.

Gradually weakening as it penetrated deeply into the North Atlantic Ocean, Hurricane Fabian (now of Category 1 intensity) underwent extratropical deepening on September 8, 2003, and essentially dissipated by the early evening hours of September 9.

Prior to 2003, the name Fabian had been applied to three North Atlantic tropical cyclones: Tropical Storm Fabian (September 15–19, 1985); Tropical Storm Fabian (October 15–17, 1991); and Tropical Storm Fabian (October 4–8, 1997). Of the three, 1985's Tropical Storm Fabian was the most intense, producing a peak minimum pressure reading of 29.35 inches (994 mb). And of the three, only 1991's Fabian made landfall—in southwestern CUBA on October 16—as a minimal tropical storm. No deaths, injuries, or property damage was reported.

Owing to the destructiveness of the fourth North Atlantic Fabian, the name has been retired from the rotating list of North Atlantic tropical cyclone names. It has been replaced on the 2009 North Atlantic naming list with Fred.

False Point Cyclone *Southern Asia, September 21, 1885* *See* INDIA.

Faye, Typhoon *Japan, September 21–26, 1957* The third of 16 western NORTH PACIFIC TYPHOONS to be christened Faye, this extremely powerful Category 4 TROPICAL CYCLONE prostrated the southernmost Japanese island of Okinawa with 146-MPH (235-km/h) winds, seven to eight inch (178–203-mm) rains, and a 12-foot (4-m) STORM SURGE on September 26, 1957. In Okinawa, where Faye's screaming winds and driving rains killed 53 people and left another 79 missing, millions of dollars in property damage were sustained by those U.S. military forces then occupying the typhoon-prone island. Coming ashore just over a year after Typhoon EMMA battered the volcanic outcropping with 140-MPH (225-km/h) winds and 22-inch (558-mm) PRECIPITATION counts on September 10, 1956, Faye's dramatic assault on Okinawa seriously impeded ongoing efforts to reconstruct its vital array of Air Force and Army installations. Aircraft hangars, barracks, administrative offices, repair facilities, and radar stations—many of them just recently rebuilt—were again unroofed or completely collapsed by Faye's strenuous passage across the center of the island, prompting exasperated officials in Washington to consider briefly abandoning Okinawa altogether, transferring air and army units to Guam or perhaps the PHILIPPINES.

The sixth tropical cyclone of the 1957 TYPHOON SEASON, Faye developed near the Mariana Islands, some 1,600 miles (2,575 km) southeast of Okinawa, on the evening of September 21, 1957. By the afternoon of September 22, as Faye's CENTRAL BAROMETRIC PRESSURE continued to fall and its associated winds and seas further intensified, storm watchers on Guam began to issue a string of warnings to allied tracking stations located around the North Pacific rim. Intended to alert shipping and shore installations to the escalating threat of a developing typhoon, these messages were in turn received by U.S. Air Force and Marine personnel on Okinawa, 420 miles (676 km) southwest of Japan's Kyushu Island.

Familiarized by the earlier Emma with the destructive consequences of powerful, late-season typhoons, U.S. military and police authorities on Okinawa hastily moved to secure the island's military bases for what appeared to be an imminent strike by still-deepening Typhoon Faye. As the Category 3 system's central barometric pressure dipped below 28.20 inches (954 mb), jacking its sustained wind speeds up to 107 MPH (172 km/h), transport and fighter aircraft were either backed into hangars, dispatched to air stations on neighboring islands, or anchored in a single row along runways edges. For the second time in slightly more than a year, storm shutters were again bolted across the windows of Okinawa, while dozens of cars, trucks, and other light equipment were driven into garages, out of reach of the overturning winds of the rapidly approaching typhoon.

In the provincial capital of Naha on the island's tip, curious citizens crowded the city's eastern beachheads, marveling at the spectacle of Faye's 12-foot (4-m) seas as they boomed ashore. Beautiful but dangerous, the cresting swells exploded against the volcanic coastline, casting arcs of spray into the hazy midday sun. Mothers came with their children, took them down to the water's edge, and made a game of outrunning the ever-rising waves. Military police, mindful of the tragedy that had overtaken 11 United States marines only a year before, soon closed all the island's beaches, posting signs in Japanese and English that warned of the presence of lurking currents, undertows that could carry an individual far out to sea in a matter of hours. Respectfully abiding by the order, the people of Okinawa remained on the breakwaters and piers, regarding the raging seas until Typhoon Faye's heralding gusts, punctuated by rain squalls and thunderbolts, forced them to return to their homes. There, tucked behind walls that had only just been rebuilt, the people of Okinawa, from commanding officers and privates to the indigenous fishers of Nago, settled down to await Faye's Category 4 arrival. Well-fitted with hurricane lanterns and bottled water, some of them no doubt listening to the creak of their roofs and pondering the history of past typhoons, asking themselves what sense it made to remain in so storm-savaged a land.

Less than 10 hours later, on the night of September 26, 1957, the EYE of Typhoon Faye crossed Okinawa's southeast coast, approximately 34 miles (55 km) northeast of Naha. Quickly penetrating inland, Faye's

160-MPH (258-km/h) wind gusts blasted neighboring military outposts, strafing runways, overturning aircraft, punching out landing light and radar systems, and toppling a control tower. Entire roofs were peeled from troop barracks, forcing army and air force personnel to seek shelter in half-built hangars. Telephone poles whipped back and forth before snapping. Three radio towers outside Naha buckled, while an empty oil storage tank near the port town of Nago crumpled beneath Faye's sustained 146-MPH (235-km/h) winds and drumming rains. Within a 14-hour period no less than 8 inches of rain fell on the greater Naha region, gorging streets with flash floods and mudslides. Hundreds of dwellings were either flooded or washed off their foundations. Thousands of people were left homeless, while an additional 70 were killed, mainly in mudslide and drowning incidents.

By the time Faye had exited Okinawa, emerging into the East China Sea, where it would spend the next two days quickly dissipating, the typhoon had claimed more than $10 million in property losses, including $2.6 million in aircraft damage to the U.S. Air Force and another $1.2 million in reconstruction costs for the Army. An additional $6.9 million in damage estimates were suffered by Naha and $600,223 by Nago. Along with the $8 million in aggregate losses incurred by Typhoon Emma, Faye's $11.3 million price tag made Okinawa, an island directly threatened by at least one tropical cyclone per year, one of the most expensive places on Earth to maintain a military installation. After casting a weary eye on nearby communist KOREA (where the Korean War had only just concluded in 1953), the United States decided that in the balance, the relentless cost of typhoons was a lesser significance than the expense of mounting any necessary residual actions against Korea from a safer but more distant base. Occupying Okinawa until 1972, the United States military units stationed there duly suffered subsequent typhoon strikes: On October 16, 1959, Typhoon Charlotte claimed 28 lives on the island, and on August 5, 1965, a TROPICAL STORM destroyed hundreds of BUILDINGS and killed 26 people.

Despite the destructive legacy of 1957's typhoon Faye, the name Faye was retained on the rotating list of TYPHOON NAMES. Indeed, between 1949 and 1995, no less than 16 tropical cyclones in the western North Pacific were identified as Faye. While the majority reached typhoon intensity, at least three were of tropical storm intensity, and one—1963's Faye—was a subtropical system that lasted only one day (August 26) in its unfavorable midwestern North Pacific environment. The most intense western North Pacific tropical system identified as Faye (18W) remained an offshore system between October 2 and 9, 1968,

but did reach Category 5 status and remain so for a 24-hour period. While barometric pressure readings are not available, estimated wind speeds within Super Typhoon Faye reached 167 MPH (269 km/h). Between July 17 and 18, 1992, Tropical Storm Faye (6W) brought tropical storm force winds to southern CHINA, injuring 24 people.

Faye, Typhoon *Japan–Korea, July 19–25, 1995*
The 16th and last North Pacific TYPHOON to be dubbed Faye, this moderately powerful midseason TROPICAL CYCLONE delivered sustained 126-MPH (203-km/h) winds, 10-inch (254-mm) rains, and a sweeping 10-foot (3-m) STORM SURGE to JAPAN's Ryukyu Islands and to the south coast of KOREA, between July 23 and 24, 1995. Regarded as one of the strongest typhoons to have struck Korea in almost a decade, Faye's 146-MPH (235-km/h) gusts and driving rains damaged or destroyed more than 22,000 BUILDINGS, leaving three times that number of people homeless. During the height of the typhoon's Category 3 fury, the 140,000-ton tanker *Sea Prince* was run aground on rock-strewn Yochon Island. Heavily laden with more than 100,000 tons of crude oil, the *Sea Prince* was immediately holed as Faye's mountainous 20-foot seas impaled its steel hull on a submerged crag. Catching fire before partially sinking, the *Sea Prince* continued to spew oil for days, eventually fouling a 40-mile stretch of the Korea Straits. At least 50 people were listed as killed or missing, while an estimated $120 million in property claims—$52 million of it for those fish and shellfish farms ruined by the *Sea Prince*'s oil spill—were submitted in Faye's wake.

The sixth typhoon of the 1995 TYPHOON SEASON, Faye was born over the hot, moist reaches of the western NORTH PACIFIC OCEAN some 1,400 miles (2,253 km) southeast of Pusan, South Korea, during the early morning hours of July 19, 1995. Steadily evolving northwest at 10 MPH (16 km/h), the slowly spinning band of thunderstorms that would soon form the core EYEWALL of typhoon Faye continued to gather, growing thicker and taller as their CUMULONIMBUS CLOUD piles reached ever higher into the sky. Enormous and gray, these clouds bands would in time touch the troposphere's underside, or the point at which the stratosphere begins, and curl over, forming the huge anvil thunderheads so characteristic of violent storm activity.

Aided by both the high and low-level winds moving in the same direction at the same speed, these towering CONVECTIVE cells were fed by the vast evaporation rates found over TYPHOON ALLEY waters that inexorably deepened during the night of July 19. Boosting the embryonic typhoon to TROPICAL DEPRESSION status, that of a cyclonic system with winds of less than 39 MPH (63 km/h), the cloud bands further organized during the early morning hours of July 20, developing both a pronounced counterclockwise spin and the crescent-shaped ring of RAIN BANDS that would encircle the EYE of the maturing TROPICAL STORM.

On the afternoon of July 21, 1995, the still-strengthening system, with winds clocked at 51 MPH (82 km/h), was christened *Tropical Storm Faye.* Maintaining its northwest course and drifting along at speeds of between nine and 11 MPH (15–18 km/h), the nascent typhoon thrashed the ocean's surface, spinning off undulating swells as it extracted every last molecule possible from the heat-charged spray. Aerial reconnaissance flights monitored by the JOINT TYPHOON WARNING CENTER (JTWC) on the island of Guam, indicated that Faye's CENTRAL BAROMETRIC PRESSURE had now steepened to 29.32 inches (993 mb) and that it was carrying significant quantities of rain.

Two days later, on the night of July 23, a now mature Typhoon Faye blazed across Japan's southernmost Ryukyu Island chain, seering the volcanic slags with 126-MPH (203-km/h) winds, warm rains, and a sweltering series of small storm surges—high waves that singed the shores of Okinawa some 220 miles (354 km) to the south. Seemingly disinterested in cyclonic irony, the second Typhoon Faye spared Okinawa a direct strike, failing to repeat its namesake's catastrophic 1957 incursion into the island in which more than 75 people were left dead and millions of dollars in property losses were incurred. By far the weaker of the two storms, the second Typhoon Faye busied itself with sinking a Japanese fishing boat, sweeping away two bridges, and dusting Suwanose Island of its trees and power lines. Farther to the north, Faye's 300-mile-wide (483-km) cloud canopy brought heavy rain squalls to Kyushu Island's south coast but otherwise caused little damage to its fishing piers and naval bases. No fatalities were reported from Faye's passage into the East China Sea, making it one of the milder storms in Japanese typhoon history.

Unfortunately the same cannot be said for Korea's southeast coast, where Faye's 130-MPH (209-km/h) winds and swamping waves came ashore near the industrial and port city of Yosu on the morning of July 24, 1995. At least 23 Korean fishers, caught at sea by the typhoon's scaling winds, were drowned when their lightly built wooden vessels capsized in cresting 20-foot (7-m) waves. Crews on larger ships fared only marginally better as Faye, moving inland over the cities of Hadong and Taegu, pounded anchorages all along a 50-mile (80-km) stretch of coastline, casting sleek freighters, supine

bulk carriers, and open-air ferry boats onto rock-punctured shores. Three sailors were drowned when huge breakers, estimated by survivors to have been 25 feet (8 m) high, overturned their launch in Posong Bay, while another two men perished near Namhae Island when a refueling barge broke apart.

On board the 1,100-foot (366-m) supertanker *Sea Prince,* battered for nearly four hours by Faye's 146-MPH (235-km/h) gusts and hammering seas, drenched crew members, who frantically labored to keep the disabled, oil-laden vessel from running aground on the pristine shores of the Hanryo Sea National Park. An aged Cypriot-registered ship, the *Sea Prince*'s engine room caught fire as the tanker was navigating the narrow, stone-studded approaches to Yosu Harbor. Now furiously burning at the stern, its tall deck-house and bridge filled with acrid smoke, the *Sea Prince* drifted wildly in Faye's rising seas, fighting to keep position in the middle of the channel until salvage tugs could arrive on the scene.

As Korean maritime police dispatched helicopters to rescue the tanker's 23 crew members, Faye's intense eyewall steadily gained on the coast, its increasing winds and blinding rains turning nearby Yochon Island's cliffs into a watery miasma that may or may not have looked as close as it really was. First surprised and then overwhelmed by the sudden onslaught of Faye's landfall, the *Sea Prince* was driven sideways by the typhoon's 20–25-foot (7–8-m) waves that slammed the vessel in a shriek of pierced steel and showering sparks onto Yochon Island's first available headland.

Within minutes another fire had broken out, this one just forward of the tanker's bridge. The *Sea Prince*'s crew, huddled beneath the long catwalk that ran the ship's length, quickly ventured out to assist the typhoon in fighting the fires but was soon repulsed by the uncontainable fury of both. Pinned to the Asian continent by a solitary pinnacle of rock, the *Sea Prince*'s burning hull wallowed in Faye's seas, groaning and rumbling as its oil-gorged innards began to split apart beneath the twisting strain.

Shortly after dusk on July 24, as the first of some 100,000 tons of crude oil began to leak into the Hanryo Sea, the storm-scrapped hulk of the *Sea Prince* settled by the stern, sank to the point that its rudder and propeller collided with the hard seabed, and shattered. Tons of oil gushed out of the stricken tanker, hardly calming the typhoon's turbulent seas as they swept the choking sludge ashore in an explosion of breaking surf. Spray-painting nearby crags and pine groves black, the *Sea Prince*'s lethal cargo would eventually darken more than 40 miles (64 km) of Korean coastline, representing one of the worst oil spills in the peninsula's history. Dozens of fish and oyster farms along the Korea Strait were fouled; patches of the tanker's oil were recorded as far north as Yosu Harbor. Although Korean environmental authorities moved quickly and efficiently to minimize the *Sea Prince*'s spill, the lingering effects of Faye's passage—two days of high seas and unusually thick fog banks—hindered their efforts to deploy containment booms and fight the fire still raging aboard the waterlogged vessel.

On the morning of July 26, as the remnants of Typhoon Faye dissipated over the Sea of Japan's chilly confines, Korean officials began the tasks of accounting for the dead and assessing the damage. Touring the countryside, civil defense personnel recorded the destruction of more than 6,000 dwellings, with another 10,000 BUILDINGS left standing but seriously damaged. Twenty-one people were listed as confirmed deaths, while another 23 fishers were posted missing at sea.

In time the *Sea Prince*, its fires extinguished and its breached hulk raised from the bottom by an expensive team of salvors, was towed away to TAIWAN and scrapped. But the ecological and economic effects of its tragic spill on the sea farms of Yochon Island were not so easily dispensed with, lingering as a reminder of the stormy interlude named Faye. The name Faye has been retired from the rotating list of TYPHOON NAMES.

Felix, Hurricane *North Atlantic Ocean, August 26–September 10, 1989* Between 1989 and 2007, four TROPICAL CYCLONES in the NORTH ATLANTIC OCEAN were identified as Felix. All four achieved hurricane intensity, with 1995's Felix reaching Category 4 status as it transited the North Atlantic, bringing some property damage to BERMUDA and eight fatalities to the eastern seaboard of the United States. In August 2007, Hurricane Felix became the second hurricane of the season (following Hurricane DEAN) to come ashore in the Western Hemisphere as a Category 5 system: It claimed some 130 lives in Nicaragua and Honduras.

The first North Atlantic tropical cyclone named Felix remained an offshore system for its entire existence. Originating near latitude 17 degrees North, 21 degrees West, on August 26, 1989, Felix reached TROPICAL STORM intensity (with a central pressure of 29.67 inches [1,005 mb]) during the evening hours of the 26th, and remained a tropical storm until downgraded to a TROPICAL DEPRESSION on the evening of August 29. By September 3, Felix's circulation had become better organized, and the system was again upgraded to a tropical storm. It fitfully continued to deepen, becoming a Category 1 hurricane (with a central pressure of 29.17 inches [988 mb]) during the early morning hours of September 5. Hurricane Felix reached its peak minimum pressure of 28.90 inches (979 mb) on September 6 before progressively weak-

ening. It underwent extratropical deepening on September 9, and by the following day had dissipated.

Felix, Hurricane *North Atlantic Ocean–Bermuda–Northeastern United States, August 8–25, 1995* The sixth named TROPICAL CYCLONE of the 1995 season, and the second North Atlantic tropical system to carry the name Felix formed north of the 14th parallel during the night of August 8, 1995. Moving to the northwest, the system quickly intensified, maturing into a TROPICAL STORM within hours of its origination. Felix remained a tropical storm until the evening hours of August 10, when it was upgraded to a Category 1 hurricane. Positioned several hundred miles to the east of the Leeward Islands, and with a central pressure reading of 29.47 inches (998 mb), Felix slowly moved to the north-northwest, gaining strength as it did so. By August 11, the system had reached Category 2 status (with a central pressure of 28.49 inches [965 mb]), and by the following morning, had been promoted to a Category 3 system, with a central pressure of 28.20 inches (955 mb). Now assuming a more northerly trajectory, Felix continued to deepen, finally reaching Category 4 intensity during the daylight hours of August 12. Felix produced its peak minimum central pressure reading later that day—27.43 inches (929 mb)—while aimed at the island of BERMUDA. Sandwiched between an encroaching high pressure ridge to the southeast and the semi-permanent anticyclone to the northwest, Felix quickly recurved northward and quickly began to weaken as cold, dry air was drawn into its circulation. On August 15, the system's EYEWALL passed just to the south of Bermuda, its central pressure of 28.46 inches (964 mb) produced 80-MPH (129-km/h) winds that caused widespread damage and disruption across the island. Between August 15 and August 20, Felix (now of Category 1 intensity) performed a complex loop-the-loop maneuver while situated several hundred miles east of NORTH CAROLINA's Outer Banks. Fearing that Hurricane Felix could still make landfall, prudent emergency management officials evacuated (*see* EVACUATION) the Outer Banks and many coastal areas in North Carolina and VIRGINIA, but by August 20, a downgraded Tropical Storm Felix was blazing to the north-northeast at nearly 25 MPH (40 km/h), and away from the eastern seaboard. On August 22, 1995, Felix underwent extratropical transitioning and by the 24th, had dissipated. All told, eight people perished—mostly in riptide and drowning incidents—during Felix's passage through the North Atlantic; three in North Carolina and five in New Jersey. Several areas from PUERTO RICO to Bermuda to the northeastern United States and Canada, suffered widespread coastal erosion from Felix's pounding seas and high surf, and a score of coastal BUILDINGS damaged or destroyed in North Carolina and Virginia.

Felix, Hurricane *North Atlantic Ocean, September 7–19, 2001* The third North Atlantic tropical system identified as Felix fortunately remained an offshore system as a portion of its Category 3 existence spanned the resource-strained days immediately following the September 11, 2001, terrorist attacks in NEW YORK, Washington, D.C., and Pennsylvania. Formed on September 7 near longitude 28 degrees West, Tropical Depression No. 7 (TD7) slowly traveled almost due west, following latitude 15 degrees North, toward the Leeward Islands. An unfavorable environment for strengthening downgraded TD7 to a TROPICAL WAVE, and it remained disorganized until the early morning hours of September 10th, when it was again upgraded to a TROPICAL DEPRESSION. On the morning of September 11, 2001, as the United States suffered the worst act of terrorist violence yet seen in history, Tropical Depression No. 7 was upgraded to Tropical Storm Felix. Steadily recurving away from landfall in the United States, Tropical Storm Felix traveled to the north-northeast, becoming a Category 1 hurricane on September 13th, and a Category 3 hurricane (with a peak minimum central pressure of 28.40 inches [962 mb]) shortly after midnight on September 14th. Pummeled by its shearing interaction with the semi-permanent BERMUDA HIGH, Felix languidly began to weaken. It was downgraded to a Category 1 system during the early morning hours of September 16, and to a TROPICAL STORM the following day. On September 19, 2001, the remnants of Tropical Depression Felix melted into history, some 200 miles (322 km/h) to the south of the Cape Verde Islands.

Felix, Hurricane *Caribbean Sea–Windward Islands–Central America, August 31–September 9, 2007* The fourth tropical cyclone identified as Felix was the most intense of the quartet, as well as the deadliest. Originating over the eastern North Atlantic Ocean from a tropical wave on August 31, 2007, Felix slowly intensified as it neared the prosperous Caribbean islands of Trinidad and Tobago, Aruba, Bonaire, and Curaçao. Upgraded to a tropical storm on September 1, the system brought 46 MPH (74 km/h) winds to Barbados, 6.2 inches (152.4 mm) of precipitation to Trinidad, and mudslides to the northern shores of Tobago. In Trinidad and Tobago, the nation's emergency management agency opened several dozen shelters and instituted evacuation plans. Passing within 100 miles (115 km) of northern South America and offloading heavy rains over Venezuela, Felix entered the Caribbean Sea on September 1. The system was upgraded to hurricane status during the early hours of September 2. As its eyewall moved westward, following a course not unlike that taken by Hurricane DEAN just

three weeks earlier, the hurricane rapidly began intensifying. By the early afternoon hours of September 2, Felix had grown to a Category 3 system, and, by the early evening hours, had achieved Category 4 status. Fueled by a favorable environment that included very low WIND SHEAR, Felix underwent explosive deepening. By midnight on September 3, it had reached Category 5 status, with a central barometric pressure reading of 27.44 inches (929 mb). Now situated over the central Caribbean Sea, Felix's sustained 165 MPH (266 km/h) winds sparked fears of a repeat of 1998's killer Hurricane Mitch over Central America. On September 4, 2007 (and after an eyewall replacement cycle that saw the system drop to Category 4 intensity before rebounding), Felix ground ashore in Nicaragua, near the lightly populated tourist hub of Puerto Cabezes. The second Category 5 tropical system to make landfall in the Western Hemisphere during the 2007 season, Felix destroyed some 9,000 buildings, sank small craft, and downed trees along the famed Mosquito Coast. In Honduras, heavy flooding swept the streets of the capital city, Tegucigalpa, claiming two lives in drowning incidents. Across Central America, Felix claimed an estimated 130 lives, most of them in Nicaragua. At the time of writing, it is believed that the United Nations' World Meteorological Organization (WMO) will retire the name Felix from the rotating list of North Atlantic tropical cyclone identifiers.

Fifi, Hurricane *Central America, September 14–22, 1974* One of the deadliest North Atlantic HURRICANES on record, Fifi's 109-MPH (175-km/h) winds and torrential rains claimed some 5,000 lives, left more than 100,000 people homeless, and wrought $500 million in property damage as it swept across the Central American nations of BELIZE and Honduras during the night of September 18 and 19, 1974. A moderately powerful, Category 2 tropical cyclone at landfall in BELIZE, Fifi's 20-inch (508 mm) PRECIPITATION counts subsequently sired dozens of landslides, capricious avalanches that erased entire villages in northern Honduras, El Salvador, and Guatemala. Transiting the isthmus and entering the eastern North Pacific Ocean on September 21, a downgraded Hurricane Fifi was renamed TROPICAL STORM Orlene.

HURRICANE FIFI CHRONOLOGY

September 14: TROPICAL DEPRESSION No. 10 begins to take shape over the southeastern Caribbean Sea, approximately 320 miles (515 km) due west of Guadeloupe island. Slowly revolving northwest at nearly 16 MPH (26 km/h), the thickening band of thunderstorms produces sustained winds of 29 MPH (47 km/h) and some light rain.

September 15: Now centered 69 miles (111 km) south of HISPANIOLA island and maintaining speedy progress northwest, Tropical Depression No. 10 steadily feeds on the warm, humid waters of the Caribbean. A rainmaker from the start, its developing circulation drops heavy showers on Haiti's southern peninsula as it brushes past during the afternoon and evening hours. Top winds measure 36 MPH (58 km/h), placing the tropical depression on the cusp of becoming a tropical storm. Emergency management authorities in nearby JAMAICA prepare the island for possible rain squalls from the grazing system.

September 16: Bearing 49-MPH (79-km/h) winds and nasty tropical downpours, newly christened Tropical Storm Fifi passes some 32 miles (51 km) south of Kingston, Jamaica, shortly after daybreak. Although Fifi's strengthening EYEWALL remains well offshore, the system's deepening RAIN BANDS deliver huge amounts of precipitation to the island's southern regions. Counts of eight inches (203 mm) and higher are observed in and around Kingston, while inundated rivers pave the crowded streets of adjoining towns with more than two feet (1 m) of water. Even though property losses reach into the hundreds of thousands of dollars, no casualties are reported.

September 17: Just before dawn, as its CENTRAL BAROMETRIC PRESSURE sinks to 29.19 inches (988 mb) and its sustained winds eclipse 74 MPH (119 km/h), Tropical Storm Fifi is upgraded to hurricane status. Situated 460 miles (740 km) east of Trujillo, Honduras, the hurricane has undergone a course change to the east-southeast during the night and is now poised to make landfall in southern Honduras, perhaps on the Nicaraguan-Honduran border, sometime in the next 24–36 hours. Despite adequate notice of the hurricane's onrushing presence, few evacuation measures if any are implemented in either country, thereby dooming thousands of people to imminent death.

September 18: Resuming its plodding northwest course a rapidly intensifying Fifi inexorably gains on the unsuspecting communities of northern Honduras and southern Belize. Upgraded to Category 2 status earlier in the day, the hurricane's expansive cloud cover brings 98-MPH (158-km/h) winds to nearby Swan Island, while a mid-afternoon curtain of rain begins to descend on Honduran banana plantations and fishing villages. Sharply sheering west during its final hours at sea, Hurricane Fifi delivers the north coast of Honduras a glancing landfall shortly after dusk. Raging past Cabo Camaron, the EYE's central pressure of 28.67 inches (971 mb) lashes the cities of Trujillo and La Ceiba with 132-MPH (212-km/h) gusts and apocalyptic rains. Fearsome flash floods

stream down timber-strewn mountainsides, battering more than 182 towns and villages into painful extinction. Almost all of Honduras's vital banana crop is ruined, the trees seemingly peeled away from the land in long furrows, their ripening green-yellow clusters squashed into rotting pulp. Power lines are downed by the mile, crowded buses wildly swerve through mountain passes before skidding into rain-swollen gorges, and thousands of buildings lose their doors and windows, their roofs and walls. Along the coast, fishing boats and other small craft are tossed against their piers by Fifi's 10-foot (3-m) seas, straining at their hawsers until broken away and driven aground. The International Red Cross will later estimate that between 800 and 1,500 people in Honduras lost their lives within 12 hours of Fifi's initial landfall in the country, indicating how effective a killing machine even the most moderate tropical cyclones can be.

September 19: Skirting Honduras's north coast, Hurricane Fifi's eye crosses into the Gulf of Honduras shortly after midnight on September 18. Its sustained winds hardly diminished by its brush with Honduras, the hurricane barrels into south Belize near the port town of Monkey River just before noon. The most powerful hurricane to have struck the country since 1961's HATTIE, Fifi continues to unleash tremendous rains as it pushes across Central America, with counts of 4 inches (102 mm) and higher being recorded as far north as Campeche, MEXICO. In Belize, where Fifi's winds are clocked at 109 MPH (175 km/h), damage to buildings and trees is quite severe: Several hundred structures become complete wreckage. In Guatemala, where Fifi's downgraded winds (tropical storm strength by late afternoon) trigger violent downpours, numerous roads and bridges are carried away in mudslides. At least 200 people are drowned in Guatemala, making Fifi the deadliest hurricane to have affected the country in almost two decades. Heavy rains also continue to blanket northern Honduras, causing sudden landslides that will over the next two days take more than 2,000 more lives.

September 20: Once again downgraded, this time to Tropical Depression, Fifi bears northwest, passing over south Mexico in what would appear to be the rainmaker's drowning demise. In those nations left reeling in its wake, however, haphazard rescue operations are hampered by residual rains, ghostly fog banks, isolated lightning strikes, and the constant roar of flowing water. Plunging through the mountain communities of northern Honduras, Fifi's 20-inch (508-mm) rains obliterate the town of Choloma in an avalanche of mud and debris shortly before dusk. Some 2,800 people unaccounted for, making it the single most tragic incident in Fifi's history.

September 21: Reaching the eastern North Pacific Ocean by midmorning, the rejuvenating remains of Tropical Depression Fifi are upgraded and renamed, transformed into Tropical Storm Orlene. Curling west-northwest, Orlene briefly reachieves hurricane strength before weakening again, making one last landfall on Mexico's northwest coast just before midnight and dissipation. In Honduras, where relief operations proceed for a third day with agonizing slowness, charges are leveled against Honduran army and government officials, alleging that they are personally appropriating the tons of foreign aid that is being sent to assist hurricane victims. Overwhelmed by the sheer magnitude of the disaster, relief organizations collect, count, and bury the bodies, paying little attention to the political infighting as Fifi's death toll rises daily by the hundreds. At first estimated to be about 10,000 dead, the tally will eventually adjust downward, ranging between 5,302 and 8,100 lost. A monster hurricane in any case, the name Fifi (which had been used once before in the North Atlantic, for a Category 1 hurricane that existed from September 4 to 12, 1958), was subsequently retired from the cyclical list of HURRICANE NAMES.

film and tropical cyclones Given the inherent drama of a TROPICAL CYCLONE strike, whether on land or at sea, it is not surprising that for almost a century, producers, directors, screenwriters, and audiences have interpreted, studied, embellished, exploited, and typified hurricanes, TYPHOONS, and cyclones through the medium of film. For large numbers of people around the globe, motion pictures, documentaries, and other visual media are the primary conduits through which tropical cyclone activity is recognized and disseminated as knowledge—and it has been that way now for decades. Cinematic portrayals of tropical cyclones and their damaging effects have been in existence since the mid-1930s, newsreel footage like that taken in the aftermath of the GREAT GALVESTON HURRICANE OF 1900 for much longer. These portrayals present hurricanes, typhoons, and cyclones as often symbolic contributions to the films they appear in, while others are purely "atmospheric" in intent, existing only to raise suspense or to provide a distant MacGuffin for a heroic character or a cowardly plotline. Almost all are included in the various literature—novels, short stories, and plays—from which many screenplays have been derived. And in the very best films, tropical cyclones serve as motivators, triggers for characters and situations that in their own drama and complexity mirror the drama and complexity of hurricanes, typhoons, and cyclones.

In 1935, MGM released *China Seas*, a stormy vehicle for characters Captain Allan Gaskell (Clark Gable), Dolly "China Doll" Portland (Jean Harlow),

Jamesy MacArdle (Wallace Beery), and Sybil Barclay (Rosalind Russell) that involves a romantic, tempestuous passage from Hong Kong to Singapore on a tramp steamship—and along the way an encounter with a severe typhoon. Unlike catastrophe films produced contemporaneously (*San Francisco*, 1936; *In Old Chicago*, 1937; *The Hurricane*, 1937), *China Seas* is not a disaster film but an action/adventure set against the pirate-infested and storm-tossed waters of the South China Sea; the typhoon appears purely as a plot device, a cinematic technique for heightening suspense, and a symbolic reflection of China Doll's grating, contentious character and her unresolved relationship with Captain Gaskell and his present romantic interest, Sybil Barclay. The black-and-white film features some interesting although antiquated special effects of the ship navigating the typhoon, and includes several relatively interesting scenes of personal peril brought on by the typhoon's fury. At one point, Captain Gaskell warns his dueling and scheming passengers that the ship is "in for a twister."

In 1936, authors James Norman Hall and Charles Nordhoff published a novel, *The Hurricane*, which in 1937 Twentieth Century-Fox released as a black-and-white feature film. Directed by John Ford, *The Hurricane* stars Dorothy Lamour as a young and beautiful Marama, Jon Hall as her new and virile husband, Terangi, Raymond Massey as the draconian Governor Eugene de Laage, and Mary Astor as his wife, Germaine de Laage, in a well-crafted tale of injustice, colonial indifference, love, rescue, and one enormous (and dramatically filmed) South Pacific cyclone. In keeping with Hollywood's "disaster film" formula of the 1930s, *The Hurricane* is not a story or film about a tropical cyclone as much as it is a set of interwoven human tales that are drawn together by the life-altering events of the climactic cyclone strike. Like *China Seas*'s substitution of the word *twister* for *typhoon*, the extraordinarily destructive tropical system in *The Hurricane* would, according to present protocols, be more accurately identified as a cyclone, as it occurs in the SOUTH PACIFIC; original authors Nordhoff and Hall were merely following a common cultural typography in the first half of the 20th century in that across the world, tropical systems were simply identified as "hurricanes." In the expert hands of director Ford, *The Hurricane* does bring new narrative techniques to the canon of tropical cyclone film, drawing upon the original novel's unfolding narration to create an atmospheric sense of impending doom through the destructive forces of nature. In one scene, as the winds begin to rise across the low-lying tropical island, the window shutters in the governor's residence blow open, and furniture is scattered; in another, palm fronds blow. The film's most remarkable scene (besides the model STORM

SURGE that washes away the village) takes place when Marama and Terangi take refuge in a large tree that subsequently becomes airborne.

Although featured in newsreels and other documentary formats, the subject of tropical cyclones essentially disappeared from U.S. movie houses during the World War II years. But in 1948, Warner Brothers released *Key Largo,* a powerful film starring Humphrey Bogart as a tormented Frank McCloud, Edward G. Robinson as the reptilian Johnny Rocco, Lionel Barrymore as a wheelchair-bound James Temple, Lauren Bacall as Temple's widowed daughter-in-law, Nora, and Claire Trevor as Rocco's boozy mistress, Gaye. Directed by John Huston and loosely based on a play of the same name penned by Maxwell Anderson in 1938, *Key Largo* is a black-and-white showcase for colorful characters (gangsters, Seminole Indians, a boatload of bootleg liquor), humid, closed rooms in a fading Florida hotel, and one extraordinarily powerful hurricane. Clearly modeled on the notorious LABOR DAY HURRICANE that swept the Florida Keys in 1935 (and which is referenced several times in the film), the unnamed but ever-present hurricane provides a symbolic and dramatic backdrop for the film's principal themes of cruelty, greed, rescue, and redemption. As with Ford's *The Hurricane,* the suspense triggers in *Key Largo* are perfectly honed, with a balanced and unfolding mix of storm references, action sequences, and scenes of personal peril and heroism that establish the film as one of director Huston's best. And like all of the tropical cyclone-related films that preceded it, *Key Largo* does not feature the hurricane as a character in of itself, but as a symbolic metaphor for the turmoil taking place in the lives of its principal victims.

Nearly a decade after Typhoon Cobra pirated three destroyers and 790 personnel from the U.S. Navy in December 1944, Hollywood launched *The Caine Mutiny,* an intensely riveting look into the nature of leadership, of loyalty, and of discipline, in the face of storm and war. Starring Humphrey Bogart as the mentally unstable Captain Philip F. Queeg and directed by Edward Dmytryk, *The Caine Mutiny* (which was closely based on the 1951 novel by Herman Wouk) largely takes place aboard the fictional U.S.S. *Caine* (DMS-18), a Navy minesweeper operating in the Pacific theater during World War II. A series of mishaps and provocations, coupled with the poisonous atmosphere of distrust and denial aboard the U.S.S. *Caine,* essentially culminates in two storms; the one that occurs on the *Caine's* bridge, the other the violent North Pacific typhoon that precipitates the final crises of overzealous duty and lost command among the ship's officers. Symbolically and thematically, the typhoon in *The Caine Mutiny* is more matured than in earlier films featuring tropi-

cal cyclones. Whereas the tropical systems in *China Seas, The Hurricane,* and *Key Largo* exist largely to provide cinematic "atmosphere" and the suspense of situational peril, the typhoon in *The Caine Mutiny* acts as a direct influence on the film's premise. If the *Caine* had never encountered the typhoon, the story and its outcome would have been vastly different.

As is somehow befitting for one of the most typhoon-ridden countries on earth, JAPAN and Japanese filmmakers of the 1950s and 1960s produced many of the screen's most notable interpretations of typhoons, generally in the guise of B-grade nuclear monsters and the storms of destruction that accompanied them. In *Mothra* (1961), released by Toho Studios, the eventual destruction of the Japanese capital by Mothra, an enormous, multicolored "moth," is directly initiated by the rescue of four seamen from a ship blown ashore on Infant Island by a typhoon. Given the significance of the typhoon—the divine wind—as a national defensive measure, it is a unique permutation in the Japanese attitude to this aspect of their popular culture and traditions that a typhoon would eventually lead not to survival, but to destruction, which they have at many times in Japanese history. Within three years, however, the symbolism had again changed. In *Mothra vs. Godzilla* (1964), the film opens with an extended typhoon sequence, one featuring an enormous storm surge and the obligatory destruction of dozens of miniature automobiles, ships, BUILDINGS, and infrastructure. The model fury also delivers to Japan a gargantuan egg, said to belong to Mothra. When that dubious boon to Japanese urban renewal known as Godzilla arrives and attempts to destroy the country and the egg, the battle of symbols is joined. Despite the film's title, it is not Mothra that eventually defeats the destructive Godzilla, but two of her offspring from within the egg that had been delivered to Japan by the divine wind of the typhoon. In the 25th film in the Godzilla franchise, *Godzilla, Mothra and King Ghidorah: Giant Monsters All-Out Attack,* released in 2001, a powerful typhoon provided cover for Godzilla's initial assault on the Bonin Islands. In this manifestation, the typhoon is little more than a popular cultural reference to present emergency practices, one indicated by television newscasts of the typhoon's landfall on the island chain.

While 1970s Hollywood, fond as it was of big-budget, symbolism-laden, special effects-spun disaster epics, never produced a disaster film (aside from a tepid 1976 remake of the 1937 classic *The Hurricane*) centered on a hurricane landfall, there were a number of lesser films with hurricane-related themes released to television. The first (and probably the finest) of these, simply titled *Hurricane,* was produced in 1974 by Metromedia Producers Corporation. A series of individual story lines drawn together by their proximity to "Hurricane Hilda," *Hurricane* is noted for its extensive use of stock footage shot during Hurricane CAMILLE's devastating landfall in 1969, as well as its thinly veiled portrayal of the inner workings of the National Weather Bureau's (now the National Weather Service) NATIONAL HURRICANE CENTER (NHC). Based on a 1972 novel of the same name by William C. Andersen, featuring a television story and teleplay by Jack Turley, and directed by Jerry Jameson, *Hurricane* stars Larry Hagman, Martin Milner, Michael Learned, Will Geer, and Patrick Duffy.

As is common with the disaster genre, several of the story lines in *Hurricane* involve extreme peril and miraculous escapes—as well as an obvious moral imperative that stresses the often ghastly consequences of not heeding preparedness and common sense lessons during times of disaster. In a waterlogged scene that could have been produced as part of a public service program for hurricane preparedness, an Army National Guardsman advises an obstinate homeowner that an EVACUATION order has been issued, only to receive the reply, "Not leaving." After explaining the imminent danger that the homeowner and his family are facing, and after receiving more or less the same response, the National Guard captain, taking out his notebook, humorlessly asks, "May I have the name of your next of kin?" After shooting the guardsman a disbelieving look, the man turns away from the door, calling out, "Mama, I guess we better pack up!" There is a further story line, no doubt inspired by a tragic example witnessed during Hurricane Camille's strike on Pass Christian, MISSISSIPPI, where a HURRICANE PARTY (rather bizarrely presided over by a messianic-like alcoholic) is crashed by Hurricane Hilda's uninvited storm surge—with predictably deadly results.

Among the more interesting themes addressed in *Hurricane* is that which exists between meteorology and storm forecasting as a pure science, and meteorology and storm prediction as a combination of science and metaphysics. Set in the offices of the "National Hurricane Warning Service" in Coral Gables, Florida, (where the actual National Hurricane Center was then located), *Hurricane*'s story line places an aged and unkempt, but very experienced, hurricane forecaster, Dr. McCutcheon (played by veteran actor Will Geer) in contrast to a young and professional, but very inexperienced, junior meteorologist, Lee Jackson (played by Michael Learned). For much of the film, their warm relationship is characterized by a debate as to where Hurricane Hilda will make landfall, and whether or not tropical cyclones are pure science, or part romance. The following exchange is illustrative of this theme:

McCUTCHEON: "Yes, siree, our lady is definitely heading north."

JACKSON: "She may be your lady, Doctor—she's not mine."

McCUTCHEON: "Well, for a woman you don't seem to be able to give Hilda a woman's due."

JACKSON: "I'll leave the romanticizing to somebody else."

McCUTCHEON: "Romanticizing, huh? Well, maybe you're right. Every hurricane has a different name, a different personality. They maybe might one day call them after some windy politician."

JACKSON: "A storm is not a woman or a dame or a lady or a politician—not to me, anyway. It's a natural phenomenon, and one I perceive in pure scientific terms."

Later, as Hilda is about to make landfall, McCutcheon (who is also the National Hurricane Warning Service's director) delivers to Jackson a now-familiar emergency management mantra about the dangers of "crying wolf" when issuing timely hurricane forecasts. "I'm the boss around here . . . it was up to me to send out the warning a little sooner . . . People listen . . . once, twice, maybe three times when they call, 'Wolf!'—but when it doesn't come, why they just shut up their ears. Balance, you know . . . there's one thing I'm really afraid of and that is that one day I won't be able to call it out soon enough, that I won't be able to call out right when I cry, 'Wolf!'"

The dangers associated with tropical cyclone activity—particularly those experienced by coastal dwellers—were highlighted in the next contribution to the tropical cyclone film canon, *Condominium*. Based on an excellent novel of the same name by John D. MacDonald and directed by Sidney Hayers, *Condominium* details the lives, loves, ambitions, and vulnerabilities of a dozen or so residents of the Silver Sands condominium complex in Florida. Released to ABC television in 1980, *Condominium* follows the primary disaster film formula of the 1970s, in which a diverse cast of characters is forced to contend with individual and collective peril, sometimes caused by natural hazards, but always exacerbated by the more human hazards of greed, mismanagement, neglect, and ignorance. In this example, the beachfront Silver Sands condominium is poorly constructed, overpriced, and not surprisingly, eventually falls to the ravages of an extravagantly powerful "Hurricane Ella." In addition to some rather interesting scenes of destruction wrought by Ella's storm surge, the film is notable for featuring an ill-fated hurricane party, and for reinforcing the message that those who fail to evacuate from the path of a dangerous hurricane are more likely than not to perish.

The use of tropical cyclones as major features in motion pictures waned in the United States during the 1980s and 1990s. This was due, in part, to a shift in the genre's formula from pure disaster and our responses to it, to an action-based adventure film in which disaster plays an incidental role. During this period, however, numerous documentaries were produced for public television, many dealing with the collective and individual histories of tropical cyclones, their meteorological workings, and their effect on economic and social stability. The rapid rise of quality cable network programming enhanced this trend, and dozens of documentaries related to tropical cyclone activity around the globe have been produced for the History Channel, the National Geographic Channel, the Learning Channel, the Discovery Channel, the Weather Channel, and others. Many of these were produced in response to the extraordinarily active, and destructive, 2005 North Atlantic hurricane season.

The disaster film genre enjoyed a brief and somewhat lukewarm resurgence in the late 1990s, with the production of *Dante's Peak* and *Volcano* (1997, volcano-centric films), *Deep Impact* and *Armageddon* (1998, meteor-centric films), and disaster-laden science fiction spectacles such as *Independence Day* (1996) and *Godzilla* (1998). Aside from the TORNADOES (which are mesocyclonic in nature) portrayed in 1996's *Twister*, tropical cyclones were largely ignored, and it was not until the first decade of the 21st century that tropical cyclones again played a major thematic and symbolic role in popular U.S. cinema. This "regeneration" of the tropical cyclone canon resulted in large part to ongoing concerns with climate change, global warming, and the advent of live television programming from areas were real-world tropical cyclones were crashing ashore.

In 2004, director Roland Emmerich and Twentieth Century Fox released *The Day After Tomorrow*, an apocalyptic extravaganza about a world beset by climatic cooling and the destructive meteorological conditions that accompany this dubious process. In strict definitional terms, a tropical cyclone appears only once in the film, as a television newscast reference to the "strongest hurricane ever recorded," Hurricane Noalani. Using actual footage taken during Hurricane INIKI's devastating landfall in Hawaii in 1992, Hurricane Noalani's role in *The Day After Tomorrow* is limited to plot and suspense enhancement. As the film progresses, global cooling creates gigantic, tropical cyclone–like systems that form over the entire Northern Hemisphere—including over the landmasses—a phenomenon of which actual moisture-fed tropical cyclones are incapable.

The Day After Tomorrow also contains some dramatic special-effects scenes of New York City being inundated by a storm surge of unprecedented magnitude, the origin of which is unexplained.

Despite the eccentricity of its scientific premise, *The Day After Tomorrow* was an enormous success with audiences around the world, and this success spawned a limited number of imitators, paramount among them *Category 6* and *Category 7*. Produced in 2004 and 2005, respectively, and released directly to television, these two films again feature a type of cyclonic activity, but not necessarily one that is tropical in nature. As seen with *The Day After Tomorrow*, an accurate portrayal of a tropical cyclone appears in *Category 7*, when "Hurricane Eduoardo" pulverizes Florida before moving northwestward to threaten Washington, D.C. The film also posits the premise that if Edouardo collides with the supercell storm system that was responsible for wreaking so much earlier destruction, the transfer of energy from one to the other could result in "the end of the world." It should be noted that at present, the SAFFIR-SIMPSON SCALE (from which tropical cyclone categories in the United States are derived) only reaches from Category 1 to Category 5; however, in the wake of destructive hurricanes Katrina, Rita, and Wilma in 2005, some observers suggested that a sixth category be added. The original Fujita Scale of Tornadic Intensity did contain an F-6 category, but as such wind speeds were considered "inconceivable," the Enhanced Fujita Scale (EFS) introduced in 2007 omits the EF6 rating.

Owing to cinema's ongoing need for dramatic base material, the Earth's growing shoreline populations, and the unpredictable nature of tropical cyclone activity, it is a safe assumption that tropical cyclones will to some degree or another continue to play a symbolic, thematic, and structural role in future film.

TELEVISION AND TROPICAL CYCLONES

In the four decades since the first RADAR images of Hurricane DIANE's rain-laden EYEWALL were broadcast to a nationwide audience in August 1955, the medium of television has profoundly altered the relationship between civilization and TROPICAL CYCLONES. Enhanced and enlivened by advances in SATELLITE communications technology and afforded an almost ritualistic predominance in the daily lives of societies around the modern world, television now enables official weather services and civil defense agencies to immediately—and with unprecedented detail—transmit vital information concerning the size, intensity, and course of approaching tropical systems to millions of people. Although a leading weapon in the arsenal of tropical cyclone forewarning and civil preparedness, television has in the process of saving countless human lives also created a popular image of the tropical cyclone that emphasizes a HURRICANE's destructive landfall at the expense of its broader climatological benefits. Played as a sort of meteorological miniseries, hurricane and TROPICAL STORM strikes in the United States frequently become the lead stories in nightly television newscasts, adroitly packaged updates that may include on-site interviews with meteorologists, historians, evacuating beachfront residents, relief personnel, and later, once the storm has passed, the survivors. When, for instance, Hurricane BERTHA was nearing landfall in July 1996, CNN dispatched no less than 45 reporters, camera operators, and technicians to NORTH CAROLINA. Some months later the Weather Channel, a 24-hour cable network dedicated solely to tracking and reporting the world's weather, reported that 75 million viewers had tuned in to follow Bertha's rare, midseason strike on the eastern seaboard.

In 1985, Neil Frank, then director of the NATIONAL HURRICANE CENTER (NHC), for the first time permitted news crews to transmit directly from the center, thereby adding even greater immediacy to both the dissemination of meteorological information and to the sense of suspense inherent in any developing event of possibly catastrophic magnitude.

Several journalists and on-air meteorologists have become minor celebrities because of their connection to tropical cyclone reporting, including retired veteran CBS News anchorman Dan Rather. In 1961, while a member of Houston's KHOU-TV news team, Rather provided some 70 continuous hours of coverage of Hurricane CARLA's intense assault on eastern TEXAS.

firefly effect   A curious phenomenon that has on occasion been observed during the landfall of a TROPICAL CYCLONE, the firefly effect occurs when billions of grains of beach sand—swept into the air by the powerful winds of a tropical system—collide while in flight, causing friction, electrostatic sparks that recall the lights of thousands of fireflies on a calm summer night.

Flora, Hurricane *Eastern Northern Caribbean, September 28–October 9, 1963*   A Category 4 HURRICANE of legendary ferocity, Flora spent 11 days of the 1963 HURRICANE SEASON treading a path of destruction through the eastern Caribbean outposts Trinidad, Tobago, and Grenada and the northern islands of JAMAICA, HISPANIOLA, and CUBA. At one point in its life as a late-season TROPICAL CYCLONE of remarkable intensity, Flora claimed between 7,186 and 8,000 lives and exacted huge property and agricultural losses on those islands most severely affected by its passage.

An unusual hurricane from the start, Flora was one of those rare Atlantic hurricanes born below latitude 10 degrees North, over the warm equatorial waters that incubate the fertile shores of the Trinidad and Tobago islands. By the morning of September 30, Flora had matured into a Category 2 hurricane, one whose 110-MPH (177-km/h) winds and six-inch (152-mm) downpours leveled scores of houses and triggered terrible mudslides in Trinidad and Tobago and on Grenada later that day. An estimated 36 people on the these islands lost their lives to the Category 3 fury, while another 469 were reportedly injured.

In Haiti, which Flora pelted with 170-MPH (274-km/h) gusts and 18-foot (6-m) seas on October 4, some 5,000 people were killed, many of them drowned when lightly constructed fishing villages along the country's long southwestern peninsula were submerged by the hurricane's 11-foot (4-m) STORM SURGE. One hundred thousand people were left homeless. Almost all of Haiti's important coffee harvest was ruined. Thousands of trees were uprooted, blocking roads and hindering rescue operations. Lagging rains dampened the remainder of the hurricane-prone country, transformed thousands of acres of mountainous farmland into a diseased quagmire of mudflats and swamps. Regarded as the most destructive hurricane to have affected Haiti during the latter half of the 20th century, Flora's strafing passage through the Jamaica Channel further set back the impoverished nation's economic refurbishment by several years, necessitating both repressive austerity measures at home and a renewed call for relief assistance abroad.

Recurving northwest during the early morning hours of October 4, Flora fortuitously weakened as it bore down on neighboring Cuba and saw its CENTRAL BAROMETRIC PRESSURE rise from 27.64 inches (936 mb) on the morning of October 3 to 28.64 inches (970 mb) by daybreak of October 4. Blaring ashore near the southern port city of Santiago de Cuba just a few hours later, Flora's 107-MPH (172-km/h) winds and torrential rains razed dozens of towns in the provinces of Oriente and Camaguey, spawning a preying band of flash floods that washed out surrounding highways, railways, railroads, and sugarcane fields. Trailing downed power lines, uprooted trees, and wind-tossed roofs behind it, Flora penetrated into central Cuba, delivering heavy rains, gale-force winds, and lightning strikes as far east as the capital, Havana. Its progress north-northwest checked by an encroaching warm front on October 5, Flora spent the next three days stalled over the island's mountainous interior. PRECIPITATION counts of between 15 and 20 inches (381–508 mm) were observed in Las Tunas and Santa Clara, while another round of floods, levee breaks, and river risings blighted 90 percent of that year's sugar and coffee stocks. Both the Cauto and Contramestre rivers reached record heights, forcing tens of thousands of Cubans to flee their vulnerable residences. More than 1,000 people in Cuba died, while another 175,000 were rendered destitute.

Despite the widespread suffering of its people—the hurricane set Castro's agrarian economy back by as much as four years—the Cuban government under the leadership of Fidel Castro rejected offers of both meteorological and financial assistance from the United States. Flora had no sooner dissipated over the BAHAMAS on October 9 than a barrage of allegations, ranging from the hypocrisy of the U.S. trade embargo on the island to the negligent operation of that year's PROJECT STORMFURY experiments, were aired over Havana radio. Unperturbed by this passionate outburst, the United States quietly retired the name Flora from the yearly list of HURRICANE NAMES.

Florence, Hurricane *North Atlantic Ocean–Southern United States, September 23–28, 1953*

Between 1953 and 2006, no fewer than eight TROPICAL CYCLONES in the NORTH ATLANTIC OCEAN and its associated basins have been identified with the name Florence. The first "F" name on the inaugural 1953 hurricane naming list, Florence has remained on the lists for over half a century, and during that time has been used to identify six hurricanes and two tropical storms. Along with the identifier IRENE, Florence is the only other name from the original naming list still being used in the North Atlantic naming system. Additionally, five tropical cyclones in the eastern NORTH PACIFIC OCEAN have been identified as Florence, including hurricanes in 1963, 1973, and 1977, and two tropical storms, one in 1965 and one in 1969. While perhaps not meteorologically significant, it is interesting to note that of the eight North Atlantic tropical cyclones identified as Florence, all but one have originated and existed within the month of September.

The first North Atlantic tropical cyclone to be so identified, Florence was one of four (observed) major hurricanes to afflict the North Atlantic basin during the 1953 tropical cyclone season. Originating in the northern CARIBBEAN SEA, near the southern coast of JAMAICA, during the midday hours of September 23, 1953, Florence remained a TROPICAL STORM as it tracked to the west-northwest and passed through the Yucatán Straits during the early evening hours of September 24. Over night, Florence deepened into a Category 3 hurricane (although "categories" as adapted from the SAFFIR-SIMPSON SCALE were not used until 1975), with its central barometric pressure reaching its (observed) peak intensity of 28.58 inches (968 mb) shortly after noontime on September 25. At that time, the system was situated less than 50 miles

from the western tip of CUBA, and Florence's tropical storm force winds and heavy seas caused considerable damage on that part of the island. Recurving into the GULF OF MEXICO and assuming a more northerly course, Hurricane Florence weakened as it drew closer to the northern Gulf coast. By the time it came ashore in FLORIDA's western Panhandle during the late-evening hours of September 26, 1953, Florence was of Category 1 intensity, with a central pressure of 29.08 inches (985 mb) and sustained, 81-MPH (130-km/h) winds. Rapidly transitioning to an extratropical storm, Florence delivered heavy precipitation and 40-MPH (65-km/h) winds to GEORGIA and SOUTH CAROLINA. Entering the North Atlantic during the early morning hours of September 28, the extratropical remnants of the first Hurricane Florence were tracked as far east as 65 degrees West, before fading into a cold front and meteorological history.

Florence, Hurricane *Gulf of Mexico–Mexico, September 11–12, 1954*

Because the rotating list of North Atlantic TROPICAL CYCLONE names was not introduced until the 1955 tropical cyclone season, the sixth named storm of the 1954 North Atlantic season was dubbed Florence. Replaced in 1955 by Flora, the 1954 incarnation—although a mature stage tropical cyclone—was not nearly so severe as 1953's Florence. It also had the shortest lifespan of the eight North Atlantic tropical cyclones thus far identified as Florence, lasting only from the early morning hours of September 11 until the early evening hours of September 12, 1954. Formed over the Bay of Campeche from a TROPICAL DEPRESSION, Florence reached hurricane intensity shortly before coming ashore near the port city of Veracruz, MEXICO, during the mid-morning on September 12. Sustained winds of 75 MPH (120 km/h) and heavy rains claimed a reported five lives, and caused several million dollars in damage to property.

Florence, Tropical Storm *Southern United States, September 17–27, 1960*

Of TROPICAL STORM intensity for only two days of its 11-day existence, the third North Atlantic TROPICAL CYCLONE christened Florence developed from a TROPICAL DEPRESSION over the extreme eastern NORTH ATLANTIC OCEAN on September 17, 1960. Situated some 300 miles to the north-northeast of PUERTO RICO, the system tracked almost due west, slowly and haphazardly strengthening as it neared the island of HISPANIOLA. Shortly after midnight on September 18, the system deepened into a tropical storm, and remained of tropical storm intensity as it passed over the southern BAHAMAS and recurved to the northwest on September 19. Trailing parallel to the northern coast of CUBA, Tropical Storm Florence steadily weakened, returning to tropical depression classification during the early evening hours of September 19. After forming an eastward loop over western Cuba between September 21 and September 23, a rainmaking Tropical Depression Florence recurved to the north-northeast and made landfall on the western tip of the Florida peninsula during the late afternoon hours of September 23. Dropping several inches of rain along the way, the depression crossed westward across FLORIDA, reentered the GULF OF MEXICO on September 25, and made eventual landfall on the Florida-ALABAMA border during the day on September 26, 1960. Hampered by unfavorable WIND SHEAR conditions for its entire existence, Tropical Depression Florence never again intensified, and although it produced heavy precipitation over Florida and the northern Gulf states, caused no major property losses or deaths.

Florence, Tropical Storm *North Atlantic Ocean, September 5–10, 1964* The fourth North Atlantic tropical system identified as Florence had a relatively long but uneventful existence over the open expanses of the eastern NORTH ATLANTIC OCEAN. Formed to the south of the Azores on September 5, 1964, the system remained of TROPICAL DEPRESSION intensity as it slowly recurved to the north-northwest, maturing into a weak TROPICAL STORM during the midnight hours of September 8. Producing peak wind speeds of 46 MPH (74 km/h) on September 9, Tropical Storm Florence continued its recurvature to the north-northeast before dissipating off the Iberian coast on September 10. An offshore system, Tropical Storm Florence caused no recorded deaths or property losses.

Florence, Hurricane *Gulf of Mexico–Southern United States, September 7–September 11, 1988* Originating from a tropical low over the extreme southern GULF OF MEXICO, the fifth North Atlantic tropical system to bear the name Florence slowly deepened from a TROPICAL DEPRESSION to a Category 1 hurricane between September 7 and September 9, 1988. First traveling on a due east trajectory, Florence briefly stalled as it was upgraded to TROPICAL STORM during the early evening hours of September 7, then proceeded to follow a course due north, one that brought it ashore in southeastern LOUISIANA on September 10. With a peak CENTRAL BAROMETRIC PRESSURE of 29.02 inches (983 mb) at landfall, Florence produced sustained, 81-MPH (130-km/h) winds and torrential rains across the greater New Orleans area. Before recurving to the northwest and dissipating over northeastern TEXAS, Hurricane Florence caused several million in property losses and one death in Louisiana.

Florence, Hurricane *North Atlantic Ocean, November 2–8, 1994* A late-season TROPICAL CYCLONE, Hurricane Florence developed over the cool waters of the mid-Atlantic Ocean from a poorly organized subtropical depression during the early morning hours of November 2, 1994. Slowly drifting to the north-northwest, Tropical Storm Florence encountered southerly WIND SHEAR that hindered its immediate development; indeed, during the morning of November 3, the system was downgraded to a subtropical depression before transitioning to a TROPICAL DEPRESSION and rapidly reintensifying. During the midday hours of November 4, as its CENTRAL BAROMETRIC PRESSURE deepened to 29.50 inches (999 mb), the system's sustained winds topped 58 MPH (98 km/h), making Florence a powerful TROPICAL STORM. By the early evening hours on the same day, as its central pressure dropped to 29.23 inches (990 mb), Florence achieved hurricane status. Steadily intensifying as it drew to the north-northwest, away from the equator, Hurricane Florence was upgraded to a Category 2 hurricane during the early evening hours of November 7, 1994. Within the next 24 hours, as the system passed to the east of BERMUDA, it achieved its (observed) maximum intensity; with a central barometric pressure of 28.70 inches (972 mb), and sustained wind speeds of 109 MPH (176 km/h), Florence was one of the most intense tropical cyclones yet observed in the North Atlantic basin during the month of November. After sharply recurving to the north-northeast on November 7, Florence quickly weakened, dropping to Category 1 intensity during the early evening hours of November 8 and, while transiting the chilly reaches of the NORTH ATLANTIC OCEAN north of 45 degrees North, dissipated. An offshore system for its entire existence, Florence caused no property damage or loss of life.

Florence, Hurricane *North Atlantic Ocean, September 10–17, 2000* Like its 1994 predecessor, the 2000 North Atlantic season's Hurricane Florence—the seventh Atlantic system to bear the name—originated from a subtropical depression, this one located over the Gulf Stream waters between FLORIDA and the island of BERMUDA, on September 11. It deepened into a TROPICAL DEPRESSION during the early morning hours of the 11th; into a TROPICAL STORM by noon, and reached Category 1 status by the early evening hours of the same day. Within a single 24-hour period, the system's CENTRAL BAROMETRIC PRESSURE had dropped from 29.73 inches (1,007 mb) to 29.29 inches (992 mb), hiking its sustained winds from 35 MPH (56 km/h) to 75 MPH (120 km/h). While such a steep pressure drop in a nascent system tends to forecast the eventual development of a robust tropical cyclone, Hurricane Florence was hampered by unfavorable WIND SHEAR conditions. Just as quickly as it had intensified, Florence weakened to a tropical storm, and remained at that classification until the late afternoon hours of September 12, when it briefly regained hurricane status. During its lifespan, Florence spent more time as a tropical storm than as a hurricane, with its sustained winds topping 74 MPH (119 km/h) only three times, and then only for several hours, at best. Tracking to the east-northeast, away from the U.S. East Coast and toward the island of Bermuda, Hurricane Florence weakened to tropical storm intensity during the midday hours of September 13. By the same time on September 15, Hurricane Florence's central pressure had risen to 29.44 inches (997 mb), and its sustained winds had fallen to 52 MPH (83 km/h), making it a moderately severe tropical storm. As it swirled closer to Bermuda, feeding on the warming waters of the Gulf Steam, Tropical Storm Florence again began intensifying, producing its peak pressure reading of 29.08 inches (985 mb) during the early evening hours of September 16. Located less than 200 miles west of Bermuda, the system generated gale-force winds and heavy rains across the island, but caused little damage and no casualties. Rapidly moving to the northeast, Florence rapidly commenced its final weakening cycle. By the early evening hours of September 17, 2000, with its central pressure rising to 29.58 inches (1,002 mb), a downgraded Tropical Storm Florence underwent its extratropical transition over the cooling waters adjacent to the Canadian Maritimes.

Florence, Hurricane *North Atlantic Ocean–Bermuda, September 3–14, 2006* The eighth North Atlantic tropical cyclone identified as Florence originated as a Cape Verde-type system, well to the east of the Lesser Antilles, on September 3, 2006. First upgraded to a TROPICAL STORM (September 5), then to a hurricane (September 10), Florence maintained a steady trajectory to the north-northwest, struggling to intensify in an unfavorable environment of WIND SHEAR and low atmospheric water vapor content. It finally reached its peak intensity on September 11, when its central pressure slipped to 28.70 inches (972 mb). By that time, the system had delivered a glancing but potent strike to BERMUDA, where 80-MPH (129-km/h) winds and more than one inch (25.4 mm) of precipitation downed power lines, uprooted trees, and damaged several houses and other structures. While a few minor injuries were reported, no deaths were recorded in Bermuda in Florence's wake. Briefly intensifying after recurving to the north-northeast, Florence rapidly weakened, and underwent extratropical transitioning on September 13, 2006, while situated over Newfoundland, CANADA. Sustained winds of 101 MPH (163 km/h) battered the province's eastern shorelines, while nearly three inches (67 mm) of precipitation fell at Salt Pond, Newfoundland. One of the most powerful extratropical cyclones yet observed

in Canada, Extratropical Cyclone Florence damaged several BUILDINGS, but claimed no lives or major injuries. Carried firmly to the northeast, the extratropical system continued to weaken and by September 15, had dissipated.

Florida An enormous peninsula that deeply penetrates the HURRICANE-prone waters of the GULF OF MEXICO and the NORTH ATLANTIC OCEAN, Florida's 1,197 miles of coastline has, over the past 498 years, withstood more TROPICAL CYCLONE strikes than any of the United States. Between 1528 (when Spanish explorers lost two ships and nearly 400 men to a hurricane in west Florida's, Apalachee Bay) and 2006, an estimated documented tropical storms and hurricanes came ashore on Florida's three principal coasts—upper gulf, east, and west—bringing heavy rains and damaging winds to their sandy barrier islands, exclusive resort communities, expansive citrus plantations, and beachside cities.

On average, Florida's coral keys and shallow inland lakes are affected by at least one tropical cyclone of varying size and intensity every year and by a major hurricane—a storm of Category 3 status or higher on the SAFFIR-SIMPSON SCALE—every five years. On at least 30 occasions between 1831 and 1995, two or more hurricanes have directly impacted the state in a single year, as partly represented by the 1835, 1837, 1844, 1848, 1870, 1886, 1903, 1916, 1928, 1933, 1946, 1964, 1995, 2004 and 2005 seasons. In 1860 Florida's west coast was struck by two hurricanes within a month. On August 11, Pensacola was afflicted with the fringe winds and expansive rains of a large hurricane that had gone ashore in LOUISIANA the night before. With winds in excess of 50 MPH (81 km/h), the tropical storm-force fury dumped more than 3 inches of rain on the coastal city. On September 14 the same area was swept by another hurricane, this one of more moderate intensity. Although heavy surf was noted all along the Panhandle region, little property damage was sustained.

During the 1837 hurricane season, no fewer than five tropical systems of hurricane strength lurched ashore between August 1 and September 26, yielding a sobering trail of destruction that reached from the Florida–GEORGIA border to the Florida Keys. A similar quintet was noted in 1886 when five mature hurricanes crisscrossed the state in just over two months' time. The years 1906, 1935, and 1948 saw four mature hurricanes visit the state per respective season, with at least six being storms of significant intensity and duration. No known hurricanes affected the state in 1889, 1891, 1893–94, 1897, 1901–02, 1904, 1907–08, 1913–14, 1918–19, 1922–23, 1927, 1930–32, 1934, 1937–38, 1942–43, 1946, 1951–52, 1957–59, 1961–63, 1968, 1969–71, 1973–74,

1976–78, 1980–84, and 1986–91. The longest lull in Florida hurricane activity stands at six years, the period between 1986 and 1991.

Florida's position between latitudes 23 and 31 degrees North presents the state with a long, very active hurricane season. Officially extending from June 1 to November 30, Florida has also endured its share of pre-and postseason tropical cyclones. In late May 1970 a dissipating tropical storm produced torrential rains over the state's northern tier but caused minimal property losses and no casualties. On June 5, 1995, Hurricane ALLISON, a relatively mild, Category 1 system, became the earliest recorded hurricane to strike the state (displacing 1966's June 9 Hurricane ALMA), while an unnamed Category 1 tempest in 1925 made landfall near Tampa on December 1. No deaths or injuries were reported in either storm, and damage from both was on the whole considered minor.

Between 1561 and 1995, Florida was affected by 12 June hurricanes, none of them particularly severe. All originated in either the Gulf of Mexico or the southwest CARIBBEAN SEA, and almost all came ashore in the Panhandle-Big Bend area where the oyster-farming city of Apalachicola enjoys the dubious distinction of being the most hurricane-ridden community in the United States. During the third week of June 1561, an early season hurricane chased Spanish explorers from nearby Pensacola when it sank several warships and speedily laid the newly built settlement to waste. On June 21, 1886, a strident hurricane lashed Apalachicola with destructive winds and flooding rains, killing three people and sinking several small watercraft; slightly more than a week later, on June 30, a second hurricane struck the city, claiming another 10 victims. The EYE-WALL of Hurricane Alma, a fairly tepid Category 2 surprise that had developed off the east coast of MEXICO's Yucatán Peninsula on June 6, 1966, brought 80-MPH (129-km/h) gusts and swarming seas to Florida's west coast on June 8 but remained on offshore menace for the rest of its existence. In 1972, Hurricane AGNES, a rainmaking upstart born near CUBA on June 15, brought 75-MPH (121-km/h) winds and TORNADOES to Apalachicola on June 19 and 19-inch (483-mm) PRECIPITATION counts to the northeast United States on June 22. Nine people in Florida were killed, and some $10 million in property damages were assessed in the historic wake of this early season storm.

Since 1715, there have been at least nine documented hurricane strikes on Florida in July, a majority of them powerful Gulf of Mexico and North Atlantic storms that left a lasting impression on those eyewitnesses who survived them. In mid-July 1715, a vengeful North Atlantic hurricane drove a heavily laden Spanish treasure fleet aground on Florida's southeast coast, wrecking seven ships and drowning scores of sailors. Scattering chests of gold and gems across the sandy seabed, the unnamed July hurricane of 1715 soon caused a financial panic in Spain and earned that rueful stretch of Florida shore the nickname "Treasure Coast." On July 19, 1886, a substantial gulf hurricane sliced ashore on Florida's northwest coast, filleting the fishing village of Cedar Key as its 85-MPH (137-km/h) winds cut across the state and emptied into the North Atlantic. Seventeen people were killed, and more than 500 houses blew down. A half-century later, the first of two hurricanes to afflict the state during the 1926 season, a minimal Category 1 system, slipped ashore just south of Jacksonville on July 27. More than $2 million in losses were tallied in the town of Indrio alone, signaling the end of the great land boom that had in previous years so shabbily (the magnitude of the weak hurricane's destruction on new construction evidences this) developed much of Daytona Beach.

July is also a month for tropical storms in Florida, cyclonic systems whose 39–73-MPH (63–118-km/h) winds do not qualify them for hurricane status but whose furious gales and sheeting rains often inflict steep damage on those counties where they come ashore. During the 1994 hurricane season, Tropical Storm ALBERTO fried much of the Panhandle with 60-MPH (97-km/h) winds and searing rains on July 2; interestingly enough, Alberto's northward course from Cuba nearly duplicated that of an unnamed tropical storm that crossed into northwestern Florida on the same date in 1919.

August, September, and October have historically been Florida's stormiest months. Fueled by the warmest water temperatures of the season and driven west by the constant trade winds, robust North Atlantic hurricanes have clouded the Sunshine State's shores at least 29 times in August, another 71 times in September, and on 55 different occasions in October. Sometimes of the CAPE VERDE STORM variety, in other instances swarming in from the southern Caribbean Sea or the western Gulf of Mexico, a majority of mid-season Floridian hurricanes have been major systems, Category 3 and 4 whirlwinds that have claimed thousands of lives and wrought billions of dollars worth of damage on BUILDINGS, roads, bridges, piers, and agricultural installations across the state. On September 16, 1928, Florida suffered what is generally considered its deadliest confirmed hurricane landfall when the mighty SAN FELIPE Hurricane blasted West Palm Beach and the shallow environs of Lake Okeechobee with 130-MPH (209-km/h) winds, blinding rains, and a merciless 20-foot STORM SURGE. As hundreds of summer cottages and lakeside farms were demolished, their screened-in front porches and flat roofs pried apart by the hurricane's wrenching gusts and cresting waves, some 1,836 people were drowned, and hundreds of thousands of dollars in losses were tallied.

The track of the intense Miami Hurricane of 1926 is shown here. One of the most severe tropical cyclones to strike southern Florida in the first half of the 20th century, the storm killed nearly 500 people.

Other important midseason hurricanes in Florida include:

A severe hurricane in September 1565 violently scattered a large fleet of French warships, forcing many to go ashore near Cape Canaveral. Only the day before engaged in a bombardment of the Spanish settlement at St. Augustine, those half-dead French survivors who washed up on the beaches were quickly butchered by the Spanish soldiers.

A hurricane on October 22, 1766, spun ashore near Pensacola, wrecking several ships, splintering wharves, and toppling a church spire. Dozens of people were said to have perished in the gulf tempest, many of them drowned by the huge storm surge that choked off the entrance to Pensacola Bay.

A particularly murderous, September 15, 1835, hurricane stymied the Florida Keys, killing more than 100 people and injuring hundreds of others. The hurricane's force was felt from Key Largo to Key West, where many boats and ships were thrown ashore, several of them becoming total losses. More than four feet (1 m) of water ran through the center of Key Biscayne, knocking houses from their foundations and depositing them in the surrounding mangrove swamps.

During the Apalachee Bay Hurricane, August 30, 1837, a 15-foot (5-m) STORM TIDE inundated Apala-chicola and the neighboring village of St. Marks with more than seven feet (2 m) of water. A schooner was cast onto the beach, and a nearby lighthouse was severely damaged, losing both its lantern and its outbuildings. A number of houses in Apalachicola were stripped of their "hurricane-resistant" slate roofs, and the resilient town's market house was collapsed. Eight people were killed and dozens more seriously injured.

A large hurricane on October 4, 1842, again battered the Apalachicola lighthouse, breaching a section of its seawall and submerging the keeper's cottage. Scores of houses in Apalachicola were unroofed; high winds as far east as Tallahassee broke windows, plucked trees from the earth, and contaminated drinking wells. Between 20 and 30 people perished in this storm, including six victims who had sought shelter in the ill-fated lighthouse.

The Tampa Bay Hurricane, September 25, 1848, delivered Category 3 winds (111–130 MPH) and an enormous, 10–15-foot (3–5-m) storm surge to the bustling coves of Tampa Bay.

The Great Middle Florida Hurricane on August 23, 1851, belted central and northern Florida with 110-MPH (177-km/h) winds and strapping seas. In the coastal town of St. Marks, the hurricane's gripping surge crushed the railroad station and then carried its splintered remains far inland. In an act that strangely mocked the tragic events in the later LABOR DAY HURRICANE OF 1935, residents in St. Marks wisely boarded their rescue train early and thus were safely evacuated to higher ground.

A Cuban-born 1906 hurricane chugged across the Florida Keys on October 18, heavily damaging the famed Overseas Railroad then being built between Key Largo and Key West for millionaire visionary Henry Flagler. Some 129 railroad laborers housed on wooden flatboats were drowned when the hurricane washed their makeshift barracks ashore, capsizing two and smashing the rest into piles of kindling.

The intense Corpus Christi Hurricane of September 6, 1919, hounded southernmost Key West with 115-MPH (185-km/h) gusts, five-inch (127-mm) rainfall counts, and wild seas while on its way to an eventual landfall in TEXAS. Caught at sea by the deepening hurricane, the Spanish steamship *Valbanera* foundered near the island of Dry Tortugas, taking 488 people to the bottom with it.

In the infamous Miami Hurricane Disaster of 1926 an estimated 243 people lost their lives to the northwesterly passage of a powerful Category 4 (27.61 inches [931 mb]) storm that brought 10-inch (254-mm) rains and a 13-foot (4-m) storm surge to large portions of the state on the night of September 17–18. One of Florida's better-known tropical cyclones, the Miami Hurricane pushed a five-masted schooner ashore near Biscayne Boulevard, collapsed hundreds of recently built cottages and bungalows in Hialeah, Coconut Grove, Homestead, and Moore Haven, and capsized a steam yacht once owned by Kaiser Wilhelm II of Germany. Bridges in Biscayne Bay were severed from the land and their steel-and-stone approaches hurled into the rising waters; downed power lines and debris-coated streets made rescue operations extremely hazardous. A number of posh oceanfront mansions in Miami Beach were destroyed by the hurricane, ending for a time the real estate boom that had, in less than a decade, turned picturesque sandy spits into overbuilt—but obviously vulnerable—small towns.

An expensive, westward-moving hurricane that traversed southern Florida on the night of September 17–18, 1947, downed power lines, swamped small boats, and destroyed more than $20 million worth of real estate. Seventeen people in the cities of Palm Beach, Fort Lauderdale, and Delray perished in the Category 4 hurricane (a central pressure of 27.97 inches [943 mb] was observed at landfall); in darkened Miami, police officers cornered and shot to death a man who had been suspected of looting.

In the so-called Gold Coast Hurricane of August 26–27, 1949, sustained 120-MPH (193-km/h) winds and relentless rains inflicted an estimated $40 million in property damage as, under the cover of night, it

passed over Palm Beach and Lake Okeechobee. Two people were drowned.

Between 1950 and 2006, the dangerous winds, flooding rains, and bulldozing surges associated with Hurricanes EASY (1950), DONNA (1960), CLEO (1964), BETSY (1965), GLADYS (1968), ELOISE (1975), DAVID (1979), ELENA (1985), JUAN (1985), KATE (1985), ANDREW (1992), GORDON (1994), and OPAL (1995), seriously affected the whole of Florida, causing scores of deaths and billions of dollars in accumulated property losses.

The last decade of the 20th century and the first decade of the 21st were no gentler to Florida's already vivid history of tropical cyclone strikes. In September 2000, Tropical Storm Gordon and Tropical Storm Helene barraged the state with heavy rains. On September 18, former Hurricane GORDON (now downgraded to a tropical storm) bluffed ashore near Florida's central Gulf coast, producing precipitation counts of eight inches (203 mm) across a broad swath. Within a week, on September 22, Tropical Storm Helene lumbered ashore in the western Panhandle, dropping an additional 10 inches (254 mm) across the northern reaches of the state.

On September 14, 2001, Tropical Storm Gabrielle (later upgraded to hurricane intensity) trumpeted ashore in central western Florida, near Venice, its central pressure of 29.02 inches (983 mb) at landfall generating considerable precipitation and widespread damage to crops and foliage. Two people in Florida died in drowning incidents.

Nearly one year later, on September 4, 2002, Tropical Storm EDOUARD waddled ashore on Florida's eastern coast, near Jacksonville, with a central pressure of 29.79 inches (1,009 mb). Traveling to the west-southwest, Edouard was downgraded to a tropical depression as it entered the Gulf of Mexico on September 5. Conditions for further strengthening were unfavorable, and the system dissipated shortly thereafter.

Between September 13 and 15, 2002, the remnants of Tropical Storm Hanna (which had made landfall in neighboring ALABAMA on September 14), delivered heavy rains to Florida, with 8.10 inches (203 mm) recorded in Chipley, Florida, and 5.05 inches (127 mm) in Marianna, Florida. Although three people were killed elsewhere, no deaths or injuries were reported in Florida following Hanna's passage.

On September 5, 2003, rains generated by a weakening Tropical Storm Henri drenched much of western Florida's St. Petersburg area. Downgraded to a tropical depression just before making landfall near Clearwater, Florida, Henri produced local downpours of up to 10 inches (254 mm) across central Florida. The southern city of Hialeah was inundated with more than nine inches (229 mm) of precipitation, and severe flooding occurred in several communities.

During the hyperactive 2004 North Atlantic hurricane season, Florida was struck by no fewer than four mature-stage tropical cyclones: Hurricanes CHARLEY, FRANCES, IVAN and Jeanne. In total, they caused some 130 deaths and $22 billion (in 2004 U.S. dollars) in damages across the Sunshine State. In addition, on August 12–13, 2004, Florida's Panhandle was pinged by Tropical Storm Bonnie, whose central pressure of 29.55 inches (1,001 mb), sustained, 55-MPH (89-km/h) winds and six-inch (152-mm) rains at landfall near Apalachicola caused some property damage and shoreline erosion. A TORNADO reportedly touched down near the eastern city of Jacksonville, and several thousand people were evacuated from low-lying counties, but only one casualty was recorded.

One day later, on August 13–14, Hurricane Charley thundered ashore near Port Charlotte, Florida, as a fairly intense Category 4 system. At the time, it was the most intense tropical cyclone to have directly hit Florida since Hurricane Andrew's Category 5 strike in 1992. With a central pressure of 27.96 inches (947 mb) at landfall during the afternoon hours of August 13, Charley produced sustained 144-MPH (232-km/h) winds across much of southwestern Florida. A durable system, Charley was downgraded to a Category 1 hurricane (with a central pressure of 28.64 inches [970 mb]) while transiting the state to the northeast, bringing wind gusts of 106 MPH (171 km/h) to the Orlando International Airport during the night of August 13–14. In Florida, Charley claimed 29 direct and indirect deaths, and caused some $16 billion (in 2004 U.S. dollars).

Less than one month later, on September 3–4, 2004, Hurricane Frances rumbled ashore near Port St. Lucie, on Florida's southeastern coast, as a powerful Category 2 hurricane. At one time boasting a central barometric pressure of 27.66 inches (937 mb), Frances weakened considerably as it slowly approached Florida, and made landfall with a central pressure of 28.34 inches (960 mb) and sustained, 104-MPH (167-km/h) winds. The largest EVACUATION yet seen in Florida's history—some 2.8 million people—was undertaken in response to Frances's approach. Widespread destruction accompanied the system's passage over the state, including severe damage to several NASA facilities at Cape Canaveral. Downgraded to a tropical storm as it spun to the northwest, Frances made a second landfall in Florida, near the Panhandle town of St. Marks, during the late evening hours of September 5. In some places, 13 inches (330 mm) of precipitation fell, causing widespread damage to citrus and other crops. Frances left over $8 billion (in 2004 U.S. dollars) in property losses, and 35 direct and indirect deaths in Florida.

Relief and recovery crews had barely settled in to remediate the destruction from Bonnie, Charley,

and Frances, when "crazy" Hurricane Ivan laid siege to the state on September 16, 2004. Charging ashore near the Alabama–Florida border as a Category 3 system (with a central pressure of 27.49 inches [931 mb]) and 127-MPH (204-km/h) winds, Ivan caused widespread coastal flooding in and around Pensacola. Sections of a major causeway over Florida's Escambia Bay were destroyed; dozens of small watercraft were cast ashore. Five days later, after performing an enormous clockwise loop across the eastern United States, and transitioning first to an extratropical depression, then an extratropical storm, then back to an extratropical depression again, Ivan slid ashore in southeastern Florida on September 21, 2004. It eventually transitioned to a tropical storm on September 23 and reentered the Gulf of Mexico. In total, 14 people in Florida lost their lives to Ivan's initial assault, while more than $9 billion (in 2004 dollars) in losses were tallied.

Like a relentless tropical conveyor belt, the North Atlantic quickly sent Hurricane Jeanne spiraling into southeastern Florida on September 25, 2004. Another tropical cyclone with a unique track, Jeanne rushed ashore near Stuart, Florida, as a fairly intense Category 3 system, with a central pressure of 28.08 inches (951 mb) at landfall. Wind speeds of 121 MPH (195 km/h) caused damage to trees, signs, and other small structures, while precipitation counts as high as 11.97 inches (279 mm) were observed in communities (Kenansville) to the southwest of the storm's landfall. Three people in Florida were killed by Jeanne; but it could have been much worse. Ten days earlier, on September 15, Category 1 Jeanne's landfall and subsequent downgrade to a rain-gushing tropical storm over the highlands of Hispaniola claimed more 3,000 lives in Haiti.

It is perhaps not surprising that Florida with its long history of multiple tropical cyclone strikes in a single season would again suffer through an active hurricane season, this the remarkable 2005 North Atlantic hurricane season. Of the 28 named storms that originated in the North Atlantic basin between June 8th and December 30th of that year, eight directly or indirectly affected the Floridian peninsula: ARLENE, CINDY, DENNIS, KATRINA, OPHELIA, RITA, TAMMY, and WILMA.

On June 11, 2005, a powerful Tropical Storm Arlene rushed ashore in the Florida Panhandle, near the city of Pensacola. The first named storm of the extraordinarily active 2005 North Atlantic hurricane season and bearing sustained winds of 70 MPH (km/h), a central pressure of 29.23 inches (990 mb), and 14-foot (5-m) seas, Arlene disrupted power supplies to more than 11,000 residences and businesses. Some eight inches (203 mb) of rain fell across northern Florida, and one direct death (in a riptide drowning incident) was recorded.

Although it had made direct landfall in Louisiana, Hurricane Cindy nevertheless dropped one inch (25 mm) of rain across the extreme western sections of the Panhandle as it recurved to the northeast on July 6, 2005. No deaths, injuries, or property damage was reported in Florida.

It was to be a short reprieve for the Florida Panhandle, however. Less than a week later, Hurricane Dennis broke ashore near Santa Rosa Island, near Pensacola, during the afternoon hours of July 10. A Category 3 hurricane, Dennis's central pressure of 27.81 inches (942 mb) produced wind speeds of 115 MPH (185 km/h) to 120 MPH (195 km/h) across much of northwestern Florida. In sections of north-central Florida, Dennis dropped more than seven inches (178 mm) of precipitation. Dennis was a deadly storm for Florida; 14 people were killed, and $2 billion (in 2005 U.S. dollars) in damage estimates were tallied.

During the morning hours of August 25, 2005, a newly upgraded Hurricane Katrina clawed ashore near Hallandale Beach, Florida. Destined to become the most expensive natural disaster yet seen in United States history, Katrina's landfall in southern Florida was relatively mild. A central pressure reading of 29.02 inches (983 mb) at landfall generated 81-MPH (130-km/h) winds and some rainfall; 14 people in Florida lost their lives to Katrina.

The first two weeks of September, 2005, delivered to Florida the heavy rains associated with an offshore Hurricane Ophelia. Seemingly tormented by its inability to complete a successful landfall, Ophelia trailed along Florida's eastern coastline between September 7 and September 10, twice intensifying to hurricane status before plunging back to tropical storm intensity. An offshore central pressure of 29.02 inches (983 mb) generated precipitation counts of between one inch (25 mm) and five inches (127 mm) along much of Florida's eastern shores. In Florida, one person was directly killed in a drowning incident caused by Ophelia's powerful surf and rip currents.

By September 20, 2005, southern Florida was again under tropical cyclone warnings, these raised in response to the approach of Hurricane Rita. Like the earlier Hurricane Katrina, Rita was essentially of Category 1 status as it came into interaction with Florida, but unlike Katrina, Rita's slowly strengthening eyewall was situated to the south of the Florida Keys. Wind speeds of 81 MPH (130 km/h) were recorded across the lower Keys. Despite a mandatory evacuation of the entire Florida Keys, two people in Florida perished during Rita's passage. Again, given Rita's strengthening history, it could have been much worse for Florida; within two days of passing through the Florida Straits, Rita had interacted with the Gulf Loop Eddy and undergone EXPLOSIVE DEEPENING. On September 22, the system produced

a central pressure reading of 26.48 inches (897 mb), making it the fourth most intense tropical cyclone yet observed in the North Atlantic basin.

On October 5, 2005, a weak Tropical Storm Tammy trundled ashore near the northeastern coastal city of Mayport, located some 15 miles (24 km/h) east of Jacksonville. Sustained winds of 50 MPH (80 km/h) lashed the U.S. naval air station in Mayport and downed trees and power lines, but otherwise caused little damage. Moving to the north at 14 MPH (23 km/h), the bulk of Tammy's three to five-inch (76.2–127-mm) rainfall was shorn over GEORGIA and SOUTH CAROLINA, although totals of one inch (25 mm) were observed across much of eastern and northern Florida.

At one time in its existence the single-most intense tropical cyclone yet observed in the North Atlantic basin, Hurricane Wilma came ashore in central-southwestern Florida on October 24–25, 2005, as a Category 3 system. With a central pressure of 28.05 inches (950 mb) at landfall near Everglades City, Wilma produced sustained winds of 115 MPH (185 km/h) across much of its landfall area. In preparation for Wilma's landfall, some 160,000 Floridians were evacuated from the Keys and other coastal areas due to storm surge predictions ranging from eight to 15 feet (3–5 m). Considerable damage was wrought by Wilma's passage across Florida, and 36 people lost their lives. Wilma's landfall in Florida made the system the eighth mature-stage tropical cyclone to have directly affected the state in less than two years.

The 2006 North Atlantic hurricane season was mercifully much less active for the Sunshine State, with only two tropical systems making landfall on the peninsula between June and August. The first, Tropical Storm Alberto, bashed ashore near Tallahassee, Florida, on June 13. Near hurricane strength when it came ashore, Alberto caused significant damage to trees, energy infrastructure, and small structures, but claimed no lives. On August 29, Tropical Storm (formerly Hurricane) Ernesto rolled ashore in southeastern Florida, bringing tropical storm–force winds and heavy rains to much of the south and eastern sections of the state. Two people were (indirectly) killed in traffic accidents due to Ernesto's passage.

Flossy, Hurricane *Gulf of Mexico–Southeastern United States, September 21–25, 1956* A moderately powerful Category 2 HURRICANE, Flossy delivered sustained 107-MPH (172-km/h) winds, five-inch (127-mm) PRECIPITATION counts, and a six-foot STORM SURGE to LOUISIANA, FLORIDA, and GEORGIA between September 23–25, 1956. Born over the southwestern GULF OF MEXICO on the evening of September 21, Flossy quickly intensified while undergoing RECURVATURE to the north-northwest. Bearing a CENTRAL BAROMETRIC PRESSURE of 28.79 inches (975 mb) at landfall just south of New Orleans on the morning of September 23, Flossy briefly returned to the Gulf of Mexico before making a second landfall at Laguna Beach, Florida, on the evening of August 24. Recurving into southern Georgia, Flossy rapidly dissipated into a string of torrential tropical downpours. One of the milder Gulf of Mexico hurricanes on record, Flossy nevertheless managed to claim 24 lives in Louisiana and Florida and to tally damage estimates to oil drilling equipment in the Gulf of Mexico in excess of $2 million. The name Flossy has been retired from the revolving list of North Atlantic HURRICANE NAMES.

Floyd, Hurricane *North Atlantic Ocean–Puerto Rico, September 3–12, 1981*   Although the name Floyd has been retired from the rotating list of North Atlantic TROPICAL CYCLONE names (and replaced with Franklin), total of three North Atlantic tropical systems carried the identifier Floyd.

The first was a hurricane that traversed the NORTH ATLANTIC September 3–12, 1981. As a TROPICAL DEPRESSION and then a TROPICAL STORM, it delivered heavy rains to the Leeward Islands and PUERTO RICO, but claimed no lives. It eventually intensified to Category 3 status in mid-Atlantic, and steadily faded thereafter.

Floyd, Hurricane *North Atlantic Ocean–Cuba–Southern United States, October 9–14, 1987*   The

second system dubbed Floyd lasted October 9–14, 1987, and along the way unfurled heavy rains and hurricane-force winds across northern CUBA and southern FLORIDA. While over the Florida Keys on October 12, Hurricane Floyd generated a pressure reading of 29.32 inches (993 mb), making it a Category 1 system.

Floyd, Hurricane *North Atlantic Ocean–Europe, September 7–13, 1993*   The third North Atlantic tropical system, also a hurricane, endured from September 7 to 13, 1993, and was meteorologically more notable as an extratropical storm than it ever was as a hurricane when it struck EUROPE with hurricane-force winds.

Floyd, Hurricane *Bahamas–Eastern United States, September 2–17, 1999*   The deadliest TROPICAL CYCLONE to have made landfall in the northeastern United States since Hurricane AGNES sloshed up the East Coast in 1972, HURRICANE Floyd's tightly wound RAIN BANDS delivered sustained 155-MPH (249-km/h) winds, torrential rains, and 19-foot (7-m) breakers to the islands of the BAHAMAS, as well as extensive flooding to the northeastern seaboard of the United States, between September 2 and 17, 1999. Some 72 people were killed, while damage estimates topped $3 billion.

Positioned 45 miles (72 km) east of Cat Island, the Bahamas, an exceptionally powerful Hurricane Floyd

Captured in three dimensions through a technological team of satellites and computers, a Category 4 Hurricane Floyd churns toward Florida in 1999. Just before it was scheduled to make landfall in the Sunshine State, a very large Floyd drew out into the Atlantic. *(NOAA)*

enveloped the low-lying island chain during the morning and afternoon hours of September 13. Boasting sustained winds of 155 MPH (249 km/h), the incipient Category 5 hurricane (powered by a central pressure of 27.29 inches [924 mb]) in fact delivered powerful winds to an enormous stretch of the NORTH ATLANTIC OCEAN. One of the largest hurricanes yet observed in the North Atlantic basin, Floyd's hurricane-force winds extended some 125 miles (467 km) from the eye, while tropical storm-force winds reached an additional 290 miles (201 km) beyond.

In the Bahamas, Floyd's 155-MPH (249-km/h) winds pried the roofs from scores of buildings, flipped dozens of automobiles, burned powerlines, and sent waves of sand scudding across empty, palm-strewn roads. On Paradise Island, situated to the southeast of Nassau, some 2,000 guests at the Atlantis hotel sought shelter in the sprawling resort's convention center. Despite widespread property damage, only one person died in the Bahamas, a miraculously low DEATH TOLL considering the severity of Floyd's passage across the coral archipelago.

Gliding to the northwest at 14 MPH (22 km/h), a sprawling Floyd became the subject of intense media attention on the afternoon of September 14, as its 175-MPH (282-km/h) gusts neared to within 100 miles of landfall on FLORIDA's eastern coastline. In the state's tourist capital of Orlando, Disney World, along with several other theme parks, closed their attractions for the first time in their histories. Touted as the largest EVACUATION in American history, some 3.1 million people, including 800,000 in SOUTH CAROLINA and another half-million in NORTH CAROLINA, boarded cars, buses, and trains and fled the beaches for relative safety inland. As far north as Savannah, GEORGIA, 75 percent of the city's 180,000 residents abandoned their houses before the very real threat of a direct hit by the most intense hurricane to threaten the United States in seven years.

Shortly after midnight on September 15, the National Aeronautics and Space Administration (NASA) ordered the evacuation of 12,500 personnel from the Kennedy Space Center in Cape Canaveral. With forecasts indicating a 75 percent chance that sustained 130-MPH (209-km/h) winds would assail the Cape by mid-morning, a skeleton crew of 102 volunteer workers was left to oversee the safety of the nation's $9 billion space shuttle fleet—four of which, *Atlantis, Endeavour, Discovery* and *Columbia*, are housed in hangars designated to withstand windspeeds not to exceed 105 MPH (169 km/h).

In southern North Carolina, severe flooding demolished 6,344 houses and rendered another 8,936 uninhabitable. As the state is one of the most hurricane-riddled in the nation, plans were soon announced to spend $600 million relocating thousands of families from those areas most vulnerable to flooding. In New Jersey, property losses reached $500 million, as more than 27,000 houses and 1,700 businesses sustained some degree of damage from Tropical Storm Floyd's passage.

Fran, Tropical Storm *North Atlantic Ocean, October 8–13, 1973* Between 1973 and 1996, four North Atlantic TROPICAL CYCLONES were identified as Fran. Of these, two were of TROPICAL STORM intensity, while the others were mature-stage tropical cyclones. The fourth Fran—which crashed ashore in the eastern United States in 1996—proved particularly severe, and it was due to the ferocity of this strike that the name Fran was retired from the list of North Atlantic tropical cyclone identifiers and replaced with Fay.

The first North Atlantic tropical system identified as Fran existed from October 8 to October 13, 1973, originated from a subtropical depression, and eventually reached Category 1 status (with an estimated peak minimum pressure of 28.88 inches [978 mb]) between October 10 and October 12, while positioned over the mid-Atlantic. On October 13, it underwent extratropical deepening and was reclassified as an extratropical storm. It shortly dissipated.

Fran, Tropical Storm *North Atlantic Ocean– Cape Verde Islands, September 15–20, 1984* The second system identified as Fran existed from September 15 to 20, 1984, and never grew beyond tropical storm intensity. It reached its peak minimum pressure of 29.35 inches (994 mb) on September 18, while situated at latitude 20 degrees North, longitude 28 degrees West, near the Cape Verde Islands. While observatories in the Cape Verde Islands recorded sustained wind speeds of 35 MPH (56 km/h), 31 people reportedly lost their lives to flash floods spawned by Fran's tropical rains.

Fran, Tropical Storm *North Atlantic Ocean– Trinidad, August 11–13, 1990* The third North Atlantic tropical cyclone identified as Fran originated on August 11, 1990, near latitude 9 degrees North, 32 degrees West, or some 1,000 miles (1,609 km) east of the Windward Islands. A short-lived system (due, in part, to its proximity to the equator, where the CORIOLIS EFFECT is at its weakest), Fran was upgraded to a TROPICAL DEPRESSION during the midday hours of August 11, but two days later, on August 13, was downgraded to a TROPICAL WAVE. Within 24 hours, the system was again of TROPICAL STORM intensity, with a central pressure of 29.79 inches (1,009 mb), and sustained winds of 40 MPH (60 km/h). Tropical Storm Fran was slowly deepening when it spun ashore on the island of Trinidad, and passed into northern Venezuela, with a central pressure of 29.73

inches (1,007 mb). A very inhospitable environment for further strengthening, Tropical Storm Fran quickly dissipated. No lives were lost and damage estimates were light.

Fran, Hurricane *North Atlantic Ocean–Eastern United States, August 23–September 10, 1996* Unfortunately, the fourth and final North Atlantic hurricane identified as Fran, this one was a fierce Category 3 system that made landfall in eastern NORTH CAROLINA with destructive and deadly effect. Existing from August 23 to September 10, 1996, Fran originated as a TROPICAL DEPRESSION while off the western coast of North Africa. It remained of tropical depression intensity until August 27, when it was upgraded to TROPICAL STORM status, with a central pressure of 29.67 inches (1,005 mb). Carried almost due westward along the 15th parallel, the system now began a shallow recurvature to the northwest and commenced strengthening, finally becoming a Category 1 hurricane (with a central pressure of 29.14 inches [987 mb]) during the late evening hours of August 29. The system was briefly demoted to Tropical Storm Fran on August 30, but by the following day was returned to Category 1 status, with a central pressure of 29.05 inches (984 mb). Continuing its RECURVATURE to the north-northwest, Hurricane Fran skirted the eastern shores of the BAHAMAS between September 3 and September 5, 1996, growing stronger along the way. On the night of September 3, Fran's central pressure dipped to 28.58 inches (968 mb) and the system was promoted to Category 2 status. Within six hours, the storm was again upgraded, this time to a Category 3 system. It reached its peak minimum central pressure of 27.93 inches (946 mb) shortly after midnight on September 5, and while situated less than 400 miles (644 km) southeast of Cape Hatteras, North Carolina. As it approached landfall in North Carolina, however, Hurricane Fran weakened, its central pressure of 28.17 inches (954 mb) at landfall near the Cape Fear area on September 5, 1996, delivering 115-MPH (185-km/h) winds to much of eastern North Carolina. A large system at landfall, hurricane force winds of 75 MPH (121 km/h) extended some 140 miles (225 km/h) from the system's center. Hurricane Fran rapidly weakened after making landfall, and by the late evening hours of September 6, was downgraded to a tropical depression while situated over the highlands of West Virginia. On September 9, as the system's remnant core scudded over the eastern Great Lakes, a once-fearsome Hurricane Fran transitioned into an extratropical depression that dropped large quantities of precipitation over NEW YORK State and NEW ENGLAND. By September 10, the system had

dissipated. Fran was a particularly severe tropical cyclone strike for the United States. In North Carolina, nearly 90 percent of the structures on North Topsail Beach were destroyed. Precipitation counts of 10 inches (254 mm) accompanied Fran's landfall in North Carolina, while rainfall totals in VIRGINIA and Pennsylvania ranged from one inch (25 mm) to five inches (127 mm). This precipitation caused dozens of flash flooding incidents, and many of the region's rivers—including the Potomac in Washington, D.C.—overflowed their banks. Flooding in the nation's capital caused some $20 million (in 1996 U.S. dollars) in damages losses alone. In total, 26 people lost their lives to Fran's assault, and damage estimates of $3.2 billion (in 1996 U.S. dollars) were tallied.

Frances, Hurricane *North Atlantic Ocean–Puerto Rico, September 30–October 10, 1961* Since 1961, no fewer than eight TROPICAL CYCLONES in the North Atlantic have been identified as Frances. Of these eight systems, only two were TROPICAL STORMS, with the remainder achieving mature-stage (hurricane-intensity) status. The eighth and final incarnation—Hurricane Frances of 2004—was so severe in FLORIDA and along the eastern seaboard of the United States that the name Frances was retired from the rotating list of tropical cyclone identifiers and replaced with Fiona for the 2010 season.

The first Hurricane Frances existed from September 30 to October 10, 1961, and delivered heavy rains to PUERTO RICO. At one point a Category 3 system, it generated a minimum central pressure reading of 27.99 inches (948 mb) shortly after midnight on October 7.

Frances, Tropical Storm *North Atlantic Ocean, September 23–30, 1968* The second system churned through the mid-Atlantic from September 23 to 30, 1968, but caused no deaths or injuries. Tropical Storm Frances reached its peak minimum central pressure (observed) of 29.55 inches (1,001 mb) on September 27.

Frances, Hurricane *North Atlantic Ocean, August 27–September 7, 1976* In 1976, the third Hurricane Frances originated on August 27 and dissipated as an extratropical storm on September 7 while near the Azores. During Frances's passage across the NORTH ATLANTIC, the system interacted with Hurricane Emmy (a Category 2 system) and an example of the Fujiwara Effect was observed on August 31. On September 3, as Hurricane Emmy's extratropical remnants were being absorbed by a rapidly weakening Hurricane Frances (then of Category 1 intensity), a Lockheed C–130 H transport aircraft belonging to the Venezuelan airforce crashed at an airport in the Azores, killing all 68 people aboard, many of whom were reportedly schoolchildren. This tragedy was indirectly linked to the stormy conditions generated by Hurricane Frances's imminent transition to a powerful extratropical storm with a central pressure of 29.29 inches (992 mb). At its most intense, Hurricane Frances produced a pressure reading (observed) of 28.43 inches (963 mb) on the morning of September 1, 1976. This made the system of Category 3 status.

Frances, Hurricane *North Atlantic Ocean, September 6–21, 1980*

The fourth North Atlantic tropical system identified as Frances did reach hurricane intensity during its existence September 6–21, 1980, but as it remained in the mid-Atlantic, tracking to the north-northeast, no deaths or property losses were incurred. It achieved its peak minimum central pressure of 28.28 inches (958 mb) during the early morning hours of September 9.

Frances, Hurricane *North Atlantic Ocean, November 18–21, 1986*

Also an offshore system, the fifth Hurricane Frances was a late addition to the 1986 North Atlantic HURRICANE SEASON, originating on November 18 and dissipating by November 21. It reached its peak intensity on November 20, when its central pressure was observed as 29.52 inches (1,000 mb), making it a fairly powerful Category 1 system.

Frances, Tropical Storm *North Atlantic Ocean–Europe, October 22–30, 1992*

During the 1992 North Atlantic HURRICANE SEASON, the sixth Frances remained an offshore system for its entire existence, but nevertheless managed to deliver heavy rains to western Spain as an extratropical depression. In existence between October 22 and October 30, 1992, Hurricane Frances reached its peak intensity on October 24, when its central pressure sank to 28.82 inches (976 mb). Banging ashore in northwestern Spain during the night of October 29–30 as an extratropical depression, the system's central pressure of 29.61 inches (1,003 mb) caused heavy downpours across northern Portugal, but claimed no lives or caused major property losses in that area.

Frances, Tropical Storm *Gulf of Mexico–Southern United States, September 11–12, 1998*

The seventh TROPICAL CYCLONE in the North Atlantic to be identified as Frances shuffled ashore in eastern TEXAS as a TROPICAL STORM September 11–12, 1998. Originating over the GULF OF MEXICO on September

8, Tropical Storm Frances slowly strengthened as it moved erratically, first to the south-southwest, then due southward before sharply recurving to the north-northeast on September 10. Tropical Storm Frances was at peak intensity at the time of its landfall near Grande Isle, LOUISIANA, on September 11, with a central pressure reading of 29.23 inches (990 mb). As is the case with many tropical storms, Frances was a rainmaker. In Matagorda, TEXAS, the storm dropped some 17 inches (432 mm) of PRECIPITATION, causing damage to road and energy infrastructure. Several dozen houses were also flooded or damaged. In southwestern Louisiana, Tropical Storm Frances dropped some 11 inches (279 mm) of precipitation, while two reported TORNADO outbreaks caused damage to structures and one direct death.

Frances, Hurricane *North Atlantic Ocean–Southern and Eastern United States, August 24–September 6, 2004*

The eighth and final North Atlantic TROPICAL CYCLONE identified as Frances was also the most powerful and destructive of the class. At one time in its existence a Category 4 system of considerable intensity, Hurricane Frances tore up the warm waters of the eastern NORTH ATLANTIC OCEAN between August 24 and September 6, 2004. Twice strengthening to Category 4 intensity (on August 29) and again between August 31 and September 2, Frances delivered widespread destruction to the BAHAMAS and parts of FLORIDA on September 1 and September 5, respectively. At the time that it passed over the Bahamas, Hurricane Frances was near its peak intensity of 27.66 inches (937 mb), with sustained winds of 144 MPH (232 km/h) firmly qualified it as a Category 4 system. Severe flooding forced the closure of the island chain's airports, and several ocean-side structures—including piers and marinas—were damaged or destroyed. The system moved west-northwest across the central Bahamas; by the afternoon of September 4, 2004, Frances had stalled about 50 miles (80 km) off the coast of Palm Beach, Florida, with wind speeds of 105 MPH (169 km/h). Frances made the second of its three landfalls in Sewall's Point, Florida, as a Category 2 hurricane on the morning of September 5, 2004, causing more than $8 billion (in 2004 U.S. dollars) in damages. Sustained winds of 105 MPH (169 km/h) accompanied this large and slow-moving system. PRECIPITATION accumulations across its Florida landfall zone topped six inches (152 mm). Frances continued to weaken as it moved west-northwest across the Florida peninsula and reemerged over the GULF OF MEXICO before making a second landfall on September 6, 2004, near St. Mark's, Florida, with tropical storm-force wind speeds of 65 MPH (105 km/h).

More than 15 inches (381 mm) of rain fell in some places as Frances tracked up the eastern seaboard toward the northeastern U.S between September 6 and September 10, 2004. In Mt. Mitchell, NORTH CAROLINA, a peak rainfall count of 23.57 inches (584 mm) was recorded, while the system produced some 123 TORNADOES in Florida, GEORGIA, the Carolinas, and VIRGINIA—the largest number yet witnessed in North Atlantic tropical cyclone history. On September 9, Tropical Depression Frances transitioned to an extratropical depression, then, while transiting the eastern Great Lakes, deepened to an extratropical storm with a central pressure of 29.55 inches (1,001 mb). Within hours, however, it had again weakened to an extratropical depression, during which time it produced rainfall counts of five inches (127 mm) over the Canadian capitol of Ottawa. In total, Hurricane Frances directly and indirectly claimed 49 lives and wrought $8.4 billion (in 2004 U.S. dollars) in property losses.

Fred, Typhoon *Taiwan–Eastern China, August 17–21, 1994* One of the most intense—and deadly—Category 4 TYPHOONS on record, Fred lashed portions of north TAIWAN and east CHINA between August 20 and 21, 1994. Brought on by a CENTRAL BAROMETRIC PRESSURE of 27.46 inches (930 mb) at landfall, Fred's 105-MPH (165-km/h) winds and 8-inch (203-mm) rains claimed three lives in Taiwan and another thousand in China's Zhejiang province. More than $1.6 billion property damage was assessed between the two nations, making it one of the most destructive storms to have struck the typhoon-ridden region in almost a quarter of a century.

The sixth storm of the 1994 TYPHOON SEASON, Fred originated in the Philippine Sea approximately 823 miles (1,324 km) southeast of Kaohsiung, Taiwan, on the afternoon of August 17, 1994. Fostered by the sun-swollen waters of the tropical NORTH PACIFIC OCEAN, the typhoon wasted little time in strengthening to SUPERTYPHOON status. By the afternoon of August 18, as fringe gales relentlessly whipped the sea into nourishing spume, Fred's core winds first escalated to Category 2 status, or 76 MPH (122 km/h), and then to Category 3, 110 MPH (177 km/h), by 8 o'clock on the same evening. By midnight on August 19, Fred had briefly achieved a Category 4 rating, with central pressures as low as 27.43 inches (928 mb) and maximum sustained winds in excess of 140 MPH (225 km/h). Weather reconnaissance flights conducted by the Typhoon Warning Center on Guam indicated that the storm was now of significant size—nearly 600 miles (966 km) across—and that it was headed in a northwesterly direction at close to 12 MPH (19 km/h).

ANALOGS based on the behavior of previous August typhoons predicted that Fred would either continue to move in the direction of Taiwan and China or else undergo RECURVATURE and come ashore somewhere in southern JAPAN.

By the morning of August 20, as Fred swiftly bore down on the Taiwanese capital of Taipei, it was clear that the typhoon was following the former scenario. All over the island, authorities issued evacuation warnings for low-lying areas and flash flood alerts for the mountain communities. Many offices and factories were closed, main roads were cleared of all nonessential traffic, and the population shuttered up their homes and waited for the wind and rain to begin.

As it turned out, Taiwan was not dealt the crippling blow everyone had come to expect from Fred. Just hours before the typhoon's EYE was set to come ashore near Taipai, its uncertain north-northeasterly course took it east of the island, where it merely grazed the coast. Although typhoon-force winds and pelting rains were felt as far inland as the city of Sinchu, a minimal amount of structural damage was suffered across the island. Three people, all of them fishers, were drowned in the opening stages of the storm, providing a small but certain hint as to the horrors the intensifying typhoon would soon inflict on slumbering China.

On Saturday morning, August 21, 1994, Typhoon Fred, with a central barometric pressure of 27.46 inches (930 mb) and weakening 105-MPH (165-km/h) winds, slammed into the curving flank of eastern China. The typhoon's eye came ashore near the city of Wenzhou, in the province of Zhejiang, bearing ferocious winds and 8-inch (203-mm) rainfalls. For 43 consecutive hours it battered the entire region, collapsing thousands of houses, washing away river embankments, and flooding rice fields, mine shafts, and railway lines. In the harbor at Wenzhou, an 11-foot (4-m) STORM TIDE, considered to have been the highest seen on that part of the coast since 1974, slammed freighters into their piers, swept dockside warehouses into the sea, and killed nearly 300 people. While at the same time that other parts of China were suffering from a prolonged drought, Fred's deluging rains swamped close to 4 million acres of farmland and destroyed more than 500,000 houses in the province. In some places the flash floods were measured at three feet and higher, with deadly consequences for the towns through which they surged. According to official news sources, no less than 8 million people were in some way affected by Typhoon Fred's passage, at least 1,000 of them fatally. The typhoon's destructive legacy, coming at a time when modern building practices and evacuation procedures are in widespread use around the world, only reemphasizes the catastrophic potential contained within each and every typhoon.

Freda, Typhoon *North Pacific Ocean–Philippines, November 12–18, 1959* An exceptionally powerful late-season TYPHOON, Freda delivered sustained, 100-MPH (161-km/h) winds and heavy rains to the northern PHILIPPINES on November 18, 1959. Proving very destructive to BUILDINGS, bridges, and roadways across Luzon Island, Freda claimed 12 lives and caused damage estimates of $7 million.

Frederic, Hurricane *Southern United States, September 7–14, 1979* The second major HURRICANE to have marched ashore in the United States during the course of the 1979 HURRICANE SEASON, Frederic hurled 130-MPH (208-km/h) winds, five-inch (127-mm) downpours, and a 20-foot (7-m) STORM TIDE at the gulf coasts of ALABAMA and MISSISSIPPI on September 12, 1979. In Alabama, where Frederic besieged Mobile Bay as a severe Category 3 TROPICAL CYCLONE, eight people were killed in BUILDING failures, automobile accidents, and drownings. Naval shipyards in neighboring Mississippi suffered extensive damage to their gantries and drydocks; flash floods spiked trees and telephone poles, knocking out electrical service to more than 100,000 households. Insurance claims quickly topped $1.7 billion, briefly making Frederic the most expensive U.S. hurricane disaster up to that time.

A textbook example of a midseason hurricane, Frederic developed over the heated waters of the southeastern GULF OF MEXICO, roughly 50 miles (80 km) northeast of Havana, CUBA, on the morning of September 7, 1979. The seventh storm of an active season—one that was then watching Hurricane DAVID unfold over FLORIDA and SOUTH CAROLINA on September 7—Frederic speedily intensified as it rode north-northwest at nearly 11 MPH (18 km/h). Reconnaissance flights conducted by the NATIONAL HURRICANE CENTER (NHC) indicated that "Frederic's CENTRAL BAROMETRIC PRESSURE had deepened to 29.54 inches (1,000 mb) by the afternoon of September 8, upgrading it to a TROPICAL STORM with sustained winds of 49 MPH (79 km/h). By the same time the next day, Frederic had matured into a Category 1 hurricane, whose 77-MPH (124-km/h) winds hinted at further strengthening.

Indeed, Frederic was close to achieving Category 4 status when it came ashore on the west side of Alabama's Mobile Bay in the vicinity of Dauphin Island just after dawn on September 12, 1979. Gusts of 145 MPH wind and ferocious rains downed power lines, uprooted hundreds of trees, drove a dozen boats

ashore, and seriously damaged some 12,000 buildings. Inundating downpours produced flash floods and eight-inch (203-mm) rain counts that transformed low-lying neighborhoods of Mobile into bayous. More than 250,000 people across the state were left without electricity, telephone, or water services; eight of them were killed, making Frederic the deadliest hurricane to have struck Alabama in more than a decade.

Fujita-Pearson Scale Introduced in 1971 by the late Professor Tetsuya Theodore Fujita and Allen Pearson, former director of the National Severe Storms Forecast Center in Kansas City, Missouri, the Fujita-Pearson Tornado Intensity Scale—known simply as the Fujita Scale—divides TORNADO intensity into seven numbered categories based on their presumed WIND SPEED, length of path, width of path, and the degree of structural damage inflicted on BUILDINGS. Numbered from 0 to 6 and preceded by the letter *F* or sometimes *FP*, the Fujita-Pearson Scale was the result of detailed studies made on more than 25,000 documented tornadoes between 1916 and 1970. F-0 tornadoes damage signs and fences; F-1 and F-2 tornadoes generally shatter windows and peel shingles from roofs; F-3 tornadoes demolish small buildings, sheds, and mobile homes; F-4 and F-5 tornadoes can flatten a substantial one- to two-story residence; F-5 tornadoes can destroy a multistory building and carry away much of its debris. A meteorological possibility, no known F-6 tornado has ever been observed.

In early 2007, the National Weather Service adopted an updated version of the Fujita Scale, known as the *Enhanced Fujita Scale*, or EFS. Designed to allow for the more accurate identification of tornado-caused damage by taking into account new structural techniques, stronger materials, and improved building codes, the EFS further includes the addition of new Damage Indicators (DI) and Degrees of Damage (DO) to allow meteorologists, emergency managers, and structural engineers to more precisely correlate different types of damage with a particular tornado's intensity and characteristics. Some adjustments have been made in the wind speed measures—the higher wind speeds of some previous F-3 and F-4 tornados would now qualify for F-5 status under the new system. Moreover, under the EFS, the as yet unobserved F-6 category has been omitted. In order to differentiate those tornadic systems classified under the previous F-scale and the new EFS configuration, the new categories are identified as EF0, EF1, and so on to EF5.

Fujiwara effect or Fujiwhara interaction Named for Japanese meteorologist Sakuhei Fujiwara (1890–1965), the Fujiwara effect describes the physical

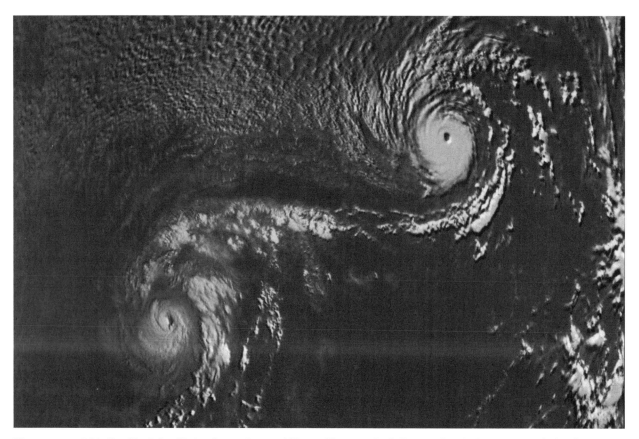

Like a vast aerial ballet, North Pacific hurricanes Ione and Kirsten illustrate the Fujiwara effect in August 1974. *(NOAA)*

processes by which two TROPICAL CYCLONES in close proximity to each other interact while over the open expanses of the NORTH ATLANTIC, NORTH PACIFIC, and Indian Oceans. First identified in 1921, the Fujiwara effect states that in order for two tropical cyclones to form what is known as a PINWHEEL CYCLONE—that is, establish a common rotational point between them—their respective EYES must be between 600 and 900 miles (965 and 1,450 km) apart. Both tropical systems must be of roughly the same size, possess CENTRAL BAROMETRIC PRESSURES that do not vary more than 20 MILLIBARS (mb) between them, and be traveling in the same direction at similar speeds. Should these criteria be met, the Fujiwara effect holds that two HURRICANES, TYPHOONS, CYCLONES, or TROPICAL STORMS will in a matter of hours begin to spin around a shared rotational point—usually located halfway between them—in a counterclockwise direction in the Northern Hemisphere, and in clockwise direction in the Southern. They may continue this pirouette for as long as three days, at which point they will have either come together as one system or split apart. In the more likely event that two tropical cyclones of different sizes and barometric pressures come into contact with each other, the Fujiwara effect main-

tains that a common rotational point will be established closer to the more intense system. In such a scenario the smaller tropical cyclone will begin to circle the larger one at increasingly higher rates of speed until it finally reaches the dominant system's speed, merges with it, or dissipates. Although documented examples of the Fujiwara effect in action over the confined waters of the North Atlantic and Indian Oceans, there have been a number of occurrences noted in both the eastern and western zones of the North Pacific Ocean, including those of Hurricanes Ione (Category 3) and Kirsten (Category 1) between August 27 and 29, 1974, and Typhoons Doug and Ellie in August, 1994. In 1976, Hurricanes Emmy and FRANCES established a Fujiwara configuration while transiting the mid-Atlantic on August 31. At the time, Emmy was the more powerful of the two, its Category 2 central pressure set at 28.85 inches (977 mb); Frances was of Category 1 intensity, with a central pressure of 29.14 inches (987 mb). Interestingly, in the end it would be Frances—eventually the more powerful of the two storms—that would consume Emmy's remnant moisture off the coast of the Azores.

In some instances, meteorologists have mistakenly predicted that a pair of tropical systems will achieve the Fujiwara effect. During the 2003 eastern North Pacific hurricane season, Hurricanes Nora and Olaf simultaneously churned toward the tip of Mexico's Baja Peninsula on October 6, with the more intense Category 2 Nora threatening to absorb the weaker Category 1 Olaf along the way. Eventually, Nora prevailed, drawing Olaf's residual moisture and energy into its own weakening circulation by October 8.

G

Gamma, Tropical Storm *Caribbean Sea–Central America, November 14–22, 2005* The 24th named TROPICAL CYCLONE of the 2005 North Atlantic HURRICANE SEASON, and the first tropical cyclone in history to bear the name Gamma (taken from the Greek alphabet), Tropical Storm Gamma delivered heavy rains and TROPICAL STORM force winds to the Central American nations of Honduras and BELIZE between November 17 and 22, 2005. Initially born over the extreme eastern CARIBBEAN SEA as Tropical Depression No. 27 (TD27) on November 14, the system briefly intensified to tropical storm intensity on November 15 (with a central pressure of 29.64 inches [1,004 mb]) but this brief period of strengthening went unrecognized by meteorological agencies at the time of its occurrence. By the early evening hours of November 16, as TD27 (Gamma) entered an unfavorable WIND SHEAR environment), it was downgraded further, to a TROPICAL WAVE with a central pressure of 29.70 inches (1,006 mb); the system more or less remained a tropical wave until November 18. During this time, the tropical wave (which had steadily been sliding to the west-north-west), trailed along the northern coast of Honduras, bringing large quantities of rain to the area. On the evening of November 18, the tropical wave that had been TD27 suddenly reintensified through its interaction with a remnant low situated near Nicaragua, and Tropical Storm Gamma was formed. It reached its peak minimum pressure the following day, with a central pressure of 29.58 inches (1,002 mb) and wind speeds of 52 MPH (83 km/h). A short-lived reprieve, Gamma was downgraded to a TROPICAL DEPRESSION on November 20, and by midnight on November 22, had largely dissipated. Although Tropical Storm

Gamma never made landfall in Central America, torrential downpours claimed at least 34 lives in Honduras, and three others in neighboring Belize. Hundreds of BUILDINGS were damaged or destroyed, while the storm's potential landfall (made more difficult by its erratic trajectory) made the EVACUATION of more than 11,000 people from vulnerable coastal areas and flood-prone to government-operated shelters that much more difficult. On November 19, for instance, a private aircraft operated by a resort in Belize crashed due to Gamma's high winds. All three people aboard the aircraft, including an American couple on their honeymoon, were killed.

Gaston, Hurricane *North Atlantic Ocean–Eastern United States, August 27–September 3, 2004* The first North Atlantic TROPICAL CYCLONE identified as Gaston, this system originated as a TROPICAL DEPRESSION some 140 miles (225 km) southeast of Charleston, SOUTH CAROLINA, on August 27, 2004. During the early morning hours of August 28, the system was upgraded to TROPICAL STORM intensity (with a central pressure of 29.67 inches [1,005 mb]), and to Category 1 status during the midday hours of August 29. At its peak intensity, Hurricane Gaston generated a pressure reading of 29.11 inches (986 mb) right before it sloshed ashore near McClellanville, South Carolina, on the 29th. Maximum sustained winds of 75 MPH (120 km/h) lashed trees and signs, but, like most tropical systems, Gaston was a rainmaker, drenching large swaths of NORTH and SOUTH CAROLINA, and VIRGINIA, between August 30 and August 31. As Hurricane Gaston weakened to a tropical storm, then to a tropical depression, Charleston, South Carolina, received eight inches (203 mm) of rain, while West End, near Richmond, Virginia, was inundated by 12.60 inches (330 mm) of precipitation. The James River topped its banks, while TORNADOES were spawned in North Carolina and Virginia. On the night of August 31, Tropical Depression Gaston reemerged into the NORTH ATLANTIC and tracked to the northeast, reintensifying to tropical storm status. As a renewed tropical storm, Gaston's peak minimum pressure was 29.50 inches (999 mb), generated on the morning of September 1, 2004. Within hours, Gaston had lost its tropical characteristics while skidding past the Canadian Maritime provinces. Sable Island, CANADA, received more than two inches (51 mm) of rain and heavy surf conditions, but no damage or injuries were reported. As is often seen in tropical systems that undergo extratropical deepening, the extratropical storm formerly known as Gaston became more intense than the tropical storm that had preceded it, generating a central pressure reading of 29.35 inches (994 mb) during the midday hours of September 2. Weakening as it shot to the northeast, the system had dissipated by September 3. In the United States, Hurricane Gaston claimed eight lives—all in Virginia—and caused $130 million (in 2004 U.S. dollars) in property losses. An interesting feature about Hurricane Gaston, it was not originally classified as a hurricane (Category 1) system at landfall, but was later upgraded to a mature-stage tropical cyclone after a more detailed review of its meteorological biography. Despite Gaston's record of flooding in Virginia, the identifier has been retained on the revolving list of North Atlantic hurricane names. Gaston is scheduled to reappear during the 2010 season.

Georges, Hurricane *North Atlantic Ocean, September 1–8, 1980* Between 1980 and 1998, two North Atlantic TROPICAL CYCLONES have been identified as Georges. Both reached hurricane status, with one of them, Hurricane Georges, wreaking enormous death and destruction during the 1998 season. The name Georges has been stricken from the list of North Atlantic tropical cyclone identifiers, and replaced with Gaston.

The first Hurricane Georges was, for a tropical cyclone, a lamb. Originating on September 1, 1980, as a TROPICAL DEPRESSION, Georges interacted with a nontropical low that aided the system's transition to a subtropical depression (with a very high central pressure of 29.94 inches [1,014 mb]) on September 5. It regained its tropical characteristics and was upgraded to Tropical Storm Georges during the midnight hours of September 7. It thereafter intensified, reaching Category 1 status shortly after midnight on September 8. A central pressure of 29.32 inches (993

mb) generated wind speeds of 81 MPH (130 km/h), but as the system was located well into the Atlantic, posed no hazard except to shipping. Within 24 hours, while rolling to the northeast past the Canadian Maritime provinces on September 8, 1980, Hurricane Georges dissipated. No deaths were reported.

Georges, Hurricane *North Atlantic Ocean–Puerto Rico–Cuba–Southern United States, September 15–October 1, 1998* Unfortunately, what was stated for the first Hurricane Georges cannot be said for the second, which, for a TROPICAL CYCLONE, was a dragon. A classic example of a CAPE VERDE STORM, Georges formed as a TROPICAL DEPRESSION at 10 degrees North, 25 degrees West, during the midday hours of September 15, 1998. Gently recurving to the northwest, the system was dubbed Georges when it reached TROPICAL STORM intensity on September 16 (with a central pressure of 29.67 inches [1,005 mb]). George matured into a Category 1 hurricane during the early evening hours of September 17, when its central pressure dipped to 29.14 inches (987 mb). At this time, Hurricane Georges was situated some 1,000 miles (1,609 km) southeast of the Caribbean island chains. If this trajectory was maintained, it would bring the system across PUERTO RICO, HISPANIOLA, and possibly CUBA. But as Hurricane Georges approached the eastern Caribbean, it suddenly began a period of intense deepening. During the midday hours of September 18, Georges was upgraded to a Category 2 hurricane with a central pressure of 28.73 inches (973 mb); it maintained Category 2 status, growing steadily more powerful, until the afternoon of September 19, when it reached Category 3 intensity (28.17 inches [954 mb]). Bearing down on the Leeward Islands, Georges continued to intensify, becoming a dreaded Category 4 system some six hours later. By the following morning, Hurricane Georges' central pressure had plummeted to 27.66 inches (937 mb), making it a very severe Category 4 hurricane on the SAFFIR-SIMPSON SCALE. Wind speeds within the EYEWALL were measured at 150 MPH (241 km/h)—and the system was now moving across the northern Caribbean islands of St. Kitts and Nevis, and Antigua. On St. Kitts and Nevis, Hurricane Georges' fury left 85 percent of the houses—along with critical infrastructure facilities such as hospitals, police stations, and schools—damaged or in ruins. The islands' airport suffered some flooding, with the terminal and control tower rendered unusable. Three people in St. Kitts and Nevis perished in Georges. On Antigua, Georges claimed two lives, and caused extensive coastal flooding that damaged piers, marinas, and cruise ship facilities.

After weakening to a Category 2 hurricane (with a central pressure of 28.52 inches [966 mb]),

Georges rolled ashore in eastern Puerto Rico—near Humacao—during the afternoon hours of September 21, 1998. Sustained wind speeds in Puerto Rico averaged near 109 MPH (176 km/h), while torrential rains washed the entire length of the island. Hurricane Georges wrought enormous damage on Puerto Rico's water and energy infrastructure networks; at one point, more than 3 million of the 3.8 million people on the island were without water and electrical service. A survey conducted by the Federal Emergency Management Agency (FEMA), indicated that 33,113 residences were destroyed, while another 50,000 suffered some magnitude of damage. Not surprisingly (given their organic nature), Puerto Rico's vital coffee and plantain crops were largely decimated by George's high winds and drowning rains; FEMA estimates place plantain losses at some 95 percent of that year's harvest, while the island's chicken-producing industry lost 60 percent of its output. One of the most severe tropical systems to have struck Puerto Rico since Hurricane DAVID in 1979, Georges caused more than $2 billion (in 1998 U.S. dollars) in property losses, and directly and indirectly claimed 12 lives.

Hurricane Georges' assault on Puerto Rico also brought heavy winds and rains to the U.S. VIRGIN ISLANDS. Although some 100 BUILDINGS were destroyed or damaged, no loss of life was reported. As with Puerto Rico, valuable crop and livestock losses were recorded, in this case to the Virgin Island's coconut and mango production.

Its eyewall once again over the open waters of the Windward Passage on the morning of September 22, Hurricane Georges quickly reintensified, becoming a Category 3 system (with a central pressure of 28.46 inches [964 mb]) shortly before barreling ashore in eastern Hispaniola later the same day. Georges' 121-MPH (195-km/h) winds and flooding downpours caused widespread damage across the Dominican Republic and Haiti. Landslides proved particularly fatal in Haiti, where Georges (downgraded to a Category 1 system with a central pressure of 29.23 inches [990 mb]) generated precipitation counts of 24 inches (610 mm) across much of Cap Haitian. Georges' rains also caused flash flooding in the Artibonite Valley and in parts of the capital, Port-au-Prince. Some 209 people were reportedly killed in Haiti, but there is some wide variance in DEATH TOLL numbers listed for the storm. In the neighboring Dominican Republic, Hurricane Georges left some 380 people dead, 100,000 people without shelter, 70 percent of the nation's bridges damaged or destroyed, and a bill for losses totaling $1.2 billion (in 1998 U.S. dollars). At one point in the recovery effort, more than 500 people were said to be missing in the Dominican Republic.

Despite the EVACUATION of more than 200,000 people preceding Hurricane Georges' landfall in Cuba on September 23, 1998, the Category 1 system (with a central pressure of 29.29 inches [992 mb]) killed six people in that country, destroyed more than 2,000 houses and caused widespread flooding across Holguin province. Georges trailed along Cuba's northern coastline between September 23 and September 25, slowly intensifying. By the time the system's core circulation reached the Straits of FLORIDA on September 25, the system had reached Category 2 status, with a central pressure of 28.79 inches (975 mb) and 104-MPH (167-km/h) winds.

Between September 25 and 26, 1998, Hurricane Georges battered the Florida Keys with 90-MPH (145-km/h) winds for nearly 10 hours, causing considerable flooding and wind-related damage to Big Pine Key, Kudjoe Key, and Key West. Nearly 1,000 residences on the Keys suffered minor damage, while another 150 were completely destroyed. In Key West, a picturesque local icon, "Houseboat Row," was decimated, with some 75 of its colorful floating treasures destroyed; for several days Key West was without electrical service.

Continuing to intensify as its recurved to the north-northeast into the GULF OF MEXICO, Georges reached its peak minimum pressure of 28.37 inches (961 mb) during the midnight hour of September 28. At this time, its eyewall was located less than 100 miles (161 km) south of Biloxi, MISSISSIPPI, where it eventually made landfall later the same day. Its core circulation disrupted, Hurricane Georges became an instant rainmaker and inundated the coastlines of eastern LOUISIANA, Mississippi, and ALABAMA with a STORM SURGE condition averaging nine feet (3 m) in height. Fading to a tropical storm, then a tropical depression, as it ground inland and sharply recurved due east on September 29, Georges delivered record-breaking rains to Mobile, Alabama (13 inches [330 mm]) and to Munson, Florida, which received 25 inches (635 mm). In Alabama, gusts of 85 MPH (137 km/h) were observed, and some 177,000 customers lost electrical service across the state. Downtown Mobile, Alabama, received considerable flood damage, and numerous businesses were disrupted. The situation was no brighter on the Florida Panhandle. Tropical Depression Georges produced several TORNADOES that damaged property and left 700,000 customers without electricity. When Georges' high waves washed away a section of Interstate 10, the U.S. Coast Guard (USCG) rescued some 200 residents who had been stranded.

By October 1, 1998, the once-formidable Hurricane Georges dissipated off the Florida-Georgia coastline. In the United States, one direct and four indirect fatalities attributable to Georges were recorded; an elderly woman who died from heat stress while being evacuated from New Orleans, and another two people who died in Florida and Louisiana in fires started by candles during power outages, while another died in a vehicular accident on a rain-slick highway near Crestview, Florida. As damage estimates in the United States alone topped $6 billion (in 1998 U.S. dollars), across the Caribbean the death toll, particularly in Hispaniola, was quickly approaching 600, making Georges one of the deadliest North American tropical cyclones in many years.

Georgia *Southeastern United States* Although Georgia's 100-mile (161-km) coastline borders on the western NORTH ATLANTIC OCEAN'S HURRICANE-prone waters, the Peach State's coastal plains, rolling hill country, bountiful orchards, and industrious seaports have during the last three centuries suffered only a moderate number of direct TROPICAL CYCLONE strikes. Between 1752 and 2006, Georgia was affected by 87 documented hurricanes and TROPICAL STORMS, with 30 of these mature systems making landfall on the state's east coast. Of these, 11 were considered major hurricanes—Category 3 and 4 systems with minimum winds of 111 MPH (179 km/h), large STORM SURGES, and mortifying DEATH TOLLS. Although the deadliest, most violent hurricanes in Georgia's history—August 26, 1881, and August 27–28, 1893—occurred on the state's North Atlantic side, the majority of Georgian storms have originated in the GULF OF MEXICO or southwestern CARIBBEAN SEA. In many cases achieving considerable size and intensity as they recurve northeast, these tropical storms and hurricanes first come ashore in northwestern FLORIDA before blasting into western Georgia, delivering damaging winds and flooding rains to the Piedmont Plateau's scenic uplands.

On average, Georgia's island-studded east shore, noted for its upscale resort communities and busy riverways, is indirectly threatened by a tropical cyclone every 3.5 years and directly impacted by a tropical storm or mature hurricane every nine years. On at least seven different occasions between 1752 and 2006, Georgia has even withstood multiple strikes in the course of a single HURRICANE SEASON, as happened in 1837, when four powerful hurricanes ravaged the state between August 1 and October 10. A similar parade of double strikes was further witnessed in 1804 (September 8 and October 4), 1852 (August 10 and August 27), 1873 (September 20 and September 24), 1885 (August 25 and September 21),

1893 (August 27 and October 23), and 1898 (October 7 and October 26). Of the two 1852 hurricanes, only the August 10 storm was of North Atlantic origin. The other, known to history as the GREAT MOBILE HURRICANE, made its initial landfall in ALABAMA before pummeling west and central Georgia with 100-MPH (161-km/h) winds and deluging rains. Two wooden bridges over the Savannah River were swept away by raging flood waters, while several towns and villages across the state were plagued by rockslides and submerged streets.

Other notable early hurricanes in Georgia include the 1752 North Atlantic tempest that on September 15 sank several merchant ships anchored at the mouth Savannah River's mouth and the powerful Carolina Hurricane of September 3–9, 1804, that, despite its name, first came ashore with great violence in Georgia on September 8. Before moving northeast where it pulverized the wharves and warehouses of Charleston, SOUTH CAROLINA, the 1804 hurricane severely battered a number of Georgia's barrier islands, washing away scores of small houses. Seventy-five slaves on Broughton Island some nine miles (14 km) south of Savannah were drowned when the hurricane's 14-foot (5-m) seas swamped the small rowboats in which they were attempting to escape. On neighboring Cockspur Island, the hurricane's bulldozing storm surge completely razed a venerable stone fortress, pushing its impregnable ramparts into the sea while at the same time scattering its 2-ton iron cannon across the island. Savannah and its surrounding countryside had no sooner cleaned up after the September 8 hurricane when another tropical cyclone (possibly of tropical-storm intensity) again raised the enclosed waters of the Savannah River on October 4. An altogether milder storm, the second 1804 hurricane tossed a multitude of small watercraft aground but otherwise spared the weary city the deadliest of assaults.

In September 1811, a mighty CAPE VERDE STORM grazed the Georgian coast on its way to a whirring landfall in Charleston, its roaring rains saturating thousands of acres of rice paddies and causing great hardship for the resilient workers who farmed the shallow banks of the state's many rivers. Similar offshore hurricanes occurred in August and September of 1822 and just two years later in August 1824. Hurried across the North Atlantic by the trade winds and then drawn northeast by the laws of RECURVATURE, tropical cyclones of this variety frequently bring heavy seas and rain squalls to eastern Georgia but little else. During the 1995 season, for instance, Hurricane Felix—a sprawling Category 4 system that developed off northern Africa's Cape Verde Islands on August 8—churned within 250 miles (402 km) of the Georgia coast before kiting northeast on August 16. Extensive beach erosion was observed from Brunswick to Savannah, and sporadic downpours spawned by the raking storm were felt as far inland as Augusta. No lives were lost nor serious injuries reported, however, making Felix one of Georgia history's weaker coastal hurricanes.

The same assessment cannot be given to the latter half of the 19th century when an epidemic of North Atlantic hurricanes withered Georgia's eastern plains, extracting steep death tolls and inflicting debilitating property damage from the developing settlements located there. Between 1856 and 1898, no less than 12 mid- to late-season hurricanes rolled into Georgia, including those of September 1, 1856, August 16–18, 1871, September 21, 1877, and August 24–25, 1885. On August 26, 1881, a gargantuan Cape Verde storm swept ashore just south of Savannah, tearing the roofs from hundreds of buildings and claiming some 335 lives. Said to have been the most vicious hurricane to have impacted the state since 1804, the 1881 maelstrom was eventually superseded by an even greater fury, the notorious SAVANNAH-CHARLESTON HURRICANE of August 27–28, 1893. In all probability the deadliest tropical cyclone in Georgia history, the Savannah-Charleston Hurricane snagged an estimated 500 lives as its swallowing seas, judged to have been 30-feet (10-m) high, inundated several riverside rice plantations. Swiftly recurring into neighboring South Carolina, the hurricane's seething STORM TIDE submerged vast sections of Charleston, drowning an additional 1,400 people.

Just over five years later, on October 7, 1898, yet another powerful hurricane passed inland at the Altamaha River's mouth, causing serious flooding and widespread crop loss; a subsequent storm, this one on October 26 of the same year, littered the Georgia-South Carolina border with broken fences, uprooted trees, leveled houses, and the carcasses of slain farm animals. Of lesser severity than the August 1893 hurricane, the second 1898 blast took less than 50 lives in Georgia, a testament to the effectiveness of those early warning procedures enacted in the wake of the Savannah-Charleston disaster.

As though in recompense for the strident horrors of the previous 40 years, the first four decades of the 20th century were a relatively quiet period for Georgia hurricanes. Between 1899 and 1940, no mature cyclonic systems crossed Georgia's Atlantic coast, while the state's western reaches withstood the gusty remains of 11 Gulf of Mexico hurricanes and tropical storms. In 1907, 1912, 1914, and 1916, low-intensity tropical storms—cyclonic systems with wind speeds of between 39 MPH and 73 MPH (62–118 km/h)—did come ashore in Savannah, Brunswick,

and Darien, but damage was on the whole considered minimal, and no lives were lost. On the night of September 17, 1928, the fading EYE of what had been one of the most catastrophic of all recorded Cape Verde storms, the infamous SAN FELIPE HURRICANE, passed some 40 miles (64 km) off the Georgia coast as it arced northeast and eventual oblivion. Earlier responsible for ending the lives of some 3,366 people in PUERTO RICO and central Florida, San Felipe's sustained, 150-MPH (241-km/h) EYEWALL winds and 30-inch (762-mm) rains had been so weakened by their fatal passage over the Florida peninsula that the former oceanic monster was barely able to produce hurricane-force gusts in downtown Savannah.

Such meteorological good fortune could not last forever. In August 1940, an unnamed Category 2 hurricane pounded Savannah and its environs with 105-MPH (169-km/h) winds, scudding rains, and a nine-foot (3-m) storm surge. Before spiraling into Beaufort, South Carolina (where it would kill 34 people), on August 11, the hurricane either damaged or destroyed 369 BUILDINGS in Georgia and left 16 people dead. Millions of dollars in insurance claims were filed, making the 1940 strike one of the costliest in state history. Seven years later, a slightly milder system with top winds of 100 MPH (161 km/h) crashed ashore at Sapelo Island, 40 miles (64 km) south of Savannah, on the afternoon of October 15, 1947. Powered by a CENTRAL BAROMETRIC PRESSURE of 28.76 inches (974 mb), this fairly weak Category 2 storm snapped tree limbs and severed power lines but caused no deaths.

On September 16, 1959, Hurricane Gracie, a moderate Category 3 system, dealt the northern Georgia coast a glancing blow as it trundled ashore just north of Beaufort. A classic coastal hurricane, Gracie's 10-foot (3-m) storm surge and lurking seas sanded away several Georgian dune ranges but claimed no lives. Damage estimates proved minimal.

On September 9, 1964, late-season Hurricane DORA killed one person and inflicted $70 million in property damage as it lashed southern Georgia with 106-MPH (171-km/h) winds and two-inch (51-mm) rains. Initially coming ashore just north of Jacksonville, Florida, Category 2 Dora haphazardly bumped its central pressure of 28.52 inches (966 mb) along the Florida-Georgia border before lurching into Georgia shortly after dusk. Hundreds of houses, schools, churches, factories, and stores were seriously damaged, prompting then President Lyndon B. Johnson (who, as a native of TEXAS, was well aware of the destructive nature of hurricanes) to declare portions of Georgia a disaster area on September 11.

Yet another coastal hurricane, DAVID, delivered 92-MPH (148-km/h) winds to Savannah's Ossabaw

Sound on September 4, 1979. Set on a course for South Carolina, David's aging fury (which in its Caribbean prime had been clocked at a sustained 155 MPH [249 km/h]), nevertheless left much of Savannah without electricity for nearly two weeks. Severe roof damage was buffered by the Savannah Civic Center, while five to six-foot (1–2-m) STORM TIDES flooded tidal creeks and neighborhoods south of the city. No lives were lost in Georgia, although statewide damage estimates reached into the tens of millions of dollars.

Seemingly blessed by both its geographical location and the arcing rules of recurvature, Georgia was, on September 21, 1989, again spared a direct strike by one of Caribbean history's most fearsome hurricanes, HUGO. Held off the Georgia coast long enough to reach devastating landfall in Charleston, South Carolina, Category 4 Hugo and its extremely low central pressure of 27.58 inches (934 mb) forced the evacuation of more than 250,000 Georgia residents from their beachfront homes. Gusts of up to 100 MPH (161 km/h) rattled the windows of Savannah, while frothing, 25-foot (8-m) waves splintered wharves, sank small boats, and breached seawalls. While no casualties were reported in Georgia, Hugo did go on to kill 28 people in South Carolina and tally up damage estimates of more than $4 billion.

Between July 5 and 7, 1994, Tropical Storm ALBERTO, an early-season, Gulf of Mexico system with 60-MPH (97-km/h) winds and cascading rains, inundated huge portions of southwestern Georgia with 22-inch (558-mm) precipitation counts and heavy gales. A lukewarm tropical storm upon initial landfall in the Florida Panhandle, Alberto's subsequent floods killed 31 people and caused nearly $200 million in property losses.

Just over a year later, intense Hurricane OPAL, following a similar course to that of Alberto, burnished western Florida and southern Georgia with 125-MPH (201-km/h) winds and killer TORNADOES on October 5–6, 1995. Nine people in Georgia perished, a majority of them in automobile accidents, while some $130 million in damage claims were filed. The arrival of the new millennium did not bring with it a reduction in direct and indirect tropical cyclone activity in Georgia. Between 2000 and 2006, Georgia was affected by the rains, winds, and waves associated with some 18 tropical systems. While a majority of these systems were of tropical depression intensity when they passed directly over or near to Georgia, they produced tremendous quantities of PRECIPITATION across the state. From the Gulf of Mexico came Tropical Depression (formerly Hurricane) GORDON, Tropical Storm Helene, and Tropical Depression (formerly Tropical Storm) Hanna in September 2000; Tropical Storms ALLISON and

Gabrielle in 2001; Tropical Depression Bill in June 2003; Tropical Depression Bonnie in August 2004 and Tropical Depressions Ivan and Jeanne in September of that year; and Tropical Depression Cindy in July 2005. From its proximity to the North Atlantic, Georgia was directly and indirectly affected by Tropical Depression Isidore in September 2002 and Tropical Storm Kyle in October 2002; Tropical Storm Henri in September 2003; Tropical Storm Alex in August 2004; and Hurricane DENNIS—an offshore system—and Tropical Storm Tammy in 2005. On September 14–15, 2002, a downgraded Tropical Storm Hanna moved northeastward over Georgia as a waterlogged tropical depression. The town of Donalsonville, in Seminole County, was seemingly inundated by some 14.59 inches (370 mm) of precipitation within a 24-hour period.

On October 6, 2005, Tropical Storm Tammy lashed southeastern Georgia with 40-MPH (65-km/h) winds, 10 inches (254 mm) of precipitation, and severe coastal flooding. Originating off the east coast of Florida on October 5, Tammy's forecasted intensification to hurricane intensity was thwarted when the system trumped ashore in southeastern Georgia. The city of Brunswick received in excess of four inches (102 mm) of rain, and electrical service to some 16,000 customers was disrupted; there were also reports that a weak tornado touched down in Glynn County. No lives were lost in Georgia, but damages totaled some $30 million (in 2005 U.S. dollars).

Gilbert, Hurricane *Eastern–Northern Caribbean– Mexico, September 8–19, 1988* The largest North Atlantic HURRICANE yet observed, lumbering Gilbert mowed a gargantuan trail of death and devastation through the Caribbean islands of GUADELOUPE and JAMAICA and across MEXICO's Yucatán Peninsula, between September 8 and 19, 1988. Extending some 500 miles (805 km) across and boasting a disproportionately small 8-mile-wide (13 km) EYE, Gilbert twice intensified to Category 5 status on the SAFFIR- SIMPSON SCALE during its 2,500-mile trek into cyclonic history, with its lowest CENTRAL BAROMETRIC PRESSURE, 26.22 inches (888 mb), being recorded off the east coast of Yucatán on the evening of September 13. Less than a day later, Gilbert scythed ashore near the Mexican resort island of Cozumel, the first hurricane to make a Category 5 landfall in the Northern Hemisphere since Hurricane CAMILLE plowed into MISSISSIPPI in August 1969. Before dissipating into a string of TORNADOES over TEXAS and Oklahoma on September 19, Gilbert left close to $5 billion in property damage in 10 countries and 318 people dead.

A quintessential late-season hurricane, Gilbert was born as a low-pressure TROPICAL DEPRESSION 375 miles (604 km) east of BARBADOS on the morning

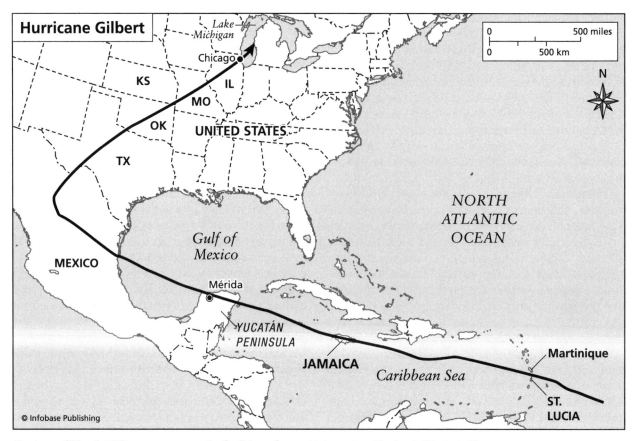

Hurricane Gilbert's 1988 rampage across the Caribbean Sea and into eastern Mexico is illustrated here.

of September 8, 1988. Carried northeast by subequatorial trade winds, Gilbert slowly intensified and was upgraded to TROPICAL STORM status while transiting Guadeloupe during the daylight hours of September 9. Although Tropical Storm Gilbert's 60-MPH (97-km/h) winds claimed no lives on the island, considerable damage was suffered by small BUILDINGS and crops.

By the evening of September 10, as Gilbert relentlessly bore down on the storm-riven isle of Jamaica, its central pressure continued to steepen rapidly. Slipping from 29.12 inches (982 mb) to 28.94 inches (980 mb) in a matter of hours, Gilbert was duly upgraded to Category 1 status. Less than a day later, as its sustained winds topped 96 MPH (155 km/h), Hurricane Gilbert was again promoted, this time to that of a Category 2 system. Now located 100 miles (161 km) south of PUERTO RICO, the hurricane remained firmly committed to its northwesterly trajectory, extracting tons of energized water vapor from the roiling surface of the CARIBBEAN SEA as its impressive CUMULONIMBUS CLOUD mass inexorably spread across the eastern Caribbean's confines.

On the afternoon of September 12, 1988, Hurricane Gilbert, with a deepening central pressure of 28.49 inches (965 mb), roared ashore in southeast

Jamaica. Traversing the island's length at nearly 12 MPH (19 km/h), Gilbert's 116 MPH (170 km/h) sustained winds and torrential downpours snapped banana trees, uprooted telephone poles, wrenched storm shutters from their hinges, stripped shingles and tiles from roofs, and blew down chimneys. On the island's south coast, where Gilbert's nine-foot (3-m) STORM SURGE inundated the crescent beaches of Morant Bay and Alligator Pond, dozens of small boats became total losses after being hurled aground, seven seafarers were drowned, and fragile coral reefs were shattered by Gilbert's powerful subsurface currents, washed ashore as a billion cutting shards that fouled tourist beaches for months to come.

Passing directly over the Jamaican capital of Kingston shortly after noon on September 12, Gilbert dropped eight-inch (203-mm) rainfalls on the city's narrow streets, spawning killer flash floods. Almost 80 percent of Jamaica's buildings suffered major structural damage, from broken windows to collapsed roofs, later necessitating a massive, multibillion-dollar reconstruction effort. Along the verdant slopes of the Blue Mountains, Gilbert's 147-MPH (237-km/h) gusts gouged vast channels through coffee and coconut plantations, leveling 19th-century houses and destroying nearly half of that season's banana harvest. All electrical, telephone, and water services were disrupted. Looters, their arms laden with food, clothing, and small appliances, scurried through the littered streets of Kingston, plundering what Gilbert did not. As a wet, windy dusk set in, the Jamaican government imposed on the capital a tenuous curfew that was immediately violated by thousands of homeless victims in search of shelter. Touted as the worst natural disaster in Jamaica's modern history, Gilbert left more than 500,000 islanders homeless, some $1 billion in property damage, and 45 people dead.

Returning to the Caribbean on the morning of September 13, Gilbert promptly underwent another cycle of rapid, terrifying intensification. A series of reconnaissance flights conducted by Florida's National Hurricane Center (NHC) indicated that during its passage over Jamaica, the hurricane's central pressure had continued to deepen, sinking to 27.56 inches (933 mb) just hours before reemerging over the open sea. Now classified as a moderate Category 4 hurricane, Gilbert's 135-MPH (217-km/h) winds whirred northwest, bound on a course that, if maintained, would eventually take it ashore somewhere in eastern Mexico or perhaps as far north as the Texas gulf coast. Caribbean civil defense authorities, appalled by Gilbert's catastrophic visit to Jamaica, wasted little time in notifying the Mexican government of the approaching hurricane's threat. As updated meteorological bulletins were passed to Mexican relief organizations, evacuation efforts in that country preceded at a brisk rate. Several vulnerable Yucatán Peninsula oceanside towns were successfully evacuated, the displaced sent inland to hillside shelters.

On September 14, 1988, Hurricane Gilbert became the first hurricane in 19 years to make a Category 5 landfall in the Western Hemisphere when it slammed ashore near the Mexican resort island of Cancun shortly after dawn. Thundering onto Cancun Beach with 218-MPH (350-km/h) gusts and a 23-foot (8-m) storm surge, Gilbert's vast cloud bands suddenly opened up, dropping eight inches of rain over the dense jungle forests of the Yucatán Peninsula. In Cozumel, hundreds of structures—hotels, apartment buildings, factories, stores, and shanties—were completely demolished, leaving more than 150,000 people without shelter. A 300-foot (100-m) Cuban cargo ship was run up on the beach and stranded. Flash floods and mudslides rushed through inland villages, killing scores of fleeing inhabitants. In the coast town of Bacalar, 50 miles north of Cancun, seven people drowned, all victims of Gilbert's unstoppable storm surge. A large tropical cyclone by any standard, Gilbert's 500-mile-wide (805-km) cloud

canopy further delivered 110-MPH (177-km/h) winds to neighboring Honduras, claiming six lives, leaving 6,000 homeless, and causing severe damage to crops and buildings.

Exiting into the southwestern Gulf of Mexico during the early morning hours of September 17, a slightly diminished Gilbert quickly regained its terrifying strength as it bore down on a sparsely inhabited section of Mexican coastline, 130 miles (209 km) south of Texas. Rushing ashore there just before dusk, the hurricane's 129-MPH (208-km/h) winds and increased 11-inch (279-mm) rains inundated hundreds of square miles of Monterrey Province, causing the Santa Catarina River to overflow its banks. Four buses being used to evacuate survivors to higher ground were quickly stalled by the rising flood waters and then overturned, killing an estimated 150 people. Scores of buildings in the area were either deroofed or simply collapsed; thousands of others had their doors, windows, porches, balconies, and other architectural details carried away. Brutal floods left more than 75,000 people in Monterrey Province homeless, trapped on roofs and dikes, awaiting aerial rescue. Moving north-northwest, Gilbert's rain-filled remains were tracked through Texas, where dozens of tornadoes killed three people on September 18. All told, nearly 300 people in seven countries perished in Hurricane Gilbert. Across the Caribbean basin, more than $2.4 billion in property damage was incurred, making Gilbert one of the most damaging storms then on record. Gilbert's 17-year record as the most intense North Atlantic tropical cyclone on record was shattered during the historic 2005 hurricane season when on the night of October 19 Hurricane Wilma generated a central barometric pressure reading of 26.05 inches (882 mb), the lowest pressure reading yet recorded in the North Atlantic basin. The name Gilbert has been retired from the rotating list of hurricane names.

Ginger, Hurricane *North Atlantic Ocean–Eastern United States, September 5–October 5, 1971*   Dancing its way into the cyclonic history books between September 5 and October 5, 1971, hurricane Ginger enjoys the rare distinction of being the third longest-lived tropical cyclone on record. For 33 windy days (20 of them spent as a mature-stage system, the remainder as a weak tropical storm), this one-time Category 3 hurricane tracked more than 5,800 miles (9,334 km) back and forth across the expansive reaches of the North Atlantic Ocean before finally dissipating. In the course of its remarkable loop-the-loop journey from the warm east waters of Florida on September 5, north to Bermuda between September 14 and 20 and then back again to the sandy shallows of NORTH CAROLINA and VIRGINIA by October 2, Ginger displayed a wide array of CENTRAL BAROMETRIC PRESSURES, ranging from 28.31 inches (959 mb) on September 16 to 28.79 inches (975 mb) on September 22. On October 1, 1971, as its central pressure deepened to 29.33 inches (993 mb), Ginger made its first and only landfall just south of Cape Hatteras, North Carolina. Quickly recurving back into the Atlantic during the early morning hours of October 2, Ginger's minimal 76-MPH (122-km/h) winds and heavy rains caused minor tree and BUILDING damage in North Carolina and Virginia but no fatalities or serious injuries. Subjected to two days of experimental seeding operations as part of PROJECT STORMFURY, Ginger retained its title as the longest-running tropical cyclone until September 1994, when the eastern North Pacific's Hurricane JOHN surpassed it. An analysis produced by the National Oceanographic and Atmospheric Administration (NOAA) in 2003 indicated that Ginger's record as the longest-lived North Atlantic tropical cyclone on record was, in fact, surpassed by the San Ciriaco Hurricane of 1899. According to NOAA's study, the San Ciriaco Hurricane first stuck PUERTO RICO on August 8, 1899, as a Category 4 hurricane; it then hit North Carolina as a Category 3 system on August 18. San Ciriaco then underwent extratropical deepening on August 21, 1899, reconstituted itself as a tropical storm on August 26 and moved through the Azores as a Category 1 hurricane on September 3, 1899, before dissipating. "It was a storm system for 33 days and a tropical storm or hurricane for 28 of those days. This ties the record with Hurricane Ginger of 1971, which was also a tropical cyclone for 28 days."

Between September 20 and October 12, 2002, Hurricane Kyle spent 22.00 days as a North Atlantic tropical system, besting 1957's Hurricane Carrie to become the fourth longest-lived North Atlantic tropical cyclone. The following is a list of the longest-lived North Atlantic cyclones:

San Ciriaco Hurricane, 1899: 28.00 days

Hurricane Ginger, 1971: 27.25 days

Hurricane Inga, 1969: 24.75 days

Hurricane Kyle, 2002: 22.00 days

Hurricane Carrie, 1957: 20.75 days

Hurricane Inez, 1966: 20.25 days

The name Ginger has been removed from the list of North Atlantic tropical cyclones.

global warming and tropical cyclones By definition, global warming concerns the warming of the atmosphere due to human-driven activities, such as the operation of internal combustion engines, automobiles, aircraft, manufacturing facilities, and even some natural phenomena, such as volcanic eruptions. Many experts fear that global warming could spark the melting of the polar icecaps, triggering a rapid rise in sea levels around the world. It is also feared that severe climatic change could occur through the cumulative effects of global warming.

Not surprisingly, considerable debate continues to surround the existence of, and the effects of, global warming. This debate extends to the generation and activity of worldwide TROPICAL CYCLONE activity.

On August 2, 2005, the science journal *Nature* published an article by Dr. Kerry Emanuel of the Massachusetts Institute of Technology (MIT) which evidenced that the duration and wind speed of hurricanes has increased by 50 percent. The National Oceanic and Atmospheric Administration (NOAA) has indicated that its most detailed hurricane records date only to 1945, so it is perhaps too early to gauge whether or not global warming is having a direct effect on tropical cyclone activity. Gerry Bell, a meteorologist attached to (NOAA) stated in a media interview that NOAA is not ". . . convinced that global warming is playing an important role yet, or it at all, in this era of increased [tropical cyclone] activity."

However, Hurricane KATRINA's stunning path of destruction during the extraordinarily active 2005 North Atlantic HURRICANE SEASON ignited continued debate on the effect global warming might be having on tropical cyclogenesis, or the formation and intensity of tropical cyclones in the North Atlantic basin. The August 2005 study conducted by MIT found that wind speeds in North Atlantic and North Pacific tropical cyclones had increased some 50 percent since the early 1970s. As evidence of the relationship between global warming and climatic change, global temperatures have risen at least one degree Fahrenheit since 1900. While North Atlantic hurricane activity has markedly increased since 1994, other experts attribute the severity of hurricanes and their increased numbers to changes in SEA-SURFACE TEMPERATURES and the salinity levels in the North Atlantic's deep water currents associated with regular 40–60 year cycles.

Gloria, Hurricane *Eastern United States, September 22–27, 1985* A fairly meek Category 2 HURRICANE, Gloria's 90-MPH (140-km/h) winds and heavy PRECIPITATION counts at landfall directly and indirectly claimed at least three lives and caused an estimated $23 million in property damage to communities in NORTH CAROLINA, New Jersey, NEW YORK, and NEW ENGLAND between September 26 and 27, 1985.

A midseason hurricane that developed off the BAHAMAS' northeast coast during the afternoon of September 22, 1985, Gloria briskly intensified as it first underwent RECURVATURE and then sped up the U.S. eastern seaboard on September 26. Aerial reconnaissance flights conducted by the NATIONAL HURRICANE CENTER (NHC) indicated that Gloria, with a CENTRAL BAROMETRIC PRESSURE of 27.82 inches (941 mb), was packing sustained winds of 130 MPH (210 km/h) and gusts of 140 MPH (225 km/h) as it grazed North Carolina's Outer Banks shortly after dawn on September 26. Although in actuality its central pressure ranked it as a Category 4 system on the SAFFIR-SIMPSON SCALE, Gloria's winds and tides, while dangerous, were not of such high caliber, thus warranting the lower classification. Aside from a few isolated power outages and some minor beach erosion, material damage in North Carolina from Hurricane Gloria's glancing passage was on the whole considered fairly moderate.

Quickly progressing up the coast in a wavering, north-northeasterly direction, Gloria's extensive spiral RAIN BANDS next grazed portions of New Jersey's east shoreline. In Atlantic City the hurricane's winds and rains brought heavy surf to the resort community's famed wooden Boardwalk, hotels, and clifflike stone jetties. More than 150,000 people in New Jersey were left without electrical power, while downed tree limbs made driving conditions extremely hazardous.

Shortly after dawn on September 27, 1985, Gloria's sprawling EYE finally came ashore on the south-central coast of New York's Long Island. Although the cooling waters of the NORTH ATLANTIC OCEAN had by this time reduced the hurricane to Category 1 status, its sustained 90-MPH (149-km/h) winds nevertheless managed to uproot hundreds of trees, knock out traffic lights and road signs, smash windows, and demolish several beachside cottages on the heavily settled island. Downed power lines left more than 750,000 households without electricity, while bayside flooding at East Massapequa left some low-lying BUILDINGS with more than three feet (1 m) of water in their basements.

Its wind and rain now considerably diminished, Gloria spiraled inland at 30 MPH (50 km/h), passing over southern and central New England during the afternoon and evening hours of September 27. Two deaths were reported in Connecticut, where TROPICAL STORM-force winds crunched trailer parks and scattered small buildings. Another individual died

in New Hampshire, where Gloria's fading EYEWALL delivered 76-MPH (122-km/h) winds to the state's granite highlands. All told, Gloria affected 13 states, leaving some $22.75 million in property losses. The name Gloria has been retired from the revolving list of HURRICANE NAMES.

Gloria, Typhoon *Taiwan, September 7–12, 1964* *See* TAIWAN.

Gordon, Hurricane *Northern Caribbean–Southern United States, November 8–21, 1994* Although late-season Gordon remained a moderate TROPICAL STORM for most of its existence, it nevertheless managed to claim 1,145 lives as it ricocheted across the northern Caribbean islands of JAMAICA, CUBA, and HISPANIOLA into southern and central FLORIDA between November 11 and 21, 1994. Some $540 million in property damage—much of it sustained by Floridian fruit and vegetable farmers—was assessed in Gordon's erratic wake, making the one-day HURRICANE the single most-expensive TROPICAL CYCLONE of the 1994 North Atlantic HURRICANE SEASON.

HURRICANE GORDON CHRONOLOGY

November 8: TROPICAL DEPRESSION No. 12 originates over the warm, humid waters of the southwestern CARIBBEAN SEA, approximately 130 miles (209 km) due east of Colorado, Costa Rica. Progressing northwest at three MPH (five km/h), the slowly developing system's top winds peak at 35 MPH (56 km/h).

November 9: Curling up the Central America's eastern flank as a loose collection of powerful thunderstorms, the as yet unnamed tropical depression deliver four to five inch rainfall counts to the inland slopes of Costa Rica. Gale-force gusts rustle palm fronds while heavy swells rock harbors as far north as the port city of Bluefields, Nicaragua. Six people in Costa Rica are killed by a seething spate of landslides and flash floods.

November 10: Bearing 31-MPH (50-km/h) winds and steady rains, Tropical Depression No. 12 comes ashore on Nicaragua's northeastern Mosquito Coast shortly after dawn. The first of five recorded landfalls that Gordon will make over the next ten days, this one occurs near the fishing village of Karawala, 200 miles (322 km) northeast of the nation's capital, Managua. Damage to trees and power lines is considered minor, and no deaths or serious injuries are reported.

November 11: After spending the previous 12 hours arcing across north Nicaragua's hilly plains in a hail of rain squalls and lightning strikes, Tropical Depression No. 12 reemerges over the Caribbean Sea as Tropical Storm Gordon. Upgraded and christened soon after its return to open waters, the budding hurricane's sustained winds are now clocked at 45 MPH (72 km/h). Moving at speeds of between five and eight MPH (8–13 km/h), Tropical Storm Gordon sharply swings northeast, setting a wavering course for Jamaica and eastern Cuba. Aerial reconnaissance missions into the nascent EYE of the system indicate that the storm's CENTRAL BAROMETRIC PRESSURE is not at present experiencing any significant deepening.

November 12: Tropical Storm Gordon remains at sea, its ill-defined EYEWALL positioned some 60 miles (97 km) north of Serranilla Bank. Its sustained winds holding steady at 46 MPH (74 km/h), the tropical storm continues its trek northeast, seemingly bound for a second landfall in central Cuba. Shortly before midnight, however, Gordon falls under the influence of a strong westerly STEERING CURRENT and begins to move due east. A light but steady rain to fall in Kingston, Jamaica.

November 13: Having passed the early morning hours bearing down on the Jamaica's southeast coast, Tropical Storm Gordon makes its second landfall just after dawn in the vicinity of Port Morant. One of the weakest tropical cyclones to have struck the hurricane-prone island, Gordon's sustained 44-MPH (71-km/h) winds, 65-MPH (105-km/h) gusts, and torrential downpours nonetheless claim two lives, and wreak more than $50 million in property damage. Serious flash floods persist for days. Gathering momentum as it recurves north-northeast at nearly 12 MPH (19 km/h), Gordon makes its third landfall right before dusk in the vicinity of Guantanamo Bay, Cuba. A fairly large tropical storm, Gordon's associated thunderstorms reach into neighboring Haiti, while gale-force gusts are observed across the lower Bahamas islands 200 miles (322 km) to the north.

November 14: While meteorologists at the NATIONAL HURRICANE CENTER (NHC) in Miami, Florida, ponder Gordon's zigzagging course across the Caribbean basin, conflicting accounts of raging mudslides and mounting DEATH TOLLS in Haiti and Cuba begin to surface in the world's newspapers. At Guantanamo Bay, Cuba, where Gordon reportedly generated an isolated, 120-MPH (193-km/h) wind gust, seven-inch (178 mm) downpours coupled with sustained 48-MPH (77-km/h) winds toppled tents and overturned water cisterns at Cuban and Haitian refugee camps. Two people were killed and another 14 were injured. In Haiti, where 14-inch (356-mm)

PRECIPITATION counts spawned a quagmire of mudslides all along the country's west coast, the death toll grows considerably higher, reaching 380 victims in a matter of hours. Recently occupied by the United States in a highly controversial MILITARY OPERATION, Haiti will diplomatically claim that only 100 to 125 people died in Gordon's passage, while United Nations observers set the figure much higher, in the vicinity of 1,122. Two hundred fifty casualties are tallied in the southern port city of Jacmel alone. Haiti soon receives $25 million in emergency aid from the United States.

The space shuttle *Atlantis,* returning from an 11-day mission to study the ozone layer's depletion is forced by Gordon's encroaching gales to scrub its planned landing at Cape Canaveral. The shuttle, complete with its cargo of 10 pregnant laboratory mice, instead safely touches down in California.

November 15: Bearing maximum winds of 50 MPH (81 km/h), Tropical Storm Gordon passes over the Bahamas' coral enclave. Spiraling west-northwest at nearly 12 MPH (19 km/h), Gordon's fringe rains continue falling across southern Florida. Tens of thousands of acres of farmland are quickly saturated as rainfall counts of between five and 20 inches (127–508 mm) are recorded. Nine-foot waves crash ashore from Key West to Palm Beach, Florida, while 30-MPH (48-km/h) wind gusts are measured as far north as Daytona Beach. A decommissioned Boeing 727, recently sunk off Miami Beach as an artificial reef, is snapped in two by Gordon's strong, subsurface currents. A 506-foot (169-m) Turkish freighter, heavily laden with fuel and steel, is run aground off Fort Lauderdale Beach; holed from a collision with one of the area's famed coral reefs, the vessel is relentlessly pounded by Gordon's high winds and heavy seas, arousing fears of an ecological disaster. Although the freighter is later salvaged without incident, its master is soundly criticized by the Coast Guard for not heeding storm warnings early enough.

November 16: Producing sustained winds of 50 MPH (81 km/h), Tropical Storm Gordon makes a midmorning landfall near Fort Myers, Florida. Accelerating northeast, the storm steadily rolls across the peninsula, bringing strong winds, flooding rains, and no less than six TORNADOES to more than forty inland counties. An elderly man in the retirement community of Barefoot Bay is killed when a twister scrambles his mobile home. Nearly a half-million people lose their electricity, 70 houses across the state are destroyed, and another are 380 damaged. A teenage surfer is drowned near Miami Beach, while another four people are killed, victims of heart attacks and automobile accidents. Millions of dollars worth of winter produce—cucumbers, tomatoes, and strawberries—are ruined, prompting dire predictions concerning rising food prices.

November 17: As the governor of Florida declares two-thirds of his state a disaster area, Tropical Storm Gordon exits the peninsula, returning its 45-MPH (72-km/h) winds to the NORTH ATLANTIC OCEAN just north of Vero Beach shortly before dusk. Firmly committed to its northeasterly RECURVATURE, Gordon races into the Atlantic, quickly feeding on the warm, undisturbed waters off the coast of GEORGIA and SOUTH CAROLINA. Just before midnight, as its central barometric pressure skids toward a low of 28.93 inches (980 mb), Gordon is upgraded to a Category 1 hurricane. Its 85-MPH (137-km/h) top winds furiously lash the sea, sending 16-foot (5-m) rollers rushing ashore as far north as NORTH CAROLINA's Outer Banks. Several oceanfront cottages, their wooden pilings quivering beneath Gordon's aquatic assault, begin to sway, threatening at any moment to topple into the clutching surf. Evacuation warnings are posted for portions of the North Carolina seashore.

November 18: One of the most unpredictable tropical cyclones on record, Hurricane Gordon commences a swift loop-the-loop maneuver during the early morning hours. First curving northeast, Gordon will spin west-northwest before shooting south-southeast by midnight of November 19. Four beach houses in Kitty Hawk, North Carolina, finally succumb to Gordon's distant but potent fury, collapsing into piles of splintered timber and shattered windows. Property losses in North Carolina and Virginia are estimated at more than $2 million. No deaths or injuries are reported, however.

November 19: Mortally wounded by upper-level wind shear brief Hurricane Gordon is downgraded to tropical-storm status hours later. Trailing south-southeast at nearly eight MPH (13 km/h), fickle Gordon eventually turns southwest and heads back to Florida as a low-level tropical system with sustained winds of 43 MPH (69 km/h). Heavy rains continue to fall 100 miles (161 km) from its center.

November 20: Its 13-day reign of terror nearly at an end, Gordon is once again downgraded, seeing its 32-MPH (52-km/h) winds fade to tropical-depression strength during the daylight hours. Languidly curling west-northwest, Gordon steadily weakens as it nears Florida's soggy east coast. Only light rain squalls are expected from Gordon's imminent demise.

November 21: After coming ashore for the last time near Melbourne, Florida, Gordon dissipates over the northern reaches of the state. Nothing more than a narrow swirl of low-altitude clouds, Gordon does not, unfortunately, die alone. Two fishers are drowned near Palm Beach, their small boat overwhelmed by Gordon's trailing 13-foot (5-m) seas. The death toll in Florida now stands at eight.

November 22: The decision is made to retain the name Gordon on the rotating list of HURRICANE NAMES. It is used again in 2000 for a Caribbean hurricane that killed 27 people in Central America and Florida; and again for a 2006 hurricane that reached Category 3 status before dissipating near EUROPE.

Grand Isle Hurricane *Southern United States, August 27–October 5, 1893* Remembered for both its unmitigated meteorological violence and its staggering 2,000-person DEATH TOLL, the Grand Isle Hurricane unleashed 137-MPH (221-km/h) winds, heavy rains, and an immense 20–22-foot (6–7-m) STORM SURGE on the southeast coast of LOUISIANA between October 4 and 5, 1893. The second killer HURRICANE to have struck the southern United States during the 1893 HURRICANE SEASON—the SAVANNAH-CHARLESTON HURRICANE had claimed some 1,900 lives in GEORGIA and SOUTH CAROLINA less than a week earlier—the Grand Isle Hurricane completely submerged the thickly settled resort spit of Grand Isle 60 miles (97 km) south of New Orleans as it stampeded ashore near the Mississippi River's mouth just before midnight. From the bayside town of Burrwood in the east to Port Sulphur in the north, and Timbalier Island in the west, the hurricane's record-breaking surge destroyed nearly every house, warehouse, customhouse, and wharf on the coast, leaving washed-out beaches littered with the remains of shattered BUILDINGS, uprooted fences, and splintered trees and shrubs. Hundreds of human corpses along with the bloated carcasses of drowned farm ANIMALS decomposed on the shore, in the gutted bayous, and in the mountains of wreckage that lined what remained of the Grand Terre Islands. Roaring north across the alluvial mudflats of the Mississippi Delta, the Category 4 whirlwind ravaged farms along the banks of Lake Borgne with 158-MPH (254-km/h) gusts and seven-foot (2-m) waves. On south Chandeleur Island 100 miles (161 km) west of New Orleans, the stone foundation of a brick lighthouse built in 1856 to keep ships from piling up on the treacherous crescent-shaped sandbar was undermined by the hurricane's clawing seas. One of its three adjoining bungalows swept away and its cracked lantern pasted together by tons of wind-driven salt, the leaning lighthouse was so severely damaged by the hurricane's passage that it would later be replaced by an iron-skeleton structure, one designed to withstand future GULF OF MEXICO storms better by allowing waves to move through its base. Before dissipating over inhospitable middle U.S. plains on October 6, the Grand Isle Hurricane had claimed between 1,800 and 2,000 lives and caused nearly $3 million in 1893 property losses to the strengthening trade and industry of Louisiana. It remains that state's deadliest confirmed hurricane strike.

Great Atlantic Hurricane *Cuba–Northeastern United States, October 11–15, 1944* A gargantuan Category 3 TROPICAL CYCLONE of considerable intensity (a CENTRAL BAROMETRIC PRESSURE reading of 28.34 inches [955 mb] was observed at landfall near Point Judith, Rhode Island, on the evening of October 14), the Great Atlantic Hurricane's 140-MPH (225-km/h) gusts, soaking rains, and engulfing seas carved a wide swath of death and destruction through western CUBA, eastern NEW YORK, and NEW ENGLAND. While the EYE of the Great Atlantic Hurricane did not come ashore in New York, its expansive 95-MPH (153-km/h) winds, augmented by a steep PRESSURE GRADIENT, rattled the towering skyscrapers of Manhattan and canceled a planned campaign speech by then President Franklin D. Roosevelt. Whirling northeast at low tide, the Great Atlantic's extreme gusts and huge waves dusted the shores of eastern Long Island but caused minimal shoreline erosion. At least two vessels—the 1,850-ton Navy destroyer *Warrington* and the Cuttyhunk lightship—foundered during the Great Atlantic's passage, taking more than 300 seamen to the bottom with them.

Another of the eastern seaboard's fast-moving tropical systems—a forward speed of 42 MPH (68 km/h) noted shortly after landfall—the Great Atlantic Hurricane downed thousands of oak, maple, elm, and pine trees, sheared the steeple from a Baptist church on Cape Cod, and washed away a steel drawbridge at West Yarmouth, Massachusetts. Local shipyards, then at the height of their World War II production levels, were severely mauled by the Great Atlantic Hurricane's flooding surf, with several unfinished wooden minesweepers and steel-hulled destroyers being tossed from their slipways and destroyed. All told, 26 people in New England perished in the storm while some $100 million in damages were assessed across the region.

Great Caribbean Hurricane of 1831 *Barbados–Southern United States, August 10–18, 1831* Occasionally referred to as the Great Barbados Hurricane of 1831, the Great Caribbean Hurricane visited cata-

clysmic devastation and gargantuan DEATH TOLLS on the Caribbean islands of BARBADOS and CUBA and on southeastern LOUISIANA between August 10 and 16, 1831. In Barbados, where the Great Caribbean Hurricane's 140-MPH (225-km/h) winds virtually leveled the capital city of Bridgetown, some 1,500 people perished, either drowned by the hurricane's 17-foot (6-m) STORM SURGE or crushed beneath the collapsed roofs of their stone houses. Judged by eyewitnesses to have been even more violent than the GREAT HURRICANE OF 1780, the Great Caribbean Hurricane's seering gusts and blistering rains stung the foliage from trees, leaving behind a tangled hive of barkless forests and what appeared to be scorched earth. First plotted in 1850 by pioneering HURRICANE scholar William Reid and subsequently published in his *An Attempt to Develop the Law of Storms*, the Great Caribbean Hurricane's northwesterly passage over Barbados bequeathed more than $7 million in property losses to the small island, a costly legacy that soon sent its weakened economy into an even steeper slump.

In Cuba, where the Category 4 EYEWALL of the Great Caribbean Hurricane smashed ashore at Guantanamo Bay on the morning of August 13, numerous merchant ships foundered along the island's southeast coast, taking an estimated 200 seamen to the rocky bottom with them. Hundreds of BUILDINGS in Havana were unroofed by the hurricane's sustained 135-MPH (217-km/h) winds, while severe rains blighted much of that year's sugarcane harvest. Scores of lives were lost, a majority of them to a suffocating series of landslides triggered by the hurricane's torrential rainfall.

Departing Cuba near the city of Matanzas and quickly transiting the GULF OF MEXICO on August 15 as a TROPICAL CYCLONE of notable size and intensity, the Great Caribbean Hurricane next blasted ashore in Louisiana, west of Last Island, during the late afternoon hours of August 16. STORM TIDES of between seven and 10 feet (2–3 m) were observed at the entrance to Lake Pontchartrain, and large hailstones beat an ominous tattoo against the lightly built roofs of a nearby army hospital. As several feet of water from Lake Borgne flowed into the hospital, the structure's roofs gave way, showering dozens of invalid soldiers with snapped beams and jagged roof tiles. An earthen levee in northeastern New Orleans was breached, allowing the wild waters of Lake Pontchartrain to flood many of the city's lakeside communities. Numerous small wharves and cottages were carried away by the hurricane's rising waters, toted inland, and deposited as piles of rubble in the surrounding bayous. In New Orleans itself, the Custom House lost part of its roof, while several merchant ships at anchor in the riverhead were pushed ashore, dragging docks and anchors behind them. An estimated 260 people in Louisiana lost their lives to the Great Caribbean Hurricane, making it one of the deadliest hurricanes in Louisiana's history.

Great Calcutta Cyclone *India, October 1–5, 1864* One of the most intense BAY OF BENGAL CYCLONES on record, the Great Calcutta Cyclone—powered by CENTRAL BAROMETRIC PRESSURE of 28.02 inches (949 mb) at landfall—delivered sustained 127-MPH (204-km/h) winds, torrential rains, and an astounding 40-foot (13-m) STORM SURGE to the northeast coast of INDIA between October 4 and 5, 1864. Some 50,000 people, many of them residents of the famous port city that gave the cyclone its name, reportedly perished during the tempest's midmorning landfall there, while an additional 30,000 soon died of starvation and cholera. More than 200 vessels, from tea and jute clippers to steam frigates and fishing boats, were smashed ashore, left holed, and dismasted by the receding surge. Nearly half of Calcutta's principal BUILDINGS were reduced to ruin, their tall windows punched out, their roofs stripped away, and their furnishings tossed into the streets. Directly resulting in expensive losses for the British East India Company, the Great Calcutta's deadly strike on the vital seaport soon prompted the company, in association with the Commission of Famine, to establish the Indian subcontinent's first telegraphic storm-warning system in 1865. A revolutionary idea at the time, the chain of transmission stations and weather observatories grew slowly, not extending to all Indian ports until 1886. In the interim, India suffered an even greater cyclone disaster, the notorious BACKERGUNGE CYCLONE of November 1, 1876. In this particular instance, more than 200,000 lives were lost, making Backergunge the most violent Bay of Bengal cyclone to have affected the region during the 19th century.

Great Colonial Hurricane *Northeastern United States, August 25, 1635* The first documented TROPICAL CYCLONE strike in NEW ENGLAND history, the Great Colonial Hurricane's strafing winds and 14-foot (5-m) STORM SURGE wrought enormous property damage and a considerable loss of life from Rhode Island's Narragansett Bay to east Massachusetts's pilgrim settlements. Chronicled by both John Winthrop, governor of the Massachusetts Bay Colony, and William Bradford, longtime governor of Plymouth Plantation, the Great Colonial Hurricane's early morning landfall near Narragansett Bay uprooted thousands of trees, razed 211 houses, peeled the roofs from countless others, demolished dozens of Indian wigwams, and drowned at least eight people. Shearing northeast, the hurricane's six-hour fury cast a 400-ton merchant ship ashore near Boston, flattened several more buildings, and twisted pliant young hardwood trees into fantastic shapes. Immediately recognized by

survivors as a HURRICANE of remarkable meteorological and historical stature, it was said that the Great Colonial Hurricane's passage up the eastern seaboard was followed by an eclipse of the Moon. Although a lack of contemporary newspaper accounts makes it difficult to compile an accurate DEATH TOLL from the Great Colonial Hurricane, it can be assumed that several dozen people perished in this truly violent storm.

Great Cyclone of 1970 *Bay of Bengal–Bangladesh, November 9–13, 1970* This Category 4 CYCLONE, viewed by a number of contemporary historians as the single deadliest TROPICAL CYCLONE in recorded history, unleashed slashing 131-MPH (211-km/h) winds, pelting, 10-inch (254-mm) rains, and an astonishing 34-foot (11-m) STORM SURGE STORM TIDE combination across the sandy south shores of BANGLADESH (then known as East Pakistan) on the terrifying night of November 12–13, 1970. Armed with a calamitously low CENTRAL BAROMETRIC PRESSURE of 27.88 inches (944 mb) at landfall, the Great Cyclone reportedly claimed 350,000 to 550,000 lives, drowned several hundred thousand head of cattle, engulfed more than a million acres of rice paddies, wheat fields, and tea plantations, demolished more than 40,000 BUILDINGS, and rendered more than a million survivors homeless. Without doubt one of the more politically tainted cyclones in modern history, the Great Cyclone has on numerous occasions been called the "cyclone that founded a nation" because of the pivotal role its catastrophic passage played in prompting East Pakistan to declare its independence from West Pakistan in March 1971. After a 10-month civil war that was nearly as bloody as the Great Cyclone itself had been, East Pakistan was granted both its sovereignty and a new name: Bangladesh.

GREAT CYCLONE CHRONOLOGY

November 9: Peeling away from the INTERTROPICAL CONVERGENCE ZONE (ITCZ), one of several low-pressure TROPICAL WAVES swiftly marshals its circulation patterns over the BAY OF BENGAL's heated waters approximately 760 miles (1,223 km) southeast of the low-lying port city of Chittagong, East Pakistan. Drifting northwest at nearly nine MPH (15 km/h), the thickening cluster of thunderstorms produces 22-MPH (35-km/h) gusts and sporadic rain showers.

November 10: Now stationed 650 miles (1,046 km) southeast of Chittagong and maintaining its meandering north-northwest course, the unnamed tropical disturbance suddenly begins to intensify shortly after daybreak. Upgraded to a TROPICAL DEPRESSION just before noon, the forward speed and course of the developing TROPICAL STORM—now generating sustained winds of about 37 MPH (60 km/h)—are haphazardly traced by radar installations in neighboring INDIA. A lack of aerial reconnaissance, however, prevents meteorologists and civil defense authorities from determining the full particulars of the young system, including its crucial barometric pressure readings. Although the depression's trajectory would seem to indicate a possible landfall on the border between northeastern India and East Pakistan sometime in the next three days, no cyclone warnings are yet posted in those threatened, very cyclone-vulnerable areas.

November 11: Photographs taken at first light by a U.S. weather satellite reveal a startling, ominous sight over the funnel-shaped waters of the Bay of Bengal. Under the steely cover of night, the compact, somewhat disorganized RAIN BANDS of the incipient tropical storm have rapidly blossomed into a sprawling, well-defined cyclone of at least Category 1 intensity. Now centered less than 500 miles (805 km) south of Dhaka, the cyclone's long anti-cyclonic tail, created by the system's OUTFLOW LAYER, extends far northeast, delivering light rain showers to the mountains of western CHINA. Within the system's mature-stage EYEWALL, sustained winds of between 85 and 90-MPH (137–145-km/h) thrash the surface of the bay into a salty haze, drawing millions of tons of water vapor into the cyclone's roiling cloud towers, where the process of condensation releases enormous quantities of heat energy. Well acquainted with the Bay of Bengal's painful history of cyclone activity, the NATIONAL HURRICANE CENTER (NHC) passes satellite data regarding the sudden severity of the storm to government weather services in India and West Pakistan, where it would seem to appear to be accepted with both humble appreciation and callous indifference.

November 12: Beating down on the Ganges River delta's alluvial lowlands at speeds of between five and seven MPH (8–11 km/h), the central pressure of the Great Cyclone steadily commences yet another inning of precipitous deepening. Although a new series of satellite images portrays the full scale of the gargantuan tropical system's CUMULONIMBUS CLOUD mass, they do not tell just how far and how quickly the storm's central pressure is falling. Meteorologists often use such data to calculate the potential size of a tropical cyclone's most destructive feature—its storm surge—by applying it to a formula for determining what is called the INVERSE BAROMETER, or the height to which the surface of the sea rises beneath the system's point of lowest atmospheric pressure. This phenomenon is of particular interest in the Bay of Bengal, where a litany of disastrous storm surges—the Great Coringa Cyclone of 1789, the BACKERGUNGE CYCLONE OF 1876, and the Bengal Cyclone of 1942—has over the

centuries left thriving cities virtually extinct and hundreds of thousands of people dead.

Tracked by radar to within 250 miles (402 km) of the Indian-East Pakistani border by midmorning on November 12, the imminent threat of the Great Cyclone's furious winds, estimated at between 115 and 130 MPH (185–209 km/h), prompts a radio station in Calcutta, India, to begin broadcasting cyclone warnings for the northeast quadrant of the country. These are, in turn, picked up by stations in Dhaka, which relay them as far as their transmitter ranges will permit. In Dhaka itself, all nonessential personnel are sent home early as the shuttered city quickly assumes the stance of a besieged fortress.

But for those teeming millions who fish and plow and harvest their lives away on the overbuilt mudflats at the Ganges River's mouths, there is little electricity and even fewer radios on which to receive storm warnings. Gathered by the thousands on secluded *chars,* narrow islands that stand no more than 15 feet (5 m) above mean sea level, the only warning entire villages will receive of the Great Cyclone's approach comes in the form of rain, in the drumming showers that accompany the gradual onset of the cyclone's leading cloud bands. As sporadic gusts of wind buffet their multistory bamboo houses, long, slow swells begin to crash ashore from the Bay of Bengal, threatening to smash or sink the hundreds of small fishing boats that ring the feathery inlets of Hatia, Sandwip, Kutubdia, and Dubla islands. Although several of these *chars* are connected to the mainland by long causeways, dikes, and bridges, very few residents heed the portentous signs and leave before the gathering cataclysm inundates their last and only chance for escape.

As dusk joins the wind and rain in darkening the Bay of Bengal's tapered shores, the distant moon—then completing its daily lunar cycle—adds the flow of its tides to the meteorological conspiracy now lurking 100 miles (161 km) to the southwest. The Bay of Bengal is known for its unusually high and low tides, and it is to be hoped that cyclones make landfall during the low, or ebb tide, thereby lessening the potential height of their storm surges. In the case of the Great Cyclone, this would mean the cyclone virtually would have to stop and sit stationary for more than a half-day to ensure that its still-strengthening eyewall came ashore during the low tide—a highly unlikely prospect considering that the cyclone had already undergone RECURVATURE and was speeding north at nearly nine MPH (15 km/h). The worst possible time for the Great Cyclone to make landfall would be around midnight when the tide was at its peak. Depending on the storm's intensity, surge conditions could reach heights of between 25 and 30 feet (8–10 m)—a nightmare scenario that could easily see the East Pakistan flatlands flooded for some 30 miles (48 km) inland.

November 13: Uncertain happenstance now tragically fulfilled, the Great Cyclone wails ashore in East Pakistan between the mouths of the Haringhata and Meghna rivers just after midnight. Sired by a central barometric pressure of 27.88 inches (944 mb), the Great Cyclone's 15-foot (5-m) surge, coupled with the above-average lunar tide, delivers a deluge of nearly unprecedented proportions to the three mainland districts of Noakhali, Barisal, and Chittagong. Nearly a dozen offshore *chars*, including Bhola, Dubla, Jabbar, and Hatia, are entirely submerged beneath 20 feet (7 m) of saltwater.

Laden with uprooted trees, the wreckage of houses and boats, and an ever-increasing number of human and ANIMAL bodies, the surge spends the early morning hours of November 13 crawling inexorably across nearly a fourth of East Pakistan's total landmass. Spreading around small hills and particularly resilient trees, it wastes little time in swamping nearly a dozen large rivers, including long stretches of the Ganges and Meghna rivers. Heralded by the dull roar of onrushing water and the snapping sound of houses and trees being devoured by its cresting teeth, the Great Cyclone's storm surge moves in concert with its torrential rains, soon covering more than a million acres of rice and jute fields with between 10 and 15 feet of brackish water. Thousands of buildings, from the pierside harbor cranes and concrete warehouses of Chittagong to the rope-lashed bamboo high-rises of the *chars,* are systematically pulverized by the cyclone's sustained 131-MPH (211-km/h) winds, 145-MPH (233-km/h) gusts, and engulfing rains. In Chittagong, a 150-ton freighter is carried ashore, followed by an armada of fishing boats, barges, interisland ferries, and even a Pakistani patrol boat.

Entire families, asleep at the time of the cyclone's landfall, perish as they are cast into the drowning maelstrom of wind and water. According to a report in the *The New York Times,* a protecting grandfather tosses his six grandchildren, aged three to twelve years, into a large wooden trunk before being swept to his death by the surge—just one of 8,000 villagers lost in the same instant. Hundreds of victims, struggling to remain afloat in almost blinding conditions, are latched onto by poisonous snakes, while thousands of others are crushed beneath collapsing buildings or else impaled by wind-thrown shards of bamboo rod. A majority of those fortunate individuals who do manage to survive do so by climbing sturdy trees. In several instances, survivors will later report that they were forced to battle frenzied snakes and other spooked wildlife for dominion over their lifesaving perches.

November 14: Downgraded to a tropical storm shortly after dawn, the Great Cyclone of 1970 quickly begins to die over the cool, jagged uplands of Bhu-

tan and Tibet. Over the course of the next week, its dissipating thunderstorms and residual moisture will merge into an existing upper-level EXTRATROPICAL CYCLONE that will eventually produce picturesque snow showers on Mount Everest, 320 miles (575 km) northwest of Dhaka.

In scores of East Pakistani communities, however, the midmorning hours of November 14 reveal anything but a pristine white landscape. Amid the muddy wreckage of breached embankments, fallen bridges, washed-out roads, flattened houses, splintered trees, and sand-blasted automobiles, hundreds of thousands of storm-dazed refugees stagger from obliterated village to decimated town in desperate search of food, water, medicine, clothing, and compassionate shelter. Literally tens of thousands of human bodies, many left stripped and bloated by the receding waters of the cyclone's surge, contaminate large portions of the countryside, providing a gruesome incubator for cholera and typhoid.

Unaware of the full dimensions of the tragedy, dozens of officials from the World Bank, the United States, Great Britain, and other nations in Western EUROPE begin to arrive in flood-savaged Barisal on a tentative fact-finding mission. Inspecting the storm-savaged district by helicopter and boat, they are visibly moved by the extent of the Great Cyclone's carnage.

November 15: Sadly accustomed to the depredations of deadly cyclone strikes on its own low-lying shores, the government of India announces that it is contributing $65,000 to relief efforts in East Pakistan.

November 16: As anguished rescuers continue to comb the wreckage in search of possible survivors, the official DEATH TOLL from the Great Cyclone rises to 55,000 people. Unofficial estimates, however, place the human toll much higher, in the vicinity of 120,000 people. Yet, in the midst of such widespread tragedy, a welcome miracle does occur: The wooden trunk that three days earlier had become an impromptu haven for the six young children washes ashore at Bhola Island, its cargo hungry and wet but otherwise unharmed.

November 17: Appalled at the extent of the devastation in East Pakistan, the United States announces that it is immediately dispatching $10 million in food, blankets, and tents to the cyclone-ravaged region. As the growing threat of typhoid and cholera outbreaks slows relief operations to a crawl, the Pakistani Red Crescent requests that the League of Red Cross Societies in Geneva, Switzerland, dispatch vaccines to the affected area.

November 20: Against a backdrop of growing international concern for the plight of those millions still left starving and homeless by the cyclone, Great Britain indicates that it will provide $7.2 million toward East Pakistani reconstruction efforts. China—a nation that has for centuries endured its own relentless round of deadly TYPHOON activity—offers an additional $1.2 million in rice, blankets, and medicine. India, putting aside decades of territorial disputes with West Pakistan, grants permission for Pakistani relief helicopters to enter Indian airspace.

November 21: Landing in two-hour intervals, squadrons of British aircraft begin to airlift troops and supplies to rain-soaked Dhaka Airport.

November 23: While officials in East Pakistan angrily charge the central government in West Pakistan with "gross neglect, callous inattention, and utter indifference" in dealing with the tragedy, the official death count from the Great Cyclone tops 168,000 people. On the sodden, debris-clogged island of Chubdia, several starving survivors are seriously injured by whirring helicopter rotors while frantically trying to collect sacks of rice and other foodstuffs.

November 24: Bearing tons of food, medicine, and building supplies, nearly 1,000 British troops begin amphibious landings on the offshore islands of Bhola, Hatia, and Dubla. Upon disembarkation, they are surrounded by hundreds of hungry children.

November 26: Completing the first leg of a 10-day journey to Asia and the Far East, a compassionate Pope Paul VI flies to Dhaka to offer his prayers and condolences to victims of the Great Cyclone.

December 7: As relief efforts continue unabated, the League of Red Cross Societies announces that a total of 200,000 people perished in the Great Cyclone of November 12–13, 1970. In February of the following year, at a time when resentment in East Pakistan toward the central government's lackadaisical reconstruction efforts had reached the point of armed insurrection, the total count of the cyclone's dead would be almost tripled, with some 550,000 men, women, and children reportedly lost.

Great Galvaston Hurricane of 1900 *Southern United States, August 31–September 10, 1900* An exceedingly violent Category 4 TROPICAL CYCLONE of legendary stature and celebrated consequence, the Great Galveston Hurricane's 140-MPH (225-km/h) winds, 9-inch (229-mm) rains, and 16-foot (3.5-m) STORM SURGE annihilated large portions of coastal and inland TEXAS between September 8 and 9, 1900. With a CENTRAL BAROMETRIC PRESSURE of 27.49

inches (931 mb) at landfall on the barrier island that gave the HURRICANE its name, the Great Galveston Hurricane not only remains one of the most intense storms to have ever afflicted the mainland United States but also the nation's deadliest. Although an exact accounting of the hurricane's DEATH TOLL has never been determined, it is believed that between 6,000 and 12,000 people perished in the tempest, making the Galveston Hurricane of 1900 the single worst hurricane disaster in U.S. history. More than one-third of all the BUILDINGS on Galveston Island, from the gingerbread Victorian mansions that graced its southern neighborhoods to the towering brick cathedrals that dominated its western enclaves and the stone-carved hotels that housed visitors to the eastern beaches, were demolished by the hurricane's remarkable fury and reduced to such complete ruin that one dazed survivor in a subsequent newspaper interview could only liken the mangled result to the handiwork of Armageddon. Damage estimates, computed with considerable difficulty in a period of weeks, soon topped $40 million (an astronomical sum in 1900), precipitating a flurry of financial donations from sympathetic communities all across the country. Tales of burials at sea, mass cremations, and heroic rescues dominated the U.S. press for days thereafter, quickly catapulting the horrific events of September 8, 1900, to the status of legend. Still one of the U.S.'s best-chronicled tropical cyclones, the Great Galveston Hurricane has been the subject of countless articles, novels, plays, and poems, as well as four major nonfiction studies, including Joseph L. Cline's *When the Heavens Frowned.*

Truly one of hurricane lore's greatest CAPE VERDE STORMS, the Galveston Hurricane originated over eastern NORTH ATLANTIC OCEAN's, temperate waters approximately 400 miles west of Africa's Cape Verde Islands, on or about August 31, 1900. Borne westward at speeds of between 12 and 15 MPH (19–24 km/h), the Galveston Hurricane, as with a majority of cyclonic systems of this type, swiftly deepened. By the afternoon of September 4, 1900, the strengthening system had progressed as far west as the CARIBBEAN SEA, bringing TROPICAL STORM-force winds and scudding rains to the northern Leeward Islands of ANTIGUA and GUADELOUPE but causing no casualties.

By the morning of September 6, the system, now touting winds of at least 74 MPH (119 km/h), had passed over PUERTO RICO and HISPANIOLA and emerged into the sun-charged confines of the GULF OF MEXICO. Steadily carried northwest by the trade winds, the Galveston Hurricane underwent another period of deepening, that elevated it to the status of a Category 3 system, with top winds clocked at 111

The City of Galveston, Texas, on the morning of September 9, 1900. As of 2007, the Great Galveston Hurricane remains the deadliest natural disaster in American history. *(NOAA)*

MPH (179 km/h), during the night of September 6. Now less than 500 miles (805 km) east-southeast of Galveston, the hurricane's rapid gain on the sandy island—then only one mile (2 km) wide, 25 miles (40 km) long, and nine feet (3 m) above sea level—was heralded by a relentless drop in barometric pressure and a concomitant lengthening of the oily swells that rolled in from the gulf throughout the day and night of September 7. Several firsthand accounts of the hurricane's approach describe what islanders called a "northern," a steady northwest wind that generally kept the boisterous waters of the Gulf of Mexico off the sandbar's southeastern beaches but, in this particular instance, did little to stop the harrowing rise of the sea. Indeed, Isaac Cline, the NATIONAL WEATHER BUREAU's observer in Galveston, dispatched a telegram to his superiors in Washington, informing them that "such high water with opposing winds" had never before been witnessed on the island. Cline, who may have had cause to question the accuracy of a central forecasting office located in distant Washington, wryly added that he suspected the phenomenon was due to the presence of a "West Indian hurricane" that the Weather Bureau had earlier tracked into the Gulf of Mexico and subsequently predicted would come ashore *north* of Galveston Island, exposing

the thriving yet vulnerable city of 30,000 people to only a glancing blow by those moderate winds and rains generally found in the storm's NAVIGABLE SEMICIRCLE.

Whistling around a central barometric pressure of 27.49 inches (931 mb) and preceded ashore by drumming downpours and tropical storm-force gale bands, the Great Galveston Hurricane's mature EYE-WALL marched into Galveston Bay on the afternoon of September 8, 1900. Shielded beneath its EYE, the hurricane's storm surge quickly piled up against the bay's shallow slopes and began to lap at piers and seep into Galveston's debris-strewn streets. All four of the island's bridges—one of them the world's longest wagon run; the other three railroad bridges—were undermined by the hurricane's 20-foot surge and submerged by the frenzied waters of Galveston Bay where sawing breakers quickly chopped them into driftwood. Grain silos, warehouses, and saloons, many of them precariously perched just a few feet above sea level, creaked and groaned as the wind's prying fingers lifted roofs, ladders, and stove-pipes high into the air and then dropped what remained of the buildings into the pounding surf.

All eight ships then at anchor in Galveston Bay were carried ashore by the hurricane's surge and

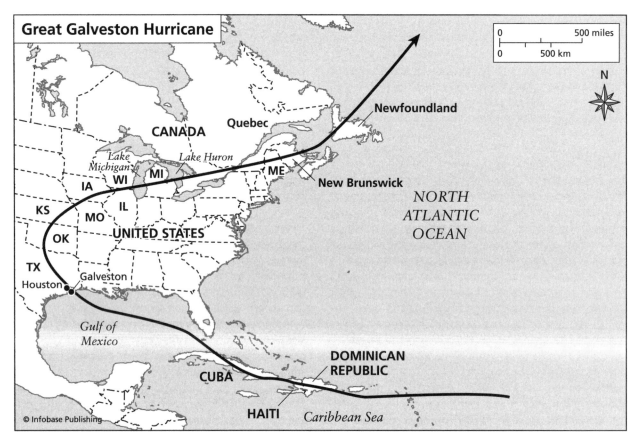

This tracking chart illustrates the course taken by the Great Galveston Hurricane of September 1900. It remains the deadliest tropical cyclone strike in American history.

unceremoniously heaved into a dented heap on the island's southeast coast where receding waters soon left them stranded. All electrical wires—Galveston was one of the first cities in the southern United States to enjoy the conveniences of electricity—were downed. Nearly every tree on the island was either uprooted or sheared off at the ground. Galveston's tallest structure, the 220-foot (76-m) brick tower atop stately St. Patrick's Church, twisted wildly against 140-MPH (225-km/h) gusts, its bells pealing out one last call to prayer as the spire slammed into the street just before dusk. A wooden span of streetcar trestle positioned over the Gulf of Mexico was caught by the oncoming surge and ridden inland with the same efficiency as a score of locomotives, running down entire rows of houses within a matter of minutes.

As traces of thunder rolled against the night, sharp flashes of lightning added a surrealistic shading to the writhing carnage below. Literally thousands of people, many of them forced from their houses by the rising waters of the gulf, fought for survival against flooded streets, wind-hurled timbers, jagged roof tiles, spooked ANIMALS, and unbridled fear, loneliness, and exhaustion. Between 6,000 and 8,000 people in Galveston died within four hours of the hurricane's

landfall, a gruesome testimonial to the cataclysmic meteorological fury possessed by the storm; 12 artillery soldiers stationed at nearby Fort Crockett were crushed to death when the roof of their barracks caved in; and 15 nuns at the Catholic Orphanage Asylum, along with 98 children under their care, drowned when the hurricane's surge completely demolished their three-story brick haven. Rescuers, coming across two-dozen corpses half-buried in beach sand, later stated that the nuns had apparently attempted to save the children by tying them together with rope. More than 50 people who had sought refuge in Galveston's impressive City Hall were likewise drowned, while only eight of the 100 souls who had fled to St. Mary's Infirmary survived the sudden collapse of the timber-framed structure. Another 2,000 people are believed to have perished elsewhere in Texas where the Great Galveston Hurricane's undiminished winds and rains shattered windows, mowed down trees, and slaughtered farm animals as far west as Odessa. Crossing the plains states of Oklahoma and Kansas on September 9, the hurricane itself came to an end over the heavily wooded forests of Minnesota and Michigan on September 10. A dangerous hurricane to the very last, its demise was characterized by strong gales and prolonged downpours that caused enormous damage to the region.

Great Havana Hurricane of 1846 *Cuba–Southern and Northeastern United States, October 6–14, 1846* A Category 5 HURRICANE of extraordinary size and strength, the Great Havana Hurricane spent four crippling days—October 10 through 14—of the 1846 HURRICANE SEASON spinning a late-season web of destruction across central CUBA, through the southern U.S. states of FLORIDA, GEORGIA, and SOUTH CAROLINA, and into NEW YORK and NEW ENGLAND. Said by contemporary accounts to have possessed an extremely low CENTRAL NORTHERN BAROMETRIC PRESSURE of 27.06 inches (916 mb) at landfall in southwestern Cuba on the morning of October 11, the Great Havana Hurricane produced terrific ADVECTION currents that scoured the island's narrow peaks, sugarcane plantations, and colonial cities at speeds of up to 175 MPH (282 km/h). In Havana, where the hurricane's fringe rains began to fall shortly before midnight on October 10, more than 85 merchant ships lying at anchor in Havana Harbor were either rammed ashore or sunk by the hurricane's 30-foot (10-m) seas; nearly every building in the capital was demolished; and hundreds of trees were uprooted and turned into impaling spears by the storm's howling 190-MPH (306-km/h) gusts. Outlying villages such as Cardenas and Santa Clara were obliterated or reduced to uninhabitable mounds of debris within a matter of a few hours. Some 50 coastal trading vessels, laboring to remain afloat in the port of Matanzas, were wrecked, taking the lives of several dozen seafarers. Before roaring into the GULF OF MEXICO as a greatly diminished Category 2 system, the Great Havana Hurricane killed at least 600 people in Cuba, making it an even deadlier storm than the Cuban Hurricane of October 5, 1844, in which an estimated 400 people perished.

Recurring northeast on the morning of October 11, the Great Havana Hurricane next battered the Florida Keys with heavy rains and a 12-foot (4-m) STORM SURGE. In Key West, where the hurricane's sustained winds topped 110 MPH (177 km/h), 14 schooners, brigs, and full-rigged ships were either sunk, dismasted, or driven ashore. A lightship standing guard over Carysfort Reef quickly foundered at its shallow, coral-bedded mooring, while twin 65-foot (22-m) lighthouses at Sand Key and Key West were undermined and collapsed, tragically drowning not only their keepers and attendant families but also 14 refugees who had sought shelter within the two white-painted brick pylons. A large naval hospital in Key West was severely damaged when a five-foot (2-m) STORM TIDE coursed through the center of the town, shearing away its stone corners and peeling apart its clay-tile roof. Another 600 BUILDINGS on the key, including Fort Taylor and a newly constructed customhouse, were utterly razed. More than $200,000 in property losses were suffered, primarily by federal installations. Streets were rendered impassable from fallen houses, stacked roofs, and jumbled timber. More than 40 lives were reportedly lost, prompting surviving eyewitnesses to deem the hurricane one of the most violent in Floridian history.

After skirting Tampa Bay on the evening of October 11, the Great Havana Hurricane's weakening EYEWALL moved inland over Cedar Key, Florida, during the early morning hours of October 12. Hundreds of spindly pine and stately oak trees were toppled, porticoed houses were gutted, and nearly 3 inches of rain fell on the state's northern tier as the hurricane barreled northeast at speeds of 12 MPH (19 km/h). Its 95-MPH (153-km/h) winds greatly diminished by its trek across the peninsula, the Great Havana Hurricane exited Florida near Jacksonville just before dawn, then turned sharply north, delivering hurricane-force winds and two-inch (51-m) rain counts to Charleston, South Carolina, later that day. Although a heavily laden schooner in Charleston Harbor was lost and many fences, chimneys, and trees in and around the port blew down, damage to the storm-prone city was considered minimal with no casualties reported by the local press.

Maintaining its recurring north-northeast swing, the Great Havana Hurricane—still boasting sustained wind-speeds of between 85 and 90 MPH (137–145 km/h)—subsequently tore its way through NORTH CAROLINA, VIRGINIA, New York, and New England. In Wilmington, North Carolina, a two-foot (1 m) storm surge crashed against the city's earthen embankments on October 13, while the Potomac River in Washington, D.C., reached its highest flood levels in 70 years on the morning of October 14. In New York City, the Great Havana Hurricane's confused seas swept away several hundred feet of the Battery's seawall. A trestle bridge in Hartford, Connecticut, was torn to pieces by 77-MPH (124-km/h) gusts, scattering railroad ties across a wide swath of track. Numerous apple orchards in Massachusetts were cored by the hurricane's dying winds, causing great financial hardship for farmers in the western half of the state. No deaths or serious injuries were reported from the Great Havana Hurricane's passage over the northeastern United States. Entering eastern Canada on October 15, the hurricane rapidly dissipated—but not before dropping more than two inches of rain on the maritime provinces.

Great Hurricane of 1780 *Eastern Caribbean–Bermuda, October 10–18, 1780* Believed by many historians to have been the single most destructive TROPICAL CYCLONE to have passed through the CARIBBEAN SEA during the 18th century, the Great Hurricane of 1780 wrought a 3000-mile swath of cataclysmic devastation through the islands of BARBADOS, St. Vincent, St. Lucia, MARTINIQUE, PUERTO RICO, and BERMUDA between October 10 and 18, 1780. In Barbados, where the Great Hurricane's 160-MPH (258-km/h) winds blew continuously for nearly 24 hours, virtually every building on the island—from the circular stone Governor's Mansion in Bridgetown to the boiling houses on the sugar plantations—was completely leveled. On tiny St. Vincent's east shores, the hurricane's enormous STORM SURGE, estimated by observers to have been some 20 feet (7 m) high, roared 90 feet (30 m) inland washing away entire villages. On French-held Martinique, the burgeoning capital of St. Pierre was virtually demolished. Nearly 1,500 buildings on the heavily fortified island, including its principal cathedral and military prison, were reduced to ruins by the Great Hurricane's 175-MPH (282-km/h) gusts and pounding storm surge. In Bermuda, where the hurricane's EYE passed 50 miles south of the island on the night of October 17–18, 50 merchant ships were run aground, many of them becoming total losses. Before dissipating over the cooling reaches of the NORTH ATLANTIC OCEAN on October 20, the Great Hurricane had claimed between 10,000 and 22,000 lives in the Caribbean. In some areas, such as those surrounding the shattered port cities of Bridgetown and St. Pierre, the devastation from wind and water was so complete that an accurate count of the dead could not be taken.

The second in a trio of remarkable storms to have coursed through the Caribbean during the month of October 1780, the Great Hurricane's exact birthplace has never been determined. Famed 19th-century hurricane scholar WILLIAM C. REDFIELD placed its origination point 400 miles (644 km) southeast of Barbados on October 10, although it is entirely possible that this remarkable storm originated off West Africa's windswept coast near the Cape Verde Islands during the opening days of October 1780. Nursed into maturity by the sun-drenched waters of the subequatorial Atlantic, the Great Hurricane quickly grew into a colossal storm whose imminent landfall in Barbados was heralded by a sky that the island's governor would later describe as being "surprisingly red and fiery." Indeed, several survivors would eventually corroborate the governor's account, recalling that the evening preceding the hurricane was noted for the stillness of the air and the ominous brilliance that accompanied the seering Caribbean sunset.

As darkness set in, a light rain began to fall. Tinged with the scent of seawater, it was the sort of rain that came with the onset of a large hurricane's outer fringe. Through the night it continued to rain, the downpours becoming heavier and of longer duration. Just before dusk, the wind began to rise from the northeast, a series of gusts that eyewitness accounts claimed rushed over the fairly level island, scattering palm fronds, sugarcane stalks, and unsecured equipment across the landscape. By ten o'clock in the morning, the wind had increased to TROPICAL STORM–intensity, with gusts approaching hurricane force. Numerous small outbuildings, drying sheds, and barns quivered and groaned as the wind pried at their gables and eaves or rushed through as though in a tunnel. In Barbados's various harbors and inlets, nearly 100 ships that had anchored themselves against the growing might of the Great Hurricane now began to part their lines, drifting helplessly toward the beaches and rocky shores. On board one such vessel, the British frigate *Albermarle,* sailors successfully saved the ship from grounding in Carlisle Bay by cutting away her foretop and painfully wearing to seaward; for many of the other craft caught in the bay, similar measures did not prove so effective. One army transport, two navy victuallers, and one ordnance vessel were driven from their anchorage and run ashore near the entrance of the bay. Quickly dismasted by the shearing winds, the wooden vessels rocked back and forth in the raging surf, opening up their seams and eventually breaking their backs. All told, more than 200 sailors were lost.

By ten o'clock on the evening of October 10, the wind in Barbados blew with the screaming intensity of a Category 5 hurricane. In Bridgetown, both the hospital and the barracks for the island's military garrison were destroyed, forcing the soldiers to seek shelter wherever they could find it. On the ramparts of the imposing fortress that guarded the entrance to Bridgetown's harbor, the Great Hurricane's gripping winds carried a 12-pound iron cannon from the south side of the battery to the north, a distance of 420 feet (120 m). Farther inland, the city's stone prison was peeled apart in pieces, necessitating the release of more than 800 prisoners—some of them French and Spanish prisoners of war—into the missile-strewn fury of the storm. In giving the order to release the prisoners, the governor (who had by that time sought refuge with his family in the cellar of his house) requested that troops be called out to prevent the miscreants from looting. Shortly thereafter, however, as first the roof and then the walls of the house collapsed, the governor rescinded his earlier request for troops. With his petrified, rain-soaked family gathered about him, he realized, as everyone else on the island now knew, that the plundering hurricane was leaving behind very little worth taking.

Shortly after midnight on October 11, the Great Hurricane's eye passed just north of Barbados, bound for the volcanic spine of nearby St. Vincent. Because the eye remained offshore the entire time, Barbados did not experience the associated lull in the storm's intensity but rather a prolonged continuation of its northeasterly fury. Hundreds of buildings near North Point, from humble bungalows to white-stuccoed plantations, were stripped of their roofs, windows, and porches and collapsed. An equal number of people, many of them slaves, were drowned as the Great Hurricane's torrential rains caused streams and rivers to overrun their banks, flooding the rubble-clogged cellar holes in which they were trapped.

On the island of St. Vincent, 100 miles west of Barbados, the Great Hurricane's burgeoning legacy began shortly after midnight on October 10–11, when estimated gusts of 172 MPH (277 km/h) and thudding rains descended on the capital city of Kingstown. More than 550 houses in the city were demolished as flash floods surged down the hillsides, boring huge channels though the fertile plantations along the way. On the island's east coast, two warships—one French, the other British—found themselves united in an ultimately hopeless struggle to keep from being thrown ashore. Of the 220 sailors on both ships, no more than 16 were eventually rescued. In nearby Grenada, 19 Dutch ships laden with sugar and other goods were stranded on a shoal and beaten to pieces by the sea, drowning several hundred more sailors.

Steadily moving northwest at approximately seven miles (11 km) per hour, the Great Hurricane next passed over St. Lucia, reaching the island in the early morning hours of October 11. Its lashing intensity hardly diminished by its earlier strikes on Barbados and St. Vincent, the Great Hurricane's thundering rains and blasting winds brought heavy seas to every anchorage on the island's southwest coast. Long wooden piers were splintered and shorn as rising waters overran the stone quays of Kingstown Harbor. Dockside warehouses collapsed, spilling their barrels of rum and molasses into the surging assault. In the harbor itself, several vessels, one of them a 74-gun British warship, rocked into each other as the hurricane's winds inexorably drove them ashore and then proceeded during the next 24 hours to break them up. What meteorological accounts survive from the Great Hurricane's direct passage over St. Lucia state that the storm's fury remained unabated until one o'clock on the morning of October 12, when the wind slowly died away in gusts, leaving just rain and a ravaged landscape littered with more than 700 dead, to signal its passage.

Seemingly growing stronger by the day, the Great Hurricane next turned its sights on Martinique, 50

miles (80 km) north of St. Lucia. In the seaside capital of St. Pierre, the hurricane's midmorning arrival brought with it a pulverizing storm surge in which 150 buildings situated along the harbor's edge were destroyed instantly. In nearby St. Pierre's Road, two French ships, the *Marquis de Brancas* and the *L'Eole,* with 150 soldiers from the regiment of Touraine aboard, were forced out to sea by the hurricane's violence and subsequently were lost with all hands. At Fort Royal, the city's cathedral, seven churches, the governor's house, the senate house, and the prison were all shattered by the hurricane's winds, while the celebrated hospital at Notre Dame with 1,600 patients in residence was riven to the ground. Although desperate efforts were made to save those patients, the nurses, and the matrons buried beneath the wreckage, a great many of them perished.

In one of the worst maritime tragedies of the 18th century, a 40-ship convoy bringing troops and provisions to Martinique from France was overtaken by the Great Hurricane while futilely searching for a sheltered anchorage on the island's west coast. Dismasted by the roiling winds, swamped by the cresting seas, those vessels that did not go to the bottom directly were eventually run ashore with an appalling loss of life. Various contemporary sources place the DEATH TOLL from this one incident at between 3,000 and 5,000 lives. On the same day, the Great Hurricane overwhelmed three British armed sloops and a supply ship belonging to Admiral Rodney's fleet as they shadowed the island's north capes. All four vessels were wrecked, adding another 72 dead to the final tally.

After belting both DOMINICA and St. Kitts with hurricane-force winds and a 22-foot (7-m) storm surge on October 12, the Great Hurricane moved north into southeastern Caribbean open waters during the early morning hours of October 13. On October 15, it passed through the Mona Passage separating Puerto Rico from HISPANIOLA, bringing high winds and flooding rains to both areas. After skirting the BAHAMAS' eastern banks on October 16, the Great Hurricane steadily underwent RECURVATURE until it was moving in a northeasterly direction that would eventually take it approximately 50 miles (80 km) south of Bermuda on October 18. In Bermuda, several dozen vessels, a majority of them British merchant ships, were driven ashore and wrecked in what contemporary observers described as one of the most severe storms to have struck Bermuda in memory. Continuing on its northeasterly course, the now fading colossus quickly entered the North Atlantic's cold, undulating waters and soon dissipated.

On those Caribbean islands directly affected by the storm's passage, however, the Great Hurricane's legacy of death and destruction lingered on for months to come. In Barbados, where an estimated 4,326 people lost their lives to the hurricane, the damage to farms and food supplies was so widespread that famine eventually set in. According to eyewitness accounts, more than 1,000 slaves perished from starvation in the first two weeks after the disaster. Although the British Parliament wasted little time in appropriating £80,000 of emergency aid for the stricken colony, their largess did virtually nothing to offset the £1.3 million in losses that had been sustained by the island's trade and property. In Martinique, where upwards of 9,000 people perished, nearly every house in St. Pierre was blown down, and more than 700,000 *louis d'or* worth of damage was assessed. In both places the lurking threat of looters and shipwreckers drew surviving soldiers away from rescue efforts, thereby increasing an already staggering death toll. In the end, an estimated 22,000 lives would be lost throughout the Caribbean to the Great Hurricane of 1780.

Great New England Hurricane *Northeastern United States, September 21, 1938* Widely regarded as the single most violent TROPICAL CYCLONE in NEW ENGLAND history, the Great New England Hurricane's 121-MPH (195-km/h) winds, 186-MPH (299-km/h) gusts, eight-inch (203-mm) PRECIPITATION counts, and cataclysmic 17-foot (6-m) STORM SURGE claimed more than 600 lives and wrought between $300 million and $400 million in property damage on NEW YORK's Long Island, Connecticut, Rhode Island, Massachusetts, New Hampshire, and Vermont.

Originating near the Cape Verde Islands on or about September 13, the Great New England Hurricane was spotted 300 miles (483 km) east of the BAHAMAS on the morning of September 19 and abreast of Jacksonville, FLORIDA, on the evening of September 20—at which point hurricane observers apparently lost track of its progress. Now deepened into a Category 4 tropical cyclone of devastating intensity (a CENTRAL BAROMETRIC PRESSURE of 27.85 inches [938 MB] was recorded aboard the Cunard White Star liner *Carinthia* on the night of September 20), the storm's distant but threatening presence was discreetly noted in the evening editions of major regional newspapers on September 20: a lukewarm HURRICANE WARNING that ultimately did little to prevent the tragic events of the following day.

Suddenly accelerating north-northwest at an average speed of 55 MPH (89 km/h), the Great New England Hurricane sideswiped NORTH CAROLINA's sandy Cape Hatteras just after dawn on September 21 and then slammed into west Long Island and southern Connecticut shortly before nightfall on the same day. Fueled by a central pressure of 27.94 inches (946 mb)

at landfall near Milford, Connecticut, the robust EYE-WALL of the Great New England Hurricane churned northwest, passing along the Connecticut Valley and into western Massachusetts with a degree of meteorological fury rarely before seen in either state. A newspaper reporter on New York's Long Island described the hurricane's onset as possessing "a steady, almost organlike note of such intensity that it seemed as if the whole atmosphere were in harmonic vibration." In Providence, Rhode Island, the waters of Narragansett Bay flooded much of the city with 13 feet (4 m) of debris, stranding hundreds of people on the upper floors of buildings and on the steps of City Hall. Vulnerable oceanside towns like Misquamicut and Napatree Point, Rhode Island, were completely obliterated by the hurricane's surge, their shingle-style cottages and finger-pier marinas reduced to mere driftwood in a matter of minutes. At Connecticut's Wesleyan University, the chapel's stone tower was blown to the ground, and in nearby Stonington five railway coaches belonging to the *Bostonian* were derailed, swept off a causeway and into a surge-swollen estuary. Despite the heroic actions of the train's crew in safely evacuating most of the 250 passengers aboard, two people were killed.

In Boston, Massachusetts, some 75 miles (121 km) east of the eyewall's path, three men drowned when their fishing trawler capsized in the harbor. At the city's airport the buckling of a 100-foot (30-m) radio tower disrupted air-traffic-control operations, while an empty American Airlines DC-2 aircraft was carried several hundred yards by the hurricane's extraordinary fringe gales and dropped in a marsh. In Springfield, 80 miles (129 km) southwest of Boston, very close to the hurricane's inland track, some 16,000 trees—many of them ancient hardwoods—were uprooted. More than 4 million bushels of apples were ruined as neighboring orchards were devoured by the hurricane's insatiable winds. Worcester, a fading textile city east of Springfield, suffered a further decline as two huge redbrick mills crumbled into dusty clouds, pulling swaying clock towers to the ground with them. In the nearby town of Hadley, the hurricane's undiminished winds collapsed a tall chimney at Northfield Seminary, fatally crushing two girls in the dormitory below. In adjacent Amherst, falling trees crushed houses, split barns, and formed impromptu dams across rain-swamped streets. A group of freshmen at Amherst College, taking part in an intelligence test during the peak of the hurricane's passage over the Pioneer Valley, reportedly achieved higher scores than any other group in the school's prior history. In distant New Hampshire, where ANEMOMETERS recorded gusts of 160 MPH (258 km/h), the base station and trestle of the scenic cog railway that chugged its way up 6,288-

The drumming rains silenced Boston's Half-Shell outdoor concert hall and resulted in flooding caused by the Great New England Hurricane of 1938. *(NOAA)*

foot (2,096-m) Mount Washington were demolished. Elsewhere in the Granite State, a smoldering rag fire in the town of Peterborough soon turned into a roaring conflagration as the hurricane's fanning winds carried flames from one wooden building to the next. Burning fiercely through the rainy, gust-punctured night, much of the town would by the following morning lay in charred ruins.

Sharply recurring northwest late in the evening of September 21, the Great New England Hurricane blasted across southwestern Vermont and New York's Lake Champlain before dissipating into 50-MPH (81-km/h) rain squalls over central CANADA on the morning of September 22. Immediately recognized as the worst natural disaster in the region's history, the Great New England Hurricane of 1938 left 600 people dead, with another 100 forever unaccounted for. Many of the 1,750 injured people were critically hurt, while 63,000 were left without shelter. Damage estimates, compiled with great difficulty over the following weeks, revealed a staggering material loss: 4,500 buildings completely destroyed; another 15,139 homes, cottages, and barns seriously damaged; 2,605 boats cast ashore, sunk, or wrecked, with another 3,369 left damaged; 1,475 ANIMALS killed, including 1,000 head of cattle and 750,000 chickens; 31,000 telephone poles plucked from the earth, forming a chaotic canopy over more than 25,000 totaled automobiles.

Great September Gale, Hurricane *Caribbean Sea–Northeastern United States, September 21–23, 1815* The most celebrated TROPICAL CYCLONE to have affected NEW YORK and NEW ENGLAND since the GREAT COLONIAL HURRICANE of 1635, the Great

September Gale pummeled Long Island, Connecticut, central Massachusetts, and New Hampshire with what one awestruck survivor described as winds of the "utmost fury."

Of possible Cape Verde origin, the Great September Gale whirled with destructive consequence through the islands of the eastern and northern CARIBBEAN SEA before tracking northward, across Long Island, and into south-central Connecticut on the morning of September 23. A truly notable hurricane in terms of its size and severity, the Great September Gale skipped ashore at Sag Harbor, Long Island, with a six-foot (2-m) storm surge in tow. Nearly every house in the settlement was flooded, with several being carried from their foundations and wrecked. The lantern atop the stone lighthouse at Montauk Point was smashed, allowing its fragile lenses to become pitted from windblown sea salt. New York City was "buffeted by very heavy rain and gales from the northeast," causing the restless waters of the harbor to submerge dozens of Manhattan's wooden piers as well as toss a heavily laden merchant ship ashore at Staten Island. As the vessel's crew labored to salvage its cargo, bursting flood waters in the nearby town of Flushing swept away a wooden drawbridge, isolating hundreds of residents.

Tearing across Long Island Sound, the Great September Gale raised an enormous storm surge in Rhode Island's Narragansett Bay that flooded much of Providence with 12 feet (4 m) of water. At least 25 ships, many of them lumber schooners and fishing sloops, were carried across the crowded anchorage by the surge and then sent careening into the long wooden bridge that spanned the head of the bay. A number of offshore islands were stripped of their trees, houses, and piers, and extensive dune and washout damage almost permanently altered the geography of Buzzard's Bay, Massachusetts. Nearly seven inches of rain fell across central Connecticut, precipitating raging flash floods and dangerously swollen rivers. Transiting the region at forward speeds of 50 MPH (81 km/h), the September Gale's winds—judged by Noah Webster to have blown "a perfect hurricane" in Amherst, Massachusetts—decimated dozens of apple orchards and cider mills, uprooted entire groves of oak, maple, and poplar trees, razed barns, gutted carriage houses, and sprayed the resulting wreckage with tons of bitter sea salt. Surprisingly, only two deaths were reported in the Great September Gale, a miraculously low number considering the obvious severity of the hurricane's passage.

Great Tokyo Typhoon of 1918 *Philippines– Japan, September 29–October 1, 1918* An exceptionally powerful Category 3 (estimated) TYPHOON, the Great Tokyo Typhoon inflicted widespread damage and huge DEATH TOLLS on the northern PHILIPPINES and central JAPAN between September 29 and October 1, 1918. The first truly great typhoon to strike Japan in the 20th century, the storm's razing winds—estimated in contemporary accounts to have been in excess of 120 MPH (193 km/h)—leveled some 200,000 houses in and around the capital city of Tokyo. Along the coastal plains of Tokyo Bay, the typhoon's heaving STORM SURGE inundated the towns of Kasai and Sunamura, flooding their narrow streets and lightly timbered houses with up to nine feet (3 m) of water. Coalfired steamships and heavy-set fishing boats were spun ashore by the dozens. Thousands of pine groves were splintered or uprooted, spread flat along once-verdant volcanic hillsides. Between 1,500 and 2,100 people lost their lives to the Great Tokyo Typhoon, making it one of the deadliest storms in Japanese history.

Like so many presatellite-era typhoons, the exact origination point of the Great Tokyo Typhoon is not known. The fact that its outer winds grazed the north Philippine island of Luzon on September 29 suggests that, like a majority of September typhoons, it most probably developed east of the 140th meridian and then steadily progressed west before undergoing RECURVATURE, or a sharp turn to the northeast, while still over the North Pacific Ocean's temperate ranges. Permitted by warm sea temperatures, favorable winds, and vast open spaces to grow into a storm of hefty size and alarming intensity, it may have deepened further as it recurved, sending it into the Japanese archipelago at an increased rate of speed shortly after dusk on September 30.

By nine o'clock on September 30, as the typhoon's EYE bore down inexorably on the main island of Honshu, heavy rains began to fall across the Tokyo and Tokaido districts. In the coastal provinces of Chiba, Ibaraki, and Kanagawa, accompanying gale-force winds rattled windows, sent farm ANIMALS scurrying into sheltered lees, and forced numerous small boats to fight for steerageway against the long, rising seas that began to roll ashore. In Tokyo, the rain's incessant hammering on the Imperial Palace's tiled roofs woke the emperor shortly before midnight. Lying there in his splendid chrysanthemum bed, the young leader stared into the darkness of the blossoming storm, instantly knowing without the aid of any weatherwise advisers that the shrieking winds and pelting rains portended a fearsome night for his country. In time, after the typhoon had worked its worst, he would magnanimously order that 100,000 yen be distributed to those victims left destitute by the storm's ruinous passage.

Shortly before 3 o'clock on the morning of October 1, 1918, the eye of the Great Tokyo Typhoon came ashore just east of the city. Those crowded fishing villages stacked along the shores of Tokyo Bay were the first to experience the typhoon's venting contagion. As wind speeds increased to a sustained 120 MPH (193 km/h), the oversailing eaves so common in Japanese house construction began to catch the wind and trap it as it coursed between the packed rows of thinly sided cottages. Eddies formed, pressurized pockets of air that would eventually pry the roof away from the exterior walls and allow the structure to fold in on itself. A set of storm shutters here, an entire section of roof there, and before long complete communities were seemingly airborne. In Kasai, directly outside Tokyo, some 10,000 houses collapsed within the space of a few hours, bulleting the air with lethal debris. Hundreds of people were struck down by flying timber and shards of clay roof tiles as they fled through the streets. Those survivors who dashed through the darkness in search of shelter were soon confronted by an even greater menace: the rising sea. As the typhoon's winds reached their peak, the shower of pounding spray along the coves and anchorages of Tokyo Bay suddenly gave way to a solid cascade of water that roared over the quays and into the streets of both Kasai and neighboring Sunamura. Rising to a height of nine feet (3 m), it swept the waterfront clean of its shattered warehouses, overturned horse carts, stray boats, and bobbing wooden casks, and then drove the jostling barricade inland. Entire families perished as the remains of their villages closed over their heads. Others—the infirm, the elderly, the very young—drowned where they lay, afterward blanketing the singed landscape with hundreds of disease-bloated corpses.

In Tokyo itself, the typhoon's rapacious winds mowed down hundreds of lampposts, fences, signs, and ornamental trees. A number of well-constructed foreign embassies in the capital sustained severe roof and window damage. Large industrial complexes, such as the Ebisu Brewery, had their tall chimneys blown down. In the case of the unfortunate brewery, a shower of bricks hurtled through the roof, weakening the entire structure; by dawn the three-story building was little more than a one-story pile of rubble. Across the city, the elegant Seiyoken Hotel, with its wide verandas and twin towers, was likewise demolished, scattering sumptuous furnishings across waterlogged streets. Although the Imperial Palace itself suffered minor damage to its roofs and decoration, its serene gardens and conservatories were blighted by the typhoon's shearing winds and ghastly rains. One of the emperor's first edicts following the passage of the storm concerned restoration of these gardens, including the hurried replanting of some 15 treasured pine trees that had been uprooted. Elsewhere in the city, some 45,000 houses were utterly destroyed, forcing tens of thousands of people to set up temporary shelters in the streets.

By midmorning on October 1, as the Great Tokyo Typhoon dissipated over the mountainous ranges of central Honshu, groups of stunned survivors emerged from their hiding places to begin the mournful duty of assessing the damage. What they found shocked them: More than 200,000 houses in and around Tokyo had been completely destroyed; more than 1,500 fishing boats were either cast ashore and wrecked or else lost at sea without a trace; hundreds of fishers were reported missing, many of them never to return. Within Tokyo city limits, rescuers retrieved 629 bodies from swollen riverbeds and flooded buildings; in the surrounding countryside, another 990 victims were pulled from vast piles of rubble. In some places the damage was so great that news of it spread throughout the archipelago, prompting giddy visitors to travel hundreds of miles to view the ruins by picturesque moonlight.

Guadeloupe *Eastern Caribbean Sea* Between 1635 and 2006, this idyllic collection of seven mountainous islands, prized for their tumbling waterfalls, sugar-white beaches, humid tropical forests, and steaming Soufrière Volcano, withstood 55 confirmed strikes from mature CAPE VERDE STORMS and eastern Caribbean HURRICANES. Composed of two major islands, Grande-Terre and Basse-Terre, Guadeloupe's position between latitudes 16 and 17 degrees North makes it greatly susceptible to those mid- to late-season TROPICAL CYCLONES that, after traveling west with the North Atlantic trade-winds, arrive on the shores of the eastern Caribbean Sea as major Category 3 or 4 systems capable of causing enormous damage and loss of life. Although Guadeloupe's record of direct hurricane strikes is not as long as those suffered by the nearby islands of ANTIGUA and MARTINIQUE, the island has nevertheless sustained serious damage from all such passing storms, making its history of hurricane activity that much more extensive. Also as with many of its Leeward Island neighbors, Guadeloupe has even resisted several powerful hurricanes in rapid succession, as witnessed in 1642 when two destructive storms made landfall within one month of each other; in 1785 when another two hurricanes spun ashore between August 25 and 31; and in 1886 when two tropical systems blasted the islands between August 20 and September 25. During the memorable 1809 HURRICANE SEASON, Guadeloupe's principal cities of Basseterre and Pointe-à-Pitre sustained heavy damage from three of

that year's most violent hurricanes, those of July 27, August 2, and September 2.

Hurricanes of historical or meteorological interest in Guadeloupe include the 1656 tempest that completely desolated the islands, sinking a number of French merchant ships at anchor in the harbor at Basseterre and in those waters surrounding the island of Maria Galante; the notorious hurricane of August 4, 1666 that overwhelmed Lord Willoughby's squadron of 17 English troop ships, sending the entire contingent to the bottom with the loss of more than 2,000 lives; the September 1713 hurricane that destroyed another 40 ships at Basseterre; the July 1762 storm that wrecked six unidentified English ships at Basseterre; the hurricane of July 31, 1765, that destroyed another 17 English merchant ships in Basseterre Harbor; the October 6, 1766, hurricane that, in addition to sinking 50 large French and English warships and traders, wrecked 12 inbound slave transports near Isle de Saints with the loss of all hands; the September 10, 1792, hurricane that drove the 44-gun French warship *Didon* aground at Pointe-à-Pitre, rendering the vessel a complete wreck; the furious Santa Ana Hurricane of July 24, 1825, that delivered a CENTRAL BAROMETRIC PRESSURE of 27.67 inches (937 mb) to the islands, decimating crops, demolishing houses, and driving numerous ships aground; the fearsome storm of September 6, 1865, that carried away 90 percent of the BUILDINGS in Basseterre; and the blistering hurricane of September 26, 1889, that killed more than 100 people on Isle de Saints. On September 12, 1928, the merciless SAN FELIPE HURRICANE claimed more than 600 lives in Guadeloupe, while the 160-MPH (258-km/h) gusts associated with Hurricane INEZ took the lives of another 90 people on September 24, 1966. Almost all of the island's $20 million banana harvest was ruined by Inez, precipitating a financial crises. On September 16, 1989, Hurricane HUGO stampeded ashore near Basseterre, killing at least five people and leaving another 12,000 homeless. In an effort to stimulate reconstruction, the French government announced it would release $5.4 million in aid and suspend all interest payments on loans held by the island's residents. Nearly all of Guadeloupe's prized coconut palms were ruined by Hugo's Category 4 passage, which severely affected the island's growing tourist trade. Finally, on September 4, 1995, the approaching winds of Hurricane Luis claimed the life of a French visitor who was swept off a pier and drowned unwisely trying to photograph Luis's dramatic seas.

Gulf of Mexico Situated between latitudes 18 and 30 degrees North, the 582,100-square-mile

($936,797$ m^2) basin of the Gulf of Mexico stands as the westernmost extension of the NORTH ATLANTIC OCEAN. Bordered to the south by MEXICO's Yucatán Peninsula, CUBA, and the CARIBBEAN SEA, to the west by Mexico and TEXAS, to the north by LOUISIANA, MISSISSIPPI, and ALABAMA, and to the east by FLORIDA, the Gulf of Mexico's position within the broad summertime parameters of the INTERTROPICAL CONVERGENCE ZONE (ITCZ) has for centuries made it one of the world's most active areas for TROPICAL CYCLONE development. Between 1528 when a powerful October HURRICANE destroyed a Spanish settlement at Florida's Apalachee Bay, and 2006, when Tropical Storm Alberto traversed its 85-degree F (29°C) waters, approximately 222 documented tropical cyclones either formed or reached maturity over the Gulf of Mexico for an average of two storms per year. While hurricanes and TROPICAL STORMS have been observed in the Gulf of Mexico during the off-season months of May and December, statistics reveal that peak periods for tropical cyclone activity in the region come in June and July and again in late-October and November or when the subequatorial TROUGH over the southwestern Caribbean Sea is at its most intense. Early season systems, which often generate over the wider expanses of the northern or western Caribbean Sea before achieving hurricane intensity over the warmer waters of the southern Gulf of Mexico, generally recurve north before coming ashore on the western Panhandle of Florida or along the coasts of northern Texas and western Louisiana. In August, September, and October, when the ITCZ flows fairly consistently over the southern Caribbean Sea, most tropical cyclone activity in the Gulf of Mexico results from those TROPICAL DEPRESSIONS and tropical storms that originate over the North Atlantic or Caribbean Sea before entering the Gulf through the Yucatán Channel or the Straits of Florida and then intensifying. Generally carried west-northwest by the increased strength of the trade winds, dozens of these systems—including the GREAT GALVESTON HURRICANE (1900), CARLA (1963), CELIA (1970), ALLEN (1980), ALICIA (1983), Katrina (2005) and Rita (2005) have made landfall on the Gulf's northeast coast and along the entire Texas coastline with deadly results. In other instances—such as September 1973's Tropical Storm Delia and October 1996's Tropical Storm Josephine—the system may be upgraded from TROPICAL WAVE to tropical depression to tropical storm within the gulf itself before coming ashore in Mississippi, Alabama, and

Florida as either a powerful tropical storm or a minimal hurricane. Other notable Gulf of Mexico hurricanes have first crossed land before entering the gulf; these include JANET (1955) in Mexico's Yucatán Peninsula and BETSY (1965) and ANDREW (1992) which moved across southern Florida. Due in large part to the above-average SEA-SURFACE TEMPERATURES normally found in the central gulf during the high summer months, the rapid deepening frequently noted in Gulf of Mexico hurricanes has long made them among the most dangerous examples of their kind. While June, July, and November tropical cyclones in the Gulf of Mexico do not typically prove as destructive as similar systems in August and September, several of them, including Hurricane AUDREY (1957), have been responsible for inflicting significant DEATH TOLLS on those low-lying land masses that ring the area. In August 1969, Hurricane CAMILLE underwent a spectacular period of intensification over the Gulf of Mexico, as did Hurricane CARMEN in September 1974. Following the passages of Hurricanes KATRINA and Rita in 2005, researchers concluded that the rapid deepening cycles experienced by both systems may have been attributable to the Gulf of Mexico's "Loop Current." Deploying sophisticated observational equipment including conductivity profilers, researchers determined that the Loop Current served to move warm waters around the Gulf, thereby keeping sea-surface temperatures high, and sparking the EXPLOSIVE DEEPENING cycles seen in Katrina and Rita.

Gulfstream IV-SP A type of aircraft used by the National Oceanic and Atmospheric Administration (NOAA) to conduct aerial research and reconnaissance missions on TROPICAL CYCLONES in the NORTH ATLANTIC OCEAN, CARIBBEAN SEA, and GULF OF MEXICO. Designed and constructed by the Gulfstream Aircraft Company and first introduced to NOAA's service in September 1996, the Gulfstream IV-SP's sophisticated ANEMOMETERS and RADIOMETERS—coupled with its twin jet-engined design, 4,692-mile range and 500 MPH air speed—allow the $33.7 million aircraft to survey a tropical cyclone's upper-level WINDS and STEERING CURRENTS comprehensively at an altitude of 40,000 feet. Presently based at MacDill Air Force Base in Tampa, FLORIDA, the Gulfstream IV-SP ably augments NOAA's existing fleet of *HERCULES WC-130* and *ORION WP-3* research aircraft in their ongoing mission of HURRICANE preparedness.

Hakata Bay Typhoon *Japan, August 15, 1281* See JAPAN.

Harriet, Typhoon *North Pacific Ocean–Japan–Korea, September 23–29, 1956* The first of two western NORTH PACIFIC OCEAN tropical systems to have been dubbed *Harriet,* this moderately powerful TYPHOON delivered 110-MPH (177-km/h) winds and torrential eight-inch PRECIPITATION counts to the island of Okinawa and to southern JAPAN on the night of September 26–27, 1956. A midseason rainmaker of considerable ferocity, Harriet's tropical downpours spawned flash floods and landslides that damaged or destroyed more than 600 BUILDINGS and claimed a total of 38 lives. Near the southern port city of Nagoya, 17 people were killed when the bus in which they were riding was swept into a rain-swollen river by a mudslide. Emerging into the Sea of Japan on the morning of September 27, a demoted TROPICAL STORM Harriet lingered over the confined waters. Briefly regaining strength during the daylight hours of September 28, Harriet was by the afternoon of September 29 again downgraded to a tropical storm. Severely weakened by upper level WIND SHEAR, Harriet finally went ashore near Pusan, KOREA, shortly before midnight on September 29. Fifteen women aboard a Korean fishing boat were drowned when the system's high seas capsized their craft. Total losses to buildings and crops in Okinawa, Japan, and Korea reportedly exceeded $50 million.

Six years later, on October 25, 1962, the second Typhoon Harriet skirted southern Japan before slamming with great violence into the southeast coast of Thailand. Fueled by a central pressure at landfall of 28.02 inches (948 mb), Harriet's 118-MPH (190-km/h) winds demolished thousands of buildings in three coastal provinces just north of the Malayan border, while pounding seas sank dozens of fishing boats. All told, the second Typhoon Harriet's incursion into Indo-China killed 769 people, left 142 missing, and seriously injured 252.

Hattie, Hurricane *Central America, October 25–31, 1961* Classified as a Category 4 HURRICANE because of its extreme winds, Hattie wrought enormous damage and hefty DEATH TOLLS on the Central American nation of BELIZE (then known as British Honduras) during the daylight hours of October 31, 1961. The eighth TROPICAL CYCLONE of an active season that had already seen one other Category 4 hurricane, CARLA, come and go in early September, Hattie's sustained 140-MPH (225-km/h) winds and ramifying 180-MPH (290-km/h) gusts claimed nearly 400 lives as its EYE swept ashore just west of the capital, Belize City, shortly after dawn. As the hurricane's scalping winds destroyed 40 percent of the BUILDINGS in Belize City, its crushing, 10–12-foot (3–4-m) STORM SURGE overran the shallow barrier islands that dot the country's heavily developed coastline. Although an honest effort had been made by officials to evacuate these islands completely, 275 people—many of them vacationers from inland communities—were drowned when Hattie's fast-moving storm surge entirely inundated both Turneffe and Caulker cays. In the fishing town of Stann Creek, west of Belize City, Hattie's fury burst through windows and doors, blasted the foliage from trees, and either destroyed or seriously damaged every last one of the town's densely stacked buildings. Aerial photographs of Stann Creek snapped after the hurricane's passage reveal a treacherous landscape of roofless boxes and toppled water towers. The grid-like purity of the town's narrow streets was marred by sudden hills of wreckage, collapsed structures that stranded the few automobiles and trucks seen parked, intact and upright, beneath crumpled car parks. Along the puddled roads, pristine picket fences pierce the uprooted foliage that has fallen around them, providing an ironic counterpoint to the severely dented kitchen appliances scattered nearby. Indeed, the damage in Stann Creek would eventually prove so cataclysmic that the town's residents renamed the new, rebuilt town in honor of its founding impetus, Hattieville. The name Hattie has been retired from the rotating list of North Atlantic tropical cyclone names.

Havana-Bermuda Hurricane *Cuba–Bahamas–Bermuda, October 14–29, 1926* An incipient Category 4 HURRICANE of extreme destructiveness, the Havana-Bermuda Hurricane delivered sustained 130-MPH (209-km/h) winds, torrential rains, and pounding seas to western CUBA, the BAHAMAS, and the entire island of BERMUDA between October 20 and 22, 1926. One of the most severe late-season hurricanes ever observed in the CARIBBEAN SEA, the Havana-Bermuda Hurricane claimed 650 lives in Cuba and another 88 in Bermuda. Ten thousand people, most of them in Cuba, were rendered homeless, while losses to property and crops exceeded $100 million, an enormous sum of money for the time.

Initially formed over the sun-charged waters of the southwestern Caribbean Sea on the afternoon of October 14, the Havana-Bermuda Hurricane rapidly deepened as it slowly recurved northeast, maturing into a Category 3 hurricane of extreme intensity as it brushed past Swan Island shortly before midnight on October 18. The first major hurricane to strike Cuba in the 20th century, the Havana-Bermuda system screamed ashore on the Isle of Youth's northern tip some five miles west of the principal trade depot of Nueva Gerona, just after daybreak on October 20. After devastating much of Nueva Gerona (where, it was said, the hurricane's winds were strong enough to drive a pine plank through a palm tree), the hurricane pushed inland at Batabano, snatching entire forests from the rain-soaked ground as it veered for Havana.

Coiled around a BAROMETRIC PRESSURE of 28.07 inches (950 mb), the hurricane's compact EYEWALL produced sustained 120-MPH (193-km/h) winds, 138-MPH (202-km/h) gusts, and 20 inches (508 mm) of rain over the capital. Destruction on a scale rarely seen even in storm-prone Cuba resulted as windows shattered, roofs took flight, and 25-foot waves easily topped the seawalls of Havana harbor. Cuban army headquarters, Camp Columbia, was virtually obliterated within a matter of hours, its legions of brick

barracks scattered across nearby parade fields like grapeshot. The ANEMOMETER at the Belen Meteorological Observatory, furiously spinning out a reading of 110 MPH (177 km/h), was suddenly carried away when the wind toppled the observatory's instrument tower. Four steel-hulled vessels in Havana harbor—one of them a Cuban warship—were either run ashore or sunk, casting their terrified crews into the stinging surf. Scores of smaller craft, from wooden schooners to an admiral's launch, were swamped by the sawing waves or splintered against the submerged embankments. One unique structural victim was the stately, eagle-festooned MONUMENT dedicated to the memory of the 266 U.S. service members who perished when the battleship *Maine* inexplicably exploded in Havana harbor one February evening in 1898. Its flanking marble columns snapped in two by the onrushing surf, the monument would in later months be restored. Elsewhere in Cuba, the 1926 tempest transformed villages from Batabano in the south, to Matanzas in the north, to Marianao in the west, into mud-swollen wastelands, uninhabitable mires of debris beneath which hundreds of victims lay trapped.

Recurving northeast and exiting Cuba during the late-evening hours of October 20, the Havana-Bermuda Hurricane slid over the northern Bahamas on the morning of October 21. Slightly weakened by its passage across western Cuba, the hurricane's maximum sustained winds topped 100 MPH (161 km/h) as it brought 15-foot (5-m) seas and several inches of rain to Great Abaco and Grand Bahama islands. In southern FLORIDA, where residents were still recovering from a direct strike by the severe Miami Hurricane of September 18, 1926, wind speeds ranged from 60 to 70 MPH (97–113 km/h), and more than two inches (51 mm) of rain fell on the city; however, there were no deaths, and damage to trees and buildings was light.

Maintaining its northeast swing, the Havana-Bermuda Hurricane swiftly gained forward momentum as it pulled away from Florida and moved ever deeper into the North Atlantic. Churning toward Bermuda at speeds in excess of 41 MPH (66 km/h), the Category 3 (a minimum pressure of 28.45 inches was noted at landfall) hurricane's arrival over that island's western shores had been accurately forecast by meteorologists but in an insufficient amount of time to save two British light cruisers at anchor in Hamilton Harbor—the *Calcutta* and the *Valerian*—from being wrenched from their moorings by the system's 136-MPH (219-km/h) winds and roaring waves. Driven across the anchorage by the hurricane's 145-MPH (233-km/h) gusts, the *Calcutta* collided with a stone breakwater but remained afloat, while the *Valerian* drifted into the roiling confines of a nearby channel and was swiftly capsized. Eighty-eight officers and sailors aboard the cruiser were drowned. Although the hurricane was a rainmaker almost from birth, the rapidity of its passage over Bermuda, while adding the strength of its STEERING CURRENTS to the system's sustained wind speeds, denied it the opportunity to linger over the island, thereby significantly reducing drowning deaths and damage from flash floods.

Seemingly committed to its northeasterly trajectory on the morning of October 23, the Havana-Bermuda Hurricane had, by late afternoon, abruptly slowed its forward speed to 22 MPH (35 km/h). Centered some 1,400 miles (2,253 km) east of Cape Hatteras, NORTH CAROLINA, and quickly diminishing in strength, the hurricane penetrated as far north as the 40th parallel before sharply altering its course *southeast* shortly after dawn on October 24. By the evening of October 27, after having described an enormous loop over the mid-Atlantic, the EYE of the Havana-Bermuda Hurricane was again located off Florida's east coast, this time nearly 2,000 miles

(3,219 km) from Miami. Turning northeast, the system passed within 700 miles of Bermuda on October 28, crossed its previous track at nearly a right angle, then shot northward at speeds of almost 30 MPH (48 km/h). Situated off Newfoundland, CANADA's northeast coast on the evening of October 29, the rain-riven remains of the Havana-Bermuda Hurricane finally dissolved over the cold reaches of the NORTH ATLANTIC OCEAN on October 30.

Hawaii *North Pacific Ocean* Although Hawaii's 130 volcanic islands span some 1,500 stormy miles (2,414 km) of the eastern NORTH PACIFIC OCEAN, the verdant chain of coral lagoons and crystalline white beaches does not have a particularly active history of HURRICANE strikes. Between 1837 and 2006, Hawaii's eight main islands—Hawaii, Maui, Kahoolawe, Lanai, Molokai, Oahu, Kauai, and Niihau—were directly affected by at least 20 recorded hurricanes and TROPICAL STORMS for an average

As Hurricane Andrew was preparing for its devastating 1992 visit to Florida, a slightly less powerful Hurricane Iniki was landing in the Hawaiian Islands. Causing destruction across the islands with its 135 MPH winds, Iniki killed six people and ran up a tab totaling more than 1 billion dollars. *(NOAA)*

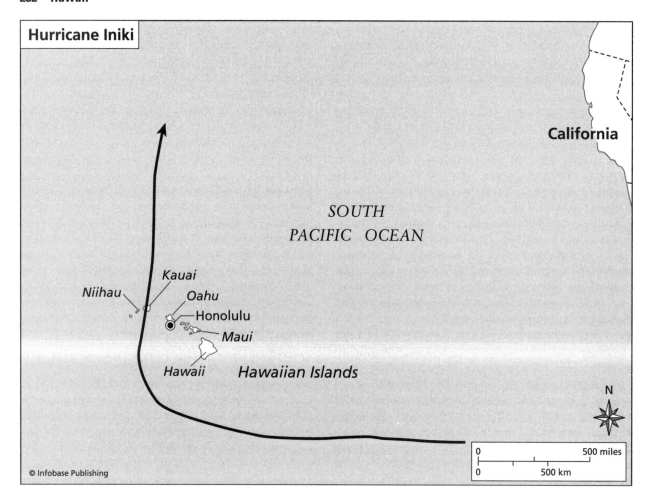

Hurricane Iniki

California

*SOUTH
PACIFIC OCEAN*

Kauai
Niihau
Oahu
Honolulu
Maui
Hawaii   Hawaiian Islands

N

0   500 miles
0   500 km

© Infobase Publishing

One of the most intense hurricanes in Hawaiian history, Iniki caused widespread property losses when it blasted ashore in 1992.

of one TROPICAL CYCLONE every eight years. Of these, only six were considered major hurricanes, storms with CENTRAL BAROMETRIC PRESSURES of 28.47 inches (964 mb) and lower, and wind speeds in excess of 111 MPH (179 km/h). In 1959 and 1960, Hawaii sustained destructive strikes from two tropical storms, Dot (August 8, 1959) and Hiki (August 12–16, 1960). The first recorded tropical cyclones to strike the archipelago in consecutive years, both storms delivered staggering 52-inch (1,320-mm) rainfalls to the island of Kauai, causing serious flash floods along Waimea River banks. One person was killed during Tropical Storm Hiki, while Dot's 60-MPH (97-km/h) winds and deluging rains destroyed much of the island's rich macadamia nut harvest.

In one respect, Hawaii's dearth of hurricane activity is rather surprising. Positioned between latitudes 19 and 29 degrees North, the scythelike spread of the Hawaiian Islands shares roughly the same parallels as storm-ridden regions of the globe, including the Caribbean basin, the Arabian Sea and BAY OF BENGAL, and those western Pacific waters north of the PHILIPPINES known as TYPHOON ALLEY. Surrounded by warm, 80°F (27°C) oceans that spawn on average

seven tropical cyclones per June-to-mid-November season, it would stand to reason that Hawaii would have as extensive a history of hurricane activity as FLORIDA, CUBA, or BANGLADESH.

But the islands do not, again partly because of their location near the center of the North Pacific Ocean. During the months (August through November) of peak hurricane generation in the eastern North Pacific, large high-pressure ANTICYCLONES develop near latitude 30 degrees North or in those waters adjacent to the Hawaiian Islands. Because anticyclones like the BERMUDA HIGH are known to influence the tracks of respective hurricanes, the dominating presence of similar anticyclones north of Hawaii has a tendency to deflect incoming North Pacific hurricanes southwest, thus sparing the popular resort islands the fury of storms that have historically been of tremendous intensity. Born in the east over the undulating feeding grounds off northern MEXICO's northern coast, most North Pacific hurricanes follow a westward course, passing several hundred miles south of Hawaii before blowing themselves out over the central Pacific. While in some cases, such as that of 1994's Hurricane JOHN, the

Hawaiian Islands have been affected by the light rains, gusty winds, and heavy surf associated with a passing hurricane's fringe, a majority of North Pacific storms leave Hawaii's fragile ecosystem unscathed.

No climatological shield, however, is entirely consistent, and it is during those early summer and out-of-season months when the Hawaiian anticyclones are at their weakest—or absent altogether—that Hawaii has endured some of its most memorable hurricane strikes. On November 7, 1837, a fierce, late-season hurricane roared into Hilo on the island of Hawaii's southeast coast, toting 114-MPH (184-km/h) winds and an enormous STORM SURGE. Hundreds of BUILDINGS in the foundling city were destroyed, leaving an estimated 730 people either dead or missing. During the first week of April 1868, a preseason hurricane at Molokai swept away the coastal whaling villages of Keauhou and Punalu. As the hurricane's furious winds scalped a nearby mission's thatched roofs, four whaling vessels that had called at Keauhou for fresh provisions were driven aground and wrecked. Hundreds of casks of whale oil were either washed ashore intact (where they were quickly looted) or smashed against the coral entrances to the harbor. Nearly 100 people died as the hurricane's storm surge—estimated by survivors to have been some 16 feet (5 m) high—completely inundated the two communities, spreading a thick pall of whale oil and wreckage for miles across the tropical countryside.

More recently, postseason Hurricane Iwa brought 110-MPH (177-km/h) winds and torrential rains to the islands of Kauai and Oahu on November 24, 1982. Although Iwa's actual wind speeds were in the vicinity of 80 MPH (129 km/h), the hurricane's 30-MPH (48-km/h) STEERING CURRENT boosted windspeeds in the storm's DANGEROUS SEMICIRCLE to more than 100 MPH (161 km/h). One person was killed and $200 million in property damage was assessed in what was considered the most severe hurricane to have struck Hawaii in 23 years.

Nearly a decade later, on September 11, 1992, the Aloha State was visited by its first "billion-dollar" gate-crasher, Hurricane INIKI. Raging across Kavai's western slopes, Iniki's sustained 130-MPH (210-km/h) winds damaged more than 20,000 buildings and left 8,000 residents homeless. Five people on the island were killed and some $1.8 billion in damage was assessed in what was then dubbed the most severe hurricane to have struck Hawaii in the 20th century.

Hawaii is the only state in the United States to be honored with its own list of HURRICANE NAMES. Officially deemed the "Central Pacific List," its six columns of eight names apiece nonetheless feature only Hawaiian male and female first names.

On September 2, 2003, Tropical Storm Jimena sped away from Hawaii at about 18 MPH (29 km/h); at this time, the system was located about 390 miles (624 km) south-southwest of Honolulu. Sustained winds were clocked at 45 MPH (72 km/h). While the system did not directly affect the Hawaiian islands, its trajectory indicates the dangers associated with tropical cyclone activity in the central North Pacific Ocean.

In August 2003, Hurricane Jimena originated as a tropical depression on August 28, and achieved hurricane intensity during the late evening hours of August 29. Generating sustained winds of 105 MPH (169 km/h), prompting hurricane watches to be posted for the Big Island of Hawaii. If Jimena were to strike Hawaii, it would be the island's first direct strike since Hurricane Iniki devastated the idyllic chain in September 1992. Although Jimena weakened into a tropical storm as it passed south of the Big Island of Hawaii on September 1, maximum sustained winds on September 1 were measured at 39 MPH (63 km/h)—with gusts to 53 MPH (85 km/h)—at South Point. As many as 1,500 people lost electricity on the Big Island and rainfall amounts ranged from 3.74 inches (95 mm) at the Hilo airport to 6.42 inches (163 mm) in the Mountain View area.

During the last week of September 2005, the tropical depression remnants of eastern North Pacific Ocean Hurricane Kenneth (at one time a Category 4 tropical cyclone) delivered torrential rains to Oahu and Kauai islands, causing flash floods and streams to top their banks.

On August 14, 2007, a weakening Hurricane Flossie brought sustained winds of 45 MPH (72 km/h) and heavy surf conditions to the big island of Hawaii. At one time a Category 4 hurricane with a lowest observed central pressure of 27.95 inches (946 mb), Flossie's approach prompted the hoisting of hurricane watches and tropical storm warnings across the southern shores of the island chain, as well as the closing of the port of Hilo. Pushed southward by a high-pressure ridge to the north, Flossie moved past the big island and steadily weakened, finally becoming a tropical depression during the late evening hours of August 15. No deaths, injuries, or property damage were recorded.

Hazel, Hurricane *Northeastern Caribbean–United States–Canada, October 5–16, 1954* One of the most eccentric—yet destructive—Category 4 hurricanes on record, Hazel spent nearly two weeks weaving an erratic path of devastation through the island of HISPANIOLA, the southern BAHAMAS, the eastern seaboard of the United States, and portions of southern CANADA between October 5 and 16, 1954. In western Haiti, where Hazel came ashore as a TROPICAL STORM on October 12, fearsome 100-MPH (161-km/h) gusts killed at least 98 people and left another 100,000 homeless. In both NORTH and SOUTH CAROLINA, where Hazel's loosely coiled EYE made landfall near the border on October 15, 130-MPH (209-km/h) winds destroyed hundreds of buildings, including a number of posh surfside cottages in Myrtle Beach and Morehead City. Nineteen lives were lost between the two states. Maintaining hurricane intensity, Hazel then progressed inland, bringing shivering winds and rains to VIRGINIA, Delaware, Maryland, Pennsylvania, and NEW YORK before crossing the border into Canada. Judged to have been one of the most severe hurricanes to have ever struck central Canada. Hazel subsequently lashed Toronto with sustained 76-MPH (122-km/h) winds and eight inches of rain. Before recurving back into the Atlantic to eventually dissipate off Norway's coast, Hazel would tally 347 dead and several hundred million dollars worth of property losses in three countries.

The eighth TROPICAL CYCLONE of the altogether remarkable 1954 HURRICANE SEASON, Hazel originated in the southeastern CARIBBEAN SEA, some 100 miles east of Grenada, during the early morning hours of October 5. Having just witnessed the deadly passage of two Category 3 hurricanes—CAROL, in August and EDNA in September—up the U.S. east coast, authorities in FLORIDA wasted little time in dispatching reconnaissance aircraft to investigate the embryonic system's behavior. What they found partly surprised them. A fairly late-season system, the fact that Hazel spent the first four days of life as a tropical storm was considered not that unusual. Although the 83-degree waters of the southeastern Caribbean were certainly warm enough to sustain rapid intensification, Hazel was being driven on an almost straight-line course west, a direction that, if maintained, would eventually take the tropical storm ashore in Central America, near the Honduran-Nicaraguan border. Although a number of previous hurricanes and tropical storms had followed a similar due-west course, Hazel's lumbering head speed of five MPH made it one of the slowest cyclonic systems ever observed. Some meteorologists believed that it could very well dissipate before it even reached land, its growth and progression was that haphazard.

But this all began to change on the evening of October 9 when Hazel's CENTRAL BAROMETRIC PRESSURE deepened slightly and the storm gradually began a sharp turn northwest. Now gusting at 29.23 inches (990 mb) of mercury, the tropical storm's wind speeds hovered just below hurricane force, seemingly defying the meteorological odds to intensify further. In addition, its forward speed had fallen to a pedestrian three MPH, almost stalling the system 300 miles (483 km) south of Haiti. Maintaining a central pressure of 29.23 inches (990 mb) throughout the day and night of October 10, Hazel reluctantly swung around, first north

by midafternoon, then northeast by the morning of October 11. As often happens with tropical cyclones, Hazel further intensified during this protracted period of RECURVATURE, with a barometric pressure reading of 29.11 inches (986 mb) taken during a midday reconnaissance flight that same day. It was further noted that the tropical storm's forward speed had increased to a respectable seven MPH (11 km/h).

Now positioned to continue moving northeast toward the Dominican Republic's rocky southwestern shores, Hazel again surprised its observers by altering its course during the early evening hours of October 11. Quickly curving back northwest, the tropical storm's central pressure slightly weakened to 29.17 inches (988 mb) as it inexorably bore down on Haiti's southern capes, eventually coming ashore over Cayes and Île à Vache shortly after midnight on October 12. Although upon landfall Hazel was still a fairly mild system, with sustained winds of less than hurricane intensity, its 100-MPH (161-km/h) gusts brought strenuous rains to the Haitian port towns of Dame Marie and Jérémie. There, as Hazel's eight-inch PRECIPITATION counts turned unpaved roads into muddy riverbeds, hundreds of lightly constructed buildings twisted and swayed beneath the groaning, two-minute gusts. From Dame Marie's civic hall to the tar-paper barrios that stepped the nearby foothills, thousands of refugees from the storm huddled in terror as entire blocks of BUILDINGS began to crash down around them. Before Hazel would pass into the Windward Passage on the morning of October 12, it collapsed 16,000 residences in Haiti, leaving some 100,000 people homeless and another 98 dead.

Seemingly unfazed by its strident assault on Haiti, Hazel lazily intensified to 29.24 inches (987 mb) as it once again changed direction on the afternoon of October 13. While busily showering the southern BAHAMAS with hurricane-force winds and torrential rains, Hazel abruptly began to curl northwest, eventually settling on a course that would bring it ashore near Savannah, GEORGIA. By midnight on October 14, however, as civil defense authorities in northern Florida and southern Georgia swiftly mobilized their evacuation units, Hazel once more swung east, setting its sights on the beachfront border between the North and South Carolina capes. The steady routine of reconnaissance flights sent into the storm's EYE now indicated that Hazel was not only traveling north-northwest at nearly 16 MPH (26 km/h) but that it had further intensified as well. Boasting a central pressure of 28.82 inches (976 mb) and falling, Hazel began a period of rapid ANTICYCLONIC strengthening that would permit it to come ashore less than 12 hours later as a hurricane of significant intensity.

The once-reluctant Hurricane Hazel made a shattering landfall at North Myrtle Beach, South Carolina, shortly after dawn on October 15, 1954. With a central barometric pressure of 27.70 inches (938 mb) and 106-MPH (170-km/h) winds, Hazel was no longer the meek tropical storm that less than a week before could barely chase shipping in the southeastern Caribbean. Quickly carried inland by a 60 MPH (97 km/h) STEERING CURRENT, the hurricane brought with it a 16.9-foot (5.5-m) STORM TIDE that caused extensive coastal erosion at Folly Beach and Pawleys Island. In nearby Garden City, 272 cottages were swept off their foundations as the sea rose nearly 14 feet, casting a clockwork series of waves ashore. In the Long Beach vicinity another 352 buildings were utterly ruined, while several small pleasure craft were left strewn across the beach. Nearly 80 percent of the exclusive oceanfront property in Myrtle Beach was either damaged or destroyed, with similar losses being recorded as far north as Cape Lookout, North Carolina. By the time Hazel had crossed into southern Virginia, 19 people were dead and $136 million in property damage had been wrought between the two Carolinas.

Because a majority of tropical cyclones weaken upon landfall, those meteorologists who had been tasked with tracking Hazel's path up the eastern seaboard naturally assumed that the hurricane would very quickly begin to dissipate into an EXTRATROPICAL system of gale-force gusts and rain-laden thunderstorms shortly after reaching Virginia's southern border. Instead, Hazel surprisingly intensified one last time as the effects of its 45-MPH (72-km/h) steering current were figured into the swirling equation. Touting 125-MPH (201-km/h) winds and five-inch rains, Hazel crashed past Richmond, Virginia, during the early evening of October 15 causing widespread power outages and extensive tree damage. Roaring into Pennsylvania just after 9 P.M. on October 15, Hazel's ripe winds shivered the fruit orchards that blanketed central Pennsylvania, sending nearly 10 million bushels of apples bouncing from their trees. In west New York, hundreds of small buildings, from open-air tobacco drying sheds to multistory cattle barns, were blasted into costly heaps of wreckage. Thousands of spindly pine trees were uprooted as Hazel effortlessly sailed across the snagging peaks of the Allegheny Mountains, thus defying all predictions that the hurricane would perish before reaching the Great Lakes' watery feeding grounds.

This would ultimately prove a tragic development, for on reaching Lake Ontario on the morning of October 16, Hazel's warm tropical core merged with an eastward-moving cold front, unleashing a fresh barrage of violent thunderstorms. In neighboring Toronto, hurricane warnings had no sooner been lifted than Hazel's 100-MPH (161-km/h) gusts descended on the startled city, bringing with them flooding rains and pinpoint lightning strikes. Encouraged by a central pressure of 28.96 inches (981 mb), the Humber River quickly overran its embankments, stranding hundreds of frantic survivors on rooftops. In the lakeside town of Kitchner, rising waters rivaled those found in the worst of southern hurricanes as numerous road and railway bridges were carried away. Dozens of houses were swiftly washed from the town's banks and shores, many with their horrified inhabitants still inside. In the end, Hazel would be responsible for the deaths of 69 people in the Toronto area alone, making it the most deadly hurricane to have struck southern Canada in recorded history. It would also prove the most costly, with some $100 million in property losses being charged to insurers.

By the afternoon of October 17, as the remnants of Hurricane Hazel drifted across the NORTH ATLANTIC OCEAN to their rainy demise off the Scandinavian Peninsula, the arduous mission of counting the dead and repairing the damage began. In the United States, then President Dwight D. Eisenhower declared North and South Carolina disaster areas, thereby making them eligible for federal relief funds. In Canada, military forces commenced a full-scale rescue effort, while the government in Ottawa appropriated several million dollars to cover those uninsured losses suffered by small businesses.

Hercules WC-130 This four-engined, turboprop aircraft was designed and built by the Lockheed Aircraft Corporation and was modified by the National Oceanic and Atmospheric Administration (NOAA) for use in TROPICAL CYCLONE research and reconnaissance missions over the NORTH ATLANTIC and NORTH PACIFIC OCEANS, CARIBBEAN SEA, and GULF OF MEXICO during the early 1980s to the present. Originally designed as a military cargo aircraft and crewed by personnel from the U.S. Air Force Reserve, the *Hercules WC-130* flies into a tropical cyclone at altitudes of between 5,000 and 10,000 feet and is equipped with the most sophisticated ANEMOMETERS, BAROMETERS, RADIOMETERS, and DOPPLER RADAR currently available. Stationed at Keesler Air Force Base in Biloxi, MISSISSIPPI, the cavernous *Hercules WC-130* (of which NOAA presently operates a fleet of 10) carries a crew of six pilots, meteorologists, and technicians and may remain aloft for up to 14 hours. Unable to fly at altitudes of greater than 10,000 feet because of its turboprop design, the *Hercules WC-130*'s upper-level reconnaissance missions were curtailed in late 1996 following the introduction of NOAA's GULFSTREAM IV-SP hurricane hunting jet.

Hilda, Hurricane *Gulf of Mexico–Southern United States, September 28–October 5, 1964* The second of two North Atlantic TROPICAL CYCLONES to have been dubbed Hilda, this powerful Category 3 HURRICANE delivered sustained 110-MPH (193-km/h) winds, six-inch (152.4-mm) PRECIPITATION counts, and a minimal two-foot STORM SURGE to the south coast of LOUISIANA on the night of October 3–4, 1964. Originating over the northern CARIBBEAN SEA some 120 miles (193 km) south of central CUBA on the morning of September 28, Hilda pursued a course to the west-northwest. Upgraded to a TROPICAL STORM on the morning of September 29, Hilda crossed the southwestern tip of Cuba and entered the GULF OF MEXICO. Although no deaths or serious injuries were reported in the wake of Hilda's passage through Cuba, $100,000 in property losses were sustained. Centered some 500 miles south of New Orleans on the morning of September 30, Hilda was given hurricane status. Undergoing a sharp course change to the north on the evening of October 1, Hilda's CENTRAL BAROMETRIC PRESSURE rapidly began to fall, reaching its lowest recorded point—27.79 inches (914 mb)—on the afternoon of September 2. Boasting sustained winds within its EYEWALL of 150 MPH (241 km/h), Hilda spent the night of September 2 quickly gaining on Louisiana's gulf coast. As has been noted in other Gulf of Mexico hurricanes, Hilda progressively weakened as it approached the shore, considerably reducing its damage potential. By the time the system's core finally made landfall slightly east of Marsh Island during the early evening hours of October 3, its central pressure had risen to 28.05 inches (950 mb) and its maximum WIND SPEEDS had decreased to 110 MPH (193 km/h). In the town of Larose, Louisiana, a large TORNADO spawned by one of Hilda's RAIN BANDS touched down on the afternoon of October 3, killing 18 people and injuring 115. Elsewhere in Louisiana, 12 people perished in building collapses, drowning, and electrocutions, while widespread crop losses were suffered by the state's sugar industry. On October 4, President Lyndon B. Johnson declared much of south Louisiana a federal disaster area and pledged $1 million in immediate emergency aid. All told, property losses in Louisiana topped $125 million, making it one of the most expensive hurricanes on record.

Nine years earlier, on September 19, 1955, the first Hurricane Hilda skirted Louisiana before coming ashore with considerable fury on the eastern gulf coast of MEXICO. Powered by a central pressure at landfall of 28.20 inches (955 mb), Hilda's 108-MPH (174-km/h) winds demolished hundreds of BUILDINGS in the port city of Tampico, while surging seas sank scores of small boats. All told, the first Hurricane Hilda's incursion into Mexico killed 166 people and left 100 missing and another 1,000 injured. The name Hilda has been retired from the repeating list of North Atlantic HURRICANE NAMES.

Himawari A series of three geosynchronous meteorological satellites placed in orbit by JAPAN between 1977 and 1984 and used to monitor TROPICAL CYCLONE activity in the western NORTH PACIFIC OCEAN. Constructed in the United States for a joint venture between the Nippon Electric Company and the Japanese National Space Development Agency (NSDA), the first of these sophisticated Geostationary Meteorological Satellites (GMS)—dubbed Himawari ("Sunflower")—was launched from Cape Canaveral, FLORIDA, on July 14, 1977. Weighing 670 pounds and standing nearly 9 feet high, *Himawari-1* was stationed at longitude 140 degrees East, where its spin-scan RADIOMETER relayed visible and infrared images of TYPHOONS, TROPICAL STORMS, and TROPICAL DEPRESSIONS to Earth at 30-minute intervals. A durable meteorological guardian, *Himawari-1* remained in service until June 30, 1989. On August 10, 1981, the second of the GMS satellites—*Himawari-2*—was fired into geostationary orbit from a launch site on the island of Tanegashima, some 450 miles southeast of Tokyo. Maneuvered into position at longitude 140 degrees East, the slightly larger *Himawari-2* transmitted nonstop TEMPERATURE and PRECIPITATION observations to receiving stations in Japan and at the JOINT TYPHOON WARNING CENTER (JTWC) on Guam until November of 1987, when its visible and infrared radiometers finally went dark. At the time of its failure *Himawari-2*'s primary tracking duties had already been assumed more than 140 degrees East by the heftier *Himawari-3* satellite, which had lifted off from the Tanegashima range on August 3, 1984. During their respective watch periods, the three *Himawari* satellites produced thousands of images of notable tropical cyclones in the western North Pacific Ocean, including Typhoons Rita (1978), Tip (1979), IKE (1984), Irma (1985) and Nina (1987).

Hispaniola *Northern Caribbean Sea* Shared by the nations of Haiti and the Dominican Republic, the 29,283-square-mile (47,126 sq km) island of Hispaniola has over the course of the past five centuries endured an extensive, often catastrophic history of TROPICAL CYCLONE activity. Situated at the CARIBBEAN SEA's northern crest and armored with a plume of 10,000-foot (3,333-m) mountain ranges, Hispaniola is the second largest of the three islands—CUBA and PUERTO RICO being the other two—that comprise the Greater Antilles group. The chain's position between latitudes 16 and 23 degrees North places Hispaniola squarely in the path of both the early season TROPICAL STORMS that sweep out of the southwestern Caribbean Sea and the mid-to-late-season HURRICANES that follow the trade winds across the NORTH ATLANTIC OCEAN from Africa's Cape Verde

Islands. The most mountainous of all the Caribbean islands, Hispaniola was directly affected by at least 97 documented tropical cyclones between 1494 and 2006. Of these, 41 were considered major hurricanes, systems with minimum BAROMETRIC PRESSURES of 28.47 inches (964 mb) or lower, sustained WIND SPEEDS in excess of 111 MPH (179 km/h), and torrential six to 12-inch (152–305-mm) rainfall counts. On at least seven different occasions between 1545 and 1928, Hispaniola suffered multiple hurricane strikes in rapid succession, as witnessed in 1545 when two destructive storms tramped ashore from August 20 to September 18; in 1851 when two hurricanes scudded across the island on August 31 and September 27; in 1878 when two mature-stage tropical systems made landfall between September 11 and October 6; and in 1901 when another two tempests battered its coastal settlements and towering inland mountain ranges between July 7 and September 12. Among the most violent of tropical cyclones in Hispaniola's history, the three mid-to-late-season hurricanes of 1751 (September 15, September 21–22 and October 8) progressively leveled the thriving port city of Santo Domingo, the present capital of the Dominican Republic, and sank dozens of warships and merchant vessels along the island's south and north coasts. Scores of lives were reportedly lost to the storms' flooding rains and swamping seas.

For centuries, Hispaniola's climatological relationship with North Atlantic tropical cyclones has often proved as paradoxical as the nature of the storms themselves. Although it is the one island in the Caribbean that even the largest hurricanes find difficult to cross and still remain intact, a statistical majority of Hispaniola's most intense tropical cyclones have nonetheless bisected the island from southeast to northwest, or parallel to its longest axis. Dominated by the forested slopes of its Cordillera Central mountain range, Hispaniola has on several occasions severely weakened intense landfalling hurricanes by shearing apart their spiral RAIN BANDS, thereby moderating much of their destructiveness before they whirled into nearby Cuba, JAMAICA, and southern FLORIDA. In September 1966, for instance, Hurricane INEZ traversed Hispaniola from east to west, its central pressure rising from 27.37 inches (927 mb) at landfall in the Dominican Republic to 28.55 inches (967 mb) upon exiting Haiti. When Hurricane DAVID grated across the eastern half of Hispaniola in September 1979, its sustained wind speeds fell from 142 MPH (229 km/h) to 85 MPH (137 km/h). In September 1987, when the first Hurricane Emily made landfall near the Barahona Peninsula, its central pressure stood at 28.29 inches (958 mb); by the time it departed the northern Dominican

Santo Domingo, the capital city of the Dominican Republic, lies in ruins following the passage of the Great Santo Domingo Hurricane of 1930. An exceptionally powerful Category 4 hurricane with sustained winds of 150 MPH at landfall, the storm killed scores of people. *(NOAA)*

Republic one day later bound for eventual landfall in distant BERMUDA, its minimum pressure had risen to 29.17 inches (987 mb) and it was downgraded to a tropical storm.

At the same time, this shearing action results in less CONVECTION within the tropical system's EYEWALL, which in turn yields tremendous quantities of rain that gush down Hispaniola's steep mountainsides, sacrificing the existence of entire villages in a matter of minutes. Not surprisingly, a survey of tropical cyclone strikes in Hispaniola reveals one of the highest DEATH TOLLS in the Caribbean, with some 23,000 of the island's residents perishing to hurricanes and tropical storms between 1930 and 2006.

In November 1994, torrential rains generated by then-Tropical Storm GORDON caused mudslides that claimed 1,122 lives in southwest Haiti and destroyed hundreds of lightly constructed BUILDINGS.

In 2004, nearly 3,000 people perished in Haiti following Tropical Storm Jeanne's passage over Hispaniola. The northern Haitian city of Gonaives (where a history of deforestation had left the soil susceptible to mudslides) suffered the worst of the casualties. All told, some 250,000 people in Haiti were left homeless.

In early July 2005, Hurricane DENNIS claimed 25 lives in Haiti, while historic Hurricane WILMA claimed an additional 12 lives in October of the same year. Tropical Storm ALPHA, the 25th named storm of the 2005 season, killed at least eight people, and left another 23 missing in the impoverished Caribbean nation. In the neighboring Dominican Republic, another three people lost their lives to flashfloods spawned by Alpha's passage over Hispaniola. When Alpha came ashore with 50-MPH (80-km/h) winds near the southern Dominican town of Barahona, it trekked to the north-northwest, destroying more than 400 buildings in Haiti. The town of Leogane, west of the capital, was particularly hard hit, with some 19 people swept away by Alpha's floodwaters.

Hispaniola's cultural connection to tropical cyclones is equally ironic in its varied dimensions and manifestations. Although Hispaniola was the last of the West Indian islands visited by Christopher Columbus on his first voyage to the New World (1492–93), it was off its southeast coast that Columbus, on his second voyage (1493–96) to the Caribbean, observed what is generally conceded to have been the beginning of documented tropical cyclone activity in the Western Hemisphere—a fairly intense early season tropical system that overtook his fleet of 17 ships sometime in mid-September of 1494 while they were exploring the Mona Passage between Hispaniola and Puerto Rico. As recounted by Samuel Eliot Morison in his *Admiral of the Ocean Sea*, Columbus was alerted to the storm's approach by the sighting of a portent—a grotesque creature swimming across the surface of the sea. Quickly shepherding his ships into a rocky cove near Saona Island, Columbus and his expedition were able to drop anchor before the system of indeterminate intensity and nature passed its high winds and rains over them.

Because a majority of early season tropical systems form over the western Caribbean Sea, there is no record of Hispaniola being struck by a tropical storm or hurricane during the month of June. In July, however, the probability of a hurricane landfall in Hispaniola sharply increases, with nine reported strikes between July 1–2, 1502, and July 24, 1926. Believed to have been the worst July hurricane in Hispaniola's history, the tempest on July 6, 1751, uprooted most of the island's flowering trees, collapsed scores of single-story houses, and drove a half-dozen French merchant ships ashore at Port-au-Prince. On July 28, 1837, a weakened system that had days earlier caused staggering destruction on BARBADOS island diagonally sliced across Hispaniola, spawning flash floods in inland valleys and claiming many lives.

It is in August when the CAPE VERDE STORMS first begin their RECURVATURE into the eastern Caribbean that the number of tropical cyclone strikes in Hispaniola escalates to 22 documented storms between 1508 and 1988. Many, such as the GREAT CARIBBEAN HURRICANE of August 12, 1831, and the so-called Antigua-Texas Hurricane of August 13, 1835, were systems of considerable size and intensity. Others, such as the St. Kitts Hurricane of August 19, 1827, were large tropical systems that brought gale-force winds and heavy rains to Hispaniola but whose eyewalls remained well offshore. Still others have

only been of tropical-storm strength as they drifted across the island, as in the example of Tropical Storm Chris which on August 25, 1988, moved northwest across northern Hispaniola, dropping tons of rain but causing no deaths.

Hispaniola's first recorded August hurricane occurred on the third of the month, 1508. By all accounts a fantastic whirlwind, the 1508 system wrecked 20 Spanish merchant ships in the harbor at Santo Domingo and took dozens of lives. Not a half-century later, on August 20, 1545, an equally strident tropical system passed over Hispaniola, mowing down acres of palm groves, demolishing houses, and destroying 12 Spanish trading vessels that had desperately sought safe anchorage on the island's east coast. On nearly the same date in 1552, another fearsome hurricane scudded across the eastern half of the island, causing severe structural damage to most of the principal buildings in Santo Domingo, as well as sinking four vessels in the harbor and drowning 130 sailors. Another August hurricane, this one on the 21st of the month, swept over the island in 1553, snapping trees at ground level, peeling the tiled roofs from houses, and throwing 15-foot seas ashore from Santo Domingo in the east to Cap-Haï-tien in the north. Sixteen Spanish merchant ships bound for Seville were lost in Santo Domingo harbor, while another four smaller vessels were wrecked elsewhere along the island's shores. One of these ships was commanded by Christopher Colón, a nephew of Christopher Columbus and a mariner who apparently did not prove as weatherwise as his illustrious predecessor. On August 14, 1680, an intense North Atlantic hurricane battered Santo Domingo, sinking 25 French warships, under the command of the Comte d'Estre, that were at anchor in the harbor, as well as several Spanish merchant ships. Hundreds of lives were lost. On August 26, 1724, the Nueva España Flota, forced to seek shelter in Samaná Bay, was overwhelmed by an intense hurricane. A half-dozen ships foundered, taking over 100 men to the bottom. Toward the end of the 18th century, 28 French ships were destroyed when a substantial hurricane barreled across southern Haiti on August 25, 1772. Contemporary reports state that for days afterward, hundreds of bodies washed up on Haiti's beaches. Just over three years later, on August 27, 1775, an equally intense hurricane slammed into Santo Domingo, causing widespread damage to buildings, piers and crops, and claiming hundreds of lives. On August 31, 1785, yet another particularly severe hurricane raided Santo Domingo, uprooting trees, flooding streets and swamping coastal villages. In what would become this particular hurricane's single worst incident, the British trader *Cornwallis*, en route to Port Royal, Jamaica, from ANTIGUA was blown ashore near Neve Bay and wrecked. Almost three years later, both Santo Domingo and Port-au-Prince were lashed by a gargantuan August hurricane, this one occurring on the 16th of the month, 1788. More than 50 ships were either wrecked or sunk by the hurricane's strident winds, many of them in the harbor at Port-au-Prince.

The latter half of the 20th century in Hispaniola has also been marked by the passage of several lethal August hurricanes, including CLEO and DAVID. On the morning of August 24, 1964, Hurricane Cleo's 135-MPH (217-km/h) gusts prostrated much of the island's forests, while blinding downpours ruined vital agricultural stocks. Some 120 lives were lost, most of them in southwest Haiti. Less than two decades later, on August 31, 1979, an even more intense Hurricane David unleashed 142-MPH (229-km/h) winds, and an 11-foot storm surge on the Dominican Republic, uprooting thousands of trees, collapsing hundreds of lightly constructed houses, shattering windows, severing power lines, and destroying much of Santo Domingo's major airport, including its air traffic control tower and terminals. Some 2,000 islanders perished in mudslides and flash floods, while millions of dollars worth of property losses were tallied.

September is the peak month for hurricane activity in Hispaniola, with at least 31 of them making landfall on the island between 1545 and 2006. Often systems of enormous size and memorable destructive consequence, these midseason furies generally travel west across the Atlantic Ocean before tearing into Hispaniola with roaring winds, pelting rains, and eroding seas. On the evening of September 18, 1545, a powerful hurricane trailed its considerable rains north along Hispaniola's east coast, triggering terrifying mudslides that obliterated a small hillside village. Ten heavily laden Spanish merchant ships were beached near the Ozama River's mouth, although their crews and cargo were eventually saved. Slightly more than two centuries later, an even greater September hurricane, this one occurring on September 9, 1737, whipped across northeastern Hispaniola. A true rainmaker, the 1737 hurricane's DANGEROUS SEMICIRCLE released enormous precipitation counts over Santo Domingo, forcing rivers to top their crude earthen embankments swiftly. Much of the capital was subsequently submerged, its narrow streets choked with mud, splintered trees, and debris from fallen houses.

In the years 1834 and 1930, Hispaniola was riven by two horrific tropical cyclones in September, the PADRE RUIZ HURRICANE and the Great Santo Domingo hurricane. The first, which spent almost two full days terrorizing the eastern half of the island with blistering winds and rocking seas between September 23 and 25, claimed several hundred victims, many of them buried beneath the rubble of their collapsed houses. The second hurricane, which stricken survivors unanimously stated was the most severe tropical cyclone to have riddled eastern Hispaniola in living memory, flattened nearly every tree; chiseled wide furrows through fields and crops; demolished thousands of houses; leveled churches, hospitals, and warehouses in Santo Domingo; sank scores of ships; and killed at least 2,000 people. Hispaniola would suffer strikes from several other notable September hurricanes, including INEZ, September 28–29, 1966; ELOISE, September 18, 1975; and EMILY, September 22, 1987. The most extraordinary hurricane to have menaced the island since 1930, Inez made landfall in the Dominican Republic as a Category 4 system of remarkable severity, one bearing a minimum pressure of 27.37 inches (927 mb) and sustained wind speeds in excess of 140 MPH (225 km/h). Sharply weakened as it bisected the mountainous length of Hispaniola, Inez brought nearly a foot of rain to southwestern Haiti during the daylight hours of September 29. All told, some 1,379 people died, a majority of them in Haiti.

Statistically, the late-season months of October and November are an active period for tropical cyclone strikes in Hispaniola. Since 1495, 18 hurricanes have come ashore on the island in October, while three have made landfall in November. Hispaniola's first October tropical cyclone, which struck the eastern half of the island during the first week of October 1495, sank seven Spanish trading vessels that had sought refuge at the Port of Isabela. During the 20th century, two notable October hurricanes pummeled Hispaniola, JÉRÉMIE (October 22, 1935) and HAZEL (October 11, 1954), both tallying death tolls in the thousands. On November 27, 1934, a mature-stage system moved *south* before striking the island's north coast, an unusual late-season example that took the lives of 38 people.

Hooghly River Cyclone *India, October 7, 1737* See INDIA.

Hugo, Hurricane *Eastern–Northern Caribbean–Southeastern United States, September 10–21, 1989* Widely considered to have been one of the 20 most-severe documented HURRICANES to have menaced the CARIBBEAN SEA and eastern NORTH ATLANTIC OCEAN during the latter half of the 20th century, Hugo delivered sustained 130–140-MPH (200–225-km/h) winds, six-to-nine-inch (152–229-mm) PRECIPITATION counts, and a 17-foot (6-m) STORM SURGE to the islands of ANTIGUA, GUADELOUPE, Montserrat, the Virgin Islands, and Puerto Rico between September 16 and 19, 1989. In the United States, where Hugo finally came ashore at Charleston, South Carolina, on September 21, powerful Category 4 winds and an inundating surge savaged much of the city, killing at least 28 people and racking up then record-setting damage estimates of $4 billion.

At one time the single most expensive natural disaster in U.S. history, Hugo developed over the temperate waters of the mid-Atlantic, 700 miles (1,127 km) east of the Leeward Islands, during the late-afternoon hours of September 10, 1989. Whisked westward at speeds of up to 10 MPH (16 km/h), Hugo, as with most hurricanes of the Cape Verde class, swiftly deepened. Satellite reconnaissance photographs taken on the afternoon of September 11 found that during the night Hugo had boldly grown into a powerful Tropical Storm, now generating eyewall winds of 49 MPH (79 km/h) and heavy, gust-driven rains.

As Tropical Storm Hugo's central barometric pressure continued to plummet, dropping from 29.00 inches (982 mb) on the evening of September 12 to 28.27 inches (978 mb) just after midnight on September 13, the system concurrently climbed the Saffir-Simpson Scale of destructive potential,

alarming weary meteorologists, civil defense personnel, and property owners along the Caribbean basin's eastern shores. Although many of them knew that it was not unusual for tropical cyclones to experience sudden drops in barometric pressure, Hugo's strident ascendency—from Category 1 on the morning of September 13 to Category 2 by midafternoon of the following day to Category 3 on the evening of September 14—had police and evacuation officials in the tiny Leeward Island nations of Antigua, Barbuda, Guadeloupe, Montserrat, and St. Kitts scrambling through the streets issuing urgent warnings for residents to evacuate their coastal homes and businesses.

Computer analogs based on the behavior of previous storms had predicted that Hugo's steady recurvature northwest would most likely take the hurricane over the northern Leeward Islands sometime during the day of September 16. Posted hurricane warnings advised that Hugo, now winding itself around a perilously low central pressure of 27.96 inches (947 mb), had sustained winds of between 125 and 130 MPH (201–209 km/h) and gusts of between 145 and 150 MPH (233–241 km/h), making it one of the most dangerous hurricanes to threaten the Leeward Islands since 1979s Hurricane David.

A swing bridge lies askew following Hurricane Hugo's road trip through South Carolina during the night of September 21, 1989. Photographs such as this attest to the destructive power of a tropical cyclone. *(NOAA)*

Well acquainted through local myth with the Leeward's extensive history of violent hurricane strikes, tens of thousands of inhabitants raced to secure their houses, boats, picturesque vacation resorts, and profitable banana plantations ahead of Hugo's approaching chaos. Closed storm shutters were braced with wooden Xs, while small outbuildings were anchored to the ground with ropes and stakes. Emergency service crews, tasked with repairing damaged power lines and water mains, were placed on standby. In the harbors and inlets of Antigua, northern Guadeloupe, and Monsterrat, all small watercraft, from sport diving vessels to fishing boats, were tightly warped alongside their piers. Some vessels, such as those with deep-water moorings, secured for Hugo's arrival by doubling up on their bow chains, rigging additional lifelines, and deploying sea anchors. By the morning of September 16, as Hugo's resonating seas sent echoing breakers into volcanic coves, the Leeward Islands stood ready.

They would need to be, for when Hurricane Hugo finally wailed ashore in northeastern Guadeloupe on the morning of September 16, 1989, it was an incipient Category 4 hurricane of extreme violence. Blasting across the French-held territory at nearly 12 MPH (19 km/h), Hugo's 130-MPH (200-km/h) winds and five-inch downpours winnowed coconut palm trees, split telephone poles, wrenched storm shutters from their hinges, stripped shingles and tiles from roofs, and toppled dozens of chimneys. On the island's southeast coast, where Hugo's 14-foot (5-m) storm surge swamped the half-moon beaches of Saint Anne and Saint François, dozens of charter boats were broken away from their moorings and cast onto rocks. Crossing directly over the principal city of Pointe-à-Pitre as it exited the island, Hugo dropped six-inch (152-mm) rainfalls on the city's gridwork streets, spawning destructive flash floods. Almost 20 percent of Guadeloupe's houses suffered structural damage from broken windows to collapsed roofs, leaving some 12,000 people without shelter. On the adjacent islands of Anguilla and Dominica, Hugo's 150-mile-wide (241-km) RAIN BANDS forced the closure of all airports and other transportation facilities, while gale-force gusts and high surf erosion caused some $43 million in crop and other property losses on St. Kitts.

Clearly obeying the rules of recurvature, Hugo maintained its northwest track, departed Guadeloupe, and entered the Caribbean Sea shortly before dusk on September 16. Immediately intensifying upon its return to open waters, Hugo's 50-mile-wide (80-km) eyewall howled toward Montserrat's green slopes, a seven-by-11-mile (11–18-km) island dominated by a 3,000-foot (1,000-m) active volcano.

Hurricane Hugo's radar signature as it appeared during the Category 4 hurricane's devastating landfall in Charleston, South Carolina. At the time this photograph was taken, Hugo's eyewall was just coming ashore. The system's eye can be seen to the lower left. *(NOAA)*

On Montserrat, where the early onset of pelting rains and tropical storm-force winds had for hours predicted Hugo's imminent landfall, the threat of the worst hurricane strike since the legendary SAN FELIPE HURRICANE of 1928 had yielded a besieged landscape of shuttered towns, abandoned streets, and gloomy gust-swept headlands. Crowded just the day before by hundreds of sun-seeking tourists, Montserrat's vacant gray-sand beaches soon disappeared beneath a thickening miasma of rain, wind, surf, and spray. Their houses and storm cellars battened into fortresses, the people of Montserrat anxiously monitored their televisions and radios, plotted Hugo's midnight approach as though it were 1928 and the great San Felipe Hurricane had returned for another 172 lives.

Seven times larger than the island itself, the Category 4 eyewall of Hurricane Hugo completely engulfed Montserrat during the early morning hours of September 17, 1989. Deep furrows of near-cataclysmic destruction were carved by 142-MPH (228-km/h) winds across the island, razing houses, uprooting ancient trees, and severely damaging the huge looms used to manufacture much of Montserrat's vital cotton exports. In the principal city of Plymouth, located on the southwest coast, Hugo's 150-MPH (241-km/h) gusts damaged or destroyed nearly every building, while gigantic waves—some estimated to be in excess of 20 feet (7 m)—hammered Plymouth harbor's 180-foot (60-m) stone jetty into extinction. Drenching seven-inch downpours turned scores of roads into brackish canals, while small landslides at the foot of Chances Peak swept 21 unlucky houses into mud-plastered heaps of wreckage. Indeed, Hugo seriously damaged or destroyed nearly 90 percent of all the houses on Montserrat, leaving 11,000 of the island's 12,000 residents homeless. Electric, telephone, and water services were disrupted for weeks, prompting a massive British and U.S. relief effort. Ten people were killed and another 89 injured. Damage estimates rapidly topped $260 million, making Hugo the costliest hurricane in Montserrat's history.

After lashing Antigua and St. Kitts with 130-MPH (200-km/h) winds and heavy rains during the day and night of September 17, Hugo passed the next two days by ravaging the Virgin Islands and eastern Puerto Rico. Sharply committed to its northwest trajectory, Hugo sailed over the Virgin Islands on the morning of September 18. Its 140-MPH (225-km/h) winds and blistering rains so severely mauled the U.S.-controlled islands of St. Thomas and St. Croix that almost every house, hotel, school, factory, and storm shelter was either damaged or utterly destroyed. The major cruise ship pier at Frederiksted, St. Croix, was partially washed away, while broken trees, severed power lines, and short-circuited electrical transformers disrupted communication systems across the entire territory. Thousands of people left destitute by Hugo's historic passage wandered through the flooded streets of St. Thomas and St. Croix, many of them looting in a panicked attempt to gather food and medicine. On September 19, the islands' governor, alarmed by reports that police officers and other defense personnel were joining the looters, requested military and financial assistance from the United States. More than 1,000 U.S. troops and $580 million in financial aid were promptly dispatched to the Virgin Islands. All told, at least six people were killed and $700 million in property losses were assessed in Hugo's wake.

Situated some 50 miles west, Puerto Rico's northeast coast was battered by Hurricane Hugo's NAVIGABLE SEMICIRCLE during the night of September 18–19, 1989. Sustained 133-MPH (214-km/h) winds, 150-MPH (241-km/h) gusts, and seven inches (178 mm) of rain left 12 people dead, another 50,000 homeless, and more than $500 million in damages to houses, hotels, hospitals, communications systems, and coffee and banana harvests. The small offshore resort islands of Culebra and Vicques, 60 miles (97 km) southeast of San Juan, were particularly devastated by Hugo, with almost all their houses and tourist hotels being swept into the sea. Much of the El Yunque Tropical Rain Forest, noted for its diverse canopy of bamboo and trumpet trees, orchid plants, and giant ferns, was either uprooted or defoliated. The governor's palace in San Juan suffered the loss of its roof and windows, while electric and water services across Puerto Rico were disrupted for weeks.

Tightly curling to the northwest, Hugo spent the better part of September 19 and 20 reintensifying. As the hurricane drew ever closer to the United States's southeast coast, its central barometric pressure teetered between 27.75 inches (940 mb) on September 19 and 27.67 inches (937 mb) by the evening of September 20, elevating it once again to the status of a Category 4 hurricane. Now firmly recognized as one of the most potentially dangerous hurricanes of recent times, Hugo's offshore trail of destruction near the eastern BAHAMAS on September 20 prompted hurricane contingency plans to be activated hurriedly from north FLORIDA to southern NORTH CAROLINA. As word passed that Hugo's central pressure was still continuing to slip (it would eventually fall to 27.58 inches [934 mb] by the evening of September 21), 300,000 people, many of them summer visitors to the barrier shores of GEORGIA and South Carolina, heeded posted HURRICANE WATCHES and began their slow withdrawal to the mainland's relative safety.

During the late-evening hours of September 21, 1989, Hurricane Hugo wailed ashore at Charleston, South Carolina, a Category 4 tropical cyclone of record-setting intensity. Fueled by a frighteningly low central pressure of 27.58 inches (934 mb), Hugo's 155-MPH (249-km/h) gusts hacked trees from the ground, shattered small BUILDINGS, blew sailboats and motor yachts ashore, and swept away several of the bridges that connected Charleston to its barrier islands. A frothing, 20-foot (7-m) storm surge rocked concrete piers and breached seawalls, flooding large sections of downtown Charleston. Nearly 80 percent of Charleston's buildings, from its antebellum town hall to its convention center to its houses and apartment buildings, lost their roofs, windows, porches, carports, chimneys, and fences. At Bulls Bay, a quiet fishing village 15 miles north of Charleston, Hugo's cresting surge carried an unfinished fishing boat five miles inland before unceremoniously depositing it in a waterlogged thicket of trees. Spanning more than 100 miles across, Hugo churned inland, its bulldozing seas plowing most of the beach houses at Folly Beach into unrecognizable driftwood. Nearby Pawley's Island was cut in two by Hugo's powerful currents, while the hurricane's encroaching wind and surf carried away nearly 100 houses on the neighboring Isle of Palms. Deemed the costliest hurricane to have struck the United States up to that time, Hugo claimed 28 lives in South Carolina, and tallied up damages estimated at more than $4 billion. Long before Hugo dissipated over the cool expanses of central CANADA on September 22, the decision had been made to retire the name Hugo from the cyclical list of HURRICANE NAMES.

hurricane *Hurricane* is the regional term given to those mature TROPICAL CYCLONES that originate over the NORTH ATLANTIC and eastern NORTH PACIFIC OCEANS. Like its meteorological cousins, the eastern North Pacific TYPHOON and the Indian Ocean CYCLONE, the hurricane is a highly organized storm system in which a warm core of low BAROMETRIC PRESSURE is surrounded by winds that rotate in a counterclockwise direction in the Northern Hemisphere. Characterized by sustained surface winds in excess of 74 MPH (119 km/h) and barometric pressures ranging from a fairly mild 29.25 inches (990 mb) to a catastrophic 27.17 inches (920 mb) or less, the hurricane generally contains enormous amounts of PRECIPITATION and may on occasion spawn TORNADOES and waterspouts. The hurricane also carries beneath its EYE a vast dome of seawater known as a STORM SURGE that in the event of a landfall can be responsible for significant property damage and loss of life.

Hurricanes originate over the warm, 84–86°F (25–27°C) waters of the North Atlantic Ocean (including the CARIBBEAN SEA and GULF OF MEXICO) and the eastern North Pacific Ocean (north of the equator and east of the international date line), every year from June to November. Although the official HURRICANE SEASON in the North Atlantic and eastern North Pacific oceans extends from June to November, there are recorded instances of mature hurricanes affecting both coasts of North and Central America and the Caribbean islands in April and December. Until the late 1980s, the generic term *hurricane* was also applied to those tropical cyclones that formed in the Indian Ocean (south of the equator and east of Madagascar) and in those waters east of AUSTRALIA (south of the equator), but this usage now appears to have been supplanted in those areas by the regional classification of *cyclone*.

In order to quantify more easily and record the size and strength of approaching hurricanes, the region's meteorological and civil defense agencies have adopted a five-point scale of hurricane intensity. Known as the SAFFIR-SIMPSON SCALE, it details the most salient highlights of hurricane generation and organization.

During the early part of the season, hurricanes form principally between latitudes 6 and 10 degrees North over the energized waters that wash the shores of the Gulf of Mexico, the western Caribbean Sea, and western MEXICO. While these early storms tend to have less intense winds than late-season hurricanes, they do carry greater amounts of precipitation. Relentlessly moving in north-northwest at speeds of five to 15 MPH (3–24 km/h), a number of them have over time brought major flood emergencies to the southern United States, CUBA, CALIFORNIA,

The fury of a hurricane in Florida. Beneath dark, rain-shadowed skies, a bayside parade of palm trees tenaciously resist a tropical cyclone's flattening winds. A dramatic photograph, the viewer can almost hear and feel the crashing of the waves against the seawalls, the howling intensity of the wind, the pelting rain that slicks the roadway. *(NOAA)*

and northern Mexico. Conversely, late-season hurricanes—those storms that primarily develop north of 10 degrees and farther out in the North Atlantic—have much higher DEATH TOLLS because of their enormous size and more organized nature. For some of them it is a 2,000-mile (3,219-km) trek across sun-drenched waters before a stormy landfall is made in ALABAMA, FLORIDA, LOUISIANA, the eastern seaboard of the United States, ANTIGUA, the BAHAMAS, BARBADOS, JAMAICA, and PUERTO RICO—plenty of time for them to grow into pressure-driven machines of enormous destructive potential.

Hurricanes are far less frequent than either Indian Ocean cyclones or North Pacific typhoons. On average, North Atlantic and eastern North Pacific hurricanes account for 27 percent of the world's annual tropical cyclone activity, whereas typhoon generation accounts for another 30 percent, and cyclone formation 43 percent. In 1969, 12 TROPICAL STORMS reached hurricane status, but only one tropical storm and no hurricanes developed in 1914, the least active year on record. The 1813 hurricane season was noted for the origination of no less than 13 storms, one of which devastated Charleston, South Carolina, on August 27–28, while the 1837 hurricane season witnessed the formation of 11 tropical storms. In terms of their trajectories, early season hurricanes that develop in the southwestern Caribbean Sea generally move north-northeast, while those that originate over the eastern North Pacific either move in a linear direction to the west or curve back toward the east and lash northwestern Mexico and southern California with high winds and flooding rains. On the other hand, late-season hurricanes will often travel great distances across their respective oceans before swinging northeast, curving in a parabolic fashion around the high-pressure ANTICYCLONES that dominate the weather patterns of the Northern Hemisphere during the summer months. Also, like a majority of typhoons and cyclones, hurricanes do not travel in a perfectly straight line but instead wobble back and forth over an imaginary track. It is not unusual for a hurricane's eye to drift 30 or more miles in one direction and then drift back to its original position a few hours later.

Some of the lowest barometric pressures ever recorded have been observed in hurricanes. On September 5, 1935, the extraordinarily powerful LABOR DAY HURRICANE produced a record-setting pressure reading of 26.35 inches (892 mb) as it came ashore on the Florida Keys. In September 1969 the now-legendary Hurricane CAMILLE blasted Pass Christian, Mississippi, with a historically low central pressure of 26.73 inches (905 mb). In August 1980 Hurricane ALLEN intensified to Category 5 status on the Saffir-Simpson Scale three times during its 1,200-mile (1,920-km) trek from the Cape Verde Islands, with its lowest central pressure, 26.55 inches (899 mb), being recorded off the northeastern tip of Mexico's Yucatán Peninsula on the evening of August 7. On September 9, 1988, Hurricane GILBERT, then 50 miles (80 km) northwest of the Cayman Islands, produced a minimum pressure of 26.22 (888 mb). During the remarkable 2005 North Atlantic tropical cyclone season, Hurricane WILMA established a new pressure reading for the Western Hemisphere of 26.04 inches (882 mb). Indeed, the 2005 season was pockmarked by extraordinarily powerful hurricanes, including DENNIS, EMILY, KATRINA, and RITA. In the eastern North Pacific Ocean, Hurricane JOHN reportedly produced a minimum barometric pressure of 27.43 inches (929 mb) during its marathon, 31-day trek across the central Pacific, while 1992s Hurricane INIKI pounded Kauai, HAWAII, with a minimum reading of 27.91 inches (950 mb).

Such remarkably low barometric pressures are responsible for the considerable size and intensity of late-season hurricanes. Although a minimal or moderate hurricane rarely exceeds 200 miles (322 km) across, a routine number of major hurricanes—those storms with barometric pressures lower than 28.47 inches (964 mb)—have widths of between 300 and 400 miles (483–644 km). Although a large hurricane's maximum winds are found directly around its tightly coiled (10–30 miles [16–48 km] across) eye, its extensive cloud mass allows for a greater distribution of the PRESSURE GRADIENT, the drop in barometric pressure over a certain distance from the eye. This brings gale-force winds, many with 74-MPH (119-km/h) gusts, to a wider area of the ocean, thereby increasing the rate of evaporation through wind-driven spray.

Those principal countries and land masses that ring the North Atlantic and eastern North Pacific basins—Barbados, BERMUDA, Cuba, GUADELOUPE, Hispaniola, Jamaica, MARTINIQUE, Mexico, Nicaragua, Puerto Rico, St. Kitts, and the United States—all have long, tragic histories of frequent hurricane strikes. A number of these storms have been of exceptional severity, making them of meteorological or historical interest.

On September 10, 1931, a small, unnamed Caribbean hurricane battered the coastal nation of Belize with 132-MPH (212-km/h) winds and prodigious rains. In the capital city of Belize, more than 1,500 people were drowned as the hurricane's brutal storm surge turned streets into sluiceways. Steamships were run ashore, houses were reduced to kindling, telephone poles were uprooted, and hundreds of steel drums that were used to store fuel oil were exploded. For months thereafter, survivors remarked of the pungent scent of chemicals that still permeated some neighborhoods. Thirty years later, on October 31, 1961, an even more powerful hurricane, HATTIE, wrought twice the damage in Belize when its sustained 140-MPH (225-km/h) winds and 180-MPH (290-km/h) gusts came ashore just before dusk. In the township of Stann Creek, hundreds of buildings were destroyed by wind action alone, leaving the narrow clay streets littered with broken windows, stray doors, and even entire walls. More than 314 lives were reportedly lost to this exceptional hurricane.

The intensely low barometric pressures found in many hurricanes have also been responsible for creating enormous, often deadly, storm surges. Caused when low central pressures allow the sea to form a literal dome of water beneath the hurricane's eye, these storm surges have frequently crashed ashore with catastrophic results. During the Labor Day Hurricane of 1935, a towering storm surge, estimated by survivors to have been 25 feet (8 m) high, swept across Florida's Lower Matecumbe Key. Rolling across the low-lying spit of land, the storm surge washed an 11-car rescue train clear off its tracks. Nearly 400 people were drowned. Other major storm surges include that of the GRAND ISLE HURRICANE, which killed an estimated 2,000 people in October 1893; Hurricane AUDREY, June 27, 1957, which claimed more than

400 lives as it deluged the delta towns of Cameron and Grand Chénier, Louisiana; and the storm surge associated with the GREAT NEW ENGLAND HURRICANE OF 1938, which leveled several coastal communities on Long Island, NEW YORK, and in south Connecticut and Rhode Island.

As for the word itself, most scholars believe that *hurricane* was initially derived from the various languages of the indigenous peoples who once populated the storm-prone shores of Central America and the Caribbean Islands. The Taino people of Puerto Rico referred to hurricanes as *huracan,* the name of the god responsible for the destructive winds that periodically ravaged their communities. The Mayan civilization christened hurricanes *hunraken* after their god of tempests, while the Galibi people of French Guiana viewed the same terrifying deity as *hyroacan.* In 1555, the Eden translation of Peter Martyr's *The Decades of the New World or West India* described a tropical cyclone the Dutch chronicler had encountered on his travels to the New World as a *furacane,* making it the first documented instance in which such a usage was seen in the English language. By the first half of the 17th century, when it came time for the great Puritan governor of the Massachusetts Bay Colony, John Winthrop, to write his account of the GREAT COLONIAL HURRICANE of August 13, 1638, the *f* in *furacane* had been supplanted by an *h,* and the rest of the word adapted to read, *hiracano.* In an anonymous account of the Carolina Hurricane of 1728, the word was rendered with only one *r,* as *huricane,* while in a colonial newspaper from the same period it was given its current usage and spelling, *hurricane.* No less august an authority on etymological matters than Noah Webster himself, creator of the dictionary that still bears his name, defined the word *hurricane* in his personal diary. Writing of the Great September Gale of 1815 as it crossed his house in Amherst, Massachusetts, Webster described the tempest that ravaged his beloved apple orchards as a "proper hurricane." In England, the word *furacane* was subsequently seen as *hurry cane, hurricaine* and, finally, *hurricane;* in France and Italy it went from *aracan* and *urican* to *ouragan* and *uragano,* respectively.

The symbol used to designate a hurricane on weather and tracking charts is as pictured left. The twin convex vanes on either side of the circle indicate the counterclockwise motion of the hurricane's winds, while the circle's solid center signifies a storm of mature strength.

Hurricane Alley The name given to the arc-shaped region of the NORTH ATLANTIC OCEAN positioned roughly between latitudes 10 to 30 degrees North and longitudes 25 to 105 degrees West, along which a majority of mid- to late-season HURRICANES originate and travel. Beginning with the Cape Verde Islands off the northwest coast of Africa, Hurricane Alley trails almost due west until it reaches the CARIBBEAN SEA's outer islands. From there, it either continues west until it terminates with the gulf coasts of TEXAS and MEXICO or else curves sharply north-northwest and follows the eastern shoreline of the United States. The phrase is also applied to those local waters, such as the western GULF OF MEXICO, over which most early season hurricanes develop. Steadily carried north-northeast, these storms have a tendency to come ashore on the FLORIDA Panhandle, an area whose extensive history of hurricane strikes reveals it to be one of the principal destinations of Hurricane Alley.

Hurricane Hunters The nickname traditionally given to military aircraft and personnel assigned to perform aerial reconnaissance on TROPICAL CYCLONES in the NORTH ATLANTIC OCEAN, CARIBBEAN SEA, and GULF OF MEXICO. Composed of squadrons of HERCULES WC-130, NEPTUNE P2V-3W, ORION WP-3, and SUPERFORTRESS B-29 aircraft crewed by pilots from the U.S. Navy and Air Force, the Hurricane Hunters service has, since its formal inception in 1944, served the meteorological community and the general public alike by providing life-saving data on hundreds of threatening HURRICANES and TROPICAL STORMS. Before the first weather SATELLITE was launched into orbit in April 1960, the Hurricane Hunters and their frequent flights through the EYEWALL and EYE of a tropical cyclone were the only practical way to obtain accurate information regarding the position and intensity of a landfalling tropical system. Also, although subsequent improvements in satellite tracking during the last quarter of the 20th century did in fact aid forecasters to predict a tropical cyclone's track and landfall more accurately, they did not render obsolete the Hurricane Hunters's original mission to obtain a tropical cyclone's CENTRAL BAROMETRIC PRESSURE, sustained WIND SPEEDS, and upper level circulation patterns with the greatest degree of accuracy possible.

hurricane names The custom of assigning a certain name to a particular North Atlantic or eastern Pacific HURRICANE or TROPICAL STORM represents a unique cultural aspect of these often memorable but virtually indistinguishable meteorological phenomena. Although scholars differ widely on exactly when and where the tradition first began, centuries of important hurricanes have been identified though the use of geographic place names, male and female first names, saint's names, different species of animals, the phonetic alphabet, holidays, and even the name of a U.S. merchant ship nearly capsized by a hurricane in August 1837. The widespread adoption of this practice in those regions affected by hurricanes has allowed all types and occurrences of TROPICAL CYCLONES to gain admittance to the meteorological discourse, while permitting others—particularly those storms of historic interest—to become part of the common folklore more easily, the shared language of memory and text.

Before the practice of granting hurricanes one-word code names began in the early 1940s, particularly notable storms were often named for the geographic location in which they made landfall. In CUBA, for instance, the 1844 hurricane that struck Havana was christened the Cuban Hurricane by one of the earliest hurricane scholars, WILLIAM C. REDFIELD, because of the enormous amount of damage and staggering DEATH TOLL it wrought on the island. Several memorable Jamaican hurricanes have been given geographic identifiers, including the SAVANNA-LA-MAR HURRICANE of October 3, 1780, in which a thunderous STORM SURGE swept away the port town of Savanna-La-Mar and claimed the lives of 763 people; and the notorious Black River Hurricane of November 18, 1912, in which 100 people lost their lives to the combined fury of wind and water that came ashore at the mouth of Negril Beach's Black River. The use of geographic names in hurricane identification further extends to both BARBADOS and MARTINIQUE, where the GREAT HURRICANE OF 1780 is simultaneously referred to as the Great Barbados Hurricane of 1780 and the Great Martinique Hurricane of 1780. In FLORIDA, a destructive spate of early hurricanes, from the Apalachee Bay Hurricane of August 30, 1837, to the Miami Hurricane of 1926 to the Lake Okeechobee Hurricane of September, 1928, have been commemorated in the names of the communities they gutted. A similar practice can be found all along the gulf coast, where ALABAMA withstood its GREAT MOBILE HURRICANE OF 1852, MISSISSIPPI suffered through its BAY ST. LOUIS HURRICANE, LOUISIANA witnessed its LAST ISLAND HURRICANE, and TEXAS endured its infamous GREAT GALVESTON HURRICANE OF 1900. In those hurricane-riddled states that line the eastern seaboard can still be overheard talk of the Charleston Hurricane of 1813, the Chesapeake-Potomac Hurricane of 1933, and the GREAT NEW ENGLAND HURRICANE OF 1938.

It was in tandem with geographic identifiers that the adjective *great* was most often applied to the names of illustrious hurricanes. Besides pegging a certain storm as one of remarkable destructive impact, the word denoted in the minds of contemporaries something of the elemental fury they had just witnessed. At a time when accurate measuring instruments were as rare as a codified naming system, eyewitness accounts of particularly violent hurricanes relied on the word *great* to vividly describe a sense of the falling barometric pressures, blazing sunsets, beating rains, sweeping storm surges, and screaming winds that periodically ravaged their lands. In such Caribbean nations as Barbados and Cuba, where hurricane activity had always been frequent and brutal, certain storms—such as the October 10, 1780, hurricane in Barbados and the October 11, 1846, hurricane that devastated Havana—were immediately accorded the sobriquet *great* because they truly were storms of extraordinary duration and severity. In the 1780 blow at Barbados, some 5,000 islanders reportedly perished, while the GREAT HAVANA HURRICANE of October 11, 1846, killed upwards of 3,000 people.

At least two recorded hurricanes have been named for individuals. The first, known to posterity as the Padre Ruiz Hurricane, slid into the Dominican Republic on September 23, 1834, bringing huge death tolls to the capital of Santo Domingo. As the earthly remains of Padre Ruiz, a priest of compassionate manner and respected memory, were being interred in the cathedral cemetery, the hurricane that now carries his name began its spirited assault on the city. No sooner had the mourners placed the padre's casket in the ground than the rising wind caused nearby church bells to begin to ring. At first quite faint, the aerial concert eventually rose to a crescendo as belfries collapsed, filling the narrow streets with crushing debris. By the time the storm had finally passed, nearly 1,756 people were dead and another name had been added to hurricane lore. Although some consternation was voiced at the association between this merciless hurricane and the kind memory of the padre, a majority of the pious considered the storm to be a sign of things to come, a portent of what life would now be like, *anno* Padre Ruiz.

The second hurricane christened for an individual, Saxby's Gale of October 4, 1869, was named in honor of a British naval officer, Lieutenant S. M. Saxby, who uncannily predicated the exact date of the storm's arrival in CANADA almost a year beforehand. A student of comets and other celestial oddities, Saxby warned that the eastern seaboard of the United States and Canada would be visited by an intense hurricane on the morning of October 4, 1869. Closely scrutinized by the world's vocally skeptical newspapers, Saxby was eventually proved correct when a hurricane of rare violence did indeed slam into north Maine and Nova Scotia on the morning of October 4, 1869. More than six inches of rain fell in Maine alone, causing widespread havoc. Although Saxby never explained exactly how it was that he arrived at his predication, the savage hurricane he seemingly conjured up was quickly dubbed Saxby's Gale in honor of his courage as well as his accuracy.

Hurricanes have also been christened for ships that they have overtaken at sea. The Calypso Hurricane, August 13, 1837, was named for the ship *Calypso,* which was nearly lost to the hurricane's spray-shrouded fury while off the coast of NORTH CAROLINA. Less than a decade later, ANTJE'S HURRICANE, September 8, 1842, was named after the schooner *Antje,* which it brazenly dismasted in the eastern CARIBBEAN SEA during the predawn hours of August 30.

In PUERTO RICO, a nation with a lengthy history of deadly hurricane strikes, storms were for decades named in honor of the saint on whose holy day the tempest came ashore. While such a tradition may have found its original impetus in the need to quantify the destructive effects of individual hurricanes, it is more likely that the ritual stemmed from a widely held belief in the ability of saints to intercede in the lives of their followers. Like the good shepherd who always keeps one weather eye turned to the sky, the saints were accorded the power to ameliorate the ruinous course of hurricanes and spare their flocks the scattering winds and deluging rains that seemed to blow right out of the Book of Revelations itself. Prayed over and sheltered against, the people's skyward plea for mercy was ritualized through church masses and names, remembrances of the first San Mateo Hurricane of September 21, 1575; the Santa Ana Hurricane, July 26, 1825, in which 374 souls were lost; the San Narciso Hurricane of October 29, 1867, in which crop damages precipitated a major famine; the San Ciriaco Hurricane in August 1899, which delivered billions of tons of rain to the island, drowning nearly 3,000 people as empty boats floated through downtown Humacao; and the second SAN FELIPE HURRICANE of September 13, 1928, in which another 300 people perished and more than $50 million in property damage was inflicted.

It was not until World War II, when a series of hurricanes threatened MILITARY OPERATIONS in the Caribbean and North Atlantic, that any effort was made by the regions' weather services to institute a systematic way of identifying and tracking these disruptive entities. Various rules ensued, such as naming hurricanes by their coordinates (which proved too unwieldy) and after observers' girlfriends (which proved equally cumbersome) before it was decided to assign tropical cyclones names from the phonetic alphabet. Already in use in the North Pacific, where TYPHOONS were frequently given single-word code

names by meteorologists who assisted the Allied forces in their strategic designs, the phonetic alphabet—Able, Baker, Charlie, and so forth—made it possible to determine where in the HURRICANE SEASON a particular storm occurred chronologically. The shorter words also made it easier and more efficient to communicate a hurricane's position, further ensuring that naval operations would not be forestalled by severe weather.

Although this system did on occasion create some bewilderment within the military—an organization that used the phonetic alphabet for purposes other than hurricane identification—it continued in widespread use until the early 1950s. During the 1950 hurricane season, for instance, Hurricane Dog brought hurricane-force winds to coastal New England, while Hurricane EASY, September 3, 1950, nearly destroyed Cedar Key, Florida, and in 1951, Hurricane Charlie killed 54 people and left another 50,000 homeless when it pummeled the island of JAMAICA. In the end it was not confusion that condemned this system to extinction but the lack of available names. The phonetic alphabet, based as it was on one set of names that could not be retired, was unable to account for major storms of the past in the same way geographic or other identifiers could. It became difficult to record the particulars of individual storms and to bring them up in discourse with absolute certainty that the same hurricane was under discussion. Various solutions, including the addition of the date after the phonetic name, were considered, but nothing could outwit the fact that the list as it now stood had lost its vitality and needed to be replaced.

To better accomplish this, meteorological services in the Caribbean and North America turned to the Pacific, where typhoon tracking agencies had adopted the use of female names as their identification system as early as 1948. In the PHILIPPINES, typhoon names were drawn from two sets of female names, with the lists alternating every other year. In a nation that is on average struck by more tropical cyclones every year than any other country on Earth, the selection of Filipino female names provided both a huge pool of possible combinations and the knowledge that such names would be recognizable to a large segment of the population, thereby increasing the effectiveness of evacuation warnings.

Such attributes were extremely appealing to Western meteorologists and civil defense authorities alike. Eager to reduce the threat of death and injury from approaching hurricanes, they recognized in the system several cultural antecedents. These included a bestselling 1941 novel by George Stewart, entitled *Storm,* in which a North Pacific hurricane swoons into the coast of California trailing a shower of epic melodrama.

Now completely convinced of the system's viability, practicality, and style, the region's weather services released the first list of female hurricane names in time for the 1953 season. Arranged in two lists of 23 names each, the lists did not include *Y* and *X* because of a lack of suitable names beginning with those letters. In addition, the names were all between one and three syllables long, thus making them easier to say and remember. The new system, which was for the most part greeted with enthusiasm by storm watchers, quickly resulted in a series of Hurricane Alices, Barbaras and Hazels, some of which were storms of violent consequence. As the rotating list was first expanded to four groups in 1960 and then to ten groups by the 1971 season, it became apparent that the historical record could be better guaranteed by the retirement of those names associated with hurricanes of significant historical or meteorological interest to a quasi "Hall of Fame." These are hurricanes which, by virtue of their considerable death tolls or expensive repair bills, will be better remembered in part by the singularity of their names. The name ANDREW, for instance, has been retired because of its then record-breaking 1992 insurance settlements, while the names AUDREY and KATRINA will always be synonymous with the seriousness of evacuation warnings and the tragic consequences of not heeding them. The name CAMILLE will forever be an archetype, just as the extraordinary Category 5 hurricane that brought 165-MPH (266-km/h) winds to Louisiana one August night in 1969 will always be viewed as a textbook example of just what unadulterated fury a hurricane is really capable of.

When an early season hurricane first originated over the energized waters of the southwestern Caribbean Sea on July 9, 1979, it had already been decided by officials at the World Meteorological Organization (WMO) that it would be christened Hurricane BOB. In response to both societal pressures and a shrinking selection of available female names, the WMO decided with the 1978 North Pacific hurricane season to incorporate male names into their alternating lists of hurricane identifiers. When this experiment proved successful with both meteorologists and the general public, a similar practice was quickly adopted for the 1979 North Atlantic hurricane season. Thus, 26 years of hurricane tradition came to an end on the morning of July 10, 1979, when the burgeoning tropical storm, then 100 miles (161 km) off the northeast tip of Mexico's Yucatán Peninsula, was officially christened Bob.

At the same time, as the practice of naming hurricanes after males and females progressed throughout the Caribbean basin, several of those nations that border the region began to view the custom as a distinctly U.S. influence, and one they desired to have some say in. In a period when such countries as Jamaica and Cuba were trying to set aside their imperial pasts in favor of a new nationalism, hurricane names began to assume a more multilingual flavor. By the 1980 season, Atlantic hurricanes were being called ALBERTO, Henri, and Klaus, thereby reflecting a diplomatic sensitivity to a cultural mosaic drawn together as a community by the common threat of cyclonic activity.

Interestingly enough, HAWAII is the only state in the U.S. union to have with its own list of hurricane names. Officially deemed the "Central Pacific List," its six columns of eight names apiece feature only Hawaiian male and female first names. Unlike the rotating lists of names used in the North Atlantic and eastern North Pacific basins, the Central Pacific List does not progress in alphabetical order, nor does it begin with the letter *A*. Dividing the list of six columns into two sets of three, the first names instead begin with *L, E,* and *P*; or *Li, Ele,* and *Peni*. In the second list of three columns, the first tropical storm of a respective season will be called Loke, Ema, and Peke, while the second will be called either Malia, Hana, or Uleki. In each list the names follow a sequential order, but this changes from column to column based upon the Central Pacific List's curious sense of symmetry. Thus the first column in the first list reads *L, M, N, O, P, U, W,* and *A*, while the second progresses *E, H, I, K, L, M, N, O;* and the third, *P, U, W, A, E, H, I,* and *K*. This pattern is also applied to the second list, where the names all begin with the same letters in the same order. While at first glance the Central Pacific List may seem somewhat confusing, an understanding of its arrangement makes it possible to identify where a particular tropical storm or hurricane occurred in a respective season.

The early years of named tropical cyclones in the North Atlantic were not without their respective confusions. In 1953, when tropical cyclones were first given female identifiers, the first named tropical cyclone was Alice, which churned across the GULF OF MEXICO in early June of that year. The season concluded in late December and early January, with the passage of Hurricane ALICE (2), which existed from December 30 to January 6, 1955. The naming of Alice 2 resulted in the first named tropical cyclone of the 1955 season being BRENDA, which existed from July 31 to August 3, 1955. The 1953 inaugural tropical cyclone naming list for the North Atlantic was: Alice, Barbara, Carol, Dolly, Edna, Florence, Gail, Hazel, Irene, Jill, Katherine, Lucy, Mabel, Norma, Orpha, Patsy, Queena, Rachel, Susie,

Tina, Una, Vicki, and Wallis. Ironically, according to records maintained by US meteorological services, six of the 14 tropical systems observed in 1953 went unnamed.

The UNNAMED HURRICANE of 1991 was the eight hurricane and the 34th tropical cyclone of either tropical storm or hurricane intensity, to remain unnamed since formal naming began in 1950.

During the 2005 North Atlantic Hurricane Season, a "W" named tropical cyclone—WILMA—first appeared, thereby exhausting the naming list for that season. Wilma was shortly followed by Tropical Storm Alpha, Hurricane Beta, Tropical Storm Delta, Tropical Storm Gamma, and Tropical Storm Epsilon. This was the first time in the history of North Atlantic tropical cyclone activity that the Greek identification system was systemically employed. During the 1972 North Atlantic hurricane season, three subtropical systems were given names from the Greek/Military naming system, Alpha (May 23–29), Charlie (September 19–22), and Delta (November 1–7). The practice of naming subtropical system was subsequently largely abandoned.

At least three eastern North Pacific hurricane names have been retired, including Adolph and Israel in 2001, and Kenna following the 2002 season.

In modern practice, the naming of hurricanes allows for greater effectiveness in identifying an incoming hurricane to both civil defense authorities and the general public, thereby increasing awareness of its potential destructiveness. It also adds a somewhat perverse persona to the nature of a hurricane, assigning to these powerful atmospheric disturbances an almost human element. Nowhere else is this more apparent than in the media, where the movements of named hurricanes are frequently identified in terms of the pronouns *he* and *she*. In the United States, where an approaching hurricane often becomes a meteorological miniseries, it is not unusual for a journalist or historian to refer to a hurricane or tropical storm with regard for the gender of its male or female name.

hurricane party

For over a century, the hurricane party has enjoyed a convivial—if somewhat reckless—cultural prominence in the history of U.S. TROPICAL CYCLONES. Arising during the early 1900s, when unforecasted HURRICANE strikes "invited" terrified coastal residents to gather hurriedly into "parties" for mutual defense, aid, and commiseration, the tradition of the hurricane party has over time evolved into an impromptu modern entertainment, a dramatic excuse for those people with ringside seats to drink and dance their way through one of Earth's most awesome meteorological events. This cultural transformation of the hurricane party from neigh-

borhood defense measure to raucous sideshow can be linked to improvements in scientific and domestic technology, to the benefits of time and preparedness that sophisticated satellite forecasting and televised warning systems now routinely provide. There is also the sense that perhaps these improvements in hurricane safety, along with the overconfidence they often foster, have overtaken the initial irony of the term *hurricane party,* a wry descriptor originally coined when hurricanes still made surprise landfalls and flimsy BUILDING codes only exacerbated the final, unamusing tragedy. In an age when television and radio media often hype a hurricane landfall into a sort of meteorological Armageddon, the once-wistful sarcasm of the phrase *hurricane party* has literally become the inspiration for an actual party, a reassuring celebration of life and endurance that continues long after the electricity has failed and the first of the candles and hurricane lamps have been lit.

Although in its more modern manifestations the hurricane party is not codified by any set rituals other than those of television, music, and beer, its early century forerunners were indeed characterized by a strictly defined etiquette of preparedness, by a collaborative effort to survive that possessed the same diligent sobriety as a traditional barn-raising. During congressional testimony given in the wake of the deadly LABOR DAY HURRICANE of September 2, 1935, Julius Stone, a federal relief administrator assigned to the Florida Keys, succinctly described the particular ritual of hurricane parties in Key West. "But Key West actually boarded up their houses. They have 'hurricane parties,' they call them, because they may be holed up for a couple of days, so they give hurricane parties. They board their places up and they get their friends together and go through the storm."

Stone's emphasis on the boarding up of houses would seem to indicate that a hurricane party of the time was principally dedicated to the securing of property, the deployment of storm shutters, or the nailing of small boards over windows. In 1935 the latter task, in particular, would have been a challenge for one person to complete alone, as plywood and other sheet lumber had yet to be developed. Teams of men and women would therefore move from house to house, efficiently battening them down with six-inch-wide pine boards before taking refuge in the largest, or sturdiest, house and waiting out the blast with a grateful toast to the calming powers of alcohol, music, and talk. In the instance of the Labor Day Hurricane, HURRICANE WARNINGS were first hoisted in Key West on Monday, September 2, at 1:30 P.M. with the storm itself eventually coming ashore 90 miles north just seven hours later. Clearly, only the cooperative enthusiasm engendered by the rituals of the hurricane party could see a city of 13,000 adequately prepared in so short a time.

In one respect the contemporary practice of issuing a call for friends and family to gather in one location under the guise of a hurricane party defies some degree of popular emergency management wisdom. To evacuate dangerous coastal areas is always advisable, but to leave one's house empty during a potentially destructive storm, or to leave it unguarded at a time when only looters venture outdoors, or to later drive home on streets strewn with snapped tree limbs and sparking power lines, all seem to be behaviors that are contrary to those of hurricane preparedness and personal safety. Most hurricane landfalls are preceded by an official state of weather emergency, making all unnecessary trips to the local convenience store (which is most likely closed to begin with) or to the beach to view the raging surf (which may be raging several miles inland) highly expensive, with possible fines ranging into the hundreds of dollars.

But in an even deeper respect, the desire for companionship before peril, for strength and solidarity and someone competent to latch the shutters, and for the reassuring hum of human conversation above the wailing tantrum of the hurricane overrules the dictates of common sense, favors instead the spectacle of 48 rain-drenched inmates breaking out of a Puerto Rican prison simply because they wanted to be with their families when formidable Hurricane Luis dashed past the Caribbean island in September 1995. It is the unifying sentiment, above all other legal, technological or safety considerations, that gives the hurricane party its cultural vitality, its popularity, its very purpose. It is the daredevil promise of a shared thrill, of an unexpected holiday, that brings people together to plot the hurricane's progress on television, to update the group occasionally on weather conditions outside, to ponder briefly—even nonchalantly—the potential for serious damage in their areas. There is a hint of the siege in the amount of canned food and bottled water that is piled high in the kitchen, a sudden recollection of a nearly forgotten time when humankind occupied a lower position in the meteorological chain of being and the withering passage of a hurricane was enough to cause a prolonged famine. There is, too, the peace of mind that comes at the terrifying height of the hurricane, the freedom of at least knowing that whatever happens to friends and family happens to all.

The practice of hosting hurricane parties persists to the present day. When Hurricane FLORENCE threatened BERMUDA in September 2006, hotels on the island that had chosen to shelter their guests in place hosted what were called "hurricane parties" for entertainment purposes and to maintain morale.

Although attendance at coastal hurricane parties is a dangerous and unwise tradition, inland hurricane parties are growing increasingly popular. Music for a hurricane party in a safe zone might include:

Deep Purple, "Stormbringer"

The Scorpions, "Rock You Like a Hurricane" "Wind of Change"

AC/DC, "Hell's Bells"

The Doors, "Riders on the Storm"

Credence Clearwater Revival, "Bad Moon Rising" "Who'll Stop the Rain"

Harold Arlen and Ted Koehler, "Stormy Weather"

REO Speedwagon, "Ridin' the Storm out"

Red Hot Chili Peppers, "Storm in a Teacup"

Blue Oyster Cult, "Divine Wind"

Frank Sinatra, "Summer Wind"

Muse, "Butterflies and Hurricanes"

Herbie Hancock, "Eye of the Hurricane"

hurricane season   The six-month period ranging from June until November of each year in which HURRICANE generation in the CARIBBEAN SEA, GULF OF MEXICO, NORTH ATLANTIC and eastern NORTH PACIFIC OCEANS is at its most active, *hurricane season* is essentially an official designation used by meteorological organizations to determine more easily the probability of a hurricanes's development in those waters during a certain time period. Hurricane season does have a basis in known meteorological behavior. During the summer months in the North Atlantic, the trade winds—those winds that move from east to west just above the equator—become stronger and more frequent. Coursing across the ocean's sun-swollen waters, they carry with them pockets of air that vary in TEMPERATURE and BAROMETRIC PRESSURE from that of the surrounding AIR MASS. Known as TROPICAL DISTURBANCES, only a fraction of these volatile pockets ever develop into mature hurricanes. However, it is during the summer months that the frequency of tropical disturbances over the Caribbean, Gulf of Mexico, and North Atlantic-Pacific oceans greatly rises, thereby proportionally increasing the likelihood that several of them will eventually develop into powerful hurricanes.

Hurricane season is also the period in which high-pressure ANTICYCLONES position themselves over the North Atlantic. These vast bodies of warm air in which the prevailing winds are moving away from the center, directly influence the course and intensity of hurricanes and TROPICAL STORMS by acting as atmospheric buffers, areas where TROPICAL CYCLONES cannot penetrate because of high barometric pressure. Because anticyclones vary from month to month within the hurricane season, early and late-season hurricanes will not only generate in different sections of the oceans but will move across them on different trajectories as well. June hurricanes tend to originate west of longitude 75 degrees and undergo RECURVATURE to the north-northeast, where they can come ashore anywhere from TEXAS to FLORIDA. July hurricanes generally develop east of longitude 75 degrees and, after experiencing recurvature, move in a north-northeasterly direction toward the U.S.'s east coast. Likewise, hurricanes in August and September form between 75 and 15 degrees East and move several hundred miles west, where they pass over the central islands of the Caribbean and into Central America. Those late-season hurricanes that originate in October and November either recurve sharply and pass back into the North Atlantic or else cross into the northeastern states of North America.

Such a varied selection of hurricane tracks further requires those nations that are prone to hurricane strikes to modify the hurricane season to fit their own needs. In Barbados, for instance, the hurricane season is five months long, extending from July to November. This is due to the fact that a large number of recorded hurricanes have struck the level island during July and November, and none in June. In Cuba, the season runs from June to November, with the highest concentration of hurricanes tearing ashore during August and September. In the United States, the hurricane season officially extends from June 1 to November 30, with Florida, Texas, ALABAMA, LOUISIANA, MISSISSIPPI, GEORGIA, SOUTH CAROLINA, and NORTH CAROLINA, susceptible to powerful strikes during the entire six-month period. The northeastern United States and CANADA, on the other hand, have a shortened hurricane season, with a peak number of tropical cyclones coming ashore during the months of August, September and October. No hurricane landfalls on the continental United States were recorded during the 2000 North Atlantic hurricane season.

The 2001 North Atlantic hurricane season was unique in that it was the first time in recorded meteorological history that three mature-stage tropical cyclones formed in the North Atlantic basin—Michelle, Noel and Olga. While still of tropical storm intensity, Michelle caused heavy rains and flooding in Honduras before intensifying to hurricane intensity over CUBA. Hurricane Olga formed in late in November, but remained an offshore system for its entire existence.

The 2002 North Atlantic hurricane season featured 12 named storms, for an annual average (1944–1996) of 10 storms per season. According to the National Oceanic and Atmospheric Administration's (NOAA) Climate Prediction Center, the 2002 North Atlantic season was slightly below average. However, the season boasted an extremely active period in September, with eight named storms and one tropical depression, which made it the most active September for tropical cyclone activity in the North Atlantic basin since record keeping began. Meteorologists attributed the relative tepidness of the 2002 season to a strengthening of the EL NIÑO phenomenon. During the 2002 season, six tropical storms struck the United States, the highest tally since 1900. Of these, four hurricanes and four tropical cyclones strikes resulted in nine deaths and some $900 million (in 2002 U.S. dollars) in property losses.

The 2003 North Atlantic hurricane season brought an early season surprise: Tropical Storm ANA, which formed more than five weeks before the official start of the North Atlantic season. Formed from a sub-tropical storm that was generated on April 20, the system was upgraded to a tropical storm on April 22. Ana's maximum sustained winds reached 50 MPH (80 km/h), but the system dissipated on April 23 without nearing landfall. According to meteorologists, Ana was the earliest Atlantic tropical cyclone since 1978, and the only tropical storm yet on record ever to originate in the Atlantic basin in April. According to NOAA's NATIONAL HURRICANE CENTER (NHC), the earliest Atlantic hurricane ever recorded was on March 7th, 1908, while the latest was on December 31st, 1954, which persisted into January 1955.

The 2004 North Atlantic hurricane season produced 15 tropical cyclones, nine of which reached hurricane status. Fourteen of these systems were named, while six reached major hurricane status, making them of Category 3 intensity or higher.

The 2005 North Atlantic hurricane season was historic in almost every respect. First, it was the busiest on record, with 31 tropical storms and hurricanes formed. The first North Atlantic tropical cyclone to be given a "V" name—VINCE—formed over the extreme eastern Atlantic on October 8, 2005; the first "W" named tropical cyclone, WILMA, originated over the northwestern Caribbean Sea in mid-October, and was followed by Tropical Storm ALPHA, Hurricane BETA, and Tropical Storm GAMMA. The 20th named storm of the season, Vince eventually grew into a Category 1 hurricane before quickly weakening to a tropical depression. The system drenched Portugal and southern Spain with heavy rains. Of the 2005 season, five hurricanes were considered major hurricanes, of at least Category 3 status or higher. On October 30, 2005, Hurricane Beta—only the second North Atlantic tropical cyclone to carry an identifier culled from the Greek alphabet—spun ashore on Nicaragua's eastern coastline. The 23rd named storm of the 2005 North Atlantic hurricane season, Beta's sustained, 90-MPH (150-km/h) winds, heavy rains, and predicted 10–15-foot (3–5-m) STORM SURGE forced thousands of Nicaraguans to seek refuge in shelters. One day earlier, Beta's Category 1 winds and rains caused considerable damage on the island of Providencia, near the Colombian coast. On November 19, 2005, Tropical Storm Gamma formed over the westernmost reaches of the Caribbean Sea. Sections of Central America were deluged by Gamma's heavy rains, killing at least 14 people. And, of course, the 2005 season featured Hurricane KATRINA, which as of 2005 remains the most destructive hurricane in recent U.S. history.

As though exhausted by its precedent-shattering activity the previous season, the 2006 North Atlantic hurricane season was a relatively mild one. Between June and October, nine tropical cyclones formed, with five of them achieving mature status. On September 18, 2006, Hurricane Helene generated a central pressure reading of 28.17 inches (954 mb), making it the most intense North Atlantic hurricane observed during the 2006 season. At the time, Helene was located in mid-Atlantic, and posed no danger to land.

hurricane warnings Perhaps the most ominous of all official weather advisories, hurricane warnings announce that a mature NORTH ATLANTIC, CARIBBEAN SEA, or GULF OF MEXICO hurricane is due to make landfall on a particular section of coastline in 24 hours or less. Issued by the National Weather Service (NWS), and superseding earlier HURRICANE WATCHES posted by that agency, hurricane warnings are designed to prompt those communities that lie in a hurricane's path to complete their preparations immediately for the arrival of 74+-MPH (119+-km/h) winds, torrential PRECIPITATION counts, and severe STORM SURGE or STORM TIDE conditions. In their strictest form, hurricane warnings appear as advisories, highlighted passages within a much larger weather bulletin that not only draw attention to a respective hurricane's size, intensity, forward speed,

and direction but also to a suggested course of defensive action. But in their cultural influence, in the unwritten power they have to mobilize effectively the evacuation of tens of thousands of people from vulnerable barrier islands and low-lying coastal neighborhoods, hurricane warnings serve as true lifesavers, expensive but necessary reminders that in 1900 a surprise hurricane landfall at Galveston, TEXAS, needlessly killed between 6,000 and 12,000 people.

When a threatening hurricane moves to within 36 hours of landfall, the following procedure for the eventual posting of hurricane warnings is activated by the National Weather Service:

- Based upon the size, strength, and forward speed of the hurricane, a probable course and point of landfall is predicted by computer ANALOGS at the NATIONAL HURRICANE CENTER (NHC) in Miami, FLORIDA. This data is then arranged in the form of a timetable, a carefully considered plan by which NHC meteorologists can determine the approximate time of the hurricane's arrival ashore.

- Additional geographical and topographical information regarding the stretch of coastline most likely to be affected by the hurricane's possible storm surge or storm tide is then figured into the equation. It is quite possible that a barrier island, headland, lake, valley, or harbor will in some way deflect a hurricane's surge, redirecting its force in such a manner that it either breaks apart and subsides or tragically inundates an unsuspecting neighboring town or city.

- The population density of endangered areas is also considered. Because many inhabited shoreside islands are accessed by a limited number of bridges, causeways, and ferry services, evacuation plans must include in their scenarios those expected delays caused by traffic jams, rain-slicked roads, the raising and lowering of drawbridges, and sheer human folly. Such intelligence is critical if hurricane warnings are to be announced in time to permit those coastal residents to evacuate their homes safely, usually a minimum of 10 to 18 hours before the hurricane is scheduled to make landfall.

- As the hurricane drifts to within 36 hours of landfall, hurricane watches are posted along jeopardized coastlines. When the hurricane, its course unchanged, slides to within 24 hours of landfall, hurricane warnings are issued for those areas most likely to be struck by the hurricane's EYEWALL, the circular core in which the system's highest wind speeds and heaviest rains are contained. Because the size of a hurricane's eyewall varies with the storm's intensity, hurricane warnings can cover at least 20 to 50 miles (32–80 km) of coastline, with hurricane watches extending outward for another 100 to 300 miles (161–483 km).

- Just to be on the safe side, National Weather Service forecasters will often "overwarn" a particular region, announcing 40-mile (80 km) and 50-mile (80 km) warnings for a system whose greatest fury lies in a 30-mile range, thereby accounting for sudden course changes just before or after landfall. While in one regard the practice of over-warning is an expensive one (it has been estimated that the cost of preparing a 579-mile [932 km] stretch of gulf coast for the October 1995 strike by 150-mile-wide [241-km] Hurricane OPAL was some $47 million), its priceless benefits are directly tallied in lower and lower DEATH TOLL statistics.

Because of the very real dangers to life and property posed by a hurricane strike, hurricane warnings are usually taken very seriously in those areas most likely to be directly affected by the storm. Local television and radio stations interrupt their scheduled programming to broadcast details regarding the hurricane's size, intensity, course, and estimated time of landfall. On televised weather maps, that section of coast targeted by the eyewall of an incoming hurricane is usually highlighted in red, the color of alarm, while adjoining stretches of coastline that have been placed under hurricane watches are rendered in yellow, the color of caution. Marinas and other shoreside installations hoist hurricane warning flags—two red squares with solid black boxes in their centers—to their mastheads, alerting stray boaters to the imminent onset of heavy seas and violent winds. Police, fire, and civil defense authorities initiate evacuation procedures for barrier islands and low-lying coastal areas; if necessary, they begin to move house-to-house to make certain residents are given sufficient time to gather important insurance, financial, and personal documents before crowding a limited number of bridges and causeways that lead to the relative safety of the mainland. Anxious last-minute shoppers form lines at supermarkets, gas stations, and lumber yards, purchasing every bottle of water, case of beer, or sheet of plywood in stock. Mobile-home parks are evacuated, their families moved to

In this 1938 photograph, an aircraft belonging to the United States Coast Guard drops a "Hurricane Warning" message to a fishing boat off the coast of Florida. The message (which is shown magnified on the right side) is blunt: it reads, HURRICANE WARNING. *(NOAA)*

community shelters. Inland homeowners close storm shutters, tape large Xs on windows, and anchor small sheds to the ground with cement blocks, steel cables, and chains. For a price, private boat yards haul expensive sail and motor yachts out of the water to protect them from the stranding dangers of the hurricane's storm surge. As in the case of Hurricane Opal, hundreds of oil company personnel are airlifted from Gulf of Mexico drilling rigs, temporarily shutting down nearly 40 percent of U.S. natural gas production in the gulf. A similar situation was observed following the passage of hurricanes KATRINA, RITA, and WILMA in 2005.

In place for the duration of a hurricane's passage, hurricane warnings are generally lifted when sustained wind speeds fall below gale-force, 44 MPH (71 km/h).

hurricane watches Announced by the National Weather Service (NWS) when a mature HURRICANE spins to within 36 hours of landfall on a certain section of coastline, hurricane watches are designed to alert those communities that lie in a hurricane's probable path to begin their preparations immediately for the possible arrival of 74+-MPH (119+-km/h) winds, pelting PRECIPITATION, and destructive STORM SURGE or STORM TIDE conditions. Superseded only by even more urgent HURRICANE WARNINGS, hurricane watches appear in weather bulletins as advisories, highlighted passages that not only describe a particular hurricane's size, intensity, forward speed, and direction but also suggest a timetable for the commencement of evacuation and security procedures. Hurricane watches are initially posted for large stretches of coastline—sometimes 400 miles (644 km/h) or more—in order to allow vulnerable oceanside villages, towns, and cities to empty their crowded tourist beaches, summer cottages, and fair-weather marinas before hurricane warnings are issued 12 hours later. Hurricane watches are also announced concurrently with hurricane warnings and are used to alert communities on both sides of the hurricane warning area to the possibility of last-minute course changes by the storm.

It has been said by emergency management officials that if hurricane warnings are a final summons to action, hurricane watches are a preliminary call to prepare. When hurricane watches are posted, residents in threatened areas should securely stow away lawn furniture, trash cans, bicycles, wheelbarrows, and any other small objects that a hurricane's powerful winds could turn into dangerous projectiles. Those people in coastal regions should consider evacuating vulnerable barrier islands and low-lying coastal areas before limited evacuation routes become crowded or impassable because of storm conditions.

I

Ida, Typhoon *Northwestern Pacific Ocean–Japan, September 21–28, 1958* Better known in JAPAN as the Kanagawa Typhoon, internationally code-named Typhoon Ida inflicted catastrophic property damage and huge DEATH TOLLS on the central Japanese island of Honshu during the night of September 26 and 27, 1958. Circling around a CENTRAL BAROMETRIC PRESSURE of 28.05 inches (949 mb) at landfall near the town of Kanagawa, some 15 miles (24 km) southeast of Tokyo, Ida's scathing 118-MPH (190-km/h) winds, 15-inch (381-mm) PRECIPITATION counts, and landslides damaged or destroyed 2,118 BUILDINGS, swept away 244 road and railway bridges, and inundated more than 120,000 acres of rice fields with two-foot (1-m) tides. The Kano, Meguro, and Arakawa rivers quickly overtopped their banks and spilled onto piers before surging into houses, stores, religious shrines, and freight depots. Of equal intensity to the North Atlantic HURRICANES Gracie (1959) and Beulah (1967), Ida claimed 1,269 lives, left another 12,000 people homeless, and racked up damage estimates of more than 50 million in 1958 dollars. Ida's landfall on September 26 is a unique meteorological coincidence in Japan; on the same date in 1954, the tragic TOYAMARU TYPHOON sailed ashore in south Hokkaido, while gargantuan Typhoon VERA went on to decimate central Honshu on September 26, 1959, one year to the day after Typhoon Ida brought such widespread fear and suffering to the densely inhabited main island.

Ike, Typhoon *Philippines–China, August 26–September 6, 1984* Regarded by the Philippine Weather Bureau as the most intense recorded TYPHOON to have struck the PHILIPPINES during the 20th century, Ike brought 137-MPH (220-km/h) winds, 11-inch (279-mm) rains, and a 13–15-foot (4–5-m) STORM SURGE to the archipelago's south and central islands and the southwest coast of CHINA between September 2 and 6, 1984. In the Philippines, where Ike touted a CENTRAL BAROMETRIC PRESSURE of 27.80 inches (941 mb) at landfall, the typhoon either damaged or destroyed more than 100,000 BUILDINGS and claimed an estimated 1,363 lives on Mindanao and Negros islands. In China, where Ike crashed ashore near the port city of Beihai on the afternoon of September 6, 13 fishers were lost at sea, drowned when their junk-rigged fishing craft were overwhelmed by the typhoon's 30-foot (10-m) waves. Before dissipating in a hail of thunderstorms over central China on the morning of September 8, Ike's lashing winds and torrential rains caused an additional 46 deaths in Guangxi and Zhuang provinces, making it one of the deadliest typhoons to have swept through the South China Sea in recent memory.

The ninth storm of a typically active TYPHOON SEASON, Ike originated over the sparkling waters of the western NORTH PACIFIC OCEAN, approximately 2,400 miles (3,862 km) east of the island of Luzon, on the morning of August 26, 1984. Churning southwest at nearly 12 MPH (19 km/h), the embryonic TROPICAL DEPRESSION swiftly sapped the ocean of its heat and moisture, maturing into a moderate TROPICAL STORM on the afternoon of August 28 and a weak, Category 1 typhoon, with a central pressure of 28.98 inches (981 mb), by midmorning on August 29. Maximum sustained winds in Ike's EYEWALL now measured 89 MPH (143 km/h).

With typhoon-force gusts around its 41-mile eyewall soon exceeding 100 MPH (161 km/h), Ike buffeted the U.S.-held island of Guam with gale-force winds and steady rains as its EYE passed some 240 miles south of the territory on the afternoon of August 29. Trying to further strengthen as it spiraled toward the weary cities and towns of the central Philippines, Ike's barometric pressure sharply seesawed between 28.35 inches (960 mb) on the evening of August 29 and 28.53 inches (966 mb) by midnight of the same day but quickly deepened thereafter. By the following afternoon, Ike's central pressure had settled to 27.93 inches (943 mb), making it a first-rate Category 3 typhoon with sustained winds of at least 130 MPH (209 km/h). Shortly before midnight on August 30, as the system's central pressure slipped even further to 27.87 inches (943 mb), it became an incipient Category 4 typhoon, one of the most intense TROPICAL CYCLONES seen in the western North Pacific in some time.

With a central barometric pressure of 27.80 inches (941 mb), Typhoon Ike stormed ashore near the port city of Surigao, the Philippines, on the morning of September 2, 1984. Gusts of 145 MPH (233 km/h) stripped buildings of their roofs, crumpled bridges, toppled thousands of trees, snapped telephone poles, and thrust before them a 16-foot (5-m) STORM SURGE, the highest seen on Mindanao's northeast coast since 1975. In the Pilar Islands, northeast of Surigao, hundreds of watercraft were driven onto volcanic shores, causing numerous drownings. Roaring across Surigao Strait's narrow waters, Ike's sustained 137-MPH (221-km/h) winds and flooding rains breached dikes and levees, washing away hundreds of houses in the hillside towns of Mainit and Tacer. Mainit, with its broad volcanic lake and winding streets, was particularly savaged when two dams burst, causing the town's immediate inundation. More than 200 people were drowned, swept away by the raging torrent that poured from the rain-swollen reservoir. Along the Negros Island's coastal lowlands nearly 60 people were drowned when Ike's towering 15-foot (5-m) storm surge rushed ashore near the town of Cebu. Bristling with splintered timbers, palm fronds, small boats, stones, and even fish that were scooped up from the seabed, the storm surge almost entirely submerged the island, decimating dozens of villages as it seemingly sought to cut the outcropping in two.

After blighting 90 percent of the rice and sugar harvest on Mindanao and leaving nearly a million people in six southern provinces homeless, Ike blasted into the South China Sea, fanning the temperate waters with 120-MPH (193-km/h) winds and renewed rains. By midnight of September 3, as

the typhoon finally began to undergo RECURVATURE to the northwest, typhoon warnings along China's southwest coast were posted from Hong Kong to Beihai. In Hong Kong, where Ike's possible landfall from the south would drive enormous waves headlong into the enclosed reaches of the harbor, deeply laden cargo ships hastily set to sea, hoping to avoid being caught at their anchorages by Ike's onrushing fury. In Zhanjiang, a major industrial settlement on the Leizhou Peninsula, sand-bagging operations continued through the night in an effort to erect a barrier against Ike's predicted 25-foot (7-m) swells. As hundreds of thousands of Chinese citizens were haphazardly evacuated from those coastal communities between Hong Kong and Zhanjiang, forecasters at the Typhoon Warning Center in Guam updated their prognosis on the storm's progress, stating that they now believed the typhoon's eye would come ashore more to the west and perhaps even striking Vietnam instead of China. In response, the Chinese government immediately widened the evacuation zone, beginning the nearly impossible task of clearing the south coastline of the most populated nation on earth.

They were still in the process of doing so when Typhoon Ike first barreled ashore on the island of Hainan, on the evening of September 6, 1984. Sweeping over the Gulf of Tonkin, Ike made its final landfall just before midnight on September 6, belting Beihai with 130-MPH (209-km/h) winds and three–four inches (76–102 mm) of rain. A mature, late-season typhoon, Ike's hefty girth ensured that while its rainfall counts remained fairly moderate, its intense Category 3 winds would extend over a greater area, causing heavy seas and thousands of building failures. Thirteen fishers, caught off the coast of Weizhou Island, were drowned as Ike's 25–30 foot (7–10 m) seas devoured their tiny boats, while sustained typhoon-force winds in Nanning, the provincial capital, uprooted trees, downed power lines, and severely damaged or ruined 13,000 houses, schools, hospitals, army facilities, and electronics plants. Before dissipating into an EXTRATROPICAL CYCLONE on September 8, Ike's remaining rain squalls claimed an additional 46 lives in the central Chinese cities of Guiyang and Chongoing. With nearly $1 billion in property losses and 1,422 deaths to Ike's credit, the World Meteorological Organization decided that despite a growing dearth of available TYPHOON NAMES, it would temporarily retire Ike's name from the rotating list of typhoon identifiers for a period of 25 years.

India *South-Central Asia*   One of the world's most densely populated nations, India's position on the northern mandible of the storm-laden Indian Ocean has long made it deadly susceptible to those early and late-summer CYCLONES that accompany the changing seasons of the Asiatic monsoon. Between 1737 and 2006, India's low-lying BAY OF BENGAL and Arabian Sea coasts were directly affected by the roaring winds, high seas, and inundating STORM SURGES of no less than 656 TROPICAL CYCLONES of varying size and intensity, for an average of two tropical systems per year. Of these documented storms, some 205 were mature cyclones of notable power and duration—Category 3, 4, and even 5 tempests with CENTRAL BAROMETRIC PRESSURES of 28.47 inches (964 mb) or below, 111–150-MPH (179–241-km/h) winds and 10–35-foot (3–10-m) STORM TIDES. Often fraught with flooding rains and TORNADOES, a majority of Indian cyclones have routinely exacted near-apocalyptic DEATH TOLLS from the shallow farming and fishing settlements of the Ganges River delta and Deccan plateaus, making the country's history of cyclonic activity one of the deadliest on record.

On October 7, 1737, India suffered what many scholars have touted as the single worst cyclone disaster in world history. Known as the Hooghly River Cyclone, this violent Bay of Bengal system delivered a thundering, 30–40-foot (10–13-m) storm surge to the mouth of the Hooghly River. Rolling across the heavily populated jute plantations that hugged the muddy delta, the surge drowned between 300,000 and 350,000 people and resulted in widespread destruction, disease, and famine. Some 20,000 vessels alone were lost, either overwhelmed by the cyclone's 35-foot (10-m) seas or thrown ashore and wrecked by crushing surf.

Fortunately, not all of India's cyclone strikes are as deadly. On the night of October 3–4, 1959, a fairly damaging Bay of Bengal cyclone pushed ashore near Chatrapur, a port town southwest of Calcutta. Possessing strident 100-MPH (161-km/h) winds, the cyclone proved to be a particularly prolific rainmaker, dropping more than 15 inches of rain over the West Bengal and Orissa districts. Only 23 people were killed—a large death toll by Western standards but relatively inconsequential by Indian statistics.

OTHER MEMORABLE INDIAN CYCLONES

The Coromandel Coast Cyclone of 1787: One of southeastern India's most capricious cyclone strikes, the Coromandel Coast Cyclone rallied ashore near Madras, a major port city southwest of Calcutta, on the morning of October 4. Ten thousand people were drowned; several thousand head of cattle were slaughtered by wind-shot debris or else trapped in rising flood waters. As has been observed in other Bay of Bengal cyclones, the Coromandel Coast Cyclone's

storm surge rushed ashore as a series of huge waves, cresting 35-foot (10-m) walls of water that progressively inundated coastal sandbanks, flowing up the Palar River before subsiding.

The Great Coringa Cyclone of 1789: First plotted in the early 19th century by cyclonic scholar HENRY PIDDINGTON, the Great Coringa Cyclone was by Bay of Bengal standards a fairly late-season system, occurring in December, with the usual severe storm-tide conditions prevailing. According to Piddington, the city of Coringa at the Godavari River's mouth southwest of Calcutta "was destroyed in a single day." Described by eyewitnesses as a succession of three waves, the Great Coringa Cyclone's storm tide drove ashore all the vessels in the anchorage, while second and third waves of even greater height simply flowed in, submerging most of the fertile fields around the delta. Some 20,000 people were reportedly killed, many of them entombed beneath the towering mounds of sand left behind by the cyclone's receding waters.

The Belle Alliance Cyclone of April 27–May 1, 1840: Named for a British merchant ship that it nearly sank on May 1, the Belle Alliance Cyclone delivered 115-MPH (185-km/h) winds, torrential rains, and a destructive 20-foot (7-m) storm tide to India's northeast coast. Finally bursting ashore on the evening of May 1, the Belle Alliance Cyclone's lowest observed barometric pressure, 28.10 inches (951 mb), unroofed scores of houses, drove several small sailing craft aground, and caused extensive flooding along Chilka Lake's lowland banks. The subject of yet another of Henry Piddington's famous meteorological reconstructions, the Belle Alliance Cyclone's PRESSURE GRADIENT was later revealed to have lost nearly an inch of barometric pressure during a two-hour period, clearly indicating the steep intensity of this particular system. Several hundred people reportedly lost their lives to the Belle Alliance storm, while damage estimates to buildings, piers, and shipping were considered significant for the time.

The GREAT CALCUTTA CYCLONE of October 5, 1864: Widely considered to have been one of India's most vicious cyclones, the Great Calcutta Cyclone claimed more than 50,000 lives as it submerged large portions of the thriving port city of Calcutta on the afternoon of October 5, 1864. Lingering with plague-like intensity, the Great Calcutta Cyclone's unsanitary legacy soon added its own horror to the carnage, seeing another 30,000 people claimed by the dehydrating ravages of cholera and dysentery. Aghast at the Great Calcutta Cyclone's economic and human scope, the British East India Company subsequently established the continent's first weather service, the India Meteorological Department. Symbolically headquartered in a rebuilt Calcutta, the service was tasked with tracking threatening Bay of Bengal cyclones through shipping reports and then telegraphing that information to vulnerable coastal areas via a comprehensive network of warning stations.

The BACKERGUNGE CYCLONE of October 31, 1876: Just one of a long line of infamous cyclones to have wobbled ashore at the Ganges River's mouths during the course of the last three centuries, the Backergunge Cyclone—complete with 147-MPH (247-km/h) winds and a 40-foot (12-m) storm surge—was responsible for one of India's most shocking natural disasters. While the devastated northeastern city of Backergunge would later rebuild itself into Barisal, BANGLADESH, the vicissitudes of nations and history could not lessen the cyclone's final toll: 1,700 well-constructed BUILDINGS reduced to rubble, 47 sailing vessels and steamships cast ashore, and a staggering, 200,000 people lost, many of them to the poststorm killers of disease and famine.

The Great Bombay Cyclone of June 6, 1882: One of few truly great Indian cyclones to have formed over the Arabian Sea, the Great Bombay Cyclone—engorged with 110-MPH (177-km/h) winds and an 18-foot (6-m) surge—reportedly claimed more than 100,000 lives when it came ashore at Bombay right before daybreak.

The False Point Cyclone of September 21, 1885: An exceptionally severe Category 5 Bay of Bengal cyclone (a minimum pressure reading of 27.08 inches was noted at landfall), the False Point Cyclone's 156-MPH (251-km/h) winds and 22-foot (7-m) storm surge rocked large sections of northeastern India just after dawn on September 22, 1885. Without doubt one of the most intense examples of its kind ever observed, the whirring eyewall of the cyclone plowed ashore at what is now known as Palmyras Point, bearing before it a single cresting wave of enormous destructiveness. Hundreds of structures—from crudely constructed wattle huts to sturdy ware- and customhouses and piers, cranes, and foundries—were entirely carried away by the surge. Scattered for miles across the False Point harbor's sodden banks, all were reduced to such unrecognizable rubble that upon returning, dazed survivors were unable to determine exactly where their property had once stood. Although an exact count of the dead does not exist, contemporary sources place the False Point Cyclone's final toll at about 10,000 people lost.

The Bengal Cyclone of October 16, 1942: A catastrophe within a catastrophe, the little-known Bengal Cyclone of 1942 beat ashore under the darkness of world war at a time when Allied MILITARY security concerns outweighed the need to alert the Calcutta lowlands to the presence of this powerful Bay of Bengal cyclone. Some 40,000 people reportedly perished in the storm; damage in and around the town of Contai was said to have been substantial.

The Unnamed Cyclone of October 27, 1949: Ripping ashore just seven years after the devastating landfall of the Bengal Cyclone, the hefty cyclone of October 27, 1949, released 105-MPH (169-km/h) winds and submerging rains on the India's southeast coast, southwest of Vishakhapatnam. Lashing the town of Machilipatnam into near extinction, the October 1949 cyclone drowned at least 1,000 people and left another 50,000 homeless. Thousands of acres of tea and jute fields were ruined, causing profound financial distress to residents and businesses alike.

The Unnamed Cyclone of June 1, 1956: Similar in size, course, and intensity to the Bengal Cyclone of 1942, the June 1956 tempest inundated much of the Midnapore district of West Bengal. Laden with vast quantities of PRECIPITATION, the 1956 cyclone rendered 20,000 people homeless, while the flood-swollen Haora River stranded another 20,000 refugees on the Assam border. An estimated 480 people perished, many of them struck down by the cyclone's attending landslides.

The Unnamed Cyclone of May 19, 1969: Following an almost straight-line trajectory across the Bay of Bengal, this deadly Category 2 cyclone battered the southeast coast of India with 100-MPH (177-km/h) winds and drenching downpours. At least 618 people in the state of Andhra perished in mudslides and flash floods, making the 1969 cyclone one of the deadliest to have affected the region in almost two decades.

The Unnamed Cyclone of October 29, 1971: Accompanied ashore by a 15-foot (5-m) storm surge, the October 1971 cyclone killed more than 75,000 cows and some 10,000 human beings in the storm-addled Orissa district. Less than two weeks later, on November 6, a second, less-intense cyclone drowned more than 100 fishers in the same region.

The Unnamed Cyclone of November 19, 1977: This monster Bay of Bengal cyclone, noted for its 115-MPH (185-km/h) winds and deluging precipitation

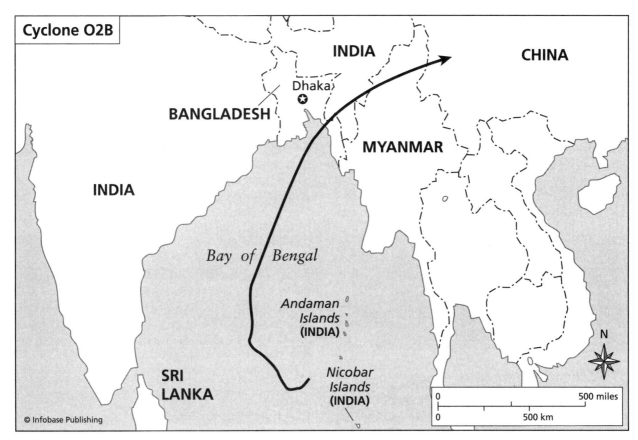

This tracking chart shows Cyclone 02B's devastating course across the Bay of Bengal and into the Indian subcontinent in 1991.

counts, devastated large portions of Devi Taluk, a coastal district located in the southeastern half of the country. Some 20,000 people were said to have perished.

The Unnamed Cyclone of May 12–13, 1979: A relatively weak Category 2 cyclone, this disorganized Bay of Bengal system and its soaking, 13–16-inch rains damaged or destroyed 1,500 villages in southeastern India. More than 350 people were reportedly killed, most of them victims of flash floods and landslides.

The Unnamed Cyclone of June 4, 1982: Bearing sustained winds of 113 MPH (182 km/h) at landfall near the eastern port city of Paradip, this moderate Category 3 Bay of Bengal cyclone claimed 200 lives, left another 200,000 people without shelter, and blighted some $107 million worth of rice stocks in central India. More than 800,000 buildings in the all-too-susceptible Orissa district were unroofed or collapsed by the cyclone's 136-MPH (218-km/h) gusts, while a 12-foot (4-m) storm surge, followed by two days of 18-foot (6-m) seas and intermittent rain squalls, sank or disabled scores of fishing boats along a stretch of coastline. A hazardous storm to endure

in any technological age, the impending arrival of the June 1982 cyclone had been well announced to the general population beforehand, allowing an early, properly organized evacuation of endangered coastal areas to occur. Like so many of the Indian cyclones that had come ashore 50 or 100 years before, the June 1982 cyclone clearly possessed the potential to become a major natural disaster, to exact an accounting of corpses that could reach the stacked underside of the storm itself. Mercifully this did not happen, a tribute to the tangible, lifesaving gains India has made in improving its cyclone early warning system.

The Unnamed Cyclone of May 9–11, 1990: The most intense Bay of Bengal cyclone to have struck southeastern India since the cyclone of November 19, 1977, came ashore at Andhra, the three-day blast between the morning of May 9 and the evening of May 11 killed 450 people, leveled thousands of homes, and ruined some $588 million worth of crops and livestock. A Category 3 cyclone of extreme severity, the 1990 cyclone's 122-MPH (196-km/h) winds, blinding rains, and flooding 21-foot (7-m) seas stripped the leaves and barks from trees, blew birds through the sides of houses, and deposited scores of boats ashore at Tamil

Nadu. Nearly a half-million people in two states were left homeless, while overall property losses mounted toward the $1 billion mark. Justifiably proud of their efforts, Indian civil defense authorities pointed out that the low death toll—450 as compared to 1977s 21,000 lost—was due to better satellite forecasting technology as well as to a more anticipatory approach to timely evacuation procedures.

Cyclone 03B, November 1, 1995: The third (03) Bay of Bengal (B) cyclone of the 1995 cyclone season, 03B veered ashore on India's southeast coast, north of Kakinada. Identified through a numerical naming system based on the Japanese model and inaugurated for use in the Bay of Bengal in 1992, 03B's 109-MPH (175-km/h) winds and 12-foot seas lashed much of coastal Orissa and Andhra Pradesh, driving fishing boats ashore and demolishing hundreds of houses. Three people were killed; damage estimates exceeded $50 million.

The Unnamed Cyclone of October 29, 1999: On October 29, 1999, an extremely powerful cyclone with wind speeds in excess of 160 MPH (257 km/h) struck the eastern Indian state of Orissa. The second tropical system to have affected the state during the month of October, the cyclone's rains and storm surge swept industrial chemicals into bathing ponds and drinking wells, causing several thousand people to suffer chemical burns. Some 10,000 lives were lost, caused by flash flooding. In the Jagatsinghpur district, some 8,119 people were said to have died, and overall, some 1,977 villages in 10 districts were directly affected. The *Hindustan Times* newspaper reported that flooding caused by the cyclone in India left livestock experiencing food and water shortages. It also noted that losses of livestock and domestic ANIMALS included 8,238 buffaloes, 78,104 sheep, 78,728 goats, 10,381 calves, and nearly 1 million broiler chickens. In addition, up to 15 million people were left homeless. In terms of its severity, the system was comparable to the 1997 cyclone that claimed thousands of lives in the neighboring Andhra Pradesh province. Media reports stated that the 1999 cyclone was the most severe to have riddled the Bay of Bengal in three decades.

Cyclone 03B of November 29, 2000: On November 29, 2000, Cyclone 03B slammed into the southeastern coast of India with 106-MPH (170-km/h) winds and a five-foot (2-m) storm surge. Dozens of lives were lost.

The Unnamed Cyclone of October 17, 2001: The 2001 Bay of Bengal cyclone season was no kinder to India. On October 17, a powerful cyclone claimed at least 38 lives when it delivered heavy rainfall to the Andhra Pradesh province. More than 50 lives were lost, and considerable property losses were tallied.

Cyclone 01B of May 16–17, 2003: On the night of May 16–17, 2003, Cyclone 01B slid across the island of Sri Lanka, causing the country's worst natural disaster in nearly half a century. Some 300 people lost their lives, and another 350,000 were left homeless. Inland and flash flooding were the primary destructive hazards associated with Cyclone 01B.

Inez, Hurricane *Eastern Caribbean–Sea– Hispaniola–Cuba–Florida–Mexico, September 22– October 11, 1966* The ninth TROPICAL CYCLONE of the very active 1966 HURRICANE SEASON, Inez remains one of Caribbean history's longest-lived HURRICANES. At its zenith, wielding a CENTRAL BAROMETRIC PRESSURE of 27.37 inches (927 mb), top winds of 145 MPH (233 km/h), and 15-inch rains, Inez scrubbed an erratic course of devastation through GUADELOUPE, HISPANIOLA, CUBA, FLORIDA, and into MEXICO between September 24 and October 10, 1966.

In Guadeloupe, where Inez's TROPICAL STORM-strength winds and rains triggered flash floods and blanketing mudslides on September 24 and 25, at least 23 people were killed and some $25 million in property losses were tallied. Almost all of Guadeloupe's staple banana harvest for that year was ruined; several small watercraft were thrown ashore by Inez's 10-foot (3-m) seas and wrecked.

In Hispaniola, where a deepening Inez delivered 135-MPH (217-km/h) gusts to the island's southeastern shores on the night of September 28–29, dozens of communities in both the Dominican Republic and Haiti sustained near cataclysmic damage. Placed by Inez's west-northwesterly track within the hurricane's DANGEROUS SEMICIRCLE, the Dominican towns of Enriquillo, Duverge, and Oviedo were completely razed, their houses, offices, schools, churches, and hospitals first unroofed and then collapsed by 140-MPH (225-km/h) winds. Nearly all of the country's sugar plantations were obliterated, the tall stalks of cane uprooted and broken and blown into long ricks at opposite ends of sodden fields. Chased from their homes by roaring floods and shifting mudslides, scores of residents took refuge in Oviedo's town hall, the only structure to survive the storm with its stuccoed stone walls still standing.

In the Haitian port towns of Jacmel and Bathet, Inez's sweeping gusts snapped palm trees, tore storm shutters from windows, turned front porches into flying carpets, and sent undulating waves crashing against towering headlands. Torrential 12-inch (305-mm) downpours quickly converted southern Haiti's mountain gorges into a watery web of deadly rivers, floods, and landslides that soon added tons of mud and wreckage to the frothing torrents below. Strewn with hundreds—possibly even thousands—of human and ANIMAL corpses, these impromptu rivers eventually emptied into the CARIBBEAN SEA, littering miles of Haiti's coastline with their gruesome cargo. All told, some 1,379 people perished during Inez's northwesterly passage over Hispaniola, 200 in the Dominican Republic and the remainder in Haiti. Another 2,200 were injured in both nations, many of them seriously.

Considerably weakened by its assault on Hispaniola, Inez slowly spiraled northwest, delivering 81-MPH (130-km/h) winds and sporadic rain showers to Cuba's eastern Oriente Province on the morning of September 30, 1966. Although several buildings and a water tank near the Guantánamo naval base were damaged, Inez's initial landfall in Cuba claimed no lives. Widespread destruction of that season's sugarcane harvest was reported, however. Struggling inland, Inez spent the next two days executing a clockwise loop-the-loop course change over the island. At one point exiting southern Cuba altogether near the city of Cienfuegos, Inez finally came full circle on October 2, at a spot that put its Category 1 EYEWALL south of Miami, Florida. Maximum winds of 79 MPH (127 km/h), violent rain squalls, and at least one confirmed TORNADO buffeted the BAHAMAS as Inez, stalled for most of October 2 by an encroaching high-pressure AIR MASS, sought a STEERING CURRENT to carry it on its way.

After killing one person in the northern Bahamas on the morning of October 3, Inez began to drift languidly west. Quickly reintensifying as it bore down on the delicate coral chain of the Florida Keys, Inez crossed into the GULF OF MEXICO later the same day as a Category 3 hurricane of extreme intensity. Still under the influence of the high-pressure ridge, Inez battered Key West and the Dry Tortugas with 125-MPH (201-km/h) winds, flooding PRECIPITATION counts, and crushing 12-foot (4-m) seas. Five Floridians perished, one of them a teenage surfer.

Churning to the west-northwest at the pedestrian speed of 6 MPH (10 km/h), it would take Hurricane Inez another eight days to reach Mexico's east coast. There, on the afternoon of October 11, 1966, the 110-MPH (177-km/h) eyewall winds of Inez lurched ashore

approximately north of the port city of Tampico. Hurricane-force winds and pounding rains uprooted trees and inundated swamps but caused no fatalities. By the afternoon of October 12, as the last of Inez's core thunderstorms dissipated over central Mexico, the storm was gone forever. Lasting a total of 18 days, claiming nearly 2,000 lives, and inflicting tens of millions of dollars in property losses, Inez proved itself to have been among the most notable of contemporary hurricanes. The name *Inez* has since been retired from the cyclical list of HURRICANE NAMES.

Iniki, Hurricane *Hawaii, September 6–11,1992*
The first "billion-dollar" HURRICANE in Hawaiian history, Iniki lashed the northwestern islands of Niihau, Kauai, and Oahu with sustained 130-MPH (209-km/h) winds, 3–6-inch (76–152 mm) rainfall counts, and 30-foot (10 m) seas on September 11, 1992. Boasting a CENTRAL BAROMETRIC PRESSURE of 27.91 inches (945 mb) at landfall on Kauai, Iniki was the first Category 3 hurricane to strike the archipelago since November 1982, when Hurricane Iwa's 111-MPH (179-km/h) winds destroyed some 2,325 BUILDINGS on Kauai and Oahu, killing one person. A more intense storm than Iwa, Iniki leveled more than 10,000 buildings on Kauai and severely damaged several high-rise hotels in neighboring Oahu. While only five fatalities were reported, Iniki's $1.8 billion price tag immediately deemed the storm the most destructive Hawaiian hurricane of the 20th century

The only TROPICAL CYCLONE of the 1992 Hawaiian HURRICANE SEASON, Iniki developed over the shimmering waters of the eastern NORTH PACIFIC OCEAN, south-southeast of the big island of HAWAII, on the evening of September 6. Steadily curling northwest at nearly 15 MPH (24 km/h), the young TROPICAL DEPRESSION gradually sapped the ocean of its heat and moisture, graving into a weak TROPICAL STORM on the afternoon of September 8 and a minimal Category 1 hurricane, with a central barometric pressure of 28.96 inches (980 mb), by midday on September 9. Maximum sustained winds in Iniki's EYEWALL now measured 82 MPH (132 km/h), while heavy rains pelted the swirling maelstrom below.

In HAWAII, where painful memories of hurricanes Iwa and Dot (1959) made subsequent central Pacific storms a matter of significant concern, forecasters promptly set to work tracking Iniki's progress toward the islands. Even though Iniki was still several hundred miles from Hawaii, comprehensive computer ANALOGS based on climatological data and the documented behavior of earlier hurricanes indicated that the tropical cyclone had every opportunity not only to reach the state but also to grow into a major storm along the way. TEMPERATURE readings taken from the central Pacific revealed that the large pools of cool water that normally surround the Hawaiian Islands were mostly absent that season, removing one of the two natural buffers that tend to moderate Hawaii's hurricane activity. The other buffer, the high-pressure ANTICYCLONE that forms just north of the islands each summer, was similarly weak, thereby allowing a hurricane of considerable size and intensity to recurve northwest instead of maintaining the trade winds' passage due west.

On the morning of September 10, as Iniki wobbled northwest, lashing the sea with 100-MPH (161-km/h) winds and moderate rains, Hawaiians commenced the long, expensive task of preparing for a major hurricane strike. Although weather analysts were still not certain exactly where Iniki's EYE would come ashore, all coastal resorts on Oahu and Kauai were nonetheless evacuated early, costing hotel, restaurant, and store operators millions of dollars in lost revenue. While a fleet of buses and cars shuttled thousands of vacationers to cramped storm shelters further inland, hotel work crews began to drain swimming pools, to stow away beach umbrellas and deck chairs, and to board up panoramic plate-glass windows. Rows of shuttered storefronts turned Honolulu's streets into empty canyons, veritable echo chambers for the rising northeast winds. In Pearl Harbor, two nuclear-powered aircraft carriers, along with ten other vessels belonging to the United States Navy, wisely recalled their liberty details, piped their personnel to General Quarters, and then steamed to sea. Unable to remain in port for fear that Iniki's 25-foot (8-m) waves would drive the ships against their piers, the hastily collected squadron escaped west-southwest, spending the storm's duration safely wallowing in the 37-foot (9 m) swells kicked up by Iniki's distant northwest passage.

With a central barometric pressure of 27.91 inches (945 mb) at landfall, Hurricane Iniki roared into the island of Kauai, north of the town of Kapaa, on the afternoon of September 11, 1992. As 145-MPH (233-km/h) gusts shivered the 5,000-foot (1,600-m) slopes of Mount Kawaikini, Iniki's 4–6-inch (102–152-mm) rains pelted roofs, carports, billboards, and palm groves, making an incongruous concert of the bangs, booms, rattles, and scrapes that now and again could be heard over the wind's dull scream. Those principal settlements located on the mountain's windward side Lihue, Koloa, and Princeville—bore the brunt of Iniki's southeasterly approach. Thousands of dwellings, from multimillion-dollar beachfront mansions to humble bungalows tucked into the forests of Kawaikini, were progressively pried apart by the hurricane's sustained 130-MPH (210-km/h) winds. More than 8,000 people,

their houses in ruins, were forced out into the storm, compelled to find refuge in cars, local police and fire stations, schools, factories, and churches. When Iniki stripped the roof from a school near Haena Point, dozens of evacuees found themselves displaced once more. Huddled on a bus, some debated how much longer the vehicle would remain upright in the buffeting winds. Because driving it to another shelter was out of the question, they decided to arrange themselves so as to balance the swaying bus against the wind and wait out the storm. They survived.

Five other Hawaiians did not, however, making Iniki the deadliest hurricane to have pounded the islands since a preseason tropical cyclone on April 2, 1868, killed nearly 100 people on Molokai. In a year that had already seen Hurricane ANDREW rampage through south FLORIDA and LOUISIANA to the record-breaking tune of $20 billion, and Typhoon Omar (August 28) destroy almost every building on the U.S., held island of Guam, Iniki's $1.8 billion in property and agricultural losses could only seem a fitting, if relatively modest, finale to such a costly storm season. Quickly responding to the tragedy, President George Bush declared Oahu, Kauai, and Niihau disaster areas on September 12; Congress drafted a record $11.1 billion relief bill for survivors of the three storms. Criticized for adding to the federal deficit, the bill was eventually signed into law on September 23, the day the World Meteorological Organization announced that the name Iniki was being retired from the alternating list of central Pacific HURRICANE NAMES.

intertropical convergence zone Sometimes referred to as the equatorial TROUGH, the intertropical convergence zone (ITCZ) is the zone where Earth's two atmospheric hemispheres (north and south) come together. Better known in layman's parlance as the doldrums, the ITCZ is also where the Northern Hemisphere's easterly trade winds meet the Southern Hemisphere's easterly trade winds, thus providing a circulatory environment favorable for TROPICAL CYCLONE generation. The trade winds themselves are a result of the clockwise-spinning, high-pressure ANTICYCLONES that dominate circulation patterns over the Atlantic, Pacific, and Indian oceans.

Satellite photographs taken of Earth clearly indicate the climatological dimensions of the ITCZ, show it to contain a fairly regular progression of cumulus and CUMULONIMBUS cloud clusters moving from east to west over Central America and across the NORTH PACIFIC OCEAN to the BAY OF BENGAL and North Africa. Determined in part by the size and position of the winter and summer anticyclones, the parameters of the ITCZ will often migrate, shifting from north to south and back again within the

This satellite photograph shows the intertropical convergence zone, or the area where the atmospheres of the Northern and Southern Hemispheres meet. As can be observed from the photograph, it is characterized by unstable air masses. Many tropical waves and even tropical depressions have generated from the ITCZ. *(NOAA)*

course of a year. During the months of August and September, the intertropical convergence zones over the NORTH ATLANTIC and North Pacific oceans can extend as far north as 30 degrees, or to the underside of their respective anticyclones. Later, with the onset of October and November, they subside south to the equator, where they remain in place until the conclusion of the southern summer.

Held to the south by the North Atlantic's BERMUDA HIGH, the central North Pacific's Hawaiian High, and the western Pacific's Ogasawara High, tropical waves—cloud-bound centers of low barometric pressure—slowly swirl across the open ocean, following the ITCZ and its slight but influential winds to possible deepening and an upgrade to TROPICAL DEPRESSION, TROPICAL STORM, HURRICANE, TYPHOON or CYCLONE. In an average year, approximately 110 tropical waves develop within the North Atlantic's intertropical convergence zone, with only 10 or so ever maturing into tropical storms or hurricanes; the numbers are slightly higher in the North Pacific's east and west quadrants, where an average year will see 20 to 25 tropical cyclones develop from 150 tropical waves.

Although the ITCZ's cradle of low-pressure is in itself conducive to tropical cyclogenesis, the ITCZ's varying high- and low-altitude winds often are not. In order for convergence, or the rapid inflow of air into a particular low-pressure area, to take place, existing air must be drawn away, usually through convective, or warming, processes. If, for instance, warm air is blocked from rising by high altitude winds that are moving at a different speed (WIND SHEAR), then convergence will not take place, and the tropical wave

will remain as it is. Should air currents in the upper atmosphere undergo ANTICYCLONIC divergence, however, and then a vertical circulation within a tropical wave can be established because the OUTFLOW of air above provides no impedance to the converging, or incoming, air below. Steadily carried westward by the ITCZ's trade wind STEERING CURRENTS, the embryonic tropical disturbance can either experience further intensification or succumb to a host of other meteorological and geographical enemies, including cool sea-surface temperatures, barricading warm and cold fronts, and, of course, landfall.

inverse barometer Proportionally linked to the PRESSURE GRADIENT of a TROPICAL CYCLONE, the inverse barometer is that rise in sea level created by a HURRICANE'S, TYPHOON'S or CYCLONE'S low CENTRAL BAROMETRIC PRESSURE. This upwelling of the sea, which meteorologists have now determined measures a half-inch for every one MILLIBAR drop in pressure, is barely noticeable while the tropical cyclone is over open waters but can, in the event of landfall, begin to shoal against the seabed, piling up into extremely dangerous STORM SURGE and STORM TIDE conditions.

Irene, Hurricane *Caribbean Sea–Southern United States, October 13–19, 1999* The ninth named tropical system of the 1999 North Atlantic HURRICANE SEASON, Irene delivered heavy rains from FLORIDA to VIRGINIA between October 15 and October 19, 1999. A Category 1 HURRICANE (with a central pressure of 29.14 inches [987 mb]) at landfall in

Florida on October 15, Irene's heavy winds and rains caused more than $100 million in agricultural losses across Florida's Miami-Dade county. Between five and 15 deaths occurred in Florida and NORTH CAROLINA, respectively, and damage estimates were estimated to exceed $800 million (in 1999 U.S. dollars).

Between 1959 and 2005, five North Atlantic tropical cyclones have bore the identifier Irene; a tropical storm in 1959; a hurricane in 1971; another hurricane in 1981; and a hurricane in 2005. The name Irene has been retained on the list of hurricane names and will be used again for the 2011 season.

Iris, Hurricane *East Caribbean–North Atlantic, August 22–September 4, 1995* At one point in its existence a Category 2 HURRICANE of considerable destructive potential, Iris spent 14 drenching days between August 22 and September 4, 1995, showering lukewarm winds and rains across the eastern Caribbean islands of GUADELOUPE, MARTINIQUE, Montserrat, and St. Lucia, and over the waters of the NORTH ATLANTIC OCEAN. Of TROPICAL STORM strength as it crossed the Leeward Islands on the night of August 26–27, Iris's 6-inch (152-mm) PRECIPITATION counts and resulting landslides killed three people and caused damage estimates of $29 million. Seventeen percent of the banana crop on the island of St. Lucia alone was ruined in the storm, the ninth TROPICAL CYCLONE of the record-setting (19 named storms) 1995 HURRICANE SEASON.

Initially a quick bloomer, Iris was first upgraded to TROPICAL STORM status on the morning of August 23, when its sustained winds topped 42 MPH (68 km/h). Iris continued to deepen throughout the afternoon of August 23 and was classified a Category 1 hurricane by early evening. Plodding to the northwest at seven MPH (11 km/h), Iris maintained its 80-MPH (129-km/h) winds until the following evening when they suddenly dwindled to tropical storm force, the victims of upper-altitude WIND SHEAR. Tweaking ANEMOMETERS at 65 MPH (105 km/h), Iris greeted its demotion with a period of further weakening that saw its 45-MPH (72-km/h) winds slowly huff across Guadeloupe and Martinique on the night of August 26–27.

On occasion producing 70-MPH (113-km/h) gusts, Iris's true threat to the Leeward Islands rested with its substantial rains and the flash floods that suddenly blossomed along the eastern Caribbean hillsides. In Martinique, where Iris's impending arrival closed down the airport and sent residents scurrying for shelter, a man and a woman were crushed to death when their mountain house was swept into a gorge by a mudslide and then buried. In Guadeloupe, Iris's six-inch (152-mm) downpours forced rivers to jump their banks, inundating several communities with more than two feet (1 m) of water. One man in Guadeloupe was drowned; several hundred houses suffered severe water damage.

On Montserrat, a tiny British dependency north of Guadeloupe, islanders faced a double threat from both Tropical Storm Iris and the restless, spewing volcano that had been rumbling beneath their houses for almost a week. Seismologists had warned that Chances Peak had a 75 percent probability of erupting sometime in the near future, a pronouncement that encouraged 3,000 of Montserrat's 12,000 residents to evacuate the island immediately by airplane and boat. Another 6,000 people, removed from their vulnerable homes on the volcano's flanks, were housed in tent cities, canvas sails that seemingly caught every last drop of Iris's rain-lashed, two-day passage. Perhaps it was the cooling effects of Iris's 4-inch (102-mm) rains that squelched the fires of Chances Peak, for within days of the storm's passage the volcano fell quiet and the islanders hesitantly, yet gratefully, returned to their homes. It was a short reprieve; in 1998, Chances Peak erupted, destroying 90 percent of the foliage and structures on the island.

Having made the last landfall of its life on Montserrat, Iris swung into the northern CARIBBEAN SEA on the morning of August 28, coming under the influence of a low-pressure trough and beginning to hike northward at nine MPH (15 km/h). Its CONVECTION cells strengthened by their return to the warm, evaporative waters of the Caribbean, Iris soon regained hurricane status. By the morning of August 29, Iris had undergone even greater deepening, with sustained winds of 85 MPH (137 km/h) now being recorded within its tightening EYEWALL. Crawling north-northeast at six MPH (10 km/h), Iris was firmly in the grasp of the low-pressure trough that would eventually draw the hurricane sharply northeast to the distant reaches of the North Atlantic, where it would dissipate on September 4—but not before Iris, the perennial tropical cyclone of many feats and surprises, would grow even larger and more powerful during the night of August 31 and September 1, 1995.

As though consciously duplicating the spectacular strengthening patterns seen during its initial transformation from tropical storm to hurricane just one week before, Iris again rapidly intensified. By the morning of September 1, as Iris suddenly stalled some 1,339 miles (2,155 km) east of Miami, FLORIDA, its winds were clocked at 105 MPH (169 km/h), giving it a Category 2 rating on the SAFFIR-SIMPSON SCALE of damage potential. By the following afternoon, however, Iris's winds had once again moderated, dropping to 100 MPH (161 km/h), or minimal Category 2 status.

Coasting within southeastern BERMUDA on September 2, Iris compelled authorities to place the island under a HURRICANE WATCH but otherwise caused little damage. The name Iris has been retired from the rotating list of North Atlantic tropical cyclone names.

Isabel, Hurricane *North Atlantic Ocean–Eastern United States, September 6–19, 2003* The fourth HURRICANE of the 2003 North Atlantic HURRICANE SEASON, Isabel reached Category 5 status twice on its long journey across the NORTH ATLANTIC before lunging ashore in NORTH CAROLINA on September 22, 2003.

Isabel originated near latitude 14 degrees North, longitude 34 degrees West, during the early afternoon hours of September 6, 2003. By the following day, as it commenced a RECURVATURE to the north-northwest, the system was first upgraded to a TROPICAL STORM, and by early evening, to a Category 1 hurricane. As it bore down on the eastern seaboard of the United States, Isabel continued to deepen, achieving Category 2 status (with a central pressure of 28.64 inches [970 mb]) on September 8, and Category 3 status later the same day. Undergoing rapid deepening, Isabel was upgraded to a Category 4 hurricane on September 9, with a central pressure of 27.99 inches (948 mb). Slashing well to the north of the Windward Islands, Isabel continued to deepen between September 9 and 11; on the night of September 11, the system's central pressure sank to 27.19 inches (921 mb), making it a Category 5 system. The system produced sustained winds of 160 MPH (257 km/h), making it an exceptionally powerful tropical system. After a brief downgrade to a Category 4 system on the morning of September 13, the system regained Category 5 intensity later that day, with a central pressure of 27.52 inches (932 mb). By September 15, 2003, Hurricane Isabel was located about 780 miles (1,255 km) south-southeast of Cape Hatteras, NORTH CAROLINA. Moving to the west-northwest at eight MPH (13 km/h), and again downgraded Category 4 Isabel generated sustained winds of 140 MPH (225 km/h) and higher gusts.

Hurricane Isabel continued its northwesterly recurvature, eventually trundling ashore in North Carolina on September 18, 2003. At landfall, Isabel was a borderline Category 2 system with a central pressure of 28.26 inches (957 mb). While 83,000 people lost electrical service, Isabel's wrath claimed three lives in North Carolina, seven in Maryland, 19 in VIRGINIA, one in New Jersey, two in Pennsylvania, and another in Rhode Island—all told, some 50 people lost their lives to Isabel's passage. In West Virginia, flooding associated with Isabel's passage caused the Potomac River to crest at 4.8 feet (2 m) above flood stage at Shepherdstown, West Virginia,

and 1.8 feet (.5 m) above flood stage at Harpers Ferry. Isabel's remnants were also responsible for an outbreak of TORNADOES in Mercer County, New Jersey, claiming no lives but causing widespread property damage. Some $3.3 billion (in 2003 U.S. dollars) in property losses were recorded.

The name Isabel—which had also been used for a tropical storm in 1985—has been retired from the rotating list of North Atlantic tropical cyclone names.

Isidore, Hurricane *North Atlantic–Caribbean Sea–Southern United States, September 14–26, 2002* The ninth named TROPICAL CYCLONE of the 2002 North Atlantic HURRICANE SEASON, Hurricane Isidore caused considerable damage in CUBA, MEXICO, and along the northern Gulf Coast of the United States between September 18 and September 26, 2002. At one point an extremely powerful Category 3 HURRICANE with a peak pressure reading of 27.58 (934 mb), Isidore originated very close to the northern coast of South America. Its proximity to the continent hindered its strengthening configuration, and the system spent the better part of four days as a TROPICAL DEPRESSION. As it recurved to the north-northwest, low WIND SHEAR and the increased CORIOLIS EFFECT allowed the system to better organize, and on September 18, 2002, it was upgraded to a TROPICAL STORM. Isabel continued to strengthen; by September 20, its sustained winds had reached 105 MPH (169 km/h) and it was now nearing the Isle of Youth on the extreme western tip of Cuba. Isabel continued to intensify, and by the following day, its central pressure had dropped to 28.55 inches (967 mb), boosting its sustained Category 3 wind speeds to 127 MPH (204 km/h). The system's passage near Cuba caused considerable flooding on the island. Isidore's brunt, however, was reserved for Mexico's Yucatán Peninsula, which the system pounded late in the day on September 22. Between 12 and 20 inches (305 mm–508 mm) of rain fell across the jungles of the Yucatán, killing two people there.

At one time a powerful hurricane, Isidore had mercifully been downgraded to tropical storm intensity by the time it came ashore near New Orleans—at Grande Isle—on the morning of September 26, 2002. Tremendous rains—on the order of 10 inches (254 mm)—deluged much of the western Gulf Coast. New Orleans was particularly hard hit; the waters of Lake Pontchartrain spilled across roads, driven in part by Isidore's sustained wind speeds of 60 MPH (97 km/h). On September 25, Isidore's cloud bands spun off a TORNADO that touched down in Lafourche parish, LOUISIANA, causing damage to several emergency vehicles. The system also generated a tornado that caused one injury and seriously damaged two dozen BUILDINGS in the western FLORIDA Panhandle. Spinning northward (dropping up to 12 inches [305 mm] of rain along the way), Isidore's rainy remnants delivered some two inches (51 mm) of precipitation to the NEW YORK CITY area on September 27, flooding subway tunnels and snarling traffic. Insurance estimates for Isidore reached some $100 million (in 2002 U.S. dollars) in losses.

Used before in 1984 for a tropical storm, in 1990 for a hurricane, and in 1996 for another mature stage tropical cyclone, the name Isidore was retired from the rotating list of North Atlantic tropical cyclone names following its 2002 passage. It was replaced on the list with the identifier Ike.

isobar By definition, *isobars* are the lines on a weather chart that link points of equal barometric pressure. Used by meteorologists to determine the size, position, and climatological character of respective AIR MASSES, isobars are derived from the hundreds of barometric pressure readings reported daily by weather observatories around the world. First sorted by position, these readings are then plotted on a map, with all points of similar pressure being joined by a line, or isobar. On any given day of the year, the isobars on a map of Earth can either appear as long, wavy lines or as smaller pockets of varying barometric pressures. Often the isobars curve, seeming to radiate from a center of high or low barometric pressure—including those particularly low readings found in TROPICAL CYCLONES. Because a HURRICANE, TYPHOON, or CYCLONE is a center of low barometric pressure around which spiral thunderstorms have banded, the isobars in such systems are circular, a series of rings that grow smaller and more closely spaced as they approach the tropical cyclone's EYE, or point of lowest barometric pressure.

Ivan, Hurricane *Windward Islands–Jamaica–Cuba–Southern United States, September 2–16 and 22–24, 2004* One of the most eccentric, and terrible, North Atlantic TROPICAL CYCLONES yet observed, Hurricane Ivan invaded the CARIBBEAN SEA and southern United States with powerful winds and deluging rains between September 2 and 24, 2004. Achieving Category 5 status at least three times during its stormy passage, Ivan tallied damage losses in the United States of at least $13 billion (in 2004 U.S. dollars) and claimed some 120 lives in Haiti and the United States. According to the NATIONAL HURRICANE CENTER (NHC), Ivan presently holds the record for the "greatest consecutive time (seven days) spent with wind speeds of 120 MPH (138 km/h) or greater for any basin."

The ninth TROPICAL DEPRESSION of the 2004 season, Ivan formed on September 2, 2004, in the eastern NORTH ATLANTIC OCEAN. On September 3,

the system was upgraded to Tropical Storm Ivan, and became a hurricane on September 5, 2004. Undergoing a very rapid deepening cycle, Ivan was upgraded to a Category 4 hurricane that very same day.

On September 7, 2004, Hurricane Ivan moved across the island of Grenada, causing widespread destruction. Once it had entered the CARIBBEAN SEA, Ivan continued to intensify, becoming a Category 5 hurricane (with a central pressure of 27.13 inches [919 mb]) during the early morning hours of September 9, 2004. Ivan's EYE passed to the southwest of JAMAICA on September 11, 2004, with sustained wind speeds of 150 MPH (241 km/h), qualifying it as a Category 4 system. As Ivan approached the Cayman Islands during the late-night hours of September 11, the system was again upgraded to Category 5 status, with an astoundingly low central pressure of 26.87 inches (910 mb). Widespread destruction was recorded in the Cayman Islands. The system remained at Category 5 intensity until September 12, 2004, when it again weakened to a Category 4 system.

Ivan regained Category 5 intensity (with a central pressure of 26.93 inches [912 mb]) on September 13, 2004. It passed the western tip of CUBA and spent the next three days moving northwestward in the GULF OF MEXICO, slowly weakening along the way.

On September 16, 2004, Hurricane Ivan roared ashore near Gulf Shores, ALABAMA, with sustained winds of 120 MPH (193 km/h), making it a Category 3 system. It tracked slowly to the northeast and weakened to an extratropical low as it moved offshore of the Delmarva Peninsula on September 19, 2004. Ivan's remnants continued to make a loop moving southwestward, then westward and crossing FLORIDA into the GULF OF MEXICO on September 21, 2004. Ivan was again upgraded to tropical storm intensity before making a second landfall over the extreme southwestern tip of LOUISIANA on September 24, 2004. The storm dissipated over eastern TEXAS later the same day. An eccentric and violent system, Ivan's destruction of petroleum producing and refining (primarily through damage of undersea pipelines) cost the American energy industry some 45 million barrels of crude oil production for the next six months. All told, 124 people were directly or indirectly killed by Hurricane Ivan, and final damage estimates soared to $19.7 billion (in 2005 U.S. dollars), making the system one of the 10 most destructive in recent memory.

Between 1980 and 2004, three tropical cyclones in the North Atlantic carried the identifier Ivan. The first was a hurricane in 1980, and the second another hurricane in 1998. Owing to the enormous destruction wrought by the 2004 system, the name Ivan has been retired from the rotating list of North Atlantic tropical cyclone names. It has been replaced by Igor and will appear during the 2010 season.

J

Jamaica *Northern Caribbean Sea* This rugged limestone plateau, famed for its invigorating mountain resorts, breezy western beaches, forested gullies, and fertile sugar plantations, has suffered an extensive history of notable HURRICANE strikes. Situated in the northern CARIBBEAN SEA, Jamaica is the third-largest of the four major islands—the others are CUBA, HISPANIOLA, and PUERTO RICO—that encompass the expansive Greater Antilles. Jamaica's position between latitudes 18 and 20 degrees North places it directly in the path of those mid- to late-season hurricanes that either spiral across the NORTH ATLANTIC OCEAN from Africa's Cape Verde Islands or originate over the south and west Caribbean's warm waters. Between 1670 and 2006, Jamaica's bauxite mines, mountain ranges, and shallow seaports were directly affected by no less than 83 recorded TROPICAL STORMS and hurricanes. Of these, 31 were major hurricanes, systems with CENTRAL BAROMETRIC PRESSURES of 28.47 inches (964 mb) or lower and winds in excess of 111 MPH (179 km/h). On at least seven occasions between 1751 and 1909, this plantation island suffered several hurricane strikes in quick succession, as happened in 1751 when three powerful storms came ashore between August 10 and October 16, and in 1772 when two hurricanes struck in just over one week's time. Similar double strikes also occurred in 1804, 1812, 1813, 1818, and 1909. On September 12, 1988, Jamaica had the misfortune to be struck by the gargantuan Hurricane GILBERT. Rated a Category 3 at landfall near Kingston, Gilbert's 122-MPH (195-km/h) winds killed 45 people, destroyed nearly 80 percent of all the structures on the island and caused damage estimates of $2 billion. Defying conventional hurricane behavior, Gilbert steadily intensified as it moved across the hostile peaks of central Jamaica, reemerging into the Caribbean as first a Category 4 and then a Category 5 TROPICAL CYCLONE before roaring ashore near Cozumel, MEXICO, on the morning of September 14, 1988.

Although a majority of early season hurricanes form in the southwest Caribbean or lower GULF OF MEXICO and move northeast, only one documented June hurricane has ever beset Jamaica. This particular hurricane, which occurred on June 7, 1692, coincided with the cataclysmic destruction by EARTHQUAKE of the notorious pirate haven, Port Royal. Considered a powerful storm by contemporary sources, the hurricane fortunately spared devastated Port Royal by directing its intense assault at Jamaica's west coast, coming ashore in the vicinity of the storm-prone township of Savanna-La-Mar on the morning of June 7. This coincidence of timing has in the past led some historians and scientists to seek a connection between the hurricane's landfall and the violent seismic activity that demolished distant Port Royal. While it is entirely possible that the June hurricane's low central barometric pressure may have been responsible in part for either the earthquake itself or the enormous tsunami that soon followed it, no uncontested data exists to confirm it.

In July, however, as the sun-baked waters of the southeastern Caribbean begin to reach the critical threshold required for hurricane formation, the probability of a hurricane landfall in Jamaica rises slightly, with 6 reported strikes between 1782 and 2006. Also it is during the hot, humid months of August, September, and October when the Cape Verde hurricanes begin to arrive at the Caribbean's east entrance that the number of strikes climbs to a hefty 23, 11, and 20, respectively. Noted midseason hurricanes in Jamaica include the violent hurricane of August 28, 1722, in which a reconstructed Port Royal suffered the loss of 26 merchant ships and the deaths of more than 400 people. Hundreds of brand-new BUILDINGS were blown down, while 25-foot (8-m) seas smashed wharves and dockyards to pieces. Just over four years later, an equally intense hurricane ravaged Jamaica's southeast coast, bringing strafing winds and rains to Port Morant, St. Anne's Bay, Port Royal, and the newly founded capital, Kingston. More than 50 vessels, from sugar-laden merchant ships to sprightly Royal Naval sloops, were caught at anchor by the unexpected tempest. Despite valiant attempts by the ships' crews to lighten their vessels by jettisoning valuable merchandise and cannon, all 50 were sent to the bottom or run onto the volcanic shore by the hurricane's prodding winds. An estimated 500 sailors perished. During a similar hurricane in October 1744, nine British warships of between 50 and eight guns, as well as 96 rum-packed merchant ships, were driven ashore in Kingston Harbour and Port Royal, taking the lives of another 400 people. A deadly hurricane during the first week of October 1818 inundated Jamaica's eastern beachheads with an 11-foot (4-m) STORM SURGE that demolished several warehouses in Port Royal, while an even greater surge during the height of the legendary SAVANNA-LA-MAR HURRICANE of 1780 demolished more than 350 houses as it rolled inland at Savanna-La-Mar, an important sugar port on the island's southwest coast, just before noon on October 3. Stated by eyewitnesses to have been 25–30 feet (8–10 m) high, the surge killed almost 200 people as it overwhelmed several square miles of the surrounding landscape. The first of three great hurricanes to traverse the Caribbean during the 1780 HURRICANE SEASON, the Savanna-La-Mar Hurricane's effects were felt as far east as Port Morant, where its residual storm surge carried three merchant ships nearly a quarter of a mile inland. Almost 1,000 lives across the island were lost to this extraordinary storm, one of the most severe to have struck Jamaica in its history.

Jamaica would later sustain similar, if only slightly less destructive, hurricane strikes in August 1781, when more than 378 people perished in Kingston; in late July 1784, when an enormous hurricane sank another 80 ships at Port Royal and Kingston and swept hundreds of buildings into the raging surf; and during the three-hour blast on August 7, 1832, when an estimated 300 lives were lost and several million dollars in damage assessed in what proved a short but intense coil of destructive energy. On September 25, 1896, a severe hurricane inundated the entire southwest coast from Savanna-La-Mar to Negril Beach, with a towering 14-foot storm surge that cast dozens of small boats ashore and ruined hundreds of dwellings. Farther inland, damage to sugar and plantain crops was widespread, precipitating a poor harvest for plantation owners. In November 1912, the sandy coves of Negril Beach were visited by a late-season storm, the notable BLACK RIVER HURRICANE of November 18. One-hundred-twenty-mile-per-hour (193-km/h) winds, blistering rains and a 30-foot storm surge killed more than 100 people and caused thousands of injurious building failures.

More recently, Jamaica was ravaged by the third tropical cyclone of the 1951 season, Hurricane CHARLIE. For nearly 24 hours between August 17 and 18, Charlie delivered 130-MPH (209-km/h) winds and more than nine inches (229 m) of PRECIPITATION to townships all over the island. More than 30,000 people were left homeless as those neighborhoods that surrounded Port Royal and Port Morant were racked by mudslides and flash floods. Extensive agricultural damage initiated a prolonged financial drought that persisted—despite the best interests of British relief efforts—through the winter of 1951–52. More than a quarter of a century later, Hurricane Gilbert brought heavy damage to the entire island. Regarded at the time as the most destructive hurricane to have ever blasted Jamaica, Gilbert's $2 billion passage served as an expensive reminder of a long history of hurricane activity that has, and will unfortunately continue to be, among the most violent in the Caribbean.

In 2004, Hurricane IVAN caused extensive damage in Jamaica, killing 14 people. During the remarkably active 2005 North Atlantic season, Hurricane EMILY killed five people in Jamaica in July; and in October of the same year, the tropical depression that would eventually grow into the most power North Atlantic hurricane yet observed—WILMA—killed one person in Jamaica, and caused flooding and mudslides in several low-lying communities. Hundreds of Jamaicans were forced to flee their residences for government shelters.

Between August 17 and 18, 2007, Category 4 Hurricane Dean passed some 50 miles (80 km) south of Kingston, Jamaica. The system generated precipitation counts of 20 inches (500 mm) and storm surges of nearly 21 feet (7 m) along the island's southern shores. Caught in the system's northern quadrant, Jamaica suffered damage to trees and small buildings, but unlike Hurricane Ivan in 2004, no deaths were reported in Dean's wake.

Janet, Hurricane *East Caribbean–Mexico, September 20–29, 1955*   One of the deadliest HURRICANES to have traversed the CARIBBEAN SEA during the second half of the 20th century, Janet's 150-MPH (241-km/h) winds and torrential rains claimed 585 lives, left more than a quarter of a million people homeless, and wrought some $100 million in property damage as it swept across the eastern islands of BARBADOS and Grenada and into MEXICO's southeast coast between September 23 and 28, 1955. An intense Category 4 hurricane of considerable size and fury at its initial landfall in Mexico, Janet's nine to 11-inch (229–279-mm) rainfalls coupled with the hurricane's 16-foot (5-m) STORM SURGE instantaneously drowned 164 people along the east coast of the Yucatán Peninsula on September 28. In nearby Belize, then known as British Honduras, an additional 61 lives were lost and 247 people injured, when Janet's speeding winds bulleted the northern port town of Corozal with flying debris and five-inch PRECIPITATION counts. In the Windward Islands, where Janet entered the southeast Caribbean as a powerful Category 2 hurricane on September 23, between 150 and 225 people were reportedly killed on Barbados and Grenada, making Janet the deadliest hurricane to have struck that part of the island chain in nearly a quarter of a century. In a tragic sidebar to Janet's violent passage, two separate plane crashes—one of a hurricane-tracking *NEPTUNE-P2V-3W* on September 27, and the other of a Mexican relief plane en route from Mérida to Chetumal on September 30—claimed an additional eleven and five lives, respectively.

The 10th storm of the remarkable 1955 HURRICANE SEASON (which had earlier seen the United States struck by no less than three major hurricanes—CONNIE, DIANE, and EDNA—since August 12), Janet originated over the simmering waters of the mid-Atlantic approximately 644 miles (1,036 km) due east of Barbados island on the morning of September 21, 1955. Steadily carried west by the equatorial trade winds, the incipient TROPICAL DEPRESSION rapidly robbed the first five feet (2 m) of the ocean's rolling surface of its heat and moisture. Maturing into a sprightly TROPICAL STORM during the early morning hours of September 21, and a well-developed Category 1 hurricane, with a CENTRAL BAROMETRIC PRESSURE of 28.89 inches (978 mb) and 86-MPH (138-km/h) winds by midnight of the same day Janet continued moving west-southwest, riding the underside of a high-pressure ridge toward the northeastern tip of South America. Meteorologists at the NATIONAL HURRICANE CENTER (NHC) in Miami, FLORIDA, speculated that if Janet maintained its present course, it could very well become one of those rare but unforgettable storms that periodically ravage Barbados before startling the islands of Trinidad and Tobago and the coast of Venezuela with hurricane-force winds and suffocating mudslides.

By the early morning hours of September 22, as Janet bore down on the lightly wooded hillsides of Barbados, weather reconnaissance flights through the hurricane's EYE indicated barometric pressures of 28.51 inches (965 mb) and intermittent wind gusts of 128 MPH (206 km/h). Early ANALOGS, based on available climatological data, predicted that the hurricane would soon turn more to the west, perhaps crossing the islands of St. Lucia and St. Vincent as a still-deepening hurricane of serious destructive intent. Once it had merged into the Caribbean, its exact course, although uncertain, ranged from a possible strike on Aruba and Curaçao, to a swift curve north-northwest. By late afternoon on September 22, as the system's central pressure strengthened to 28.49 inches (964 mb), it became clear to storm watchers that Janet was indeed turning west, beginning a 250-mile (402-km) arc that would carry it across Barbados, past St. Vincent and the Grenadines Islands, and into Grenada, the southernmost island in the chain.

Roaring across Barbados just before midnight on September 22, 1955, Janet's sustained 105-MPH (169-km/h) winds and 4-inch (102-mm) rains tore into the island's ripening sugarcane fields, uprooting trees, leveling warehouses, and killing 24 people. In the capital city of Bridgetown, hundreds of BUILDINGS—from the stone hall that housed the island's government to the narrow rows of storefronts that lined the embankments of Bridgetown Harbor—were either destroyed or heavily damaged. Seven thousand people were evacuated as Janet's seven-foot (2-m) STORM SURGE lashed low-lying towns, sweeping away entire houses. All electrical and telephone services across the island were interrupted, while dozens of small boats were wrenched from their moorings and cast ashore, choking volcanic inlets with waves of wreckage. Several million dollars in damages were sustained, making Janet the costliest hurricane to have struck Barbados since 1928.

Maintaining its southwest curve, Janet moved back into the Atlantic, churning up tons of heat-laden spindrift as it spiraled toward the nearby volcanic pylon of Grenada. In Grenada, where a steady drizzle had for hours predicted Janet's approach, evacuation warnings had yielded an overcast landscape of closed spice plantations and abandoned, gust-riven beaches. Earlier crowded with hundreds of harborside refugees, the terraced roads of Grenada's capital, St. George's, were now cleared, ready to become deadly spillways for Janet's heavy rains and spontaneous floods. Tucked within their red-tiled houses, their storm cellars, and their mass shelters, the people of Grenada anxiously listened to their radios, monitoring with a palpable sense of dread the slew of damage reports and DEATH TOLLS then being issued by overwhelmed relief agencies in Barbados.

Completing its curl west, Janet roared past Grenada, approximately 42 miles (69 km) north of the island, on the morning of September 23, 1955. Although in many respects Janet's assault on Grenada was of a more offshore nature, the hurricane's 106-MPH (171-km/h) EYEWALL winds, extending outward for at least 50 miles (80 km), brought heavy rains and hurricane-force gusts to most of the island. Passing the north shores of the island at nearly 12 MPH (19 km/h), Janet's strafing winds and six-inch (152-mm) rains mowed down palm trees, uprooted telephone poles, wrenched storm shutters from their hinges, stripped shingles and tiles from roofs, and blew down chimneys. On Grenada's north coast, where Janet's nine-foot (3-m) storm surge inundated the beaches and bungalows of Levera Bay, seven men were drowned when their private yacht was broken away from its mooring and driven onto the rocks. Farther inland, Janet's steady downpours spawned serious flash floods along the 2,757-foot (919-m) slopes of Mount St. Catherine. More than half of Grenada's commercial and residential communities, including dozens of warehouses filled with nutmeg and coconut stores, suffered some degree of structural damage, later necessitating a year-long reconstruction effort. Between 125 and 200 people were reportedly killed on Grenada, the victims of floods, building failures, and mudslides. Many of the dead were forever buried beneath the mountainous floes, making it difficult for authorities to determine a more accurate death count.

Entering the southeast Caribbean Sea for the first time on the evening of September 23, 1955, Janet promptly underwent another rapid period of intensification. An increased round of reconnaissance flights indicated that following its passage through the Windward Islands, the hurricane's central pressure had deepened to 28.46 inches (962 mb), spiking its sustained wind speeds at between 111 and 115 MPH (179–185 km/h). Now classified as a low-grade Category 3 hurricane, Janet inexorably whirred northwest, bound on a course that if maintained would eventually take it ashore somewhere in eastern Mexico or perhaps as far north as the gulf coast of TEXAS. Caribbean civil defense authorities, appalled by Janet's staggering toll in Barbados and Grenada, immediately notified the Mexican government of the hurricane's potential threat. As updated meteorological data were passed to Mexican relief organizations, evacuation efforts in that country proceeded at a competent rate. Several vulnerable oceanside towns on the Yucatán Peninsula were successfully evacuated, the displaced sent inland to hillside shelters.

Just before midnight on September 26, 1955, as Janet's plodding trek to the northwest took it within 450 miles (724 km) of the Yucatán Peninsula's east coast, the U.S. Navy dispatched one of its hurricane-hunting Neptune P2V–3W aircraft to penetrate the storm's dangerous EYEWALL to record the wind velocities and barometric pressures found there. Departing from the Naval Air Station at Jacksonville, Florida, the Neptune headed southwest, droning on through the night as its crew of nine calibrated their instruments, the ANEMOMETERS, BAROMETERS, thermometers, and RADARS needed to complete this mission successfully. Also on board were two civilian photographers for one of CANADA's largest newspapers. Permitted to join the flight on the condition that a copy of every picture they took of the hurricane would be added free of charge to the weather service's photo library, the two journalists excitedly chatted with the crew as they loaded film into their cameras and then settled down in the tail of the plane to await the Neptune's arrival at Janet's position.

Dawn was hovering just over the horizon when the aircraft first encountered Janet's fringe cloud bands on morning of September 27, 1955. Eventually jolted awake by turbulence, the two photographers were disappointed to find that in the course of their sleep, the night had gone from black to gray and that the clear skies over the BAHAMAS had given way to a universe completely shrouded in heavy, seemingly indistinguishable clouds. Assured by a crew member that the view would greatly improve once the aircraft had entered the hurricane's eye, the journalists occupied their time by photographing the crew as they busily began to record Janet's vital statistics.

While anemometers recorded a host of wind speeds, from 65 MPH (105 km/h) near the system's outer edge to 150 MPH (241 km/h) around the eyewall, radar screens pictured crescent-shaped bands of rain, stark white spaces that in time the Neptune-P2V would deliberately, and at its own peril, enter. As the first series of observations were immediately radioed to the newly founded National Hurricane Center in Miami, the aircraft's pilot, Lieutenant Commander G. B. Windham, added that he was entering the eyewall at an altitude of 700 feet (233 m) in order to record the still-deepening hurricane's central barometric pressure.

What exactly happened next has never been determined, in part because no trace of the Neptune-P2V or of the 11 men aboard was ever found. Later inquiries into the aircraft's loss theorized that a faulty altimeter may have been responsible, but no conclusive verdict was ever reached. In this scenario, the intensely low barometric pressures found in Janet may have caused the Neptune's altimeters—which are essentially barometers—to provide a false reading. Commander Windham, believing that his aircraft was in fact much higher than it really was, would have maintained his assumed altitude until the plane simply flew into the sea. The loss of the Neptune-P2V, along with the subsequent crashes of three Air Force typhoon-tracking aircraft in the NORTH PACIFIC OCEAN because of false readings, prompted the eventual development of the radar altimeter.

Besides representing the first loss of a HURRICANE HUNTERS' aircraft in U.S. history, the crash of the Neptune-P2V denied meteorologists crucial insights into the strengthening configuration subsequently undertaken by Hurricane Janet. By the morning of September 27, 1955, as the hurricane came within 150 miles (241 km) of the Yucatán Peninsula, its central barometric pressure had dropped to a portentous 27.69 inches (937 mb), catapulting Janet into the status of a moderate Category 3 storm with sustained winds of 120 MPH (193 km/h). Because the previous day's pressure readings had gone to the bottom with the scattered remains of the Neptune-P2V, the evolutionary chain that had seen Janet's barometric pressure go from 29.17 inches (987 mb) on September 25 to 27.69 (937 mb) on September 27 was broken, leaving storm watchers with only speculative theories as to the rate at which the hurricane had deepened. Such information is necessary if the climatological influences of cool water pools and venting ANTICYCLONES on the system are to be synchronized with its growth, timed in such a manner that their connection to a tropical cyclone's respective maturation cycles is clearly defined.

As it was, Janet did not disappoint those meteorologists tracking it, for on the night of September 28, 1955, as the hurricane made its first landfall on the central east coast of the Yucatán Peninsula, its minimum pressure had further steepened to 27.19 inches (920 mb), rating it as a Category 4 hurricane of extreme intensity. The city of Chetumal, capital of the Quintana Roo territory on the Yucatán Peninsula, was nearly completely leveled in what several survivors later likened to the explosion of an aerial bomb. Just before midnight on September 28, as Janet thundered ashore with sustained winds of 150 MPH (241 km/h), its spiral cloud bands suddenly opened up, dropping nine-inch rains over the dense jungle forests of the peninsula. In Chetumal, hundreds of structures, from the missionlike city hall to the art-deco theater, were seriously damaged, while blacks upon blocks of apartment buildings, factories, stores, and shanties were completely demolished. Flash floods and mudslides rushed through inland villages, killing dozens of fleeing inhabitants. In the coastal towns of Xcalak, Sacxan and Bacalar, 164 people died, victims of Janet's 14–16-foot (4–5-m) storm surge. A large tropical cyclone by any standard, Janet's 130-mile-wide (209 km) eyewall delivered 146-MPH (235-km/h) winds to neighboring Belize, claiming 61 lives and injuring 247 others in the northeastern town of Corozal.

Exiting into the southwestern GULF OF MEXICO during the early morning hours of September 29, a slightly diminished Janet quickly regained its terrifying strength as it bore down on the storm-prone city of Tampico, some 250 miles (402 km) south of Brownsville, TEXAS. Rushing ashore there just after dawn on September 29, the hurricane's 129-MPH (208-km/h) winds and increased 11-inch rains inundated nearly 75 percent of the city. Hundreds of buildings were either deroofed or simply collapsed; thousands of others had their doors, windows, porches, balconies, and other architectural details carried away. Flash floods left more than 22,000 people in Tampico homeless, trapped on roofs and dikes, awaiting their eventual rescue by units of the U.S. Navy. Stalled over the city by an entrenched high pressure ridge to the north,

Janet proceeded to spend the next five days dissipating into a stream of downpours, precipitation counts that remain among the highest ever observed in northern Mexico. As the United States mounted extensive MILITARY OPERATIONS in an effort to assist victims of the hurricane, Janet's lingering rains spawned flash floods, mudslides, and levy breaks. Snakes, snapping turtles, and other ANIMALS swarmed out of their swamped habitats, attacking with terrifying frenzy any unwary survivor who stumbled into the flooded streets. All told, nearly 100 people in the greater Tampico area perished in Hurricane Janet, while across the Caribbean basin more than $100 million in property damage was incurred, making it one of the most damaging storms then on record. The name Janet has subsequently been retired from the rotating list of HURRICANE NAMES.

Japan *North Pacific Ocean*   With 3,500 volcanic islands spread over a 2,000-mile (3,219-km) range of the NORTH PACIFIC OCEAN, Japan presents a formidable mountain barrier between the Korean Peninsula and incoming TYPHOONS. Since 1610, Japan's four major islands—Hokkaido, Honshu, Kyushu, and Shikoko—have been struck by no less than 1,022 recorded typhoons. On average, its coastal plains and shallow bays are inundated by three to four mature or formative-stage typhoons a year. In some instances, such as the 1955 season, nine or more typhoons have directly affected the island nation in a single year. A majority of Japanese typhoons are storms of significant intensity and duration, with barometric pressures of less than 28.10 inches (959 mb) and wind speeds in excess of 120 MPH (193 km/h). Some, such as the 1281 Hakata Bay typhoon that fortuitously destroyed the invasion fleet of Kublai Khan, have even entered the national folklore, assuming a cultural importance as *kamikaze,* or the "divine wind."

By tradition, the Japanese recognize two classes of typhoons. Those storms that come ashore in the spring and autumn months are called *ame taifu,* or rain typhoons, because of their enormous PRECIPITATION counts; those in summer are known as *kaze taifu,* or wind typhoons. Because Japan's lengthy typhoon season extends from May to December and encompasses the climatological vagaries of all four seasons, the differentiation between the two types of storms is not without justification. For instance, the rain typhoons, which strike Japan between May and July and November through December, tend to be part of larger, EXTRATROPICAL storm systems. Generally, these typhoons are in the final, or decaying, stage of their existence, a period characterized by both a steep rise in barometric pressure and the formation of fronts. Moving north-northwest, rain typhoons deliver enormous quantities of precipitation to the country's north regions and in many cases cause severe flooding. On the other hand, wind typhoons occur during the height of the typhoon season, during the hot, humid months of August and September. By far the more intense of the two classes, wind typhoons tend to move north-northeast and come ashore in either their deepening or mature stages. Although the heaviest rainfall is generally restricted to a 20-mile (32-km) radius directly surrounding the EYE, wind speeds can reach 200 MPH (322 km/h), and STORM SURGES of 15–20 feet (3–7 m) are not uncommon.

The Japanese typhoon season's height coincides with that of the North Atlantic HURRICANE SEASON, or during the months of August and September, when the westward-flowing trade winds are at their strongest. Like the powerful CAPE VERDE STORMS that frequently bring devastation to the shores of North America and the CARIBBEAN SEA, Japan's most intense typhoons generally make landfall during September. A comprehensive survey of recorded typhoon strikes conducted by the Japan Meteorological Agency indicates that an above-average number of storms have come ashore on either the 16–17th or the 25–26th of September and that these dates must represent singularities, or days on which typhoons are statistically more prone to strike than others. Indeed, many of Japan's most memorable typhoons—storms noted for both their catastrophic loss of life and enormous amounts of property damage—have occurred on these two dates. Between 1883 and 1963, for instance, no less than 12 major typhoons came ashore on September 16–17, while another 15 made landfall on September 25–26. Estimated DEATH TOLLS for the two periods stand at 16,500 and 10,000, respectively. It should also be noted that Japan has had its share of early to late season typhoons. One particularly damaging preseason storm struck on April 11, 1910, while another recorded typhoon occurred on November 24, 1892.

Because a majority of typhoons develop between latitudes 31 degrees North and 5 degrees South, Japan's sprawling position between latitudes 46 degrees North and 31 degrees South naturally makes for an uneven distribution of typhoon strikes on its various land masses. During the summer months, when the northward movement of typhoons is dictated by the rules of RECURVATURE, typhoons have a tendency to make landfall more frequently in the country's southern districts. The island of Okinawa, for example, lies 500 miles (805 km) southwest of Kyushu—right in the heart of TYPHOON ALLEY—and averages at least one direct typhoon strike per year. Conversely, the island of Hokkaido, which reaches far into the Northern Pacific, receives a typhoon every seven years and then only in the spring and autumn months when the *ame taifu,* or rain typhoons, are prevalent. Major metropolitan areas like Osaka and Tokyo average a typhoon strike every two to three years and a particularly destructive strike

every five to six—frequently enough for vulnerable Osaka Bay to be equipped with one of the most sophisticated storm-surge control systems in the world.

Not surprisingly, such a record of relentless and intense typhoon activity is not without its share of disastrous storms. On many occasions, Japanese typhoons have been events of terrifying fury, of wind-swept mountainsides and submerged coastal villages, of death tolls that reached into the tens of thousands. Memorable Japanese typhoons include the Nagasaki Typhoon of September 17, 1828, in which a towering storm surge in the Ariake Sea drowned nearly 15,000 people; the infamous VERA, also known as the Ise Bay Typhoon, which killed 5,159 people on September 26, 1959; Typhoon KATHLEEN, September 16, 1947, in which 2,360 people lost their lives; and both the Muroto and Muroto II typhoons of September 21, 1934, and September 16, 1961, respectively. During Typhoon HARRIET, September 27, 1956, 17 people were lost when a mudslide washed their bus into a rain-swollen river near the industrial port city of Nagoya. Japanese history also tells of the 1281 Hakata Bay typhoon which, in a single night of stinging vengeance, sank 90 percent of the Mongol conqueror, Kublai Khan's, 2,200-ship invasion fleet. Although both the passage of time and national pride have made it difficult to say with certainty how many Mongolian and Korean soldiers perished in this particular typhoon, most reputable sources place the death toll between 45,000 and 65,000.

Despite such a long litany of tragedy and death, Japan has come to accept, and indeed depend upon, the presence of typhoons for an estimated 4 to 8 percent of its annual rainfall count. Because a fair-sized typhoon can deliver between 150 and 500 billion tons of precipitation in a single strike, Japan's lush forests, picturesque inland reservoirs, and bountiful rice fields routinely rely on typhoons for their seasonal sustenance and renewal. Although a number of these storms often bring death-dealing floods to the volcanic shores of the archipelago, the alternative—drought—could possibly extract an even greater, more unfortunate loss of life and livelihood.

To quantify and record the size and strength of incoming typhoons more easily, the Japan Meteorological Agency has adopted a five-point scale of typhoon intensity. Similar in design and purpose to the SAFFIR-SIMPSON SCALE (the system used to define the destructive potential of North Atlantic hurricanes), the Japanese scale details the most salient highlights of typhoon generation and construction. Because the Japanese define a *typhoon strike* as the passage of the storm's eye within a 62-mile (100-km) radius of a certain location, the Japanese scale of typhoon intensity includes not only the force of the winds—Weak, Moderate, Strong, Severe, and Extraordinarily Severe—but also the size of the storm itself; that is, Very Small, Small, Medium, Large, and Very Large. In addition, the Japanese typhoon scale lists the radius of the 1,000 mb ISOBAR lines (in other words, how far a barometric pressure of 1,000 mb extends outward from the eye), as well as those areas of the storm in which wind speeds are above 70 MPH (113 km/h). It should also be noted that in terms of barometric pressure, the typhoon scale begins at 29.00 inches (990 mb), whereas the Saffir-Simpson scale starts at 28.94 inches (980 mb). For a reproduction of the typhoon intensity chart, *see* TYPHOONS.

On August 21, 2001, Typhoon Pabuk crashed ashore in central Japan but was downgraded to a tropical storm as it moved inland. Six people were killed, with another 50,000 evacuated from low-lying coastal areas. In places, more than 12 inches (300 mm) of rain were reported.

On July 10, 2002, Typhoon Halong, (Vietnamese for a scenic bay in Vietnam), passed just south of Guam, producing heavy rains and gale-force winds. Typhoon Halong weakened into a tropical storm before crossing the island of Honshu on July 16 with maximum sustained winds near 45 MPH (75 km/h). The tropical storm forced the closure of hundreds of schools in Tokyo and seriously disrupted transportation and electrical networks. Typhoon Halong was the second typhoon to strike the country within one week's time. A swift-moving storm, Halong dropped between two and three inches (40–70 mm) of rain, destroying nearly 117 buildings.

On September 5, 2002, Typhoon Sinlaku swept over Japan's Okinawan islands, causing widespread flooding and disrupting electrical systems. Sustained winds were clocked at 89 MPH (137 km/h). Five sailors from the PHILIPPINES were lost and another 29 people injured. Sinlaku later struck CHINA.

On October 1, 2002, Typhoon Higos swept coastal areas in eastern Japan (Kanto and Tokoku regions) with 90-MPH (150-km/h) winds. Ports and highways were closed, hundreds of flights were cancelled and tens of thousands of homes were left without electricity. Four people in Japan lost their lives to Typhoon Higos, while another 62 were injured. Coming ashore at Tomakomai, Hokkaido, with sustained winds of 56 MPH (90 km/h) and a central pressure of 28.93 inches (980 mb), Typhoon Higos was the first mature-stage tropical cyclone to have struck Hokkaido in 23 years. It destroyed at least 92 houses in the province. In Ibaraki prefecture, nine steel towers used to carry high-tension electrical wires were toppled by the typhoon's gusting winds.

On August 12, 2003, Japan was struck by Typhoon Etau (Palauan for "storm cloud"), but after the system had been downgraded to a tropical storm. It moved across Japan's northernmost island of Kokkaido, and

weakened further to tropical depression intensity before moving into the Pacific. Ten people were reported as missing after flooding incidents. Twelve people died in Japan, all told. The storm had earlier caused seven deaths and 79 injuries in Japan. Up to 16 inches (400 mm) of precipitation fell across parts of Hokkaido, and wind speeds topped 90 MPH (114 km/h). More than 1,000 homes were flooded.

On October 21, 2004, Typhoon Tokage roared ashore near the Japanese port city of Fushiki. A powerful typhoon, Tokage capsized the Russian passenger/car ferry *Antonina Nezhdanova*, while the vessel rested at its pier, rendering the vessel a total loss and causing considerable damage to the quay.

Jérémie, Hurricane *Northern–Western Caribbean Sea, October 17–25, 1935*   Sometimes referred to as the Hairpin Hurricane, this deadly Category 2 (estimated) HURRICANE delivered 107-MPH (172-km/h) winds, terrific rains, and heavy surf to the northern CARIBBEAN SEA islands of JAMAICA, HISPANIOLA, and CUBA between October 21 and 24, 1935. Initially dubbed the Hairpin Hurricane by contemporary historians because of its wildly eccentric track, the storm has in recent years been more commonly identified as Jérémie, after a seaport in southwestern Haiti that it tragically inundated on October 22. In that particular incident, some 2,000 people, many of them agricultural workers, were drowned when the system's spiral RAIN BANDS released between 10 and 12 inches (254–305 mm) of PRECIPITATION over the jagged peaks of the Massif Peninsula, triggering enormous flash floods and suffocating landslides.

Originating over the southwest Caribbean Sea some 100 miles north of Colón, Panama, on the morning of October 17, Jérémie quickly intensified as it swept north-northeast at 12 MPH (19 km/h). As maximum wind velocities within its EYEWALL exceeded 74 MPH (119 km/h), Jérémie slid across Navassa Island (located 115 miles [185 km] south of Guantánamo Bay, Cuba) on the evening of October 21. Always the rainmaker, Jérémie's tropical downpours turned the U.S. held island into a poisoned quagmire of downed trees, collapsed BUILDINGS, and flooded streets. A deeply laden schooner, overtaken by Jérémie's furious seas while navigating the hazardous shallows off Cuba's Port Antonio, was dismasted and then sunk, taking its entire 31-member crew to the bottom with it.

Sharply recurving northwest over southern Cuba on October 23—where the port city of Santiago de Cuba endured considerable damage and loss of life at the hands of Jérémie's clutching rains—the hurricane then speedily rounded the northern tip of Jamaica during the daylight hours of October 24. More than $2 million in damages were assessed on the island, mostly in losses to houses and crops. Now firmly headed in a southwesterly direction, a much weakened Jérémie returned to the south Caribbean Sea in time for a blustery landfall on the northeast coast of Honduras on the morning of October 25. While no longer as severe as they had been three days earlier, Jérémie's gale-force gusts and pelting rains killed some 150 people at Cape Gracias a Dios, and caused enormous losses among the nation's banana and coffee harvests.

Jerry, Hurricane *Southern United States, October 10–15, 1989* A moderately powerful Category 1 HURRICANE, Jerry's 80-MPH (130-km/h) winds, six-inch rains, and seven-foot STORM TIDE battered large portions of southeastern TEXAS on October 15, 1989. With a CENTRAL BAROMETRIC PRESSURE of 29.02 inches (982 mb) at landfall near the port city of Galveston, Jerry's relatively light assault on the storm-prone community nevertheless resulted in the deaths of three people and the destruction of between $5 million and $8 million worth of property and agricultural stocks.

A fairly late-season TROPICAL CYCLONE, Jerry originated over the southwest CARIBBEAN SEA's, warm waters some 77 miles (124 km) northeast of Honduras on the morning of October 10, 1989. The 10th storm of an exceptionally turbulent HURRICANE SEASON—which less than a month earlier had witnessed the Category 4 rampage of Hurricane HUGO through the northern Leeward Islands and into the SOUTH CAROLINA coast—Jerry leisurely intensified as it steadily progressed north-northeast. Confined to the east by CUBA's long shores and to the west by the boxlike mass of MEXICO's Yucatán Peninsula, Jerry was denied the large, open seas that most favor hurricane intensification. Developing with a near fatal hesitancy, Jerry was slowly upgraded to a TROPICAL STORM on the afternoon of October 11 with sustained winds of 41 MPH (66 km/h) and a mature, if somewhat disorganized, Category 1 hurricane two days later.

Indeed, Jerry was still intensifying, on its way to becoming a low-level Category 2 hurricane, when it eventually came ashore on the gulf coast of Texas just before dusk on October 15, 1989. Driven into Galveston Bay by a high-pressure ridge then positioned over central FLORIDA, Jerry pounded the barrier island through the night, its 100-MPH (161-km/h) gusts and torrential rains disrupting electrical and telephone services to more than 75,000 people. Jerry's four-foot (1-m) STORM SURGE, coupled with the onset of high tide, treated a significant eight-foot (3-m) storm tide, that exploded into spray as it crashed against Galveston's famous concrete seawalls. Causeways to the mainland were closed as the storm tide completely surrounded the island, flooding both bridge approaches. Dozens of BUILDINGS sustained damage to their roofs and sid-

ing, while at least five dwellings were crushed by falling trees. Road signs, billboards, and street lights were all bent to the ground or sent tumbling through the air, smashing with an audible thud into rain-washed intersections. Numerous mobile homes in trailer parks farther inland were either overturned or wrenched apart like tinfoil, scattering tufts of yellow insulation across a splintered landscape of faux walnut paneling and dented kitchen appliances. Three people were killed. One man in Galveston was electrocuted, while two on the mainland died in rain-slick road accidents. Pressing inland, Jerry quickly dissipated, collapsed into a series of thunderstorms whose gale-force winds and steady rains caused isolated power outages and minor flash floods outside of Houston on the night of October 16.

Although between $5 million and $8 million in property damage was assessed in the wake of Hurricane Jerry, the decision was made to retain the name Jerry on the rotating list of future HURRICANE NAMES. Even more lukewarm the second time around, Jerry reappeared during the record setting 1995 hurricane season as a rain-soaked tropical storm whose sustained winds never exceeded 42 MPH (68 km/h) but whose intense five to eight-inch (127–203-mm) rains brought widespread flooding to Florida, GEORGIA, and NORTH and South CAROLINA between August 23 and 29, 1995. Seven people in the Carolinas were killed, principally in drowning incidents.

Joan, Hurricane *Caribbean Sea–Central America–North Pacific Ocean, October 11–27, 1988* One of the most unusual NORTH ATLANTIC HURRICANES to have appeared during the latter half of the 20th century, late-season Joan inflicted widespread property losses and ramifying DEATH TOLLS on the southwest Caribbean coasts of Venezuela, Colombia, Costa Rica, Honduras, and Nicaragua, between October 16 and 23, 1988. The strongest TROPICAL CYCLONE to have ever made a recorded landfall on the Central American nation of Nicaragua, Joan's sustained 145-MPH (233-km/h) winds, seven-inch (178-mm) PRECIPITATION counts, and heaving 15-foot (5-m) STORM SURGE claimed 256 lives in that country, most of them through drowning and landslide incidents. Some 12,000 BUILDINGS along Nicaragua's southeast coast were pounded to pieces by the hurricane's Category 4 winds and breaking seas; extensive flood damage to inland crops forced loss estimates past the $500 million milestone. Nearly 300,000 people across the southern Caribbean basin were left homeless for months to come. All told, Joan claimed 326 lives in five countries and left property losses of close to $900 million. On October 23, a spent Hurricane Joan transited the Central American isthmus and entered the eastern North Pacific Ocean. Ceremoniously renamed Miriam, the powerful tropical storm became only the third documented North Atlantic tropical cyclone—1974s Hurricane Fifi was another—to have accomplished this notable feat.

HURRICANE JOAN CHRONOLOGY

October 10: What will eventually become the fifth hurricane of the record-setting 1988 hurricane season (eleven named storms, five of which reach hurricane-force, three of these rated at Category 4 or better), receives its start over the North Atlantic Ocean's subequatorial waters some 1,100 miles (1,770 km) east of the island of Trinidad, just after midday. Gently swirling west-northwest at nearly 17 MPH (27 km/h), the newly organized tropical depression originates below latitude 10 degrees North, an area where the formation of late-season tropical cyclones is a rare though not unprecedented occurrence.

October 11: Now stationed approximately 975 miles (1,569 km) east of the island of Grenada and maintaining its speedy progress west-northwest, the burgeoning tropical depression comfortably feeds on the Atlantic's warm, humid waters. A rainmaker from birth, its brimming showers spend the better part of the morning lashing the ocean's surface, while sustained 35-MPH (56-km/h) winds place it on the threshold of becoming a mild tropical storm. By midafternoon, as its top winds spike at 43 MPH (69 km/h), the tropical depression is upgraded to tropical storm and given the next available identifier on the alphabetized list of hurricane names: Joan.

October 12–14: Its spiral cloudbands now bearing 47-MPH (76-km/h) winds and soaking downpours, Tropical Storm Joan leisurely spends the next three days at sea. Initially arcing to the northwest at 16 MPH (26 km/h) on the evening of October 12, the midlevel system, held to the south by the subsiding Bermuda High, patiently shears southwest during the day and night of October 13. The velocity of its winds virtually unchanged by this abortive attempt at recurvature, Tropical Storm Joan plods westward at speeds of between 14 and 17 MPH (23–27 km/h) and seemingly targets the southern Windward Island of Grenada as it draws to within 200 miles (322 km) of the Caribbean Sea on the afternoon of October 14. In Barbados and Grenada, the onset of night is further darkened by the posting of tropical storm warnings, terse weather bulletins that advise that a system with sustained winds of between 39 and 73 MPH (63–118 km/h) and torrential precipitation is due to make landfall sometime in the next 24 hours or less.

October 15: Armed with 51-MPH (82-km/h), 64-MPH (103-km/h) gusts, and blasting rainfalls, Tropical Storm Joan commences its assault on Grenada during the early morning hours. Precipitation counts of six inches (152 mm) and higher are observed in and around the capital city of St. George's, while swollen streams transform the deserted streets of neighboring villages into gushing sluiceways. Although damage to houses and crops across the island exceeds several hundred thousand dollars, no fatalities or serious injuries are reported. Entering the southeastern Caribbean Sea at dawn, Joan pursues its westward course at 17 MPH (27 km/h). Scantly diminished by their passage over Grenada, Joan's maximum winds spend the remainder of October 15 burnishing the sea at speeds of between 45 and 48 MPH (72–77 km/h).

October 16: Centered less than 150 miles (241 km) north Caracas, Venezuela, Tropical Storm Joan lashes the Los Roques Islands with 56-MPH (90-km/h) winds and heavy rains during the early morning hours of October 16. Scores of palm trees are shorn of their foliage; gunning rapids shoot 11 people across the shallow coral islands to their deaths. As news of Joan's first confirmed casualties is flashed to the world, the slowly deepening tropical storm relentlessly maintains its trek due west, bound, it would appear, for the popular tourist islands of the Netherlands Antilles.

October 17: The first tropical cyclone to affect the islands of Curaçao and Aruba directly in more than a quarter of a century, Tropical Storm Joan neatly bisects the former during the midmorning hours of October 17. Unaccustomed to cyclonic activity, residents on both islands scurry for safety as blocks of untested buildings succumb to Joan's 69-MPH (111-km/h) gusts and stinging rains. Dozens of excursion boats, from manually powered paddlewheelers to well-fitted diving craft, are beached by the storm's five-foot (2-m) seas. Some 100 miles southwest of Aruba, Colombia's Guajira Peninsula—though positioned in Joan's milder navigable semicircle—is deluged with tons of rain during the afternoon and early evening hours. Hulking floods, swollen rivers, and treacherous mudslides soon smother another 25 lives, making Joan the deadliest tropical system to have struck the northern tip of South America since 1933.

October 18–21: Just before daybreak, as its central barometric pressure skids to 29.19 inches (988 mb) and its sustained winds eclipse 74 MPH (119 km/h), Tropical Storm Joan graduates to hurricane status. Now located some 60 miles (97 km) north of Santa Marta, Colombia, the strengthening eyewall of Hurricane Joan finally begins its long, unhurried recurvature northwest, a course change that considerably slows its forward momentum. Between midday on October 18 and the late-evening hours of October 21, Joan's speed across the extreme southwestern Caribbean Sea rarely tops nine MPH (15 km/h). Meteorologists later speculate that it is this pedestrian pace that allows Joan to deepen over the warm, confined waters of the Caribbean so quickly. Promoted to Category 2 status on the afternoon of October 19 and to Category 3 status during the early morning hours of October 20, Joan's expansive cloud mass delivers 118-MPH (190-km/h) winds and ribbons of rain to the northeast shores of Panama and Costa Rica. Fearsome floods rush down timber-clad mountainsides, causing the narrow Terraba River to overrun its banks. As the villages of Neily and Cortes are submerged beneath impromptu lakes, scores of shrimp boats and other small craft along the coast are wrenched from their moorings by Joan's 15-foot (3-m) seas and driven aground. More than 1,000 buildings in Costa Rica, including several major hospitals, schools, and canneries, are gutted. All told, four people lie dead in Panama and another 18 in Costa Rica by nightfall on October 21.

October 22: Its sustained winds seemingly invigorated by its brush with Panama and Costa Rica, Hurricane Joan plows into southeastern Nicaragua, just north of the industrial port city of Bluefields, at noon. The most intense tropical cyclone ever recorded at this latitude (12 degrees North), Joan's central pressure of 27.52 inches (932 mb) generates wind speeds of 145 MPH (233 km/h), gusts of 160 MPH (258 km/h), and violent Category 4 rains. Of average size for a late-season tropical cyclone, Joan's circular cloud bands reach from the Honduran–Nicaraguan border to the north to the Costa Rican–Panamanian border in the south, a distance of nearly 400 miles (644 km). On Great Corn Island (Isla del Maiz Grande), 50 miles (80 km) northeast of Bluefields, all 7,500 buildings are demolished leaving the slight coral atoll littered with tons of splintered wood, corrugated tin roofing, and dozens of trapped victims. On the Nicaraguan mainland, almost all of the structures in Bluefields—from the city's colonial-style civic hall to its solid concrete jetties and embankments—are destroyed by the hurricane's vindictive winds and pummeling breakers. Ninety percent of Bluefield's 60,000-odd residents are left without food, water, electricity, and shelter, compelling Nicaragua's Sandinista government to declare a month-long state of emergency. Churning northwest at 12 MPH (19 km/h), Joan's weakening eyewall brings 100-MPH (161-km/h) gusts and dangerous downpours to the Nicaraguan capital of Managua during the late-evening hours of October 22. Capricious flash floods coupled with seething mudslides buckle numerous bridges and wash away miles of

roads, hampering relief efforts. At least 137 people in Nicaragua are killed outright; another 119 are listed as missing. More than a quarter of a million Nicaraguans are rendered homeless, making Hurricane Joan the worst natural disaster in the country's history.

October 23: Its once-formidable winds never again to exceed 60 MPH (97 km/h), a downgraded Tropical Storm Joan enters the eastern North Pacific Ocean during the early morning hours. Rechristened Tropical Storm Miriam, it will last until October 28, when it dissolves into an EXTRATROPICAL system some 869 miles (1,399 km) northwest of Acapulco, MEXICO. The name Joan has been retired from the repeating, six-year list of HURRICANE NAMES.

John, Hurricane/Typhoon *North Pacific Ocean, August 11–September 10, 1994* A marathon Category 5 (estimated) HURRICANE, John enjoys the unique distinction of being the longest-lived TROPICAL CYCLONE on record. For 31 stormy days between August 11 and September 10, 1994, John traveled more than 8,000 miles (12,875 km) across the expansive NORTH PACIFIC OCEAN. In the course of its westward journey, it crossed the international date line, was rechristened TYPHOON John, then recrossed the date line a few days later, reverting to its previous status as a hurricane. On August 26 it overtook Johnston Island, a small atoll 700 miles (1,127 km) southwest of HAWAII and lashed it with 110-MPH (177-km/h) winds and a 20-foot (7-m) STORM SURGE. Although no casualties were reported on the island, damage to both its shoreline and its various military installations was considered severe.

The fifth North Pacific hurricane in a season noted for its spate of powerful storms (Hurricane Emilia, July 17, and Hurricane Gilma, July 22, both with winds in excess of 160 MPH), John originated 410 miles (660 km) south-southeast of Acapulco, MEXICO, on August 11, 1994. For more than a week John, with 40-MPH (64-km/h) winds and heavy quantities of PRECIPITATION, lingered in the eastern Pacific as a low-grade TROPICAL STORM. Moving north-northwest at approximately 15 MPH (24 km/h), John steadily gained strength from the warm 84°F (29°C) waters, becoming a full-fledged hurricane on the morning of August 19. Although its powerful winds were clocked at more than 90 MPH (145 km/h), meteorologists gave the hurricane, now located in the vicinity of 16 degrees North, 134 degrees West, little chance for long-term survival. It was expected that its steady drift west would bring it into contact with a deep pool of relatively cooler water that was then present off the south coast of the Hawaiian Islands, seriously hindering any opportunity for further intensification.

But on the morning of August 23, as Hurricane John crossed into the Central Pacific, its BAROMETRIC PRESSURE rapidly began to drop. By midafternoon John's central pressure was measured at 27.38 inches (929 mb), one of the lowest barometric readings ever recorded in a North Pacific hurricane. As its 144-MPH (232-km/h) winds kicked up vast quantities of spray, heavy surf warnings were posted for the southern Hawaiian Islands, now slightly less than 500 miles (805 km) north of the storm's turbulent EYEWALL. While the hurricane's steady north-northwesterly direction indicated that it would not in all likelihood pose a direct threat to the popular resort beaches of Waikiki and Oahu, there was always the chance that the storm would suddenly double back on itself and strike HAWAII with the devastating fury of another Hurricane Iwa (1982) or INIKI (1992).

Continuing north-northwest at close to 15 MPH (24 km/h), John now leveled its sights on the small, sparsely inhabited pinnacle of Johnston Island. Located 700 miles (1,127 km) southwest of Honolulu, Johnston Island was home to more than 1,000 United States military personnel and their families. As the hurricane, with winds in excess of 110 MPH (177 km/h), bore down on the atoll, a squadron of government helicopters moved in to begin a mass evacuation. While John's strident winds were of considerable concern to authorities, an even greater threat was presented by the hurricane's 15–20-foot (5–7-m) storm surge. Because Johnston Island housed an incinerator used to destroy chemical weapons, officials feared that should a major storm surge sweep over the island, stockpiled chemical agents could be released into the maelstrom with catastrophic results.

On the afternoon of August 26, 1994, John crashed ashore at Johnston Island. One-hundred-fifty-mile-per-hour (241-km) wind gusts uprooted palm trees, sheared the roofs from barracks and houses, and knocked out power lines. A 20-foot (7-m) storm surge, judged to be the highest ever to strike the atoll, flooded the island's principal harbor, seriously damaging a dockside supply depot. While property damage on Johnston Island reached into the millions of dollars, none of the 100 military personnel left on the island as a skeleton crew were killed or injured, and the chemical weapons incinerator remained intact.

Its winds barely diminished by its spirited assault on Johnston Island, Hurricane John continued to spiral north-northwest across the Central Pacific's undulating expanse. On the afternoon of August 29, 1994, as its erratic winds continued to oscillate between a hefty 100 MPH (161 km/h) and a dangerous 120 MPH (193 km/h), John officially crossed the international date line and entered the western North Pacific Ocean. Now formally rechristened Typhoon John, the storm made it as far west as longitude 178 degrees East before an abrupt about-face forced the storm to head back toward the eastern Pacific. On September 3, after executing a loop-the-loop maneuver, John's CENTRAL BAROMETRIC PRESSURE began to rise steadily and its winds speed dropped to a rather tame 34 MPH (55 km/h). No longer considered either a hurricane or a typhoon, John's status was downgraded to that " of a TROPICAL DEPRESSION. But by the morning of September 4, John had again changed course, this time in a north-northwesterly direction that took it as far north as latitude 30 degrees. There it rapidly swung back east and recrossed the international date line during the evening hours of September 5. It also regained tropical storm grade as its winds strengthened to 59 MPH (95 km/h).

Now christened Tropical Storm John for the second time in its unusually long life, the storm was quickly approaching that point in its existence when it would enter the record books as the most durable tropical cyclone in known history. Sharply moving northeast, John's barometric pressure again began to drop, bringing with it a commensurate rise in wind speed. On September 7, 1994, with 65-MPH (105-km/h) winds roaring around its loosely defined EYE, John broke the 23-year record, held by Hurricane GINGER (September 5–October 1, 1971) for the most durable cyclone on record. As though reinvigorated by its newly honored status, John's winds again strengthened to hurricane force on the evening of September 9. Firmly maintaining its northeasterly course, John crossed the 40th parallel and swiftly entered the cool seas that wash the shores of Alaska's Aleutian Islands. Cut off from the heat-laden waters of the tropics, John quickly began to dissipate. By midnight on September 10 it was gone forever.

Joint Typhoon Warning Center (JTWC) Located in Pearl Harbor, Hawaii and staffed by meteorologists and technicians from the U.S. Navy and Air Force, the Joint Typhoon Warning Center is responsible for the tracking and forecasting of all TYPHOONS, CYCLONES, TROPICAL STORMS, and TROPICAL DEPRESSIONS in the western and southwestern NORTH PACIFIC OCEAN. Established by the commander in chief of the U.S. Pacific Fleet on May 1, 1959, the JTWC is similar in design and mission to the NATIONAL HURRICANE CENTER (NHC), except that the JTWC's jurisdiction extends from longitude 180 degrees West to the coast of CHINA, and from latitude 60 degrees North to 60 degrees South. Tasked with assigning TYPHOON NAMES and cyclone designators from a list prepared by the World Meteorological Organization (WMO), the JTWC annually compiles and publishes an Annual Tropical Cyclone Report. Originally located in Guam, it was moved in 2000.

Josephine, Tropical Storm *South–Eastern United States, October 4–10, 1996*   Although its 65-MPH (105-km/h) gusts were not of HURRICANE intensity when it first came ashore at Apalachee Bay, FLORIDA, on the night of October 7–8, 1996, TROPICAL STORM Josephine and its torrential, five–six-inch (127–152-mm) rainfalls later caused widespread flooding in northwestern Florida, GEORGIA and SOUTH and NORTH CAROLINA between October 8–10.

The 10th-named TROPICAL CYCLONE of the 1996 HURRICANE SEASON, Josephine originated off the east coast of MEXICO some 100 miles (161 km) northeast of Tampico on the afternoon of October 4. Carried east-northeast and away from the Bay of Campeche at forward speeds of between five to 12 MPH (8–19 km/h), Josephine spent nearly one week lazily intensifying over the warm SEA-SURFACE TEMPERATURES of the central GULF OF MEXICO before being upgraded to a tropical Storm during the late-evening hours of October 6.

Centered some 425 miles (684 km) southwest of Tallahassee, Florida, and pursuing its meandering trajectory east-northeast, Josephine's sustained winds topped 45 MPH (72 km/h), causing large waves to radiate outward from beneath its slowly organizing EYEWALL. Breaking ashore in southern LOUISIANA, the tropical storm's pounding seas—coupled with the presence of above-normal STORM TIDES associated with the gale-force winds of a large high pressure system concurrently positioned over the northeastern United States—swamped the two-lane highway that connected the barrier island of Grand Isle to the mainland, compelled the city of New Orleans to close its famous floodgates, and forced the evacuation of 3,000 people from St. Bernard and Plaquemines parishes.

Shortly after midnight on October 7, with computer ANALOGS indicating a midweek landfall somewhere along the Florida Panhandle, TROPICAL STORM WATCHES were posted from Apalachicola to Venice. As emergency officials evacuated 4,500 gulf coast residents from mobile homes and beachfront cottages, Josephine continued to intensify, its maximum sustained winds climbing to 60 MPH (97 km/h) during the midmorning hours of October 7. Quickly gaining forward speed, the tropical storm pulled to within 75 miles (121 km) of landfall near the town of St. Marks just after midday, causing storm watches to be superseded by TROPICAL STORM WARNINGS and a new round of evacuations and business closings to begin.

Sliding northeast at nearly 23 MPH (37 km/h) and carrying beneath it a hulking, six-foot (2-m) STORM SURGE, Tropical Storm Josephine rushed ashore between Apalachicola and Cross City along Florida's sparsely populated Big Bend coast just before midnight on October 7. Swiftly churning inland, the tropi-

cal storm's spiral RAIN BANDS began to splinter into downpours, gust-whipped sheets that when added to the three inches (76 mm) of rain already received by the state from the earlier high pressure system, caused widespread flooding, numerous tree falls, and the further evacuation of residents from nine counties. In Jacksonville, which Josephine deluged with 9.08 inches (254 mm) of rain in a 48-hour period, more than 80 streets were completely flooded, while downed power lines threw 100,000 houses into darkness. In Putnum County, Josephine spawned a substantial TORNADO that damaged 131 mobile homes but caused no deaths or injuries. Elsewhere in central and northern Florida, Josephine dropped four inches (102 mm) of rain on Gainesville and triggered another 12 reported tornado outbreaks, one of them responsible for cutting a $400,000 swath of destruction through Naples but none causing serious injuries.

Scudding northeast at nearly 41 MPH (66 km/h) during the morning hours of October 8, Tropical Storm Josephine briefly entered the NORTH ATLANTIC OCEAN before cresting ashore just north of Charleston, South Carolina, later that afternoon. Its sustained winds reduced to 52 MPH (84 km/h), the tropical storm swung inland over South Carolina, rapidly began to undergo EXTRATROPICAL deepening as it sheared across North Carolina's eastern banks and collided with an existing low pressure system then moving southeast. Disintegrating into water-laden thunderstorms, Josephine's tattered remains dropped more than four inches (102 mm) of rain across Wilmington, North Carolina, before delivering blustery rain squalls to NEW YORK, NEW ENGLAND, and eastern CANADA on October 9. While no lives were reportedly lost to Tropical Storm Josephine, property damage estimates across six states exceeded $3 million.

The third of four North Atlantic tropical cyclones to have been designated Josephine, the 1996 tropical storm was preceded in 1990 by Hurricane Josephine, a Category 1 system that developed over the mid-Atlantic on September 21 and lasted until October 6. With a CENTRAL BAROMETRIC PRESSURE of 28.93 inches (980 mb) and maximum winds of 85 MPH (137 km/h), Hurricane Josephine remained an offshore system for its entire existence.

Juan, Hurricane *Southern United States, October 21–November 9, 1985*   The third "billion-dollar" HURRICANE to have made landfall in the United States during the course of the 1985 HURRICANE SEASON, Juan delivered 85-MPH (135-km/h) winds and eight-inch (203-mm) rains to the gulf coast of LOUISIANA and staggering nine to 18-inch (203–457 mm) PRECIPITATION counts to FLORIDA, ALABAMA, MISSISSIPPI, TEXAS, VIRGINIA, and Pennsylvania between October

20 and November 9, 1985. In Louisiana, where Juan initially came ashore as a moderate Category 1 hurricane on October 26, 15 people were killed in BUILDING failures, automobile accidents, and drownings. In northeastern United States, where Juan's dispersing RAIN BANDS dropped 18 inches of rain on Virginia's Blue Ridge Mountains between November 2 and 4, more than 40 people lost their lives to raging flash floods and mudslides. Insurance claims quickly topped $1.6 billion, making Juan one of the most expensive hurricanes to have struck the United States up to that time.

A relatively late-season TROPICAL CYCLONE, Juan originated over the tepid waters of the southwest GULF OF MEXICO, roughly 500 miles (805 km) south of New Orleans, Louisiana, on the afternoon of October 21, 1985. The 10th storm of an active season—one that had earlier seen Hurricane Danny strike Louisiana and Texas on August 15, Hurricane ELENA belt Mississippi on September 2, and Hurricane GLORIA roar up the eastern seaboard and into NEW ENGLAND on September 26–27—Juan rapidly strengthened as it churned north-northwest at nearly 11 MPH (18 km/h). Reconnaissance flights conducted by the NATIONAL HURRICANE CENTER (NHC) indicated that Juan's CENTRAL BAROMETRIC PRESSURE had deepened to 29.54 inches (1,000 mb) by the afternoon of October 23, upgrading it to a TROPICAL STORM with sustained winds of 49 MPH (79 km/h). By the same time the next day, Juan had matured into a Category 1 hurricane, sustaining 77-MPH (124-km/h) winds hinting at further intensification.

In fact, Juan was well on its way to achieving Category 2 status when it came ashore on Louisiana's gulf coast some 100 miles west of New Orleans just before dawn on October 26, 1985. One-hundred-five-mile-per-hour (169-km/h) gusts and ferocious rains downed power lines, uprooted hundreds of trees, drove a dozen shrimp boats ashore, and seriously damaged some 10,000 buildings. Inundating downpours spawned flash floods, while eight-inch (203-mm) rain counts turned the low-lying neighborhoods of Morgan City and Houma into bayous. More than 250,000 people across the state were left without electricity or telephone service. Fifteen people were killed, making Juan the deadliest hurricane to have struck Louisiana since Hurricane BETSY claimed 61 lives there in September 1965.

In a move reminiscent of Hurricane Elena's erratic trek across the Gulf of Mexico just seven weeks earlier, Juan had no sooner crossed the coast, penetrating as far inland as the city of Thibedaux before it abruptly stalled, slightly weakened, and then slowly pulled back to the nourishing waters of the gulf on October 27. Held off the coast by an equally erratic cold front, a downgraded Tropical Storm Juan spent the next four days haphazardly careening along the north gulf coast of the United States. Seemingly unable to decide just where exactly it intended to come ashore, Juan rushed inland at Biloxi, Mississippi, on October 29, delivering 50-MPH (81-km/h) gusts to the shipyards and steelworks of the city, only to retreat once more to the gulf the next day. Drifting almost due east, Juan rolled past Alabama, bringing heavy rains and swollen tides to Mobile Bay on October 30. Before coming ashore on the border between Alabama and Florida for the third and last time on the afternoon of October 31, Juan further weakened as its sustained winds dropped to 38 MPH (61 km/h), TROPICAL DEPRESSION stage. Although torrential 8-inch (203-mm) rains caused several outbreaks of localized flooding, no deaths or major damages were reported from Juan's passage across the Florida Panhandle.

Unfortunately the same could not be said of Juan's final disintegration over the inland mountain ranges of Virginia and West Virginia between November 2 and 7, 1985. In what would become the most serious flood emergency in the region since 1972's Hurricane AGNES, Juan dropped nearly 19 inches (483 mm) of rain on the Blue Ridge Mountains in just two days. From November 2 to 4, Virginia, West Virginia, and portions of Maryland were transformed into veritable sluiceways, rain-soaked states that were quickly isolated by bridge collapses, road washouts, and widespread power and communications outages. Creeks in the Allegheny Mountains overran their banks, sweeping unwary leaf watchers and their vehicles into raging gorges. Twenty-one people were killed in Virginia, a majority of them the victims of fast-moving flash floods. Another 22 perished in West Virginia, five of them in a tragic mudslide outside Lewisburg on November 4. Property losses in Virginia and West Virginia eventually topped $441 million and $200 million, respectively.

While an estimated $1.6 billion in damage and 61 deaths were eventually tallied in the wake of Hurricane Juan, the decision was made to keep the name Juan on the repeating list of future HURRICANE NAMES.

Kate, Hurricane *Northern Caribbean–Southern United States, November 15–22, 1985* The first November HURRICANE to have struck the mainland United States in more than a half-century, Kate remains one of the CARIBBEAN SEA's most destructive late-season storms. Displaying a CENTRAL BAROMETRIC PRESSURE of 28.87 inches (977 mb) and top winds of 108 MPH (174 km/h), Kate blazed a 1,000-mile (1,609-km) trail of devastation through JAMAICA, CUBA, FLORIDA, and GEORGIA between November 18 and 22, 1985. In west Jamaica, where Kate's TROPICAL STORM–force rains spawned flash floods and mudslides on November 18 and 19, seven people were killed and some $5 million in property losses were tallied. In Cuba, where a strengthened Kate delivered 125-MPH (201-km/h) gusts to the island's northwest shores on the night of November 19 and 20, nearly 200,000 people were evacuated from vulnerable Havana and sent inland to the relative security of mountainside storm shelters. Ten people in Cuba nevertheless lost their lives to Kate, a majority of them in drowning accidents, while some 2.5 million acres of the nation's staple sugarcane harvest were uprooted by the hurricane's sustained 105-MPH (169-km/h) winds and saturating rains. So widespread and severe was Cuba's economic loss that the United Nations was compelled to provide nearly $2 million in emergency relief funds to those towns and villages most seriously impacted by Kate's passage. In Panama City, Florida, where Kate's 100-MPH (161-km/h) EYEWALL winds came ashore on the afternoon of November 21, damage to coastal residences, small buildings, water craft, power lines, road signs, and trees was fairly severe, an estimated $892 million in losses being assessed in three states. Five people in Florida were killed, four of them in the tense, evacu-ation-driven hours immediately preceding the hurricane's landfall. Downgraded to tropical storm status during the early morning hours of November 22, Kate brought 55-MPH (89-km/h) rain squalls to southwestern Georgia, precipitating isolated power outages, flash floods, and the deaths of another two people.

Kate (Sening), Typhoon *Philippines, October 1–15, 1970* The second of four deadly SUPER TYPHOONS to have slammed into the PHILIPPINES during the remarkable 1970 TYPHOON SEASON, Kate crushed 12 provinces on the north islands of Mindanao and Luzon beneath its 120-MPH (193-km/h) EYEWALL winds, 10-inch rains, and splintering 11-foot (4-m) STORM SURGE between October 14 and 15, 1970. More commonly known in the Philippines by its local name, Sening, Category 3 Kate was preceded ashore by Typhoon GEORGIA on September 15 and followed by Typhoons JOAN (Titang) and PATSY (Yoling) on October 16 and November 19, respectively. Bearing gusts of 171 MPH (275 km/h) just hours before landfall near the city of Virac, Catanduanes, Typhoon Kate either damaged or destroyed some 51,000 BUILDINGS, displaced a quarter of a million people, and claimed 583 lives. Millions of dollars in losses were sustained by the nation's rice and sugar crops, while scores of villages remained submerged for several weeks following the passage of the typhoon into the South China Sea on October 16. Filipino President Ferdinand Marcos had no sooner declared the entire island of Luzon a disaster area when an even more intense Typhoon Joan rushed across the central and southern islands of the storm-prone archipelago on October 16 and 17, flattening another 42,000 houses and taking an additional 526 lives in 12 southern provinces.

Kathleen, Typhoon *North Pacific Ocean–Japan, September 9–16, 1947* An exceptionally powerful Category 3 TYPHOON, Kathleen inflicted widespread property damage and catastrophic DEATH TOLLS on the central Japanese island of Honshu between September 15 and 16, 1947. The last truly great typhoon to strike JAPAN in the first half of the 20th century, Kathleen's scything winds—clocked at more than 115 MPH (185 km/h)—razed some 9,928 BUILDINGS in and around the capital city of Tokyo. Along the coastal havens of Tokyo Bay, Kathleen's heaving STORM SURGE, brought on by a CENTRAL BAROMETRIC PRESSURE of 28.34 inches (960 mb) at landfall, raided a number of low-lying communities, tragically drowning their narrow streets and lightly timbered houses beneath several feet of water. Between 1,692 and 2,320 people perished in Kathleen, making it one of the deadliest typhoons in modern Japanese history.

Katrina, Hurricane *Southern United States, August 23–31, 2005* The first truly catastrophic TROPICAL CYCLONE to strike the continental United States in the 21st century, Hurricane Katrina remains one of the most devastating natural disasters in U.S. history. At one point in its existence a Category 5 HURRICANE of record-setting intensity, Katrina clawed a 1,200-mile (1,931-km) trail of stunning death and destruction through southern FLORIDA, MISSISSIPPI, LOUISIANA, and ALABAMA, between August 25 and August 30, 2005. In the BAHAMAS, where Tropical Depression Twelve (TD 12), then an upgraded Tropical Storm Katrina, produced between six and 12 inches (15–30 cm) of rain on August 24, no deaths or injuries were reported. In southeastern Florida, where a Category 1 Katrina delivered sustained 80 MPH (129 km/h) winds and flooding rains on August 25, some 11 people were killed and nearly $1 billion in property losses were recorded. But in Louisiana, Mississippi, and Alabama, where Katrina roared ashore on August 29 with a central barometric pressure of 27.11 inches (918 mb), sustained winds of 145 MPH (227 km/h), and a 30-foot (10-m) STORM SURGE, a score of coastal parishes and counties were nearly obliterated. Built on swampland between the east bank of the Mississippi River and vast Lake Pontchartrain to the north, the largest city in the region, New Orleans, initially survived Katrina's Category 3 assault, but the hurricane's torrential rains and 18-foot (16-m) storm surge inundated Lake Pontchartrain, causing its waters to rise more than eight feet above sea level. Sections of the vast network of levees constructed over the past century by the U.S. Army Corps of Engineers to protect the low-lying city from flooding, failed under these extreme conditions. Some 80

percent of New Orleans was consequently flooded; 200,000 residences and commercial buildings were inundated beneath 20–25 feet (7–8 m) of brackish, contaminated water. More than 500,000 people—the entire population of New Orleans—had to be evacuated (*see* EVACUATION), as well as tens of thousands of others left homeless elsewhere in Louisiana and in neighboring Mississippi. In what would become one of the worst natural disasters in U.S. history, much of the City of New Orleans was rendered uninhabitable as the DEATH TOLL across the region topped 1,836 people, making Katrina the deadliest North Atlantic hurricane in more than a century. Amid an enormous relief and recovery operation, property losses from Katrina were estimated at $82 (in 2005 U.S. dollars) billion.

The tropical cyclone that would eventually make the third-most intense hurricane landfall in U.S. history originated some 175 miles (282 km) southeast of Nassau, the BAHAMAS, during the mid-afternoon hours of August 23, 2005. The 12th tropical depression of the exceptionally active 2005 North Atlantic HURRICANE SEASON, Tropical Depression 12 (TD12) slowly tracked to the northwest near eight MPH (13 km/h). Appearing to SATELLITES as a disorganized splay of clouds with a band of deep convection building to the east of the system's center, TD12, with an estimated minimum CENTRAL BAROMETRIC PRESSURE of 29.74 inches (1007 mb), produced maximum sustained winds of 35 MPH (55 km/h) and higher gusts as it delivered between three and six inches (7–15 cm) of PRECIPITATION to the Bahamian city of Georgetown and Great Exuma Island.

In a unique clarification issued by the NATIONAL HURRICANE CENTER (NHC) during its first discussion on Tropical Depression Twelve, it was explained that TD 12 had been designated TD 12 instead of TD 10 because it was in essence a combination of earlier Tropical Depressions 10 and 11. Although TD 10 had essentially dissipated by August 19, a remnant of TD 10 that remained tucked in the middle of the atmosphere was absorbed into a weak TD 11 on August 20, three days before TD 12 formed.

During the night of August 23, TD 12 continued to strengthen as it drifted to the west-northwest at just over six MPH (10 km/h). Steered through interaction with the BERMUDA HIGH to the northeast, TD12 drew increasing amounts of moisture from the warm NORTH ATLANTIC waters and delivered 23-MPH (37-km/h) winds to the Bahamas.

By mid-morning on August 24, the National Hurricane Center, using satellite imagery, DOPPLER RADAR data and aircraft reconnaissance data, determined that TD 12's sustained winds were in excess of 40 MPH (64 km/h) on the surface, warranting its

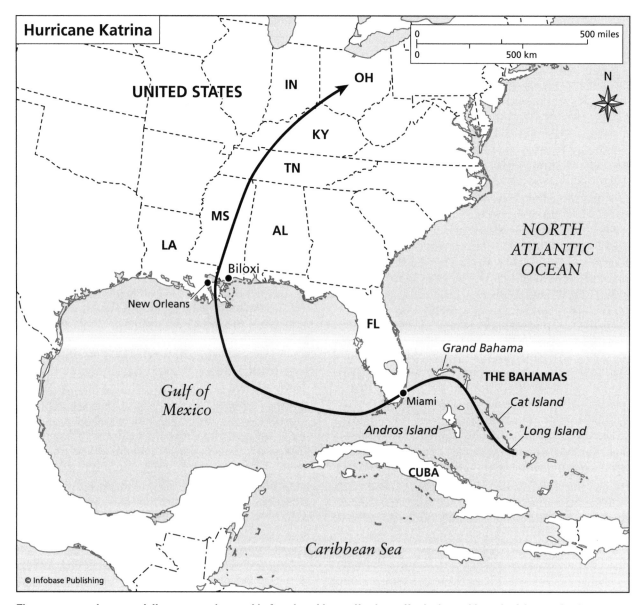

Hurricane Katrina

The most expensive natural disaster yet witnessed in American history, Hurricane Katrina's notable path of destruction is rendered in this tracking chart.

upgrade to TROPICAL STORM status. In accordance with National Weather Service (NWS) naming criteria, the system was dubbed Katrina.

As the system developed a symmetrical upper-level outflow pattern and WIND SHEAR conditions remained weak, Katrina continued to strengthen during the afternoon and early evening hours on August 24. Now located some 165 miles (266 km) east of the Florida coast, Katrina's forward speed had increased to eight MPH (13 km/h), and its sustained winds, spread around a deepening banding structure, had increased to 46 MPH (74 km/h). Shortly before midnight, as it fell under the influence of a subtropical ridge to the north, Tropical Storm Katrina turned almost due west, and its sustained winds topped 52

MPH (84 km/h), prompting a hurricane warning to be posted throughout southeastern Florida.

Although warm SEA-SURFACE TEMPERATURES (SST) provided a ready supply of moisture, Tropical Storm Katrina's intensification slowed during the early morning hours of August 25, as dry air and some wind shear from the system's interaction with the low-level ridge to the north disrupted its circulatory patterns. Early morning forecasts issued by the NHC indicated that Katrina would make landfall in southeastern Florida before weakening and reemerging into the GULF OF MEXICO, where low wind shear and high SST's portended intensification.

Katrina continued to deepen during the mid-morning hours of August 25. While the system's forward speed had decreased to five MPH (8 km/h), its sustained winds had increased to 58 MPH (93 km/h). By the early evening hours on August 25, Katrina's central pressure (as observed by a hurricane hunter reconnaissance flight) had dropped to 29.09 inches (985 mb), and its sustained winds had increased to 78 MPH (126 km/h) at the surface. Its forward speed remained at five MPH (8 km/h) as it continued to interact with its steering currents. Owing to the presence of a upper-level trough that had developed northward from the CARIBBEAN SEA, Katrina's trajectory had turned more to the southwest.

With a central barometric pressure of 29.09 inches (985 mb) and sustained winds of 80 MPH (129 km/h), Hurricane Katrina pounced ashore in southeastern Florida—between Hallandale and North Miami beaches—during the early evening hours of August 25. Across southern Florida, schools and airports were closed, and some emergency shelters were opened. Several of Florida's barrier islands had previously been evacuated, thereby lessening the potential for casualties from storm surge flooding. While a highway underpass that was under construction in Miami collapsed during Katrina's passage, overall property damage in the state was relatively light, despite rainfall totals of 18 inches (46 cm) in some southeastern locations. Eleven people in Florida lost their lives to Katrina, many in accidents related to the hurricane's landfall. Katrina was the third tropical cyclone to affect Florida during the 2005 North Atlantic HURRICANE SEASON, with Tropical Storm ARLENE and Hurricane DENNIS striking the state earlier in the year.

Now under the steering influence of a mid-level high positioned to the northwest, Hurricane Katrina's trajectory turned more to the southwest, and its forward speed increased to seven MPH (11 km/h), taking its core across the moist flatlands of the Florida Everglades. After four hours over the Florida peninsula, Katrina's sustained winds were still clocked at 75 MPH (121 km/h), and the system was producing between one and two inches (3–5 cm) of rain every hour as it purred across southeastern Florida.

Although computer models used by the National Hurricane Center (NHC) had as early as August 24 predicted that Katrina would eventually strike Louisiana and Mississippi as a very powerful hurricane, the system's rapid and often harrowing intensification following its emergence into the Gulf of Mexico on August 26, confirmed for meteorologists how favorable an environment with low vertical wind shear, high SST's, and a strong upper level (200 mb) anticyclone can be to a hurricane's intensification cycles. During the late night hours of August 26, large clusters of deep convection developed within

Katrina's regenerating EYEWALL, and surface wind speeds rose to 104 MPH (167 km/h), making the system a powerful Category 2 storm. Moving to the west-southwest at seven MPH (11 km/h) and producing 12-foot (4-m) seas in the Gulf of Mexico, Katrina—if the latest dynamical model runs were correct—would grow into a Category 4 hurricane with sustained winds of 132 MPH (212 km/h) during the next 24 hours.

Positioned some 400 miles (644 km) southeast of New Orleans, Louisiana, Hurricane Katrina continued to rapidly intensify during the early morning hours of August 27. During the night the system's EYE had become visible to infrared satellite imagery. An air force reconnaissance flight conducted shortly before dusk observed a minimum pressure of 27.90 inches (945 mb), giving Katrina surface winds in the 115 MPH (186 km/h) range. Upgraded to Category 3 status, Katrina wobbled (a common trait in many large, intense tropical cyclones) to the west-southwest, and continued to intensify.

As Hurricane Katrina drew to within 300 miles (483 km) of the Gulf of Mexico's northern coast during the early morning hours of August 28, an air force reconnaissance aircraft operating in the eyewall's northwestern quadrant observed flight level wind speeds of 158 MPH (254 km/h)—or 144 MPH (232 km/h) at the Earth's surface—and a minimum barometric pressure of 27.61 inches (935 mb). This prompted the NHC to issue a special advisory indicating that Katrina would "perhaps even reach Category Five status sometime during the next 36 hours," and that the system would make landfall with sustained winds of at least 150 MPH (241 km/h).

Shortly after dawn, Hurricane Katrina was upgraded to a Category 5 hurricane. Now moving at 10 MPH (16 km/h), the recent most survey conducted by the HURRICANE HUNTERS had indicated winds of 180 MPH (290 km/h) at flight level, which translated into sustained surface winds of 161 MPH (260 km/h). Despite fluctuations caused by an impending eyewall rejuvenation cycle, Katrina was "expected to be a devastating Category Four or Five hurricane at landfall."

Katrina continued to intensify during the afternoon hours of August 28. Prompting recollections of past Gulf of Mexico hurricanes that had grown into meteorological terrors in the course of a day—CAMILLE (1969), FREDERIC (1979), OPAL (1995)—Katrina's central barometric pressure had slid to 26.78 inches (907 mb), and sustained surface winds of 173 MPH (278 km/h) lashed the Gulf of Mexico. According to an advisory issued by the NHC, "Katrina is comparable in intensity to Hurricane Camille of 1969 . . . only larger." Katrina continued

Hurricane Katrina
1445 UTC 29 August 2005
GOES-12 visible channel
over a MODIS true-color background

GOES Project NASA-GSFC

The now iconic Hurricane Katrina as it smashed ashore in Louisiana and Mississippi in late August 2005. At one time a Category 5 system, Katrina had weakened considerably by the time it made landfall. Its inundating rains, however, undermined the elaborate levee system surrounding the city of New Orleans, flooding much of the historical metropolis and prompting one of the largest evacuations in American history. *(NOAA)*

its recurvature to the north, following an opening caused by a weakness in a subtropical ridge to the north of the system.

Hurricane Katrina reached its meteorological peak shortly before dusk on August 29, 2005. As aircraft reconnaissance missions recorded central barometric pressures of 26.63 inches (902 mb)—the fourth lowest on record for the North Atlantic basin behind GILBERT (1988), the LABOR DAY HURRICANE (1935), and ALLEN (1980)—and sustained winds at the Earth's surface in excess of 167 MPH (269 km/h), Hurricane Katrina churned northward at 11 MPH (18 km/h), bound for landfall in the shallow coastal plains of Louisiana and Mississippi. One of the largest hurricanes yet observed in the Gulf of Mexico with a cloud cover 230 miles across, Katrina was

forecast to deliver hurricane force winds (74 MPH and higher [119+ km/h]) at least 140 miles (225 km) inland.

As forecasts indicated that Katrina could very well become only the fourth Category 5 hurricane to strike the United States, behind ANDREW (1992), Camille (1969), and the LABOR DAY HURRICANE (1935), numerous precautions and safeguards were implemented. Oil companies evacuated dozens of nonessential personnel from 21 of their Gulf of Mexico drilling rigs and platforms, while the reactor at the Waterford nuclear power station at Killona, Louisiana, was shut down as a precaution. In New Orleans and other coastal communities, thousands of people heeded evacuation orders and recommendations, and moved far inland.

But just as quickly as Hurricane Katrina had deepened, it began to weaken. The onset of a 12–24 hour eyewall replacement cycle, as well as the infiltration of dry air from the west, provided the first meteorological checks on Katrina's harrowing intensification. During the early morning hours of August 29, as the Category 4 system drew to within 60 miles (97 km) of the Gulf coasts of Louisiana and Mississippi, its central barometric pressure rose to just under 27.01 inches (915 mb). Moving due north at 13 MPH (21 km/h), and with sustained surface level wind speeds of 150 MPH (241 km/h), Katrina was forecast to produce storm surges conditions of 28 feet (9 m) along the Louisiana and Mississippi coasts.

The third-most intense hurricane yet observed to have made landfall in the continental United States, Hurricane Katrina roared ashore in Plaquemines Parish, on the extreme southern tip of Louisiana, shortly after dawn on August 29, 2005. Driven by a central pressure at first landfall of 27.11 inches (918 mb), Katrina's sustained, 140-MPH (225-km/h) winds, blinding rains and 18–30-foot (5–9-m) storm surge caused widespread coastal destruction in Louisiana and Mississippi. In Biloxi, Mississippi, Katrina produced a 30-foot (9-m) storm surge that inundated much of the city; it was the largest storm surge yet measured in the United States. After briefly crossing the mouths of the Mississippi River, Katrina—now generating sustained winds of 125 MPH (201 km/h)—made a second landfall on the Louisiana-Mississippi border just before midday. Air Force reserve reconnaissance flights observed 138-MPH (222-km/h) winds at 5000 feet (1,667 m) above the ground, and sustained surface winds of about 127 MPH (204 km/h) at the surface.

Still moving due north, Katrina's forward speed increased to 14 MPH (23 km/h) as it ground through inland Mississippi and Louisiana with widespread destructiveness. By midnight on August 29, Katrina's sustained winds had fallen to 58 MPH (93 km/h) and the system was moving rapidly to the north-northeast at 19 MPH (31 km/h).

As it continued to churn inland over the forests and highlands of the southern United States, Katrina spawned numerous TORNADOES of F0, F1 and F2 intensity on the Fujita-Pearson Scale of Tornadic Intensity. In northern and central GEORGIA, trees were uprooted, numerous houses and commercial properties were damaged, and at least one person was killed. Three of the tornadoes to strike Georgia were of F2 intensity. Tornadoes associated with Katrina's dying circulation also touched down in three Pennsylvania counties on August 31, damaging trees and BUILDINGS, but claiming no injuries

This satellite photograph shows the magnitude of the flooding suffered by the city of New Orleans during Hurricane Katrina's passage into meteorological history. The flooded areas are to the lower right-hand side of the photograph. Street after street, block after block, the lives and communities damaged by Katrina may take generations to recover. *(NOAA)*

or lives. National Weather Service (NWS) observers pegged these tornadoes at between F0 and F1 on the FUJITA SCALE. Tornadic touchdowns associated with Katrina were also reported in VIRGINIA and Alabama.

While Katrina was being downgraded to a tropical depression near Clarksville, Tennessee, during the late afternoon hours of August 30, an enormous local, state, and federal relief and recovery operation was being launched in response to the hurricane's unprecedented destruction. By September 3, the United States Coast Guard (USCG) had performed 9,500 rescues, double the entire number of rescues undertaken in 2003 alone. Some 70,000 National Guard and federal troops were deployed to Louisiana and Mississippi, and in the following weeks and months a Herculean task awaited the United States as it rebuilt one of its most historical cities and attempted to restore the many communities affected by Hurricane Katrina. As of 2007, some 1,836 people were said to have been killed by the direct and indirect effects of Hurricane Katrina, with more than 1,000 dead in Louisiana, more than 200 lost in Mississippi, 11 in Florida, and two each in Alabama and Georgia. Although state and local officials had initially predicted that "probably thousands were dead in the city [of New Orleans]" subsequent tallies sharply reduced the projected final death toll.

Hurricane Katrina's ferocity and graphic destruction ignited the ongoing debate concerning the effect GLOBAL WARMING was having on tropical cyclogenesis, or the formation and intensity of tropical cyclones in the North Atlantic basin. An August 2005 study conducted by the Massachusetts Institute of Technology (MIT) found that wind speeds in North Atlantic and North Pacific tropical cyclones had increased some 50 percent since the early 1970s. As evidence of global warming (which is attributable to the by-products of internal combustion engines), global temperatures have risen at least 1°F ($^-$17°C). While North Atlantic hurricane activity has markedly increased since 1994, other experts attribute the severity of hurricanes and their increased numbers to changes in sea-surface temperatures and the salinity levels in the North Atlantic's deep-water currents associated with regular 40–60 year cycles.

2005's Hurricane Katrina was the third and final North Atlantic tropical cyclone to bear the name Katrina. Between November 3 and 7, 1981, the first Hurricane Katrina delivered Category 1 winds and rains to Cuba and parts of the Bahamas; between October 28 and November 1, 1999, Tropical Storm Katrina made landfall in Central America and brought torrential rains to MEXICO. By virtue of the enormous loss of life and destruction caused by 2005's Hurricane Katrina, the name Katrina has been permanently retired from the rotating list of North Atlantic tropical cyclone names.

Kim, Typhoon *Philippines, November 3–15, 1977* A fairly intense Category 2 TYPHOON, Kim's sustained 102-MPH (164-km/h) winds, 127-MPH (204-km/h) gusts, and torrential rains seriously battered large portions of the northern PHILIPPINES between November 13 and 14, 1977. On Luzon Island, Kim's seven-inch (178-mm) downpours caused treacherous flash floods, raging rams of water that swept several mountainside villages into an early-morning oblivion. At least 30 people were killed, while another 51,000 were left homeless and destitute by as many as 20 separate landslide incidents. In the wind-riven capital of Manila, widespread power outages caused by the typhoon's strident fury soon triggered a dangerous fire in a seven-story harbor-front hotel. Caused by a lit candle and fanned by the typhoon's strenuous winds, the fire quickly engulfed the landmark structure, trapping scores of terrified guests on its upper floors. Although rain counts of up to seven inches (178 mm) were observed in downtown Manila, Kim's aerial fire brigade did little to contain the choking spread of the flames. Forty-seven people eventually perished in the fire, indirectly making Kim one of the most tragic Filipino typhoons in recent memory.

Kit, Typhoon *Japan, June 20–28, 1966* The second North Pacific TYPHOON to be christened Kit, this relatively weak, early season rain typhoon delivered 78-MPH (126-km/h) winds and astonishing 20–30-inch (508–762-mm) PRECIPITATION counts to the east coast of JAPAN's Honshu Island on June 28, 1966. In the town of Koga, 40 miles north of Tokyo, spontaneous floods and roaring mudslides swept away several hundred BUILDINGS, claiming the lives of 12 people. In the city of Nako, 75 miles (121 km) northwest of Koga, another spate of mudslides added 20 more lives to the DEATH TOLL, making Kit one of the deadliest recorded *ami*, or rain typhoons, in contemporary Japanese history.

Korea *Northwestern Pacific Ocean* A broad, 600-mile peninsula that deeply penetrates the TROPICAL CYCLONE–ridden waters of the western NORTH PACIFIC OCEAN, Korea has during the past three centuries repulsed a veritable ambuscade of TROPICAL STORMS and TYPHOONS. Although Korea is positioned to the north of TYPHOON ALLEY beyond the reach of most early and late-season typhoons, its shallow coastal plains, brace-work of islands, and cluttered sea and fishing ports have on average been affected by at least one typhoon or tropical storm per year. Partially protected by the volcanic barrier of the Japanese islands, Korea's peak period for tropical cyclone activity occurs in July and August when the first of the North Pacific's SUPERTYPHOONS begin to recurve north-northwest from their breeding grounds west of the PHILIPPINES. While not as frequent as the rain-laden *BAGUIOS* of the Philippines or as deadly as the immense wind typhoons that yearly haunt the coasts of TAIWAN and CHINA, a notable number of Korean typhoons have been systems of substantial size and intensity upon landfall. In some instances responsible for inflicting enormous DEATH TOLLS and widespread property damage on the peninsula, these storms nonetheless provide Korea with a fascinating cyclonic chronology:

KOREAN STORM CHRONOLOGY

July 18, 1930: A gargantuan early season rain typhoon careens across the Korea Strait from Japan. Loaded with tons of PRECIPITATION, the unnamed monster spawns dozens of flash floods and landslides over the southeastern mountain cities of Ulsan and Taegu. Some 3,400 BUILDINGS are swiftly devoured by the typhoon's 101-MPH (163-km/h) winds and rains. All along Korea's south shores, flotillas of small boats are thrown ashore and wrecked, drowning two dozen fishers. As damage estimates top $34 million, the death toll climbs to 219 dead and 15 others missing.

August 28, 1936: A midseason wind typhoon drifts ashore near the southwest trading station of Kohung during the early morning hours. The first known super typhoon to have struck Korea in the 20th century, the unnamed tempest claims 1,104 lives and leaves another 1,028 injured.

August 5, 1937: Another midseason typhoon, this one brimming with rain, deluges the northern port town of Wonsan, some 60 miles (97 km) north of Seoul, with 11-inch (279-mm) precipitation counts. One-hundred-thirty people perish, a majority of them in drowning incidents.

July 10, 1940: Scudding out of the East China Sea, an unnamed rain typhoon barrels into southwest Korea just after midnight. Triggering enormous rains, the tempest leaves an estimated 52 people dead, 100 injured, and 900 others without shelter. Washed-out bridges severely hamper relief efforts in the storm-ravaged region.

December 8, 1949: One of only six recorded typhoons to have struck the Korean peninsula during the month of December, this fearsome late-season surprise flattens hundreds of houses, uproots acres upon acres of trees, and inundates rice paddies all along the east coast. Believed to be among the deadliest of Korean typhoons, the unnamed maelstrom of 1949 reportedly kills a staggering 2,000 people.

July 20, 1950: Yet another July typhoon makes landfall near the southern island city of Kohung, some 200 miles (322 km) south of Seoul. Torrential rains drown 56 people.

June 6, 1953: A rare June typhoon that had originated 1,200 miles due west of the south Philippines on May 26, recurves through Taiwan and east China before slamming into the central highlands of the peninsula at dawn. Hundreds of buildings on Kang-

hwa Island are demolished by the typhoon's 106-MPH (171-km/h) winds and sawing 15-foot (5-m) seas. Forty men, women, and children perish.

September 10, 1956: Typhoon EMMA's 140-MPH (225-km/h) gusts and nine-inch (229-mm) rains cause several million dollars in damage to the southeastern city of Pusan. In a related incident, 16 American service members are lost when their Boeing RB-50, assigned to measure the typhoon's intensity, disappears over the Sea of Japan.

September 17, 1959: One of Korea's most monstrous midseason typhoons, Sarah makes an early evening landing on the Peninsula's southeast shores. Bearing sustained winds of 115 MPH (185 km/h), Sarah spreads more than ($43 million dollars worth of destruction across much of Kyongsang and Namdo provinces. Some 669 people perish, while another 171 are posted as missing. In what will prove to be Sarah's most tragic incident, 1,200 men are drowned when 46 fishing boats, caught off Pusan by Sarah's raging winds and swamping seas, founder.

August 26, 1960: The 120-MPH (193-km/h) winds associated with Typhoon Carmen claim 24 lives in southern Korea.

July 9, 1967: One of Korea's most memorable July typhoons, Billie kills 347 people and leaves another 2,000 homeless. An early season interloper from Japan, Billie's towering waves drown several dozen Korean fishers by overturning their boats on the Yellow Sea.

July 11, 1974: One of the most intense typhoons to strike Korea during the 1970s, Gilda kills 108 people, levels 12,000 buildings, and tallies more than $300 million in property and agricultural losses.

August 25–26, 1979: Typhoon Judy drops copious rains on much of southwestern Korea. Severe flooding, conjoined by rumbling landslides, carries away villages, roads, and bridges. Sixty Koreans die, while another 20,000 are left destitute.

July 1, 2002: Typhoon Rammasun crossed the Philippine Sea with maximum sustained winds near 125 MPH (200 km/h). The system battered coastal China with flooding rains, causing at least six deaths. Downgraded to a tropical storm, Rammasun made landfall in South Korea on July 5, with no major casualties or property losses. On September 3, 2002, Typhoon Rusa (Malaysian for "deer") pounded the Korean peninsula, claiming at least 200 lives, and making it one of the deadliest typhoons in recent memory. Rusa was also the most powerful typhoon to have struck Korea since Typhoon SARAH in 1959, which killed more than 840 people. The east coast province of Gangwon was severely battered, and the regional capital, Gangneung, was swamped with high waters after some 36 inches (89 cm) of rain fell across the region. Sustained winds were pegged at 127 MPH (160 km/h) while flooding consumed 12,621 acres (5,110 hectares) of farmland. Rusa's tearing gusts destroyed some 7,800 electrical poles. The typhoon was the worst to strike South Korea since 1959, causing 200 deaths and dumping between 11.8 and 19.7 inches (300 and 500 mm) of rain on parts of the country in less than 12 hours. Rusa damaged more than 20,000 BUILDINGS and destroyed at least 200 bridges and highways.

September 14, 2003: Super Typhoon Maemi (Korean for Cicada chirps) blasted the Korean port city of Pusan. Towering dockside cranes crumpled, cars were tossed into heaps, and a large cruise ship capsized at its pier as Maemi's sustained, 135 MPH (175 km/h) winds and 18-inch (240-mm) rainfalls overwhelmed the Korean Peninsula. Damage estimates topped $1 billion (in 2003 U.S. dollars) as 87 people were confirmed dead with another 28 listed as missing. At one point in its existence an extremely powerful super typhoon with maximum sustained winds of 170 MPH (280 km/h), spawned mudslides and floods that claimed at least 117 deaths, and forced 25,000 to evacuate their homes (see evacuation). Estimated property damage was set at $4.1 billion (in 2003 U.S. dollars). Some 5,000 homes were destroyed and another 13,000 damaged.

L

Labor Day Hurricane of 1935 *Southern United States, August 29–September 3, 1935* A NORTH ATLANTIC HURRICANE of record-shattering intensity and remarkable meteorological violence, the Labor Day Hurricane remains one of the most extraordinary TROPICAL CYCLONES in U.S. history. The first of only three Category 5 hurricanes to have ever stampeded ashore on continental United States, the Labor Day Hurricane's extremely low CENTRAL BAROMETRIC PRESSURE at landfall of 26.35 inches (892 mb), blistering 200-MPH (322-km/h) winds, 225-MPH (354-km/h) gusts, stinging rains and inundating 25-foot (7-m) STORM SURGE rampaged across the FLORIDA Keys' shallow coral chain on the night of September 2, 1935. Small, tightly coiled, and particularly deadly, the Labor Day Hurricane's 20-mile-wide (32-km) EYEWALL pushed the seething waters of the North Atlantic nearly 25 feet (7 m) above sea level at Matecumbe and Long Keys, two densely inhabited outcroppings some 90 miles (145 km) southwest of Miami. Much of the Florida East Coast Railroad, its concrete spans and stone embankments undermined by the hurricane's pounding breakers, was washed into the sea. An 11-car rescue train, dispatched from Miami to evacuate residents of the low-lying keys, was derailed by the hurricane's rampaging storm surge, leaving only the locomotive and tender standing on tracks. The American Red Cross later determined that 408 people perished in the Labor Day Hurricane, with 244 known dead and another 164 posted as missing.

Lake Okeechobee Hurricane *Southeastern United States, September 16, 1928* *See* SAN FELIPE HURRICANE.

Last Island Hurricane *Southern United States, August 9–12, 1856* Sometimes referred to as the Isle Derniere Hurricane, this celebrated TROPICAL CYCLONE unleashed pulverizing 115-MPH (185-km/h) winds, torrential rains, and a deadly nine–12 foot (3–4 m) STORM SURGE on LOUISIANA's south and east coasts between August 10 and 11, 1856. Named for a low-lying resort island it inundated on August 10, the Last Island Hurricane claimed between 140 and 160 lives and was responsible for hundreds of thousands of dollars worth of property damage to BUILDINGS, piers, and shipping from New Orleans to Morgan City. Regarded by many eyewitnesses as one of the most intense HURRICANES to have struck the United States during the second half of the 19th century, the Last Island Hurricane's cataclysmic destruction of the popular summer community of Isle Derniere gained it widespread notoriety, including a central role in Lafcadio Hearn's 1889 fictionalized account of the tragedy, *Chita: A Memory of Last Island.*

As is the case with so many pre–20th century tropical cyclones, the Last Island Hurricane's exact place of origination has never been determined. Because past investigations into the Last Island Hurricane's provenance have failed to reveal any record of the storm's landfall in any of the many islands that dot the CARIBBEAN SEA, it is entirely possible the system formed in either the southwestern or eastern reaches of the GULF OF MEXICO. In other respects, its midseason status, coupled with its considerable violence at landfall in Louisiana, would seem to imply that it was of the Cape Verde variety, a cyclonic system that formed off the Africa's northwest coast before moving west-northwest and simply missed the major islands along the way. Contemporary weather reports indicate that the hurricane's onset in New Orleans on August 10, 1856, was heralded by a shift in wind direction from east to southeast, the sort of circulatory progression found in those counterclockwise tropical systems that are traveling northeast.

Reared over sun-smote waters, the Last Island Hurricane underwent rapid intensification, growing into a storm of significant size and destructive potential within a matter of hours. On the morning of August 10, as the hurricane moved within 50 miles (80 km/h) of the Louisiana coast, a full-rigged ship overtaken by its EYE recorded a CENTRAL BAROMETRIC PRESSURE reading of 28.20 inches (954 mb), making the Last Island Hurricane a storm of firm Category 3 status. In the log, the vessel's British master added that the barometric pressure had dropped nearly two inches (67 mb) during the past 24 hours, indicating a precipitous increase in the system's PRESSURE GRADIENT and a proportional increase in its ferocity. Heavy rains whipped across the gulf as wind gusts within the hurricane's EYEWALL escalated to 130 MPH (209 km/h), forcing the vessel to heave-to in broaching seas and screeching winds. Within hours, it was a dismasted wreck.

In New Orleans, some 250 miles (402 km) northeast of the eye, light rains accompanied by intermittent gales began to assail the city just after daybreak on August 10. Completely unexpected by the citizens of New Orleans, the hurricane's fringe furies caught several Sunday morning churchgoers in a virtual deluge of gust-driven rain and portentous winds. Hats, veils, and umbrellas were scattered across the muddy streets of New Orleans, soon to be followed by awnings, tree limbs, and chimney pots. By midmorning, as the sky turned citrine yellow, the winds increased, becoming more sustained as the hurricane's intense core drew closer to Louisiana's south coast. Heavy rains began to roil the expansive waters of Lake Pontchartrain, casting waves of ever-increasing height and frequency against the lake's wooden embankments. Street lamps swayed as those horse-drawn carriages still out in the tempest were overturned or blown sideways into trees and storefronts. Several property owners still in the process of shuttering their buildings found themselves showered with broken glass; along the New Orleans riverbank, steamboat operators fought rising waters to keep their laboring vessels from being run aground.

On Caillou Bay's resort island of Isle Derniere, 110 miles (177 km) southwest of New Orleans, the situation was far more severe. By 10 o'clock on the morning of August 10, the wind was already gusting to TROPICAL STORM strength, driving the steady rains it carried against the 20 small summer cottages, hotel, restaurant, and ferry depot of Last Island.

Only 22 miles (35 km) in length and standing less than 10 feet (3 m) above sea level, Last Island was by midmorning confronting the horrifying spectacle of an encroaching Gulf of Mexico, of a sea that had risen two feet (1 m) in one hour. Immediately recognizing the significance of such a rapid rise in sea level, the 413 people then on the island hastily made preparations to abandon it. A few of the hardier ones actually made it to the ferry landing on the southern tip of the island, only to discover that the swollen waters of the gulf had already pounded the wooden docks to pieces.

Retreating to their lightly timbered cabins, the alarmed vacationers listened to the creak of their roofs and the intensifying howl of the wind. As candles and hurricane lamps flickered against a darkening noonday sky, watch parties, struggling to remain upright against 60-MPH (97-km/h) winds, recorded an ominous rise in Caillou Bay's water level. Surging five feet (2 m) in just two hours, the burgeoning waves sent the watch parties scurrying back to their families with the news that the end was about to be seen of Last Island and that there was nothing anyone could do to prevent it. While several residents did eventually seek shelter in the two-story hotel and houses, it was with a very real understanding that in time the hurricane's storm surge would overwhelm the island, washing it clean of everything but its summer memories.

Just before four o'clock on the afternoon of August 10, 1856, as sand-laden winds blinded those terrified victims seeking shelter from the storm, the hurricane's 13-foot (4-m) surge suddenly overtook three-quarters of Last Island. Rolling ashore as an endless series of ever-rising breakers, the hurricane's onrushing surge stood five feet (2 m) higher than the island's lowest point. It blasted through the gridwork of streets that connected the various buildings, knocking cottages off their pilings, uprooting trees, shrubs, and fence posts, and tearing at the stout walls of the two-story Muggah Billiard-House. A few of the cottages, their rooms echoing with the screams of trapped residents, briefly floated, crazily rocking beneath the winds until their joints opened up and they sank. Other structures on the island simply collapsed upon impact, seemingly folded into one as the surge graded most of the island, removing nearly two feet (1 m) of sand from its topography.

One hundred forty of the 413 men, women, and children on Last Island that afternoon were drowned. All but five cottages on the island's east shore were utterly demolished. Clutching at floating doors, furniture—even entire sections of roof—bands of survivors miraculously rode the gargantuan maelstrom of swirling wreckage that trailed across the island for most of the evening. In time, 265 of them would find shelter on the 620-ton steam-packet *Star,* en route from Bayou Boeuf to Last Island when it was cast ashore. Completely dismasted, its funnel gone, its anchors dropped in a last-minute attempt to keep it off the land, the *Star* came to rest near the dwindling remains of the Billiard-House. Forced out into the storm by the collapse of the Billiard-House, several dozen survivors saved themselves from almost certain drowning by climbing aboard the grounded wreck of the steamer.

All told, between 140 and 160 people perished in the Last Island Hurricane, a majority of them on the island that gave the storm its name. Damage was felt as far east as Pensacola, where heavy rains lashed the bay there. The Opelousas Railroad, which connected New Orleans with Morgan City and Last Island, suffered numerous track washouts and bridge collapses, requiring weeks of work before regular service could be resumed. Newspaper reports recorded in lurid detail how thieves had brazenly taken to robbing the corpses left on Last Island by the receding storm surge, pulling multicarat diamond rings off the fingers of dead society ladies.

Such stories, whether or not they were true, eventually gained the Last Island Hurricane considerable notoriety throughout the hurricane belt. Moralizing tales and editorials about the hurricane leveling justice on Last Island's "privileged" residents quickly entered popular literature and became subject to subsequent retellings of greater and greater exaggeration. At a time when lowbrow melodrama proved a profitable genre in homespun theaters all over frontier America, the Last Island Hurricane's apocalyptic coincidences served to foster a gruesome legend that persisted well into the latter part of the 19th century. Propagated through newspaper articles, plays, and eventually the Hearn novel, the lessons of the Last Island tragedy endured the Civil War, the shock of Reconstruction, and the rise of the U.S. industrialist to become a cautionary exemplum on the evils of social arrogance and financial negligence.

Thus it is not surprising to discover that Hearn's highly embellished account of the Last Island disaster was reissued in 1889, 33 years after the hurricane occurred to a readership that had just recently been exposed to the ethical horrors of the May 31, 1889, flood at Johnstown, Pennsylvania. In that particular incident, more than 3,000 people in the steel town of Johnstown were killed when a dam, owned by the exclusive South Fork Fishing and Hunting Club, burst during heavy rains, sending millions of tons of water crashing through mountain valleys. Subsequent investigations into the dam's failure proved that the club, whose members included steel mag-

nates Andrew Carnegie and Henry Clay Frick, had neglected to maintain the dam adequately, thereby precipitating its final demise. When placed against the cultural foil of Hearn's sensationalized remembrances, the horror of the Johnstown Flood was further augmented by the realization that it could have been avoided and that its outcome was foretold three decades earlier in the legendary course of the Last Island Hurricane.

latent heat   First defined in 1750 by Scottish scientist Joseph Black as the opposite of sensible heat, latent heat describes the physical processes in which heat is initially absorbed by the action of EVAPORATION, is later released through the action of CONDENSATION, and does so in each instance without registering any rise in internal TEMPERATURE. In a TROPICAL CYCLONE, latent heat is collected from warm SEA-SURFACE TEMPERATURES through evaporative processes before being returned to the atmosphere at altitudes above that of the CONDENSATION LEVEL. In this way, latent heat aids in tropical cyclone development and maturation by adding enormous quantities of energy to the system's CONVECTION cells.

Late Season Hurricane of 1770 *Northeastern United States, October 20, 1770*   Following a similar trajectory to the SOUTHEASTERN NEW ENGLAND HURRICANE of October 23–24, 1761, the Late Season Hurricane's strident northeasterly winds and 3-inch (76 mm) rainfall counts battered coastal towns and villages from Rhode Island to Maine. In New London, Connecticut, two merchant ships were run ashore and then pounded to pieces by the hurricane's surf. In Boston, where hail accompanied the hurricane's inland passage, one of the highest STORM TIDES in a half-century overflowed piers, submerged several streets, and swept away the approaches to a principal drawbridge. Standing guard over a small island in New London's harbor, Fort William surrendered its sentry boxes, its timber stocks, and several chimneys to the hurricane's conquering winds, while scores of small boats were sent spinning against the fort's stone ramparts and were smashed. Producing an early CENTRAL BAROMETRIC PRESSURE reading of 28.96 inches (980 mb), the Late Season Hurricane maintained much of its strength as it blew northward, through southern New Hampshire and into what is now eastern Maine. Damage was considerable, with numerous houses, barns, and sheds demolished at Portsmouth and unusually high tides measured at Portland, Maine.

Late Season Hurricane of 1888 *Northeastern United States, November 25–27, 1888*   One of the largest HURRICANES ever observed in NEW ENGLAND, the Late Season Hurricane of 1888 spent no less than three days bringing 75-MPH (121-km/h) winds, drenching rains, and tumultuous seas to Cape Cod, eastern Massachusetts, and southern New Hampshire. As thousands of trees, small buildings, and fences were damaged by the hurricane's powerful offshore winds, 15 merchant ships at anchor in Boston harbor were thrown ashore by huge waves and wrecked. The spectacular effect this particular TROPICAL CYCLONE had on the waters of Massachusetts Bay is further illustrated by reports that spray overtopped the stone lighthouse on Minot's Ledge, some 114 feet above the sea.

Louisiana *Southern United States* With some 397 miles (639 km) of coastline gracefully fanning out into the hurricane-prone GULF OF MEXICO, Louisiana has endured a long, often violent history of HURRICANE activity. In the 291-year span between 1715 and 2006, Louisiana's fragile delta plains, shallow barrier islands, and riverbank cities were directly affected by no less than 67 recorded hurricanes. Of these, 27 were regarded as major hurricanes, Category 3 systems whose CENTRAL BAROMETRIC PRESSURES produced readings of less than 28.47 inches (964 mb) and sustained wind speeds greater than 111 MPH (179 km/h). Since 1715, 22 hurricanes of varying degrees of intensity have come ashore within the densely populated limits of New Orleans, causing hundreds of deaths and millions of dollars in property damage. On the night of August 17, 1969, eastern Louisiana had the rare misfortune to be struck by one of only three Category 5 hurricanes yet to make a documented landfall in the United States, Hurricane CAMILLE. And on August 29, 2005, much of coastal Louisiana (including the city of New Orleans) was devastated by Hurricane KATRINA, whose Category 3 landfall eventually claimed 1,577 lives in Louisiana, and an additional 259 lives elsewhere in the United States.

On average, the state's shallow coast, noted for its tranquil summer havens and vast shrimp fleet, is indirectly threatened by a TROPICAL CYCLONE every two-and-a-half years and directly impacted by one every five years. On various occasions Louisiana has even withstood multiple strikes in the course of a single HURRICANE SEASON, as happened in 1860, when three dangerous hurricanes struck the Louisiana coastline in the space of seven weeks. The first, which came ashore near New Orleans on August 11, raised the brooding waters of Lake Pontchartrain and drowned about 13 people. The second of the trio, which struck on September 14, lashed the delta towns of Balize and Pass-a-l'Outre, claiming 10 lives between them, while the last of the series, on October

2, generated considerable wind and five-inch (127-mm) PRECIPITATION counts at New Orleans, but a minimal STORM SURGE elsewhere along the coast. A similar series of triple strikes was further witnessed in 1860 when three large hurricanes came ashore in Louisiana between August 11 and October 3. Double strikes took place in 1812, 1834, 1885, and 2005.

Louisiana's position on the northwestern curl of the Gulf of Mexico generally presents it with a prolonged, frequently vulnerable hurricane season. Between 1715 and 1993, Louisiana was struck by at least three June hurricanes, all of them tropical cyclones of considerable fury and duration. The first, known as the June Storm of 1812, cast a seven-foot storm surge into the farming town of Franklin, demolishing dozens of houses and killing entire herds of cattle on the night of June 11–12, 1812. In 1934 an unnamed June hurricane killed six people as it crossed Louisiana's southwest coast, battering Morgan City and Jeanerette with barometric pressures as low as 28.52 inches (966 mb) on June 16. The third June hurricane, christened AUDREY, proved to be an even deadlier storm when it crashed ashore at Cameron Parish, Louisiana, on June 27, 1957. Five hundred eighteen people perished as Audrey's 20-foot (7-m) storm surge completely inundated thousands of acres of low-lying bayou-country. One thousand nine hundred BUILDINGS were totally destroyed; another 19,000 were seriously damaged, making Audrey one of the most destructive early season hurricanes in Louisiana history.

There has been only one documented hurricane strike on Louisiana in July, and it was a storm of memorable size and intensity. Christened the BAY ST. LOUIS HURRICANE OF 1819, its fleeting passage across the lower Mississippi Delta on the night of July 27–28 delivered 74-MPH (119-km/h) winds to New Orleans and heavy seas to the mouth of the Mississippi River. A number of farms along the shores of Lakes Borgne and Pontchartrain were washed away, while two schooners and a brig were run aground. Between 16 and 35 lives were lost in Louisiana, a relatively mild DEATH TOLL when placed against the dozens of lives claimed by the storm when it later burst ashore in neighboring MISSISSIPPI.

Yet it is in August, when the first of the late-season hurricanes begins to transit the Gulf of Mexico, that the number of confirmed strikes in Louisiana rises to a brisk 16. Famous midseason tropical cyclones in the state include the hurricane of August 31, 1794, in which an enormous storm surge—estimated by some historians to have been 15–20-feet (5–7-m) high—engulfed the lower Mississippi Delta. Roaring upriver for nearly 80 miles (129 km), the surge eventually inundated the towns of Balize and Plaquemines, forcing the flight of hundreds of local residents. At nearby Fort St. Philip, 17 British soldiers who had been put to work building the garrison were drowned when the surge swept them into its seething torrent and then leveled the entire outpost.

Less than a decade later, another August hurricane, this one on August 19, 1812, spun ashore southwest of New Orleans. Known as the Great Louisiana Hurricane of 1812, its vast storm surge caused the gulf waters to swell more than six feet (2 m). Flooding past New Orleans and into the confined basin of Lake Pontchartrain, it initiated one of the largest inundations ever experienced by the city and its environs. Coursing across the rice and cotton plantations downriver from New Orleans, the Great Louisiana Hurricane's surge flattened buildings, uprooted crops, and killed an estimated 60 people, many of them slaves.

In what would become one of the most celebrated storm disasters of the late 19th century, the LAST ISLAND HURRICANE of August 10–11, 1856, claimed between 140 and 160 lives when it submerged the Louisiana resort island of Isle Derniere. After demolishing almost every structure on the island, the hurricane moved inland, lashing New Orleans with seesawing tides and 90-MPH (145-km/h) winds. More than $350,000 worth of 1856 property damage was assessed, making the Last Island Hurricane one of Louisiana's most notable August hurricanes.

On August 17, 1969, Hurricane Camille, a Category 5 storm of extraordinary ferocity brought sustained 190-MPH (306-km/h) winds, 250-MPH (402-km/h) gusts, and a 20–25 foot storm surge to Louisiana's Boudreau Islands while on its way to a catastrophic landfall at Pass Christian, Mississippi. Piling up against the narrowing approaches to Lake Borgne, the surge eventually sheared its way into the lake, forcing its tides to rise 10 feet (3 m) above normal against miles of concrete and stone seawalls. Never before tested by a hurricane of Camille's calibre, a number of these walls failed, were overtopped and then breached. Nine people in Louisiana perished as Camille's overflow knocked cottages from their foundations and washed out highway bridges.

As recently as 1992, Louisiana was struck by an August hurricane of Category 4 status. Hurricane ANDREW, the midseason upstart that grew to inflict $15 billion in insured losses on southern FLORIDA and southeastern Louisiana, pummeled the lower Mississippi Delta with 140-MPH (225-km/h) winds and TORNADOES on the morning of August 26. Although only one person was reportedly killed, Andrew's monetary losses in Louisiana totaled more

than $5 billion, making it the single most expensive hurricane in state history.

In September, when the Gulf of Mexico and the CARIBBEAN SEA actively begin to host a banquet of CAPE VERDE STORMS and local hurricanes, Louisiana's history of strikes jumps to 31 hurricane landfalls between 1722 and 2006. One of these September storms, the Great Hurricane of September 23–24, 1722, devastated the newly founded capital of New Orleans. Preceded ashore by a seven to eight-foot (2–3-m) storm surge, the hurricane's 100-MPH (161-km/h) EYEWALL winds scoured the young city, completely leveling 34 crudely constructed dwellings. Almost every public building in New Orleans, from the hospital to the cathedral, was either unroofed or totally ruined. Ten flatboats at dock were holed and sunk; four other vessels—one of them a floating gunpowder magazine—were carried ashore and grounded.

In September 1909, an unnamed hurricane of extreme intensity killed approximately 350 people as it churned ashore east of Houma, some 48 miles (77 km) southwest of New Orleans, on the afternoon of September 20. Inundating the coastline from Gulfport, Mississippi, to Louisiana's Grand Isle with a 14-foot (5-m) storm surge, the hurricane's 138-MPH (222-km/h) winds, brought on by a pressure reading of 27.49 inches (931 mb), later gutted the state capitol building at Baton Rouge. Six years later another Category 4 hurricane of equal barometric pressure nearly stalled while crossing the Mississippi River's mouth on the afternoon of September 29, 1915. Toting a 16-foot (5-m) storm surge beneath its eye, the hurricane swamped Lake Pontchartrain as it slowly passed inland, lashing greater New Orleans with hurricane-force winds for over seven hours. Almost every structure along the lake was swept into wreckage, claiming the lives of 275 people.

In mid-September 1947, yet another Category 4 hurricane—this one boasting a central barometric pressure of 27.75 inches (940 mb)—barreled ashore just south of New Orleans on September 19. Eighty-four people along Louisiana's gulf coast were reported dead or missing; more than $4 million in property damages were assessed in New Orleans and Baton Rouge.

More recently, Hurricane Flossy's passage into New Orleans on September 25, 1956, wrought more than $2 million in damage on oil drilling rigs positioned in the Gulf of Mexico; Hurricane BETSY either damaged or destroyed more than 25,000 houses when it made landfall east of Louisiana's Marsh Island on the night of September 7–8, 1965. Before crossing into northern Mississippi and being downgraded to a tropical storm, Betsy killed 61 people in

Louisiana and left another 60,000 homeless. Nearly $700 million in damage was assessed, making Betsy the most expensive tropical cyclone to have then affected the state.

Come October, when the BERMUDA HIGH's autumn subsidence to the south-southeast restricts Atlantic hurricane activity to the western Caribbean Sea and the Gulf of Mexico, Louisiana's record of cyclonic strikes decreases to seven documented landfalls between 1715 and 2006. Of these seven hurricanes, four were storms of substantial size and intensity. One of them, the GRAND ISLE HURRICANE of October 1–2, 1893, still remains the deadliest confirmed tropical cyclone in Louisiana history, with 1,800 to 2,000 deaths to its credit. Treading ashore near the Mississippi River's mouth as a tightly coiled Category 4 monster, the hurricane inundated the densely populated barrier spit of Grand Isle with a colossal 20–22-foot (7–8-m) storm surge. Entire villages disappeared as first the hurricane's 137-MPH (221-km/h) winds, followed by its surge, wrenched cottages from their pilings, plowed over barns, hammered shoreline hotels into driftwood, and sawed restaurants down to the bone. Hundreds of vacationers on Grand Isle were drowned, overwhelmed by the high waves, or pushed under by the floes of wreckage that scudded across the gust-whipped surface of the Gulf of Mexico.

Before the Grand Isle Hurricane of 1893, the last major October storm in Louisiana memory occurred in 1837, during the first week of the month. Known as Racer's Hurricane (in honor of the British armed sloop *Racer*, which the tempest nearly sank on October 1), the storm came ashore near Louisiana's Sabine River on the afternoon of October 6. Enormous precipitation counts drove Lake Pontchartrain over its embankments, washing away nearly two miles (3 km) of the Pontchartrain Railroad's track, while four steamboats were driven against the breakwater at Port Pontchartrain and wrecked. A brick lighthouse located on the bayou side of the lake was toppled, claiming the lives of the keeper and his family. Several other lives were lost in Port Pontchartrain, where 17 buildings were washed into a swamp.

Such continual hurricane activity has over time served to reshape several major geographical features of the Louisiana shoreline. Many of its low-lying peninsulas, spits, and alluvial islands have been repeatedly swept by the rushing winds and deluging surges of passing hurricanes, causing incalculable damage to the area's finely tuned ecosystem.

In August 2002, Tropical Storm BERTHA—which had been born over the northern Gulf of Mexico as a tropical depression on August 4—tracked northwestward across Louisiana before recurving to the southwest. Bertha reemerged over the Gulf of Mexico and briefly strengthened before lumbering ashore in Texas on August 9, 2002.

On October 3, 2002, Hurricane Lili delivered a Category 2 strike to Louisiana's Vermillion Bay. One hundred mile per hour winds (161 km/h) lashed coastal regions, but caused little damage.

On June 30, 2003, an early season Tropical Storm BILL rapidly developed in the Gulf of Mexico before slicing ashore in southeastern Louisiana with sustained winds of 60 MPH (97 km/h). Electrical service to thousands of residences was disrupted and Bill's enormous rainfall amounts caused flooding across several southeastern states. At least eight inches (203 mm) of rain fell in Pascagoula, Mississippi (on the border with Alabama) and tornado outbreaks were reported in Mississippi, ALABAMA, Tennessee, Georgia, NORTH CAROLINA and VIRGINIA.

Formed from a tropical wave in the southwestern Gulf of Mexico on October 6, 2004, Tropical Storm Matthew pounded ashore in central Louisiana on October 10, delivering as much as 10 inches (230 mm) of precipitation. A fairly mild tropical system, Matthew dissipated over Arkansas on October 11, 2004.

During an early morning landfall on July 6, 2005, a weak Hurricane CINDY delivered 75-MPH (161-km/h) winds and a four to six foot (1–2 m) storm surge to the southern coasts of Louisiana and Mississippi. Bearing a central barometric pressure of 29.47 inches (998 mb) at landfall, Cindy dropped between four and six inches (102–152 mm) of rain and prompted the posting of tornado watches as it moved inland across Louisiana, Mississippi, Alabama, and northwestern Florida at 13 MPH (21 km/h). Louisiana's most populous city, New Orleans, suffered some minor street flooding and downed trees, but Cindy's landfall and inland passage to the east of the city otherwise sparred it.

In 2005, Hurricane Katrina devastated southern Louisiana, causing unprecedented death and destruction in the historic city of New Orleans. Less than two weeks later, on September 23, Hurricane RITA delivered some 15 inches (381 mm) of rain to the city while on its way to a landfall in Texas, prompting the failure of a hastily repaired levee, and the reflooding of several parts of the city with almost 12 feet (4 m) of brackish water. One person was reportedly killed.

Ludlum, David M. (1911–1997) U.S. meteorologist, historian, author, and founder of *Weatherwise* magazine, Ludlum received his Ph.D. in history from Princeton University in 1938 and in 1940 joined the U.S. Army where he undertook training as weather forecaster. Commissioned a captain in the Army Air Force at the start of World War II, Ludlum operated a mobile weather station in EUROPE, where he aided in the Allied assault on Monte Cassino, Italy, in May 1942 by preparing weather charts and forecasts in advance of the MILITARY OPERATION. In honor of his thoroughness as well as his accuracy, the mission to liberate Monte Cassino was officially dubbed Operation Ludlum. Promoted to lieutenant colonel in 1944, Ludlum resigned his commission in the Army following the conclusion of the war and returned to New Jersey in 1946 to found a weather instruments and consulting firm. In an effort to provide a larger advertising forum for his company, Ludlum founded *Weatherwise* magazine in 1948. Between 1948 and 1978, Ludlum served as *Weatherwise*'s editor in chief, transforming the homespun enterprise into one of the U.S.'s premier historical and meteorological publications. Following the magazine's sale in 1978, Ludlum served as a "Weatherwatch" columnist until his retirement in 1994. A gifted writer and weather historian, Ludlum was the author of numerous important meteorological texts, including *Social Ferment in Vermont* (1938), *Early American Hurricanes* (1963), *Early American Winters* (1966), *Early American Winters II* (1968), *Early American Tornadoes* (1970), *The American Weather Book* (1982) and the *Audubon Society Field Guide to North American Weather* (1991).

M

Madagascar *Indian Ocean* A large, mountainous island approximately 400 miles (644 km) off the southeastern coast of Africa, Madagascar is frequently struck by cyclones of varying intensity during the Indian Ocean tropical cyclone season. Officially extending from mid-November to mid-April of each year, the Madagascan cyclone season has on several occasions produced TROPICAL CYCLONES of considerable intensity and destructiveness.

On February 9, 2003, for example, Tropical Cyclone Gerry developed off the northeastern coast of the island of Madagascar. By February 13, the system was located to the north of the Mascarene Islands and produced sustained winds of 115 MPH (185 km/h) and some seven inches (178 mm) of PRECIPITATION of Réunion Island.

On February 26, 2003, Tropical Cyclone Japhet generated off the west coast of Madagascar. On February 27, the following day, the system was producing TROPICAL STORM force (70-MPH [110-km/h]) winds across the Mozambique Channel, and delivered heavy rains to northern Madagascar. No deaths or injuries were reported.

The 2003 Indian Ocean cyclone season was an active one for Madagascar. On May 9, Tropical Cyclone Manou made landfall along the eastern coast of Madagascar on with maximum sustained winds of 85 MPH (140 km/h) and wind gusts of 125 MPH (200 km/h) and higher. One of Madagascar's deadliest tropical cyclones, Manou claimed some 265 fatalities in the Vatomandry district. Nearly 85 percent of the district's buildings were destroyed and Manou's lingering cloud remnants dropped several inches of rain on the mountainous island, causing severe flash flooding.

On the night of December 9–10, 2003, Tropical Storm Cela scudded across the northern reaches of Madagascar, its 50-MPH (85-km/h) winds generating torrential rains that claimed some five lives.

The most severe tropical cyclone to have affected Madagascar during the last decade, Tropical Cyclone Gafilo struck the northeastern reaches of the island on March 7, 2004, killing 172 people, injuring another 879, and leaving some 214,000 without shelter. In one instance, 111 people were reportedly lost in the sinking of a ferryboat. During the 2004 Indian Ocean tropical cyclone season, two systems—Elita and Gafilo—pounded Madagascar, causing widespread damage to clove, vanilla and ylang-ylang crops, depressing an already struggling agricultural economy.

Martinique *Eastern Caribbean Sea* This rugged volcanic island, noted for its black sand coves, fertile slopes, and verdant rain forests, has had a venerable, often devastating history of HURRICANE strikes. Located in the southeastern CARIBBEAN SEA, French-held Martinique is the largest of the five major islands—Martinique, St. Lucia, St. Vincent, the Grenadines, and Grenada—that comprise the expansive Windward Islands chain. The archipelago's position between latitudes 12 and 15 degrees North places it directly in the path of the mid- to late-season hurricanes that spiral across the Atlantic Ocean from Africa's Cape Verde islands. Between 1623 and 2006, Martinique's stone *communes,* Lamentin plains, and notoriously active volcano, Mont Pelée, were directly affected by 60 recorded hurricanes. Of these, 22 were major hurricanes, with BAROMETRIC PRESSURES of 28.47 inches (964 mb) or lower and winds in excess of 111 MPH (179 km/h). At times this plantation island has even suffered several hurricanes in quick succession, as happened in 1775, when two destructive storms came ashore within one month of each other and in 1887, when two hurricanes struck in just over one week's time. Similar double strikes also occurred in 1642 and 1765.

Since 1623, there is no record of Martinique having been struck by a June hurricane. Most early season hurricanes tend to form in the southwestern Caribbean or lower GULF OF MEXICO and move northeast, thereby sparing the distant island their developing fury. In July, however, as the sun-baked waters of the southeastern Caribbean begin to reach the critical 80–84-degree F (27°–29°C) threshold needed for hurricane generation, the probability of a hurricane landfall in Martinique dramatically increases, with nine reported strikes between 1623 and 1980. Also it is during the hot, humid months of August, September, and October when the Cape Verde hurricanes begin to arrive at the Caribbean's east door that the number of strikes jumps to a hefty 22, 10, and 12, respectively. Famous midseason hurricanes in Martinique include the violent hurricane of August 3, 1680, in which 20 French men-of-war and two British merchant ships were driven ashore in Cul-de-Sac Bay, claiming more than 400 lives. In September 1776, an enormous convoy of more than 100 French and Dutch merchant ships, laden with sugarcane, pineapples, and bananas bound for EUROPE, was overtaken by a ferocious hurricane at Point Pitre Bay. Despite a strenuous effort by the vessels' crews to lighten the ships by jettisoning their valuable cargo and cannon, 60 percent of the convoy was nonetheless sent to the bottom or run onto the volcanic shore by the hurricane's prodding winds. An estimated 2,500 sailors perished. During a similar hurricane in October 1695, several French warships were driven ashore near the town of Bellefontaine, claiming another 600 lives. A deadly hurricane in August 1788 inundated Martinique's eastern beachheads with a 14-foot (5-m) STORM SURGE that swept away an entire village on the Caravelle Peninsula, while an even greater surge during the height of the legendary GREAT HURRICANE of October 12, 1780, demolished more than 150 houses in the then capital of St. Pierre. During the same storm, a fleet of 40 troop transports that had just arrived from France was completely destroyed near St. Pierre's Road, drowning between 3,000 and 5,000 soldiers. Nearby, two French frigates, the *Marquis de Brancas* and the *L'Eole*, with 150 soldiers from the regiment of Touraine aboard, were driven from the harbor by the violence of the storm and subsequently lost without a trace. Further inland, the Great Hurricane's

180-MPH winds destroyed Fort Royal's principal cathedral, seven churches, the governor's house, the senate-house, and the main prison. More than 1,000 people perished in the capital alone, while the estimated DEATH TOLL across the island would eventually exceed 9,000. Martinique would later suffer similar if only slightly less destructive hurricane strikes in October 1817, when more than 200 people perished; in late July of 1825, when the Santa Ana Hurricane killed another 450 people; and during the four-hour blast on August 18, 1891, when 700 lives were claimed and more than $10 million in damage was assessed. More recently, Martinique was directly affected by the fourth tropical cyclone of the 1970 season, Hurricane Dorothy. For 17 hours between August 20–21, Dorothy brought 100-MPH (161-km/h) winds and nearly 12 inches (305 mm) of PRECIPITATION to the hillside communities surrounding the capital, Fort-de-France. Mudslides surged into the city's center killing at least 42 people. Less than a decade later, on August 30, 1979, Hurricane DAVID caused heavy damage to the northern resort towns of Basse Pointe and Grand Riviere as it passed through the Martinique Passage on its way to PUERTO RICO and HISPANIOLA. Although David's 150-MPH (241-km/h) winds killed 37 people on the neighboring island of DOMINICA, no fatalities were reported in Martinique. Just over a year later, on August 5, 1980, Hurricane ALLEN closely followed in David's path, leaving 20,000 people homeless in Dominica and claiming one life in Martinique.

mercurial barometer   A type of BAROMETER that uses the movement of mercury to measure BAROMETRIC PRESSURE. First developed in 1643 by the Italian inventor Evangelista Torricelli, the standard mercurial barometer of the present day remains essentially unchanged from its Renaissance prototype; a glass tube between 35 and 40 inches long, sealed at the top but left open at the bottom, that is partially immersed—open end down—in a container of mercury. The weight of the air outside the tube pushes down on the mercury, forcing it to rise inside the tube; conversely, any decrease in atmospheric pressure relieves the outside pressure on the mercury, allowing it to slide downward in the tube. These changes in the height of the mercury column are then observed against the engraved calibrations on the tube as barometric pressure in inches or MILLIBARS. Most mercurial barometers intended for domestic use are of the one-tube variety, with a range of between 31.00 and 28.50 inches (1,039–965 mb). More sophisticated mercurial barometers employed are of the two-tube design, that not only allows for greater sensitivity and accuracy but also possesses a much wider range,

As fascinated onlookers gaze in wonder, a waterspout churns beneath a clearly defined mesoscale cyclone in this photograph taken in 1896 in Vineyard Sound, Massachusetts. The result of convective activity, they do not actually "draw up" the sea. Their shape and color is the result of condensation from the low pressures in the spout's column. *(NOAA)*

from 31.99 to 26.00 inches (1,083–880 mb). As a forecasting device, the mercurial barometer is most accurate when the direction and degree of its rise and fall over a defined time period is observed. By classifying the characteristics of a mercurial barometer's movement as steadily falling, rapidly falling, falling then rising, and steadily rising and rapidly rising, a meteorologist or weather watcher can determine the onset of HIGH and LOW pressure AIR MASSES, EXTRATROPICAL CYCLONES, and TROPICAL CYCLONES. In the case of the latter, a mercurial barometer in the path of a mature-stage tropical cyclone will register steadily falling and then rapidly falling before rapidly rising with the passing of the system's CENTRAL BAROMETRIC PRESSURE.

Mesoscale Convective Complex (MCC)   TROPICAL CYCLONES do not always generate from TROPICAL WAVES. Sometimes a large collection of thunderstorms centered in a nontropical low-pressure environment can mature into a tropical system. Hurricanes MITCH (1998) and VINCE (2005) received their start as mesoscale convective complexes.

Mexico *North America*   Mexico, the vast crescent of a nation that straddles the northern half of the Central American isthmus, has suffered a long, continuous history of powerful TROPICAL CYCLONE activity. With 1,756 combined miles (2,826 km) of coastline bordering on two hurricane-prone bodies of water—the NORTH PACIFIC OCEAN to the west and the GULF OF MEXICO and CARIBBEAN SEA to the east—Mexico is particularly vulnerable to mid- and late-season HURRICANE strikes on both its shores.

Between 1552 and 2006, 92 mature hurricanes and TROPICAL STORMS came ashore on Mexico's coasts, bringing torrential rains and flailing winds to its marshy lagoons, arid highlands, and Sierra Madre ranges. On average, Mexico's gulf-coast plains and shallow Pacific basins are likely to be inundated by at least one hurricane every three to six years, and by a major hurricane—storms characterized by BAROMETRIC PRESSURES of less than 28.50 inches (965 mb) and WIND SPEEDS in excess of 110 MPH—every five to seven years. In some instances, three or more hurricanes have directly affected the nation in a single year, as happened during the 1898 season, when three hurricanes inflicted damage on the Yucatán Peninsula and during the 1993 season, when five North Pacific tropical cyclones came ashore on the northwest coast, two of them bearing winds of hurricane intensity. On September 17, 1988, Mexico had the dubious honor to be struck by one of only four recorded Category 5 hurricanes to ever make landfall in the Northern Hemisphere, the extraordinarily destructive Hurricane GILBERT.

Early accounts of Mexican hurricanes are largely culled from the country's east coast, where the Spanish were among the first of the European conquistadores to record the number of men and ships, and the amount of gold extracted from them by the avenging fury of North Atlantic storms. During the latter half of the 16th century, a veritable ambuscade of hurricanes struck in and around the major seaport of Veracruz, causing the Spanish to ponder both God's displeasure and their mounting losses. In September 1552, a hurricane overtook the *Flota de Nueva España* while it was en route to Havana, sinking 12 treasure-packed ships. In 1567, three ships from that year's *flota* were driven ashore as an October hurricane pounded Veracruz with blinding winds and surging seas. Less than a decade later, a July 1574 hurricane scattered 27 ships belonging to the same *flota*, sinking five of them and claiming the lives of hundreds of sailors. On September 12, 1600, an intense hurricane overwhelmed the *Flota de Nueva España* as it tried to seek refuge in the harbor at Villa Rica, some 5 miles north of Veracruz. Seventeen ships were wrecked, more than 1,000 men were drowned, and 10 million pesos worth of gold, silver, mercury, and spices were lost. Three stormy decades later, 19 more galleons were run aground and broken up when a hurricane in late October 1631 again laid waste to Veracruz.

On August 18, 1835, the 28-hour Antigua Hurricane came ashore at Matamoros, Mexico, a town located near the mouth of the Rio Grande. Boasting intense winds and a seven-foot (2-m) STORM TIDE, this hurricane not only ravaged shipping in the neighboring harbor of Brazos Santiago but also swept more than 300 houses into the sea. Only four lives were reportedly lost, although it is entirely possible that the final count was much higher. Two years later, Matamoros would again be impacted by a deadly hurricane, this one known as Racer's Hurricane. During the course of this three-day (October 1–4, 1837) storm, all vessels lying at anchor in Brazos Santiago were cast ashore or sunk, while the town's new customs house was swept away. A similar hurricane again struck Brazos Santiago on September 12–13, 1849, likewise running a number of small fishing and supply craft ashore.

More recently, a thrashing hurricane in 1951 claimed 115 lives as its 160-MPH (258-km/h) gusts blazed into Tampico during the night of August 22. Of the 115 people who were killed, 42 perished when floods swamped a dam near Cardenas on August 23, and 50 more were drowned when the Guayalijo River subsequently flooded the town of La Paloma. A particularly vicious hurricane, this same storm had just a few days earlier struck JAMAICA with 130-MPH (209-km/h) winds, killing an 162 people there. Just over four years later, the first Hurricane Hilda also came ashore at Tampico on September 19, 1955; 166 people were killed and more than 1,000 were injured in the resulting floods. Finally, on September 21, 1993, Hurricane Gert killed another 76 people when it came ashore at Tampico, bearing 100-MPH (161-km/h) winds and pelting rains.

Mexico's west coast, which is susceptible to strikes from those North Pacific hurricanes that curl back on themselves, suffered a paralyzing storm in late October 1957. On October 21, a hurricane with 130-MPH (209-km/h) winds struck scenic Mazatlán, driving fishing boats ashore and killing eight prisoners when the roof of the city's jail was carried away. Between August 31 and September 2, 1967, Hurricane Katrina demolished several neighborhoods in San Felipe, leaving 2,500 people homeless; Hurricane Bridget, June 17, 1971, brought 100-MPH (161-km/h) winds and heavy seas into the harbor at Acapulco, killing 40 people and sinking the flagship of the Mexican Navy. In October 1975, Hurricane Olivia's 100-MPH (161-km/h) winds left 29 dead in Mazatlán. Just under a year later, Hurricane Liza drowned at least 630 people in La Paz when its torrential rains and 130-MPH (209-km/h) winds collapsed an earthen dam on the Cajoncito River. More recently, Hurricane Tico laid waste to portions of Mazatlán on October 19, 1983. Bearing down from the northwest, Tico's 125-MPH (201-km/h) winds wrecked nine ships and left some 25,000 people homeless. In the first week of July 1993, early season Hurricane Calvin brought a 15-foot (5-m) STORM SURGE to the streets of Acapulco, flooding posh vacation resorts and killing 34 people.

After careening across the Baja Peninsula, Hurricane Ismael wailed ashore on mainland Mexico's northwest coast, some 410 miles (660 km) south of the Arizona border, on the night of September 14–15, 1995. Said to have been the fifth-deadliest North Pacific hurricane on record, Israel's 100-MPH (161-km/h) gusts and 30-foot (10-m) seas harpooned several dozen shrimp and fishing boats in the Gulf of Baja, sinking 32 of them and drowning 100 people; 150 other fishers were left marooned on waterlogged sandbars, their wooden-hulled livelihoods either sunk or hopelessly stranded. In Mexico's Sinaloa and Sonora provinces, Ismael's central pressure of 29.02 inches (983 mb) at landfall produced sustained winds of 80 MPH (129 km/h), torrential 8-inch (203-mm) rains, and roaring breakers. For more than six hours, Ismael lashed the coastal village of Topolobampo, splintering palm trees, flooding streets, and destroying more than 5,000 lightly timbered BUILDINGS. Five people in Topolobampo were killed and another were 60 injured; 9,000 Mexicans were left without shelter, while damage estimates to crops and farms totaled nearly $10 million.

In the days following Ismael's deadly demise, the Mexican National Meteorological Service was bitterly criticized for not providing adequate warnings and updates to those shoreline communities most likely to be affected by Ismael's growing menace. Relying on erratic data from offshore weather balloons and satellite photographs, Mexican meteorologists believed the burgeoning hurricane to be bearing west-northwest, when in reality Ismael was quickly swinging north-northeast toward the staple fishing grounds of the Baja. Some 18 bulletins and weather advisories issued over a two-day period were therefore not enough to prevent the loss of 105 lives to Ismael's surprise arrival—an indication, perhaps, of the importance that accurate, timely information plays in the art and science of tracking a tropical cyclone.

In the eastern North Pacific Ocean, the 1996 hurricane season was an unusually active one for western Mexico. Between June 24 and October 2, Mexico's northwest coast and the Baja Peninsula were struck by seven tropical cyclones, four of them mature hurricanes. On June 24, the 100-MPH (161-km/h) winds associated with Hurricane Alma came ashore near Playa Azul, a small town west of the port of Ciudad Lazaro Cardenas, some 185 miles (298 km) southwest of Mexico City and 170 miles (274 km) northwest of Acapulco. Twelve-inch precipitation counts triggered violent landslides, while three people were reportedly killed when their house collapsed in the village of Chucutitla, some 20 miles (32 km) northwest of Lazaro Cardenas.

Less than two weeks later, Hurricane Boris lumbered ashore in west central Mexico, about 60 miles (97 km) northwest of Acapulco, during the midmorning hours of June 30. A powerful Category 1 hurricane, Boris's 90-MPH (145-km/h) winds and five to 10-inch (127–254-mm) rainfall counts uprooted trees and road signs, downed power lines, and flooded the lobbies of several luxury hotels. A number of fishing boats were thrown against a seawall and wrecked. A destructive hurricane, Boris left at least 10,000 people homeless. Four people were listed as killed, including an elderly woman and a six-year-old boy who was crushed by a falling beam. Another 70 people were injured.

On the cusp of becoming a mature hurricane, Tropical Storm Cristina made landfall on Mexico's Oaxaca province during the midmorning hours of July 3. A poorly organized system, Cristina's 60-MPH (97-km/h) winds quickly dissipated into gusty rain showers, wind-splattered downpours that peppered windows with the velocity of bullets. In what initially seemed a horrific repeat of 1995's Hurricane Ismael, 62 shark fishers from the village of Puerto Madero were listed as missing at sea, their small boats reportedly overwhelmed and sunk by Cristina's rocking seas. An extensive air-search mission soon found the fishers alive, their craft intact, on their way back to Puerto Madero. No lives were lost to Tropical Storm Cristina, while damage estimates barely topped $1 million.

On August 1, 1996, an eight-foot (3-m) wave generated by offshore Hurricane Douglas swept four U.S. tourists into the sea at Cabo San Lucas, drowning two and seriously injuring the others. The third hurricane to have threatened Mexico's east coast in five weeks, Douglas was one of those rare Pacific systems that had originally developed in the Caribbean Sea before crossing Central America and being renamed. Formerly Hurricane César, Douglas was at the time of the drowning incident an intense Category 2 system with maximum sustained winds of 110 MPH (177 km/h). The hurricane's outer RAIN BANDS dropped considerable quantities of precipitation on the southwestern provinces, of Colima, Guerrero, and Michoacan, with three- and four-inch (76–102 mm) counts being measured in downtown Manzanillo. The most powerful hurricane yet seen in the 1996 Pacific hurricane season, Douglas eventually recurved northwest, making its two-person DEATH TOLL in Mexico much lower than it had been in Venezuela and Costa Rica. Still tallying the damage from César on August 7, the day a fading Tropical Depression Douglas finally disappeared over the cool waters between Mexico and Hawaii, authorities in Central America stated that at least 70 people had died in Venezuela, Nicaragua, Costa Rica, and El Salvador.

On September 4, 1996, Tropical Storm Elida's 50-MPH (81-km/h) winds and seven-foot (3-m) swells forced the temporary closure of seaports on Mexico's Baja Peninsula. Situated some 175 miles (282 km) west of Cabo San Lucas, Elida haltingly intensified as it churned northwest at three MPH (5 km/h), drawing its scattered gales and isolated thunderstorms away from east Mexico. By the following morning, with clear skies on the horizon, HURRICANE WATCHES along the Mexican coastline were discontinued.

Like alphabetical clockwork, the next hurricane to terrorize Mexico—this one christened Fausto—emerged from the southeastern North Pacific on the afternoon of September 9, 1996. Swiftly developing into a Category 3 heavyweight with top winds of 115 MPH (185 km/h), 12-inch rains, and 10-foot (3-m) storm surge, Fausto's meteorological fists hammered the Baja Peninsula throughout the day and night of September 13. In the resort cities of La Paz and Cabo San Lucas, snapped utility poles knocked out electrical and telephone service; fallen trees buried cars and small buildings. Half a dozen yachts were washed ashore on Baja's famed beaches. Some 1,700 homes in neighboring La Paz were either damaged or destroyed, reducing 2,500 people to seek refuge in Red Cross shelters. One person, a 26-year-old U.S. tourist, was killed when a power line fell onto his trailer. Transiting the Gulf of California, Fausto made its final landfall near Topolobampo, the same city trounced by Hurricane Ismael a year before, on the morning of September 14. Its sustained, 105-MPH (169-km/h) winds quickly diminished upon landfall, fell to 85 MPH (137 km/h) as the center of circulation pushed farther inland later that day. By midnight the first of Fausto's tropical moisture would be felt across western Texas and southern New Mexico, an arid region some 400 miles (644 km) from the Pacific that does not normally see much rain. One of the most intense hurricanes to have made landfall in northwestern Mexico in a decade, Fausto left one person dead and some $25 million in property losses.

On the afternoon of October 2, the fifth mature hurricane of the 1996 Pacific hurricane season to affect Mexico directly, Hernán, chugged ashore near the coastal trading post of Punta Tejupan. A moderately powerful Category 1 hurricane, Hernán's 85-MPH (137-km/h) winds uprooted hundreds of trees, and scudding rains spawned numerous flash floods and road washouts. Coming ashore in a sparsely inhabited section of Mexico, Hernán claimed no lives and left no long-term ecological consequences.

On Mexico's east coast, where the 1995 hurricane season had been one of the busiest in years, one whimsical hurricane, one rain-swollen tropical storm, and a powerful tropical depression killed a total of 39 people and wrought nearly $1 billion in damage. On August 11, Tropical Storm Gabrielle made landfall just north of Tampico, Mexico. Formed in the western Gulf of Mexico on August 9, Gabrielle's central pressure of 29.17 inches (988 mb) soon serenaded northeast Mexico and south Texas with 70-MPH (113-km/h) winds, heavy rains, and at least one documented TORNADO. An estimated six people in Mexico were killed; another 100 were injured.

Between September 27 and September 29, Tropical Depression OPAL, a precipitation-driven upstart from the southwest Caribbean Sea, delivered such heavy rains to Mexico's Yucatán Peninsula that it provided itself the evaporative means to continue strengthening overland. On the morning of September 30, while still stationed over the sodden jungles of the peninsula, Tropical Depression Opal was upgraded to a tropical storm. As its sustained winds topped 45 MPH (72 km/h), 12–15-inch (305–381-mm) downpours inundated north central Yucatán, uprooting entire forests, carving hillside rifts, and drowning 19 people. Sliding into the Gulf of Mexico on October 1, Tropical Storm Opal diligently intensified as it sped northward, bound for a stunning strike on northwest Florida on the evening of October 4.

The rain-swept towns and cities of the Yucatán had no sooner commenced their recovery than an even more powerful tropical system, Hurricane Roxanne, targeted the resort island of Cozumel on October 9. Armed with 110-MPH (177-km/h) winds and six-inch (152-mm) rains, the North Atlantic's first *R* hurricane first forced the mass evacuation of some 16,000 tourists and residents and then invaded Cozumel from the east at dusk on October 10, shattering windows, toppling lampposts, and collapsing small sheds and hotel cabanas. While Roxanne inflicted $3 million in damages on the hotels and houses of the island, no lives were lost, making it one of Cozumel's milder hurricane strikes.

In September 2002, Hurricane ISIDORE delivered Category 3 winds to Mexico's Yucatán Peninsula, forcing the evacuation of at least 70,000 people and killing two people.

On October 25, 2002, Hurricane Kenna whisked ashore on Mexico's central Pacific coast, just north of Puerto Vallarta, with maximum wind speeds of 140 MPH (230 km/h). At the time the third- most intense tropical cyclone to have struck Mexico's Pacific coastline, Kenna inundated several shoreside hotels in Puerto Vallarta. At one time a Category 5 hurricane that generated sustained winds of 160 MPH (257 km/h), Kenna's forecasted arrival was preceded by the evacuation of 10,000 fishers and residents from some 30 coastal communities. In Puerto Vallarta, 20-foot (7-m) waves topped the harbor's breakwaters, killing at least one man and injuring another 32.

On Mexico's east coast, Tropical Storm Erika originated over the eastern Gulf of Mexico on August 14, 2003. Boasting sustained winds of 70 MPH (110 km/h) at landfall some 45 miles (70 km) southeast of Brownsville, Texas, Erika delivered high winds and rains to northern Mexico. Widespread damage to foliage, crops, and small BUILDINGS were reported in northern Mexico.

On August 25, 2003, western Mexico was affected by Hurricane Ignacio, which devastated the Baja Peninsula's grape vineyards. Sustained winds of 90 MPH (150 km/h) and a six-foot (2-m) storm surge delivered large quantities of precipitation to Cabo San Lucas—in some places 20 inches (508 mm) were observed.

On October 5, 2003, Tropical Storm Larry slammed southern Mexico's Tabasco state. Downgraded to a tropical depression shortly after landfall near the town of Coatzacoalcos, Larry's 50-MPH (85-km/h) winds spawned heavy rains and some damage to oil production and distribution facilities, but caused no deaths or injuries.

The 2005 eastern North Pacific and Atlantic Ocean HURRICANE SEASONS resulted in two tropical systems coming ashore on the nation's two coasts simultaneously. Hurricane OTIS delivered high winds and rains to western Mexico's Baja Peninsula on October 1, 2005. Otis was the 15th tropical cyclone of the 2005 eastern North Pacific Ocean hurricane season, and the seventh mature-stage tropical cyclone. Flooding was observed in and around the tourist city of Cabo San Lucas, on the southern tip of Mexico's Baja Peninsula. At the same time, western Mexico was affected by Hurricane Marty. A Category 2 system, Marty made landfall along Mexico's Baja Peninsula on October 22 near San Jose del Cabo with maximum sustained winds near 100 MPH (160 km/h). Marty weakened as it tracked northward across the Gulf of California, bringing gusty rains to the arid expanses of Arizona on October 24. The second tropical cyclone to trundle ashore in Baja in one month's time, Marty killed 10 people.

On August 25, 2003, western Mexico was affected by Hurricane Ignacio, which battered the Baja Peninsula. Sustained winds were 90 MPH (150 km/h) with higher gusts. Ignacio was the first hurricane of the eastern North Pacific Ocean season in 2003. Forecasts called for a four to six foot storm surge. Areas like Cabo San Lucas and elsewhere were affected. Ignacio was forecast to deliver residual moisture to Arizona and New Mexico. The system lingered off the coast of the Baja on August 26, 2003, and weakened rapidly. Maximum sustained winds decreased to near 60 MPH (95 km/h). Rainfall counts of 20 inches were recorded in some areas of the Baja Peninsula and its grape-growing vineyards. Most of the grapes used for the Mexican wine industry are grown in the Baja area.

The extraordinary 2005 North Atlantic hurricane season brought a number of tropical systems to eastern Mexico, including the most intense July hurricane on record, EMILY. On July 18, Emily plowed into the Yucatán Peninsula. Bearing sustained winds of 115 MPH (185 km/h), Emily caused widespread damage. On July 21, Emily made its second Mexican landfall near San Fernando. A moderately powerful Category 3 hurricane, Emily's sustained, 125 MPH (201 km/h) winds, and 3–5-inch (76–127-mm) rainfalls, lashed northeastern Mexico, destroying small buildings, uprooting trees, and causing numerous flash floods. In the mountain city of Monterrey, more than four inches (102 mm) of rain fell in a six-hour period.

Tropical Storm (later Hurricane) STAN, the 20th tropical depression of the 2005 North Atlantic hurricane season, delivered between five and 10 inches (127–254 mm) of rain to the Yucatán Peninsula on October 2, 2005. Soon entering the Bay of Campeche, Stan made a second Mexican landfall near the port city of Veracruz on October 4, causing no known deaths or injuries.

Between them, tropical cyclones Opal and Roxanne were responsible for nearly $1.5 billion in losses, making the 1995 hurricane season one of the most expensive in Mexico's history.

On August 23, 1996, residents of the greater Tampico area met Hurricane Dolly, a moderately powerful Category 1 system that had been gate-crashing cities and towns across the Yucatán Peninsula for much of the previous week. Bursting ashore in a flurry of 80-MPH (129-km/h) winds and six–12-inch (152–305-mm) downpours, Dolly's visit to southeast Mexico forced the port of Tampico to suspend docking operations; 2,300 people were evacuated from low-lying coastal areas. Extensive street flooding soon paralyzed traffic in the city, hampering efforts to find shelter for those 1,500 people already left homeless by Dolly's crushing gusts. As some 500,000 other Mexicans suffered without electrical, telephone, and water services, two people died—a woman who perished when a tree fell on her house in Pueblo Viejo on the outskirts of Tampico and a man, drowned when he was swept away by flood-waters in the northern indus-

trial city of Monterrey. Forty-two other people were injured in Hurricane Dolly, two of them seriously.

microbarograph A model of ANEROID BAROMETER used to observe and record BAROMETRIC PRESSURE and fitted with two vacuum chambers (known as sylphons) arranged vertically inside a corrugated metal cylinder, the microbarograph is similar in design and operation to the standard aneroid barometer except that it is larger in size, more sensitive to changes in barometric pressure, and can through the presence of a scroll drum and pen arm produce a printed record of barometric activity called a MICROBAROGRAPH TRACE. Accurate to atmospheric pressures within 0.01 of an inch (0.34 mb), the basic microbarograph contains a series of adjustable timing gears that govern the speed at which the scroll drum revolves, thereby allowing the instrument to record pressures during a one-day or one-entire-week period. Although subject to the same technical limitations as the basic aneroid barometer, the microbarograph's ability to provide meteorologists with a detailed record of changes barometric pressure over a certain geographical location within a particular time span has made it an indispensable tool in the tracking of TROPICAL CYCLONES. When powered by a battery pack, the microbarograph can be deployed in remote locations and then left unattended during the landfall of a tropical cyclone's dangerous EYEWALL to track its core CENTRAL BAROMETRIC PRESSURE. Securely anchored to a table or platform, numerous microbarographs on oil rigs, buoys, and radio towers have figured prominently in the study of tropical cyclones by providing scien-

Now an artifact from an earlier age of meteorology, this microbarograph nevertheless shows the general principle behind a microbarograph's operation. The microbarograph trace can be seen on the printed roll to the left. While surpassed by modern technology, the microbarograph and its observations provide a valuable record of past tropical cyclones. *(NOAA)*

tists and meteorologists with data that is as valuable in scope as it is in detail.

microbarograph trace This name is given to the record of BAROMETRIC PRESSURE readings as observed and printed by a MICROBAROGRAPH. Much like a medical electrocardiogram (EKG) in appearance, the microbarograph traces and its jagged lines similarly record for detailed diagnosis the PRESSURE GRADIENTS that lurk at the heart of a TROPICAL CYCLONE's powerful wind systems. Measuring six inches (152 mm) high by 12 inches (305 mm) long and imprinted with a grid that is scaled vertically for barometric pressure and horizontally for time (with each day of the week divided into two-hour segments), most microbarograph traces have a basic sensitivity range of 28.50–31.00 inches (960–1,050 mb) and can extend in length of record from one to seven days. During periods of relatively tranquil weather, changes in barometric pressure generally produce a microbarograph trace that contains very few peaks and valleys and that rises and falls fairly slowly. But during the onset and subsequent passage of a tropical cyclone's EYEWALL, trace signatures begin to drop, gradually at first and then with increasing steepness as the system's CENTRAL BAROMETRIC PRESSURE draws closer to the instrument's location. A seven-day microbarograph trace taken at Kings Point, NEW YORK, during the passage of the GREAT ATLANTIC HURRICANE, for instance, began on October 11 with a six-hour trace at 30.18 inches (1,022 mb), followed by a two-and-a-half day span in which barometric pressures steadily lost 0.03 inches (1.02 mb) in just over 36 hours. Then, shortly after noon on October 14, the microbarograph trace started to fall very quickly, reaching its lowest point of 29.07 inches (984 mb) some 10 hours and 0.08 inches (2.7 mb) later. By midnight of October 17, however the Great Atlantic Hurricane's microbarograph trace had regained 1.41 inches (4.7 mb) in 72 hours.

military operations and tropical cyclones It was during a debate on the strategic value of meteorological stations in the CARIBBEAN during the Spanish-American War that President William C. McKinley most adroitly defined the relationship between military operations and TROPICAL CYCLONES. "I am," he said, "more afraid of a West Indian hurricane than I am of the entire Spanish Navy."

McKinley's regard for the destructive potential of the HURRICANE was not misplaced; the history of global military operations is fraught with instances in which the mightiest battle squadrons and most disciplined legions have been soundly defeated by a combined assault of wind and water. At best, hurricanes, TYPHOONS and CYCLONES have brought reduced visibility and moderate rain cover to advancing armies and fleets, while at their worst they have dismasted, driven ashore, or sunk warships, prostrated entire armies, and nullified even the most skillfully planned of military maneuvers. In September 1281, the Hakata Bay typhoon saved JAPAN from invasion when it destroyed 90 percent of Mongol conqueror Kublai Khan's 2,200-ship invasion fleet. Spain's efforts to establish hegemonic control over the Caribbean basin during the 16th and 17th centuries were in large part thwarted by the costly depredations of hurricanes against her treasure-laden silver fleets. During the American Revolution, a spate of hurricanes disrupted British naval operations along the U.S. eastern seaboard, hampering efforts to provide logistical support to their armies fighting on the mainland. On September 12, 1818, the colorful career of Jean Lafitte was brusquely curtailed when a surprise hurricane wrecked four of his pirate ships as they lay at anchor in Galveston Bay, TEXAS. Admiral Perry, returning from his historic 1853 visit to Japan, was nearly denied his laurels when his flagship, the steam frigate *Mississippi,* bucked into a powerful NORTH PACIFIC typhoon and was badly damaged. On December 17, 1944, Admiral William "Bull" Halsey's Third Fleet, then participating in the liberation of the PHILIPPINES, suffered the loss of three destroyers and nearly 800 men in Typhoon COBRA. Cobra's 115-MPH (185-km/h) winds and 70-foot seas simply overwhelmed the fast but slight vessels, sending them to the bottom in a matter of minutes. The typhoon's siege was sufficiently strident enough to delay by several days the Third Fleet's scheduled mission of launching offensive strikes against Japanese positions on the island of Luzon.

It was the 16th- and 17th-century Spanish conquistadores on their paramount quest for gold, God, and glory in the Caribbean who first encountered the avenging fury of the NORTH ATLANTIC hurricane. In September 1494, Christopher Columbus, then on his second voyage to the New World with a fleet of 17 ships, sought refuge from a TROPICAL STORM on the southeastern tip of HISPANIOLA. Protected from the rising winds by the rocky lee of Saona Island, Columbus's fleet of high-sided *naos* and caravels bobbed and churned beneath the incipient hurricane but sustained little damage. During his fourth and final exploration of the New World in 1502–1504, Columbus encountered a mature, midseason hurricane on July 10, 1502, in the Mona Passage between Hispaniola and PUERTO RICO. Although Columbus again found suitable shelter for his ships, a nearby *flota* of 30 galleons, caravels, and *naos,* many of them deeply laden with gold and silver plate for

Spain's coffers, was ravaged by the convulsive tempest. In such relentless winds, the vessels' towering sterncastles and webbed top hampers served like the bantam tail of a weathervane, spinning the ships into the tossing broadsides of enormous waves. Overwhelmed by the high seas or driven onto the craggy, spray-shrouded shore, 26 of the 30 ships that had so grandly set sail from Santo Domingo two days before were lost, and with them, several tons of mortgaged treasure and more than 500 lives.

The tragic destruction of Admiral Antonio de Torres's fleet that July night in 1502 was only the first in a long litany of such hurricane-borne disasters to afflict Spain's efforts to establish military and economic dominance in the Caribbean region. Despite a change in the *flota*'s annual sailing date to coincide with the end of the HURRICANE SEASON, on at least 40 different occasions between 1527 and 1766, hurricanes in the Caribbean Sea and GULF OF MEXICO did overtake the treasure-stacked convoys bound for Spain. In some instances, such as a 1551 hurricane in the FLORIDA Keys, only a single, 480-ton galleon was separated from the fleet and wrecked, while in others, like the September 6, 1622, hurricane between CUBA and Florida, two entire *flotas,* the Armada de Tierra Firme and the Flota de Tierra Firme, of 32 and 24 ships respectively, were battered into almost complete oblivion. As many of the *flota,* including the 1622 convoy, carried gold, precious gems, mercury, and other commodities valued at several million pesos per voyage, their cumulative loss to hurricanes and tropical storms was directly tallied in Spain, where the government was forced with alarming frequency to rely on the usurious assistance of European banks and merchant houses to finance its overseas ambitions. As late as 1766, long after Spain had ceded her Caribbean monopoly to British, French, and Dutch raiders, the remaining *flotas* were still racked by the occasional hurricane. On September 4, 1766, a hurricane in the Gulf of Mexico fell upon five Spanish treasure galleons en route from Veracruz to Havana and dashed them ashore near Bay St. Bernard on the Florida Panhandle. One of the vessels was later refloated, while the other four were stripped of their cannon and fittings and burned.

During the same period, Spain's naval incursions into the Caribbean, made in a lukewarm effort to halt the roving influence of Dutch and English traders, proved no less immune to the dangers of hurricanes than did her ungainly *flotas.* In August 1779, a powerful hurricane in MISSISSIPPI temporarily delayed a planned Spanish assault on British bayou holdings in LOUISIANA. Hundreds of small boats and a frigate that were to be used in the upriver expedition were smashed in place on the beaches,

necessitating an additional week of preparation. Just over a year later, in October 1780, another Spanish operation against the British—this one to take place in Florida—was similarly thwarted by the arrival of an intense hurricane. Much larger than the 1779 foray into Louisiana, the 1780 affair was nothing less than a full-scale naval invasion, with an armada of 64 heavily armed warships, transports, supply vessels, and scouts being used to ferry 4,000 soldiers from Havana to Pensacola. Placed under the dual command of Admiral Antonio Solano and General Pedro Galvez, the fleet was overtaken in the Gulf of Mexico on October 17 by the third in a trio of extraordinary hurricanes (the others being the SAVANNA-LA-MAR HURRICANE of October 3 and the GREAT HURRICANE of October 10) to cross the Caribbean that season. Known as Solano's Hurricane, the five-day blast of wind, rain, and shearing seas first reduced several of the admiral's ships to dismasted hulks and then scattered the storm-weary survivors as far north as New Orleans and Mobile, ALABAMA. Subsequently unable to regroup their far-flung, jury-rigged forces, both commanders chose to return to Havana at once. It was there, after a mournful consideration of the humiliating mauling they had just received from the skies, that the Spanish canceled the operation completely.

Perhaps their decision was prompted in part by a slew of reports then arriving in Havana about how severely the English and French forces in BARBADOS, JAMAICA, and MARTINIQUE had been beaten by the two other storms to strike the region earlier that month; and of how the governor of Barbados, Major-General Cunningham, released his Spanish prisoners-of-war at the height of the storm because the jail they were being held in was about to collapse and he feared they would be buried beneath it; or of how the same governor, in a letter to his superiors in London, later praised the Spanish soldiers for the friendly manner in which they conducted themselves during the difficult days following the storm when DEATH TOLLS on the island daily mounted by the hundreds; and of how the governor would in return "show them every indulgence in my power." There were similar stories in Martinique, where the French military governor had declared a general amnesty following the Great Hurricane's passage over the island on October 11. In setting free his British prisoners, the governor magnanimously stated that in the midst of such a cataclysm, "all men should feel as brothers."

This 18th-century sentiment regarding the ennobling virtues of hurricanes did not, however, extend as far north as the American colonies, where the sanguinary struggle between the rebellious colonists and

Great Britain was entering its final stages. During the years of the American Revolution (1775–1782), 30 recorded hurricanes either moved through the Caribbean basin or passed up the eastern seaboard of the colonies. A number of these storms, including the twin hurricanes of September 2 and September 9, 1775, and the hurricane of August 10, 1781, eventually made landfall in areas where some of the Revolution's heaviest fighting was then taking place. While damage from the September 9. 1775, hurricane was largely confined to the coastal regions of NEW ENGLAND and CANADA, the September 2 storm, known as the Independence Hurricane of 1775, killed nearly 200 people in NORTH CAROLINA and VIRGINIA. This same hurricane decimated a significant portion of the year's corn harvest in eastern North Carolina, forcing the colony's cash-strapped legislature to authorize additional funds for the purchase of supplies for its gathering militias. During the later hurricane of August 10, 1781, British forces occupying Charleston, SOUTH CAROLINA, were pelted with 85-MPH (137-km/h) winds and heavy rains. Two British ships, the *Thetis* and the *London,* sank at their piers; a number of other vessels were either dismasted or driven aground. Although another year would pass before the British finally evacuated Charleston, the hurricane of 1781 clearly indicated to their military planners one of the more hazardous conditions that would have to be overcome if further victories in the region were to be assured.

Other hurricanes of the Revolutionary period, such as the offshore storm of August 11–13, 1778, principally affected naval operations along the colonies' coast. In the case of the 1778 hurricane, a fleet of French warships that had been sent to assist the Continental Army in its prolonged assault on British forces entrenched at Newport, Rhode Island, was in part prevented from doing so by the passage of an intense two-day hurricane over Narragansett Bay. Although the French commander, Admiral D'Estaing, had initially sailed his 16-ship fleet out of Newport in the hopes of engaging a smaller British squadron that had been sent north from NEW YORK to challenge him, the hurricane's escalating winds and raging seas prohibited any opportunity for actual combat. For more than two days, both fleets found themselves limited to a crippling battle for self-preservation as the tempestuous winds and seething seas sprung masts and seams, carried away bowsprits, spars, and stern galleries, swept boats and other deck gear over the side, and in the instance of the 90-gun French flagship *Languedoc* shattered her steering gear. By the time the hurricane had moved northwest on the morning of August 14, the opposing fleets had been so widely scattered by the hurricane's fury that no

In October 1854, following the conclusion of his second visit to compel the Japanese empire to open its borders to the West, Commodore Matthew Perry's flagship, the USS *Mississippi,* encountered a powerful typhoon. As this drawing illustrates, the vessel was nearly lost. *(NOAA)*

battle was ever joined. The British fleet returned to New York, and the French slowly regrouped for a renewal of the bombardment of Newport.

Despite the numerous innovations and improvements made to naval architecture by the industrious 19th century, the seemingly predatory tropical cyclone still proved a viable hazard to military operations around the globe. Although the invention of the steam engine, the screw propeller, and the iron hull had certainly served to make vessels less vulnerable to the driving whims of the wind, these advancements could not overcome both a lack of proper meteorological forecasting or the sheer force of a tropical cyclone's fury. On more than one occasion during the empire-building century, the mightiest steam-powered frigates, cruisers, and ironclads afloat found themselves overtaken by a surprise hurricane, typhoon or cyclone; with their sails tightly furled, their auxiliary coal-fired boilers spewing black smoke into the storm's ripping winds, their crews could do little but try to keep the ship's bow into the circular winds, praying that the tons of water that sluiced across the decks every hour would not dampen the fires and leave the ship powerless and completely assailable in the midst of cresting 40-foot seas. In such a situation, with the vessel nearly dismasted, the rudder smashed, the iron hull plates groaning beneath the strain of the twisting seaway, the greatest concentration of firepower in existence would have proved as useless a defense as launching a broadside into the EYE of the storm itself.

During the first year of the U.S. Civil War, the powerful Union Hurricane of November 2, 1861, seriously disrupted a planned Northern assault on Confederate shore installations in South Carolina. Touted as "the largest fleet of warships and transports ever assembled," the Union armada was to simultaneously seize Port Royal, located between Charleston, South Carolina, and Savannah, GEORGIA, and establish an all-encompassing blockade of the mid-Atlantic coast. Four days after departing Chesapeake Bay on August 29, the fleet, comprised of 18 iron-hulled cruisers, 12 steam frigates, and numerous smaller transports, supply, and scout vessels, encountered the hurricane as it unexpectedly roared around the Carolina Capes. Laboring to keep his command from being scattered by the hurricane's bucking winds, the rear-admiral in charge ordered the entire fleet to beat a course to the open sea and avoid being caught too close to the shoaling sandbanks of the Carolina coast. With the threat of collision hovering over the entire maneuver, the Union fleet successfully accomplished this, although not without casualty; two ships—one a frigate, the other a transport—were sunk, claiming the lives of some 125 Union sailors. While the hurricane's voracious assault did delay the fleet's mission by almost a week, it did not in the end scuttle it. Regathered and repaired by midweek, the Union triumphantly captured the fortress at Port Royal on November 7, 1861.

In other regions of the globe, military forces from both Great Britain and Germany found their colonial pursuits equally harassed by 19th-century tropical cyclones. Like an elusive army of guerrillas, a steady round of BAY OF BENGAL cyclones tempered England's continuing rush through the Indian subcontinent. In early November 1839, a formidable cyclone struck Coringa, India, killing nearly 300,000 people and sinking several thousand ships in the harbor. At the storm's height a phenomenal 30-foot (10-m) STORM SURGE lurched ashore, destroying the Royal Naval dockyards situated there. The second disaster of such magnitude to occur in less than a half-century, the Coringa Cyclone prompted the British to shift a larger fraction of their fleet operations to Calcutta, a port city that eventually proved no less storm-swept than Coringa. On October 5, 1864, for example, an immense cyclone smashed into bustling Calcutta, killing in excess of 50,000 people and driving every vessel, from armored cruisers to jaunty paddle steamers, onto the quays. Likewise, British inroads into CHINA were often derailed by those cyclopean North Pacific TYPHOONS that periodically blasted down TYPHOON ALLEY. One such typhoon, in mid-July 1841, choked the harbor at Hong Kong with so much wreckage that those warships still afloat were unable to move.

Germany, too, suffered heavy losses to tropical cyclones. On March 16, 1889, her punitive efforts to establish a Pacific foothold in Samoa were dashed to oblivion when a predawn cyclone struck the island of Apia Upolu. In a place where imperial tensions between Great Britain, the United States, and Germany had recently escalated into shelling raids and raucous landing parties, the intense cyclone's devastating siege proved far more effective in diffusing the situation than did the seven heavily armed warships then anchored in the harbor. For their part, six of the seven vessels, including the 2,169-ton German cruiser *Olga* and the 3,900-ton American corvette *Trenton*, were either sunk or run ashore, claiming the lives of an estimated 147 sailors. The one British ship in the contingent, the 2,770-ton *Calliope*, spent several hours steaming to seaward and escape. Sensational photographs of the catastrophe were published in EUROPE and the United States, showing the capsized wreck of German steam frigate *Adler* on the beach at Apia. Her keel snapped, her propeller gone, her rigging a thicket of shattered spars and twisted lines, the glaring loss of the 884-ton *Adler* chided officials in Berlin into seriously reconsidering their Pacific ambitions. Germany finally peacefully relented, leaving the sprawling archipelago to be divided between its British and U.S. competitors.

Although, on numerous occasions during World War II, hurricanes and typhoons did indeed influence the design and success of military operations in the Atlantic and Pacific theaters, their greatest significance can be found in those improvements made to weather forecasting procedures. Although a 1944 landing of Marines on Okinawa was seriously dampened by a typhoon's rains, and Admiral William Halsey's Third Fleet lost nearly 1,000 men to the swamping effects of Typhoons Cobra and Viper, modifications in typhoon tracking techniques—including the installation of a codified naming system and the use of radar in prediction ANALOGS—nevertheless made it much easier for military leaders to plan their offensive strategies. Nearly every vessel in the Allied navies was assigned a weather officer, and observation stations were established on Allied-held islands, thereby forming a comprehensive network of relay and forecasting offices throughout the region. There can be little doubt that the increasing amounts of data available on hurricanes and typhoons greatly contributed to the success of individual operations, thereby hastening Allied victory and peace.

It was also during the war years that the fearsome cultural reputation of hurricanes and typhoons was conscripted for the purposes of morale enhancement and product identification. Nazi Germany's climactic assault on Moscow in October 1941 was christened

Operation Taifun, or Operation Typhoon, in the hope that it would somehow infuse the winter-weary invaders with the destructive power of a true aerial cataclysm. In a similar vein, the British developed two classes of fighter aircraft that they pointedly identified as the Hurricane and the Typhoon. Though made of wood both planes evoked the vengeful spirit of the winds as they bravely formed an airborne bulwark against the relentless squalls of incoming Axis aircraft.

Not all military operations associated with tropical cyclones have been of an offensive nature, however. On more than one occasion, military units from any number of the world's nations have been deployed in rescue efforts following the injurious passage of a hurricane, typhoon, or cyclone. When in September 1938 the GREAT NEW ENGLAND HURRICANE isolated the region by washing out both railway lines and roads, the battleship USS *Wyoming* was enlisted to carry the mail from New York City to Boston. When Hurricane JANET struck Tampico, Mexico, on October 4, 1955, the United States Navy not only provided the Mexican government with meteorological updates on the storm's violent pro-

gression across the gulf, but it also dispatched squadrons of helicopters from the USS *Saipan* to rescue the hundreds of people left marooned on roofs and hilltops by the hurricane's extensive flooding. The Navy would later reinforce the *Saipan*'s humanitarian efforts by sending a cargo vessel with doctors, food, and medicine aboard to the storm-stricken port city. In March 1982, air and naval units from AUSTRALIA and NEW ZEALAND were deployed to the South Pacific nation of Tonga to assist in reconstruction after the devastating April 9 passage of Cyclone Isaac. In November of the same year, the fast-attack submarine, USS *Indianapolis* was ordered to the Hawaiian Islands to provide electrical service to those communities ravaged by the passage of Hurricane Iwa. According to media sources, more than a week after the intense storm had made landfall, tens of thousands of people on the island of Kauai were still without electricity, making the *Indianapolis*'s mobile nuclear-power plant of crucial importance to reconstruction efforts. Following the ruinous passage of the April 1991 cyclone in BANGLADESH, eight U.S. ships, 7,550 U.S. Marines and sailors, and tons of water-purification tablets, tents, powdered milk, and

The U.S. military assists civilians in the wake of a hurricane disaster. A subject for heated debate, the use of military resources in responding to a natural or other large-scale disaster is sometimes necessary. *(NOAA)*

body bags were sent to assist the Bangladeshi armed forces in the herculean cleanup. Equipped with transport helicopters and hovercrafts, the squadron was further augmented by a supply ship and two helicopters from the British Royal Navy.

When Typhoon Louise struck Okinawa on October 9, 1945, with 92-MPH (153-km/h) winds, a central pressure of 28.61 inches (969 mb) and 30–35-foot (10–12-m) waves, a total of 12 ships and small watercraft belonging to the United States Navy (and assigned to Operation Olympic, or the planned invasion of JAPAN) were grounded or severely damaged along the shores of Buckner Bay. Personnel casualties were 36 killed, 47 missing, and 100 seriously injured. More than 60 aircraft on the island were damaged. Four tank landing ships, two medium landing ships, a gunboat, and two infantry landing craft were lost.

When a tropical cyclone threatens to make landfall near one of its many installations, the U.S. Navy generally sends its ships to sea, and moves its aircraft and other land-based equipment well inland. On June 10, 2005, for instance, several ships homeported at the Naval Station in Pascagoula, Mississippi, were sent to sea in preparation for Tropical Storm ARLENE's passage across the northern coast of the Gulf of Mexico. By deploying its vessels in this fashion, the navy avoids a repeat of the 1889 disaster in the Samoan Islands, where several warships broke away from their moorings and were either driven ashore, or into one another, causing considerable damage and loss of life.

When Supertyphoon Yoyong struck the PHILIPPINES on December 3, 2004, up to 30 percent of the Philippine air force's aircraft and helicopters were deployed for relief operations, while 48 military teams were mobilized for relief and rescue operations. The U.S. military also deployed five heavy-lift helicopters to assist the Filipino military in relief operations.

The United States Navy, United States Coast Guard, the National Guard, and the Civil Air Patrol all performed heroically during and after Hurricane KATRINA's record-shattering landfall on the northern Gulf coast in August of 2005. The Coast Guard airlifted hundreds of stranded people from roof and treetops, while the United States Army provided tents and other logistical supplies needed to care for the hundreds of thousands of displaced residents. The United States Navy dispatched the hospital ship, USNS *Comfort,* to New Orleans, thereby providing emergency medical facilities for those hospitals and other medical facilities damaged or destroyed by Katrina's heavy strike and the subsequent flooding caused by the failure of several levees. The military's domestic response to Hurricane Katrina remains one of the largest of its kind in American history.

millibar (mb) A unit of BAROMETRIC PRESSURE used on the SAFFIR-SIMPSON SCALE and by meteorological services. A millibar, as the *milli* implies, is one thousandth of a bar, or an inch of mercury. It has become a popular unit of measurement because sea-level barometric pressure is nearly 1,013 millibars, making it easy to gradate the CENTRAL BAROMETRIC PRESSURES of TROPICAL CYCLONES. Often abbreviated *mb.*

miniswirls First identified in 1992 by Professor Tetsuya Theodore Fujita of the University of Chicago, miniswirls are small tornadic vortices that originate within the EYEWALL of a mature-stage TROPICAL CYCLONE. Discovered during ground-and-aerial surveys conducted in the wake of Hurricane ANDREW's devastating passage across southern FLORIDA, miniswirls are believed responsible for the narrow swaths of total destruction caused to BUILDINGS and trees over a distance of several thousand feet by this remarkable HURRICANE. Although research into the mechanics of miniswirl activity is still incomplete, comprehensive studies of Andrew's damage path indicate that WIND SPEEDS in the leading edge of a miniswirl can exceed 40 MPH (64 km/h). When this factor is added to those sustained wind speeds normally found in a tropical cyclone's EYEWALL, maximum miniswirl wind speeds can reach as high as 200 MPH (322 km/h).

Mississippi *Southern United States* Although its fairly sheltered 65-mile (105-km) coastline is one of the shortest of the five southern states—ALABAMA, FLORIDA, LOUISIANA, and TEXAS—that border on the hurricane-prone GULF OF MEXICO, Mississippi has had an extensive, often tragic history of HURRICANE activity. Between 1722 and 2006, the state was threatened by no less than 53 documented hurricanes, with 24 of these being direct landfalls in the highly industrialized port areas of Bay St. Louis, Biloxi, and Pascagoula. Since 1722, 12 major hurricanes—Category 3 storms with BAROMETRIC PRESSURES of less than 28.47 inches (964 mb) and 9–25-foot (3–4-m) STORM SURGES—have come ashore in Mississippi, causing hundreds of deaths and millions of dollars in property damage. On August 17, 1969, Mississippi had the rare misfortune to be struck by one of only three Category 5 hurricanes to ever make a recorded landfall in the United States, the now legendary CAMILLE. On average, the state's low-lying coast, noted for its picturesque resort beaches and thriving shipyards, is threatened by a hurricane every 2.5 years and directly affected by one every seven years. At times it has even received several hurricanes in succession, as happened in 1860 when three powerful TROPICAL CYCLONES, those of August 11, September 14–15, and October 2–3, came ashore in less than two months' time. Indeed, as recently as 1985, Mississippi sustained minimal damage from two of that year's passing gulf storms: Hurricane Danny (August 15, 1985) and Hurricane JUAN (October 26, 1985). Hurricane ELENA, August 31, 1985, caused major damage as it hammered the coast from Pascagoula to Gulfport with 125-MPH (200-km/h) winds and a six-foot (2-m) STORM SURGE. Four people were killed; official damage estimates eventually topped $1 billion.

Although the fanning spread of the Mississippi River Delta does provide Mississippi with a degree of protection from incoming hurricanes, the state's location at the northern apex of the Gulf of Mexico nevertheless affords it a long and vulnerable HURRICANE SEASON. Between 1722 and 2006, Mississippi was struck by at least two July hurricanes, both of them of significant intensity. The first, the BAY ST. LOUIS HURRICANE of July 27–28, 1819, cast a six-foot (2-m) storm surge into the farming village of Pass Christian, destroying a number of houses and killing several hundred head of cattle. Farther east, the Mississippi shoreline from Rigoulets to Mobile Bay was utterly ravaged by the hurricane's searing winds and pulverizing surf. At Cat Island's west end, the 150-ton hulk of the naval vessel *Firebrand* was later found in the shallows, dismasted and overturned. When searchers finally cut their way into the hull, they discovered the decomposing bodies of 39 of its 47-man crew. The second July hurricane, which came ashore between Pascagoula and Mobile, Alabama, on July 5, 1916, was of equal if not greater intensity than the 1819 storm. Although it raised an 11-foot (4-m) storm surge in Mobile Bay and brought 107-MPH (172-km/h) winds to that city, its effects just over the border in Mississippi were generally limited to uprooted trees and some minor beach erosion.

It is in August and September when the first of the late-season hurricanes begin to arrive in the gulf that the number of strikes in Mississippi rises to a hefty three and seven, respectively. Significant midseason hurricanes in Mississippi include the August 15, 1901, storm that pelted the ironworks of Biloxi with 100-MPH (161-km/h) winds and drenching rains; the destructive hurricane of September 15, 1855, that raised a 10–15-foot (3–5-m) storm surge along the coast between Louisiana's Lake Pontchartrain and Mississippi's Bay St. Louis; and the September 29, 1915, hurricane that caused several million dollars' worth of damage when it swept away both an amusement park on Deer Island and most of the eastbound lane of US Highway 90 linking Pass Christian with Gulfport. In September 1906, the

Pascagoula–Mobile Hurricane left 78 people dead in Mississippi and caused several thousand 1906 dollars in damage to shipping in the Mississippi Sound. More recently, the hurricane of September 19, 1947, thudded into Harrison County, Mississippi, with sustained winds of 132 MPH (213 km/h) and a staggering 15-foot (3-m) storm surge at Bay St. Louis. Twenty-two people were killed, and $17.5 million in damage—$3.4 million in Biloxi alone—was assessed. A rather unusual hiatus in hurricane activity during the 1950s ended in September 1960 when Hurricane Ethel came ashore between the barrier islands of Petit Bois and Horn Island on September 15. With sustained winds of 85 MPH (137 km/h) and a meager storm surge, Ethel was one of the weakest hurricanes to ever afflict the Mississippi shoreline. Its rain-shorn demise over the state's heavily timbered interior ushered in another lull in hurricane activity, one disturbed only by the relatively tame Hurricane Hilda of October 3, 1964, and the slightly more intense Hurricane BETSY, September 7–8, 1965. Both storms initially made destructive landfalls to the east of Louisiana's Marsh Island before crossing into northern Mississippi as TROPICAL STORMS. Aside from some significant structural damage to farm and resort communities along the coastline, both Hilda and Betsy were as fair in their passage over inland Mississippi as the earlier Ethel had been.

Tragically, the same cannot be said of the extraordinary hurricane that, on the night of August 17, 1969, ended the quiet decade in a furious maelstrom of coiling winds and surging seas. Without question, Hurricane Camille was one of the most intense recorded hurricane to strike Mississippi. With a CENTRAL BAROMETRIC PRESSURE of 26.84 inches (909 mb), sustained winds of 180 MPH (290 km/h), and a 15–25-foot (5–7-m) storm surge observed from Pass Christian to Biloxi, it was, in fact, one of the two most powerful hurricanes to ever strike mainland United States. Although Camille was slightly less intense than the earlier Category 5 storm, the LABOR DAY HURRICANE of 1935, its toll in terms of property damage and human lives was no less significant. At Pass Christian, a small resort community in the Long Beach area of Mississippi, 143 people were killed when Camille's cresting storm surge swept away most of the town. Later, after Camille had been downgraded to tropical storm status, it continued to move northeast to the high-sided Appalachian Mountains, where its deluging rains and crushing mudslides killed an additional 113 people.

Such relentless, often destructive hurricane activity has served to reshape many of the principal geographical features of the Mississippi coastline. Several of its low-lying peninsulas, spits, and lagoons have been repeatedly scoured by the torrential winds and inundating tides of passing hurricanes, causing incalculable damage to the area's fragile ecosystem. The barrier island of Petit Bois, which now lies like a marshy sentinel across the east entrance to Mississippi Sound, has been swept by so many hurricanes in the past century that it has literally migrated several hundred feet west. On a number of stormy occasions it has also been breached. Nearby Sand Island, once the site of a towering lighthouse, almost completely disappeared in the 1947 hurricane, while the susceptible, shores of Bay St. Louis, located 27 miles (43 km) west of Petit Bois, have become crowded and cold with dozens of concrete embankments and 20-foot (7-m) seawalls. At least 15 times in the past 200 years, the Bay, once heralded as the best-sheltered harbor on the Mississippi coast, has been inundated with destructive, often deadly storm surges, some of them reaching heights of 10–12 feet (3–4 m). Although state and federal authorities have undertaken a multitude of measures to better protect Mississippi's coastal integrity, it is unlikely any of them will in the long run prove an adequate match for the steady cycle of gulf and Caribbean hurricanes that will continue to prune the shores of the Magnolia State.

On September 26, 2002, Tropical Storm ISIDORE delivered 10-foot (3-m) waves to Biloxi, Mississippi, that caused damage to several beachfront casinos. The system later went ashore near New Orleans; in that city, sustained winds were clocked at nearly 55 MPH (89 km/h).

The now legendary Hurricane KATRINA visited historic devastation to Mississippi on August 29, 2005. Said to have been the most destructive and intense tropical cyclone to have affected the state in recent memory, Katrina claimed 238 lives in Mississippi, and left another 67 people missing. Sustained winds of 120 MPH (195 km/h) winds, eight to 10-inch (200–250-mm) rainfall counts, and a whopping 27-foot (18-m) storm surge caused widespread coastal destruction. In what seems a cruel irony, the town of Waveland, Mississippi, was obliterated by a relentless barrage of sea waves—some reportedly as high as 50 feet (17 m)—as well as a record-setting storm tide. Several bridges in the state were destroyed, while 11 TORNADOES touched down in Mississippi, one of the highest hurricane-related tornadic tallies in the state's history. All told, more than $3 billion (in 2005 U.S. dollars) in property losses were recorded, making Katrina the state's costliest natural disaster.

Mitch, Hurricane *Central America–Southern United States, October 22–November 5, 1998* Said by contemporary sources to have been the second-deadliest NORTH ATLANTIC TROPICAL CYCLONE on record, the strongest documented October HURRICANE, and the fourth-strongest hurricane ever observed in the Caribbean Basin, Hurricane Mitch spent six harrowing days between October 27 and November 3, 1999, charting an almost unrivaled course of death and destruction through the Central American nations of Nicaragua, Honduras, and Guatemala. At one point a Category 5 hurricane of equal intensity to the legendary Hurricane CAMILLE Mitch's final DEATH TOLL surpassed that of the earlier storm by the thousands. In one Nicaraguan town alone, some 2,000 people perished beneath a boulder-laden landslide, while 25 percent of the people in Honduras were rendered homeless. All told, Hurricane Mitch claimed between 9,000 and 12,000 lives in Central America, as well as $5.5 billion in property damage.

A tropical cyclone that would come to be defined by its unfortunate superlatives, Hurricane Mitch was born over the southwestern CARIBBEAN SEA, some 50 miles (80 km) northwest of Venezuela, on the evening of October 20, 1998. According to a later analysis of this remarkable hurricane's development, the embryonic TROPICAL DEPRESSION that would eventually become Mitch was formed from the union of a weak tropical wave and what is known as a MESOSCALE CONVECTIVE COMPLEX (MCC), or a dense banding of intense but noncyclonic thunderstorms. Drifting slowly to the northwest, this combination of low BAROMETRIC PRESSURE and intense CONVECTION activity quickly drew additional strength from the warm, relatively insulated waters of the southwestern Caribbean. By the evening of October 22, the two-day-old depression had become the newly minted TROPICAL STORM Mitch.

During the mid-morning hours of October 25, 1998, Mitch matured to Category 5 status, the first hurricane of such intensity to transit the Caribbean Sea since 1988's Hurricane GILBERT. And the system's CENTRAL BAROMETRIC PRESSURE continued to plummet. By dusk on October 26, Mitch's central pressure had sunk to 26.74 inches (905 mb), vying Hurricane Camille's August 17, 1969, record. On October 27, while positioned some 100 miles north of the eastern coast of Honduras, Mitch's 180-MPH (290-km/h) winds, powered by a central barometric pressure of 26.73 inches (905 mb), overwhelmed the four-masted schooner *Fantome*. Once owned by Greek shipping magnate Aristotle Onassis, the 282-foot (94-m) sailing ship, now operated as a cruise ship, was overtaken by the hurricane's 50-foot-high (17-m) swells as it attempted to seek safe harbor on the eastern coat of Nicaragua. All 31 crew members were lost.

A terrific rainmaker, Hurricane Mitch produced rainfall counts of between 50 and 75 inches (1,270–

1,905 mm). In one six-hour period, 25 inches (635 mm) of rain fell. In northwestern Nicaragua, Mitch's flooding rains spawned numerous landslides, including one particularly savage instance on October 30 in which a mudslide traveled 13 miles (21 km) downslope of the Casitas Volcano, burying 10 communities and claiming the lives of some 2,000 people. Northern Honduras, in the vicinity of the Ulúa River, was inundated. Several large banana plantations were destroyed.

Emerging into the Bay of Campeche on November 3, 1998, Mitch was again upgraded to a tropical storm. Lashing Florida as a rain-laden tropical storm during the morning of November 5, Mitch caused no deaths or injuries, but $40 million in property damages. Maximum sustained winds were 60 MPH, while an extreme gust of 73 MPH (117 km/h) was recorded near Miami. The name Mitch has been retired from the rotating list of HURRICANE NAMES.

monuments and tropical cyclones Erected in memory of those unfortunate individuals who lost their lives to the depredations of HURRICANES, TYPHOONS, and CYCLONES, monuments stand as revealing dolmans on the cultural landscape of TROPICAL CYCLONES. In their aesthetic, in the way they are architecturally composed, monuments indicate something of the cultural context in which they were raised, the views on beauty and remembrance that prevailed during the age that built them. In the way they were commissioned, executed, and paid for can be found the rituals of commemoration, the behaviors—political and otherwise—that moved a people to justify their respect for both the natural invincibility of the hurricane and the lives of those human beings it claimed. Some monuments, such as the one on the FLORIDA Keys that marks the mass grave of some 423 men, women, and children who perished in the extraordinary LABOR DAY HURRICANE of 1935, are quite bold, almost defiant, as they rise in art-deco splendor not more than 100 feet (31 m) from the Atlantic Ocean. Others are more modest: A simple pair of black iron plaques bolted to the corner of a building in downtown Providence, Rhode Island, shows how high the water rose during the GREAT SEPTEMBER GALE OF 1815 and the GREAT NEW ENGLAND HURRICANE OF 1938. Whether they are dedicated to the memory of those who have died or to the heroism of the community that has survived, monuments define the very language of tropical cyclones, providing a textual cross-section of the society on which a particular hurricane made cataclysmic landfall.

When one considers the enormous amount of negative publicity that accompanied the Labor Day Hurricane of September 2, 1935—the subsequent

This tablet, which resides in the chapel at the Mare Island Navy Yard in California, memorializes the men lost during the Samoan Cyclone of March 16, 1889. *(NOAA)*

congressional hearings into why a government-contracted rescue train did not arrive on the keys in time to save more than 200 Civilian Conservation Corps road workers and Ernest Hemingway's eloquent rebuttal to the inevitable whitewash—the choice of an art-deco sunset, of an outspoken tribute in coral rock and stainless steel, becomes readily apparent. This monument, directly off the Overseas Highway near Islamorada, is nothing less than a gravestone that recalls not only the plight of the victims but that of the perpetrators as well. Raised by public subscription during the Great Depression, its great size and modern outlook reflects both a distrust of New Deal government and a celebration of individual humanity. A majority of the victims from this hurricane were displaced veterans, survivors of the 1932 Bonus Army march on Washington, D.C., and it was deemed appropriate by the committee that oversaw the memorial's construction to honor the downtrodden struggle of these ignored patriots by turning their deaths into a political statement. Well taken at the time, the protest has since been digested by a plethora of social programs, and the Hurricane Memorial, as it is generally known to locals, has subsequently fallen into disrepair. This, in turn, may say something about the priorities of our own age, although only time and the next deadly hurricane strike will tell.

Less than a year after a late-season hurricane swept up the NEW ENGLAND coast on October 3, 1841, the residents of Truro, Massachusetts, a small fishing village on the northeastern edge of Cape Cod, gathered on the town's highest headland to dedicate a simple yet elegant monument to 57 of their lost members. A white marble obelisk, unornamented except for a brief inscription, was poignantly unveiled, then reverently prayed before. Overlooking the sea, it commemorates the heroism of those citizens of Truro who were drowned when seven of their Georges Bank fishing schooners were overwhelmed by the OCTOBER GALE, one of the most intense Atlantic hurricanes then on record. Couched in the religious iconography of the obelisk, of the many ancient columns that rise before the most notable churches and cathedrals, the memorial's architectural purity reflects not only the famed New England penchant for practicality but also the devotional simplicity evidenced by the first line of the engraved dedication: "Sacred to the Memory of . . ." As sacred in lesson and example as the 57 lost, many of them youths from the same families, a community of grief comes together to remember and heed the varied warnings found in the arrival of unexpected hurricanes.

Hurricane monuments mourn the living as well as the dead—or perhaps *celebrate* and *proclaim* are better descriptors of what one might feel as one views the two black plaques, firmly bolted to the corner of the 17th-century brick Market House in Providence, Rhode Island, on which are emblazoned thick lines and a series of impressive measurements. Descriptive captions explain that the first plaque was so placed to commemorate the fact that on September 23, 1815, the Great September Gale drove the waters of Narragansett Bay so far into the heart of the city that the sea rose to "the level of this line," and the building still survived. The second plaque, which rests even higher than that of 1815, makes certain that the viewer understands the hierarchy of hurricanes by touting the fact that during the Great New England Hurricane of September 21, 1938, the waters of the bay rose in this spot to be "One-Foot-Eleven and Three Quarter Inches Above the Old Mark." Almost a footnote really, the inscription nonetheless implies a historical standard, a reference to an earlier, and perhaps even more intense, hurricane that stormwise New Englanders still tell tales about. Neither of the plaques mention the hundreds of lives lost in each storm, so they cannot be considered memorials in the strictest sense of the word. Rather, the twin monuments in Providence impart more a sense of civic pride and less a pall of subsequent fear and regret. Their textual emphasis on "Wind Driven Waters" and "In the Great Gale" hint at a greater fascination with the endured fury of the storms than they do a citywide concern over the enormous amounts of damage wreaked by both. Here, hurricanes have become record-setting entities that the human-fashioned walls of the colonial Market House have twice withstood, eliciting a heroic response to the unending struggle between humankind and the great lingering storms of monumental memory.

N

Nagoya Typhoon *North Pacific Ocean–Japan, September 19–27, 1953* The sustained 127-MPH (204-km/h) winds, 10-inch (254-mm) PRECIPITATION counts, and 10-foot STORM SURGE of this extremely powerful Category 3 TYPHOON battered large portions of central JAPAN on the night of September 25–26, 1953. Possessing a CENTRAL BAROMETRIC PRESSURE of 27.99 inches (948 mb) at landfall in the port city of Nagoya some 60 miles southwest of Tokyo, the Nagoya Typhoon demolished 8,604 BUILDINGS, carried away numerous roads, bridges, and piers, flooded beachheads, submerged more than 50,000 acres of rice paddies, and left nearly a million people without shelter. Of equal intensity to 1965's Hurricane BETSY in the NORTH ATLANTIC OCEAN, the Nagoya Typhoon claimed 478 lives across Honshu Island before spinning itself into extinction over the Sea of Japan on September 27, making it one of the deadliest TROPICAL CYCLONES to have affected the densely populated volcanic chain during the stormy decade of the 1950s.

Nancy, Typhoon *North Pacific Ocean–Japan, September 10–19, 1961* More commonly known in JAPAN as the Muroto II TYPHOON, internationally code-named Typhoon Nancy inflicted widespread property damage and steep DEATH TOLLS on the central Japanese island of Honshu during the early morning hours of September 16, 1961. Possessing a CENTRAL BAROMETRIC PRESSURE of 27.49 inches (931 mb) at landfall near the burgeoning port city of Osaka, 90 miles (145 km) southwest of Tokyo, Nancy's sustained 135-MPH (217-km/h) winds, nine-inch (229-mm) PRECIPITATION counts, and 13-foot (4-m) STORM SURGE utterly destroyed 15,258 BUILDINGS,

washed away scores of roads, bridges, and piers, uprooted thousands of trees, and left more than 700,000 people without shelter. The second of two gargantuan TROPICAL CYCLONES to have besieged Osaka during the 20th century, Nancy's voluminous rains and inundating storm surge caused Osaka's Tosabori River to top its concrete embankments, flooding many of the city's residential and commercial districts with several feet of brackish seawater. Similar in size and intensity to 1961's Hurricane CARLA in the NORTH ATLANTIC OCEAN, Typhoon Nancy snagged 202 lives across south Honshu Island before dissipating over the Sea of Japan on September 19. One of modern Japan's most expensive natural disasters, Nancy's $500 million price tag prompted Osaka to construct an extensive series of flood gates, pumping stations, and drainage canals, an ambitious project that was completed in 1981.

National Hurricane Center (NHC) Located just outside Miami on the campus of Florida International University in West Dade, the National Hurricane Center (NHC) is primarily responsible for the tracking and forecasting of all TROPICAL CYCLONES in the NORTH ATLANTIC OCEAN, CARIBBEAN SEA, GULF OF MEXICO, and eastern NORTH PACIFIC OCEAN. A unit of the National Weather Service, the National Hurricane Center's state-of-the-art satellite communication systems, coupled with its experienced staff of scientists, meteorologists, analysts, and technicians, have in recent years made the NHC a familiar, somewhat reassuring media destination during storm emergencies, thereby expanding its mission to include educating the public on the unique hazards associated with TROPICAL STORMS and HURRICANES and the need for coastal residents to obey promptly any posted evacuation warnings in their areas. The National Hurricane Center's extensive library has further established it as one of the world's preeminent research centers into the scientific and historical nature of tropical cyclones, while a number of individuals connected with the center have significantly contributed to a greater understanding of cyclonic systems through their respective technical publications. In southern FLORIDA itself, the National Hurricane Center is a popular local landmark, a fascinating place where schoolchildren go on field trips and where visitors can watch hydrogen-filled WEATHER BALLOONS be launched twice a day.

Established in its present form in 1955, the NHC collects, assesses, and disseminates a broad array of meteorological data regarding the formation and behavior of tropical storms and hurricanes. Sophisticated satellite communication networks allow it to receive climatological and storm data from a host of technological sources, including two National Weather Service SATELLITES in orbit around the equator; a DOPPLER RADAR network reaching from the gulf coast of MEXICO to southern CANADA; a string of satellites that fly from pole to pole every two hours; and the United States Air Force HERCULES WC-130 aircraft, stationed in Biloxi, MISSISSIPPI, that fly reconnaissance missions into and around tropical cyclones. On an average day, when the tropics are devoid of cyclonic activity, the NHC spends its satellite time engaged in prediction, in a search for those climatological elements—from the speed and direction of upper level winds over the central Pacific to the temperature of the AIR MASSES drifting west through Africa to the evaporation rates found above the warm waters of the subequatorial Atlantic—that favor hurricane generation. Interpreted by both comprehensive computer ANALOGS and an experienced team of weather researchers and theoreticians, this meteorological intelligence is then transmitted to the world, sent out as daily bulletins to the military, newspapers, radio stations, and civil defense organizations in Hawaii, North and Central America, and the Caribbean basin.

During those critical periods when a tropical cyclone is in existence, the NHC becomes a veritable fortress, a castle in which the weapons of vigilance and forewarning are kept. As the NHC's twin-jet GULFSTREAM IV-SP aircraft joins the National Oceanic and Atmospheric Administration's (NOAA) two ORION WP-3 turboprop planes in sending data regarding the tropical cyclone's upper-level winds back to the NHC, endless computer programs begin to dissect the information, placing it within the greater context of sea-surface temperatures, wind speed and direction, position of summer ANTICYCLONES, and those weather patterns then dominating the inhabited regions of the Western Hemisphere. Further computations, including a comparison with the documented hurricanes of history, yield a series of predictions, forecasts of the storm's track that are then transmitted to the Federal Emergency Management Agency (FEMA) in Washington and the American Red Cross and to state and local civil defense authorities across the U.S. southern and eastern seaboards. If (as many often do) a mature hurricane moves into the Caribbean Sea or Gulf of Mexico, reconnaissance flights supervised by the National Hurricane Center increase in their frequency and are conducted every hour as a potentially dangerous system moves closer to land. Round-the-clock shifts are established among NHC staff as a constant watch is placed on the system, designed to detect major changes in the storm's course and intensity within an hour of their occurrence.

In the event of a tropical storm or hurricane's imminent landfall in the United States, the National Hurricane Center, in its capacity as the country's principal forecasting office, becomes the center of intense media interest—particularly on the part of those regions most likely to suffer from the hurricane's strike. Beneath the glare of klieg lights and television monitors, the NHC's director reassuringly takes time from his busy schedule to grant interviews and offer updates regarding the storm's statistics and to urge further that all evacuation warnings in afflicted areas be heeded. Before costly evacuation and other storm-preparedness measures are initiated, however, officials in threatened areas consult with the more anonymous members of the NHC, hearing firsthand from the experts how serious a particular storm is liable to prove. Because years of study into the tracking of cyclonic storms has yet to result in an accurate prediction beyond 24 hours of landfall, entire sections of coastline can either be needlessly evacuated or left vulnerable as a powerful hurricane, defying the favored forecast, makes a surprise landfall there. In either scenario, only last-minute information can avoid a major economic or human disaster, requiring the NHC to accumulate a ready pool of data continually from which to derive its vital forecasts.

Given the unpredictable behavior of tropical storms and hurricanes, the National Hurricane Center has in the past been remarkably accurate in forecasting just where a certain cyclonic system is going to make landfall. The meteorological difficulties inherent in determining just where along a 100–4,000-mile (161–6,437-km) coastline a hurricane's 25–60 mile-wide EYE (40–97-km) is poised to come ashore, coupled with the often tragic ramifications associated with an inaccurate prediction, places great pressure on the NHC to seek ever improved methods for the acquisition and analysis of storm data.

But even the most sophisticated of analogs can sometimes be wrong, and it is during such times that the National Hurricane Center finds itself subjected to largely unfair criticism by angry property owners and conflict-driven reporters. When in June 1957 Hurricane AUDREY claimed 518 lives in LOUISIANA, the newly christened National Hurricane Center was roundly criticized for not issuing updated bulletins that forewarned of the Category 4 storm's progress toward the Louisiana coast. Although subsequent investigators revealed that the urgency of the NHC's ample warnings had been diluted at the local level, the resulting controversy tacitly compelled the organization to develop even bolder methods of hurricane prediction and moderation, including PROJECT STORMFURY, the moderately successful effort conducted during the mid-1970s to weaken tropical cyclones by sprinkling their cloud masses with silver iodine crystals. As recently as the 1995 HURRICANE SEASON, residents of Florida's Panhandle accused the National Hurricane Center of misforecasting the path of Hurricane Erin, leaving tens of thousands of people trapped on flooded barrier islands when the weak Category 2 system came ashore near Pensacola on August 4. While in truth the NHC's well-publicized predictions had included Pensacola in their lists of possible landfall sites, curious tourists and nonchalant residents in the Panhandle, fooled into a false sense of security by Erin's earlier lukewarm assault on southern Florida, either chose to ignore these forecasts or else construed them in such a way that mandatory evacuation warnings somehow became an invitation to a HURRICANE PARTY. Such unwise behavior eventually provoked a stinging rebuke by state civil defense authorities, and an entirely new campaign of hurricane-awareness programs were promptly conducted across the state.

Although no organization enjoys negative publicity, the National Hurricane Center does recognize the significant role that criticism—warranted and otherwise—has played in the history of its mission. When the National Hurricane Center's ancestor, the United States Weather Bureau, was founded in 1890, it was partly in response to the destructive 1875 hurricane that had unexpectedly leveled the burgeoning port town of Indianola, TEXAS. Charging that the military authorities then responsible for the issuance of hurricane forecasts were not only incompetent but negligent when it came to civilian weather interests, the public created such a clamor for accurate meteorological forecasts that Congress mandated the first weather service in U.S. history. Very quickly, hurri-

cane forecasting stations were constructed in Kingston, JAMAICA, in 1892 and later in Havana, CUBA, in 1899, establishing the art of hurricane prediction as an international effort.

The 1890s were, however, hurricane-prone years for the United States. Several powerful hurricanes, such as the 1893 tempest that claimed more than 2,000 lives in GEORGIA and SOUTH CAROLINA, lashed coastal towns and major seaports, tallying millions of dollars in combined property losses. After the GREAT GALVESTON HURRICANE killed between 6,500 and 12,000 people at Galveston, Texas, on September 8, 1900, and vituperative allegations that no warnings had preceded the Category 4 hurricane's landfall were aired across the nation, the principal hurricane forecasting office in Havana was moved to Washington, D.C. At the time, authorities believed that a bureaucratically centralized weather office in the heart of the country's capital would effectively correct what was essentially a scientific, rather than an organizational, problem. Not surprisingly, a spate of dangerous hurricanes, including the 1926 Miami Hurricane and the deadly SAN FELIPE/Lake Okeechobee hurricane of September 16, 1928, soon revealed that only improvements in storm tracking, coupled with an updated warning network, could lessen the nation's hurricane problem.

To this end, Congress decentralized the UNITED STATES WEATHER BUREAU in early March 1935. Replacing the Washington office with forecasting stations in Jacksonville, Florida; New Orleans, Louisiana; and San Juan, PUERO RICO, the federal government was clearly perplexed when less than six months after its establishment, the Jacksonville hurricane center proved unable to prevent the loss of more than 400 lives to the extraordinary LABOR DAY HURRICANE of September 2, 1935. Although this remarkable Category 5 hurricane had been well tracked, with regular updates as to its progress toward the Florida Keys broadcast to threatened areas, preestablished evacuation procedures were not carried out as planned, leaving hundreds of hapless road workers and civilians to the mercy of this cataclysmic storm. Five years later, in the spring of 1940, another hurricane forecasting center was organized in Boston, Massachusetts. Inspired by concerns voiced in the aftermath of the GREAT NEW ENGLAND HURRICANE of September 21, 1938, the Boston office was tasked with the mission of ensuring that never again would 680 people perish in a storm that a majority of them knew was in existence long before it made its unexpected landfall at Long Island, NEW YORK and southern NEW ENGLAND.

During World War II, the nation's primary hurricane forecasting office at Jacksonville was transplanted to Miami, where a joint hurricane observation corps was formed between the Army Air Corps and the United States Navy. One year later the first aerial reconnaissance flights into tropical storms and hurricanes were implemented, placing the Miami office at the forefront of efforts to revamp hurricane prediction methods radically. As subsequent improvements were made to the aerial reconnaissance system, it became apparent that the forecasting centers in Louisiana, Puerto Rico, and Boston were quickly becoming redundant. In 1955 they were finally abolished, and their staff and equipment was shifted to the newly created National Hurricane Center in Miami.

In 1965, long after the first weather satellite was launched into orbit and a coastal radar network extending from Maine to Texas had been installed, the National Hurricane Center was again moved, this time to the Miami suburb of Coral Gables, where it was housed in a 12-story office building. Here the NHC flourished, supervising a number of highly visible research projects designed to moderate the destructive effects of hurricanes during the stormy 1970s and 1980s. On August 24, 1992, Hurricane ANDREW badly damaged the National Hurricane Center as it passed over south Florida. Andrew's Category 5 gusts, clocked at more than 160 MPH (258 km/h), tore radar domes from the building's roof, carried away ANEMOMETERS, and caused the air-conditioning system to malfunction. Although the building was in no immediate danger of collapsing, the damage it sustained during the height of Andrew seriously impinged on the NHC's ability to continue its vital mission successfully at such a crucial hour. The loss of air-conditioning, for instance, disrupted climate control parameters in the building, threatening to short out much of the NHC's fragile computer equipment. Shortly thereafter, the decision was made to construct a new building for the National Hurricane Center that would be designed from the foundation up to withstand any future Floridian furies.

On May 31, 1995, a brand-new $5 million home for the National Hurricane Center was dedicated in Miami. Nicknamed "Bob's Bunker" in honor of former NHC director Dr. Robert C. Sheets, who supervised its construction, the new center is a long, low, white-painted building designed to resist sustained wind speeds of 134 MPH (216 km/h). Shuttered windows and a steel-reinforced concrete roof 10 inches thick ensure that in the likely event of another hurricane, the building's structural integrity is not compromised. The building's placement on a 14-foot (5-m) embankment guarantees that only the most implausible of STORM SURGES could ever flood it, while emergency generator and diesel stocks always remain ready in the event of a

power outage. Thus equipped, the National Hurricane Center stands with necessary funding poised to carry its life-saving mission of hurricane forecasting and prediction effectively forward.

The National Hurricane Center maintains a close relationship with the Tropical Prediction Center (TPC). While federal funding for hurricane research has increased since the destructive 1995 North Atlantic hurricane season, overall funding for TPC operations and early warning systems has remained unchanged, challenging the TPC's ability to complete its mission.

navigable semicircle The term, now somewhat obsolete, given to that part of a HURRICANE, TYPHOON, or CYCLONE in which the circular winds and rains are less intense. Primarily used by mariners, the term is derived from the tradition of dividing a TROPICAL CYCLONE into navigable and DANGEROUS SEMICIRCLES, or right and left halves based upon the system's forward motion. In the Northern Hemisphere, where tropical cyclones spin in a counterclockwise direction, an observer facing into the winds of an approaching hurricane will find the navigable semicircle on the left, or western side, and the dangerous semicircle on the right, or eastern side. The opposite is applicable, of course, to the Southern Hemisphere, where the clockwise spinning of a cyclone will present a navigable semicircle on the right, or eastern side, and a dangerous semicircle on the left, or western half. While in truth both halves of a tropical cyclone are extremely dangerous, the navigable half, which is not strengthened by both the forward speed of the system's STEERING CURRENT and the storm's own forward velocity, will possess considerably weaker winds and lower seas. Mariners who divided an oncoming hurricane, typhoon, or cyclone into dangerous and navigable semicircles were better able to steer their vessels into the storm's "calmer" regions, thus greatly increasing their chances of surviving its fury.

Neptune P2V-3W A twin-engined aircraft designed and built by the Lockheed Aircraft Corporation, this was used by U.S. Navy HURRICANE HUNTERS for TROPICAL CYCLONE reconnaissance missions over the NORTH ATLANTIC OCEAN, CARIBBEAN SEA, and GULF OF MEXICO during the 1950s and early 1960s. Painted navy blue and flying through a tropical cyclone at altitudes of between 300 and 1,500 feet (100–500 m), the Neptune P2V-3W's bank of ANEMOMETERS and ANEROID BAROMETERS collected data on surface WINDS and SEA-SURFACE TEMPERATURES and then radioed this information to the NATIONAL HURRICANE CENTER (NHC) in Miami, FLORIDA. During the 1960s the Neptune P2V-3W, a reliable aerial workhorse, was replaced on the hurricane hunting circuit by a modified version of the Super Constellation. A Neptune P2V-3W carrying 11 people on a reconnaissance mission to Hurricane JANET was lost in September 1955.

New England *Northeastern United States* This six-state region of the United States—Rhode Island, Connecticut, Massachusetts, New Hampshire, Vermont, and Maine—has over the course of the past three centuries witnessed a rigorous procession of notable TROPICAL CYCLONE strikes. Between 1635 and 2006, New England's NORTH ATLANTIC shores, beloved for their rocky coves, antique fishing villages, and dune-laced beaches, were directly and indirectly affected by at least 53 documented HURRICANES and TROPICAL STORMS. Of these tropical systems, 22 were considered major hurricanes, systems with BAROMETRIC PRESSURES of 28.50 inches (965 mb) and lower, sustained wind speeds of 111 MPH (179 km/h) and higher, and drenching PRECIPITATION counts.

On average, New England is likely to be inundated by at least one tropical cyclone of varying size and intensity every 5 years and by a major hurricane every 25 years. In some instances, two or more tropical cyclones have directly impacted the region in a single year, as seen during the 1893 HURRICANE SEASON when three storms—all former hurricanes possessing top winds of between 55 and 71 MPH (89–114 km/h)—transited central and eastern New England from August 24 until October 14; during the 1888 season, when another three systems streaked across Massachusetts, New Hampshire, and Maine between September 17 and November 28; and during the remarkable 1954 season when two mature hurricanes—CAROL of August 31 and EDNA, September 11—killed 71 people and wrought nearly $500 million in property damage.

On September 21, 1938, New England suffered what is generally considered its deadliest confirmed hurricane landfall when the GREAT NEW ENGLAND HURRICANE raced ashore in southwestern Connecticut, releasing sustained 121-MPH (195-km/h) winds, 186-MPH (299-km/h) gusts, and eight-inch rains from Rhode Island to Vermont. Preceded ashore by an unprecedented 17-foot (6-m) STORM SURGE, the Great New England Hurricane of 1938 obliterated dozens of coastal communities in Rhode Island and Connecticut, claiming the lives of some 600 people.

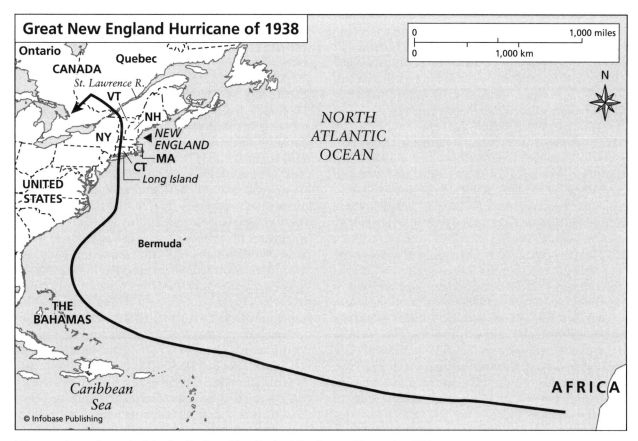

This map shows the track of the deadly Great New England Hurricane of September 1938.

This riveting photograph of high surf slamming against a seawall illustrates how, in part, the Great New England Hurricane of 1938 was so destructive to the region. *(NOAA)*

New England's position north of the 43rd parallel (where SEA-SURFACE TEMPERATURES rarely reach the minimum 79°F [26°C] threshold required for tropical cyclone development) extends the region a brief, relatively mild hurricane season. Ostensibly ranging from June 1 to November 30, New England's hurricane season historically peaks in August and September, when the first of the CAPE VERDE STORMS and eastern CARIBBEAN SEA hurricanes begin to recurve north-northeast and trail around the western edge of the BERMUDA HIGH before crossing with considerable fury into northeastern United States.

Since 1825 New England has been struck by only one mature June hurricane, a powerful anomaly from the northern Caribbean that on June 5 sideswiped Cape Cod, uprooting trees on Nantucket Island and throwing seven-foot seas ashore at Provincetown. Nearly a century and a half later, on June 14, 1966, fading Hurricane ALMA, an early season surprise that had initially made landfall in west FLORIDA on June 9, dropped almost two inches of rain on the outer banks of Cape Cod while recurving northeast. On June 22, 1972, the deluging remnants of Hurricane AGNES besieged southwestern Connecticut and western Massachusetts with 37-MPH (56-km/h) gales and torrential downpours. Earlier responsible for spawning record-breaking flood emergencies in Pennsylvania and NEW YORK, Agnes's toll in New England was mercifully much lower, with no deaths or injuries suffered and only sporadic incidents of property damage sustained.

In 1916 New England's most significant July strike—from an unnamed tropical storm that had originated over the eastern Caribbean Sea on July 12—delivered 40-MPH (64-km/h) winds and sweeping rains to Martha's Vineyard on July 21; loosely held together by a fairly mild central pressure of 29.37 inches (995 millibars), the 1916 tropical storm swamped several small boats on the island but caused no death or injuries.

During the summer months of August, September, and October, when the growing anticyclonic swirl of the Bermuda High increasingly dominates whether patterns over the North Atlantic, New England's susceptibility to hurricane and tropical storm strikes rises sharply. Quickly carried northward by recurvature and kept to the west by the expanding high-pressure barrier over the mid-Atlantic, many Cape Verde and eastern Caribbean Sea tropical cyclones are able to retain their intensity as they move across the cooler, less-evaporative waters of the northeast coast and in turn deliver powerful, often deadly strikes to New England's south and east shores.

Indeed, several of New England's most violent hurricane disasters have come from fast-moving Category 3 storms, well-developed tropical tempests that have torn northward at speeds of 30 MPH (48 km/h) or higher. Denied the opportunity to linger and weaken over colder waters, and with much of their terrific forward speeds added to the sustained winds of their DANGEROUS SEMICIRCLES, hurricanes such as the Great September Gale of 1815, Great New England Hurricane, the GREAT ATLANTIC HURRICANE of September 15, 1944, Carol (August 31, 1954), and GLORIA (September 27, 1985) have provided New England with the sort of large-scale weather disasters usually seen on Caribbean islands and across the southern United States. The 1938 Hurricane, still the greatest of all New England hurricanes, remains the sixth-deadliest tropical cyclone in U.S. history, its 600-person DEATH TOLL only exceeded by the GREAT GALVESTON HURRICANE OF 1900, the Lake Okeechobee Hurricane of 1928 the twin inundations of 1893 in SOUTH CAROLINA and LOUISIANA and hurricane Katrina, 2006. Notorious midseason hurricanes in New England include:

Hurricane DIANE, August 17–19, 1955. Its Category 1 EYEWALL reduced to rain-laden tatters by its earlier passage over northeastern United States, Tropical Depression Diane dropped more than 20 inches (508 mm) of rain on Connecticut, Rhode Island, and Massachusetts. Widespread flooding of near-record proportions across southern New England killed 82 people and caused $800 million in damages.

Hurricane DONNA, September 12, 1960. At one point in its existence a Category 4 hurricane of immense destructive potential, Donna's greatly diminished central pressure of 28.55 inches (967 mb) delivered 95-MPH (153-km/h) winds and lashing seas to Connecticut, Rhode Island, Massachusetts, and Maine. Seven people were killed, thousands of trees were uprooted, and $30 million in lost apple and potato harvests were assessed.

Hurricane Esther, September 22–26, 1961. A last-minute loop-the-loop maneuver carried Esther's 83-MPH (134-km/h) eyewall winds to within 400 miles (644 km) of Cape Cod. Falling under the influence of recurvature, Esther soon trained northwest, bisecting Cape Cod before crashing ashore near Rockland, Maine, on September 26. Between seven and eight inches (178–203 mm) of rain fell over Connecticut and Rhode Island, causing serious flash floods. No lives were lost, however, and damage estimates proved minimal.

Hurricane Ginny, October 29–30, 1963. A Category 3 hurricane of considerable intensity as its center whirred 125 miles (201 km) southeast of Cape Cod, Ginny produced sustained 65-MPH (105-km/h) winds on Nantucket Island and 100-MPH (177-km/h) gusts along the rock-sculpted coast of Maine. Rapidly intensifying as it recurved northward, Ginny's central pressure deepened to 27.99 inches (948 mb) while approaching New England, making it comparable in strength and destructive potential to 1965's Hurricane BETSY in FLORIDA and LOUISIANA. Colliding with a cold front moving down from central Canada, Ginny's outer RAIN BANDS produced heavy snowfalls over northern New England. Some 18 inches (457 mm) of snow were observed in central Maine, eight inches (203 mm) in New Hampshire, and a record-setting 48 inches (1,219 mm) on the gust-scaled summit of Mount Katahdin. No deaths or injuries were reported. Damage estimates did not exceed $100,000 1963 dollars.

Hurricane BELLE, August 10, 1976. A mild Category 1 hurricane, Belle's 74-MPH (119-km/h) winds and torrential rains destroyed fruit trees in Connecticut, downed power lines in Massachusetts, and spawned raging flash floods in the mountains of Vermont. Two people in New England were killed, and $6.5 million in agricultural claims were filed.

Hurricane GLORIA, September 27, 1985. Three people—two in Connecticut, one in New Hampshire—died when this former Category 3 hurricane blustered ashore in eastern Connecticut. Considerably weakened by their trek across Long Island, Gloria's 50-MPH (81-km/h) winds caused $16 million in property damage, making the storm one of the meekest hurricanes in New England history.

Hurricane BOB, August 18, 1991. Generated by a central pressure of 28.91 inches (979 mb) at landfall near Providence, Bob's 125-MPH (187-km/h) gusts and six-foot (2-m) storm surge caused extensive pier damage along the shores of Narragansett Bay. On Cape Cod several houses were toppled into the sea. Progressing north at speeds of 30 MPH (48 km/h), Bob dropped 10 inches (254 mm) of rain over central and eastern Maine, claiming the lives of two people and causing $21 million in crop losses.

Hurricane BERTHA, July 14, 1996. Only the 16th tropical cyclone to have made landfall in the United States since 1900, Bertha lumbered ashore in NORTH CAROLINA on the afternoon of July 12, a powerful Category 3 hurricane. Two days later, its 115-MPH (185-km/h) winds mortally weakened by its sodden passage over Virginia and New York, Bertha's fading eyewall blustered across southeastern New England. The highlands of western Massachusetts received nearly five inches (127 mm) of rain, while 70-MPH (113-km/h) gusts shivered the summer palaces of Cape Cod, prompting the closure of beaches and the cancellation of concerts. Although Bertha's sustained, 35-MPH (56-km/h) winds downed power lines and damaged trees, no deaths or injuries were reported in the wake of its rare July visit to New England.

Hurricane EDOUARD, September 1, 1996. At one point in its existence a Category 4 hurricane of alarming intensity, Edouard's sustained 140-MPH (225-km/h) winds had weakened to 90 MPH (145 km/h) by the time it lashed southeastern Massachusetts and Rhode Island with tropical storm-force winds and rains. Passing some 80 miles (129 km) southeast of Nantucket at 12 MPH (19 km/h), Edouard's sustained 50-MPH (81-km/h) winds, 80-MPH (129-km/h) gusts, and four-foot (1-m) seas flooded waterfront streets with more than 12 inches (305 mm) of water. On island beaches, 15 small boats were broken from their moorings and cast ashore, many becoming total losses. In the town of Hyannis, a fire station and several cottages lost their roofs to Edouard's five-inch (127-mm) rain squalls, tallying several hundred thousand dollars in property damage. Elsewhere across eastern Massachusetts, limb-severed power lines threw 65,000 people into an early afternoon darkness that persisted throughout most of the rain-swept night. The second tropical system to have affected New England in as many months, Hurricane Edouard claimed no lives and caused minimal structural damage.

Hurricane ISABEL, September 21, 2003. A large wave swept a man into the sea from a beach in Rhode Island, drowning him.

Tropical Storm Hermine, August 31, 2004. An extratropical system that crossed eastern Massachusetts with sustained winds of 35 MPH (56 km/h). After originating to the west of BERMUDA on August 29, Hermine slid northward, its sustained winds increasing to 50 MPH (80 km/h). The system's interaction with the cooler waters north of latitude 30 degrees North caused the system to undergo extratropical deepening on August 31. Crossing eastern Massachusetts on August 31, the diminishing system brought sustained winds of 35 MPH (56 km/h) to the Bay State.

Hurricane WILMA, late October 2005. Raced up the eastern seaboard at speeds approaching 55 MPH (89 km/h), remaining at sea, but delivering 20-foot (7-m) waves and moderate coastal flooding to northeastern New England. During Wilma's passage, a large extratropical cyclone (known in local parlance as a nor'easter) formed over northern New England, aided by moisture drawn from passing Wilma. Several inches of rain fell across the New England states, and wind gusts of between 27 and 35 MPH (43–56 km/h) were recorded at Boston's Logan International Airport.

New London Hurricane *Northeastern United States, August 30, 1713* The first major TROPICAL CYCLONE to strike NEW ENGLAND during the 18th century, the New London Hurricane, like so many of its predecessors, initially made landfall in southern Connecticut before skating northward through Massachusetts and New Hampshire. Scores of BUILDINGS in the burgeoning port city that gave the HURRICANE its name were razed, while thousands of trees, many of them fruit trees, were so badly blasted by the hurricane's winds that one incredulous observer likened the withered result to the effects of an early season frost.

New York *Northeastern United States* The snagging presence of Long Island, the 118-miles (190-km) pincer of land that extends New York's eastern boundaries well into the NORTH ATLANTIC OCEAN has, over the past three centuries, clinched for the Empire State one of the most active TROPICAL CYCLONE strike records in the northeast. Between 1693 and 2006, 43 documented HURRICANES and TROPICAL STORMS directly affected New York State, with at least 26 of these systems making landfall on Long Island's south or east shores. Of the 40 tropical cyclones that have lashed New York's diverse terrain, 15 were considered major hurricanes, Category 3 and 4 whirlwinds with CENTRAL BAROMETRIC PRESSURES of less than 28.47 inches (964 mb) and minimum wind speeds of 111 MPH (179 km/h). While on average, New York's beaches, seaports, and cities are inundated by one tropical cyclone of varying size and intensity every seven years, hurricane strikes have in several instances occurred every two or three years, as seen in 1783, 1785 and 1788; and in 1854, 1856, 1858, and 1861. During the extraordinary 1954 HURRICANE SEASON, Long Island and upstate New York were battered by two intense hurricanes, CAROL (August 31) and HAZEL (October 15). Less than a year later, two more tropical cyclones dropped extreme rains on eastern and central New York State—CONNIE (August 12, 1955) and DIANE (August 19, 1955), making the 1954–1955 seasons the most active in the state's modern history.

New York sustained what is widely regarded as its deadliest confirmed hurricane landfall on the afternoon of September 21, 1938, when the GREAT NEW ENGLAND HURRICANE—better known in New York as the Long Island Hurricane of 1938—barrelled ashore near Patchogue, Long Island. Measuring some 240 miles (386 km) across and possessing a record-setting pressure of 27.94 inches (946 mb), the Long Island Hurricane released 130-MPH (209-km/h) winds, 186-MPH (299-km/h) gusts, and a 21-foot (7-m) STORM TIDE on the entire southwest coast of the island. Dozens of communities from Babylon to Moriches to Westhampton to Sag Harbor were virtually obliterated, their shoreside summer cottages, porticoed churches, and shingle-style mansions reduced to sodden piles of timber, brick, and stone. At least 78 New Yorkers perished in the Long Island Hurricane, 19 of them in Westhampton and another 10 in the New York metropolitan area.

New York's location, north of the 40th parallel and adjacent to the chill waters of the NORTH ATLANTIC OCEAN, affords the state a brief, relatively mild hurricane season. Officially extending from June 1 to November 30, New York's hurricane season traditionally crests in August, September, and October, when late-season CAPE VERDE STORMS and eastern CARIBBEAN SEA hurricanes begin to follow the western edge of the clockwise-spinning BERMUDA HIGH into northeastern United States. Since 1825, New York has been struck by at least three June tropical cyclones, with only the June 4, 1825, system being of mature status. Known to history as the Early June Hurricane, the 1825 tempest glided across the eastern tip of Long Island, uprooting trees, carving inlets through the sandy dunes, and sinking a U.S. schooner. In New York City, some 120 miles west, fierce outer gales whipped the harbor into a frenzy, damaging piers and causing a Colombian warship to founder. Nearly a century later, on June 29, 1902, the rain-laced remnants of an early season hurricane that had initially come ashore in MEXICO and TEXAS before recurving northeast brought heavy PRECIPITA-TION to central and western New York, inundating the towns of Elmira and Ithaca with more than five inches of rain as it moved eastward across the state and into the North Atlantic Ocean. In late June 1972, then President Richard Nixon was forced to declare 14 New York counties disaster areas after the water-laden remains of New York's third June cyclone, AGNES, cascaded ashore at New York City on June 22. Its 37-MPH (56-km/h) squalls and 11-inch (279-mm) rains penetrating far inland, Agnes spawned severe flash floods in Elmira, Olean, Salamanca, and Rochester. The city of Corning was particularly hard hit, with its famed glassworks museum left awash in four feet (1 m) of water. Seven people in New York State perished, while property losses from Agnes soared past the $50 million mark.

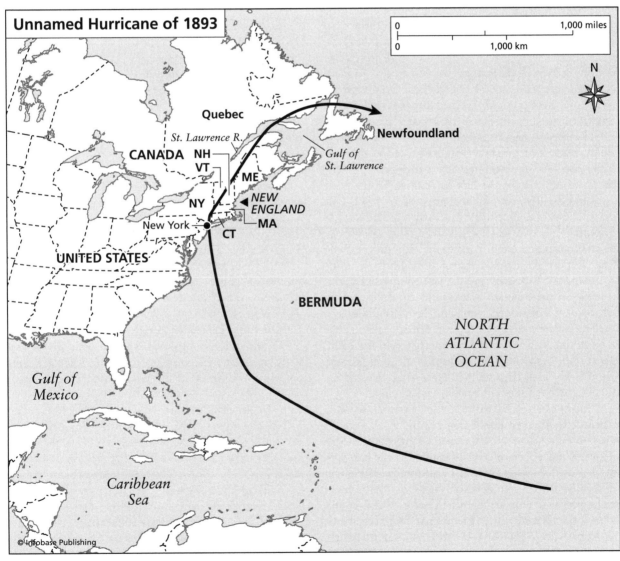

This hurricane passed directly over New York City in late August 1893. Of Category 1 intensity at the time, it is one of the few mature-stage tropical cyclones to have directly affected the city in its history.

There is no record of New York State being directly struck by a July hurricane. On July 21, 1916, however, the loosely organized EYEWALL of an unnamed tropical storm did send five-foot (2-m) rollers ashore on Long Island as it passed some 90 miles (145 km) east of the island. Bearing northeast, the weakening system would eventually make a grazing landfall in extreme eastern NEW ENGLAND on July 23—but not before sinking several small watercraft at Montauk Point and Block Island.

It is during the months of August, September, and October when the blossoming anticyclonic sweep of the BERMUDA HIGH increasingly dominates weather patterns over the North Atlantic that New York's vulnerability to strikes from mature hurricanes and powerful tropical storms rises. Swiftly borne northward by recurvature and held west by the impenetrable high-pressure cell over the mid-Atlantic, most Cape Verde and eastern Caribbean Sea tropical cyclones are able to retain their intensity as they move across the cooler, less-evaporative waters of the northeast coast and can, in turn, deliver powerful, often deadly strikes to all sections of New York State.

On September 17, 1903, central and western New York were pounded by the remnants of a northwestward-moving hurricane that had initially made landfall in New Jersey. Carrying its fairly mild central pressure of 29.25 inches (990 mb) over Rochester, the hurricane's diminishing squalls dropped 3 inches (76 mm) of rain on the city and its environs. Less than a decade later, on September 15, 1912, the watery bones of a hurricane that had originally made landfall in LOUISIANA before sweeping northeast, crossed southern New York State, downing power lines in New York City and causing flash flooding elsewhere in the state. The city of Syracuse took a beating on the night of August 19–20, 1915, when the tropical storm-force winds of the Second Galveston Hurricane trekked northeast. Earlier responsible for claiming the lives of 275 people in Texas, the 1915 hurricane's 120-MPH (193-km/h) winds had greatly faded by the time they reached New York, producing only 32-MPH (52-km/h) gales over Syracuse.

During the record-setting 1933 hurricane season, far-western New York was deluged by an exceptionally powerful Category 3 hurricane (a pressure reading of 27.98 inches was noted at landfall) that had initially come ashore in NORTH CAROLINA on the night of August 22–23. Known to history as the Chesapeake-Potomac Hurricane, this 113-MPH (182-km/h) behemoth released heavy downpours over west New York. In New York City, 45-MPH (72-km/h) winds scaled the summit of the recently completed Empire State Building, while a four-foot (1-m) storm tide was observed at the Battery. On Long Island, 60-MPH

(97-km/h) gusts toppled trees and small buildings but caused no fatalities or serious injuries.

Other mid- to late-season tropical cyclones of meteorological or historic significance in New York State include:

The Great Storm of October 29, 1693. The first documented tropical cyclone in New York history, this "most violent storme" of uncertain origin trampled the northeast coastline from VIRGINIA to Long Island, sinking ships, unroofing houses, and causing widespread tree and crop damage. Largely remembered for its raging seas, the Great Storm of 1693 apparently inundated several low-lying spits and barrier islands, its swirling, subsurface currents swiftly eroding new channels and inlets all along the Virginia shore. New York's Fire Island, located five miles (2 m) south of Long Island, was reportedly cut in two, an event later duplicated by the Great Long Island Hurricane of 1938, whose tides split the sandy beach between Fire Island Inlet and Shinnecock Bay. As with many early American tropical cyclones, no record of deaths or injuries survives from the 1693 storm.

The Unnamed Hurricane of October 14, 1706. This late-season rainmaker of indeterminate intensity caused "mighty Floods" in eastern and central New York State. The fledgling town of Albany was particularly devastated, with steady rains of falling over a three-day period.

The October Hurricane of 1749. After raising a 15-foot storm tide along the shores of Maryland and New Jersey, this "violent gale" sent powerful seas rolling into New York harbor on the night of October 18–19, 1749. Before carrying its considerable fury over Long Island Sound and into southern New England, the October Hurricane drove scores of small sloops, water lighters, and launches ashore at the Hudson River's mouth.

The Unnamed Hurricane of October 9, 1783. The first of two October tropical cyclones to have raced up the eastern seaboard during the 1783 hurricane season, the October 9 hurricane pummeled New York City and western Long Island with abnormally high tides and stiff rains. In the harbor itself, rising waters carried away wooden piers and flooded nearby cellars, causing considerable financial distress to local merchants.

The Equinoctial Storm of September 23–24, 1785. So named by eyewitnesses because its landfall coincided with that year's summer equinox, the Equinoctial Storm was particularly severe in New York City and western Long Island. In New York harbor a Swedish warship was thrown ashore on Governor's Island and dismasted, while a Spanish merchant ship, wallowing in heavy seas off Long Island, was eventually pushed ashore there and wrecked.

The Enigma Hurricane of August 19–20, 1788. A meteorological riddle of uncertain classification, this small, fast-moving disturbance visited staggering destruction on New York City and the Hudson River valley. Called a "most violent hurricane" in one survivor's newspaper account, the Enigma Hurricane's recorded characteristics are not always in keeping with known cyclonic behavior. In New York City, for instance, the Enigma's winds blew with "incredible fury" for some 23 minutes and then suddenly abated; the waters of the harbor, risen to a "very great height," had submerged much of Manhattan's Front and Water streets. The western ramparts of the Battery were knocked into the harbor. Dozens of vessels, blown into a tangled dam at the Hudson River's outlet, formed a net for the ominous tons of debris—tree trunks, sections of houses, casks, crates of apples and pumpkins, even the carcasses of dead ANIMALS—now beginning to float downriver.

In Poughkeepsie, a Hudson River valley town some 60 miles (97 km) north of New York City, the Enigma's surprising onset lasted for almost an hour with fearsome winds and flooding rains that uprooted "the largest oaks in the woods," swept away bridges, and moved down many acres of orchards and cornfield. In the nearby settlement of Hillsdale, a number of sheep and oxen were killed, and fences, mill dams, potash works, and other sites suffered severe damage. According to an account of the storm in the *Hudson Weekly Gazette* of August 26, 1788: "The day had passed . . . attended by showers, the air remarkably thick and sultry, the wind at southeast, when it chopped around, as it were in an instant, and blew a most violent hurricane from the northwest, attended by an amazing deluge of rain; it could hardly, however, with propriety be called rain, it rather wore the appearance of large rivers precipitated from huge mountains, and driven through the atmosphere by the irresistible force of the wind."

Soon crossing into western New England, the Enigma Hurricane left the populous banks of the Hudson River in gloomy, rain-washed ruin. Although contemporary newspaper accounts did not report any known fatalities, enormous amounts of tree and structural damage did occur along a 75-mile-wide (121-km) swath of central and eastern New York. A meteorological mystery of enduring fascination—the question of whether the Enigma Storm was a tightly coiled hurricane or a TORNADO of extraordinary size and duration—may remain unsolvable until a similar disturbance retraces its remarkable path.

The Carolina Hurricane of September 8, 1804. The first of two tropical cyclones to affect New York State that year, the Carolina Hurricane grazed the eastern mandibles of Long Island while recurving northeast. Breaking seas and hurricane-force gusts tossed watercraft ashore and toppled trees but caused no reported deaths.

The Snow Hurricane of October 9, 1804. Following the Carolina Hurricane ashore by nearly a month, the Snow Hurricane not only produced a pressure reading of 28.87 inches (977 mb) in New York City, but nearly eight inches (203 mm) of snow over large sections of western New York State. Highlighted by two-inch (51-mm) rains and a rapid TEMPERATURE drop to 42 degrees F (6°C) at dusk on October 9, the Snow Hurricane's "considerable flight of snow" did not linger long, however, in many places melting "as it fell."

The GREAT SEPTEMBER GALE September 23, 1815. This immense Cape Verde hurricane, perhaps the most powerful tropical cyclone to have struck Long Island since colonial times, wrought considerable damage across the entire New York City area on the morning of September 23. At Sag Harbor, Long Island, where the Great September Gale skipped ashore with a six-foot (2-m) storm surge in tow, nearly every house was flooded, with several being carried from their foundations and wrecked. The lantern atop the stone lighthouse at Montauk Point was smashed, its fragile lenses becoming pitted from windblown sea salt. New York City was "buffeted by very heavy rain and gales from the northeast," causing the restless waters of the harbor to submerge dozens of Manhattan's wooden piers, as well as toss a heavily laden merchant ship ashore at Staten Island. As the vessel's crew labored to salvage their cargo, bursting flood waters in the nearby town of Flushing swept away a wooden drawbridge, isolating hundreds of residents.

The Long Island Hurricane of September 3, 1821. Only marginally weaker than the Great

September Gale of 1815, this fast-moving tropical cyclone touched down in southwestern Long Island, near the present location of Kennedy Airport, on the morning of September 3, 1821. Passing over Nassau County and into southern New England, the Long Island Hurricane splintered piers along the East River, inundated much of the Battery, uprooted scores of hardwood trees, and killed 10 cows on Broadway by collapsing their barn. As a ferryboat at Whitehall Dock vainly struggled to remain afloat, the hurricane's heaving waves sank 10 merchant ships at anchor in the Quarantine and another 12 near the Hudson River's entrance.

The GREAT HAVANA HURRICANE of October 14, 1846. At one point in the existence of this Category 5 hurricane of spectacular fury, the Great Havana Hurricane's 175-MPH (282-km/h) winds had mercifully dwindled to 82 MPH (132 km/h) by the time it reached New York City on October 14. A system of durable destructiveness, the Great Havana's rocking seas and scooping surf soon undermined the Battery's vulnerable seawall, sending it tumbling into the harbor.

The SAVANNAH-CHARLESTON HURRICANE of October 29, 1893. Recurving northeast after making twin landfalls in GEORGIA and SOUTH CAROLINA, this dangerous hurricane unrolled heavy rains and tropical storm-force winds over central New York State on October 29, 1893. One of the deadliest tropical cyclones in U.S. history, the Savannah-Charleston Hurricane's approach to New York was unsurprisingly marred by tragedy. On August 23, the 240-ton Lightship No. 37, moored at Five Fathom Bank off the entrance to Delaware Bay, was capsized by a large wave generated by the hurricane. Four crew members perished.

The Unnamed Hurricane of September 29, 1896. One of New York State's deadliest hurricanes, this powerful eastern Caribbean hurricane first sliced inland near Cedar Key, FLORIDA, and then moved northeast. Knifing into western New York during the early morning hours of September 29, the system's hearty rain squalls hemorrhaged over Buffalo and Rochester, claiming some 100 lives and causing $7 million in property losses.

The Unnamed Hurricane of September 16, 1903. Of only tropical storm-force as it traced through eastern and central New York State, this former Category 1 hurricane had originally come ashore at Atlantic City, New Jersey. Combined with torrential rains, 50-MPH (81-km/h) winds caused $2 million in damages near Syracuse and Oneida.

The *Morro Castle* Hurricane of September 8, 1934. Named for the ill-fated luxury liner *Morro Castle*, this mature Category 2 hurricane transited eastern Long Island on the afternoon of September 8, 1934. Producing a barometric pressure reading of 28.56 inches (966 mb) at landfall, the hurricane's sustained 105-MPH (167-km/h) winds uprooted trees and demolished several oceanside cottages near Montauk but caused no injuries. Fanning 20–30-MPH (32–48-km/h) winds on the hurricane's west fringes seriously hindered efforts to rescue 435 passengers and crew from the *Morro Castle*, then burning furiously some 15 miles (24 km) off the New Jersey coast. Eventually pushed west by the distant hurricane's seas, the charred *Morro Castle*, containing the bodies of 171 of its passengers and crew, soon grounded at Asbury Park.

The GREAT ATLANTIC HURRICANE of September 14–15, 1944. While the EYE of this hurricane did not come ashore in New York, its expansive 95-MPH (153-km/h) winds, augmented by a steep PRESSURE GRADIENT, rattled the towering skyscrapers of Manhattan. Brushing to the northeast at low tide, the Great Atlantic's extreme gusts and huge waves dusted the shores of eastern Long Island but caused minimal shoreline erosion. At least two vessels—the 1,850-ton Navy destroyer *Warrington* and the Cuttyhunk lightship—foundered during the Great Atlantic's passage, taking more than 300 sailors to the bottom with them.

Hurricane CAROL, August 31, 1954. An intense Category 2 at first landfall on central Long Island, Carol's 125-MPH (201-km/h) gusts, scouring rains and 12-foot (4-m) seas toppled huge elm trees in East Hampton, washed out roads on the west end of the island, and stranded 4,500 people in the town of Montauk. An extreme gust of 135 MPH (217 km/h) was noted on Block Island, some 20 miles east of Long Island. Two people perished, both of them in New York City.

Hurricane HAZEL, October 15, 1954. One of the most powerful tropical cyclones to have ever traveled over central New York, Hazel delivered record-setting 73-MPH (118-km/h) winds and torrential rains to the western reaches of the state. On the nearby coast of New Jersey, Hazel's passage saw a three-story hotel crash into the sea; 527 beach houses were completely destroyed.

Hurricane CONNIE, August 12, 1955.   Although the precipitation-swollen eyewall of this memorable hurricane only sideswiped New York, its excessive rains—estimated to have measured 12.20 inches (305 mm) at La Guardia Airport—soon caused massive flood emergencies in New York City and western Long Island.

Hurricane Doria, August 28, 1971.   Following a similar course to the Long Island Hurricane of 1821, a much-weaker Doria made its Category 1 landfall near Kennedy Airport on the morning of August 28, 1971. Eighty-MPH (129-km/h) winds forced the closing of the facility but caused only minimal property losses.

Hurricane BELLE, August 10, 1976.   A weak Category 1 hurricane at landfall in western Long Island, Belle's 105-MPH (169-km/h) winds killed one person in New York City.

Hurricane GLORIA, September 27, 1985.   A considerably more intense hurricane than Belle, Category 3 Gloria blasted the communities of western and central Long Island with 125-MPH (201-km/h) gusts, showering rains, and severe high tides. The Long Island town of East Massapequa was flooded with four feet (1 m) of water, while skidding 50-MPH (81-km/h) winds killed two people in New York City.

Hurricane BOB, August 19, 1991.   Passing east of Long Island, over Block Island, and into Narragansett Bay, Bob's sustained 54-MPH (87-km/h) winds, 100-MPH (161-km/h) gusts, and whirring tornadoes caused extensive property damage on Fire Island.

Hurricane BERTHA, July 14, 1996.   After leaving 10 people dead in North Carolina, Hurricane Bertha's 105-MPH (169-km/h) eyewall winds continued their inland recurvature, passing over Washington, D.C., and Virginia before dousing southeastern New York with 50-MPH (81-km/h) gusts and two-inch (51-mm) rainfall counts. Downgraded to a tropical depression shortly before arriving in New York, Bertha was the first July tropical cyclone to have directly affected the state since the offshore hurricane of July 21, 1916. Nine storm shelters were opened in New York City, while 58,000 residences on Long

Island were left without electricity. One death—that of a young girl who was killed in an automobile accident on Staten Island—was reported. Minimal damage in New York State, however, was assessed in the wake of this $194 million early season surprise.

Tropical Storm Floyd, September 16, 1999.   Dropped between three and five inches (76–127 mm) of rain across New York City; Brewster, New York, received an inundating 13.7 inches (330 mm). In preparation for the system's arrival, the New York City Mayor's Office of Emergency Management (OEM) opened 24 emergency shelters, but owing to the tepidness of Floyd's passage, they were sparingly used.

Tropical Storm Isidore, September 29, 2002.   While centered over the Ohio Valley, the remnants of the storm delivered heavy rains to the New York City area.

Hurricane Isabel, September 19, 2003.   At one time a Category 5 hurricane, this system passed close to New York State as it moved inland after coming ashore in North Carolina. Although no deaths, injuries, or major property damage was recorded in the Empire State, one person was killed in New Jersey by a falling tree.

Extratropical Storm Barry, June 3, 2007.   The former tropical storm dropped 3.9 inches (0.762 mm) in Central Park, causing widespread ponding conditions across the five boroughs.

New Zealand   One of the world's most beautiful and scenic nations, New Zealand's seemingly primordial forests, mountain ranges, and active volcanoes do not have a particularly active history of TROPICAL CYCLONE activity. Although located in the Southern Hemisphere, most tropical cyclone activity occurs between 4 degrees South and 22 degrees South. While New Zealand lies beyond this range, the country's verdant islands are often affected by some degree of tropical cyclone activity, even if most of that activity is limited to systems of TROPICAL STORM intensity, or to powerful extratropical cyclones that generate from the remnants of southward-moving tropical systems.

As a review of data maintained by the Joint Typhoon Warning Center (JTWC) indicates, New Zealand's history of tropical cyclone activity between 1945 and 2005, is largely determined by EL NIÑO patterns over the southern Pacific and southern Indian Oceans.

No tropical cyclone activity affected New Zealand between 1945 and 1950; and between the years 1952–54; 1957–60; 1965–66; 1969–73; 1977–78; 1981–88; 1989–96; 1998–2005.

Between 1908 and 2005, New Zealand was struck by only two mature-stage tropical cyclones. The first, which struck on March 19, 1918, produced a pressure reading of 28.64 inches (970 mb); the second, which trundled ashore on February 1, 1936, also produced a pressure reading of 28.64 inches (970 mb). Both of these systems were of Category 1 status.

New Zealand's history of cyclonic activity is, like the geography of the island nation itself, varied and cyclical. Between 1908 and 1936, New Zealand was affected by at least 15 known tropical cyclones. Between July 4 and 7, 1908, New Zealand was struck by a fairly powerful out-of-season tropical system that generated a pressure reading of 28.93 inches (980 mb) as it came ashore on the North Island. Just over two years later, a second out-of-season tropical storm—this one bearing a pressure reading of 29.08 inches (985 mb)—dashed ashore on the extreme western tip of the North Island. And during the first week of August 1916, a powerful tropical storm delivered a pressure reading of 29.38 inches (995 mb) and tropical storm force winds to much of the North Island.

Two of New Zealand's most severe weather events, while closely connected to tropical cyclone activity in the South Pacific and southern Indian Oceans, were not actually tropical cyclones. Although they originated as tropical cyclones, the two systems had undergone extratropical deepening by the time they reached the islands of New Zealand. The first battered New Zealand on April 9–10, 1968. Known officially as Storm No. 32 and unofficially given the name Giselle by French meteorologists as the system bore down on the French Caledonian capital of Noumea, Giselle originated near the Solomon Islands before tracking to the south-southeast and crossing Cape Reinga (on the extreme northwestern coast of the North Island) on April 9, 1968. Merging with a potent extratropical cyclone that had earlier originated over the frigid waters of Antarctica, Giselle slammed large sections of the North and South Islands with wind speeds of between 99 MPH (160 km/h) and 171 MPH (275 km/h). Torrential rains caused the Avon River in Christchurch to top its banks, while tearing gusts lifted the roofs from more than 98 houses near the capital of Wellington. On the morning of April 10, 1968 (Good Friday), the 8,948 ton interisland ferry *Wahine* attempted to enter Wellington Harbor during Giselle's fury and was driven ashore on Barrett's Reef. Holed below the waterline, the 488-foot (149-m) *Wahine* eventually capsized. Of the 610 passengers and 125 crew members aboard, 44 passengers, six crew members, and one stowaway lost their lives to one of the most perplexing maritime disasters in modern memory. In addition to the 51 people lost on the *Wahine,* another three people were killed in BUILDING collapses and accidents indirectly linked to Giselle's landfall.

The second event, which occurred on March 7, 1988, involved an extratropical cyclone that had originated from Cyclone Bola as that system has tracked to the south-southeast over the tepid waters of the southern Indian Ocean, and trundled ashore on the North Island's eastern coast. As sustained winds of 62 MPH (100 km/h) uprooted trees, deroofed buildings, and downed powerlines, torrential 36-inch (916-mm) rains flooded sections of Tolaga Bay, Gisborne, Dargaville, and Te Karaka. Landslides destroyed transportation networks as hundreds of acres of topsoil were lost to Bola's flash flooding. Three people in New Zealand lost their lives, while damage estimates topped $100 million NZ dollars.

On January 1, 2004, Cyclone Heta devastated the island of Niue, located between Tonga to the west and the Cook Islands to the east. An independent state under New Zealand protection, Niue and its 1,800 or so permanent residents were battered by sustained wind speeds in excess of 155 MPH (250 km/h) as Heta tracked to the west some 43 miles (69 km) from the volcanic outcropping. Struck by cyclones in 1990 (Ofa), 1979 and 1960, Niue's tiny capital, Alofi, was riven by towering waves, torrential precipitation counts, and Category 5 winds as Cyclone Heta blasted past the island at forward speeds reaching 19 MPH (30 km/h). One person died as well more than 95 percent of the buildings on the island were destroyed.

Nicole *Sub-Tropical Storm North Atlantic Ocean, October 10–11, 2004* Sub-Tropical Storm Nicole formed on October 10, 2004, near BERMUDA from an extratropical low pressure system. Nicole was short-lived; on October 11, 2004, the system was absorbed by a large extratropical low pressure system about 345 miles (555 km/h) south-southwest of Halifax, Nova Scotia.

Nimbus A series of nine meteorological research SATELLITES placed in orbit by the United States between 1964 and 1978, these were used to study TROPICAL CYCLONE activity in the NORTH ATLANTIC, NORTH PACIFIC, and Indian Oceans. Designed and constructed by General Electric on behalf of the National Aeronautics and Space Administration (NASA) and the U.S. WEATHER BUREAU, the first of these sun-synchronous research satellites—designated *Nimbus I*—was launched on August 28, 1964, from CALIFORNIA's Vandenberg Air Force Base. Weighing 830 pounds and standing just under 11 feet high, *Nimbus I* carried a trio of one-inch, black-and-white television cameras and a radio transmitter. Powered by 10,500 solar cells, *Nimbus I* elliptically orbited Earth in concert with the Sun's cycles at an altitude of 263 miles (423 km). Before the malfunction of its solar panels knocked the satellite out of commission on September 23, 1964, *Nimbus I* had managed to capture and transmit nearly 27,000 weather-related images, including several of HURRICANES ETHEL (September 4–16) and Gladys (September 13–25).

Primarily intended to serve as a test platform for an array of newly developed meteorological instruments, *Nimbus I*'s successful but brief mission was followed by the launch of the 912-pound *Nimbus II* satellite on May 15, 1966. Placed into a sunsynchronous orbit of 684 miles, *Nimbus II* carried a Medium Resolution Infrared Radiometer (MRIR), a highly sensitive heat-measuring device that enabled scientists and meteorologists to observe a 2,000-mile-wide (3,219 km) section of Earth's atmosphere for carbon dioxide levels and the distribution of water vapor (humidity). In service until January 18, 1969, *Nimbus II* relayed to Earth more than 200,000 visible and infrared images, including several of TYPHOON IRMA, May 15, 1966, when the powerful early season system was positioned 400 miles (644 km) to the southeast of the northeastern PHILIPPINES. *Nimbus II* also visually tracked gargantuan Hurricane Beulah as it lumbered across the CARIBBEAN SEA and the GULF OF MEXICO from September 8 to September 20, 1967.

The remaining seven *Nimbus*-class research satellites were rocketed in orbit between April 14, 1969 (*Nimbus III*) and October 24, 1978 (*Nimbus VII*). *Nimbus B*, launched on May 18, 1968, was stranded at an unusable altitude and soon destroyed. The pioneering *Nimbus III*, however, successfully completed nine lengthy research trials, including the first-ever measurement of the vertical TEMPERATURE differentials in the TROPOSPHERE, as well as soundings of Earth's ozone layer. On August 16, 1969, *Nimbus III* photographed one of the most extraordinary of all North Atlantic hurricanes, CAMILLE, as its tightly wound EYE and EYEWALL slid across the western tip of CUBA. Launched on April 8, 1970, the 1,366-pound *Nimbus IV* satellite photographed both Typhoon Rose as it neared to within 100 miles (161 km) of Hong Kong on August 16, 1971. Between September 13–30, 1971, *Nimbus IV*'s image dissector camera system tracked Hurricane GINGER as the marathon tropical cyclone crisscrossed the western North Atlantic Ocean.

Norfolk–Long Island Hurricane *Bahamas–Northeastern United States, September 1–4, 1821* Said to have been only marginally weaker than the GREAT SEPTEMBER GALE of 1815, this fast-moving TROPICAL CYCLONE delivered powerful winds and torrential rains to the eastern banks of NORTH CAROLINA, VIRGINIA, southeastern NEW YORK State, and southern NEW ENGLAND between September 1–3, 1821. Of probable Cape Verde origin, the Norfolk–Long Island Hurricane was first spotted some 325 miles (523 km) east of Turks Island, the BAHAMAS, on the morning of September 1. Rapidly undergoing RECURVATURE northwest, the system's EYEWALL passed just east of Grand Bahama Island on the night of September 1, its NAVIGABLE SEMICIRCLE delivering heavy gales and pounding seas to the sandy anchorages of Eleuthera and Great Abaco islands. By midmorning of September 2, the EYE of the Norfolk–Long Island Hurricane was churning less than 300 miles (483 km) east of Charleston, its radiating currents and breaking surf creating above-normal STORM TIDES all along the SOUTH CAROLINA coastline. Its recurvature northwest had in the last 12 hours been sharply checked to the east, placing the system on a north-northeasterly trajectory. Racing ashore near Beaufort, North Carolina, just after noon on September 2, the eyewall of the Norfolk–Long Island Hurricane sliced across eastern North Carolina, uprooting trees, collapsing barns, and demolishing houses. Passing directly over eastern Virginia with considerable destructive consequence to the ships and piers of Norfolk, the hurricane continued its course swing northeast during the early afternoon hours of September 2. Swiftly transiting the mouth of the Chesapeake Bay before trailing along the eastern shores of Maryland, Delaware, and New Jersey, the Norfolk–Long Island Hurricane touched down in southwestern Long Island near the present location of Kennedy Airport shortly before dusk. It is believed to have been the only mature-stage tropical cyclone to have passed directly over New York City in almost four centuries of record-keeping. The Norfolk–Long Island Hurricane splintered piers along New York's East River, inundated much of the Battery, uprooted scores of hardwood trees, and killed 10 cows on Broadway by collapsing their barn. It then crossed through Nassau County and into southern New England just after midnight on September 3. As a ferryboat at Whitehall Dock vainly struggled to remain afloat, the hurricane's heaving waves sank 10 merchant ships at anchor in the Quarantine and another 12 near the entrance to the Hudson River. No human fatalities were reported.

North Atlantic Ocean A 14,000,000-square-mile chevron of water that reaches from the Arctic Circle in the north (latitude 66 degrees North) to the equator (latitude 0 degrees) in the south and is bound to the east by the continents of EUROPE and Africa and to the west by the eastern seaboard of North America and the CARIBBEAN SEA islands, the North Atlantic Ocean has for centuries existed as one of Earth's premier breeding grounds for TROPICAL CYCLONES. During the years 1494–2006, 2,062 documented HURRICANES and TROPICAL STORMS formed over the mid to lower latitudes of the North Atlantic valley, for a mean average of five tropical systems per year. In 1933, 1953, 1969, 1990, 1995, and 2005, the sub-equatorial (between 6 degrees and 30 degrees North) waters of the North Atlantic produced an above-average number of mature stage tropical cyclones, with 11 of 21 recorded storms developing over the North Atlantic in 1933; eight out of 14 in 1953; nine of 17 in 1969; seven of 14 in 1990, eight of 19 in 1995, and 7 of 31 in 2005. The remaining hurricanes and tropical storms during those record-setting years either originated over the enclosed basins of the Caribbean Sea or the GULF OF MEXICO.

While tropical cyclones have been observed in the North Atlantic Ocean during the off-months of January, May, and December, statistics indicate that the peak season for hurricane and tropical storm formation over the sun-seared waters of the Atlantic comes in August, September, and October—or during those periods when the oscillating influences of both the BERMUDA HIGH and the INTERTROPICAL CONVERGENCE ZONE (ITCZ) on air circulation patterns over the North Atlantic are predominant.

In the early summer months of June and July, the clockwise-spinning Bermuda High is relatively small and is positioned close to the equator, where SEA-SUR-FACE TEMPERATURES during the preceding northern winter have been at their warmest. Any easterly or TROPICAL WAVES drifting west with the trade winds between latitudes 5 and 30 degrees North are thus prevented from completely crossing the North Atlantic by the blocking presence of the high-pressure dome and soon dissipate into harmless rain showers. For this reason hurricanes and tropical storms very rarely form over the North Atlantic during June and July, instead confining their strident winds and furious rains to the southwestern Caribbean Sea or southern Gulf of Mexico.

But come the mid-to-late-summer months of August and September, as the Bermuda High undergoes its seasonal expansion and migration north, the Intertropical Convergence Zone—also known as the easterly TROUGH—settles between latitudes 10 and 30 degrees North, thereby providing incubating tropical systems with an unobstructed path across the North Atlantic. Referred to as CAPE VERDE STORMS by meteorologists, a majority of these howling North

Atlantic systems begin to organize into TROPICAL DISTURBANCES and TROPICAL DEPRESSIONS while still centered near the Cape Verde Islands, some 400 miles (644 km) off the North Africa coast. Diligently following the trade winds westward, Cape Verde storms are granted more than 2,000 miles (3,219 km) of open water over which to strengthen into capacious tropical storms and colossal hurricanes before an often-violent landfall is made in the eastern and northern Caribbean, Central America, or southern or eastern United States. Achieving an intensity rating of Category 3, 4, or even 5 on the SAFFIR-SIMP-SON SCALE along the way, mid-to-late-season North Atlantic hurricanes of this variety include such memorable tempests as CHARLIE (1951), DONNA (1960), CAMILLE (1969), DAVID (1979), ALLEN (1980), HUGO (1989), Katrina, Rita and Wilma (2005).

With the onset of the northern autumn, however, tropical cyclone formation over the North Atlantic in October and particularly November reverts to its early season parameters. Falling sea-surface temperatures in the central Atlantic force BAROMETRIC PRESSURES within the Bermuda High to drop several MILLIBARS, thereby sharply reducing the ANTICY-CLONE's size and intensity. Slowly retreating toward the equator in preparation for its annual hibernation, the Bermuda High restricts hurricane and tropical storm activity to the Caribbean Sea and Gulf of Mexico for the remainder of the North Atlantic season.

North Carolina *Eastern United States* Displaying some 301 miles (484 km) of tidewater plains, scenic barrier islands, lacework river mouths, and snagging capes to the HURRICANE-churned waters of the NORTH ATLANTIC OCEAN, North Carolina has over the past four centuries routed a veritable legion of TROPICAL CYCLONE strikes. Between 1586, when the English adventurer Sir Francis Drake encountered a powerful early season tropical system off the Outer Banks, and 2005, when no less than two mature-stage hurricanes affected in North Carolina, the Tar Heel state was directly affected by at least 82 recorded hurricanes and TROPICAL STORMS. Of these 82 tropical systems, 35 were landfalling hurricanes and tropical storms, meaning that they came ashore somewhere on North Carolina's east coast. The remaining systems either made landfall in GEORGIA and SOUTH CARO-LINA before recurving northward into North Carolina or crashed ashore in northwest FLORIDA, ALABAMA, MISSISSIPPI, and LOUISIANA before plowing in a hail of TORNADOES and flooding PRECIPITATION northeast, where they frequently delivered widespread destruction and loss of life to the inland hills of the Piedmont region. Between 1853 and 2006, an additional 31 hurricanes and tropical storms sideswiped North Carolina's landmark Cape Hatteras and broad Pamlico and Albemarle Sounds while passing north-northeast. Often storms of significant size and intensity, the high winds and eroding seas associated with these offshore tempests have on numerous instances caused serious pier and jetty damage to North Carolina's principal trading depots of Wilmington, Morehead City, New Bern, and Nags Head.

On average, North Carolina's east shores are indirectly threatened by a tropical cyclone every two years and are directly impacted by a tropical storm or mature hurricane every five years. On at least nine different occasions between 1822 and 2006, North Carolina has endured multiple strikes in the course of a single HURRICANE SEASON, as seen in 1822 when three notable hurricanes chugged into the state between August 12 and September 28. Other multiple strikes were noted in 1827 (July 30 and August 28), 1874 (September 9 and November 5), 1933 (August 22 and September 16), 1954 (August 30 and October 15), 1955 (August 11, August 17 and September 19), and 1996 (July 12 and September 5). All 20 of these tropical systems were of North Atlantic or eastern CARIBBEAN SEA origin.

Since 1586, North Carolina has been directly affected by at least two documented June tropical cyclones, both of which proved memorably destructive along the Atlantic shore. The first, a slow-moving tempest that raged between June 23 and 26, 1586, so seriously mauled Roanoke Island—later a part of Virginia—that the first English settlement in America—founded on the island the year before by Sir Walter Raleigh—had to be temporarily abandoned. Producing thunder, heavy rain, and hailstones "as big as hen's eggs," the 1586 system broke anchors and carried much of Roanoke Colony's shipping out to sea. There, according to an awestruck account by the colony's governor, Ralph Lane, the jostled ships encountered "great spouts at the seas," a description that at first glance suggests the existence of waterspouts but upon closer scrutiny probably refers to the enormous waves and copious quantities of spray driven into the air by the storm's hurricane-force gusts. Bucking into the 1586 whirlwind as it sheared across the Outer Banks, Sir Francis Drake's five-ship expedition spent two days navigating Pamlico Sound and the Roanoke River basin, only to arrive at a destination littered with flattened BUILDINGS, uprooted trees, broken fences, and smashed boats. On June 29 the inhabitants of Roanoke Colony boarded Drake's ships and set sail for England, few of them ever to return. Sir Francis, however, would return to North Carolina in July 1587, only to duel again with an intense Atlantic hurricane, this one blasting ashore with some destructiveness near Roanoke Island on August 31.

North Carolina's second June hurricane strike came in 1825, when a severe tropical cyclone of Caribbean Sea origin walloped Charleston, South Carolina, before recurving northeast on the night of June 3–4. Known to history as the Early June Hurricane, the 1825 system—complete with a 14-foot (5-m) STORM TIDE—leveled houses, toppled church spires, shattered piers, and flooded inlets all along North Carolina's coastline, causing no reported casualties in the state but significant damage to trees and crops.

During the month of July, when SEA SURFACE TEMPERATURES north of latitude 30 degrees rise to just below the 79-degree (26°C) minimum required to sustain a tropical cyclone's CONVECTION currents, North Carolina was challenged by seven documented tropical cyclone landfalls between 1788 and 2006. Of these, one was a moderately powerful tropical storm (July 10, 1901); three were fairly mild hurricanes that caused minimal coastal damage; and three were intense, mature-stage hurricanes that ravaged houses, crops, and shipping and killed dozens of people.

North Carolina's unique tradition of July tropical cyclones was inaugurated on the night of July 23–24, 1788, when a severe hurricane most probably born east of the Bahamas marched ashore near Ocracoke Inlet. Dubbed George Washington's Hurricane in honor of the U.S. president who noted the storm's harrowing passage over his home in VIRGINIA, the July 1788 system evidently spent several days stalking shipping along the west coast of BERMUDA before recurving north-northwest. Seventeen American and British merchant ships at anchor on Portsmouth Island's northeast coast were either dismasted and beached or swamped by the hurricane's 25-foot (8 m) breakers and then sunk. At least 29 people were killed, most of them sailors.

A far more intense early season tropical cyclone, known to history as the North Carolina Hurricane of 1842, passed its violent offshore EYEWALL over Ocracoke Inlet as it skirted northwest on the night of July 11–12, 1842. Judged in contemporary narratives to have been the most severe hurricane to have hit North Carolina since the notably destructive Eastern Carolina Hurricane of September 7, 1769, the North Carolina Hurricane of 1842 obliterated a substantial town on Portsmouth Island, either drove ashore or wrecked some 43 vessels at Cape Hatteras, Ocracoke Inlet, and Diamond Shoals; demolished a bridge at Albemarle Sound; and severed the railroad link between coastal Wilmington and inland Raleigh. In the towns of Edenton, Washington, and New Bern, thousands of trees were uprooted; broad swaths of surrounding tobacco and cornfields were deeply furrowed by the hurricane's gorging winds.

Some 100 people, mostly seafarers, reportedly lost their lives to the storm; in one tragic incident, seven men, attempting to retrieve the cargo from a wrecked merchantship on Diamond Shoals, were swept away by the hurricane's huge breakers and drowned.

On the morning of July 30, 1908, a Category 1 hurricane skipped ashore just north of Cape Hatteras. Possessing a CENTRAL BAROMETRIC PRESSURE of 29.18 inches (988 mb), the unnamed hurricane generated 75-MPH (121-km/h) winds and gusty downpours but caused no deaths or injuries. Property damage was relatively minor. Just over a half-century later, on July 24, 1959, North Carolina's sixth July tropical cyclone, a tame Category 1 hurricane dubbed Beth, wandered ashore just north of Beaufort, South Carolina, before suddenly sprinting northward. Slightly weaker than the July 1908 system (29.32 inches), Beth snipped foliage, rocked boats, and severed power lines across the state but took no lives and caused few injuries.

Hurricane Bertha, the most powerful July tropical cyclone to have directly hit eastern North Carolina in 154 years, lumbered ashore near Wrightsville Beach, just east of Wilmington, during the midafternoon hours of July 12, 1996. Fortified by a central barometric pressure of 28.55 inches (966 mb) at landfall, Bertha's 35-mile-wide (56-km) eyewall generated sustained 105-MPH (169-km/h) winds, an extreme gust of 108 MPH (174 km/h) at Jacksonville, North Carolina, and isolated reports of hail. Deeply intensifying as it crossed the warmer waters of the gulf stream, Category 2 Bertha caused considerable damage all along the shores of Onslow Bay, from Wilmington in the south to the United States Marine Corps base at Camp Lajeune to New Bern in the north. At Carolina and Kure beaches some 15 miles (24 km) south of Wilmington, Bertha's 15-foot (5-m) breakers buckled fishing piers, smashed storefront windows, tore the roofs and porches from a half-dozen houses, and rolled a carnival Ferris wheel off its foundation blocks. The southern tip of nearby Topsail Island was inundated by Bertha's nine-foot (3-m) STORM SURGE, a sand-muddied, frothing-white invasion of the North Atlantic that carried away 40 feet (13 m) of a concrete fishing pier, demolished 40 oceanfront cottages, and washed out a principal bridge to the mainland, marooning 15 families who had unwisely defied earlier orders to evacuate. Wet, shaken, but alive, they were later plucked from their rooftops by police and Coast Guard helicopters and carried inland to safety. In the city of New Bern at the mouth of the Neuse River, Bertha's flooding seas splintered three hotel marinas, casting a jumble of small fiberglass sailboats onto lawns and terraces. Strands of pine trees in the Croatan National Forest were uprooted or lost many

of their branches. Passing over largely rural farming country shortly before midnight on July 12, Bertha's downgraded 75-MPH (121-km/h) winds downed hundreds of miles of power lines, plunging some 400,000 people across the state into darkness. Pounding four-inch (102-mm) rainfall counts ruined 17,000 acres of corn and 3,000 acres of tobacco in 10 southeastern counties, costing North Carolina farmers some $155 million in losses. Before scudding into Virginia on the morning of July 13, Bertha spawned a number of tornadoes over central and western North Carolina, one of which touched down with some destructiveness in the state capital at Raleigh. According to figures later released by the American Red Cross, Bertha's $60 million in property damages included damage to nearly 5,800 houses, with 180 completely destroyed and another 900 rendered uninhabitable. The most costly July hurricane in North Carolina's history, Bertha killed four people and tallied $194 million in cumulative losses.

It is during the months of August and September, when the first of the CAPE VERDE STORMS begins to enter the west portals of HURRICANE ALLEY, that the number of North Carolinian hurricanes escalates. The state's vulnerability to strikes from early to mid-September hurricanes in particular is evidenced by the fact that an above-average number of North Atlantic tropical cyclones have come ashore in North Carolina on either the 1–3rd, 14–15th, or 16–18th days of the month; and that these dates may represent singularities, or days on which tropical cyclones are statistically more likely to make landfall than others. Indeed, many of North Carolina's most notorious hurricanes—mature tropical systems marked by a catastrophic loss of life and steep property-damage estimates—have occurred on these dates. On September 2, 1755, an intense hurricane inundated much of eastern North Carolina, sinking five British and colonial merchant ships and stranding a dozen other vessels at Portsmouth Island. Another tropical cyclone on the same day in 1775 ravaged the shore-side town of Bar, demolishing numerous buildings, blighting much of that year's corn harvest, and killing some 150 people. Between September 1 and 3, 1772, a tenacious tropical system whipped its destructive winds and eroding seas across the northern trade post of Edenton; 15 deeply laden vessels, several of them British warships, were wrecked. Fatalities numbered at least 50. On September 1, the first of two severe hurricanes to batter eastern North Carolina during the 1893 hurricane season sliced ashore at Onslow Bay; five fishing boats were capsized, claiming ten lives. In 1913, a Category 1 hurricane that had originated north of PUERTO RICO passed over the Cape Lookout lightship on the morning of September 3. Producing sustained 74-MPH (119-km/h) winds at Cape Hatteras, the 1913 hurricane stripped the supine lightship of its lifeboats, davits, and deck-rails before riding ashore with diminishing intensity near Morehead City.

After ripping ashore near Charleston, South Carolina, a major hurricane sharply recurved northward into the greater Wilmington area on the morning of September 17, 1713. Scores of coastal residents were killed, many of them buried beneath the wreckage of their collapsed dwellings. A peak-season system of remarkable intensity, the 1713 hurricane produced a storm surge that reportedly swept a small sloop three miles (5 km) inland before setting it down in a clump of trees. On September 17, 1906, a Category 3 hurricane skated ashore just south of Wilmington. Boasting a central pressure at landfall of 27.96 inches (947 mb), the 1906 hurricane's 129-MPH (208-km/h) winds and steady rains flooded much of the Cape Fear area, crushing houses, toppling church spires, yanking trees from the sodden ground, and beaching scores of small boats. Seven people in North Carolina perished, while damage estimates quickly topped 2 million 1906 dollars. Twenty-seven years to the day later, a slightly less intense Category 3 hurricane burst ashore at Cape Hatteras on the afternoon of September 17, 1933. Brought on by a central pressure of 28.25 inches (956 mb), the system's 113-MPH (182-km/h) winds caused widespread structural and crop damage in New Bern. Twenty-one North Carolinians lost their lives, while property losses mounted to more than $1 million.

Other historic August and September hurricanes in North Carolina include the twin hurricanes of 1590, which on August 19 and August 28 twice prevented a British expedition from landing men and supplies at Roanoke; the ferocious hurricane of August 26, 1591, that raised enormous seas in Pamlico Sound; the tremendous tempest of September 30, 1752, that wailed ashore between Cape Fear and Cape Lookout, submerging barrier islands, demolishing houses, driving eight ships aground on Ocracoke Bar, and killing at least seven people; the Eastern Carolina Hurricane of September 7–8, 1769, that thundered over the coastal towns of New Bern, Edenton, and Brunswick, destroying the Brunswick County courthouse, uprooting "hundreds of thousands of the most vigorous trees in the country, tearing some up by the roots, others snapping short in the middle," and undermining no less than 20 sawmill dams; the exceptionally severe Independence Hurricane of September 2–3, 1775, that momentarily subdued the rebellious Carolina countryside with oppressive winds and invading seas, claiming nearly 150 lives on the Outer Banks and another 13

elsewhere in the region; two destructive storms (the latter of which an eyewitness described as resembling "a West Indian hurricane") between August 3 and August 13, 1795, that spawned extensive flooding in the Piedmont region and washed out bridges, mills, and corn harvests along the Cape Fear coast; the particularly damaging hurricane of August 23, 1806, that so devastated Wilmington that the city's newspaper declared it to have been the "most violent and destructive storm of wind and rain ever known here;" the legendary Norfolk and Long Island Hurricane, whose vicious eyewall, on September 3, 1821, tore northwest across Cape Fear, Portsmouth Island, Edenton, and Kitty Hawk—hundreds of houses were completely demolished, and an unspecified number of lives were lost; the unnamed hurricane that on August 18, 1879 produced a sustained wind speed of 138 MPH (222 km/h) over Cape Lookout; the particularly deadly hurricane of September 11, 1883, that sank more than 50 ships in Pamlico and Albemarle Sounds; the severe North Atlantic hurricane that on September 13, 1889, so seriously mauled the Five Fathom Bank Lightship No. 24 that the stalwart vessel was later sent to the shipbreakers; the 107-MPH (172-km/h) whirlwind that ravaged Cape Hatteras on the night of August 17–18, 1899, killing five people and irreparably altering the cape's shoreline; the unnamed Category 1 hurricane that slogged ashore near the mouth of the Cape Fear River on September 22, 1920, capsizing small boats, submerging piers, and claiming one life; the Chesapeake-Potomac Hurricane, August 22–23, 1933, that initially unraveled its central pressure of 27.98 inches (947 mb) and 113-MPH (182-km/h) winds over Cape Hatteras before recurving into Virginia; Hurricane CAROL, a deepening Category 1 system that on August 30, 1954, lunged ashore at Morehead City with 100-MPH (161-km/h) winds before ricocheting back into the North Atlantic over Elizabeth City; the second hurricane of the 1954 season to affect North Carolina, EDNA, whose 103-MPH (166-km/h) eyewall winds stormed past the Outer Banks on September 10; Hurricane CONNIE, August 13, 1955, whose central pressure of 28.40 inches (962 mb) at landfall generated sustained 119-MPH (192-km/h) winds at Holden Beach—27 people in North Carolina perished; an even more-intense Hurricane Ione, that on September 19, 1955, bounced ashore at Morehead City, its central pressure of 28.34 inches (960 mb) manufacturing 119-MPH (192-km/h) winds and record-breaking precipitation counts of nearly 49 inches (1,245 mm) over the town of Maysville—five people were killed, some $100 million in property losses assessed; Hurricane DONNA, September 11, 1960, whose 110-MPH (177-km/h) gusts claimed six lives and demolished scores of buildings along the Outer Banks; Hurricane DIANA, September 11–13, 1984; Hurricane GLORIA, September 26, 1985; Hurricane HUGO, September 21, 1989, that shivered the inland city of Charlotte with 100-MPH (161-km/h) winds and caused stunning damage on the southeast coast; offshore Hurricane EMILY, August 31, 1993; Tropical Storm Jerry, August 28, 1995, whose torrential rains claimed four lives in North Carolina, including that of a firefighter who was washed away while attempting to rescue a stranded motorist; and FRAN, September 5, 1996, a Category 3 hurricane whose 115-MPH (185-km/h) winds and 12-foot (4-m) storm surge took 12 lives and caused more than $1 billion in property losses.

In October and November, when much of the North Atlantic's tropical cyclone activity becomes concentrated in the Gulf of Mexico and southwestern Caribbean Sea, North Carolina's number of direct strikes between the years of 1749 and 2006 decreases. During the early morning hours of October 18, 1749, North Carolina's first recorded October hurricane, christened the October Hurricane of 1749, strafed Cape Hatteras with "very violent" winds, a 15-foot (5-m) storm tide, and torrential rains. Contemporary accounts state that 11 ships at anchor near Ocracoke Bar were either sunk or stranded, a majority of them becoming total losses. Less than a half-century later, on October 8, 1783, another October hurricane unleashed furious winds from Wilmington to Winston-Salem, overturning barns, splitting massive trees, and breaching fences. The mid-nineteenth century was marked by the passage of the GREAT HAVANA HURRICANE through western North Carolina on October 13, 1846, and the tragic encounter off Cape Hatteras between a squadron of Union warships and the Expedition or Union Hurricane on November 1, 1861, during which two warships foundered, taking 200 lives to the bottom with them. During the evening hours of October 14, 1944, the Great Atlantic Hurricane—a mighty Category 3 system that had only a day earlier enveloped much of central CUBA—offloaded 129-MPH (208-km/h) winds and raging seas to Cape Hatteras. Arcing northward at 30 MPH (48 km/h), the Great Atlantic Hurricane's central pressure of 27.96 inches (947 mb) caused extensive coastal flooding and beach erosion but no casualties. The worst of North Carolina's October hurricanes, HAZEL, darted ashore near Swansboro on October 15, 1954. Recurving north-northwest at 30 MPH (48 km/h), Hazel's deadly eyewall scoured the oceanfront communities of Long Beach and Holden Beach with sustained 106-MPH (170-km/h) winds and a pulverizing 17-foot storm tide. Hundreds of beachside cottages were destroyed,

while large boats were carried several hundred yards inland and then stranded. Hazel's toll was steep for North Carolina—19 people dead and some $136 million in property losses assessed. On the night of November 17–18, 1994, late-season Hurricane GORDON brushed its central pressure of 28.93 inches (980 mb) across the Outer Banks, its 85-MPH (137-km/h) winds causing extensive beach erosion and some damage to coastal BUILDINGS but no fatalities.

Although very rare, there are documented instances of tropical cyclone activity occurring in North Carolina during the postseason month of December. Between 1888 and 1925, two tropical systems—one a mature hurricane, the other a tropical storm—transited the dunes of Cape Hatteras on the same date—December 2. Neither storm was particularly destructive, and no casualties were reported in either event, making the December systems two of the most fascinating, most benign examples of North Carolinian tropical cyclones.

On September 18–19, 2003, Hurricane ISABEL trundled ashore in North Carolina. Maximum sustained winds were clocked at 105 MPH (169 km/h) and 45-foot (15-m) high seas that caused extensive coastal erosion.

North Pacific Ocean Divided into *eastern* and *western* zones by the international date line (longitude 180 degrees West), the 30-million-square-mile expanse of the North Pacific Ocean—confined to the east by the continents of North and South America and to the west by the continent of Asia—has for centuries served as one of Earth's most hospitable environments for TROPICAL CYCLONE formation. Including that midocean zone commonly referred to as the *central North Pacific* (between longitudes 140 degrees West and 180 degrees West), statistical averages based on the frequency of documented tropical cyclones in the North Pacific Ocean during the years 1900–2006 indicate that between the years 1281 (when the Hakata Bay Typhoon saved JAPAN from invasion) and 1996 (when 10 HURRICANES and TROPICAL STORMS blasted across the eastern North Pacific), some 25,000 TYPHOONS, hurricanes, and tropical storms developed over the undulating blue folds of the North Pacific Ocean, for an average of 34 tropical systems per year.

Influenced by the seasonal north-south migration of the INTERTROPICAL CONVERGENCE ZONE (ITCZ) and by the relative strength and position of the North Pacific's two anticyclonic high-pressure areas, the frequency of tropical cyclone generation in the eastern and western halves of the North Pacific is generally active but uneven, with the west producing nearly twice as many tropical systems each systems each year as the east. On average, 15 percent of Earth's tropical cyclones form over the eastern North Pacific Ocean, while another 30 percent originate over the western North Pacific. In 1952, for instance, 14 hurricanes and tropical storms developed in the eastern North Pacific, while another 21 typhoons and tropical storms originated in the western North Pacific; in 1962, 24 tropical systems took root in the western North Pacific, whereas only 11 blossomed in the eastern North Pacific. In 1996, at the same time that 27 typhoons and tropical storms were swirling across the western Pacific, 11 hurricanes and tropical storms were churning up the 80°F (27°C) waters of the extreme eastern Pacific. In 2006, 25 tropical cyclones crossed the Eastern North Pacific Ocean.

While the relationship between the North Pacific Ocean's varied climatological parameters and tropical cyclone generation can be described in terms of one circulatory model, the North Pacific's gargantuan size, coupled with its seemingly endless assortment of atmospheric genealogies, permits each of its three zones to possess their own unique meteorological characteristics. These zonal singularities, in turn, shape and define the life cycles of their respective tropical cyclones, determining nearly every factor in typhoon and hurricane frequency, genesis, trajectory, and intensity. They are, from east to west across the North Pacific.

EASTERN NORTH PACIFIC OCEAN

In some texts referred to as the southeastern North Pacific Ocean, this stretch of the North Pacific spreads from the west coast of the United States, MEXICO, and northwestern South America to longitude 140 degrees West; and from the Arctic Circle (latitude 66 degrees North) south to the equator. Although the eastern North Pacific's climatological HURRICANE SEASON begins on June 1 and continues until November 15, there are documented instances of mature-stage tropical cyclones appearing in May and December. During the months of January through April, or when the northern winter is at its coldest, the ITCZ generally swings northward over the extreme southeastern North Pacific, extending its nearly windless influence between latitude 3 degrees North and latitude 15 degrees North. Often delivering a brood of embryonic TROPICAL WAVES to the lukewarm incubator of 72°F waters and slow-moving, low-altitude winds off the central coast of western Mexico, the ITCZ's winter migration neatly bisects the 15-degree (latitude 10 degrees North to latitude 25 degrees North) range of ocean within which a statistical majority of eastern North Pacific tropical systems have in the past originated. Reaching some 900 miles (1,200 km) to the west, this crucial formation range is overlaid for most of the winter months by the blocking presence of a diminished ANTICYCLONE—the Pacific High—whose outward-spinning winds and resulting WIND SHEAR severely restricts tropical cyclone development over the area. Elsewhere in the eastern North Pacific Ocean, the polar-driven Aleutian Low is at its midwinter height, its counterclockwise winds sending dozens of PRECIPITATION-filled storms into the cold shores of western CANADA and the United States—all climatological occurrences that are not conducive to tropical cyclone formation during the months of January through April in the eastern North Pacific Ocean.

The climatological guard over the eastern North Pacific does change, however, during the spring, summer, and fall months. By the end of April, the cold-core Aleutian Low has usually disappeared, replaced in ever-widening circles by the northward-moving warmth of the Pacific High. Caused by the sun's heating of the eastern North Pacific's surface, the clockwise-spinning Pacific High will spend much of the next six to seven months dominating wind circulation patterns over the region. The ITCZ completes its southward retreat by the middle of May, hovering just north of the equator as the first of its fragile tropical waves begins its tenuous voyage westward. Over the center of the tropical cyclone formation range, determined by scientists to be positioned some 360 miles (625 km) due south of the Baja Peninsula, warm winds course in from the northeast, the result of a burgeoning thermal low that forms seasonally over the western United States and Canada. As SEA-SURFACE TEMPERATURES in July and August approach the 84° Fahrenheit threshold required for widespread evaporation, hurricanes and tropical storms are born over the eastern North Pacific Ocean. Often ranking as Category 2, 3, and even 4 systems on the SAFIR-SIMPSON SCALE, most tropical cyclones that form in the eastern North Pacific are harmlessly carried west-northwest by the westward-flowing trade winds, away from densely inhabited Mexico and CALIFORNIA. Unfortunately for hundreds of people, this has not always been the case with all eastern North Pacific tropical cyclones, as the history of deadly strikes along Mexico's west coast will evidence.

Although very rare, there are recorded examples of tropical depressions being formed in the eastern North Pacific Ocean before recurring *eastward* across Central America and into the CARIBBEAN SEA or GULF OF MEXICO. Between June 11 and 18, 1965, an unnamed tropical depression developed in the eastern North Pacific, some 300 miles southeast of the city of Tapachula, Mexico, and then traveled through Mexico to the Gulf of Mexico where it was upgraded to a tropical storm on the evening of June 13. Bearing a CENTRAL BAROMETRIC PRESSURE of 29.68 inches (1,005 mb) and maximum winds of 60 MPH (97 km/h), the system finally made landfall on the Florida Panhandle on the morning of June 15. Notable for reasons other than destructiveness, the early season Pacific interloper claimed no lives and caused only minor property damage.

CENTRAL NORTH PACIFIC OCEAN

Encompassing the entire 1,500-mile spread of the Hawaiian Islands, the Central North Pacific Ocean extends from longitude 140 degrees West to longitude 180 degrees West—or where the international date line separates the eastern North Pacific from the west—and from the Arctic Circle (latitude 66 degrees North), south to the equator. While at first glance a seemingly arbitrary designation, the zone officially deemed the Central North Pacific does, in fact, have its own history of tropical cyclone activity apart from that of the eastern North Pacific. Between 1832 and 2006, at least 130 hurricanes and tropical storms affected the area, with more than half originating west of longitude 140 degrees. A number of these systems, such as hurricanes IWA (1982) and INIKI (1992), have grown into storms of enormous destructive potential as they bear down on HAWAII and its neighboring islands, sometimes making landfall with deadly consequence. While the Central North Pacific's climatological hurricane season runs from June 1 to October 31, there are recorded examples of mature-stage tropical cyclones taking shape in May and November.

Circulation patterns over the central North Pacific during the winter and summer months are similar to those found for the same periods in the eastern North Pacific, with several notable exceptions. In the winter the ITCZ—which tends to arc northward in the vicinity of longitude 160 degrees West—forces seedling tropical waves against the underside of the Aleutian Low, where counterclockwise-spinning winds send them recurving northeast into the frigid waters of the Gulf of Alaska, where they dissipate into snow showers before ever reaching cyclonic maturity. With the Pacific High safely lingering many hundreds of miles southeast, some of these Central North Pacific disturbances are able to slip eastward with the prevailing winds of March and April, undergoing EXTRATROPICAL deepening before eventually delivering great quantities of rain to the northwestern United States.

During the prime months of August and September, tropical cyclone activity in the Central North Pacific is tempered by both a large pool of cooler sea-

water that habitually drifts just south of the Hawaiian Islands and the strength and position of the Pacific High, sometimes referred to as the Hawaiian High, which overlies most of the region from early July to late October. A vast high-pressure cell, the Pacific High's hurricanes and tropical storms westward, away from Hawaii and the Line Islands. At times when the anticyclone has been weak or absent (its average BAROMETRIC PRESSURE ranges between 30.38 inches [1,029 mb] and 30.30 inches [1,026 mb]), severe Central North Pacific hurricanes have been able to recurve northward, as seen with postseason Hurricane Iwa in November 1982.

In late May 2001, Adolph became the strongest May hurricane on record in the eastern Pacific basin with sustained winds of 145 MPH (233 km/h) at peak intensity. Because of opposition from pressure groups, the name Adolph was later retired from the list of rotating North Pacific HURRICANE NAMES.

On September 26, 2001, Hurricane Juliette generated the second-lowest CENTRAL BAROMETRIC PRESSURE yet recorded in the eastern North Pacific Ocean: 27.25 inches (923 mb), making the system of firm Category 4 status.

On August 24, 2004, Hurricane Frank reached a peak intensity of 86 MPH (138 km/h) while churning northward near Cabo San Lucas. Frank's interaction with MEXICO's western coastline made it a short-lived system. Moving over cooler waters, the system dissipated on August 26.

WESTERN NORTH PACIFIC OCEAN

Beginning at longitude 180 degrees West—the international date line—and flowing westward to the heavily settled coasts of CHINA, JAPAN, KOREA, the PHILIPPINES, and TAIWAN, the western North Pacific Ocean has long held the distinction of being Earth's premier region for tropical cyclone activity. Although there is no set TYPHOON SEASON in the western North Pacific Ocean—the occurrence of typhoons and tropical storms has been documented in every month of the year—there are a series of climatological conditions upon which the remarkable spike in tropical cyclone generation most often seen during August and September is contingent. As with the eastern North Pacific and central North Pacific zones, circulation patterns over the western North Pacific are dictated by the actions of a high-pressure anticyclone, the Ogasawara High. Sometimes a part of the Pacific High, sometimes its own autonomous cell, the Ogasawara High vanishes from the western North Pacific during the winter months and then reappears during the summer heat, reaching a northernmost position between latitudes 30 degrees North and 35 degrees North. Dependent upon sunlight reflected by the ocean's surface for its size and strength, the Ogasawara High tends to spend the principal months of July through November alternately wandering north and south, forcing some westward-traveling typhoons into the northern Philippines, while allowing others to slide across the central islands of Japan. Almost all of the mid-to-late-seasons typhoons and tropical storms that menace TYPHOON ALLEY are borne westward by the anticyclone's clockwise-turning winds, sent tearing ashore in China and Vietnam with relentless regularity.

The ITCZ's northward transit, which usually commences around May, further defines the principal areas of tropical cyclone formation in the western North Pacific. During the winter and spring months of January to April, any tropical cyclone formation in the western North Pacific is generally limited to between latitudes 3 degrees North and 10 degrees North. As the ITCZ continues its seasonal progression northward during the month of June, principal typhoon and tropical storm breeding ranges extend from 4 degrees North to 12 degrees North. When the ITCZ achieves its northernmost configuration, in the vicinity of the 30th parallel, during the months of July and August, typhoon development spans latitudes 9 degrees and 25 degrees North. With the onset of fall in the Northern Hemisphere, both the ITCZ and the Ogasawara High begin their ritual withdrawal toward the equator. In September, October, and most of November, tropical cyclone formation zones shrink to between 10 degrees North and 20 degrees North before returning to their winter confines of latitudes 3 degrees North and 12 degrees North in December. It is during these waning-season periods, when typhoon trajectories are kept from recurving to the northeast by both the subsiding ITCZ and the dwindling Ogasawara High, that the Philippines Islands have traditionally endured their most devastating tropical cyclone strikes.

Norton, Grady (1895–1954) U.S. meteorologist, one-time director of the U.S. WEATHER BUREAU's primary TROPICAL CYCLONE tracking office in Jacksonville, FLORIDA, and celebrated HURRICANE forecaster, Grady Norton was said by his contemporaries to have possessed the ability to "smell" an approaching hurricane. Born in northern ALABAMA, Norton joined the Weather Bureau on graduating from college in 1913. In 1919, after a period of advanced study at Texas A&M University, Norton joined the U.S. Army as a junior meteorologist. Always a precocious observer of tropical cyclone behavior, it was not until 1928 and a trip to Florida following the devastating September passage of the SAN FELIPE/Lake Okeechobee Hurricane that Norton reportedly decided to become a professional hurricane forecaster. After traveling to New Orleans, LOUISIANA, in 1929 to receive further training in tropical cyclone forecasting, Norton was assigned to the Weather Bureau's Jacksonville office in 1935. Equipped with short-wave radio transmitters and a recently developed teletype network, the Jacksonville station provided Norton with the technology that in time helped him to become the most trusted hurricane forecaster of the 1930s and 1940s. Linked by microphone to 25 radio stations across Florida, his reassuring voice—characterized by a slight southern cadence—became a familiar, reassuring accompaniment to evacuation and other storm preparedness rituals. In an age when SATELLITE tracking systems and live television communications did not exist, Norton provided his listeners with timely, accurate information regarding the intensity and course of threatening hurricanes and tropical storms. In 1936, Norton published an important work on tropical cyclones entitled *Florida Hurricanes*. Appointed director of the Jacksonville tracking station in 1942, he supervised its subsequent transfer to Miami in the spring of 1943. In 1944, following JOSEPH DUCKWORTH's historic flight through the EYE of a hurricane, Norton oversaw the establishment of the HURRICANE HUNTERS aerial reconnaissance service. Responsible for saving hundreds, perhaps thousands, of lives during two decades, Grady Norton died of a heart attack while tracking Hurricane HAZEL's progress through HISPANIOLA on October 11, 1954. He was succeeded as director of the Miami forecasting office by his assistant, Gordon E. Dunn.

October Gale, Hurricane *Northeastern United States, October 3, 1841* One of NEW ENGLAND's deadliest TROPICAL CYCLONES, the October Gale killed at least 85 people as it knifed past Nantucket, Martha's Vineyard, and the outer banks of Cape Cod. Slashing its way northward over Cape Cod Bay and along the coast of Maine, the powerful EYEWALL of the October Gale tragically overwhelmed the Georges Bank fishing fleet as it departed Truro, Massachusetts, for the rich mackerel grounds to the east. Nine schooners, fore-and-aft rigged vessels of about 53 tons burthen, were quickly capsized by the HURRICANE's howling gusts and rocking seas. Eighty-five men and boys—19 of them between 11 and 18 years of age—perished in the storm, a tally of lost youth that a contemporary newspaper account poignantly stated caused "many a heart to wring with anguish." In the town of West Harwich, 11 miles (18 km) north of Truro, 15 sailing vessels were blown ashore, while scaling winds unroofed houses, tore large trees up by the roots, mowed fences down at ground level, stripped orchards of their fruit, and dropped nearly 18 inches (505 mm) of snow over central Connecticut. Several tons of salt hay that had been harvested and stacked in the marshes were swept away by the October Gale's seven-foot (2-m) STORM SURGE. Elsewhere along the shores of Cape Cod Bay, enormous saltworks—tall wooden windmills used to manufacture salt through solar evaporation—were first undermined by the hurricane's heavy seas and then collapsed. At the time the core industry on the Cape, the loss of the saltworks coupled with the lingering effects of the 1837 bank panic soon worsened an already grave economic depression across the area, creating a long winter of hardship and deprivation for the resilient people of northern Cape Cod. Undaunted, they would

in July 1842 erect a MONUMENT in Truro inscribed to the courageous memory of the 57 husbands, fathers, and sons from the town who were lost at sea during the "memorable gale of October 3, 1841."

Offshore Hurricane of 1924 *Northeastern United States, August 26, 1924* Although the EYEWALL of this powerful midseason HURRICANE remained over the NORTH ATLANTIC OCEAN's open waters for most of its existence, the Offshore Hurricane's TROPICAL STORM–force gales and 6-inch (152-mm) PRECIPITATION counts caused widespread property losses and crop damage across the southeastern NEW ENGLAND states of Rhode Island and Massachusetts. Scores of small watercraft were forced ashore at Cape Cod and the islands; downed power and telephone lines cast much of eastern New England into rain-soaked darkness. No fatalities or serious injuries were reported, however, making the Offshore Hurricane of 1924 one of New England's more merciful TROPICAL CYCLONES.

Olive, Typhoon *Philippines, June 20–27, 1960* The first of four major TROPICAL CYCLONES to strike the PHILIPPINES during the 1960 TYPHOON SEASON, Olive crushed the northern island of Luzon beneath its 115-MPH (185-km/h) winds, eight-inch (203 mm) rains, and powerful 10-foot (3-m) STORM SURGE on June 27 and 28, 1960. Preceded ashore by TROPICAL STORM Lucille on May 24 and followed by TYPHOONS KIT (October 7) and LOLA (October 13), Olive claimed 404 lives, demolished some 32,000 BUILDINGS, and left more than $30 million worth of property damage behind. In Cabanatuan City, 50 miles northeast of Manila, Olive's unrelenting rains caused oily mudslides that trapped hundreds of people in their houses or on their rooftops; several ships, among them a patrol boat belonging to the nation's coastal defense force, were either sunk or run aground by Olive's 24-foot (7-m) seas. Said by the Philippine Weather Bureau to have been one of the most violent early season typhoons on record, the storm's legacy was subsequently honored by the retirement of the name Olive from the World Meteorological Organization's list of TYPHOON NAMES.

Opal, Hurricane *Mexico–Southern United States, September 24–October 6, 1995* The first Caribbean TROPICAL CYCLONE to be christened with a name beginning with the letter O, Opal claimed 44 lives and inflicted some $2.5 billion in property damage as it swept through eastern MEXICO, the FLORIDA Panhandle, and central ALABAMA between September 29 and October 5, 1995. In Mexico, where TROPICAL DEPRESSION No. 17 (Opal) dusted Yucatán Peninsula's north coast with 35-MPH (56-km/h) winds and 10-inch (254-mm) rain counts, hundreds of BUILDINGS in Campeche and Tabasco provinces were demolished. Twenty-five people, among them an infant boy and three fishermen, were killed, while another 472 were reportedly injured. In Florida, where Opal made landfall as a powerful Category 3 HURRICANE on the evening of October 4, a stretch of coastline from Pensacola Beach to Panama City was left battered and waterlogged, an overwashed landscape of stranded sailboats, gutted cottages, fallen power lines, and denuded trees. Two people in Florida were killed, and more than 200 buildings were left utterly ruined. Pressing inland, Opal's flagging winds brought tropical storm-force rain squalls to Alabama and GEORGIA, spawning deadly TORNADOES and leaving more than 1.5 million people without electrical and telephone services. Six people in Alabama perished, two of them crushed when a tree toppled onto a mobile home. In Georgia, nine people were killed in storm-related traffic accidents, while another two victims were tallied from Opal's dying passage over NORTH CAROLINA.

The 15th named TROPICAL CYCLONE of the wholly remarkable 1995 hurricane season, Opal originated as a tropical depression east of Piacer, Mexico, on the afternoon of September 24, 1995. Drifting northwest at nearly three MPH (5 km/h), the depression, christened No. 17, landed on Yucatán Peninsula's east coast, south of the port town of Punta Allen, just before midnight on the 25th. For the next four days, Opal zigzagged across the Yucatán's northern plateaus. Swinging almost due north on September 26, the system slowly progressed to historic Chichen Itza, where on September 27 it sharply turned northwest and stalled. Its sustained winds erratically rising and falling,

Tropical Depression Opal suddenly gained a southeasterly STEERING CURRENT on the afternoon of September 27, swiftly sending it across the peninsula to Mérida, where it again stalled on the following day.

Finally emerging into the southwest Gulf of Mexico's rejuvenating waters on September 29, Opal again shifted its course, shearing southwest at nearly 7 MPH (11 km/h). By the afternoon of September 30, the strengthening EYE of the now upgraded TROPICAL STORM was located to the southwest of Mérida, deep within the Bay of Campeche, and headed northwest once again. Stalled over the bay's heated waters, Tropical Storm Opal was prevented from making landfall in hurricane-prone Veracruz by yet another sudden course change on the night of October 1, this one taking it northeast at 21 MPH (34 km/h).

Now located northwest of Mérida, Tropical Storm Opal spent the greater part of October 2 rapidly intensifying. Hiking its sustained winds to 100 MPH (161 km/h) in a matter of hours, the powerful tropical storm was first upgraded to a Category 1 hurricane by noon of October 2 and a Category 2 hurricane, with winds of between 96 and 110 MPH (155–177 km/h) by six o'clock the same evening. Tropical storm-force winds and pounding rains reached 230 miles (360 km) from the hurricane's tightening eye, walloping the Yucatán Peninsula's northwest coast with 70-MPH (113-km/h) gusts and frenzied rains. Isolated flash floods raged through the outskirts of Mérida, washing houses from their foundations and sweeping unwary travelers into swollen rivers and streams. Tens of thousands of people in Yucatán were left without electricity and running water, while collapsed bridges and blocked roadways restricted efforts by relief agencies to enter those areas most severely effected by the grazing tempest. More than 10,000 people were left homeless.

By the morning of October 3, as Opal neared to within 400 miles of the Florida Panhandle, HURRICANE WATCHES were posted along a 550-mile (980-km) stretch of gulf coastline, from Grand Isle, LOUISIANA, to Panama City, Florida. The scene of Louisiana's worst hurricane disaster, Grand Isle itself was evacuated, its affluent vacationers and year-round residents moved to the mainland before Opal's potential fury flooded bridges and causeways. Offshore oil drilling platforms in the GULF OF MEXICO were abandoned by helicopter, while public works departments in Louisiana, MISSISSIPPI, Alabama, and Florida removed traffic lights, inspected storm drains, and prepared to unseal emergency water supplies. The New Orleans Levee Board closed the 31 floodgates that protected the delta city from destructive hurricane surges. Thousands of gulf-coast residents, familiarized through awareness programs and legend with the region's violent history of hurricane activity, raided supermarkets and lumber yards for ice, flashlight batteries, bottled water, kerosene, and plywood. Although it was still too early to tell when and when Hurricane Opal would make landfall, the people of the U.S. storm belt knew that hurricane preparedness was a choice best made early, a prompt decision that in the next 24 to 48 hours could quite possibly change the course of their lives.

By now there were clear signs coming from the gulf that any contact with Hurricane Opal could indeed be a matter of life or death; during the early morning hours of October 4, the hurricane underwent a harrowing period of deep and organized intensification. At dusk on October 3, Opal's CENTRAL BAROMETRIC PRESSURE of 28.74 inches (973 mb) produced maximum wind speeds of 109 MPH (175 km/h) and gusts as high as 127 MPH (204 km/h); by midnight that night, Opal's central pressure had plummeted to 28.37 inches (960 mb), lifting its howling 125-MPH (201-km/h) winds to solid Category 3 status. By dawn of October 4, as the hurricane slid to within 200 miles (400 km) of landfall in the Fort Walton Beach–Pensacola area of Florida, its barometric pressure fell again, this time to 27.32 inches (925 mb), making its sustained 150-MPH (241-km/h) winds some of the most powerful ever observed in the Gulf of Mexico.

Awaking to overcast skies and light drizzle on the morning of October 4, 1995, gulf-coast residents turned on their radios and televisions only to discover with horror that the Category 2 baby they had put to bed last night had now matured into an oncoming Category 4 monster. As thousands of beach dwellers, business owners, and emergency personnel watched and listened in apprehensive wonder, meteorologists explained that Opal's sudden strengthening was due to warmer-than usual SEA-SURFACE TEMPERATURES in the northern Gulf of Mexico. Uncannily similar to killer hurricanes CAMILLE, which blasted Mississippi and Louisiana during the equally active 1969 season, and KATRINA and RITA that ravaged nearly the same area in 2005, Opal had developed a small, 11-mile-wide (20-km) eye, around which spun a 125-mile EYEWALL band. Within the band, sustained wind speeds of 150 MPH (241 km/h)—just five MPH (8 km/h) short of Category 5 status—were being recorded by aerial ANEMOMETERS. Barometric readings, while indicating that Opal's precipitous drop in central pressure had now stabilized, revealed that the hurricane could indeed further deepen as it moved, at nearly 22 MPH (35 km/h) over the warm northern shallows of the gulf. A minimum 18-foot (6-m) STORM SURGE was predicted for the western shores of the Florida Panhandle.

More than 100,000 people in northwestern Florida promptly fled the coast, crowding highways and bridge-heads in an effort to escape the fast-moving siphon of life behind them. As factories and offices closed for the day, the people who remained behind shut their remaining storm shutters, filled their bathtubs with water and their hurricane lamps with kerosene, and settled down in front of the television to track Opal's progress. Schools were transformed into storm shelters; civil defense brigades built sandbag fortifications around low-lying neighborhoods in Florida and Alabama. At Cape Canaveral, NASA postponed by one day the launch of the space shuttle *Columbia*. In Louisiana, where Opal's gale-force onslaught brought down tree branches, power lines, and road signs, a deputy sheriff was seriously injured when the huge U.S. flag he was helping to furl caught the wind like a sail, whisking him 30 feet into the air and then dropping him to the pavement.

Shortly before 6 o'clock on the evening of October 4, 1995, Hurricane Opal made a burnishing landfall on the Florida Panhandle. Coming ashore east of Pensacola, in the vicinity of the U.S. Air Force's Hurlburt Field installation, the hurricane's sustained 125-MPH (201-km/h) eyewall winds had been downgraded to that of a Category 3 system, although its 144-MPH (232-km/h) gusts still qualified for Category 4 status. Mercifully weakened during its final hours over the Gulf, Opal nevertheless raced inland with great destruction, spawning tornadoes, tidal surges, and swiftly moving floods. At Panama City, Florida, some 100 miles (161 km) east of where Opal's eye came ashore, huge waves pounded the pilings of a recently completed fishing pier, collapsing several hundred feet of its 1,500-foot (300-m) concrete span into the raging surf. In Okaloosa County, Florida, an elderly woman was killed when a tornado exploded her mobile home. Park benches, streetlights, billboards, and road signs along a 120-mile (315-km) stretch of coastline were wrenched out of the ground and then hurled several hundred feet though the air. Two hundred homes in Bay County, Florida—from million-dollar resort cottages to quaint bungalows—were utterly demolished, their walls, roofs, and entire floors thrown down like playing cards across the sodden green landscape. Another 1,000 structures—restaurants, hotels, arcades, bars, gas stations, and churches—were seriously damaged.

Radiating out from the eye, Opal's destructive web ensnared almost all of northern Florida. At Fort Walton Beach, swank motor yachts and sailboats were carried ashore by Opal's 20-foot (7-m) waves and then left stranded on highways and median strips as the roiling waters receded. Large segments of US 98, the region's principal freeway, were washed out, requiring months of repair work. In Navarre Beach, a barrier island located between Fort Walton and Pensacola, dozens of beachfront cottages were hammered into driftwood, while Destin suffered the loss of its high school gymnasium. With the roof peeled off by Opal's 142-MPH (229-km/h) gusts, the walls of the sports complex soon folded in on themselves, crashing down on the rows of bleachers and dunking backboards through the wooden floor. A giant roller coaster in an amusement park in Panama City Beach shivered and creaked beneath Opal's peaking 120-MPH (193-km/h) winds but remained intact.

Blazing inland at nearly 21 MPH (34 km/h), Opal's weakening 79-MPH (127-km/h) winds raided central Alabama and northwestern Georgia, plucking utility poles from the ground, uprooting trees, and overturning mobile homes, before passing into the Ohio Valley on the morning of October 5. Now downgraded to tropical-storm status, Opal's 45-MPH (79-km/h) gales

brought heavy rains to Tennessee and Kentucky and death to two people in NORTH CAROLINA. In Maryland, nearly 120 residences were damaged or destroyed by three of Opal's tornadoes on the night of October 5. Although no casualties were reported, some $5 million in burst windows, toppled chimneys, and missing roofs were assessed. The name *Opal* has been retired from the cyclical list of HURRICANE NAMES.

Ophelia, Hurricane *Southeastern–Eastern United States, September 6–18, 2005* The 15th named storm and the eighth mature stage TROPICAL CYCLONE of the remarkable 2005 North Atlantic HURRICANE

Although a tropical system of considerable power, a lingering Hurricane Ophelia came close to making landfall several times, but was ultimately unsuccessful in crossing the eastern seaboard. Ophelia did, however, leave drowning rains and three people dead. *(NOAA)*

SEASON, Hurricane Ophelia spun a crazy course along the eastern seaboard of the United States, lingering for nearly two weeks, and shifting between hurricane and TROPICAL STORM intensity. In addition, it made no fewer than three complete loops during its star-crossed trajectory. A copious rainmaker, Ophelia dropped up to 10 inches (254 mm) of precipitation across NORTH CAROLINA's Outer Banks, forcing the temporary closure of the historic Cape Hatteras lighthouse. In preparation for Ophelia's forecasted landfall in North Carolina, some 30 shelters were opened and 350 National Guard troops were deployed to the state. All told, Ophelia claimed three direct and indirect fatalities, and caused some $70 million (in 2005 U.S. dollars) in property losses. The bulk of Ophelia's rage was directed at coastal erosion.

Orion WP-3   A four-engined, turboprop aircraft designed and built by the Lockheed Aircraft Corporation, from the early 1970s to the present it has been used by the National Oceanic and Atmospheric Administration (NOAA) for TROPICAL CYCLONE research and reconnaissance missions over the NORTH ATLANTIC and NORTH PACIFIC OCEANS, CARIBBEAN SEA, and GULF OF MEXICO. Originally intended to track Soviet submarines during the 1960s and 1970s, the Orion WP-3 flies into a tropical cyclone at altitudes of between 1,500 and 24,000 feet (300 m–800 m) and is equipped with the most sophisticated ANEMOMETERS, BAROMETERS, and RADIOMETERS available. Stationed at MacDill Air Force Base in Tampa, FLORIDA, the spacious Orion WP-3 can carry between seven and seventeen crew members, meteorologists, and technicians and may remain aloft for up to 12 hours. Unable to fly at altitudes of greater than 24,000 feet because of its turboprop design, the Orion WP-3's upper-level reconnaissance missions were curtailed in late-1996 following the introduction of NOAA's new GULFSTREAM IV-SP hurricane hunting jet.

Otis, Hurricane *Eastern North Pacific Ocean-Mexico, September 28–October 3, 2005*   The 15th tropical system of the 2005 eastern North Pacific HURRICANE SEASON, Hurricane Otis originated from Tropical Depression No. 15 (TD15) on the morning of September 28. On October 1, it achieved its peak intensity of 28.64 inches (970 mb) and produced sustained winds of 105 MPH (169 km/h). Although forecast to slice into MEXICO's Baja Peninsula, Otis quickly weakened, remained an offshore system, and finally dissipated on October 3 without claiming any lives or causing significant property damage.

outflow   This term is given to that meteorological process by which rising air currents in the EYE of a TROPICAL CYCLONE are vented, or released, from the pinnacle of the system. Essentially serving the fluelike eye as a high-altitude pump, the outflow of a HURRICANE, TYPHOON, or CYCLONE counteracts its low-level INFLOW by drawing spent air out of the top of the eye and then ejecting it away from the fragile heart of the storm at speeds of up to 60 MPH (97 km/h). A necessary component if a tropical cyclone is not to strangle on its own intake of warm, humid air, the outflow often manifests itself as anticyclonic action, moving in a clockwise direction in NORTH ATLANTIC hurricanes and western NORTH PACIFIC typhoons, and counterclockwise in Australian and Indian Ocean cyclones. Its presence generally accompanies cyclonic storms of mature strength, and results in the formation of an extensive OUTFLOW LAYER, or high-elevation canopy of cirrus and CIRROSTRATUS CLOUD types. As with a majority of a tropical cyclone's elements, the outflow performs several tasks at once, including directly assisting in a system's intensification by delivering cool, dry air to the outer fringes of a storm; once there, the cool, sinking air acquires new quantities of heat and moisture before being drawn into the burgeoning tropical cyclone once again.

outflow layer   Sometimes referred to as the *outflow canopy*, the outflow layer of a TROPICAL CYCLONE represents the summit of the storm, the final plateau atop a five to six-mile mountain of cumulus and CUMULONIMBUS CLOUD bands. The thinnest of a mature HURRICANE's, TYPHOON's, or CYCLONE's horizontal cloud stacks, the outflow layer's wispy collection of icy cirrus and CIRROSTRATUS CLOUDS, ranging from 3,000 to 5,000 feet (1,000–2,000 m) in height, is formed when rising air currents in the storm's EYE, now stripped of most of their warmth and moisture, are expelled into the cold confines of the upper atmosphere. Fanning across the underside of Earth's troposphere in concert with the system's OUTFLOW circulation, what moisture that has been left in the air currents, quickly freezes, forming a veritable roof of ice clouds over the entire tropical cyclone. Often under the influence of localized anticyclonic activity, the outflow layer's high-elevation clouds spin in a clockwise direction in the Northern Hemisphere and counterclockwise in the Southern (southwest and west in hurricanes, typhoons, and BAY OF BENGAL cyclones; east and northeast in the cyclones of the Southern Hemisphere).

In its earliest stages the outflow layer can obscure a tropical cyclone's developing eye, denying satellites and other aerial observers a clear view of the system's blossoming and potentially dangerous EYEWALL. But as the hurricane, typhoon, or cyclone further intensifies, witnessing a concomitant increase in its low-level INFLOW rates, the upper-level outflow layer begins to better organize itself, serving as a more efficient exit for the vast quantities of heated air now being siphoned through the eye. Drifting to the edges of the storm's vortex, the outflow layer gradually assumes the eye's circular shape, trailing off in one direction or another like a long, balancing tail of thick, densely packed clouds. Its expanse contained only by WIND SHEAR, or the presence of high-altitude winds moving in different directions at different speeds, both the outflow layer's position and growth configuration can indicate to storm watchers the relative health of a respective tropical system.

P

Padre Ruiz Hurricane *Eastern–Northern Caribbean, September 18–24, 1834* Named for a deceased Roman Catholic priest whose funeral services coincided with the hurricane's landfall in the Dominican Republic, the Padre Ruiz Hurricane visited cataclysmic devastation and gargantuan DEATH TOLLS on the Caribbean islands of DOMINICA and HISPANIOLA between September 20 and 23, 1834. In Dominica, where the hurricane's winds virtually leveled the capital city of Roseau, some 230 people perished, either drowned by the hurricane's 12-foot (4-m) STORM SURGE or crushed beneath the collapsed timbers of their houses. In northeastern Hispaniola, where the Padre Ruiz Hurricane's EYEWALL raced ashore on the afternoon of September 23, searing gusts and blistering rains stung the foliage from trees, leaving behind a tangled hive of barkless forests and what appeared to be scorched earth. First plotted in 1850 by pioneering HURRICANE scholar William Reid and subsequently published in his *An Attempt to Develop the Law of Storms,* the Padre Ruiz's north-northwesterly passage over Dominica and Hispaniola bequeathed steep property losses to the islands, a costly legacy that soon catapulted their ravaged economies into a prolonged depression. In the Dominican Republic, where the Category 3 eyewall of the Padre Ruiz Hurricane smashed ashore at Santo Domingo, numerous merchant ships foundered at the Ozama River's mouth, taking an estimated 170 sailors to the rocky bottom with them. Hundreds of BUILDINGS in Santo Domingo were unroofed by the hurricane's sustained, 135-MPH (217-km/h) winds, and severe rains blighted much of that year's sugarcane harvest. Scores of lives were lost, a majority of them to a suffocating series of landslides triggered by the hurricane's torrential rainfall counts.

Patsy (Yoling), Typhoon *Philippines–Vietnam, November 12–22, 1970* Judged by the Philippine Weather Bureau to have been the most severe Category 3 TYPHOON to have passed over Manila in 106 years, Patsy delivered 124-MPH (200-km/h) winds and 12-inch (305-mm) rains to the southern and central provinces of the PHILIPPINES and the northeast coast of Vietnam between November 19 and 22, 1970. Better known in the Philippines by its local name, *Yoling,* Patsy destroyed 32,000 BUILDINGS and claimed an estimated 1,108 lives in the Greater Manila area. Before dissipating over northern Vietnam on the afternoon of November 22, Patsy's slicing winds and furious rains caused an additional 30 deaths, making it one of the deadliest typhoons to have swept through the South China Sea in recent history.

The 16th and final storm of a most extraordinary TYPHOON SEASON (one which had seen the Philippines struck by no less than three other major typhoons—Georgia, KATE (Sening) and JOAN (Titang)—since September 15, Patsy originated over the western NORTH PACIFIC OCEAN northeast of Guam on the morning of November 12, 1970. Positioned almost directly over the 15th parallel, the young TROPICAL DEPRESSION steadily churned southwest at 17 MPH (27 km/h), smartly sapping the ocean of its heat and moisture as it quickly grew into a weak TROPICAL STORM on the afternoon of November 13 and a minimal Category 1 typhoon with a CENTRAL BAROMETRIC PRESSURE of 29.00 inches (982 mb) by midday on November 15. Maximum sustained winds in Patsy's EYEWALL now measured 82 MPH (132 km/h). After progressing southwest direction until nightfall on November 14, Patsy suddenly swung northwest, heading for a return to its former course along the rigid track of latitude 15 degrees North.

With typhoon-force gusts now eclipsing 95 MPH (153 km/h), Patsy reached the 15th parallel on, of all days, November 15, a coincidence that seemingly set the typhoon on a rare, straight-line track west across the North Pacific Ocean for most of its remaining life. Between November 15, when Patsy rifled through the Mariana Islands, and November 19, when it finally came ashore on southern Luzon, the typhoon traveled 1,300 miles (2,400 km) on a course nearly parallel with that latitude 15 degrees North. Lashing Guam with gale-force winds and heavy rains as it passed 100 miles north of the island on the afternoon of November 15, Patsy inexorably trod straight to the west, further strengthening as it spun toward the slumbering cities and towns of the Philippine Islands.

With a central barometric pressure of 27.17 inches (920 mb), Typhoon Patsy blasted ashore near the port town of Jomalig, the Philippines, on the morning of November 19, 1970. Gusts of 145 MPH (233 km/h) stripped buildings of their roofs, crumpled billboards, toppled trees, snapped telephone poles, and gushed a 12-foot (4-m) STORM SURGE, the highest seen on Luzon's southeast coast since 1921. In the Polillo Islands, 83 miles east of Manila, hundreds of watercraft, from ungainly barges to refrigerated fruit carriers, were driven ashore, causing numerous fatalities. Lathered waves pitched a disabled coastal ferry, laden with 380 people, onto a sandbar near Agta Point. Filipino naval personnel courageously fought 30-foot (10-m) seas in order to rescue the ferry's entire complement successfully and to secure the stricken vessel against the brutal hammering of the ocean.

Steadily moving across the narrow waters of the Polillo Strait, Patsy's sustained 124-MPH (200-km/h) winds and considerable rains breached dikes and levies, washing away 1,600 houses in the mountain towns of Antipolo, Taytay, and Marikina. In Manila, as government-owned television and radio stations suspended regular broadcasting in favor of civil defense alerts and patriotic calls for strength and calm, the electricity abruptly failed. Almost one-third of Luzon Island was plunged into a sort of noonday darkness as Patsy's clipping winds and shorting rains undermined much of the National Power Corporation's service grid. At the Presidential Palace, rising flood waters in the streets began to overflow into the basement, endangering a vital government communications center located there. As technicians worked to install a temporary switchboard higher up in the palace, Patsy's prodigious rains continued to sweep over Manila, eventually inundating 90 percent of the capital's streets. Ordering the closure of all banks, stock exchanges, schools, manufacturing plants, and Manila's International Airport (where, shortly after 10 o'clock in the morning, Patsy's strongest sustained windspeed of 124 MPH [200 km/h] was recorded), then President Ferdinand Marcos commandeered all food stocks and private transportation facilities available in the southern provinces of Luzon Island, enlisting them in the coming relief effort by warning that all food merchants and private transportation owners who defied his orders were subject to immediate arrest.

By early evening, as the EYE of Typhoon Patsy lurched into the warming shoals of the South China Sea, the rain and wind finally began to subside in Manila. Under a leaden sky tinged with drizzling rains and bantering gusts, weary survivors forged through the capital's swamped streets, simultaneously fascinated by and appalled at the devastation wrought by Patsy's day-long fury. In Manila's splendid cathedral, swelling floodwaters rapidly turned the marble floor of the nave into a lake bed, compelling those devout individuals who had sought mercy and shelter in the cathedral to either climb onto wooden chairs or scramble for the high-ground safety of the ornate altar. In Manila's high-rise office buildings and hotels, first-story lobbies were mired in dark, mud-thickened torrents of rainwater, stranding hundreds of people on upper floors. Just to the east of Manila, rescue crews toiled into the night, sifting through the tombs of wreckage that once housed the towns of Antipolo, Taytay, and Marikina. With the threat of more flash floods from Patsy's lingering rains hanging over their heads, the teams struggled through muddy bogs and swamped coconut groves to save hundreds of victims trapped beneath fallen houses. Although in some instances the effort was spectacularly successful, it came too late to save the lives of some 900 residents of the three towns. During the subsequent week, when all schools and businesses remained closed and electricity snapped on only infrequently, rescue operations in Manila itself would yield another 206 casualties, making Patsy one of the 10 most deadly typhoons to have struck the archipelago in its recorded history.

After wreaking nearly $40 million in property damage on the Philippines, Patsy continued to move almost due west across the South China Sea, fanning the temperate waters with 100-MPH (161-km/h) winds and renewed rains. By midday of November 20, the typhoon's course had finally begun to undergo its long-anticipated RECURVATURE, or an abrupt shift northwest. Crossing the 15th parallel for the last time on November 20, Patsy steadily progressed west-northwest, bound for the Indochinese

Peninsula's east shores. Patrolling the coastal waters of the then-divided Vietnam, units of the U.S. Navy quickly prepared themselves for battle with a much greater foe than the Vietcong. As General Quarters was sounded throughout the fleet and thousands of sailors hastened to secure their respective ships against the coming heavy winds and seas, Patsy inexorably churned toward the 17th parallel, the border between North and South Vietnam. In both nations, military officials considered the strategic and propaganda value of a typhoon strike on the enemy's territory and then diligently tracked the typhoon in a hail of anxious hopes and destructive dreams. Dismissing the nightmare possibility that Patsy might indeed strike their own respective countries, North and South Vietnam continued to hurl the typhoon at each other long after it came ashore near the fishing village of Co Lieu, North Vietnam, during the late evening hours of November 21, 1970. In South Vietnam, the press pointed out that not only had the U.S. fleet successfully weathered Patsy's passage across the South China Sea, but that the typhoon had also killed 30 people and destroyed several hundred houses in the North. All this was a journalistic estimation, of course, because North Vietnam never officially recognized that Patsy had come ashore there and so never released any casualty or damage reports. In view of Patsy's severity in the Philippines, it is entirely possible that the typhoon proved quite serious in North Vietnam, prompting the government to withhold embarrassingly high DEATH TOLL counts to prevent their use in South Vietnamese propaganda efforts.

Pauline, Hurricane *Mexico, October 8–10, 1997* On October 8–10, 1997, Hurricane Pauline thrashed southwestern MEXICO, killing 272 people.

During the 1998 NORTH PACIFIC HURRICANE SEASON, two moderately powerful tropical cyclones affected Mexico's western coast. In early September, Hurricane Isis proved the only tropical system of HURRICANE strength to make landfall during the 1998 season when it pounded the northwestern cities of Colima and Guadalajara with high winds and rains between September 2 and 3. With a CENTRAL BAROMETRIC PRESSURE of 29.14 inches (987 mb) at landfall, Isis's 75-MPH (121-km/h) winds and intense rains claimed eight lives and wrought extensive structural damage.

Between September 6 and 12, 1998, torrential rains spawned by offshore Tropical Storm Javier caused churning rivers in Mexico's Chiapas region to surmount their banks, isolating nearly a half-million people and claiming the lives of another 100. Topping 16 inches (406 mm) within a two-week period,

tropical downpours precipitated huge mudslides that destroyed bridges and blocked mountain passes for several weeks.

During the 1999 North Atlantic hurricane season, intense rains associated with Tropical Depression No. 11 caused widespread flooding over Mexico's eastern Tabasco state on October 7. An embryonic system that would eventually grow into Tropical Storm Harvey, Tropical Depression No. 11's passage along the Gulf coast was felt the hardest in the city of Villahermosa, where dozens of bridges and roads were blocked or destroyed.

During the 1999 hurricane season in the eastern North Pacific Ocean, one hurricane in particular, Greg, wrought considerable death and devastation along the northwestern Mexican coast. A powerful Category 2 hurricane, Greg's 113-MPH (182-km/h) winds and seven inch rains lashed the resort city of Cabo San Lucas during the afternoon hours of September 8. The most destructive hurricane of the 1999 eastern North Pacific season, Greg's roaring landslides, engulfing flash floods, and rampaging seas killed nine people and caused property losses of $30 million.

Philippines *North Pacific Ocean* With some 7,000 volcanic islands, shallow coral lagoons, and rockbound passages strewn along a 1,200 mile (3,000 km) range of the western NORTH PACIFIC OCEAN, the Philippines suffers the dubious distinction of being the most TYPHOON-ridden nation on Earth. Positioned by an accident of nature between latitudes 5 and 22 degrees North and at the western terminus of the stormy stretch of Pacific Ocean known as TYPHOON ALLEY, the Philippines have over the centuries been directly affected by so many typhoons and TROPICAL STORMS that an exact count of their number is nearly impossible to compile. Records maintained by the Philippine Weather Bureau indicate that between 1948 and 2006, the archipelago's 780 inhabited islands were struck by no less than 248 typhoons, or an average of four storms per year. In some instances, such as the 1972 season, six or more typhoons have affected the chain's principal islands of Luzon, Mindanao, Samar, Mindoro, and Leyte in a single year. Of the storms, 80 were considered major typhoons, systems with CENTRAL BAROMETRIC PRESSURES of below 28.10 inches (959 mb) and wind speeds greater than 120 MPH (193 km/h). During the same time period, 9,104 lives were reportedly lost (2,517 in the 1970 season alone), while upward of $790 million in property damage was assessed, making the 46-year storm span one of the most devastating in Philippine history.

In practice, the Philippine Weather Bureau recognizes two classes of typhoons, or *BAGUIOS*, as they are known locally. Those early-season TROPICAL CYCLONES that make landfall during the spring months are called *bean*, or rain, typhoons because of their considerable PRECIPITATION counts, while those storms that come ashore in summer and autumn are known as wind typhoons. Because the Philippines' extensive TYPHOON SEASON ranges from April to December and includes the climatological characteristics of all four seasons, the segregating of the two typhoon types is not without some justification. For example, the rain typhoons that generally strike northern Luzon and the Bataans between April and August tend to be part of larger, EXTRATROPICAL storm systems. Often these typhoons are in the final, or decay, stage of their existence, a phase noted for both a steep rise in barometric pressure and the formation of fronts. Moving in a north-northwesterly direction, Philippine rain typhoons deliver prodigious quantities of precipitation to the country's northern regions. In June and July 1972, three rain typhoons—Konsing (June 23–24), Bebeng (June 26–July 2), and Gloring (July 10–25) dropped an astounding 52 inches (1,260 mm) of rain within three weeks, causing record-breaking flood emergencies in the mountain communities north of Manila. All told, 413 people were killed and some 70,679 houses completely destroyed in this remarkable triple assault.

Conversely, wind typhoons occur toward the end of the Philippine typhoon season, during the hot, humid months of October, November, and December. By far the more intense of the two classes, wind typhoons tend to move on a west-northwesterly course and come ashore in the Bicol, Visayas, and northern Mindanao regions in either their deepening or mature stages. Although a wind typhoon's heaviest rainfall is generally limited to a 15–25-miles (25–40-km) radius directly surrounding the EYE, WIND SPEEDS in Philippine wind typhoons can reach 160 MPH (215 km/h) and are often accompanied by 15–20-foot (5–7-m) STORM SURGES.

The height of the Philippine typhoon season roughly coincides with that of the North Atlantic HURRICANE SEASON, the months of September through November, when the westward trade winds over both the North Atlantic and North Pacific oceans are at their strongest. As with the powerful Cape Verde HURRICANES, which sometimes bring late-season destruction to the shores of North America and the Caribbean, the most intense typhoons generally make landfall in the Philippines during the months of October and November. Indeed, a survey of Philippine typhoon strikes for the years 1948 to 1994 reveals that October and November storms are often not only deadlier than their early and late-season counterparts but that they also frequently come ashore in pairs, generally within a week of each other.

In November 1928, two unnamed typhoons struck different regions of the Philippines within a matter of days: The first, which caused extensive damage to the central islands of Samar, Panay, and Negros, crashed ashore on November 24, killing more than 50 people; three days later, a second typhoon struck the island of Leyte, claiming 200 lives and leaving another 10,000 homeless. In early November 1937, twin typhoons ravaged Manila and the surrounding Bulacan Province within a week of each other, killing 277 people and destroying hundreds of valuable coconut groves. During the 1956 season, two December typhoons, Orchid and Polly, swept through the Greater Manila area between the 2nd and the 13th, respectively. Orchid, which was barely of typhoon strength when it came ashore on the Polillo Islands, did little damage to the capital as it lightly crossed into the South China Sea and blossomed two days later, killing 300 people in eastern Java. Polly, on the other hand, roared inland with 98-MPH (158-km/h) winds and 10-inch (254-mm) rains, spawning flash floods and BUILDING collapses that claimed 79 lives. In October 1960, Typhoons KIT (October 7) and Lola (October 13) drowned 92 people when they spiraled ashore in northern Luzon. A decade later, on October 14, 1970, Typhoon KATE (known as *Sening* in the Philippines) struck central and southern Luzon, killing 583 people; less than a day later Typhoon JOAN (*Titang*) roared ashore in the southern Philippines, claiming another 526 lives.

Other late-season Philippine typhoons that did not travel in packs include the unnamed tempest of November 6, 1885, in which 10,000 people perished as the typhoon's storm surge rushed ashore along the Tiburon Peninsula; the storm of November 14, 1934, in which 300 people were killed when the township of Mauban was obliterated by 111-MPH (179-km/h) winds; and the powerful Christmas Typhoon of December 25–26, 1947, during which a minor EARTHQUAKE and tsunami accompanied the storm's landfall in southeastern Luzon—at the height of this typhoon's fury, the Danish motorship *Kina* was driven ashore on the Calagua Islands, Northern Samar, and wrecked, claiming the lives of 34 of its 63 passengers and crew. In mid-November 1970, Typhoon PATSY's 124-MPH (200-km/h) winds and 12-inch (305-mm) rains claimed 1,108 lives and destroyed some 32,000 buildings, prompting the Philippine Weather Bureau to declare Patsy the most severe typhoon to have passed over Manila in 106 years.

Such a record of relentless and intense typhoon activity further extends to those early or out-of-season storms that have frequently mauled the Philippines with deadly effect. On May 11, 1913, a vicious typhoon struck the province of Mayon, drowning an estimated 827 people as its 16-foot (5-m) storm surge inundated several coastal fishing villages. In June 1925, central Luzon was ravaged by an enormous typhoon that claimed another 1,000 lives and left tens of thousands of people homeless. Half a decade later, on May 27, 1929, the storm surge associated with an unnamed typhoon submerged six small villages on the southern island of Leyte, killing 119 people. During the first week of April 1935, a major typhoon thrashed parts of Luzon and Samar islands, killing 70 people, while another, on December 26 of the same year, claimed an additional 39 lives in northern Luzon. On June 27, 1960, Typhoon Olive, the second of four powerful typhoons to strike the Philippines that season, crashed ashore in southeastern Luzon, killing 604 people and left 60,000 others homeless. Damage estimates from Olive's 170-MPH (274-km/h) gusts approached $30 million, making it the most expensive typhoon to have struck the country up until that time. In late June 1964 Typhoon Winnie slammed across southern Luzon, killing 43, leveling 24,000 buildings and earning itself notoriety as the most powerful typhoon to have struck that part of the country since 1882. On April 26, 1971, Tropical Storm Wanda delivered a wailing, preseason strike on the central islands, killing at least 125 people and causing widespread damage to crops and villages. On July 22, 1983, the relatively mild Typhoon Bebeng (international code-named Vera), besieged Bataan with 80-MPH (129-km/h) winds and a 12-foot (4-m) storm surge. One hundred eighty-two people perished as Bebeng's massive flooding inundated at least 10 villages on Manila Bay's western banks, including the major fishing port of Pantalan Luma; no fewer than 466,465 people were displaced at some 49,000 houses were utterly destroyed in eight provinces. One survivor later stated that Bebeng was worse than what befell the Philippines during World War II. "We lost everything," he mourned, "in a flash."

On September 2, 1984, Typhoon IKE clobbered the southern and central islands of the Philippines with 137-MPH (220-km/h) winds, 11-inch (279-mm) rainfall counts, and a deadly, 15-foot (5-m) storm surge. Viewed at the time as one of the most powerful typhoons to have impacted the archipelago during the 20th century, Ike damaged or destroyed more than 100,000 buildings and claimed upward of 1,400 lives.

Less than a year later, Typhoons Hal and Irma delivered 128-MPH (206-km/h) winds, 15-foot (5-m) storm surges, and extensive flooding to the islands between June 1 and July 1, 1985. Some 190 people were killed; another 25,000 were left homeless. On November 28, 1987, President Corazon Aquino was forced to declare a state of emergency after Typhoon NINA's 120-MPH (193-km/h) winds and torrential rains decimated seven densely populated provinces in the country's central region. More than 1,200 people reportedly lost their lives to the depredations of the storm, while another 250,000 were left homeless.

The Philippines had no sooner dug itself free of Nina's rubble than Typhoon RUBY, on October 24, 1988, blasted northern Luzon with sustained 130-MPH (209-km/h) winds and heavy rains. Sweeping away bridges, fields, even entire towns and villages, Ruby killed 540 people and tallied more than $150 million in property losses.

On August 30, 1995, powerful Typhoon Kent unleashed heavy rains as it transited Luzon Island in a hail of 100-MPH (161-km/h) winds and pinpoint lightning strikes. In the town of Bacolor, north of Manila, swelling flash floods hurled a torrent of volcanic ash down the slopes of Mount Pinatubo, demolishing many of the town's principal buildings. Five people were lost, and millions of dollars in damages were accrued. During the very active 1995 typhoon season, the second Tropical Storm Nina guided further rains to the area on September 2, triggering another round of mudslides. Hundreds of people were stranded on rooftops for days, one woman giving birth to a boy while awaiting rescue. Another 105,000 fled their homes, crowding hotels and government shelters as far south as Manila.

Struck by no less than 14 tropical cyclones during the course of the 1995 season, the Philippines would later lose more than 100 people to Typhoon Sybil on October 2, another 160 people to Tropical Storm Zack on October 30, and an estimated 740 victims to the stunning ravages of SUPER TYPHOON ANGELA on November 3. The most violent typhoon to have ravaged the nation since Ruby, Angela's 141-MPH (227-km/h) winds, 12-inch (305-mm) rains, and 30-foot (10-m) seas leveled thousands of buildings, knocked out telephone and electrical service to 30 provinces, and left 640,000 people without shelter. Cast into complete darkness, Manila suffered severe damage to its closed stock exchange and international airport, and acres upon acres of surrounding sugarcane fields were submerged beneath five feet of water, precipitating months of financial hardship for thousands of plantation workers. Stunned by allegations that the DEATH TOLL had been elevated by lax evacuation measures, then President Fidel Ramos announced on November 8 that local officials who had not adequately prepared for Angela's Category 4 landfall would be prosecuted.

Despite such a long succession of tragedy and death, the Philippines have come to accept, and indeed rely upon, the destructive presence of typhoons for at least 40 percent of its yearly rainfall requirements. Because an early season *bean* typhoon can deliver between 75 and 175 billion tons of precipitation in a single strike, the Philippines' mountainside lumber plantations, coconut groves, and rice paddies depend on typhoons for their seasonal growth and renewal, and although a number of these storms frequently bring deadly floods to the archipelago's volcanic bays, the alternative—drought—could possibly extract an even greater, more unfortunate loss of live and livelihood. The torrential rains of Typhoon Bebeng, for instance, broke a prolonged, eight-month drought over the southern provinces of Luzon, permitting that year's coconut and sugar crop to later be harvested successfully.

Three typhoons affected the Philippines on or about September 8, 1935; none of these were direct strikes, and wind speeds never exceeded 30 MPH (56 km/h); but like many Filipino tropical systems, they were slow-moving and dropped a great deal of PRECIPITATION. In Baguio, some 44 inches (1,118 mm) of precipitation fell from August 2 to 4, 1935. Forty bridges were washed out in the northern Philippines, and at least three people were drowned.

In October 1998, Super Typhoon ZEB inundated the northern Philippines; just over a week later, Super Typhoon Babs struck the central Philippines, inundating several neighborhoods in Manila.

On November 3, 2000, Typhoon Bebinca struck the Philippines, leaving 28 people dead and forcing the closure of government offices, schools, and financial institutions. The Filipino National Disaster Coordinating Council reported that more than 10,000 individuals were evacuated from their homes as a result of the increasing water levels. The storm also spawned landslides in the Manila suburb of Antipolo, and several deaths reportedly occurred in the Cagayan and Nueva Vizcaya provinces northeast of Manila. All telecommunications and electrical networks around the capital were also disrupted.

On July 22, 2003, Super Typhoon Imbudo paralyzed the northern island of Luzon with sustained winds of 150 MPH (240 km/h). Said to have been the most intense typhoon to strike the islands since 1998, Imbudo's unwelcome visit killed at least 10 people and caused nearly $20 million (in 2003 U.S. dollars) to crops and other agricultural facilities. A durable system, Imbudo churned westward through the South China Sea and eventually made landfall near the Chinese city of Yangjiang on July 24. At landfall, Imbudo's sustained winds topped 105 MPH (165 km/h) and claimed some 20 lives in CHINA. According to the Hong Kong Observatory, Typhoon Imbudo was the most intense typhoon to strike Guangdong province since Typhoon Sally riddled the area in 1996, killing 123 people and injuring an additional 4,300 people.

On November 24, 2004, Tropical Storm Muifa (Winnie; Filipino naming list) slammed into central and southern Luzon, killing at least 407 people and leaving another 174 missing and at least 100 injured. Downgraded from a typhoon that at one time was of Category 4 status and had sustained winds of 132 MPH (213 km/h), Muifa dropped torrential rains and caused tremendous flooding.

On December 2, 2004, Super Typhoon Nanmadol made landfall along the northeast coast of the Philippines with sustained winds of 115 MPH (185 km/h), and 138-MPH (220-km/h) gusts. More than 400 people were listed as dead, and another 560 listed as missing. Quezon Province was most heavily struck; in the town of Real, more than 100 people were killed, and large mudslides buried much of the town as well as two nearby fishing villages. One day later, on December 3, 2004, Typhoon Yoyong struck the eastern Philippines, near the town of Aurora, killing one person and downing power lines. Bearing winds of 115 MPH (185 km/h), Yoyong (the 24th tropical system to affect the Philippines during the 2004 North Pacific Ocean typhoon season), pounded the island of Catanduanes, killing at least one person.

On March 16, 2005, Typhoon Roke struck the Philippines and weakened into a tropical storm. No lives were reportedly lost.

One of the deadliest Filipino typhoons in a decade, Typhoon Durian pounded the Philippines with 125-MPH (201-km/h) winds and deluging rains on December 1–2, 2006. More than 500 fatalities were recorded, with another 398 listed as missing. The fourth major typhoon to have struck the Philippines during the 2006 typhoon season, the disaster forced then president Gloria Macapagal Arroyo to declare a state of "national calamity" with massive flooding and volcanic mudslides. Durian's rains caused widespread flooding and volcanic mudslides in the province of Albay. According to the Red Cross, rescue craft were used to convey some 8,900 families to 305 EVACUATION shelters. In addition, an astounding 66,719 people were left without shelter.

In order to better quantify and record the size and strength of incoming typhoons, the Philippine Weather Bureau has adopted a five-point scale of typhoon intensity. Modeled on the SAFFIR-SIMPSON SCALE (the system used to define the destructive potential of North Atlantic hurricanes), the Filipino scale details the most important elements of typhoon generation, including wind force—Weak, Moderate, Strong, Severe, and Extraordinarily Severe—and also storm size itself—Very Small, Small, Medium, Large, and Very Large.

Piddington, Henry (1797–1858) A former captain with the British East India Company and later curator of the Museum of Economic Geology in Calcutta,

INDIA, Henry Piddington remains one of early cyclonic history's greatest scholars. The first to apply the word *cyclone* to tropical systems in the Indian Ocean and BAY OF BENGAL and the pioneering inventor of the famed horn cards that allowed mariners to easily chart wind patterns in approaching TROPICAL CYCLONES, Piddington spent the latter half of his life investigating and compiling the tracks of more than 400 Indian Ocean CYCLONES. Between 1839 (when the East India Company tapped him to organize their newly founded meteorological service) and 1855 (when he went blind), Piddington published no fewer than 40 treaties on Bay of Bengal and Indian Ocean cyclones in the *Journal of the Asiatic Society of Bengal*. One of the first systematic efforts to record the characteristics and behavior of Indian Ocean storms, Piddington's writings are still consulted on a regular basis today, although some of his assessments—such as the theory that tropical cyclone activity in the Indian Ocean was due to "Sumatran Volcanoes" and "traceable [to] the volcanic centers of Cheduba"—can only seem quaint, even absurd when judged in light of modern scientific knowledge.

pinwheel cyclones   This term is used to describe the meteorological condition that ensues when two TROPICAL CYCLONES closely approach one another and then begin to spin—or pinwheel—around a shared rotational point. First identified in the early 1920s as occurring in TYPHOONS and subsequently defined worldwide as the FUJIWARA EFFECT, pinwheel cyclones are as rare to spot as they are spectacular to watch; on average, weather SATELLITES observe pinwheel cyclones in the NORTH ATLANTIC and Indian Oceans once every 15–20 years, and over the stormy expanses of the NORTH PACIFIC OCEAN once every three to five years. Continuously tracked by visible and infrared imagining technology, the gliding whirl of pinwheel cyclones in motion has reminded some meteorologists of an atmospheric ballet or perhaps an aerial waltz. In point of fact, however, the physical interaction between two pinwheeling tropical cyclones is closer in character and consequence to a boxing match in which both mature-stage systems slowly circling each other until dominance by one is asserted and they either drift apart or merge. Initially drawn to within 600 miles (1,200 km) of one another by a string of meteorological and climatological factors (including different forward speeds and the presence of conjoining STEERING CURRENTS), a pair of counterclockwise-spinning tropical cyclones in the Northern Hemisphere will develop into a pinwheel circulation when their respective upper and lower-level wind patterns begin to collide, and turbulently attempt to meld. It is over this point of ocean, where the circula-

tory leverage of one system's DANGEROUS SEMICIRCLE comes into contact with the circulatory resistance of the NAVIGABLE SEMICIRCLE of the other, that a common rotational center is formed. Satellite photographs of eastern North Pacific pinwheel cyclones indicate that the clockwise-direction OUTFLOW LAYERS of both systems will further join, forming a sort of anticyclonic umbilical cord that serves to strengthen the venting of both EYES. The presence of this upper-level linkage, which extends from eye to eye, is often indicated by a long S-shaped line of CIRROSTRATUS CLOUDS or by the frozen ice crystals that are formed at high altitudes. Depending on the size and intensity of the individual tropical cyclones involved, this cyclosynchronistic existence can last only a matter of hours or as long as three or four days. Tropical cyclones with comparable meteorological characteristics have a much greater chance of creating a lengthy pinwheel period because the resistance of one system is less likely to be immediately engulfed by the leverage of the other. In the event that two tropical cyclones of differing size and intensity come into close proximity—such as in the example of a large, fast-moving HURRICANE overtaking a slower, much smaller TROPICAL STORM—the weaker system's CUMULONIMBUS RAIN BANDS will generally be pulled into orbit around the stronger one and quickly subsumed before any pinwheel cycle can begin.

Polly, Typhoon   *North Pacific Ocean–Philippines, December 7–14, 1956*   A relatively weak TYPHOON, Polly released sustained 100-MPH (161-km/h) winds and flooding 11-inch (279-mm) PRECIPITATION counts across the northern PHILIPPINES on December 13, 1956. Raging flash floods accompanied by mudslides and BUILDING collapses killed 79 people on central Luzon Island and tallied damage estimates in excess of $25 million.

precipitation   This is the meteorological process wherein moisture in the air cools and condenses and then falls back to earth as rain, snow, or sleet. Understood by scientists, historians, and emergency relief authorities to be one of a TROPICAL CYCLONES's most notable physical traits, precipitation counts in TROPICAL STORMS, HURRICANES, TYPHOONS, and CYCLONES are often recorded in inches (centimeters), or even feet (meters) and generally prove very destructive to those areas over which they are measured. During an average year, hundreds of lives around the world are lost to the flooding rains associated with tropical cyclones, and millions of dollars worth of structural damage—from washed-out bridges and roadways to submerged farms and towns—are all too frequently assessed in their wake. Spawned by the immense, water-laden CUMULONIMBUS CLOUDS that comprise most of a tropical system's heavy cloud mass, precipitation counts are determined in part by both the particular stages of a tropical cyclone's development (Formative, Deepening, Mature, and Decay) and by geographical features such as mountains, valleys, and plains through which the soggy system passes. Although even the most sophisticated of rain gauges sometimes finds it difficult to accurately measure wind-driven downpours, documented rainfall observations in all classes of tropical cyclones have nonetheless indicated captured amounts as little as 0.5 inches or veritable cascades of 45 inches (1,224 mm) and higher. On at least three separate occasions, North Atlantic hurricanes over NEW ENGLAND have even produced snow, triggered by a sudden infusion of Arctic air into a hurricane's fringe circulation. One such instance, the famed Snow Hurricane of October 9, 1804, saw portions of central Massachusetts blanketed by 10 inches (254 mm) of drifting snow, while a later example, on October 3, 1841, forced residents in one Connecticut town to attend their town meeting in a sleigh.

On a most basic level, precipitation in a tropical cyclone is a by-product of a storm's heat mechanism, a tangible result of the relentless intake of warm, moist air that characterizes the inner workings of a tropical system's immense, wind-scaled CONVECTION cells. Positioned over the humid waters of Earth's tropical zones, mature and developing cyclones of all classes feed on the high evaporation rates usually found in such aquatic environments, drawing dozens of moisture-laden air currents through the EYE of a respective system before exhausting them into the atmosphere's upper reaches. Steadily cooling as they rise, these currents of water vapor eventually condense into rain clouds, folding into the towering EYEWALL of a hurricane, typhoon, or cyclone as gray-black cumulonimbus cloud piles. These, in turn, will continue to deepen, growing darker and more unstable as millions, even billions (2.6 billion in the case of the August 1899 hurricane in PUERTO RICO) of tons of water vapor further accumulate within them. When, in time, these cumulonimbus clouds become so saturated that they can no longer contain any additional moisture, they will begin to drop their cargo as rain, heavy downpours that have tended to be the most torrential ever witnessed on the planet. During the course of Saxby's Gale, on October 4, 1869, two inches of rain fell on western Massachusetts in two hours; the colossal SAN FELIPE HURRICANE of 1928 dropped more 30 inches (762 mm) of rain on the island of Puerto Rico in the course of a single afternoon.

On a more specialized level, however, an understanding of precipitation in a tropical cyclone needs to take into account a tropical system's symbiotic nature, those subtle meteorological processes by which hurricanes, typhoons, and cyclones simultaneously sustain and grow. As warm, humid air currents within the eye and eyewall rise toward the tropopause and cool, their latent heat energy is released through condensation. Because tropical lows and their attendant thunderstorms thrive on warm air, this heat immediately adds itself to the powering of a tropical cyclone's convection cells, or the giant thunderstorm circulations caused by the rise of warm air currents and the fall of cool ones. At the same time, the water vapor created through condensation immediately adds itself to the cumulonimbus cloud mass that forms the physical structure of a tropical cyclone, thereby increasing the height of the cloud towers that ring the system's fluelike eye. This, in turn, aids in establishing an OUTFLOW circulation by more efficiently venting the eye, allowing greater quantities of warm, moist air to be ingested by the system's intake. In this way the saga of precipitation in a hurricane is like that of smoke in a chimney: Rising because it is hot, the smoke over time coats the inside of the chimney with several layers of tar, in a sense adding to its existing mass. Moisture in a tropical cyclone does much the same thing, only the twin processes of evaporation and condensation further drive the very circulation of the system, permitting it to feed, build, and sustain.

For this reason the largest quantities of precipitation in a tropical cyclone are most often found in its eyewall, within that thick ring of clouds that immediately surrounds the eye. Just as a tropical cyclone is always in flux, its enormous cloudbanks continually merging with one another, areas of heavy precipitation will similarly shift, gathering to the right or left of its forward direction before sliding back and spreading out, or dividing into smaller, localized downpours. Sometimes in the case of large, powerful Category 3 and 4 tropical cyclones, the greatest amounts of precipitation will be observed beyond the eyewall, on the outer fringes of the storm. In other instances Category 3 and 4 systems will carry their heaviest precipitation right around the eye, seemingly using it to feed even more moisture into the vortex.

Although the reasons for such anomalies are not fully known, it is fairly well understood that precipitation counts in tropical cyclones are in some way linked to the intensity and organization of a particular storm at a certain time. For years tropical meteorologists were astounded to find that tropical storms, those cyclonic systems with sustained WIND SPEEDS of between 39 and 73 MPH (63–118 km/h), dropped on average more rain than a mature cyclone did and that, based on this observation, it could be held that the higher the wind speed, the smaller the rainfall counts would be. Surmising that precipitation levels were somehow determined by the efficiency of the convective cells, intensive aerial and ground studies of rainfall in tropical cyclones indicated that while certain standards could be applied to rainfall measurements between different classes of storms, no ready-made formula for correlating a cyclonic system's CENTRAL BAROMETRIC PRESSURE or wind speed with its potential precipitation counts existed. There were simply too many extraneous factors—the forward speed of the storm, the velocity of its winds, the accuracy of the rain gauge—to ever know for certain just how much precipitation a typical tropical cyclone contained.

A review of history's best-documented cyclonic strikes, however, does at least give some idea of the severity of the precipitation counts that can be expected from certain types of storms. At times driven horizontal by winds of 111 MPH (179 km/h) or higher, rain in tropical cyclones has trickled through stone walls 14 feet (5 m) thick, bombarded the foliage from trees, and felt like grapeshot against the naked skin of those unfortunate victims forced to counter the fury head-on. During the Coastal Hurricane of 1806, between 18 and 36 inches (458–945 mm) of rain reportedly fell on Cape Cod; the GREAT SEPTEMBER GALE of 1815 carried so much rain inland that several freshwater streams across New England were temporarily turned into poisoned salt baths. The infamous Cuban Hurricane of 1844 swept Key West, FLORIDA, with some 15 inches (372 mm) of rain over a three-day period. An Indian Ocean cyclone on February 21, 1896, deluged the island of Mauritius with up to 47 inches of rain, while a waterlogged 1913 typhoon at TAIWAN dropped 81.5 inches (2,050 mm) of rain on the volcanic outcropping in just two days' time. The very severe Corpus Christi Hurricane of September 14, 1919, inundated the TEXAS port city with nearly 14 inches (350 mm) of precipitation, causing extensive damage to streets and piers, and in August 1960 Hurricane Hiki battered HAWAII with 52-inch (1,300 mm) cloudbursts, sudden downpours that caused Kauai's Waimea River to quickly overflow, bringing catastrophic flooding to the rich macadamia nut plantations that shared its fertile banks. Australia's Mackay Cyclone produced precipitation counts of 56 inches (1,411 mm) in January 1918. This caused severe flooding along the banks of the Pioneer and Fitzroy Rivers. Several hundred residences in Rockhampton were destroyed, and six lives lost.

Cyclone TRACY (1974) dropped some 7.67 inches (195 mm) of precipitation on Darwin, Australia, in less than nine hours, a record for the time.

pressure gradient   This term is given to that incremental decrease in barometric pressure that occurs over a certain distance of Earth's atmosphere. In a TROPICAL CYCLONE, where extremely low BAROMETRIC PRESSURES are observed in the EYE, or center of the system, the pressure gradient is measured as the difference in air pressure between the outer perimeter of the storm (usually set at the 1,000 MILLIBAR—29.53 inches—range) and its distant eye. Primarily responsible for the creation of ADVECTION currents, or wind, the steepness of a tropical cyclone's pressure gradient can range from 0.5 inches to 2.5 inches of mercury or more, with a proportionate increase in wind velocity being the direct result. Because the average HURRICANE, TYPHOON, or CYCLONE measures between 300 and 400 miles (600–790 km) across, the pressure gradient, following the shape of the circular ISOBARS, will only extend through 150 to 200 miles of its total cloud mass. However, within this relatively short distance, the pressure gradient can become quite steep, rapidly falling an inch or more over the space of just a few miles. In an effort to equalize this sharp drop, surrounding air currents with higher pressures will quickly move in to fill the gradient, producing ever-increasing wind speeds as they race along the barometric slide toward the center of the storm. Just as a ball rolls faster down a steep hill, the wind speed will continue to follow a tropical system's pressure gradient throughout its lifetime, slowing when the gradient rises, plummeting when it abruptly tumbles. Microbaragraph traces of mature tropical cyclones have revealed a funnellike convergence that becomes almost vertical as the pressure gradient progresses toward the storm center. Slowly tapering as it moves inward from the cyclone's fringes, the pressure gradient suddenly shears off as it enters the confines of the EYEWALL, coming together to form an asymmetrical V. It is along this course that the hurricane's fiercest winds are to be found, roaring away until the two halves of the gradient come together within the calm limits of the eye.

Project Cirrus   Conceived in 1946 by scientists from the General Electric Corporation and implemented during the 1947 HURRICANE SEASON by the National Weather Bureau, Project Cirrus sought to weaken dangerous HURRICANES by seeding them with dry ice. A radically new approach to the perennial threat of TROPICAL CYCLONES, Project Cirrus's attempt to moderate the destructive potential of cyclonic systems by reducing their PRECIPITATION counts and wind speeds while they were still harmlessly positioned over the open sea, represented a bold change in the meteorological and civil defense strategies of the time. Assuming an anticipatory, rather than a reactionary, stance to the hurricane problem, Project Cirrus was based on the premise that a hurricane's fury would be lessened if its violent thunderstorm bands could be made to release their moisture before landfall. Although preliminary experiments had indicated that silver iodide crystals might work just as well—if not better—than dry ice in prompting a hurricane's EYEWALL thunderstorms to precipitate, scientists hoped that a dry ice seed would lower an eyewall's temperature to the point that its cloud banks would condense into rain or snow, thus dismantling its convective circulations and shutting down the entire system.

RELATION OF PRESSURE CHANGE TO TROPICAL CYCLONE DISTANCE

Hourly rate of fall (inches)	Distance to center (miles)
0.02–0.06	250–150 miles
0.06–0.08	150–100 miles
0.08–0.12	100–80 miles
0.12–0.16	80–50 miles

On October 13, 1947, the third hurricane of that season—a fairly powerful Category 2 system (28.76 inches, 974 mb), then off the coast of northern FLORIDA—was seeded with 80 pounds of dry ice with no measurable effect other than an abrupt course shift from northeast to southwest. While scientists speculated that the experiment may have failed because the dry ice evaporated too quickly, the unfazed tempest delivered 100-MPH (161-km/h) winds and torrential rains to GEORGIA on the afternoon of October 15 and a resultant stormy backlash of criticism to those individuals and organizations associated with Project Cirrus. While no lives were reportedly lost to the hurricane's passage over Savannah, its expensive course change was incorrectly attributed to the influence of the failed dry ice experiment, dampening all enthusiasm for further continuance of the Cirrus project. It would not be until late August 1958 when a swarm of specially modified aircraft pollinated Hurricane Daisy with several canisters of silver iodide crystals that vital seeding experiments on tropical cyclones would be resumed. In 1961 these ongoing missions were redefined as PROJECT STORMFURY.

Project Stormfury A second, far more ambitious program of HURRICANE moderation than the earlier PROJECT CIRRUS, Project Stormfury was a series of federally funded experiments conducted between 1961 and 1983 in the hopes of diminishing deadly hurricanes by seeding their CUMULONIMBUS CLOUD towers with silver iodide crystals. A collaborative effort by the United States Weather Services and the United States Navy, Project Stormfury aimed to stall or even turn back the development of a TROPICAL CYCLONE

by abruptly reconfiguring its PRESSURE GRADIENT and thus expanding the diameter of its EYEWALL. Because studies had suggested that the most intense hurricanes usually possessed small, tightly wound eyewalls, scientists theorized that by releasing flares and canisters filled with silver iodide into the eyewall's cloud tops, they could force the dangerous ring to collapse, reforming at a much greater distance from the EYE. This would, in turn, lessen the steepness of the system's pressure gradient, or the distance over which BAROMETRIC PRESSURE decreases, thereby ameliorating the hurricane's high winds.

Although the theory's preliminary test run in August 1958 proved somewhat of a disappointment—Hurricane Daisy's Category 2 eyewall survived unscathed, not at all affected by the bombardment of silver iodide pellets that had been delivered to it—the decision was subsequently made to seed the fifth storm of the 1961 HURRICANE SEASON, Esther, with 1,048 pounds of silver iodide while the September hurricane was still 1,500 miles (2,700 km) southeast of FLORIDA. This accomplished, it was observed that while a segment of Esther's eyewall was in fact broken down, its sustained 130-MPH (209-km/h) winds reduced to 119 MPH (192 km/h), the wound was hardly mortal. Indeed, within two hours of the 18,000-foot (6,000-m) seeding, the hurricane had not only recovered but began to intensify further as well. Within two days, Hurricane Esther's maximum winds were topping 150 MPH (241 km/h), leading scientists to again ponder the well-publicized difficulties inherent in any seeding operation. Brushing up the eastern seaboard of the United States as a Category 4 hurricane of significant intensity, Esther eventually blew itself out over the frigid fields of the NORTH ATLANTIC—but not before stalling Project Stormfury in its tracks for another two years.

On September 13, 1963, Hurricane Beulah (the first of three tropical cyclones to bear that name) was seeded with silver iodide crystals while safely located over the western Atlantic Ocean. A large Category 1 system, Beulah's eyewall resisted the dozens of silver iodide flares fired into it by military aircraft, prompting Stormfury scientists to deem this initial assault a failure. A resilient hurricane in any number of respects, Beulah was still in existence when the decision was made two weeks later to recommence Stormfury's seeding experiments. This time the results were more encouraging. Beulah's 19-mile (25-km) eyewall widened to 32 miles (48 km), while the hurricane's sustained winds dropped from 89 MPH (143 km/h) to 80 MPH (129 km/h). Although the changes were only temporary, it was clear to those involved that silver iodide seeding could indeed bring about measurable modifications in hurricane characteristics, thereby providing the Stormfury program with a much needed boost in morale—and in its crucial federal funding.

Despite both growing support for the continuance of Stormfury's mission and the spate of treacherous hurricanes that sliced across the North Atlantic Ocean and the GULF OF MEXICO during the intervening years, no seeding experiments were carried out between 1964 and 1968. Always mindful of the criticism that had been heaped upon earlier unsuccessful attempts to modify hurricanes, Project Stormfury's guardians had devised a narrow series of meteorological and geographical standards by which candidates for future seeding were to be judged. Fearful that a seeded hurricane might drop even larger quantities of PRECIPITATION than an unseeded one, scientists agreed that modification experiments could only be conducted on those TROPICAL STORMS or hurricanes that had no more than a 10 percent chance of approaching within 50 miles (78 km) of a densely populated shoreline. They further decided that only weaker, less-organized Category 1 and 2 systems would be seeded, thus bettering the chances of delivering a fatal blow to a storm's deepening configuration the first time out. One did not have to be a public relations wizard to realize that a single success of this magnitude would not only prove the viability of the mission once and for all but also protect Stormfury's tenuous funding from possible future budget cuts.

So for the next five years, Project Stormfury quietly waited for the perfect deep-water hurricane to appear, weathering the inevitable whirlwind of speculation and concern on the part of the public that its self-imposed rules had inspired. One of the most fascinating meteorological experiments of the age, Project Stormfury soon sired a well-publicized brood of ideas, plans, scams, and schemes for the moderation of cyclonic systems. Ranging from the useless detonation of a nuclear device within a hurricane's eye to the impractical spreading of plastic sheets across the sea and the bombardment of a storm's eyewall with tons of ice cubes, a majority of the suggestions were so outlandishly sincere that they could only be viewed as an indication of how greatly the public feared tropical cyclones and of their desire to see the destructive effects of such threats diminished. Fully cognizant of the stakes involved, Project Stormfury scientists and meteorologists cautiously scrutinized each and every one of the 45 tropical cyclones (28 of them mature hurricanes) that cropped up, eventually discounting such noted tempests as CLEO, Dora, HILDA, and Isabell (1964) because of their proximity to Caribbean islands, 1965's BETSY and 1966's ALMA due to their predicted landfalls in Florida, and the second Hurricane Beulah (1967) because of its sheer size.

It was not until the remarkable 1969 hurricane season, when no fewer than 18 tropical cyclones originated over the North Atlantic, CARIBBEAN SEA, and Gulf of Mexico, that Project Stormfury found the right system to seed: Hurricane Debbie, some 1,400 miles (2,200 km) southeast of Miami, Florida, and moving east-northeast, away from the principal islands that dot the region. A moderately powerful hurricane with sustained wind speeds of 100 MPH (161 km/h), Debbie was first seeded on August 18, 1969, with more than 1,000 pounds of widely dispersed silver iodide pellets. Considered the most successful of all the Stormfury experiments, Debbie's top winds later fell to 70 MPH (113 km/h), while its thickening eyewall suffered numerous breaches, causing a near-mortal rise in the storm's central pressure.

On August 19, as an unexpected course change to the northwest brought Debbie within 1,000 miles (1,600 km) of landfall on the United States's east coast, the decision was made at the NATIONAL HURRICANE CENTER (NHC) in Miami to again seed the hurricane. Although Debbie's 125-MPH (183-km/h) winds and landward drift made it an undesirable hurricane to seed, it was not a politic time for Project Stormfury to fall back on old rules or appear timid or disinterested; just three days earlier, the third hurricane of the season, CAMILLE, had grown into a Category 5 legend while traversing the northwestern Gulf of Mexico. Raging ashore in LOUISIANA and MISSISSIPPI on the night of August 17, Camille's 200-MPH (322-km/h) winds, driving rains, and horrendous STORM SURGE killed more than 100 people, making it one of the worst hurricane disasters in living memory.

At one time a serious candidate for seeding operations, Camille was the sort of tragedy that left Stormfury scientists playing a rueful hand of retrospection. Perhaps an effort to seed Camille would have succeeded, sufficiently disrupting its sophisticated yet fragile construction enough to moderate its most violent aspects. Concurrently, there was the ongoing fear of negative publicity, of the pundits who would ask how it was that the second most powerful hurricane in U.S. history could come ashore at a time when Project Stormfury was allegedly in full effect. In order to protect the legitimacy of the Stormfury mission, it was imperative that Debbie, then well on its way to becoming another Camille, be met with every weapon in the Stormfury arsenal.

Accordingly, the second seeding of Hurricane Debbie took place on the morning of August 20, 1969. Continuing well into the night, squadrons of land-based jets released nearly 2,000 pounds of silver iodide crystals into the Category 3 system. Although little change was noted in the hurricane's CENTRAL

BAROMETRIC PRESSURE, its sustained winds did decrease to 110 MPH (177 km/h), and a temporary expansion of its eyewall was observed. Within a matter of hours, however, Debbie had fully recovered its fury and would thereafter maintain its strength until the dissipating expanses of Newfoundland, CANADA, put an end to it on August 25.

Stormfury's government funding, which was slated for renewal at the close of the 1971 hurricane season, permitted scientists to conduct two more seeding runs on marathon hurricane GINGER on September 26 and 28, 1971. One of the longest-lived tropical cyclones on record, Ginger spent 31 days wandering across the mid-Atlantic, 20 of them as a mature hurricane. Seeding of its RAIN BANDS occurred as the rainmaker neared the coast of NORTH CAROLINA and was largely geared toward experimental goals. As it was, neither Ginger's eyewall nor its central barometric pressure registered any change, and there was only a slight decrease in its 75-MPH (121-km/h) wind speeds.

Its findings deemed inconclusive, the mission to seed Hurricane Ginger was the last ever conducted by Project Stormfury. Officially ended in 1983, Stormfury did not fade quietly but, in keeping with its tumultuous history, perished beneath a seeming barrage of political criticism. On July 1, 1983, MEXICO's foreign minister announced that U.S. hurricane reconnaissance aircraft would be prohibited from entering Mexican airspace until certain allegations surrounding Project Stormfury's 22-year-old experiments had been investigated. Believing that a prolonged drought then in progress over northern Mexico had been caused by misguided experimentation with rainfall patterns, the misinformed foreign minister created such a furor with his statements that the dormant but controversial operation was summarily cancelled. While only the gullible would ever accept the accuracy of the charges, the minister's outburst did raise an interesting issue regarding a possible side effect of hurricane moderation—that of reduced precipitation counts in those areas that depend upon tropical cyclones for much of their annual rainfall needs.

Puerto Rico *Northeastern Caribbean Sea* An island of ancient volcanic mountains, steep valleys, and fertile coastal plains, Puerto Rico has endured a long, sometimes catastrophic history of TROPICAL CYCLONE activity. Located on the northeastern crown of the CARIBBEAN SEA, Puerto Rico is the smallest of the three major islands—CUBA and HISPANIOLA being the other two—that comprise the Greater Antilles group. The chain's position between latitudes 16 and 23 degrees North situates easterly Puerto Rico directly in the path of the mid- to late-season HUR-

RICANES that follow the trade winds across the North Atlantic Ocean from Africa's Cape Verde Islands. Between 1515 and 2006, Puerto Rico's sandy beaches, 4,000-foot (1,300-m) mountain ranges, and historic stone fortresses were directly affected by 80 documented TROPICAL STORMS and hurricanes. Of these, at least 41 were major hurricanes. On six different occasions between 1530 and 1899, Puerto Rico even withstood several hurricanes in quick succession, as happened in 1537 when two destructive storms came ashore within one month of each other, and in 1772 when two hurricanes struck in less than three weeks' time. Other historic double strikes include those of July 23 and August 21, 1812 and the twin October tempests of 1891. Between July 26 and August 31 of 1530, Puerto Rico was struck by three hurricanes, the last of them a storm of significant intensity. More than half of the island's houses were leveled, and several hundred head of cattle were drowned. Seven years later, another trio of hurricanes battered Puerto Rico within two months' time. Of even greater severity than the 1530 strikes, the three 1537 storms ran a dozen merchant ships onto the rocks, inundated seaside villages, and temporarily halted Spanish development of the island. Hundreds of people reportedly perished, many of them enslaved Taino workers.

Since 1780, Puerto Rico has been affected by at least one June hurricane, although this particular storm, which occurred on June 13th, was of fairly minor intensity and caused little damage except to fragile crops. In July, the probability of a hurricane landfall in Puerto Rico steeply increases, with seven reported strikes between 1515 and 1952. One July hurricane, in 1515, killed several hundred of the island's indigenous Taino people, while another—christened *Santa Ana* after the Saint's Day on which it blasted ashore—wrought enormous damage across the entire length of the island on July 26, 1825. Nearly 7,000 BUILDINGS were reportedly destroyed, while 374 people lost their lives to the raging winds and rains of the Santa Ana storm, considered among the worst to have ever affected the island during the 19th century.

But it is in August and September, when the first of the Cape Verde hurricanes begins to arrive at the western terminus of HURRICANE ALLEY, that the number of Puerto Rican hurricanes jumps to a fearsome 15 and 19, respectively. Notorious midseason hurricanes include:

CHRONOLOGY OF NOTORIOUS MIDSEASON HURRICANES
The First San Mateo Hurricane of September 21, 1575: The 10th documented hurricane to have lashed Puerto Rico since 1515, San Mateo's fierce

winds and thudding breakers hounded shipping, uprooted trees, flattened houses, and turned inland valleys into rain-gorged sluiceways. Scores of lives were said to have been lost, many of them sailors on Spanish treasure ships at sea.

The Unnamed Hurricane of 1720:   Although few historical details survive to commemorate the passage of this storm (including the day and month on which it occurred), it is known that a 50-gun Spanish warship was driven ashore on Puerto Rico's southeast coast and wrecked, drowning more than 500 seafarers.

The Second San Mateo Hurricane, September 21, 1804:   The second Puerto Rican hurricane to bear the name, San Mateo II visited enormous destruction on plantations and seaports, collapsing buildings, splintering piers, and breaching stone jetties. More than ten merchant ships were overwhelmed by the hurricane's huge waves and strident winds while rounding the island's rock-strewn west coast and sunk, taking several dozen sailors to the bottom with them.

The Los Angeles Hurricane of August 2–3, 1837:   One of the most destructive Caribbean hurricanes on record, Los Angeles's CENTRAL BAROMETRIC PRESSURE of 28.00 inches (948 mb) and sustained 120-MPH (193-km/h) winds razed scores of houses, drove several ships aground at San Juan, and took an estimated 296 lives.

The Santa Elena Hurricane of August 18, 1851:   A relatively minor hurricane, Santa Elena brought 85-MPH (137-km/h) winds to San Juan and heavy rains to Arroyo, on the northwestern corner of the island. No casualty figures from this storm exist, indicating that the DEATH TOLL, if any, was small.

The San Ciriaco Hurricane of August 8, 1899:   An intense Category 4 system (a central pressure reading of 27.75 inches (939 mb) was observed at landfall in the port town of Arroyo), San Ciriaco's 135-MPH (217-km/h) winds and 13-foot STORM SURGE claimed more than 3,000 lives on the island, while some $20 million in property losses were assessed.

The Second San Felipe Hurricane, September 13, 1928:   Generally regarded as the most violent North Atlantic hurricane to have struck Puerto Rico during the first half of the 20th century, San Felipe's sustained 150-MPH (241-km/h) winds delivered 30-inch (750-mm) PRECIPITATION counts to the island's central highlands in just under 12 hours. More than 19,000 buildings, from residences and hotels to factories, schools, and churches, were utterly demol-

ished by flash floods and mudslides, leaving some 284,000 people homeless. More than $50 million in losses were submitted, while 374 people (a conservative estimate) perished.

The San Ciprian Hurricane, September 26, 1932:   Roaring ashore just four years after the devastating passage of the San Felipe Hurricane, gargantuan San Ciprian's 120-MPH (193-km/h) winds, 11-inch (279-mm) rainfall counts, and flooding storm surge took some 225 lives, injured another 3,000 people, and left at least 75,000 others homeless. More than $30 million in damages were tallied, making San Ciprian the second most-expensive hurricane to have affected the island up to that time.

Hurricane Lili, December 19, 1984:   The fifth documented tropical cyclone to have developed in the North Atlantic after the official hurricane season's end on November 30, Hurricane Lili originated 780 miles (1,200 km) southeast of Puerto Rico on the afternoon of December 12. Quickly borne northwest by the trade winds, Lili entered the confined waters of the southeastern Caribbean Sea and rapidly intensified. Generating 79-MPH (127-km/h) winds and five-inch precipitation counts on reaching southeastern Puerto Rico on the morning of December 19, Lili downed hundreds of trees, toppled a radio tower outside of San Juan, and collapsed several dozen houses. Fifteen-foot seas pounded the port towns of Guayama and Ponce, flooding waterfront warehouses and casting small watercraft aground. While no lives were lost to Lili's postseason flowering, some $25 million in property damage did result.

Hurricane HUGO, September 18, 1989:   One of the most severe Category 4 hurricanes in Puerto Rican history, Hugo's 133-MPH (214-km/h) winds raked the northeast coast of the island, killing one person and causing nearly $2 billion in damage to houses, hotels, and banana and coffee harvests. The small offshore communities of Culebra and Vicques were particularly ravaged by Hugo's 150-MPH (241-km/h) gusts, with 90 percent of the buildings on both islands swept into the sea.

Hurricane LUIS, September 6, 1995:   The most powerful hurricane to have transited the Caribbean Sea since 1989, Luis's offshore EYEWALL passed 100 miles (161 km) north of Puerto Rico, pounding its northeast shores with tropical storm-force winds and heavy rains. Bearing down on the island with 125-MPH (201-km/h) winds on September 5, Luis's potential fury so frightened inmates at a minimum-security prison near San Juan that 48 of them launched a successful escape. Later recaptured, they stated that they had only wanted to be with their families when the dreaded hurricane finally came ashore. Two fatalities were reported.

Hurricane Hortense, September 10, 1996: The eighth-named tropical cyclone of the 1996 HURRICANE season, Hortense is better remembered in Puerto Rico for its 28-inch (700 mm) rainfall counts than its fairly tame 80-MPH (129-km/h) winds. A classic Cape Verde hurricane, Hortense received its start as a tropical depression 240 miles west of the Cape Verde Islands on September 1. Feeding on the mid-Atlantic's warm waters as it slowly spiraled westward, Tropical Storm Hortense socked the British and U.S. VIRGIN ISLANDS with 70-MPH (113-km/h) winds and torrential rain showers on the afternoon of September 8. Recurving northwest at 10 MPH (16 km/h), Hortense fitfully intensified as it crossed the northeastern Caribbean basin on the morning of September 9. Upgraded to minimal hurricane status while still centered 110 miles (230 km) southeast of Puerto Rico, Hortense continued its lukewarm recurve northwest, sometimes stalling in its tracks, other times crawling forward at speeds of between four MPH and seven MPH (6 km/h–11 km/h). As the new hurricane's central pressure continued to drop, as its maximum winds were boosted to 80 MPH (129 km/h) by late-afternoon on September 9, HURRICANE WARNINGS were posted across Puerto Rico. Almost immediately, some 7,000 people took refuge in Red Cross shelters, converted schools, and sturdy government buildings. The San Juan International Airport was closed, its incoming flights promptly diverted to Miami. Hours before Hortense's Category 1 eyewall even came ashore in southwestern Puerto Rico, widespread power outages were being reported across the southeastern reaches of the island, with 8,000 families in the greater Guayama area suddenly cast into the thickening darkness by Hortense's leading rain squalls. Creeping northwest at five MPH (8 km/h), Hortense's 470-mile-wide (640-km) mass threw 30-foot (10-m) breakers against Puerto Rico's volcanic coves and concrete seawalls, causing a coastal road on the offshore island of Vieques (Crab Island) to be washed out shortly before midnight on September 9. Slicing across the southwestern tip of the island on the morning of September 10, Hortense's hurricane-force winds, relentless rains, and four-foot (1-m) storm surge brought enormous destruction to Guayama and Ponce: 2,000 houses were destroyed, a majority of them flooded with more than five feet (2 m) of water. An additional 2,419 buildings were seriously damaged, their windows shattered and their corrugated-tin roofs lifted by Hortense's

100-MPH (177-km/h) gusts. More than 24 inches (610 mm) of rain over a 48-hour period turned Ponce's network of streets into a single lagoon reefed with flooded cars and barricading mounds of debris. In the capital of San Juan, highways were gradually transformed into raging canals by Hortense's accumulating rains; at least five overpasses were carried away by the torrent. Along the slopes of the Cordillera Central mountain range, dozens of small villages were besieged by mudslides and landslips, soupy mires triggered by Hortense's hourly deluge. Five railroad bridges were swept away in a matter of hours. Some 1,000 people in the San Juan suburb of Dorado, driven from their homes by rising floodwaters, found tenuous refuge on rooftops and in trees while they awaited rescue by military helicopters. Off the eastern town of Humacao, the cargo ship *Isabella*, sluggishly wallowed in Hortense's 30-foot (10-m) seas. As a Coast Guard helicopter braved Hortense's 55-MPH (89-km/h) gales, the vessel's 11-man crew clad in life jackets huddled on deck. After several hours of grueling, danger-fraught work by the helicopter and a swimmer, all 11 crew members were safely rescued. Fourteen other men, women, and children in the U.S. commonwealth, most the victims of drowning, lost their lives, and there were some $300 million in insured property losses caused by Hurricane Hortense from structural damage to waterlogged coffee and plantain harvests, making it the most expensive tropical cyclone to have affected the island since 1989's Hurricane Hugo.

Come October and November, when the BERMUDA HIGH's seasonal subsidence to the south-southeast confines Atlantic tropical cyclone activity to the western Caribbean Sea, Puerto Rico's 500 year span of hurricane landfalls yields eight October hurricanes and two more in November. Of the documented October storms, three were systems of substantial size and intensity. One of them, the San Narciso Hurricane of October 29, 1867, remains one of the deadliest hurricanes in Puerto Rican history, with between 1,000 and 1,200 casualties to its credit. Jolting ashore near the town of Caguas as a tightly wound Category 3 killer, San Narciso inundated the harbor at San Juan with a ruinous 15-foot (5-m) storm surge. Hillside neighborhoods vanished as first the hurricane's 127-MPH (204-km/h) winds followed by its intense rains, stripped buildings from their foundations, flayed the foliage from trees, plowed under coffee and banana plantations, and spawned terrifying floods. San Narciso went on to claim an additional 600 lives in the VIRGIN ISLANDS on October 30. On August 12, 2003, tropical storm Odette's outer bands brought heavy rainfall and flooding to Puerto Rico, sending dozens of people to shelters.

Q Owing to a shortage of available male and female names beginning with the letter *Q*, there have been no *Q* HURRICANES, TYPHOONS, CYCLONES, or TROPICAL STORMS christened in the world's principal storm zones. When the practice of naming NORTH ATLANTIC tropical cyclones began in 1953, the initial lists did include the name *Queen* (1953), *Queen* (1954), and *Queena* (1955). None of these names were used, however, and by the start of the 1956 Atlantic hurricane season they had been omitted.

R

radar   In the broadest sense of the definition, *radar* is a system or device by which radio waves are first transmitted to an object, are in turn reflected by it, and then are received and measured to gauge their respective length and speed. In its meteorological sense, *radar* refers to the scores of radars and radar systems installed in coastal regions and on board aircraft between 1946 and 2006 and used to study and monitor TROPICAL CYCLONE activity over the NORTH ATLANTIC, NORTH PACIFIC and Indian OCEANS. Some six decades have passed since the first radar image of a tropical cyclone's spiral RAIN BANDS was obtained in 1945. During that time, progressively more sophisticated radar system have not only provided meteorologists and scientists with vast amounts of new visual data on the circulatory characteristics of tropical cyclones but also established the groundwork for unprecedented advances in forecasting accuracy and civil defense preparedness. Supplanted as a tracking tool by the early 1960's development of the first meteorological SATELLITES, radar soon found a new niche in the meteorological research field. Principally used to observe the size and density of a tropical cyclone's constituent cloud banks, radar technology enabled scientists and meteorologists to unlock many secrets regarding one of a tropical cyclone's most destructive yet little understood attributes, its PRECIPITATION distribution.

In September 1945, Major Harry Wexler of the U.S. Air Force Weather Service published the first radar images of a tropical cyclone's CUMULONIMBUS rain bands. Quickly recognizing radar's advantages as a forecasting device, the U.S. Navy transferred 25 surplus radar systems to the U.S. WEATHER BUREAU in early 1946. After modifying the aircraft-borne radars for ground-based applications, the Weather Bureau founded a chain of TORNADO tracking stations in the U.S. Midwest. In 1951, spectacular radar images were obtained by aircraft flying through the EYE of TYPHOON Marge off the coast of JAPAN. Following the destructive North Atlantic HURRICANE SEASON of 1954, the Navy installed an even more powerful, long-range radar station at Cape Hatteras, NORTH CAROLINA, in early August 1955. Established in time to track that year's spate of huge HURRICANES, including CONNIE, DIANE, and IONE, the well-publicized success of the Cape Hatteras station encouraged the U.S. Congress to appropriate additional funds for the procurement of additional radar systems specifically designed for meteorological applications. The first of these new-generation radars, dubbed WSR-57, was installed at the NATIONAL HURRICANE CENTER (NHC) in time for the start of the 1959 hurricane season. Able to survey a 150-mile (261-km) circle of ocean—but ultimately limited in distance by Earth's curvature—the land-based WSR-57 was soon adapted for deployment in HURRICANE HUNTER aircraft, permitting scientists to scan an additional 200,000 square miles of Earth's surface. Within the next 10 years, an additional 55 WSR-57 systems were positioned from TEXAS to NEW ENGLAND at 200-mile (310-km) intervals, thereby placing the entire gulf and Atlantic coasts of the United States under continuous meteorological surveillance. The invention of DOPPLER RADAR in the early 1970s, combined with improvements in computer design and imaging technology, saw the aging WSR-57 radar network replaced by the WSR-88D radar system during the early 1990s. Designed to scan a 250-mile-wide section of coastline, the WSR-88D

Doppler radar remains at the leading edge of tropical cyclone research.

radiometer An instrument that measures the emitted and reflected thermal energy of Earth's surface and, in essence, a sophisticated electronic camera that uses visible and infrared wavelengths to create photographs, the radiometer is most often deployed on weather reconnaissance aircraft and in meteorological SATELLITES to observe the amount of solar heat reflected by Earth, carbon dioxide levels, water vapor distribution layers, and SEA-SURFACE TEMPERATURES. First developed in the early 1960s, radiometers use different wavelength bands to measure individual climatological factors. For instance, when the first of the *Nimbus*-class satellites was launched into orbit in 1966, it carried a five-band Medium Resolution Infrared Radiometer (MRIR) that allowed for measurements to be taken at five levels of the TROPOSPHERE. In October of 1978, the TIROS-N satellite employed line-scan radiometers that produced visible (photographic) and infrared images with a resolution of 1,900 feet. When the GOES-8 satellite went into service in November 1994 it was equipped with a Visible and Infrared Spin-Scan Radiometer (VISSR) that enabled it to observe temperature and humidity levels at 19 different points of the atmosphere.

radiosonde Most often carried up into Earth's TROPOSPHERE by a WEATHER BALLOON, a radiosonde is an instrument that simultaneously observes and transmits meteorological data concerning BAROMETRIC PRESSURE, WIND SPEED, and air TEMPERATURE. In the past radiosondes have been adapted for use on small rockets and fired into TORNADOES in an effort to study their complex circulation patterns; they have also been dropped by HURRICANE HUNTER tracking aircraft into TROPICAL CYCLONES in the hopes of shedding light on PRESSURE GRADIENTS, CENTRAL BAROMETRIC PRESSURE, construction of the EYEWALL, the movement of air currents within the system's EYE, and PRECIPITATION counts. Radiosondes are also deployed on meteorological buoys.

rain bands The name *rain bands* describes the circular bands of CUMULUS and CUMULONIMBUS CLOUDS that radiate outward from the EYE of a TROPICAL CYCLONE. Often referred to as *spiral rain bands*, the PRECIPITATION-producing cloud bands of a tropical cyclone generally measure from three to 25 miles (8–40 km) in width, and between eight and 300 miles (12–483 km) in circumference. Determined in large part by the intensity of a particular tropical system, rain bands can extend from 10,000 to 45,000 feet (3,048–13,716 m) in height and may

As a Hurricane Hunter aircraft carries out its lifesaving mission, the spiral rain bands of an unidentified tropical system are clearly visible beyond. *(NOAA)*

number from three (in a small TROPICAL STORM) to six or seven in a moderately powerful HURRICANE to twelve in a large TYPHOON or CYCLONE. The heaviest rain bands in a mature-stage tropical cyclone are observed around the eye, where their densely packed mass of CONVECTION cells constitutes the windy parameters of the EYEWALL. Beyond the eyewall, spiral rain bands become thinner and spaced farther apart. Their tops grow less lofty, their bases develop at higher altitudes of the TROPOSPHERE, and they may form smaller pockets of precipitation that sometimes appear on meteorological RADAR screens as individual cloud clusters of an ancillary nature. Between the rain bands sustained WIND SPEEDS decrease to 30 MPH (48 km/h) or less and rainfall is light and sporadic. The cirrus and CIRROSTRATUS CLOUDS of a fully developed OUTFLOW LAYER often shield rain bands from view in SATELLITE images and in those photographs obtained by the space shuttle, making them difficult for meteorologists to trace and observe. Unknown before the application of meteorological radar in the early 1950s, the size, position and precipitation potential of a tropical cyclone's rain bands are presently monitored by a network of DOPPLER RADAR stations and satellite-borne RADIOMETER systems.

recurvature This term is given to that abrupt change in forward direction that occurs when a TROPICAL CYCLONE falls under the influence of the upper-level, high-pressure ANTICYCLONES that dominate circulation patterns over the NORTH ATLANTIC, NORTH PACIFIC, and Indian OCEANS. In the Northern Hemisphere, where CAPE VERDE STORMS and SUPER TYPHOONS generally spend several days swirling westward from their places of origin before encountering the southwestern edge of the clockwise-spinning BERMUDA HIGH and western Pacific Ogasawara High, recurvature is marked by a long, often parabolic shift in course to the north, northeast, or northwest, away from the equator and into the cooling confines of the extreme northern Pacific Ocean. In the Southern Hemisphere, BAY OF BENGAL and Australian CYCLONES—while initially carried west by their STEERING CURRENTS—soon recurve south, southeast, or southwest and either burst ashore in Madagascar and AUSTRALIA or else fade away over the southern Indian Ocean's inhospitable climes.

Although the exact point at which any one tropical cyclone commences recurvature varies from storm to storm and between individual HURRICANE SEASONS, a statistical compilation of previous tropical cyclone tracks indicates that a clear majority of mid- to late-season North Atlantic HURRICANES have begun their recurvature shortly before or after entering the eastern reaches of the Caribbean basin or within that square of ocean that extends from latitudes 10 to 35 degrees North and longitudes 50 to 75 degrees West. In the eastern North Pacific Ocean, TROPICAL STORMS and hurricanes tend to recurve less frequently than their North Atlantic counterparts but have most often done so while positioned to the west of the Hawaiian Islands. Across the international date line (180 degrees West), tropical storms and TYPHOONS in the western North Pacific Ocean commence their recurvature between latitudes 12 and 40 degrees North and longitudes 150 and 110 degrees West—a course change that eventually sends them bounding ashore in the northern PHILIPPINES, eastern CHINA, and south-central JAPAN.

In the life of a tropical cyclone, recurvature is a period of uncertain transition, of meteorological and climatological changes that can pose any number of positive and negative challenges to the future size and intensity of the system. In some instances, such as Hurricanes BETSY (1965) and ELENA (1985), the forward speed of a hurricane or typhoon can significantly drop during recurvature and stall from 12 MPH (19 km/h) to five MPH (8 km/h) or less in a matter of hours. In others, a tropical cyclone's momentum can sharply increase during recurvature, accelerating to between 18 and 20 MPH (29–32 km/h), depending on how quickly recurvature is accomplished. Some tropical systems even cease moving forward altogether, sitting stationary while their strenuous winds and gathering gusts whip the surface of the sea into a miasma of energy-laden spray. Some stationary hurricanes, such as the BLACK RIVER HURRICANE of 1912 and BELLE (1976), have boldly intensified while experiencing recurvature, while other examples have sustained weakening, even dissipation, during their recurvature periods. Once recurvature is completed, however, a tropical cyclone's forward speed can substantially increase, jumping to between 30 and 60

MPH (48–97 km/h), as witnessed in the GREAT NEW ENGLAND HURRICANE of 1938, Hurricanes CAROL (1954) and GLORIA (1985), and the western North Pacific typhoons FAYE (1957) and VERA (1959).

Redfield, William (1789–1852)

A Connecticut-born shopkeeper with a keen, seemingly instinctive understanding of atmospheric physics, William Redfield stands as one of 19th-century's most accomplished meteorological theorists and historians. The first individual to have defined systematically the circular nature of those wind patterns found in the TROPICAL CYCLONE, the self-educated Redfield devoted much of his adult life to the pioneering practice of retracing the tracks of notable NORTH ATLANTIC and CARIBBEAN SEA HURRICANES by studying the logbooks of ships that had encountered these hazardous—yet wildly misunderstood—storms at sea. First published in 1831, Redfield's voluminous biographies of past hurricanes, including the three remarkable storms of 1780 (see GREAT HURRICANE, SAVANNA LA MAR HURRICANE, and MILITARY OPERATIONS), as well as the GREAT CARIBBEAN HURRICANE of 1831, enjoyed an admiring readership during the 1840s and 50s and exerted a profound influence on the principal meteorological thoughts and practices of the time.

The story of Redfield's initial insight into the workings of hurricanes has over time achieved the quality of myth. In mid-October of 1821 while traveling across northwestern Connecticut in the wake of the devastating Long Island Hurricane of September 3 (see NEW YORK), Redfield became intrigued by the varying directions in which large trees had been toppled by the hurricane's sustained winds. In the town of Cromwell, Connecticut, where Redfield lived over his small dry-goods store, trees uprooted by the hurricane all fell to the northwest, indicating that the storm's winds had originated from the southeast. But, as Redfield was surprised to discover, less than 100 miles northwest, all the downed trees had fallen to the southeast, evidencing a prevailing wind from the northwest.

Now convinced that a hurricane was, in fact, one colossal storm wherein winds blew in a counterclockwise direction, Redfield spent the following decade constructing the first practical model of a tropical cyclone. Employing many of the same resources and techniques used to generate modern hurricane ANALOGS, Redfield published in July 1831 his first treatise on cyclonic circulation in the prestigious *American Journal of Science*. Entitled, "On the Prevailing Storms of the Atlantic Coast," Redfield's brilliant study—controversial from the moment it blew across the scientific horizon—quickly revolutionized the world's understanding of cyclonic systems. Expounding on subjects from the probable height of a hurricane's CUMULONIMBUS CLOUD towers to the proven presence of a calm EYE in its center to a tropical cyclone's forward speed and circulation, Redfield's astounding investigations subsequently inspired a veritable school of meteorological theory, one later expanded and refined by the works of William Reid and HENRY PIDDINGTON.

Rita, Hurricane *North Atlantic Ocean–Caribbean Sea–Cuba–Southern United States, September 18–26, 2005* The 18th named TROPICAL CYCLONE of the extraordinarily active 2005 North Atlantic HURRICANE SEASON, Hurricane Rita ticketed the BAHAMAS, CUBA, FLORIDA, and the southern Gulf states with Category 3 winds, rains, and heavy seas. At one point in its existence (September 21–22) a Category 5 HURRICANE of remarkable intensity (a central pressure reading of 26.48 inches [897 mb] was observed), Rita was at the time the fourth most intense North Atlantic tropical cyclone on record.

On September 23, as American Red Cross personnel were preparing to open 14 shelters in the Houston area, Rita was located about 220 miles (354 km) southeast of Galveston, TEXAS, and 210 miles (337 km) southeast of Port Arthur, TEXAS. Moving to the northwest at near 10 MPH (16 km/h), Rita's sustained winds lashed the warm waters of the Gulf at 135 MPH (217 km/h), making the system of firm Category 3 status. Rita was also a large storm; at its peak, the system measured some 350 miles (560 km) across.

During EVACUATIONS for Hurricane Rita near Dallas, Texas, a commuter bus carrying 45 elderly evacuees from a nursing home burst into flames and exploded, killing 24 people. Response officials speculated that the fire began in the vehicle's braking system, then spread to the oxygen containers onboard, sparking the deadly explosions.

Before Rita even neared New Orleans, its destructive effects were being felt in a city that had yet

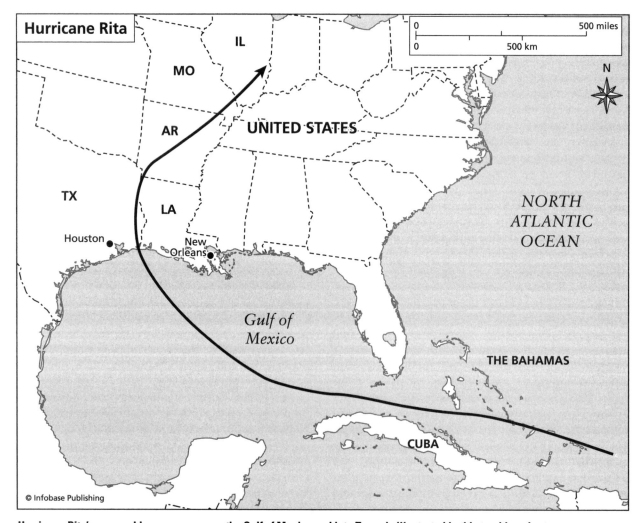

Hurricane Rita's memorable passage across the Gulf of Mexico and into Texas is illustrated in this tracking chart.

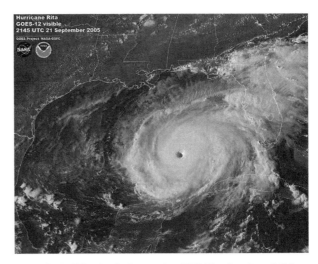

Two views of Hurricane Rita as the record-setting storm drove across the Gulf of Mexico in September 2005. The first is a visible image of Rita on the outside; the second is an infrared version of the same. A colorized IR version will allow meteorologists to evaluate a tropical cyclone's strength by the height of its cumulonimbus cloud towers—the darker the color, the more intense the convective activity. In this case, an inside look at Rita indicates that strongest winds are located directly around the system's eye. *(NOAA)*

to recover from Hurricane KATRINA's catastrophic strike on August 29, 2005. At the time measuring 250 miles (353 km) across, Rita's expansive cloud cover placed New Orleans within the system's tropical storm force quadrants. Rita's heavy seas raised an eight-foot (3-m) STORM SURGE that rose over one of the failed but temporarily repaired levees, causing localized flooding in the city's devastated 9th Ward.

Hurricane Rita struck LOUISIANA and Texas (near Sabine Pass, Texas) on October 24, 2005, as a Category 3 system with sustained winds of 120 MPH (193 km/h). More than 2.5 to 3 million people were evacuated. In Houston, 350,000 customers were without electricity, while nine people lost their lives. In Lake Charles, Louisiana, the storm smashed BUILDINGS and filled streets with water. In Galveston, Texas, which was spared a direct hit from Rita, high winds of 70 MPH (113 km/h) caused a blizzard of blowing embers as two residences and a commercial building went up in flames. In MISSISSIPPI, one person was killed when a TORNADO spawned by Rita's remnants ripped through two housing divisions. On September 25, 2005, Rita's sustained winds dropped to 20 MPH (32 km/h), and the system was downgraded to a TROPICAL DEPRESSION while centered near Hot Springs, Arkansas. During this time, nearly 2,400 National Guard troops were redeployed from Hurricane Katrina rescue efforts in New Orleans to those of Rita in western Louisiana and eastern Texas. In total, Hurricane Rita directly and indirectly claimed 120 lives, and tallied $10 billion (in 2005 U. S. dollars) in property losses.

Between 1948 and 1978, the identifier Rita was also used for 10 tropical cyclones in the west-

ern NORTH PACIFIC OCEAN. Of these, seven were typhoons and the remaining three of tropical storm intensity. The last of the western North Pacific typhoons dubbed Rita was also one of the most intense yet observed in the North Pacific basin. On October 23, 1978, Super Typhoon Rita generated a central pressure reading of 25.92 inches (878 mb), making it considerably more powerful than its 2005 North Atlantic namesake. After weakening, Typhoon Rita hammered the PHILIPPINES and Hong Kong, between October 26 and October 29, injuring three people in Hong Kong but claiming scores of lives in the Philippines. The identifier Rita has been retired from the rotating list of North Atlantic tropical cyclone names.

Rosie, Typhoon *Japan, July 25, 1997* Raking much of the Japanese islands during the morning of July 25, 1997, Rosie, a moderately powerful early-season rain TYPHOON, caused widespread power failures, landslides, and flash floods. As Rosie's 90 MPH winds and heavy rains forced the suspension of air and rail services throughout the country, two shellfishes were drowned when the rain-swollen river in which they were harvesting overflowed its banks, sweeping them to their deaths. Emerging into the Sea of Japan on the morning of July 26, a downgraded Tropical Storm Rosie carried its 56 MPH (90 km/h) winds northward to oblivion.

Roxanne, Hurricane *Caribbean Sea-Mexico, October 7–21, 1995.* Surging inland, its weakening RAIN BANDS lashing the heavy foliage and isolated villages of east Yucatán, Hurricane Roxanne spent the next two days transiting the peninsula, treading a westward path of waterborne damage from Cozumel in the east to Mérida in the west. Sustained 78-MPH (126-km/h) winds, accompanied by five-inch (127-mm) downpours, undermined numerous river embankment, causing flash flooding and the destruction of several hamlets. More than 20,000 people were driven into shelters, while nearly three-quarters of the valuable lobster traps in the state of Quintana Roo were lost. Nine Mexicans perished.

Shortly after reaching the Bay of Campeche at midnight on October 11, Roxanne was downgraded to a tropical storm. As though angered by its demotion, the still strident tropical storm spent the next two days chugging northwest before regaining hurricane status on October 14. Sharply veering southeast, Roxanne fitfully strengthened as it slowly struggled across the Bay of Campeche, seemingly bound for a return visit to the Yucatán Peninsula's northwestern tip. Always a slow mover, Roxanne's top winds raced to 85 MPH (137 km/h) at the same time that its forward momentum had nearly stalled. Between October 15 and 18, Roxanne's changing drift to the northwest saw it move only 20 miles (32 km) in a 24-hour period, making it one of the slowest hurricanes ever observed.

On the evening of October 16, a pipe-laying barge with 248 people aboard sank in heavy seas generated by Hurricane Roxanne. Two offshore supply boats and two construction tugs quickly pulled 222 people from the frothing waters of the bay but had to discontinue the rescue effort because of deteriorating conditions. The barge was returning personnel to offshore oil platforms after they had been evacuated because of Roxanne's near miss some seven days earlier. Bravely withstanding Roxanne's 95-MPH (153 km/h) gusts and 30-foot (10-m) seas, several survivors clung to the barge's derricks—which thankfully remained above water when the vessel went down—until they were eventually rescued. In all, five of the barge's passengers were drowned.

Downgraded to a tropical depression on October 19 Roxanne headed southwest toward Mexico's east coast. On October 20, 30-MPH (48-km/h) gales and roaring seas submerged abandoned streets in the capital of Campeche beneath two feet (1 m) of water, the worst inundation the busy port city had seen in 70 years. Neighboring Ciudad del Carmen was also battered, its waterfront left littered with sunken fishing boats and scattered oil drums.

Ruby, Typhoon *Marianas Islands–Philippines, October 18–24, 1988* An extremely powerful Category 4 TYPHOON, Ruby remains one of Filipino history's most destructive late season TROPICAL CYCLONES. With a CENTRAL BAROMETRIC PRESSURE at landfall of 27.52 inches (931 MILLIBARS) and sustained winds of 130 MPH (209 km/h), Ruby mined a 2,100-mile (5,100-km) path of devastation through the Marianas Islands and the northern PHILIPPINES between October 22 and October 24, 1988. In the U.S.-held Marianas, Ruby's 100 MPH (161 km/h) winds, six-inch (152-mm) rains, and 12-foot (4-m) STORM SURGE, brought enormous property damage to the archipelago's southern and central cays and to the main island of Guam. In the Philippines, where Ruby's 42-mile-wide EYE thudded ashore on the principal island of Luzon, more than 52,000 BUILDINGS were either damaged or destroyed, leaving an estimated 110,000 people homeless. On the afternoon of October 24, 1988, the 2,846-ton coastal ferry *Dona Marilyn* with more than 500 passengers and crew aboard bucked into Ruby's cresting 35-foot (10-m) seas and was capsized. Three hundred eighty-nine men, women, and children drowned, while another 140 spent several harrowing hours in the raging waters of the Visayan Sea before being rescued. Before rapidly dissipating in a shower of thunderstorms over the South China Sea on the afternoon of October 25, the most powerful typhoon to strike the northern Philippines in more than 18 years had claimed 540 lives and wrought upward of $150 million in damage to dwellings, roads, bridges, railway lines, and agricultural stocks in the northern half of the nation.

The 17th storm of a very active TYPHOON SEASON, Ruby formed over the iridescent waters of the western NORTH PACIFIC OCEAN, 2,000 miles (4,800 km) east of Luzon island, on the afternoon of October 18, 1988. Beating southwest at nearly 13 MPH (21 km/h), the young TROPICAL DEPRESSION swiftly liberated the ocean of its heat and moisture, maturing into a powerful TROPICAL STORM on the morning of October 21 and an intense Category 2 typhoon with a central pressure of 28.88 inches (977 mb) by the early morning hours of October 22. Maximum sustained winds in Ruby's EYEWALL now exceeded 105 MPH (169 km/h).

With gusts around its 38-mile-wide (59-km) eye soon reaching 120 MPH (193 km/h), Ruby lashed Guam with typhoon-force winds and heavy rains as it passed south of the territory on the afternoon of October 22. Undergoing further intensification as it spiraled toward the storm-weary cities and towns of the central and northern Philippines, Ruby's barometric pressure sharply deepened, sinking from 28.88 inches (977 mb) on the afternoon of October 22 to 28.53 inches (966 mb) by midnight of the same day, to 27.91 inches (945 mb) by daybreak the next day. Maximum winds at the system's core now measured between 115 and 121 MPH (185–195 km/h). Shortly after one o'clock on the afternoon of October 23, as aerial reconnaissance flights from the Joint Typhoon Warning Center (JTWC) on Guam penetrated the roiling updrafts of the typhoon's eye, Ruby's central pressure slid even further to 27.85 inches (943 mb), making it a Category 4 typhoon of enormous destructive potential.

Bearing a central barometric pressure of 27.88 inches (944 mb), Typhoon Ruby swept ashore near the bayside town of Siniloan, the Philippines, on the morning of October 24, 1988. Gusts of 140 MPH (225 km/h) stripped buildings of their thatched roofs, uprooted thousands of palm trees, downed power lines, and pushed before them a 16-foot (5-m) storm surge, the greatest seen on the southeast coast of Luzon since Typhoon Patsy delivered a 12-foot (4-m) inundation to the area in November 1970. In the Polillo Islands, east of Manila, two tornadoes destroyed more than 100 houses near Agta Point in a matter of minutes, while nine-inch (229-mm) rainfalls—some extending as far south as the island of Panay—spawned raging flash floods and mudslides that washed out precipitous mountain roads and flimsy wooden bridges. In the central province of Antique 35 people were drowned when a vicious gust blew their bus off a bridge and into the rain-gorged Sibalom River. Eyewitnesses on the banks of the river quickly moved to assist those survivors who were able to escape the submerged vehicle, but the whitewater fury of the torrent, coupled with Ruby's 100-MPH (161-km/h) winds, tragically nullified their efforts.

Farther to the north, where Ruby's sustained 130 MPH (209 km/h) winds and considerable rains hammered seawalls along the Polillo Strait's shallow confines, several 150-ton fishing boats were cast onto Luzon's volcanic beachheads, drowning 32 people. A fairly small storm by typhoon standards, Ruby's 200-mile-wide (322-km) cloud bands nonetheless managed to kick up record seas throughout the Philippine Islands, 30–40-foot (10–12-m) waves that quickly disabled and then sank all small watercraft caught in the typhoon's path. Two steel-hulled fishing boats transiting the Verde Island Passage were battered for nearly an hour before sinking with all hands; a coastal oil tanker in ballast was snapped in two near the entrance to Manila Bay.

Another such vessel, the 2,846-ton ferry *Dona Marilyn*, was overwhelmed while on an overnight voyage from Manila to Tacloban, the provincial capital of Leyte Island. Never designed for deepwater service, the *Dona Marilyn*'s three decks, left open to allow passengers a proper promenade as the ferry navigated through the country's picturesque islands, were by midmorning awash with frothing fingers of the Visayan Sea. Within the pitching ferry, 529 passengers and crew—many of them shoppers returning from a weekend trip to Manila—either huddled in fear or diligently worked to save the ship.

As Typhoon Ruby wore on into early afternoon, the *Dona Marilyn* developed a serious list to starboard, one which took her lowest deckrail within feet of the sea. Unable to shift the ferry's ballast or even to operate the pumps, the captain had no sooner issued the order to abandon ship than the swamped vessel lurched onto her starboard side and began to sink stern-first. Shortly before two o'clock on the afternoon of October 24, 1988, as Ruby's airborne fury raged on unabated, the *Dona Marilyn* finally went down west-northwest of Leyte. One hundred forty survivors either clung to those life rafts that had floated off the ferry when it went down, or drifted aimlessly in the roaring seas, straining to keep breathing against the smothering spray. Quite remarkably, 13 survivors managed to swim to a nearby island, while the remaining 127 found relative safety on Marapipi Island, their rafts having been blown ashore there by the typhoon's westerly winds. All told, 389 people perished in the tragedy, prompting then president Corazon Aquino to direct the Philippine minister of transportation to conduct a full investigation into the incident.

After destroying nearly 75 percent of the rice and sugar harvest on Luzon, and leaving nearly 765,000 people in 35 provinces with shattered homes, families, and lives, Ruby exited into the South China Sea, fanning the temperate waters with 94 MPH (151 km/h) winds and diminishing rains. By late evening on October 24, as the typhoon began its recurvature northwest, it suddenly lost much of its intensity, returning to tropical storm status just before dawn the next day. By the same time on October 26, Ruby had disintegrated into a roving series of thunderstorms and lingering showers that by nightfall had dissolved into clear skies. With nearly $220 million in property losses and 540 deaths to Ruby's credit, the World Meteorological Organization decided that, despite a growing dearth of available typhoon names, it would temporarily retire the name *Ruby* from the rotating list of typhoon identifiers for a period of 15 years.

Saffir-Simpson Damage Potential Scale The official scale used by the National Weather Service (NWS) to rate the size, intensity, and destructive capabilities of NORTH ATLANTIC and western NORTH PACIFIC HURRICANES and developed in 1973 by construction engineer Herbert Saffir and then-director of the NATIONAL HURRICANE CENTER (NHC) Robert Simpson, the Saffir-Simpson Scale separates mature hurricanes (TROPICAL CYCLONES with winds of 74 MPH or greater) into five ascending classes, or what are called categories. Ranging from 1 to 5, these categories are determined by such meteorological factors as the hurricane's CENTRAL BAROMETRIC PRESSURE, its maximum sustained WIND SPEEDS, and the potential height of its dangerous STORM SURGE. Primarily used by the Weather Service to alert civil defense agencies to the destructive potential of a hurricane that lies within 72 hours of landfall, the Saffir-Simpson Scale has over time gained widespread popularity with the general public, becoming an easy, reliable way for storm watchers, historians, media outlets, and homeowners to quantify and qualify the severity of threatening hurricanes quickly.

Adapted for use by weather services in JAPAN and AUSTRALIA, the Saffir-Simpson Scale's universal effectiveness springs in part from its correlation of the various degrees of hurricane intensity with the types of structural damage that can be reasonably expected from a hurricane of definable intensity. **Category 1** hurricanes, for instance, have a central pressure of greater than 28.94 inches (980 MILLIBARS), sustained winds of between 74 and 95 MPH (119–153 km/h), and a likely four to five-foot (1–2-m) storm surge. Damage potential for Category 1 hurricanes is minimal, with some tree and crop damage, downed power lines, and minor BUILDING damage (broken shutters, buckled antennas, busted fences) being suffered.

The central barometric pressure of **Category 2** hurricanes ranges from 28.91 to 28.50 inches (979–965 mb). Maximum wind speeds extend from 96 to 110 MPH (155–177 km/h), while storm surges of six to eight feet (2–4 m) are generally recorded in storms of this class. Damage potential is often moderate, with the brunt of the hurricane's fury being borne by trees, shrubs, crops, road signs and traffic lights, small boats, and toppled sheds. Some window, roof, and chimney damage is also likely.

Category 3 hurricanes are considered the first of the major hurricanes, storms with barometric pressures of between 28.47 and 27.91 inches (964–945 mb), winds of 111–130 MPH (179–209 km/h), and an anticipated nine–12-foot surge. Category 3 systems cause extensive damage to property, from the uprooting of large trees to the crumpling of mobile homes, small aircraft, and coastal cottages.

In **Category 4** hurricanes, major cyclonic systems where barometric pressures slide from 27.88 inches (944 mb) to 27.17 inches (920 mb), damage potential is often widespread and extreme. Sustained winds roar across the surface of Earth at speeds of between 131 and 155 MPH (211–249 km/h), pushing before them an enormous 13–18-foot (4–6-m) storm surge. Under such conditions the roofs of small houses, restaurants, shopping malls, and condominium complexes begin to fail, tearing off in pieces or lifting into the air virtually intact. Along the edges of harbors, coves, bays and river deltas, Category 4 hurricane's raging seas splinter piers, breach seawalls, and drive small vessels—fishing boats, oil barges, pleasure yachts, and coastal freighters—aground.

Category 5 hurricanes, those rare but recorded events of seemingly indescribable meteorological violence, possess spiraling central pressures of 27.17 inches (920 mb) or below, screaming winds of greater than 155 MPH (249 km/h), and a pulverizing storm surge that measures at least 18 feet (6 m) in height. Damage potential in hurricanes of this intensity is always catastrophic, with entire land and cityscapes being reduced to rubble in a matter of hours.

Sam, Typhoon *China, August 22, 1999* During the 1999 western NORTH PACIFIC tropical CYCLONE SEASON, a moderately powerful Typhoon Sam trundled ashore just to the east of Hong Kong on the afternoon of August 22, 1999. Sam's maximum sustained 81-MPH (130-km/h) winds uprooted trees, buckled billboards, flooded streets, and smashed piers. Attempting to land during the height of the typhoon's passage over the territory, a Chinese jetliner with 315 passengers and crew aboard crashed and burned at the newly inaugurated Hong Kong International Airport. Two people aboard the MD-11 aircraft were killed, another 206 injured. Although the crash was later attributed to low-altitude WIND SHEAR caused by Typhoon Sam's winds, airport authorities stated to news services that it was standard operating procedure to keep the technologically advanced airport operating in all meteorological conditions, including typhoons.

Samoan Cyclone of 1889 *South Pacific Ocean– Samoan Islands, March 15–16, 1889* An exceptionally destructive TROPICAL CYCLONE, the Samoan Cyclone of 1889 wrought considerable damage to warships belonging to the United States, Great Britain, and Germany that were at the time moored in Apia harbor in the Samoan Islands.

Although lost to the mists of meteorological history, the approximate track followed by the Samoan Cyclone of 1889 can be reconstructed from the paths of known tropical cyclones that have subsequently affected the Samoan Islands. A majority of Samoan cyclones originate to the north-northwest of the archipelago before tracking to the south-southeast, bringing destructive wind, rain, and seas to the northern coasts of Savai'i and Upolu islands. Between January 13 and 20, 1946, for instance, a TROPICAL STORM was born to the west of the islands, over Indonesia, then tracked east, passing just to the west of Savai'i and Upolu. Between February 1 and 4, 1990, Cyclone Ofa rattled the Samoan Islands, killing three people, damaging or destroying nearly 3,000 structures, and wreaking enormous devastation across the islands' breadfruit and banana plantations. Less than one year later, the sixth named

The German warship SMS *Olga* lies stripped and aground in Apia Harbor following its encounter with the March 1889 tropical cyclone in the Samoan Islands. *(NOAA)*

cyclone of the 1991 South Pacific cyclone season, Val, lumbered into Western Samoa on December 7. Of Category 4 intensity at landfall on Savai'i's northern coast, Cyclone Val deepened as it passed over the relatively small Samoan Islands, its sustained winds rising from 132 MPH (213 km/h) to 144 MPH (232 km/h) within a matter of hours. Stalled for nearly a day over Savai'i and Upolu, Val's EYEWALL produced average wind speeds of 104 MPH (167 km/h) and 150-MPH (241-km/h) gusts. In the capital of Apia, the library and post office were destroyed. Nearly all of the residences on Savai'i were damaged, while up to 80 percent of those on Upolu lost windows, roofs, porches, and outbuildings. One of the deadliest cyclones in Samoan history, Val killed at least 12 people and left dozens of others with injuries. On January 26, 1998, Tropical Storm No. 16 (later to become Cyclone Tui) passed directly over the eastern part of the island of Savai'i in Western Samoa, causing only minor structural damage, but electrocuting a young boy. Tropical Storm No. 16's minimum pressure in Samoa of 29.26 inches (991 mb) produced maximum sustained winds of 46 MPH (74 km/h) and gusts up to 69 MPH (111 km/h). On January 31, 1993, Cyclone No. 15 tracked to the southeast of its origination point and passed directly over Samoa. While the system eventually deepened into a Category 2 cyclone on February 1, 1993, it was of tropical storm intensity as it passed over the islands, delivering 60 knot winds and heavy rains. On January 5, 2004, an extremely powerful Cyclone Heta passed less than 100 miles (161 km) to the west of Samoa. While sustained winds within Heta's eyewall topped 161 MPH (259 km/h), the system's offshore passage delivered 105-MPH (169-km/h) winds to much of Savai'i and Upolu, knocking out power, uprooting trees, and causing extensive roof damage. A rainmaker for much of its existence, Cyclone Heta deluged the volcanic Samoan islands with tons of precipitation, much falling at a rate of two inches (51 mm) an hour.

Had it made landfall in Samoa, 2005's Cyclone Olaf would have most closely mirrored the trajectory and effects of the 1898 cyclone. Born to the northwest of the islands on February 13, 2005, Olaf deepened to a Category 5 cyclone with sustained winds of 167 MPH (269 km/h) and an observed gust of 201 MPH (324 km/h), as it approached Apia harbor. On February 15, however, as it drew to within 50 miles (80 km) of the capital, Olaf abruptly recurved to the east of Savai'i and Upolu, and weakened into a tropical storm. Sustained winds of 161 MPH (259 km/h) tattered roofs and foliage in the northern coastal towns of Asau and Avao. Less than two weeks later, Cyclone Percy—also of Category 5 intensity—struck American Samoa and the neighboring Cook Islands with 150 MPH (241 km/h) winds on February 28, 2005. During the 2005 cyclone season, Cyclones Meena and Nancy—both of which were of Category 4 intensity—further trounced the islands on March 5 and March 15, respectively, making the 2005 season one of the most active in recent Samoan memory.

The Samoan Cyclone struck during the night of March 15–16, 1889. At the time identified as a *tropical hurricane,* it is now referred to as a tropical cyclone under the new identification nomenclature for those tropical systems that generate south of the equator.

The U.S. Navy's Pacific Station flagship, USS Trenton, and the smaller warships USS *Vandalia* and USS *Nipsic* were countered in force by the German corvette SMS *Olga* and the gunboats SMS *Adler* and SMS *Eber.* The British also had a vessel in the harbor at Apia, the corvette HMS *Calliope.*

Although the weather was visibly problematic in the hours before the storm, the senior officer, Rear-Admiral Lewis A. Kimberly, decided to keep his vessels at Apia, "basing his decision on local opinion that the hurricane season was already past." There had been March cyclones in Samoa, like the March 17–25, 1946, system that originated to the east of Samoa, tracked in a parabolic course to the southwest, but spared the islands a direct strike.

SMS *Eber* was blown onto the reef and completely destroyed. Nearly all of her crew was killed; SMS *Adler* and USS *Vandalia* also suffered terrible casualties and were effectively left total wrecks; the USS *Trenton* had its steam plant extinguished through water entering the vessel's hawse pipes, and dragged her anchors; the SMS *Olga* and USS *Nipsic* were run aground, but later hauled off and repaired.

More than 50 U.S. sailors and about 90 German sailors lost their lives. Some sources list 200 lives lost. Seventy-six men were lost in the wreck of the SMS *Eber.* The ship's decorated bow section landed on the beach very near to where the *Eber* struck the reef.

Two merchant ships were also destroyed. They were anchored in an estuary to the east of the harbor. In total, six merchantmen, ranging in size from 25 to 500 tons, and a number of smaller craft in the harbor, were destroyed.

At the height of the cyclone, the USS *Trenton* dragged its anchors along the reef; water entered the hawse pipes, put out her fires, and left the vessel drifting in the wind and high seas. *Trenton* had laid down four riding cables, with the last one extending 90 feet (30 m) before breaking loose. According to RADM Kimberly's memoirs, it was "the swift and tremendously strong under tow that was a great factor in keeping the USS *Trenton* from pounding down on the sharp reefs as each wave threw her upon them."

The USS *Vandalia* sank to her main deck right beside the USS *Trenton.* Lines were rigged from the *Vandalia*'s rigging to the *Trenton* in order to facilitate the rescue of those aboard the *Vandalia.*

The SMS *Olga* was beached on the eastern side of Apia harbor, Upolu. The USS *Nipsic* was beached near the sunken USS *Vandalia,* and the battered but afloat USS *Trenton.* The *Trenton*'s starboard quarter gallery was almost entirely torn away, and the vessel was sunk to the gun deck. After the cyclone had passed, all the food and provisions were removed from the USS *Trenton* to be used for feeding sailors ashore. USS *Trenton* had broken her back in two places, a hole in the hull, and the propeller and rudder broken away. She was eventually broken up where she lay, but after all guns and other items were salvaged. The guns were of the Dahlgren type, with iron carriages.

SMS *Adler* was a complete loss; stripped of weapons and valuable fittings, the wreck was left in place, and could still be seen well into the 20th century. The *Adler*'s wreck came to rest on the western edge of Apia Harbor, Upolu. The USS *Nipsic* was beached, but eventually refloated. Her propeller was bent beyond repair, her rudder and rudder post torn away. Her funnel collapsed, but was replaced with the funnel from the sunken USS *Vandalia.* USS *Nipsic* departed Apia on April 17, 1889, but had to return because she bucked into a gale, had her jury rigged rudder smashed, and was forced to return to Apia. She did not depart Apia until May 15, 1889, bound for Honolulu and repair. Constructed in 1879, she lasted until 1913. During the cyclone, her master, Captain Mullan, attempted to save his vessel by keeping steam up by using pork fat from the ship's rations and using them in the boilers to keep up steam. During this operation, and before the *Nipsic* was grounded on the eastern reef, she collided with and sank the anchored schooner *Lily,* killing two of the schooner's crew.

San Felipe Hurricane *Caribbean Sea–Southern United States, September 5–16, 1928* Better known in the United States as the Lake Okeechobee Hurricane of 1928, this fierce Category 4 TROPICAL CYCLONE remains one of the deadliest CAPE VERDE STORMS on record. Driven by harrowingly low CENTRAL BAROMETRIC PRESSURES of 27.76 inches (940 MILLIBARS) and below, the San Felipe Hurricane claimed some 3,366 lives as it careened across the eastern Caribbean Islands of GUADELOUPE, St. Kitts, and Montserrat; the northeastern territories of PUERTO RICO and the VIRGIN ISLANDS; and into central FLORIDA, between September 12 and 16, 1928. In Guadeloupe, where the hurricane's sustained 135-MPH (217-km/h) winds razed nearly every BUILDING on the island, more than 400 people were killed on September 12. The second hurricane of the same name to have struck Puerto Rico since September 1876, San Felipe claimed nearly 1,000 lives when it deluged the island with 30-inch rainfalls and 150-MPH (241-km/h) winds on September 13. Soaring over the BAHAMAS and into southern and central Florida on September 15–16, the San Felipe Hurricane raised a 15-foot (5-m) STORM SURGE along the shallow shores ta Lake Okeechobee, breaching dikes and inundating the farming and resort towns of Okeechobee and Belle Glade. An estimated 1,836 men, women, and children perished, making the San Felipe/Lake Okeechobee Hurricane the third-deadliest hurricane disaster in U.S. history.

Of equal or greater caliber than later Hurricanes, DAVID (1979), DONNA (1960), HUGO (1989), and LUIS (1995), the San Felipe Hurricane developed over the North Atlantic Ocean's eastern reaches, most likely in the Cape Verde Island's vicinity, during early September 1928. Poorly tracked for the first week of its life, the hurricane was eventually spotted in the mid-Atlantic on September 10, by which time it had matured into a colossal Category 4 HURRICANE of awesome intensity, with sustained winds peaking at 135 MPH (217 km/h). A vigorous rainmaker for most of its existence, San Felipe's torrential downpours flooded the sea surface, adding immense quantities of moisture and latent heat to the system's siphoning EYE as it spun inexorably toward the Leeward Islands chain.

The San Felipe Hurricane first came ashore at Guadeloupe southeast of the island's commercial center of Pointe-à-Pitre, during the late-morning hours of September 12, 1928. A Category 4 hurricane of near-catastrophic intensity, its central pressure of 27.76 inches (940 mb) produced wind gusts of between 160 and 170 MPH (256–274 km/h), gigantic seas, and 11–12-inch (279–305-mm) PRECIPITATION counts. Hundreds of buildings on Guadeloupe were completely razed and blasted across the sodden landscape with the same destructiveness as a small nuclear device. Entire palm groves were sawed down at ground level. Coastal roads disappeared, dragged into the sea by terrifying 30-foot (10-m) waves. Considered one of the deadliest hurricanes in Guadeloupe's history, an estimated 520 people perished, a majority of them drowned in the raging flash floods that spilled across the island's famous eastern plains.

Not at all weakened by its brutal passage over Guadeloupe, a recurving San Felipe steadily continued northwest, gathering strength as it entered the CARIBBEAN SEA just before dusk on September 12. Bound for Puerto Rico, the hurricane spent the remainder of the night roaring northward, its DANGEROUS SEMICIRCLE showering darkness and death on the nearby islands of Montserrat and St. Kitts. Some 100 people were killed on the two eastern islands, while hundreds of building failures reduced both to mere shadows of their former prosperity and beauty.

Less than 12 hours later, San Felipe, now deepened into a hurricane of even greater size and ferocity, rushed ashore on the southeast coast of Puerto Rico. Named for the Saint's Day on which it occurred— and, ironically enough, the second hurricane to bear the name—San Felipe's still-stinking central pressure of 27.50 inches (931 mb) made landfall near the port city of Arroyo, southeast of San Juan, shortly after dawn on September 13. Far more severe than the earlier San Felipe Hurricane of September 13, 1876, San Felipe II sheared northwest, its gargantuan EYEWALL diagonally bisecting the island as it traveled from Arroyo and Guayama in the southeast to the northwestern town of Isabela. In San Juan, ANEMOMETERS maintained by the U.S. WEATHER BUREAU recorded sustained wind speeds of 150 MPH (241 km/h), while rain gauges in the central highlands of the island caught up to 30 inches (762 mm) of precipitation over a 12-hour period.

Cataclysmic devastation on a scale rarely before seen in Puerto Rico ensued. More than 19,000 buildings were utterly demolished, leaving some 284,000 people without shelter, food, clothing, or medical assistance. In the town of Cayey, southwest of San Juan, airborne sheets of tin roofing decapitated scores of fleeing townspeople. The unfinished dome of a nearby church, pried from its piers by the hurricane's clawing winds, quickly collapsed into the church's nave, crushing to death 14 parishioners who

Another of tropical cyclone history's most famous images, this photograph of a palm tree that has been pierced by a 12-foot (3.6-m) long plank was taken in Puerto Rico in the wake of the ferocious San Felipe/Lake Okeechobee Hurricane of 1928. There are reasonable explanations for why this photograph remains so celebrated. On one level, it portrays the power of this Category 5 hurricane's passage over Puerto Rico, while on another it contains potent religious imagery and symbolism. *(NOAA)*

had gathered at the rain-soaked altar to pray. A 10-foot (3-m) pine plank, thrown like a javelin by San Felipe's 180-MPH (290-km/h) gusts, pierced the thick trunk of an otherwise unscathed royal palm tree, later providing a local sightseer with one of hurricane lore's most famous photographs. Dozens of small boats, many surprised at anchor by San Felipe's sudden onset, were thrown into rocky coves all along the island's south and east coasts, battered into mere chipwood by San Felipe's 19-foot (6-m) storm surge and spray-shrouded seas.

Farther inland, record-setting precipitation counts initiated serious flood emergencies in the mountainside communities of Caguas and Adjuntas. Rumbling landslides sent tons of mud and rock hurtling down the slopes, erasing both villages in a

matter of minutes. Shallow valleys and creekbeds in central Puerto Rico quickly filled with natural and humanmade debris, creating impromptu dams that held back floodwaters for miles. Temporary lakes began to form, swelling cauldrons of brackish water dotted with broken trees, stalled automobiles, resilient cottages, and the bloated remains of people and ANIMALS. These reservoirs' slow draining would in subsequent weeks spawn dam breaks and further flash floods, sudden cascades of water that kept weary survivors from returning to their ravaged lands. All told, some 1,498 people in Puerto Rico died during the San Felipe Hurricane. Damage estimates of $50 million, a figure that today would inflate to billions of dollars lost, attests to the material severity of the hurricane's disastrous passage over the island. Not surprisingly, the memory of San Felipe continues to endure in Puerto Rico, inspiring a veritable canon of stories, plays, poems, ballads, and folk songs concerning the unforgettable events of September 13, 1928.

Spinning northwest at 16 MPH (26 km/h), a slightly diminished San Felipe soon quit wounded Puerto Rico completely, spending much of September 14 and 15 delivering 119-MPH (192-km/h) winds and torrential rains to northern HISPANIOLA and the southern Bahamas. In Florida, where tree and dune damage from the deadly Miami Hurricane of September 18, 1926, was still widely visible across southeastern portions of the state, hurricane forecasters vainly sought to predict just where—if at all—along the Atlantic coast San Felipe's rejuvenating core would come ashore. If the hurricane maintained its steady progression northwest, it could be expected to make landfall in the vicinity of West Palm Beach sometime during the evening hours of September 16. Slicing diagonally across midland Florida, San Felipe would no doubt bring terrific winds and flooding rains to the muddy banks and crude earthen dikes of Lake Okeechobee before exiting into the Gulf of Mexico on or about the 17th. On the other hand, the storm could fall under the influence of a high pressure AIR MASS, then slowly spreading east-southeast across LOUISIANA and MISSISSIPPI, hopefully finding itself pushed out into the North Atlantic by the encroaching high just before or after its landfall in southeastern Florida; if the high arrived earlier, growing stronger as it moved across the sunbelt states, then there was even the possibility that most of San Felipe's dangerous eyewall would be kept over open waters, exposing Florida's east coast to the somewhat weaker furies of the hurricane's NAVIGABLE SEMICIRCLE.

But by midmorning of September 16, as an intensifying San Felipe's 135-MPH (217-km/h) eyewall winds neared to within 200 miles (322 km) of West Palm Beach, it became apparent to observers that the weather waiting game was over. Steady fringe rains were already falling in Fort Lauderdale, while enormous breakers—some 15 feet (5 m) high—splashed ashore at Pompano Beach. Storm warnings were belatedly hoisted from Miami to Jacksonville, reflecting the hurricane's projected northeasterly swing inland over the peninsula. Along the fertile rim of vast, brooding Lake Okeechobee, thousands of people—from the shanty-dwelling vegetable farmers who tilled the lake's alluvial lowlands to the seasonal vacationers who crowded its pink-painted hotels and lantern-lit ballrooms—began their hasty retreat toward safer ground. Several hundred women and children anxiously sought refuge on large party barges moored near the towns of South Bay and Belle Glade, while scores of others in Moore haven and Pelican Bay climbed trees or gathered in multistory buildings away from the lake's fatal shores.

Pirouetting around a central barometric pressure of 27.34 inches (929 mb) and preceded ashore by swirling downpours and uncoiling lightning bolts, the San Felipe Hurricane's Category 4 eyewall danced ashore at West Palm Beach on the evening of September 16, 1928. Powered by the lowest barometric pressure observed in the United States up until that time (a record that would stand until the LABOR DAY HURRICANE pulverized the Florida Keys with a central pressure of 26.35 inches in September 1935), San Felipe's sustained 148-MPH (238-km/h) winds stitched a blanket of unimaginable destruction from Boynton Beach to Lake Park. Building roofs were unzipped from walls and folded end over end until reduced to needlelike splinters. Thousands of palm trees were hemmed at ground level, their frond pompoms plucked and tossed by San Felipe's 160-MPH (258-km/h) gusts. Parked automobiles were driven through brick walls. An 11-foot (4-m) storm surge, one of the highest ever noted in southern Florida, washed out large spans of coastal highway A1A, the principal road linking Palm Beach with the neighboring city of Delray. Nearly 100 oceanfront homes, mission-style mansions with red tile roofs, concrete seawalls, and private yacht marinas, were razed as the surge swamped Lake Worth Inlet, tallying some $11 million in 1928 property losses. Nearly 700 people reportedly perished in Palm Beach County, a majority of them drowned by the surge's passage over the densely inhabited barrier island on which much of scenic Palm Beach is tenuously situated.

Swiftly curling, as had been predicted, north-northeast, San Felipe's fearsome eyewall bounded across the northern Everglades, penetrated as far inland as Orlando before sharply recurving toward GEORGIA and the North Atlantic just after midnight on September 16. By then, however, the worst had been suffered by those delicate towns ringing the shallow basin of Lake Okeechobee—Belle Glade, Pahokee, Port Mayaca, and Moore Haven—mauled so brutally by San Felipe's harrowingly low barometric pressures, 135-MPH (217-km/h) winds, and 20-foot (7-m) storm surge. Taming the wind-wild waters of the 35-mile-long (53-km) lake into an enormous aquatic mane as it passed overhead, the counterclockwise hurricane carried the surge to Okeechobee's southern banks before unleashing it as waves—shoaling claws of water that breached dikes—submerged hundreds of acres of cropland, left 25-foot (8-m) marks on trees, and instantly drowned 200 fleeing refugees on the lake road between Belle Glade and Pahokee. Hundreds of buildings—from farmhouses, barns, and roadside stands to summer cottages, restaurants, and dance halls—were swept into extinction. Carried ashore at South Bay by the surge, two rain-soaked barges loaded with 274 women and children were safely set down on the beach, left stranded by San Felipe's receding waters. Likewise, another 300 people survived the hurricane's wrath by seeking sanctuary in sturdy buildings by climbing to the top stories of nearby hotels.

But for 1,836 people in Florida, San Felipe was the end. Later fished out of the lake by the boatload, their bodies were either buried in mass graves or cremated on roaring funeral pyres. Another 1,879 people were reportedly injured, many of them seriously. Presently ranked as the third-highest hurricane DEATH TOLL in U.S. history, San Felipe's human carnage along the shores of Lake Okeechobee has been topped only twice, during the GREAT GALVESTON HURRICANE of 1900 and the GRAND ISLE HURRICANE of 1893.

The United States government eventually reconstructed Lake Okeechobee's dikes and embankments, spending seven years and $5 million fashioning an ambitious network of drainage canals and stone levees to protect the rebuilt towns of Moore Haven and Belle Glade from future hurricanes. Designed to empty a storm's overflowing surge into the nearby Atlantic Ocean, the project (which included a stone seawall some 85 miles long and 38-feet high) was one of the most significant engineering achievements of its time and did much to restore confidence and prosperity to the Lake Okeechobee region.

Sarah, Typhoon *Japan–Korea, September 11–20, 1959* A particularly deadly western North Pacific TYPHOON, Sarah delivered 112-MPH (180-km/h) winds and flooding, nine-inch (229-mm) PRECIPITATION counts to southern JAPAN and KOREA between

September 17–19, 1959. A peak-season system of unusual severity, Sarah's scouring winds and tropical downpours triggered flash floods and landslides that damaged or destroyed more than 3,000 BUILDINGS and claimed a total of 1,869 lives in both nations. In southern Japan, where Sarah's 123-MPH (198-km/h) gusts and heavy seas uprooted trees and shattered scores of piers on the afternoon of September 17, 104 fishers were drowned when their small boats were overwhelmed by the typhoon's wind-driven swells and capsized. Within weeks, Japan would be pummeled by an even greater typhoon, VERA. In southern Korea's Kyongsang and Namdo provinces, which Sarah blasted with 110-MPH (177-km/h) winds and scudding rains on September 19, hundreds of houses were unroofed; some 432 people were killed and another 171 were posted as missing. Damage estimates in Korea alone mounted to $43 million. Several hundred miles off the southern coast of Korea, 46 fishing boats that had departed from Oerarodo Island on August 28, 1959, were caught at sea during Sarah's passage on September 18 and foundered; some 1,200 people were reportedly lost, making Sarah one of the deadliest typhoons to have traversed the western North Pacific Ocean in decades.

satellites Broadly defined, a satellite is any object designed and built by humankind and placed into semipermanent orbit around Earth; in a narrower meteorological sense, the dozens of satellites launched into orbit between 1960 and 1996 used to study and monitor TROPICAL CYCLONE activity over the NORTH ATLANTIC, NORTH PACIFIC and Indian oceans.

Nearly four decades have passed since the United States sent the first operational prototype meteorological satellite (*TIROS*) into space on April 1, 1960, a period in which progressively more sophisticated satellites not only provided meteorologists and research scientists with volumes of new visual data on the physical nature of tropical systems but also laid the groundwork for unprecedented improvements in forecasting accuracy and civil defense preparedness. Powered by solar panels, stabilized with gyroscopes and booster rockets, and equipped with advanced RADIOMETERS, magnetometers, and other atmospheric sensors, civilian and military meteorological satellites possess the ability to observe and transmit visual and infrared photographs of air TEMPERATURES, SEA-SURFACE TEMPERATURES, humidity levels, and carbon dioxide readings from across many spectrums of Earth's TROPOSPHERE. Their advanced communications technology further enables them to receive and transmit data regarding BAROMETRIC PRESSURE and WIND SPEEDS from remote weather stations on mountaintops and ocean buoys, thereby

exponentially increasing the scope of surveillance. During the course of an average day, U.S. based receiving stations obtain 1,500 meteorological readings and 120 photographs from orbiting satellites.

Satellites have also changed the culture of tropical cyclones by permitting the near instantaneous dissemination of meteorological information to television and radio stations, as well as to those federal and state emergency organizations tasked with implementing evacuation procedures and conducting relief operations. One of the most invaluable of all forecasting tools, satellites now allow millions of people—from those individuals residing in the path of a storm to those living thousands of miles distant—to follow a system's progress ashore visually, participating in some way in its innate, natural drama.

Since the mid-1970s, two types of meteorological satellites have been used for tropical cyclone reconnaissance. The first, a *polar-orbiting satellite*, circles Earth longitudinally—from pole-to-pole—at an altitude of between 450 and 550 miles. Designed to provide photographs with a resolution of between two and 10 miles and principally used for tropical cyclone research, previous examples of this class include *TIROS*, *NIMBUS*, and *TIROS-N*. Satellites of this type are generally sunsynchronous in orbit, meaning that their solar cells and instrumentation packs are always turned toward the Sun. The second type of meteorological satellite, a *geostationary satellite*, orbits Earth latitudinally, or parallel to the equator, and is most often employed in weather forecasting. Positioned above a certain point of Earth's surface, geostationary meteorological satellites travel at the same speed as Earth (approximately 18,000 MPH [28,962 km/h]) and snap photographs every 18 minutes or so, thereby permitting them to maintain continual weather surveillance over particular geographical areas. Spinning at 100 revolutions per minute, most geostationary meteorological satellites, such as the *GOES* and *HIMAWARI* classes, operate at altitudes of between 22,000 and 23,000 miles (35,398–37,007 km), enabling them to survey a 1,700-mile-wide section of Earth between latitudes 60 degrees North and South.

A brief chronological history of the development of meteorological satellites follows.

CHRONOLOGY OF METEOROLOGICAL-SATELLITE DEVELOPMENT

1947: Equipped with a black-and-white camera and two rolls of 16-mm film, an Aerobee sounding rocket is launched from New Mexico's White Sands Proving Grounds on September 25. Reaching an altitude of 81 miles, the Aerobee's cameras manage to capture the image of a previously unidentified tropical system in the western GULF OF MEXICO. Quickly recognizing the forecasting potential of Earth-orbiting weather stations, meteorologists and scientists begin preliminary development of a storm-tracking satellite system for the U.S. military.

1949: The U.S. Navy launches its first *Viking* atmospheric research rocket from White Sands Proving Ground on May 3. Intended to observe the workings of the upper troposphere but unable to reach a usable orbit after launch, the first *Viking* rocket is followed by seven others by midsummer of 1951.

1958: The Radio Corporation of America (RCA) unveils its design for the world's first civilian meteorological satellite in early February.

1959: Constructed by RCA and named *TIROS*, the first weather satellite is placed under the operational authority of NASA's Goddard Space Flight Center on April 13 and readied for its successful launch on April 1, 1960.

1960: In orbit for fewer than ten days, the TIROS meteorological satellite photographs an intense Coral Sea CYCLONE several hundred miles off the east coast of AUSTRALIA on the morning of April 10. In a midafternoon press conference, Francis W. Reichelderfer, director of the U.S. WEATHER BUREAU, triumphantly displays the world's first satellite images of a tropical cyclone and announces that the development of meteorological satellites now means "that hurricanes spawned anywhere over the vast oceanic areas can be detected much earlier than ever before possible."

1961: The lifesaving benefits of tracking tropical cyclones by satellite are boldly evidenced during Hurricane CARLA's landfall in eastern TEXAS on the afternoon of September 10. An intense Category 4 HURRICANE, Carla's 46-person DEATH TOLL could have been much higher had adequate HURRICANE WARNINGS—made possible by TIROS III's surveillance between September 7 and 10—not been issued beforehand.

1963: A feasibility study for the world's first geostationary weather satellite, christened Synchronous Meteorological Satellite (SMS), is commissioned from a U.S. aerospace contractor by NASA on February 1.

1964: Primarily designed as an atmospheric testing station, the first of the *NIMBUS*-class satellites—designated Nimbus I—is launched from Vandenberg Air Force Base in CALIFORNIA on August 28.

1966: *Essa II,* the first meteorological satellite to be equipped with the new Automatic Picture Transmission (APT) system, is launched from Cape Canaveral, FLORIDA, on February 28. On December 7 of this year, the first *Applications Technology Satellite (ATS-1)* is sent into geosynchronous orbit from Cape Canaveral.

1967: Formation of the World Weather Watch takes place in Geneva, Switzerland, on March 23. Directed by the World Meteorological Organization, an agency of the United Nations and drawing upon the technological resources of the United States, the Soviet Union, and Australia, the World Weather Watch is intended to enhance the effectiveness of weather satellite reconnaissance by establishing computerized data banks and 24-hour global forecasting centers.

1969: *Meteor 1,* the former Soviet Union's first operational meteorological satellite, is launched into orbit from the Russian outpost of Plesetsk on March 26. Situated at an altitude of 443 miles, *Meteor 1* is fitted with two cameras as well as infrared imaging technology. Between 1969 and 1981, a total of 32 *Meteor*-class satellites are deployed by the Soviets and used for tropical cyclone tracking over the western North Pacific Ocean.

1970: The first of the *TIROS-M* class of meteorological satellites in launched into orbit from Vandenberg AFB on January 23.

1972: The first *TIROS-M* meteorological satellite to be fitted with a spin-scan radiometer, *NOAA-2,* is placed in orbit on October 15.

1975: *GOES-1,* the first of a new class of geostationary meteorological satellites, is launched into orbit by the United States on October 16.

1977: *Himawari,* the first of a new class of geostationary satellites, is placed into orbit over the western North Pacific Ocean by JAPAN on July 14.

1978: The first of the *TIROS-N* meteorological satellites is rocketed into orbit from Vandenberg on October 13.

1981: Europe's *Meteosat 2* weather satellite is placed into geostationary orbit above the equator on June 19. Equipped with television cameras and a radiometer, *Meteosat 2* is designed to monitor tropical cyclone generation over the North Atlantic and Indian oceans.

1994: The most advanced satellite of its kind, *GOES-8* is launched into geostationary orbit on April 13; less than one year later, in March 1995, it is followed into orbit by *GOES-9*.

GOES is an acronym for Geostationary Operational Environmental Satellite, GOES is a series of 13 geosynchronous meteorological SATELLITES launched into orbit by the United States between 1975 and 2006 and used to monitor TROPICAL CYCLONE activity in the Atlantic, Pacific and Indian Oceans. Designed and constructed under contract to the National Oceanic and Atmospheric Administration (NOAA), the nine GOES-class satellites were the first to be equipped with the Synchronous Meteorological Satellite (SMS) operational system and were operational successors to the ESSA satellites of the late 1960s. Able to remain positioned over a fixed geographical point by orbiting at the same speed as Earth's rotation, the GOES series not only provided the world's meteorological and scientific communities with their first continuous-surveillance weather satellite network, but also served as a prototype for subsequent generations of geostationary weather satellites, including EUROPE's Meteosat class, INDIA's Indiasat line and JAPAN's HIMAWARI series. Extraordinarily reliable despite their small size and fragile components, the GOES satellites have for two decades successfully photographed and transmitted to Earth hundreds of thousands of weather-related images, many of them featuring some of the most notable North Atlantic HURRICANES and North Pacific TYPHOONS on record.

The first of the GOES-class satellites—christened *GOES-1*—was launched from Cape Canaveral, FLORIDA, on October 16, 1975. Measuring some eight feet (3–2 m) high by six feet in diameter and weighing 650 pounds, pill box-shaped *GOES-1* carried a visible and infrared spin-scan RADIOMETER, a magnetometer, and an instrument used to observe thermal and water vapor levels in the TROPOSPHERE called an Atmospheric Sounder (VAS). Positioned over longitude 98 degrees West and fueled by a bank of solar panels, *GOES-1* circled Earth at 18,000 MPH at an altitude of 22,300 miles, transmitting images with a resolution of five miles to the NATIONAL HURRICANE CENTER (NHC) and the National Weather Service (NWS) every 30 minutes.

Operated by NOAA's National Environmental Satellite, Data, and Information Service, the remaining eight GOES-series satellites were placed in service between June 16, 1977 (GOES-2) and April 13, 1994 (GOES-8). On June 16, 1978, *GOES-3* was launched into orbit at longitude 135 degrees West. Less than two years later, on September 9, 1980,

GOES-4 was carried into orbit above 98 degrees West by a Delta rocket. Substantially larger than its predecessors (12 feet [4 m] high, and having a deployment weight of 975 pounds), *GOES-4* possessed technology that allowed it to collect meteorological information automatically from ocean buoys and remote weather stations and then retransmit that data directly to receiving dishes at NHC and NWS. In many respects a flagship satellite, *GOES-4* was followed into orbit by *GOES-5* on May 22, 1981, and *GOES-6* on April 28, 1983. *GOES-G*, launched on May 3, 1986, and intended to become *GOES-7*, was destroyed by remote control after the propulsion unit on its launch-rocket failed shortly after liftoff. Its successor, *GOES-7*, was successfully launched into an orbit above longitude 83 degrees West on February 26, 1987. During the early morning hours of August 24, 1992, *GOES-7*'s infrared photography captured Hurricane ANDREW's catastrophic rampage across southern Florida.

Constructed at a cost of nearly $220 million and placed into regular service in November of 1994 after nearly a year of trials, the 2,161-pound *GOES-8* satellite possessed the ability to obtain temperature and humidity levels electromagnetically over 19 wavelength regions of the troposphere, as well as to record day-and-night cloud-cover measurements and SEA-SURFACE TEMPERATURES with unprecedented accuracy and detail. Powered by 2,622 solar cells arranged along a 70-foot-long (23-m) panel, *GOES-8* was also fitted with sensors that allowed it to receive radio distress calls from foundering ships and lost aircraft. Commissioned in time to observe the very active 1995 North Atlantic HURRICANE SEASON, *GOES-8* (also known as *GOES-NEXT*) provided NHC forecasters with several hundred high-resolution, visible and infrared images of that season's tropical cyclones, including one spectacular shot taken on the morning of August 30 that shows no less than five tropical systems—three hurricanes and two TROPICAL STORMS—swirling across the North Atlantic. Between 1995 and 2006, five additional satellites were launched, GOES 9–13.

Savannah-Charleston Hurricane *Southeastern United States, August 18–30, 1893* This remarkable Category 3 HURRICANE, long considered one of the deadliest TROPICAL CYCLONES to have occurred during the last decade of the 19th century, brought 130-MPH (209-km/h) winds, five-inch (127-mm) rains, and a violent 12-foot (4-m) STORM SURGE to the barrier islands and principal cities of GEORGIA and SOUTH CAROLINA between August 27 and 28, 1893. In Georgia, where the hurricane's 30-foot seas flowed nearly 20 miles inland, almost 500 people perished in the inundation of riverside rice plantations. In South Carolina, where a raging STORM TIDE submerged almost a quarter of the city of Charleston, another 1,400 people were drowned, their bodies swept into swamps and flooded inlets. Millions of dollars in property damage were assessed as ships were run aground, church steeples toppled, streetcar lines were downed, and the steamboat *City of Savannah* was wrecked on Hunting Island. Although all 30 passengers and crew on board were eventually rescued, the 2,100-ton vessel, pounded into pieces by the hurricane's enormous waves, was declared a total loss.

Like many midseason NORTH ATLANTIC tropical systems, the Savannah-Charleston Hurricane originated near the Cape Verde Islands off the northwest coast of Africa on the evening of August 18, 1893. Borne on a long, arcing course northwest and having missed all the principal islands of the CARIBBEAN SEA along the way, the intensifying hurricane blasted the northern BAHAMAS on August 26 with estimated 113-MPH (182-km/h) gusts and swirling 19-foot (7-m) seas. By the early morning of Sunday, August 27, the hurricane's calm EYE, accompanied by roaring EYEWALL winds most probably in excess of 125 MPH (201 km/h), was bearing down on the Georgia coast, where a haphazard series of storm warnings were passed by both word-of-mouth and official, but unpublished, weather bulletins.

Its CENTRAL BAROMETRIC PRESSURE of 28.81 inches (975 MILLIBARS) accompanied by a train of hurricane-force gusts and pelting rains, the Savannah-Charleston Hurricane rushed ashore between the two cities on the afternoon of Sunday, August 27, 1893. The Georgia dock town of Beaufort, northeast of Savannah, was ravaged; nearly every building in the prosperous town was demolished. The fury of the wind and waves was of such intensity that several 4,500-ton iron weights used to anchor buoys in a nearby channel were rolled 100 yards (300 feet; 10 m) inshore. The steamboat *Pilot Boy* was driven aground near the mouth of the Beaufort River, while neighboring Paris Island, where the United States military was in the process of building a naval dockyard, was inundated by the hurricane's terrific storm surge. Some 20 people, many of them former slaves, were drowned as their crude shanties melted into the raging surf. A contemporary newspaper account of the tragedy stated that in many instances, people were so frightened by the sudden violence of the hurricane's onset—the tide, for instance, rose 3 feet per hour for most of the storm's duration—that they did not leave their houses.

To the south, where the timber-trading city of Savannah lay within the hurricane's NAVIGABLE SEMICIRCLE, the Weather Bureau's ANEMOMETERS recorded sustained wind speeds of 70 MPH (113 km/h) before being carried away by a rapid series of screaming gusts. The hurricane's shoaling surge, rolling inland across the Tybee Sound before spilling into the Savannah River, completely submerged all the wooden wharves belonging to the Savannah, Florida and Western Railroad companies. Several barkentines and sloops anchored at Savannah's Quarantine were washed ashore and left stranded, sometimes dismasted, on the bubbling marshes that embraced the city. Strafing gusts downed telephone and telegraph lines, toppled the Savannah Brewing Company's brick smokestack, peeled the tin roof from the Market Building, and dropped the steeple of the First African Baptist Church into the street. Dozens of rice plantations along the sodden banks of the Savannah River were virtually obliterated by the hurricane's rearing seas, claiming not only one of the most bountiful rice harvests in years but also some 500 lives, many of them freed slaves who had remained on the plantations after the Civil War (1860–1865). Surprisingly, only one person in Savannah itself—a barber electrocuted by a fallen power line—reportedly perished; a remarkably low DEATH TOLL, considering both the severity of the hurricane and the huge number of casualties it claimed elsewhere in the region.

The same cannot be said for the border city of Charleston, South Carolina, north of where the hurricane's 140-mile-wide (210-km) eyewall came ashore at Beaufort. Positioned within the hurricane's DANGEROUS SEMICIRCLE, Charleston suffered far greater damage than Savannah, with nearly a quarter of the city being deluged by a remarkable 17-foot (6-m) storm tide on the afternoon of August 27. As the height of the hurricane's surge was added to the coincidental peak of the high tide, the North Atlantic Ocean began to flow across the western half of the Charleston Peninsula, filling streets, immersing gas meters, and plunging the city into complete darkness as the stormy night progressed. Eventually reaching a height of 10 feet (3 m), the flood joined both the Cooper and Ashley rivers, adding their combined flow to the pooling devastation that was once the Charleston waterfront.

Causeways and embankments were overrun and then breached. The recently constructed Ashley Drawbridge, a steel-and-stone marvel between Charleston and Ashley Island, was first buckled by the wind and then sent twisting into the river by the hurricane's tides. Charleston's famous Battery fortress was similarly razed, its sheer brick walls and wooden railings first undermined by the roaring seas and then engulfed. The roof of a prominent downtown hotel was sheared off; 14 boxcars on a siding were blown off the track and overturned. Every one of the 73 churches in Charleston received some degree of structural damage, ranging from the loss of windows to the snapping of steeples to complete building failure. Thousands of valuable timber trees were plucked from the ground, hurled like battering rams through the sides of buildings. The lightship at Rattlesnake Shoals, just outside Charleston Harbor, was run aground and wrecked, as were 30 other vessels in Charleston Harbor at high tide. Three of the lost vessels, elegant, clipper-bowed steam yachts, were slammed ashore and broken up when the hurricane's surge carried away the entire Carolina Yacht Club—piers, ornamental lighthouse, and all. One schooner was lifted well inland and came to rest in the middle of a city street. A large portion of Charleston's cotton crop was riven while still in the field, picked, and scattered by the wind; the baled remainder, awaiting shipment in waterfront warehouses, was dumped into the sea and ruined. Six people in Charleston reportedly perished, one of them an old woman who was said to have died of fright. Nearly a million 1893 dollars was assessed in property damage, making the Savannah-Charleston Hurricane one of the most costly strikes in the city's history.

On the morning of August 29, as Charleston's fire department began to use its pumper trucks to clean out contaminated wells and cisterns, the people of the ravaged city emerged to gawk, pray, and wonder at the destruction around them. Streets were carpeted with leaves, while fallen trees and limbs made travel slow and hazardous. Rivers and streams, clogged with muddy debris, occasionally backed up, spawning levee breaks and flash floods. By midmorning, reports began to filter into the city concerning the devastation on Sullivan's Island, a sandy barrier outpost near the entrance to Charleston Harbor. Its surface topography completely changed by the storm tide's inundation, the island's 32 houses were gutted, its church was collapsed, and its resort hotel was lightened by two stories. A heavy, steam-powered pile driver, used to build piers and bridges, was carried from one end of Sullivan's Island to the other, a weighty testament to the power of this particular hurricane. Two people on the island were drowned.

But it was not until August 30, when the morning editions of the newspapers (which had themselves suffered damage to presses and copy rooms) were delivered to the streets, that the true dimensions of the tragedy began to emerge. In places like Beaufort, Georgia, and Port Royal, South Carolina, where the hurricane's eyewall fury had been at its most intense, little remained of the thriving farming, fishing, and trading settlements that once clustered there. Hundreds of houses, cottages, and shanties serv-

ing as storm shelters had been washed away. Shallow barrier islands—among them St. Helena, Ladies' Island, Wawthas Island, Dawtha Island, and Coosaw Island—were thoroughly submerged by the hurricane's storm surge—storm tide combination. Before it was all over, between 775 and 1,400 people would die, some so gruesomely that their identities would remain forever unknown. Another 7,000 survivors, many of them impoverished former slaves, were displaced by the hurricane's destruction and forced to endure a biting winter of hardship and neglect while they sought to rebuild their communities.

In the meantime, the Savannah-Charleston Hurricane, apparently maintaining a great deal of its strength as it blustered inland, delivered considerable damage to portions of South and NORTH CAROLINA on August 29. In the town of Ridgeland, South Carolina, three mules were crushed to death when the hurricane's 110-MPH (177-km/h) gusts leveled a barn. More than 100 buildings in the town of Kernerville, North Carolina, were likewise flattened, and a nearby brick church suffered the loss of its steeple, portico, roof, windows, and pews. One woman was killed, and some $60,000 in 1893 property damage was tabulated.

Savanna-La-Mar Hurricane *Northern Caribbean Sea, October 1–5, 1780* The first of three severe HURRICANES to have whipped through the CARIBBEAN SEA during the memorable month of October 1780, the Savanna-La-Mar Hurricane released 125-MPH (201-km/h) winds, streaming rains and a rampaging 20-foot (7-m) STORM SURGE on the south coasts of JAMAICA and CUBA between October 3 and 4, 1780. Named for a Jamaican port town it obliterated at landfall on October 3 and followed into hurricane history by both the GREAT HURRICANE of October 10–18 and Solano's Hurricane of October 17, the Savanna-La-Mar's midafternoon tempest demolished hundreds of wooden BUILDINGS, cut down entire sugarcane plantations, smashed warships like teacups, and even (if eyewitness accounts are to be believed) triggered not one but two accompanying EARTHQUAKES. Some 3,000 people—over half of them sailors on ships at sea—reportedly perished during the hurricane, making the Savanna-La-Mar storm one of the deadliest natural disasters in Jamaican history.

A late-season killer born over the reaches of the Caribbean Sea north of Santa Marta, Colombia, on or about October 1, the Savanna-La-Mar Hurricane rapidly deepened as it recurved north-northeast, seemingly targeting the towns and villages of southwestern Jamaica for an imminent assault by its Category 3 EYEWALL winds and rains. Winding past the mouth of the

Black River during the late-morning hours of October 3, the vicious core of the hurricane was finally ashore by 3 o'clock that afternoon, tearing its way through a thriving rum-and-sugar depot named Savanna-La-Mar. Crowding onto a 15-foot (5 m) headland in order to gaze in wonder at the fury of the sea, several hundred of the town's residents were said to have been swept to their deaths while observing the onslaught of the hurricane's tsunamilike storm surge. Forty others who had taken shelter in the town's courthouse perished when that building's roof collapsed, pinning them beneath tons of timber and roof shingles. In the neighboring parishes of Westmoreland, Hanover, St. James's and Elizabeth's, nearly every house—even those few built of stone—was left in ruins, causing great hardship and misery for thousands of dazed survivors. At Blue Castle, a large sugar plantation outside the town of Cambridge, 200 slaves were crushed to death when the boiling house in which they had taken refuge came apart beneath the hurricane's brutal winds. In the port town of Lucea, situated north of Savanna-La-Mar, all but two houses were completely razed, leaving not a tree, shrub, or walk to mark the outline of its former streets. Four hundred people in Lucea were killed, while the gravely injured numbered in the thousands. In Montego Bay, 363 lives—250 of them slaves—were lost to wind-shot debris, structural failures, and drownings. Damage estimates in Westmoreland alone reached nearly 1 million English pounds.

At sea, the Savanna-La-Mar Hurricane's DEATH TOLL was higher still. The British transport HMS *Monarch*, en route to Kingston with hundreds of Spanish prisoners aboard, foundered in Savanna-La-Mar's enormous seas on October 1, leaving no survivors. On October 4, as the hurricane's diminishing EYE relentlessly bore across the southeastern midlands of Cuba, the British frigate HMS *Phoenix*, 44 guns, was carried ashore at Cabo de la Cruz, Cuba, and wrecked. Two hundred of its crew perished, drowned in smothering seas, their bodies battered into a bloodless pulp against the jagged rocks that guard the infamous southern cape. Near Montego Bay, the HMS *Ulysses*, violently buffeted by the hurricane's confused seas, lost both its main and mizzenmasts; they toppled over the side in a treacherous tangle of twisted rope, torn canvas, and broken spars. Fearing that his ship was in danger of capsizing, the *Ulysses*'s captain ordered all upper-deck guns thrown over the side, a procedure that did, indeed, save the valiant vessel and its exhausted crew. In the harbor of Savanna-La-Mar itself, sailors aboard HMS *Princess Royal* were given quite a surprise when one of the hurricane's 140-MPH (225-km/h) gusts suddenly snatched the ship's fifth anchor from

the foredeck and dropped it over the side. As though adding insult to injury, the hurricane eventually finished off the *Princess Royal* completely, driving the dismasted, gutted warship ashore where, according to a contemporary newspaper account, it was set on an even keel by an earthquake and used by displaced survivors as a shelter.

sea-surface temperature This term is frequently used to describe the average seasonal TEMPERATURE of the *Ekman Layer*, or the first 300 feet of any given ocean surface. Because TROPICAL CYCLONES depend upon the processes of EVAPORATION and CONDENSATION for their creation and sustenance, sea-surface temperature readings can serve as a reliable indicator of both the potential for tropical cyclone development in a particular storm zone and the chance for future deepening or weakening in those HURRICANES, TYPHOONS, CYCLONES, TROPICAL STORMS, and TROPICAL DEPRESSIONS already in existence.

Nearly three decades of comprehensive research into the connection between sea-surface temperature and tropical cyclone activity has indicated that water temperatures must be 80° F (27°C) or higher and must extend at least 200 feet (70 m) down before the vast rates of evaporation required to feed the condensing RAIN BANDS of tropical cyclones can be generated. Because Earth's oceans are laced together by a network of independent currents, tidal eddies, and shallow coastal pools—all with their individual thermal characteristics—sea-surface temperatures can sharply vary by as much as 6 degrees F (⁻14°C) over the distance of just a few miles. Meteorologists have linked the sudden periods of intensification and weakening observed in some previous tropical cyclones to the presence of these individual sea-surface conditions and their rising or falling evaporation rates.

During the peak summer months of August and September, when tropical cyclone activity is at its most active in the NORTH ATLANTIC and NORTH PACIFIC OCEANS, semienclosed bodies of water such as the CARIBBEAN SEA, the GULF OF MEXICO, and the Gulf of California tend to possess a band of average 80°F (27°C) sea-surface temperatures along their coastlines, with a large pool of above average (2–3 degrees higher) sea-surface temperatures lying farther offshore. In the instances of the narrow Gulf of California and the tapered BAY OF BENGAL, it is not unusual for sea-surface temperatures to range 4 and 5 degrees above average, a result of the slow-moving currents that faintly circulate through their confined headwaters. Along the eastern seaboard of the United States, average sea-surface temperatures during August and September are equal to those of the North Atlantic in general, except for the stretch of coastline that extends from VIRGINIA to southern NEW ENGLAND, where average sea-surface temperatures reach between 2 and 3 degrees higher. Similar above-average thermal conditions can also be found along the west (particularly southwest) coast of FLORIDA, the western tip and islands of CUBA, the south and east coasts of HISPANIOLA and PUERTO RICO, and across the entire Leeward and Windward Island chains of the eastern Caribbean Sea.

While not every tropical cyclone that has either strengthened or weakened has done so because of an encounter with changing sea-surface temperatures, there is sufficient meteorological data to establish that in passing over a particular stretch of ocean, several mature-stage tropical cyclones have significantly lowered sea-surface temperatures, in some cases by as much as 10 degrees. Studies conducted in 1977 on Hurricane ANITA revealed that sea-surface WIND SPEEDS within the Category 4 system's EYEWALL had swiftly churned the first 100 feet (31 m) of the Gulf of Mexico into deep radiating swells; 100 feet (31 m) beneath that, the hurricane's mounting INVERSE BAROMETER developed a series of upward-spiraling currents that drew cooler water to the surface in its wake. Sea-surface temperatures recorded following Anita's passage were 8 degrees below what they had been 24 hours earlier. Other tropical cyclones where variances in sea-surface temperatures proved a leading factor in development include DIANE (1955), DONNA (1960), TRACY (1974), and OPAL (1995).

Meteorological SATELLITES equipped with infrared technology measure sea-surface temperatures through the use of an instrument called a RADIOMETER.

On July 5, 2000, Typhoon Kai-Tek passed over the South China Sea, lingering for four days before traveling northward over TAIWAN. Based on data taken from NASA's *Quikscat*, a satellite that measures WIND SPEEDS over water, sea surface temperature measurements made by the joint U.S.–Japanese Tropical Rainfall Measuring Mission satellite showed a 16° drop in the area where the counterclockwise-spinning storm had stalled. Colder water, drawn upward by the typhoon, most probably caused the drop in SSTs.

While a majority of tropical cyclones tend to leave colder SSTs in their passage, there have been observed examples where SSTs will actually rise slightly following a tropical cyclone. On July 12, 2005, an analysis prepared by the University of Miami for the NATIONAL HURRICANE CENTER (NHC) indicated that when Hurricane DENNIS passed through the northeastern Caribbean Sea during the first two weeks of July, its circulatory winds essentially covered its SST tracks by moving warm

waters around and behind the system. This, in turn, allowed the 5th named tropical cyclone of the 2005 season—EMILY—to maintain its intensity as it basically followed Dennis's track during the third week of July 2005.

Second Colonial Hurricane *Northeastern United States, September 7, 1675* While perhaps not as powerful or large a storm as the GREAT COLONIAL HURRICANE of August 25, 1635, the Second Colonial Hurricane nevertheless pilloried much of the NEW ENGLAND seaboard with intense winds and crushing seas. In New London, Connecticut, numerous merchant ships were either driven aground or wrecked. In Boston, some 90 miles (145 km) northeast, apple trees were plucked of their fruit, hay and corn stocks were laid over in ruin, and several ships and piers in the bustling harbor were destroyed. Damage estimates for Boston alone totaled 1 thousand pounds, a princely sum in colonial New England. As is the case with so many pre-19th century hurricanes, no complete count of the dead or missing exists.

South Atlantic Ocean A 10,000,000-square-mile span of the Atlantic Ocean that reaches from the Antarctic Circle in the south (latitude 66 degrees South) to the equator (latitude 0 degrees) in the north and is bound to the east by the continent of Africa and to the west by that of South America, the South Atlantic Ocean has a limited history of tropical cyclone activity. While the South Atlantic is subject to the furies of large, powerful EXTRATROPICAL CYCLONES, only five documented tropical cyclones have ever originated over or been sustained by its relatively tepid (mean average temperature is 75 degrees F (24°C)) waters. Confirmed by two decades of nearly continuous SATELLITE surveillance, this climatological curiosity has given rise to a number of hypothesis, including the presence of cooler SEA-SURFACE TEMPERATURES across the South Atlantic, a northward-positioned INTERTROPICAL CONVERGENCE ZONE (ITCZ) that does not migrate below the equator, and a dearth of seedling, low-pressure TROPICAL WAVES from central and southern Africa. South Atlantic tropical cyclones were reportedly observed in 1991, January 2004, March 2004 (Catarina), and two tropical systems in February and March 2006.

South Carolina *Southeastern United States* Presenting some 187 miles (301 km) of coastal plains, palmetto-lined barrier islands, and feathery river mouths to the HURRICANE-roiled waters of the NORTH ATLANTIC OCEAN, South Carolina has over the last four centuries repulsed a veritable broadside of TROPICAL CYCLONE strikes. Between 1686 and

2006, South Carolina's principal seaports, Charleston and Georgetown, were affected by 88 documented hurricanes and TROPICAL STORMS, with 41 of these mature systems making landfall on the state's east coast. Of the latter, 20 were considered major hurricanes, Category 3 and 4 systems with minimum winds of 111 MPH (179 km/h), nine–18-foot (3–6-m) STORM SURGES, and sizable DEATH TOLLS. Although the deadliest hurricanes in South Carolina's history—September 27, 1822, and August 27–28, 1893—occurred on the state's North Atlantic side, a number of South Carolinian storms have originated over the GULF OF MEXICO or southwestern CARIBBEAN SEA. In many instances achieving considerable size and intensity as they recurve northeast, these tropical storms and hurricanes first come ashore in northwestern FLORIDA before tearing though central GEORGIA and western South Carolina, releasing earth-moving rains and drilling TORNADOES over the Piedmont Plateau and southern Blue Mountains.

On average, South Carolina's east coast, noted for its upscale golf resorts and active riverways, is indirectly threatened by a tropical cyclone every 2.5 years and directly impacted by a tropical storm or mature hurricane every 5.5 years. On at least six different occasions between 1728 and 1959, South Carolina has even withstood multiple strikes in the course of a single HURRICANE SEASON, as occurred in 1728, when two severe hurricanes slammed into the state between August 11 and September 14. Other double strikes were noted in 1752 (August 15 and September 15), 1822 (August 11 and September 27), 1873 (September 20 and September 24), 1898 (October 7 and October 24), and 1959 (July 7 and September 16). Both 1873 hurricanes were of Gulf origin and first passed over Florida and Georgia before causing widespread damage to houses, piers, and shipping along South Carolina's Atlantic coast.

Since 1761, eastern South Carolina has been directly affected by at least three June tropical cyclones, all of them storms of damaging consequence. The first, which stranded five merchant ships on Charleston's Rebellion Road, may in fact have only been of tropical storm—strength when it passed over that city on June 1, 1761. A poorly documented event, it is entirely possible that the June 1 system developed over the Gulf of Mexico, migrated to the northeast, crossed northern Florida as a weak tropical storm (or perhaps a healthy TROPICAL DEPRESSION), and first entered the North Atlantic near the Florida–Georgia border. Maintaining its RECURVATURE northeast, the early season storm would then have been in a position to trail along the South Carolina coast, off-loading its reputed fury on the blossoming industry of Charleston Harbor.

Less than a decade later, on June 6, 1770, another early season tropical cyclone of indeterminate origination and strength dusted Charleston, its low-pressure sweeps lifting small roofs, scattering fences, and plucking many of the city's feathery palmetto trees from the ground. Similar in many respects to the 1761 storm, the 1770 system, too, was most likely born over the temperate waters of the Gulf of Mexico, retracing the earlier storm's track northward before inundating Charleston's vulnerable harbor with terrific rain squalls and a five-foot storm surge.

Conversely, the 1761 and 1770 storms could have wandered in from the southeastern Caribbean Sea, following the path taken by South Carolina's third—and most powerful—June hurricane, that of June 3, 1825. Ship's logs place the sprawling tropical cyclone, later christened the Early June Hurricane of 1825, at Santo Domingo on May 28 and over CUBA's east coast by June 1. By the morning of June 3, the system was abreast of Charleston, its powerful EYEWALL flaying the sea's surface into long swells that curled ashore as far north as Norfolk, VIRGINIA. Traveling very slowly north-northeast, the Early June Hurricane spent the next 24 to 30 hours crippling the city with pummeling 5-inch (127-mm) PRECIPITATION counts, limb-snapping winds, and a 14-foot (5-m) STORM TIDE. A number of small boats foundered; nearly every pier along the waterfront was splintered. Further inland, hundreds of small pine trees were either uprooted or shaved off at ground level, an untidy yet welcome windfall that later went toward replenishing timber stocks depleted by rebuilding efforts. Despite their apparent severity, no deaths or serious injuries were reported in the aftermath of South Carolina's three recorded June hurricanes.

In July, when SEA-SURFACE TEMPERATURES north of latitude 30 degrees still hover just below the 80 degree F (27°C) minimum required to sustain a tropical cyclone's CONVECTION currents, South Carolina has been tested by two confirmed hurricane strikes, both occurring during the 20th century. On the evening of July 14, 1916, an unnamed Category 1 hurricane slipped ashore just south of Myrtle Beach. Possessing a CENTRAL BAROMETRIC PRESSURE of 28.9 inches (980 mb), the hurricane produced 85-MPH (137-km/h) gusts and sporadic downpours but caused no deaths or injuries. Property damage was light. Not a half-century later, on July 23, 1959, South Carolina's second July hurricane, a mild Category 1 system named Abby, blew ashore just north of Beaufort. Much weaker than the July 1914 hurricane, Abby lashed the foliage from trees and severed power lines across the state but claimed no lives.

It is during the months of August and September, when the first of the CAPE VERDE STORMS begin to knock at the west gates of HURRICANE ALLEY, that the number of South Carolinian hurricane visits rises to 13 and 27, respectively. The state's susceptibility to strikes from September hurricanes in particular is evidenced by the fact that an above-average number of North Atlantic tropical cyclones have come ashore in South Carolina on either the 14–15th or the 16–17th of the month and that these dates must represent singularities, days on which hurricanes are statistically more likely to make landfall than others. Several of South Carolina's most notorious hurricanes—mature systems characterized by a catastrophic loss of life and enormous property damages—have occurred on these dates. On September 14, 1728, a howling hurricane inundated Charleston harbor, sinking eight merchant ships and stranding twenty-three others, while another, on the same day in 1844, demolished four waterfront warehouses, a large pier, and part of a recently reinforced stone seawall. On September 15, 1752, a hurricane of considerable intensity razed nearly every two-story BUILDING in Charleston, sank 20 cotton-laden English merchant ships at anchor in the harbor, and pushed ashore three Royal Navy frigates near Sullivan's Island. Reported to have produced a storm tide of at least 10 feet (3.1 m), the September 15, 1752, hurricane claimed more than 100 lives, many of them slaves who were trapped in their shanties on Sullivan's Island by the arrival of the sweeping tide. At the time it was said in South Carolina that the hurricane's shearing winds had caused thousands of apple and peach trees to flower a second time and many to bear a second harvest of fruit. On September 16, 1700, a major hurricane trounced greater Charleston, killing scores of residents and driving a Scottish merchant ship onto Charleston's treacherous sandbar. Thirteen years later, an even more violent hurricane struck the city on September 16, with submerging storm tides and rain-splattered winds wrecking 34 small boats and ships, washing beachheads away in nearby Port Royal, and toppling an 80-foot (25-m) brick lighthouse on Sullivan's Island. For several days following the hurricane's passage, the port of Charleston was closed to all shipping while work boats trawled tons of wreckage—snapped spars, entangled masts, and hundreds of wooden casks—from its rain-swollen, mud-colored environs.

On September 17, 1906, a strapping, Category 3 hurricane belted eastern and inland South Carolina with 129-MPH (208-km/h) winds and torrential rains. Three people were drowned when their steam launch was overturned in Charleston harbor. Twenty-two years to the day later—the night of September 17, 1928—the dying EYE of what had been one of the most catastrophic of all recorded Cape Verde storms, the celebrated SAN FELIPE HURRICANE, passed northward along the South Carolina coast. Responsible earlier for ending the lives of some 3,366 people in PUERTO RICO and Lake Okeechobee, Florida, San Felipe's 110-MPH (177-km/h) winds and seven-inch (178 mm) rains claimed another five lives in South Carolina and caused millions of dollars in property damage.

Other historic midseason hurricanes in South Carolina include the 1686 tempest that in September of that year prevented Spanish raiders from driving English settlers out of Charleston; the hurricane of August 9, 1781, which destroyed six ships at Charleston harbor, two of them the English warships *Thetis* and *London;* the blustery hurricane of September 10, 1811, during which a storm-generated TORNADO gouged a broad swath of destruction through downtown Charleston; the hurricane that on August 27, 1813, submerged much of South Carolina's northeast coast, driving five ships ashore at the mouth of the Edisto River and drowning between fifteen and twenty people on Sullivan's Island; the particularly deadly hurricane of September 27, 1822, which killed 300 people in Winyah Bay and another 300 on North Island as it blasted ashore at Cape Romain; the severe hurricane that, on August 26, 1881, claimed nearly 100 lives as it recurved into South Carolina from Georgia; the 125-MPH (201-km/h) tempest that ravaged Charleston on the night of August 24–25, 1885, killing 21 people and irreparably damaging the city's famed lighthouse; Hurricane Gracie, a moderate Category 3 system that lunged ashore just north of Beaufort on September 16, 1959, its 10-foot (3-m) storm surge and rough seas destroying a number of beachfront buildings but claiming no lives; Hurricane DAVID, September 4, 1979, which delivered 92-MPH (148-km/h) winds to Hilton Head Island, causing huge washouts and much tree damage; and the immense, much earlier Carolina Hurricane of September 7–8, 1804, that first came ashore in northeastern Georgia before skating up the coast and into Beaufort on the morning of September 8. Referred to in some histories as the Antigua-Charleston Hurricane of 1804, the Carolina Hurricane decimated the piers and warehouses of Charleston, washing away scores of small houses in the process. As the seemingly impregnable stone ramparts of Fort Moultrie surrendered to the sea's undermining assault, 20 houses on nearby Sullivan's Island disappeared, casting their unfortunate inhabitants into the maelstrom. Five vessels caught by surprise in Charleston Harbor were sunk, while another eleven were dismasted. Some 380 people reportedly perished in the Carolina Hurricane of 1804, making it one of South Carolina's deadliest recorded hurricanes.

In all probability the deadliest tropical cyclone in South Carolina's history, the SAVANNAH-CHARLESTON HURRICANE ended an estimated 1,400 lives as its encroaching seas, said by eyewitnesses to have been 30 feet (10 m) high in places, flooded a number of riverside rice plantations and the port of Charleston on August 28, 1893. A hurricane of remarkable size and ferocity upon arrival in Charleston, the Savannah-Charleston Hurricane's destructive impact on the city would remain unsurpassed for 96 years, until the early evening landfall of Hurricane HUGO, September 21, 1989. Held off the Georgia coast long enough to reach devastating landfall in Charleston, Category 4 Hugo and its ominously low central pressure of 27.6 inches (934 mb), forced the evacuation of more than 300,000 South Carolina residents from beach-front resorts. Gusts of up to 155 MPH (249 km/h) yanked trees from the ground, flattened small buildings, swept sailboats and motor yachts ashore, and carried away several of the bridges that connected Charleston to its barrier islands. A frothing 17-foot (5-m) storm surge rocked concrete piers and breached seawalls, flooding large sections of downtown Charleston. Deemed the costliest hurricane to have struck the United States up to that time, Hugo claimed 28 lives in South Carolina and tallied up damage estimates of more than $4 billion.

In October and November, a time when the North Atlantic's heaviest hurricane activity generally shifts to the Gulf of Mexico and southwestern Caribbean Sea, South Carolina's number of direct strikes subsides to a fairly manageable four hurricanes in October and none in November. South Carolina's first documented October hurricane struck on October 13, 1894; the second came ashore at Charleston on October 7, 1898. Both systems were relatively minor, causing few deaths and minimal property damage. On October 11, 1947, an unnamed Category 2 hurricane toted its central pressure of 28.76 inches (974 mb) ashore at Myrtle Beach, downing power lines and snapping tree limbs but claiming no lives. The worst of South Carolina's October hurricanes, HAZEL, barreled ashore at North Myrtle Beach on October 15, 1954. Sustained winds of 106 MPH (170 km/h) at landfall, along with a 17-foot (6-m) storm tide, made Hazel one of the Palmetto State's most destructive hurricanes. One man died, while some $27 million in property losses were assessed.

On October 11, 2002, Tropical Storm Kyle came ashore in South Carolina. Heavy rains were reported in near Edisto Beach State Park, some topping 6.35 inches (152 mm). Some coastal flooding was recorded, and tides were one to two feet (.5–1 m) above normal.

In 2004, South Carolina suffered two direct hurricane strikes, the first time the state had been directly affected by multiple storms since 1959. After pounding Florida, CHARLEY reemerged into the Atlantic, and struck South Carolina as a Category 1 hurricane. Less than three weeks later, Hurricane GASTON made landfall a few miles from where Charley came ashore in South Carolina.

In 2005, Hurricane OPHELIA brought flooding rains and coastal erosion to South Carolina. Seesawing between hurricane and tropical storm strength, Ophelia lingered off the coasts of North and South Carolina for nearly one week, falling below hurricane intensity on September 12, 2005, but regaining hurricane classification within the next day. On September 12, Ophelia was located about 145 miles (233 km) south of Wilmington, North Carolina, and about 140 miles (225 km) east-southeast of Charleston, South Carolina. Although Ophelia remained a largely offshore tropical system, its tropical storm force winds extended as far as 160 miles (257 km) from its center. Up to 10 inches (254 mm) of rain was forecast. Voluntary EVACUATIONS were called for oceanfront and riverside areas in northeastern South Carolina and North Carolina's Outer Banks.

After rolling shore on the Florida–Georgia border on October 5, 2005, Tropical Storm Tammy dragged northward, delivering six inches (152 mm) of precipitation and high winds to the resort community of Hilton Head on October 6, 2005.

Southeastern New England Hurricane *Northeastern United States, October 23–24, 1761* This intense TROPICAL CYCLONE, judged by eyewitnesses to have been the "most violent storm in 30 years," unraveled great destruction on the coastal and inland communities of southeastern NEW ENGLAND. After collapsing a wooden bridge over Rhode Island's Narragansett Bay and toppling the wooden spire of Newport's Trinity Church, the Southeastern New England Hurricane traipsed northeast, uprooting thousands of trees and causing damaging tides in Portsmouth, New Hampshire, and Casco Bay, Maine.

Stan, Hurricane *Caribbean Sea–Central America, October 1–5, 2005* Hurricane Stan, the 10th hurricane and 18th named storm of the 2005 North Atlantic HURRICANE SEASON, made landfall on MEXICO's Yucatán Peninsula and subsequently at Veracruz, on October 4, 2005. Bearing sustained winds of 80 MPH (130 km/h), Stan forced the closure of all three of Mexico's primary Gulf coast ports (Coatzacoalcos, Cayo Arcas, and Dos Bocas) and delivered pounding rains to Oaxaca state. A copious rainmaker, Stan dropped between five and 10 inches (127–254 mm) of rain fell across the jungles of the Yucatán and in neighboring BELIZE. In Veracruz, three people were killed and another seven injured. Quickly weakening to a TROPICAL DEPRESSION, Stan's remnants drifted across Central America for nearly five days, inundating the region with heavy PRECIPITATION falls. As many as 2,000 people died as Stan ground across Central America, spawning mudslides in El Salvador and Guatemala. In Guatemala, 1,400 Maya Indians were killed, while 400 fatalities were recorded elsewhere in Central America. In El Salvador, more than 40,000 people fled their homes, while landslides claimed at least 62 lives. As of October 6, 2005, landslides and marine accidents spawned by Stan were blamed for nine deaths in Nicaragua, one in Costa Rica, four in Honduras, and six others in southern Mexico. The total DEATH TOLL was 796 people (with many missing) as of October 19, 2005.

steering current This term is given to those surrounding air currents that influence, or "steer," a TROPICAL CYCLONE's direction. Because a HURRICANE, TYPHOON, or CYCLONE possesses only minimal means for self-propulsion, its spiraling mass must depend upon adjoining AIR MASSES to determine what track it will follow during its lifespan. These climatological features can range from high-pressure ANTICYCLONES to the westward flow of the trade winds to barricading cold fronts.

storm surge One of two meteorological terms (the other being STORM TIDE) given to that rise in stillwater sea level that accompanies the landfall of a TROPICAL CYCLONE, the storm surge has been long recognized by scientists, historians, and civil defense authorities as the single most destructive aspect of a HURRICANE, TYPHOON, or CYCLONE. Laden with seaweed, rocks, shattered trees, automobiles, pieces of buildings, and even ships, it has on numerous occasions been responsible for substantially adding to a respective storm's DEATH TOLL. Variously appearing to surviving eyewitnesses as a rapid rise in water levels, a towering wall of water, or a series of enormous waves, the type and severity of individual surges varies from storm to storm and between the assorted kinds of coastlines found in those regions of the world most subject to cyclonic activity. Directly linked to changes in a tropical cyclone's CENTRAL BAROMETRIC PRESSURE, surges can range from four to five feet (1–2 m) in a Category 1 hurricane, to 9 to 12 feet (3–4 m) for a Category 3 typhoon, to 18 feet (6 m) and above in a Category 5 cyclone. When, by a coincidence of timing, the storm surge comes ashore during high tide, the resulting condition is known as a storm tide; in this case, the height of the

tide is added to that of the surge. In either scenario, the consequences are often severe for those BUILDINGS, ships, and residents situated along the afflicted shore. Storm surges of between 20 and 35 feet (7–10 m) have often been recorded in modern hurricanes, typhoons, and cyclones, while several reliable narratives from the 17th and 18th centuries document surges of even greater heights, 40–45 feet (12–14 m). Rolling inland like a road grader, the titanic weight of storm surges bearing down on the earth has further been connected to seismic activity—those EARTHQUAKES that many eyewitness have claimed coincided with the landfall of a tropical cyclone.

Years of concerted study into the nature and behavior of the storm surge have yet to yield a complete picture of this remarkable meteorological phenomenon. During the 16th and 17th centuries, when barometers and other devices for measuring a tropical cyclone's pressure were still in their developmental stages, the storm surges that accompanied particular cyclonic systems were solely attributed to the influence of long-fetch winds on the sea's surface. A contemporary analysis of the devastating SOUTH CAROLINA Hurricane of September 16–17, 1713, stated that during the storm's peak, "the winds raged so furiously that it drove the sea into Charlestown, damaging much of the fortifications whose resistance it is thought preserved the town. . . ." Less than a half-century later during the intense hurricane of September 15, 1752, Charleston was again partly inundated by a storm surge whose "unexpected and sudden fall . . . [of] five feet in the space of 10 minutes . . ." an observer later ascribed to a shift in the wind from northeast to southwest.

In September 1759, the northeasterly winds of a mighty Floridian hurricane "so greatly impeded the current of the Gulf Stream" that the waters of the GULF OF MEXICO were, according to a contemporary source, forced over the Tortugas Islands, completely inundating them. In December 1789, a storm surge associated with an extraordinarily powerful BAY OF BENGAL cyclone rushed ashore as a series of three large waves at Coringa, INDIA. Estimated by survivors to have ranged from 25 to 40 feet (7–13 m) in height, the waves were said to have resulted from an accumulation of seawater at the head of the bay, blown there by the furious northwesterly winds of the cyclone. As late as 1821, long after barometers had become widely available, eyewitness accounts of the NORFOLK–LONG ISLAND HURRICANE of September 2, 1821, described the storm surge as "a perpendicular wall some five feet high driven by the wind when it changed to northwest. . . ."

In the mid-19th century, however, as investigations conducted by early cyclonic scholars WILLIAM

C. Redfield and Henry Piddington revealed a number of anomalies in the characteristics of documented storm surges, a new model for surge formation began to take shape. Based on a more holistic approach to the causes of storm surges, studies into their development began to include such climatological and geological factors as size and speed of the tropical cyclone, current strength and direction, and coastline configuration. Redfield, drawing upon data gleaned from years of pioneering study into the nature of Caribbean hurricanes, believed that many of the huge, tsunamilike storm surges observed in Barbados, Cuba, Jamaica, Martinique, and Puerto Rico were primarily due to the shoaling effects of shallow seabeds on wind-driven waves. Inspired by accounts of the cresting 20–30-foot (7–10-m) surge reportedly witnessed during the 1780 Savanna-La-Mar Hurricane at Jamaica and of the series of battering waves that inundated parts of Barbados during the gargantuan hurricane of August 10, 1831, Redfield speculated that the height of the storm surges that came ashore in these places was increased by both a gradual rise of the sea bottom and the sheer rock faces that embrace many of the Caribbean's harbors.

In a similar vein, Piddington, whose contributions to the field were largely statistical in nature, argued in his famed treatises on Bay of Bengal cyclones that storm surges were in essence large "storm waves" that the rotating winds of the cyclone pushed up into a cone beneath its eye and then carried ashore with it. A later student of Indian Ocean cyclones, John Eliot refined Piddington's thesis, positing instead that water currents over the entire area of the storm follow its circular motion and in doing so are liable to pile in on each other as they enter the narrowing shores of the northern Bay of Bengal. Forming a type of aquatic dam, they continue to accumulate until sheer inertia overcomes remaining resistances, leaving an unimpeded wave of water to swamp the bay with often tragic results for those low-lying nations, such as densely populated Bangladesh, that ring it.

But Eliot's model of surge formation, based as it was on the behavior of respective storm surges in the Bay of Bengal and Indian Ocean, was not always able to provide a plausible explanation for certain surge activities in other regions of the world. Meteorologists studying storm surges along the southeast coast of the United States, for instance, discovered that more than half of the surges reported in North and South Carolina were not fast-moving waves at all but rather gradual inundations, a swelling of the sea that transpired over a period of hours. During the Eastern Carolina Hurricane of September 1769, the tide rose twelve feet above highwater in "a few hours," while a hurricane at Charleston, South Carolina, on August 27–28, 1813, added nearly five feet (2 m) to an incoming tide within the space of two hours. Just 9 years later, twelve people were drowned when Charleston was struck by a similar hurricane on September 27–28, 1822. In this particular case, however, the surge moved much faster than it had before, rising and falling six feet (2 m) in 45 minutes.

Along the gulf coast of the United States, an area that Eliot believed most closely resembled the geographical outlay of the Bay of Bengal, similar incongruities between individual storm surges were also being observed. During the Last Island Hurricane of August 10–11, 1856, the waters of the Gulf of Mexico rose eight feet (3 m) in six hours, inundating most of Isle Derniere, a barrier resort island on the southeast coast of Louisiana. At least 140 people were drowned when the storm surge, which had been steadily encroaching on the island at a rate of nearly two feet (1 m) per hour, suddenly rose six feet (2 m) in one hour, sweeping over a dozen buildings into the roiling surf. Toward the end of the century, the second Indianola Hurricane of August 19, 1886, delivered an enormous surge to the gulf coast of Texas, one which gradually rose before swiftly spiking and then receding. The town of Indianola was never rebuilt. During the Great Galveston Hurricane of September 8, 1900, I. M. Cline, custodian of the United States weather office in Galveston, witnessed a four-foot (1.5-m) rise in the Gulf of Mexico that he stated took place in four seconds. Cline further reported that the waters that submerged a fair portion of the island were rapidly moving from east to west, or roughly parallel to the direction of the hurricane's winds, seemingly lending credence to Piddington's theory of spiraling surface currents and their cumulative effect on the formation of storm surges.

Between 6,500 and 12,000 people perished in the Great Galveston Hurricane, a majority of them by drowning. Along with the twin inundations of 1893 that killed between 2,600 and 3,300 people in Louisiana, Georgia, and South Carolina, the Great Galveston Hurricane clearly reaffirmed a need not only for more seawalls but also for a comprehensive model of storm surge formation that addressed the many unanswered questions raised during the studies of the past century. It was now apparent to observers like Cline that while wind speed certainly played a role in deciding the force and direction of circular surface currents and in powering the relentless series of surface waves that batter afflicted coastlines, it did not lie at the heart of storm-surge formation. Nor did the topography of respective seabeds, bays, harbors, and inlets—though favorable to the augmentation of incoming surges—solely cause adjacent bodies of water to gather themselves suddenly into a mound or cone before coming ashore. There remained another component in the origination of storm surges, one that had been suggested by Piddington as early as 1856 and toward which many of the 20th century's most ambitious meteorological investigations would be directed.

It was not until the mid-1920s, however, that a connection between a tropical cyclone's central barometric pressure and the development of its storm surge was conclusively drawn. Made possible by ongoing improvements in the design and deployment of barometers, this model allowed meteorologists to account for the known behavior of surface currents and coastal geography on the severity of respective surges, while at the same time providing a viable reason for a surge's initial generation. Under this model, the intensely low barometric pressures found in many hurricanes, typhoons, and cyclones cause the waters beneath such systems to literally rise, forming a bulging dome that appears as a gentle swell while still at sea but begins to shoal, violently piling up as it approaches land.

Known as the INVERSE BAROMETER, this increase in sea-level water height is closely linked to a tropical cyclone's PRESSURE GRADIENT, or the steepness of the drop of its barometric pressure over a certain radius of its mass. A round of pressure experiments executed by meteorologists during the 1950s revealed that the inverse barometer measures as much as three feet (1 m) per every 2.953 inches (100 mb) of pressure drop, or about a half-inch (1 cm) for every millibar of pressure that the sea's surface gains, thus providing a numerical standard by which the category and degree of cyclonic systems can be observed or even predicted. In this way, those scientists and civil defense authorities tracking a tropical cyclone can fairly easily translate a central barometric pressure reading of 28.50 inches (965 mb) into an anticipated eight-foot (3-m) storm surge or a pressure reading of 27.17 inches (920 mb) into a potential 17–18 foot (5–6 m) inundation.

Trailing some 10–15 miles (16–24 km) behind a tropical cyclone's eye, the storm surge, which can range in size from 30 to 50 miles (48–80 km) across, occurs in the right, or east and north quadrants, of hurricanes and typhoons in the Northern Hemisphere, and in the left, or west and north quadrants, of cyclones in the Southern Hemisphere. So positioned within the DANGEROUS SEMICIRCLE, or that part of a tropical system in which the winds are at their fastest, the surge's complex array of surface and deep-water currents gradually assumes the major characteristics of the storm itself, including a swirling motion that follows the clockwise or counterclockwise direction of the storm's winds. In some of the most powerful tropical cyclones, these currents spiral 200 to 300 feet (70–100 m) downward, much deeper than either Piddington or Eliot ever imagined. Upon reaching their maximum depth, these currents then flow outward, carrying the cooler seawater they have churned up away from the surface. Were they to spiral in the opposite direction and instead carry cooler waters to the surface, they would quite possibly kill the tropical cyclone by impeding the warm-water evaporation rates on which it feeds.

The inverse barometer model of storm-surge generation has also refined tropical meteorology's understanding of how surge activity is affected by the shape of a particular coastline, the depth of its shoaling waters, and the topography of its principal bays, lakes, and rivers. Because a westward-moving hurricane's storm surge is positioned to the right and rear of its eye, any bay, inlet, or estuary to the right will be flooded when that system finally makes landfall. Should the hurricane come ashore near the mouth of a river or shallow cove, its storm surge will push existing waters back in on themselves, forcing them to overrun their banks in a gushing quest for space. In the past, several large lakes, including Pontchartrain in Louisiana and Okeechobee in Florida, have sustained destructive surges of this type, while a number of river outlets—from the Mississippi River to England's River Thames to India's engorged Ganges—have not only been choked off but driven 80 miles (129 km) back inland.

As most lakes and rivers are bordered by fairly low-lying plains, the destructive consequences of such surges are usually severe, with dozens of deadly flood emergencies being registered in Bangladesh, JAPAN, and the United States. Notable lake inundations include the 1928 SAN FELIPE HURRICANE, in which a storm surge on Lake Okeechobee killed between 1,800 and 2,000 summer vacationers in Florida, and the Nagasaki Typhoon of September 17, 1828, in which a towering storm surge at Ariake Bay drowned nearly 15,000 people. Bangladesh, built as it is on the shifting sands of the Ganges River delta, is acutely susceptible to those Bay of Bengal surges that frequently turn the huge river on itself, ramming its waters upstream until they transform fertile jute fields into killing grounds for tens of thousands of people. During the 18th and 19th centuries the equally shallow fan of the Mississippi River delta was inundated by similar surges in August 1794, July 1819 (the BAY ST. LOUIS HURRICANE), and during the Last Island Hurricane of August 1856, necessitating the construction of an elaborate network of levees and embankments to protect neighboring New Orleans.

Faced with the fact that at one time 90 percent of all deaths in tropical cyclones are caused by storm surge activity, those nations of the world most vulnerable to cyclonic strikes developed a number of different methods for dealing with this terrifying menace. In the United States, the National Weather Service has devised several computer ANALOGS specifically for the purpose of predicting degrees of surge severity along the coastal regions of North America and HAWAII. Commonly known by the acronyms SPLASH (Special Program to List Amplitudes of Surges from Hurricanes) and SLOSH (Sea, Lake, and Overland Surges from Hurricanes), these analogs analyze barometric pressure readings from an incoming tropical cyclone, compute the potential height of its storm surge, and then compare that information with topographical information from those areas most threatened by the system.

Color coded to show different water depths, these analogs, which require nearly a billion computations to complete, not only permit the NATIONAL HURRICANE CENTER (NHC) to determine just how far inland a certain surge will penetrate but also how it will interact with such geographical features as lakes, valleys, harbors, and mountain ranges. It is entirely possible that the destructive potential of a surge may be deflected by the presence of a hill or seawall, its force redirected in such a manner that it either breaks apart or submerges an unsuspecting nearby town or city. This data is in turn employed by civil defense agencies as they set up an efficient timetable for the implementation of evacuation procedures and other security measures. Because haphazard evacuations can result in both needless expense and overcrowded storm shelters, the versatile SLOSH and SPLASH systems with their preconfigured data banks save lives and money as they simultaneously provide scientists with vital statistics on the physical meteorological characteristics of individual storm surges.

In JAPAN, where direct strikes from North Pacific typhoons average between two and nine per TYPHOON SEASON, storm-surge control measures have taken a decidedly more mechanized course. At the bustling port city of Osaka on the main island of Honshu, an extensive web of sliding flood gates, electrically operated barriers, and pumping stations capable of moving 67,320 gallons of water per second, protects the densely populated industrial center from the mid- to late-season wind typhoons that frequently flood the shallow confines of Osaka Bay. Designed and built between 1961 and 1981, the formidable array of 130 embankments, 33 steel barriers, 80 pumping stations, and 6 deep spillways was constructed in response to a devastating history of surge activity that included the Muroto and Muroto II typhoon strikes of September

21, 1934, and September 16, 1961, respectively. In the first Muroto typhoon, 3,036 people drowned, and some 43,000 buildings in and around Osaka were destroyed when an 18-foot (6-m) storm surge swept into the city; in the second typhoon, 202 people lost their lives, and another 15,000 houses were ruined by the flooding onslaught of a 13-foot (4-m) surge.

Engineered with an eye toward containment, the Osaka control system does not prevent the surge from entering Osaka Bay itself but rather keeps it from rushing into the three principal rivers that flow through the city, backing them up until a serious flood emergency ensues. Since the project's completion in 1981, no less than five typhoons—two of them major systems with surges in excess of 10 feet (3 m)—have come ashore near Osaka, testing the system's structural and design integrity to the highest degree. Deployed in less than an hour's time on each occasion, the massive barrier had held, not only reducing the death rate by 90 percent but also saving property owners and insurance companies millions of dollars in damage claims. In this way, Osaka's $600 million project, one of the most ambitious of its kind ever undertaken, will justify the cost of its construction in fewer than 10 storms.

Culture, too, has over the centuries taken its precautions, protecting its vitality by portraying the storm surge as an entity of mythical or legendary proportions. Reminiscent of the Great Deluge, of the universal flood that an angry God set forth upon Earth, the storm surge has found its way into *literature* and myth, assuming a normative import that far exceeds its pure meteorological state. Imbued with a spiritual, military, or even coincidental significance, the storm surge and its catastrophic effects have at different times and by different cultures been viewed as a foretelling of the Apocalypse by NEW ENGLAND Puritans and Southern Baptists; as a sign of the *kamikaze,* or of the saving workings of the divine wind that in 1281 brought a pulverizing storm surge to Hakata Bay, Japan, and to where the invasion fleet of Kublai Khan fatally bore its brunt; as a moment of displeasure for the earth gods who punished the indigenous people of the Carolinas with a surge that "raised the water over the tops of the trees where the town now stands . . . ;" and as an alleged message to the voodoo believers of Haiti from their recently departed "father," François "Papa Doc" Duvalier, who at the very moment he was being laid to rest in April 1971 supposedly sent a hurricane and its destructive storm surge ashore at Cap Haitian. Although Category 5 Hurricane Edith did indeed brush past Haiti's Cap Haitian in September 1971, any connection between the feared dictator's burial and the powerful hurricane that slightly damaged the country five months later is purely cultural, a popular reaction to the surge's devastating sense of timing, as well as its symbolic implications.

In August 2005, Hurricane KATRINA delivered a 30-foot (10-m) storm surge to the Louisiana–Mississippi border, some five feet greater than that observed in 1969's Hurricane CAMILLE (24.3 feet), and setting a storm surge record in modern meteorological history. Dozens of fatalities were attributed to Katrina's storm surge, which seemingly countered recent trends that indicated that the largest killer in tropical cyclones was inland flooding.

storm tide   This meteorological term is used to describe the destructive rise in sea level that occurs when the STORM SURGE of a TROPICAL CYCLONE coincides with the high lunar tide.

Superfortress B-29   This twin-engined aircraft was designed and built by the Boeing Aircraft Corporation and used by U.S. Air Force HURRICANE HUNTERS for TROPICAL CYCLONE reconnaissance missions over the NORTH ATLANTIC and NORTH PACIFIC OCEANS, CARIBBEAN SEA, and GULF OF MEXICO during the 1950s. Painted silver and flying through a tropical cyclone at speeds of between 190 and 290 MPH (306–468 km/h), the *Superfortress B-29* carried a crew of 10 as well as an array of ANEMOMETERS, RADIOSONDES, DROPSONDES, and RADAR screens. Up forward were situated the weather observer, pilot, copilot, engineer, radar operator, and navigator. In the rear of the aircraft were seated the crew chief, radio operator, radiosonde operator, and dropsonde operator. In essence a flying meteorological laboratory, the *Superfortress B-29* collected data on upper level WIND SPEEDS and STEERING CURRENTS and then radioed this information to the NATIONAL HURRICANE CENTER (NHC) in Miami, FLORIDA. On November 3, 1949, a *Superfortress B-29* stationed at Kindley Field in BERMUDA was lost while on hurricane hunting duty off the island's coast. On November 1, 1952, a *B-29* departed Guam for Typhoon Wilma, then located off the coast of Samar Island in the PHILIPPINES, and was lost with all hands. A reliable workhorse, the *Superfortress B-29* was also used in relief operations following the passage of particularly severe HURRICANES and TYPHOONS. On September 17, 1952, a *Superfortress B-29* delivered 300 gallons of water and 2 tons of food rations to Wake Island following the devastating transit of Typhoon OLIVE's sustained 120-MPH (193-km/h) winds. During the late 1950s, the *Superfortress B-29* was replaced on hurricane hunting duty by the *Boeing B-50,* which could not only fly 10,000 feet higher than the *Superfortress B-29* but also carried an oven, thereby allowing its crew hot meals.

supertyphoon   This popular, official designation is used by meteorologists, journalists, and historians to describe those dreaded western NORTH PACIFIC typhoons that in the span of their short lives attain great size and exceptional intensity. Sometimes spelled *super typhoon,* the term originated during the early 1940s when Allied weather reporters, in a systematic effort to improve the accuracy and effectiveness of their military forecasts, first began to classify threatening North Pacific typhoons according to their strength and potential severity. Drawing upon a list of familiar descriptors, meteorologists began to refer to relatively mild, early season rain typhoons as "bean" or "midget" typhoons, while large, powerful, late-season wind systems were known as supertyphoons.

Although this early, somewhat interpretive typhoon ranking system was eventually superseded by a regional adaptation of the five-point SAFFIR-SIMPSON SCALE, the term *supertyphoon* nonetheless remained in common usage around the world. Sustained in part by its innate sense of drama, it is now most often applied to those cyclonic systems that are of firm Category 4 status or better on the graduated Scale of Typhoon Intensity. (*See* TYPHOON for a reproduction of this scale.) Ranging from 450 miles to 700 miles across, containing minimum BAROMETRIC PRESSURES of 27.88 inches (944 mb), and possessing sustained winds of 130 MPH (209 km/h) or higher, the three to four supertyphoons that occur during the course of an average TYPHOON SEASON are often responsible for inflicting widespread property damage and sacrificial DEATH TOLLS on those tempest-racked nations—CHINA, the PHILIPPINES, JAPAN, and TAIWAN—that ring the western terminus of TYPHOON ALLEY.

On September 26, 1959, for instance, Supertyphoon VERA slammed into central Japan, killing 5,159 people, while Supertyphoon Oscar, September 17, 1995, came within 60 miles of Tokyo before veering northwest and the open sea. In September 1984, Supertyphoon IKE claimed 1,363 lives as it besieged the southern Philippines with 137-MPH (221-km/h) winds and heavy seas, while 1995s Supertyphoon ANGELA blasted central Luzon Island with 155-MPH (249-km/h) gusts and torrential rains. More than 800 people reportedly perished, making Angela the deadliest typhoon to have struck the country since Supertyphoon Nina snuffed out 1,000 lives on the archipelago in late November of 1987.

Curiously enough, use of the adjective *super* in conjunction with powerful North Atlantic HURRICANES became a comparatively rare tradition in the West. While on occasion a U.S., Canadian, or British newspaper may have made a passing reference to a *super hurricane,* the term instead found a similar, considerably more widespread manifestation in *major hurricane.*

T

Taiwan *Northwestern Pacific Ocean* Formerly known as Formosa, the mountainous, deeply forested island of Taiwan has for centuries repulsed a veritable armada of North Pacific TYPHOONS. Positioned at the west end of TYPHOON ALLEY, 122 miles (196 km) off the coast of central CHINA, Taiwan's extensive plains and industrious seaports of Taipei, Kaohsiung, and Taichung have on average been directly affected by between one and four typhoons per TYPHOON SEASON. Although Taiwan has suffered through its share of early and late-season typhoons and TROPICAL STORMS, its peak period of TROPICAL CYCLONE activity occurs in August and September, when the first of the North Pacific's SUPER TYPHOONS begins to recurve north-northwest and into the volcanic islands of JAPAN. While not as frequent as rain-making Filipino *BAGUIOS,* or as deadly as those wind typhoons that yearly stalk the coasts of Japan and China, a number of Taiwanese typhoons have been systems of considerable size and intensity upon landfall. In some instances responsible for inflicting steep DEATH TOLLS and immense damage estimates on the island, these typhoons nevertheless provide Taiwan with an intriguing storm history.

On December 18, 1867, a large, late-season wind typhoon, accompanied by a 15-foot (5-m) STORM SURGE, inundated the busy harbor at Keelung, Taiwan. Several vessels at anchor in the northern port, including a coastal passenger liner, were thrust ashore, many becoming total wrecks. Triggered by the typhoon's 11-inch (279-mm) PRECIPITATION counts, shuddering landslides quickly swallowed two mountainside villages, killing an estimated 500 people.

In late July of 1896, an equally powerful typhoon crossed southern Taiwan, sinking numerous ships and leveling hundreds of BUILDINGS in the towns of Pingtung and Hengchun. Before slamming with great violence into Shantung, China, on July 24, the typhoon claimed 379 lives in Taiwan and left thousands more homeless.

On September 16, 1912, the celebrated Taito Typhoon rolled ashore in northern Taiwan, blighting rice fields, sinking the island's fishing fleet, and razing to the ground 91,400 buildings in and around the port city of Taito. Said by survivors to have been one of the most violent typhoons to have crossed the island in living memory, the titanic Taito Typhoon of 1912 killed 107 people and left another 293 injured.

Less than a quarter of a century later, a particularly deadly typhoon sliced into eastern Taiwan, just south of the town of Kuangfu, on the morning of August 16, 1936. Generating sustained winds of 107 MPH (172 km/h), the typhoon's whirring EYEWALL cut a broad swath of destruction across the center of the island, uprooting thousands of trees and spawning horrific flash floods. Some 720 people reportedly perished, making the August 1936 typhoon one of Taiwan's deadliest weather disasters.

On September 29, 1940, southern Taiwan was again battered by the wind typhoon that demolished 5,000 buildings and left 50 people dead.

More than 100 people were killed when an unnamed typhoon lashed both the Pescadores Islands and mainland Taiwan with 107-MPH (172-km/h) winds and 7-inch (178-mm) precipitation counts on the night of September 26–27, 1951.

Boasting top winds of 115 MPH (185 km/h), an unnamed Category 3 typhoon trounced southwestern Taiwan on the afternoon of November 14, 1952. A thousand buildings were destroyed and 67 people, many of them fishers, were killed, while another 500 lay injured.

Not a decade later, on June 27, 1957, 86 people died in an early season strike by Typhoon Virginia, a relatively small rain typhoon that nevertheless initiated capricious mudslides near the hillside town of Toufen.

On July 17, 1958, midseason Typhoon Winnie, a Category 3 rainmaker from the Mariana Islands, stomped ashore in eastern Taiwan. Twenty-four people perished, most of them victims of Winnie's sudden downpours.

During the course of the 1959 typhoon season, Taiwan was struck by two immense midseason typhoons in less than 10 days' time. On August 20, 1959, a funereal Typhoon Iris unfolded over southern Taiwan, mowing down trees, houses, telephone poles, and road signs before exiting into the Taiwan Strait just before midnight. Swiftly penetrating the Chinese coastline, Iris was soon snuffing out lives all over Fukien Province—2,334 of them, according to official reports. Much kinder to Taiwan than China, Iris claimed only six lives on the island, permitting military and civil defense authorities a collective sigh of relief as they quickly moved to lift storm warnings. By the morning of August 30, however, Taiwan was back on alert, awaiting the arrival of Typhoon Joan. A voluminous typhoon with 112-MPH (180-km/h) winds and heavy rain squalls, Joan closely followed Iris's track, loyally blasting Taiwan's south coast before dissipating in a shower of thunderstorms over an already saturated Fukien Province. Eleven people in Taiwan were killed, and several million dollars worth of property damage was assessed in the wake of Typhoon Joan.

On August 2, 1960, Typhoon Shirley trampled ashore at Taitung, a small fishing village on Taiwan's southeast coast. Producing sustained winds of 110 MPH (177 km/h), Shirley's eyewall rains and headlong gusts left 50,000 people homeless and another 104 dead, buried beneath tons of fallen debris. Ninety-five fishers were also lost at sea, their lightly built junks broken apart by Shirley's grinding seas.

Typhoon Pamela, a Category 4 super typhoon that had originated over the humid waters of the eastern North Pacific Ocean nearly a week before, shot across northern Taiwan on the night of September 12–13, 1961. One hundred six people, many of them children, perished, while another 913 were reportedly injured.

Shortly after Supertyphoon Gloria streamed ashore at Keelung on the night of September 11–12, 1964, four Taiwanese meteorologists were arrested and charged with dereliction of duty for failing to predict properly Gloria's eventual landfall on the island's north coast. A Category 4 typhoon of extreme ferocity, Gloria's 132-MPH (212-km/h) winds and 14-foot (5-m) storm surge killed 330 people and wrought some $17.5 million in 1964 property losses, making it one of modern Taiwan's most catastrophic typhoon strikes. The meteorologists, widely viewed as nothing more than political scapegoats, were eventually released.

Barely of typhoon intensity as it traversed the southern foothills of Taiwan on June 18, 1965, Typhoon Dinah's early season rains and resulting landslides left 2,531 houses in ruins and another 4,355 seriously damaged. Thirty-one people perished in what experts deemed one of the most destructive spring typhoons in Taiwanese history.

Just two years later, on October 19, 1967, Typhoon Carla, a fairly substantial Category 3 wind typhoon from the PHILIPPINES, grazed Taiwan's southeast coast while recurving north-northeast. Seventy-three people died.

The 1969 typhoon season would see another two typhoons strike Taiwan; Elsie, on the morning of September 27, and Flossie, on the afternoon of October 3. The weaker of the two systems, Category 2 Elsie killed 47 people and injured another 66 as it drifted across the southern half of the island, toppling radio towers and collapsing 200 small buildings. Less than a week later, Typhoon Flossie, an extreme Category 3 system with sustained winds of 113 MPH (182 km/h), moved ashore at Taipei. A huge rainmaker, Flossie's flash floods soon submerged large portions of the capital, killing 75 people and rendering another 31,000 homeless.

A similar double strike took place during the 1977 typhoon season, when two powerful 120-MPH (193-km/h) wind typhoons—Thelma, July 25, and VERA, July 31—howled across Taiwan within one week of each other. Thelma riddled the southern port city of Kaohsiung with 150-MPH (241-km/h) gusts and stinging rains, causing shore cranes to buckle and six cargo ships to sink at their moorings. To the north, Vera pulverized the concrete quays of Keelung, knocking out all electrical and telephone service across half of the island. All told, 39 people died in both typhoons, while damage claims mounted into the millions.

On October 21, 1981, Typhoon Gay's fringe rains flooded two neighborhoods in Taipei, claiming no casualties but inflicting large property losses on the prosperous capital.

In August 1994, Typhoon FRED, on course for a deadly midseason strike on the Chinese port city of Wenzhou, killed three Taiwanese fishers as it passed 75 miles (121 km) northeast of Taipei.

During the notably active 1995 typhoon season, Taiwan's hill-ringed harbors and pine-clothed beaches endured the deleterious effects of three mature typhoons and one tropical storm. On August 25, a westward-moving Typhoon Irving, complete with 110-MPH (177-km/h) winds and 15-foot (5-m) seas, slunk along the south coast of Taiwan before falling upon Hong Kong the following day. A large tropical cyclone, Irving's rains were felt as far north as the city of Chiai. Two days later, a deepening tropical storm, Janice, released gale-force gusts and light drizzle over Taipei, flooding several streets and uprooting dozens of trees before moving northwest to the open waters of the North Pacific Ocean. Swiftly intensifying, Typhoon Janice eventually came ashore with great fury in KOREA on August 30, killing 53 people and leaving 110,000 others homeless. Closely trailing Tropical Storm Janice, Supertyphoon Kent sent gargantuan breakers crashing against Taiwan's southeast coast on September 1. Bound, like the earlier Irving, for landfall near Hong Kong, Typhoon Kent drowned six people in China's Guangdong province and caused extensive surge damage to the coastal communities of Hainan Island. On September 29, Supertyphoon Ryan killed two people in Taiwan as its 135-MPH (217-km/h) gusts brushed the island's northeast coast. Several dozen fishing sampans in Suao were smashed to splinters by Ryan's 15-foot (5-m) seas, causing financial distress for scores of families.

The 1996 typhoon season was no kinder to Taiwan than the 1995 season had been, with one enormous typhoon—Herb—smoking the towns and villages of central and southern Taiwan with sustained 121-MPH (195-km/h) winds and nine-inch (229-mm) rains on the night of August 1. At one time a supertyphoon with top winds approaching 150 MPH (241 km/h), Herb had weakened considerably before rolling ashore near the city of Taoyuan 50 miles (80 km) south of Taipei. Gusts of 130 MPH (209 km/h), combined with torrential rain showers, produced a bumper crop of local calamities, from a flood-damaged dam in the north to submerged streets in the west to rumbling rockslides in the south and east. More than half of Taiwan's low-lying counties were declared disaster areas after Herb's 12-foot (4-m) storm surge inundated thousands of houses and treacherous rockslides marooned 20,000 people in isolated mountain communities. Only a 24-hour airlift of food, medicine, and clothing kept the rain-soaked multitude alive. Considered the most severe typhoon to have affected Taiwan in three decades, Herb left 46 people dead, 10 missing, 580 others injured, and more than $800 million in property losses.

In August 2000, Super Typhoon Billis rousted Taiwan with high winds and seas. Widespread flooding claimed dozens of lives across the island.

In early 2001, Typhoon Toraji killed 200 people in Taiwan. Months later, in September, Typhoon/Tropical Storm Nari caused Taipei's worst flooding on record and killed 100 people. Inundating northern Taiwan, Nari flooded sections of the capital and caused enormous disruption to financial markets and critical infrastructure. Like many tropical systems, Nari was an inveterate rainmaker; in some places, 32 inches (813 mm) of precipitation fell.

On September 2, 2003, Typhoon Dujuan skirted southern Taiwan and swirled toward China's Pearl River estuary. One of the most powerful tropical cyclones in recent memory, Dujuan created a unique "double eyewall" configuration. Typhoon Dujuan was a powerful Category 3 hurricane when it first hit Taiwan with sustained winds of 120 MPH (200 km/h). Downgraded to a tropical storm as it sped westward at 21 MPH (34 km/h), Dujuan bore down on the Chinese mainland's Pearl River Estuary. One person was listed as missing in Taiwan following the typhoon's passage.

On July 18, 2005, the first typhoon to strike Taiwan during the 2005 North Pacific typhoon season—Haitang—made landfall on the island's eastern coast, killing at least seven people, overturning buses and trucks, and sinking a cargo vessel near the southern port of Kaohsiung. One of the most intense typhoons in contemporary Taiwanese history with sustained winds of 119 MPH (191 km/h) and 145-MPH (234-km/h) gusts at landfall, Haitang forced the closure of Taiwan's financial markets, and dropped nearly three feet (1 m) of rain over inland mountainous regions. Media sources placed damage estimates to crops and other property at more than $40 million, and indicated that some 1.5 million people were left without electricity.

Typhoon Longwang (Chinese for "Dragon King") struck the eastern Taiwanese city of Hualien in October 2005, grounding a cargo ship, destroying seven buildings, disrupting electricity to 500,000 residences, and canceling air travel. A historic temple was also destroyed by Longwang's thumping winds. Two people were killed and 46 others suffered minor injuries. After its passage over Taiwan, Typhoon Longwang crossed the Taiwan Straits and struck the Chinese mainland.

temperature   Long a subject of spirited debate among meteorologists and cyclonic scholars, the exact relationship between the presence of a TROPICAL CYCLONE and an observable rise or fall in surrounding air temperatures has yet to be completely defined. Possibly linked to the existence of descending air currents in a tropical cyclone's EYE and contingent upon respective humidity levels at different altitudes, recorded temperatures in mature HURRICANES, TYPHOONS, and CYCLONES have been found to rise, fall, and remain unchanged. During the Great September Gale of September 23, 1815, recorded air temperatures at Yale University registered 48 degrees F (9°C) at sunrise and then rose to 65 degrees (18°C) by early afternoon, just as the hurricane's eye was making landfall in southern Connecticut. A severe, late-season hurricane that lashed Galveston, TEXAS, on October 1–3, 1867, produced a two-day average temperature of 80 degrees F (27°C) and a sharp rise to 85 degrees as the eye of the immense storm passed directly over a weather observatory on the night of October 3. On October 20, 1882, the midmorning passage of a typhoon's eye over Manila, the PHILIPPINES, generated temperature readings of 75 degrees F (24°C) on the eye's leading edge, 88 degrees (31°C) in its center and 75 degrees F (24°C) following its transit. When plotted on a microbarograph, compared with both the Manila Typhoon's CENTRAL BAROMETRIC PRESSURE and wind speed measurements, the series of temperature readings revealed that as the typhoon's eye approached the observatory and its central pressure continued to decrease, air temperatures within the eye increased proportionally.

Texas *Southern United States* Presenting some 367 miles (591 km) of shallow coastal plains, sandy barrier islands, and flowing river mouths to the HURRICANE-prone waters of the GULF OF MEXICO, Texas has for the last 236 years routed a veritable invasion of TROPICAL CYCLONES. Between 1766 and 2006, Texas's principal port cities of Galveston, Corpus Christi, and Brownsville were affected by 100 documented hurricanes and TROPICAL STORMS. Of these 100 storms, 50 were judged major hurricanes, Category 3 and 4 systems with minimum winds of 111 MPH (179 km/h), nine to 18-foot (3–6-m) STORM SURGES, and monstrous DEATH TOLLS. Since 1766, at least 15 hurricanes of varying degrees of intensity have come ashore within the heavily populated environs of Galveston Bay, causing thousands of deaths and millions of dollars in property damage to the famed seaport located there. On the afternoon of September 8, 1900, Texas had the rare misfortune to be struck by the deadliest of all recorded American hurricanes, the GREAT GALVESTON HURRICANE. Between 6,000 and 12,000 people reportedly died during the seven-hour tempest, making it one of the most infamous natural disasters in U.S. history.

On average, Texas's wide shores, noted for their peaceful summer havens and gushing oil refineries are directly impacted by at least one TROPICAL DEPRESSION, tropical storm, or hurricane every year. On at least 12 occasions between 1854 and 2006, Texas even endured multiple strikes in the course of a single

HURRICANE SEASON, as witnessed in 1854, when two hurricanes came ashore between September 16 and 19; in 1871, when two June hurricanes (June 4 and 9) and one October hurricane (October 2–3) made landfall near Galveston; in July and September of 1874; in 1886, when four hurricanes struck the state between June 14 and October 13; again, just two years later, when two early season hurricanes lashed the north Texas coast on June 17 and July 5; and in 1933, when four more tropical cyclones swept the state between July 6 and September 15.

Texas's position on the western and northwestern curl of the Gulf of Mexico provides it with a full June to October hurricane season. A majority of June and July hurricanes, born over the southwestern CARIBBEAN SEA before wandering northward, have made landfall on the northwest coast of Texas between Galveston and Port Arthur. Although many early season Gulf of Mexico hurricanes were storms of moderate intensity—the June 4, 1871, hurricane at Galveston produced a fairly mild barometric pressure reading of 29.51 inches (999 mb) at landfall—some, such as the July 4, 1874, blast at Indianola, a small town 40 miles (64 km) northeast of Corpus Christi, blew down fledgling settlements and sank ships, thereby seriously hindering coastal trade. Another hurricane in July 1909 produced a pressure reading of 28.29 inches (958 mb) as it trundled ashore near Beaumont, uprooting trees and shattering windows.

It is in August, September, and October, when the first of the monster CAPE VERDE STORMS begins stomping its way across the Caribbean Sea and Gulf of Mexico, that Texas becomes most vulnerable to damaging hurricane strikes. Celebrated midseason Texas hurricanes include:

- The September 4, 1766, hurricane that delivered an enormous storm surge to Galveston Bay, driving five Spanish treasure ships aground on the island and uprooting hundreds of trees. Although the five vessels were later condemned as total wrecks, most of their treasure and metal fittings were salvaged.

- The hurricane of September 11, 1818, that sent more than four feet of water scudding across Galveston Island, collapsing dozens of wooden BUILDINGS, smashing piers, undermining bridges and tossing numerous ships and small boats onto the beach. According to contemporary accounts, hundreds of bodies were left scattered across the island, while wreckage 10 feet (3 m) thick in places hampered efforts to rescue trapped survivors. Among those vessels that foundered in Galveston Bay were four sloops belonging to the notorious pirate

Jean Lafitte. The virtual dictator of Galveston Island since 1817, Lafitte had turned the sandy spit into a pirate fortress, one capable of supporting 20 ships and more than 1,000 cohorts. Although seriously damaged by the September 1818 hurricane, Lafitte's lair endured until 1821, when another hurricane of sorts, the United States Navy, finally drove Lafitte into exile and death.

- The hurricane of October 1, 1837, known as Racer's Hurricane, named for the British sloop-of-war HMS *Racer*. This immense hurricane spent nearly one week savaging the greater Galveston area with wailing winds, torrential rains, and a 12-foot (4-m) STORM TIDE. Several dozen buildings on Galveston Island, including its new customshouse and warehouse, were demolished. Eleven ships, one of them a naval schooner and another a brig, were carried ashore and then stranded. One of the more severe hurricanes to have slammed into Texas during the 19th century, Racer's Hurricane was felt as far inland as Houston, where low-lying northern districts were flooded with four feet (1 m) of water on October 6. One death was reported in Galveston.

- The hurricane of September 17–18, 1842, whose early evening landfall snipped the spire from Galveston's recently constructed Episcopal Church and demolished several smaller buildings around the burgeoning city. No lives were lost, but damage estimates rose as high as $10,000, a steep sum in 1842.

- Twelve years later to the day, the tightly coiled Matagorda Hurricane of September 17–18, 1854, that ground ashore in central Texas between Matagorda and Lavaca bays. A terrific rainmaker, the Matagorda Hurricane released heavy rains over Columbus and Houston, while its six-foot (2-m) storm surge was felt as far north as Galveston Island. There, newly built warehouses, laden with cotton bales and sugarcane, were washed from their foundations, causing considerable financial hardship for the storm-weary city.

- The first Galveston Hurricane of October 3, 1867, a small, intense storm that inflicted more than a quarter of a million dollars worth of property damage on the island city. Accompanied ashore by a seven-foot (2-m) surge, the hurricane's 100-MPH (161-km/h) winds sliced the top two floors from a brick hotel, while

heavy rains melted huge cones of distilled sea salt into stinging rivers. Much of Galveston's railway system was submerged, the trestle bridge leading across the Bay was washed out and all mail and passenger transport to the mainland was disrupted. One man perished in a building collapse.

- The hurricane of August 16, 1869, that came ashore on Texas's lower coast, between the cities of Rockport and Corpus Christi. The Episcopal church in storm-riven Indianola was demolished, its bell tower cast into the hurricane's swirling surge by shearing gusts. Between 10 and 17 people were reportedly drowned when their houses were pulled into the sea and broken up by the hurricane's roaring surf.

- On September 16, 1875, that again trampled Indianola. This hurricane, a 100-MPH (161-km/h) whirlwind, carried beneath its eye an unusually large storm surge. Piled up against the sloping banks of Matagorda Bay, the surge eventually rolled ashore at Indianola, reducing three-quarters of the town's 2,000 buildings to splinters in a matter of hours. One hundred seventy-six townspeople died, making the Indianola Hurricane of 1875 one of the deadliest Texan hurricanes on record.

- A massive late-season hurricane, one far greater in intensity than either the 1869 or 1875 strikes, that just over a decade later, on October 12, 1886, brought the town of Indianola itself to an end, sweeping almost all of the rebuilt town into Matagorda Bay. Littering the sandy landscape with broken carriages, half-buried suitcases, a keyless upright piano, and the carcasses of hundreds of drowned farm ANIMALS, the hurricane's subsiding waters left 250 townspeople dead, their bodies snagged beneath the mangled ruins of a once-prosperous trade center that would never rise from its tidal grave. Migrating westward to Port Lavaca, southward to Corpus Christi, and northward to Freeport, Indianola's survivors established new lives and new professions elsewhere, leaving their hurricane-haunted peninsula to the ghosts of memory and nature.

- The Second Galveston Hurricane of August 16, 1915, during which Galveston's recently completed, 8-mile-long (13-km) seawall was breached in several places, flooding the city's business district with more than 6 feet (2 m) of water. Brought on by a central pressure of 27.9

inches (945 mb), the hurricane's 120-MPH (193-km/h) winds, 19-inch (483-mm) PRECIPITATION counts, and 21-foot (7-m) storm tides killed 275 people in Galveston and wrought some $50 million in property damage. In San Antonio, 250 miles (402 km) due west of Galveston, tropical storm-force winds and drenching rains destroyed much of that season's valuable cotton harvest. Fifteen years and one massive seawall earlier, the vulnerable barrier island had been virtually obliterated by the Great Galveston Hurricane of September 8, 1900. Between 6,000 and 12,000 lives were lost in the 1900 disaster, graphically illustrating the effectiveness of the $9 million in improvements subsequently made to Galveston's storm surge defenses.

- The mammoth Corpus Christi Hurricane of September 14, 1919, that remains one of the most intense Texas hurricanes on record and was bolstered by a central barometric pressure of 27.37 inches (927 mb) at landfall on the central gulf coast. Joined by a 16-foot (5-m) storm surge, 122-MPH (192-km/h) winds decimated the port city that gave the hurricane its name, demolishing hundreds of buildings, sinking 10 ships, and claiming the lives of 287 people. The neighboring town of Port Aransas, 25 miles (40 km) north of Corpus Christi, was almost completely inundated by the hurricane's surge and saw many of its principal buildings pulled into the gulf by the rush of receding waters. A killer from birth, the Corpus Christi Hurricane had just two days earlier claimed nearly 500 lives off the FLORIDA keys when it sank the Spanish passenger liner *Valbanera,* drowning all aboard. An estimated $20 million in damages were suffered by Corpus Christi and Port Aransas, making the 1919 strike one of the region's most expensive.

- Hurricane Gladys, a Category 3 whirlwind whose 113-MPH (182-km/h) winds and five-foot (2-m) storm surge brought widespread destruction to Corpus Christi on September 6, 1955. Four people were killed and another 56 were injured.

- Hurricane CINDY, which just two years after Hurricane CARLA ravaged Port Lavaca with Category 4 winds and tides, drifted into Galveston Bay on the morning of September 17, 1963. A loosely organized Category 1 system, Cindy's central pressure of 29.41 inches (996 mb) generated 18-inch (457-mm) downpours and flash floods that cost the state and its industry more

than $12 million. No deaths or serious injuries were reported, however, making Cindy one of the kindest of all Texas hurricanes.

- One of the largest of all recorded Gulf of Mexico hurricanes, BEULAH, that on September 20, 1967, lumbered ashore at Brownsville, a prosperous industrial city just north of the Texas–Mexico border. A Category 3 (28.05 inches of mercury) storm of moderate intensity at landfall, Beulah's 10-inch (254-mm) downpours and 100 observed TORNADOES spent the next five days ravaging the Rio Grande Valley, damaging or destroying 10,000 buildings and leaving 38 people dead. Precipitation counts of nearly 28 inches were collected in south-central Texas, while the flood-swollen San Antonio River topped its banks in a record-setting inundation.

- The third tropical cyclone of the 1970 hurricane season, Celia, that during the midafternoon hours of August 3, 1970, screamed ashore between Corpus Christi and Aransas Pass as an incipient Category 4 hurricane of considerable intensity. A central barometric pressure of 27.90 inches (945 mb) at landfall produced sustained 146-MPH (235-km/h) winds and 161-MPH (259-km/h) gusts but only moderate six-inch (152-mm) rains. Downtown Corpus Christi was nearly obliterated, with 90 percent of its office towers, stores, schools, and fire and police stations either damaged or destroyed. Dozens of shrimp boats were snatched from their moorings and then cast aground at Rockport, many of them becoming total losses. Howling inland over San Antonio and Odessa, Celia's powerful winds wrested trees and telephone poles from the ground, crushed automobiles, fleeced thousands of acres of cotton fields, and left 30,000 people homeless. Twelve Texans were killed, and damage estimates of $500 million catapulted Celia into the aristocracy of hurricane strikes, making it one of the state's most expensive natural disasters.

- Tropical Storm Delia, a midseason wanderer from the Gulf of Honduras, that on the morning of September 4, 1973, made the first of two halfhearted landfalls between Galveston and Freeport. Quickly returned to the gulf by an expanding high-pressure ridge on the afternoon of September 5, a dissipating Delia and its central pressure of 29.12 inches (986 mb) went ashore near Freeport later that night. Coupled with 25-inch (635-mm) precipitation counts, 69-

MPH (111-km/h) winds claimed five lives and caused some $15 million in agricultural losses.

- One of several documented instances where a tropical cyclone from the eastern North Pacific Ocean made a deviant landfall in southwestern Texas. Tropical Storm Paul crossed the west coastline of MEXICO to deliver gale-force winds and 15-inch (381-mm) rainfall counts to the Rio Grande Valley on September 26, 1978. Some 7,000 acres of Texas farmland were inundated as the Rio Grande surged to more than 12 feet (4 m) above flood stage. A major railroad bridge linking the state with Mexico was swept away, disrupting rail service for a week. Although no deaths or injuries were reported in the wake of Tropical Storm Paul, nearly $20 million in agricultural losses, washed-out roads, and collapsed bridges were assessed. Ironically, western Texas would, on September 29, 1982, suffer a second strike from a storm named Paul, this one a powerful Category 2 hurricane that had originated off the coast of El Salvador on September 19. Recurving northeast on September 22, Paul's tropical storm-force gusts and heavy rains killed more than 1,000 people in Guatemala before returning to the energized waters of the North Pacific on September 23. Deepened into a mature, 100-MPH (161-km/h) fury, Paul slammed into Los Mochis, Mexico, on the morning of September 29, killing eight people and leaving several thousand others homeless. Its dissipating remains over El Paso on September 30, Paul's heavy rains caused serious localized flooding but no casualties.

- On July 24, 1979, another tropical storm, this one dubbed Claudette, that meandered ashore near Matagorda Island. A more violent tropical storm than. Delia, Claudette's 40-MPH (64-km/h) winds and 25-inch (635-mm) deluges killed seven people and wrought $408 million in property losses across central portions of the state.

- On August 18, 1980, severe Hurricane ALLEN that burnished Brownsville and neighboring San Padre Island with 160-MPH (258-km/h) gusts and 10-foot (3-m) storm tides. One of the most dangerous Gulf of Mexico hurricanes ever experienced in Texas, Allen killed 13 people and rang up nearly $1 billion in losses.

- Another interloper from the eastern North Pacific Ocean, Hurricane Norma that gate-crashed southwestern Texas on October 12, 1981. First making landfall near Mazatlán, Mexico, Category 2 Norma later released 25 inches (635 mm) of rain over the Texas cities of Bridgeport and Gainsville, spawning roaring flash floods and multimillion-dollar damage estimates. In one heart-rending incident, Norma's encircling waters pounced on the Gainsville Zoo, cornering several howling animals in their cages before drowning them. It was said that the zoo's wily elephant, later found trapped in a large tree, had survived Norma's 12-hour passage by holding its trunk above water. Festooned with 13 tornadoes, Norma's late-season romp through Texas and Oklahoma claimed seven people and left another 67 injured. Damage claims topped $150 million.

- Hurricane ALICIA that three years to the day after Hurricane Allen sliced ashore at Brownsville, hammered Galveston and Houston with 115-MPH (185-km/h) winds and torrential rains on August 18, 1983. Twenty-one people were killed, and nearly $1 billion in damages accrued.

- In mid-October 1989, Hurricane JERRY that lashed Galveston with 80-MPH (130-km/h) winds and heavy rains. Three people perished in the late-season $15 million strike.

- On July 27, 1995, a strengthening Tropical Storm Dean that blustered ashore at Galveston with 45-MPH (72-km/h) winds and much-needed rains. While some local flooding was reported on Bolivar Peninsula, damage from Dean's five-foot (2-m) tidal surge was considered minimal. No deaths or injuries resulted.

- Between August 23 and August 26, 1996, the downgraded remains of Hurricane Dolly brought intense thunderstorms and welcome four to eight-inch (102–203-mm) rain counts to southeastern Texas, thereby tempering a nine-month drought that had hardened the cattle-grazing folds of the West Gulf Coastal Plain. Generated by Dolly's unstable cloud bands, a fairly substantial tornado touched down just south of Brownsville. Several houses, a garage, and a small store were seriously damaged, but no injuries were reported. Although Dolly's insured losses in Texas totaled nearly $2 million, they were minimal when compared to the $2.4 billion in losses suffered by the state's agricultural sector during the preceding drought.

- The remnants of Tropical Storm Fay dumped heavy rains and spawned tornadoes in Texas on September 7, 2002, as it came ashore between Houston and Corpus Christi. Up to one foot (.5 m) of rain fell over Freeport and West Columbia in coast Brazoria County, while five to eight inches (127–203 mm) of precipitation were observed in Matagorda and Wharton counties.

- On the morning of July 15, 2003, Tropical Storm Claudette dripped ashore near Port O'Connor, Texas, with maximum sustained winds of 75 MPH (121 km/h). Born in central Caribbean Sea on July 8, 2003, about 415 miles (670 km) east-southeast of Kingston, Jamaica, the system subsequently entered the Gulf of Mexico to become the first hurricane of the 2003 North Atlantic season. A fairly destructive system, Claudette killed two people in Texas and left thousands without power along the central Texas coast before being downgraded to a tropical depression on July 16. Extreme gusts of 88 MPH (142 km/h) were observed at the Wadsworth nuclear power station. A 92-foot-long shrimp boat was also lost, requiring that the assistance of the United States Coast Guard (USCG) in rescuing its crew.

- Texas's South Padre Island was affected by 30–40-MPH (48–64-km/h) winds and high seas as Tropical Storm Erika sailed into northern Mexico on August 16, 2003.

- In August of 2003, Tropical Storm Grace slowly and poorly organized over the southeastern Gulf of Mexico before making landfall at Port O'Connor, Texas, on August 31. Maximum sustained winds of 40 MPH (65 km/h) were observed, while two to five-inch (51–127 mm) rains splashed across large sections of the state's Gulf coast.

- On July 20, 2005, Hurricane Emily rolled ashore near San Fernando, Mexico, with sustained, 125-MPH (201-km/h) winds and heavy rains. In southern Texas, Emily's rolling seas eroded dunes and damaged foliage, but caused little permanent damage. On July 21, at least eight tornadoes touched down in southern Texas. Generated by Hurricane Emily's constituent thunderstorms, the tornadoes destroyed mobile homes and houses near the town of Alice, and one person was injured. Brownsville, Texas, was buffeted by high winds and rains, but sustained only minor damage and no injuries.

- At one point a record-breaking Category 5 hurricane, Hurricane Rita had cooled to Category 3 intensity when it crashed ashore in northeastern Texas on September 24, 2005. Armed with a central pressure of 27.58 inches (934 mb), Rita's 121-MPH (195-km/h) winds caused considerable damage in the state's northeastern quadrant. Some 113 lives were lost, while more than $2 billion (in 2005 U.S. dollars) in property losses were assessed. Because of Rita's landfalling location, many Texan oil refineries and distribution facilities were either shut down or damaged, causing a temporary spike in domestic gasoline prices.

No tropical cyclones directly affected Texas during the relatively tepid 2006 North Atlantic hurricane season.

In mid-August 2007, a disorganized Tropical Storm Erin—the fourth North Atlantic system to bear the name—came ashore near Lamar, Texas. Dragging its 40 MPH (65 km/h) winds ashore on August 16, Tropical Storm Erin dropped some 11 inches (279 mm) of precipitation across much of the state's eastern and northern counties. Seventeen people were killed in the state and damage estimates were in the hundreds of millions of dollars.

TIROS (Television Infra-Red Observation Satellite)   This pioneering class of civilian weather satellites was launched into orbit by the United States between 1960 and 1965 and was used to monitor tropical cyclone activity in the North Atlantic Ocean, Caribbean Sea and Gulf of Mexico. Commissioned' by the U.S. Weather Bureau and the National Aeronautics and Space Administration (NASA) and constructed by RCA, the 10 first-generation TIROS satellites—TIROS I through TIROS X—signaled the start of a new age in tropical cyclone tracking and prediction. Equipped with television cameras and infrared technology, the TIROS satellites not only provided the world's meteorological and scientific communities with unprecedented views of global weather circulations but also served as operational prototypes for subsequent generations of weather satellites, including the Japanese Himawari series, the European *Meteosat* line and the U.S. GOES-class. Remarkably durable despite their small size and fragile components, not one of the 10 TIROS satellites remained in service for longer than five years, and yet in that time they had managed to photograph and transmit to Earth successfully no less than 649,077 images, many of them atmospheric portraits of some of the greatest North Atlantic HURRICANES on record.

The first of the the TIROS-class satellites, TIROS I, was launched from Cape Canaveral, FLORIDA, on April 1, 1960. Weighing 270 pounds and measuring some 19 inches high and 42 inches in diameter, drum-shaped TIROS I carried two television cameras, a radio transmitter, and two small solid-fuel rockets used to steer the satellite and keep its black-and-white cameras continually directed at Earth. Powered by solar-charged batteries, TIROS I circled Earth every 99 minutes at an altitude of 435 miles. Relayed every hour to newly constructed receiving stations in Fort Monmouth, New Jersey, and Kaena Point, HAWAII, TIROS I's relatively simple photographs—although sometimes made hazy by the satellite's ineffectual stabilizers—provided meteorologists with an entirely new visual archive with which to study the world's complex weather systems. Before the malfunction of its wide-angle camera took the satellite out of commission on June 17, 1960, TIROS I had managed to capture and transmit nearly 23,000 photographs. Intended by the U.S. Weather Bureau to provide "an operational system for viewing the atmosphere regularly and reliably on a global basis," TIROS I and its prodigious output proved that both goals were not only attainable but imminently so.

On November 23, 1960, the second of the TIROS satellites—dubbed TIROS II—was launched into orbit from Cape Canaveral. Slightly larger than its predecessor, TIROS II not only carried two black-and-white cameras but also a RADIOMETER, an instrument that allowed the satellite to observe and record such factors critical to tropical cyclone generation as SEA-SURFACE TEMPERATURE and anticyclonic activity. Steadied in orbit by a magnetic coil, TIROS II lasted until March 1961 when its cameras finally went dark. Amongst the more than 22,000 photographs that TIROS II recorded are several views of the enormous EXTRATROPICAL CYCLONE that brought 40-inch snowfalls to western NEW YORK on February 4, 1961.

The eight remaining TIROS-class satellites were placed in orbit between May 18, 1961, and July 2, 1965. During the first two weeks of September 1961, TIROS III and its visible/infrared photography was credited with saving hundreds of lives in TEXAS by assisting meteorologists at the NATIONAL HURRI-CANE CENTER (NHC) with the tracking of Hurricane CARLA. A tropical cyclone of record-setting intensity, Carla's 46-person DEATH TOLL in Texas would have been much higher had TIROS III not provided forecasters and civil defense authorities with timely, accurate data regarding Carla's size, forward speed,

and direction. On August 8, 1963, TIROS VI photographed a weak Hurricane ARLENE as it moved 300 miles to the southwest of BERMUDA, allowing storm watchers to study the system's interaction with a cold front to the north. TIROS VII, launched on June 19, 1963, remained in continuous operation the longest, for just under five years. Before ceasing transmission in March 1968, TIROS VII added 125,331 images to meteorology's satellite library. TIROS X, the final satellite in the first-generation series, was launched on July 2, 1965, in time to assist NHC meteorologists in tracking formidable Hurricane BETSY as it battered the BAHAMAS, Florida, and LOUISIANA during the first two weeks of September. Several of the satellite's finest shots were of Betsy's 40-mile-wide EYE and Category 3 EYEWALL.

TIROS-M   Variously referred to as either the ITOS (for Improved TIROS Operational System) class, or the NOAA 1–5 class, TIROS-M was a series of eight meteorological SATELLITES rocketed into orbit by the United States between 1970 and 1976 and used to track HURRICANES, TYPHOONS, CYCLONES, and TROPICAL STORMS in the NORTH ATLANTIC, NORTH PACIFIC, and Indian Oceans: Constructed by RCA for the National Aeronautics and Space Administration (NASA) and the National Oceanic and Atmospheric Administration (NOAA), the first of these sunsynchronous satellites was launched from Vandenberg Air Force Base in CALIFORNIA on January 23, 1970. Weighing 682 pounds and measuring some 40 inches high by 40 inches wide, box-shaped TIROS-M (ITOS-1) was fitted with one wide-angle and one telephoto television camera; a radio transmitter, a RADIOMETER, and several instruments intended to measure humidity, SEA-SURFACE TEMPERATURE and the altitudes of respective cloud-tops. Fueled by two five-foot-long solar panels, TIROS-M circled Earth in tandem with the cycles of the sun at an altitude of 890 miles and inclined toward the equator at 102 degrees. Its infra-red and visible photographs, transmitted every 12 hours to a dozen receiving stations around the globe, providing meteorologists with double the amount of visual data they had previously received from the then-operational NIMBUS meteorological satellite series. After a six-month test period, TIROS-M began regular transmission on June 15, 1970.

Renamed NOAA upon subsequent transfer of their operation to that agency, the remaining seven TIROS-M (ITOS A-H) satellites were blasted into orbit between December 12, 1970 (NOAA-1) and July 29, 1976 (NOAA-5). Launched on October 21, 1971, and intended for redesignation as NOAA-2 once in orbit, ITOS B was stranded about 200 miles above Earth and rendered unusable, ITOS D, also known as

NOAA-2, was the first meteorological satellite to carry a spin-scan radiometer, a development that on August 17, 1973, allowed its infrared technology to monitor Hurricane Brenda's generation over the southeastern CARIBBEAN SEA; and on August 23, 1974, to record in comprehensive detail the thermal dynamics of Hurricanes Ione and Kirsten as the pair established a PINWHEEL CYCLONE configuration over the eastern North Pacific Ocean. ITOS E, launched on July 16, 1973, was also stranded in an unusable orbit, while ITOS F (NOAA-3), launched on November 6, 1973, captured deadly Hurricane FIFI on film as it washed across Central America between September 21 and 23, 1974. On August 10, 1976, a newly launched ITOS H (NOAA-5) visually tracked Hurricane BELLE's Category 1 passage into the northeastern United States.

TIROS-N A series of nine meteorological SATELLITES placed in orbit by the United States between 1978 and 1991, and used to monitor TROPICAL CYCLONE activity in the NORTH ATLANTIC, NORTH PACIFIC, and Indian Oceans. Commissioned by the National Aeronautics and Space Administration (NASA) and the U.S. Air Force Space and Missiles Systems Organization (SAMSO) and constructed by RCA, the first of these polar-orbiting satellites—designated TIROS-N—was launched from Vandenberg Air Force Base in CALIFORNIA on October 13, 1978, Weighing 1,000 pounds, standing just over six feet high and fitted with thermal control louvers, TIROS-N carried two television cameras, a transmitter, a highly sophisticated RADIOMETER, and an array of instruments designed to observe humidity and cloud conditions. Powered by 10,000 solar cells, TIROS-N circled Earth every 90 minutes at an altitude of 503 miles. Its infrared and visible pictures, relayed every 15 minutes to three ground stations, had a resolution of 1.7 miles, providing meteorologists with visual data regarding high- and low-altitude cloud over that was unprecedented in its scope and detail. Essentially an operational prototype for a planned third-generation class of TIROS satellites, TIROS-N was the first example of its kind to possess the ability to collect meteorological information automatically from ocean buoys and remote weather stations and then to retransmit that data directly to the NATIONAL HURRICANE CENTER (NHC) in Miami, FLORIDA. On November 17, 1979, while positioned 1,150 miles east of Madagascar, TIROS-N photographed its first tropical cyclone; an extremely intense Indian Ocean CYCLONE then bearing down on the island.

Complete with a change in designation the remaining eight TIROS-N class satellites rocketed in orbit between June 27, 1979, as NOAA-6; and May 14, 1991 as NOAA-12. NOAA-B, launched on May 29, 1980, was stranded at an altitude of 155 miles and destroyed upon reentry less than a year later. When the geostationary GOES-West satellite over the central North Pacific Ocean malfunctioned on November 25, 1982, NOAA-7 was redirected to track Hurricane Iwa as the destructive late-season system bore down on HAWAII. Launched on March 28, 1983, NOAA-8 lasted until December 1985 when its battery pack exploded; during its relatively brief service, however, it did manage to photograph Hurricanes ALICIA (August 16–19, 1983) and GLORIA (September 24–27, 1985).

tornado A distant meteorological cousin of the TROPICAL CYCLONE, a *tornado* is defined as a tightly organized column of rapidly rotating winds centered about a vortex, or counterclockwise spinning funnel, containing an exceptionally low point of barometric pressure. Although much smaller in size and shorter in lifespan than a mature HURRICANE, TYPHOON, OR CYCLONE, a tornado's writhing mesoscale winds, clocked in some rare instances at more than 250 MPH (402 km/h), are capable of enormous, if somewhat limited, destructiveness. Formed when powerful thunderstorms link updrafts of warm, humid air with Earth's CORIOLIS EFFECT and then swiftly intensified by vacuumlike contact with either land or water, a majority of the 700 or so tornadoes that touch down in the United States each year are between 54 feet and 165 feet (20–50 m) wide, possess maximum wind speeds of between 55 MPH and 112 MPH (89–180 km/h), and trace paths that range from a few hundred feet to one or two miles (2–3 km) across. Some 3 percent of tornadoes observed in the U.S. Midwest have been considered truly violent outbreaks, howling tubes of air—turned black, gray, and sometimes red by collected dust and debris—that boast maximum wind speeds of between 207 MPH and 318 MPH (333–512 km/h), and paths of between six and 30 miles (10–48 km) long. Not surprisingly, at least 68 percent of all deaths attributable to tornadic activity are due to tornadoes of this class, making the early detection and tracking of such developing storms of paramount concern to meteorologists and civil defense authorities alike.

Generated by the towering CUMULONIMBUS CLOUD circulations that form the better part of a tropical cyclone's cloud mass, tornadoes have on numerous documented occasions accompanied the landfall of a hurricane, typhoon, cyclone, or TROPICAL STORM. Indeed, studies into the connection between tropical cyclones and tornadic activity indicate that some 59 percent of the hurricanes that affected the United States between 1948 and 1986 spawned a minimum of one tornado, while at least

10 percent of tropical cyclones observed during the same period produced two or more tornadoes. On a more regional basis, 70 percent of all hurricanes that came ashore on the GULF OF MEXICO—particularly those that made landfall in northeastern TEXAS, southwestern LOUISIANA, and the west coast of FLORIDA—yielded tornadoes, whereas only 40 percent of North Atlantic hurricane strikes contained any confirmed tornadic outbreaks. Meteorologists ascribe this disparity to the fact that Gulf of Mexico hurricanes are generally more intense at landfall than their North Atlantic counterparts, thus lending statistical credence to the theory that stronger tropical cyclones are more likely to propagate tornadoes than weaker, less-organized systems.

Tropical cyclone-spawned tornadoes are relatively small, fleeting phenomena, wiry offspring less than 500 feet (150 m) wide that travel for a mile or so across the land before vanishing into their parent clouds. Containing wind speeds of between 50 and 100 MPH (81–161 km/h), they most frequently form beyond a tropical cyclone's EYEWALL in those outer RAIN BANDS situated 150 to 250 miles (241–402 km) forward and to the right of the system's EYE.

For this reason, most recorded tornadic outbreaks precede a tropical cyclone's landfall, touching down hours before the brunt of the storm even nears the coast. The imminent arrival of an immense hurricane in Charleston, SOUTH CAROLINA, on September 10, 1811, was heralded by a tornado of unprecedented size and severity that devastated several of the city's neighborhoods just before midday. On October 3, 1964, a rapacious tornado hatched by onrushing Hurricane Hilda claimed 18 lives in Larose, Louisiana, and wrought nearly $1 million in property damage on BUILDINGS and crops. A little over two weeks later, Hurricane Isbell's October 15 assault on Florida generated a tornado that killed one person and injured another 22. Between September 8 and 9, 1965, Hurricane BETSY's Category 3 winds stitched together six flagging tornadoes, three of which occurred only hours before the hurricane made landfall in southern Florida. The remaining three, produced while Betsy neared its final landfall in Louisiana, affected neighboring ALABAMA and MISSISSIPPI. In early September 1979, Hurricane DAVID crafted no less than 34 deadly tornadoes as it arced across southeastern Florida on the 3rd of the month, killing one person and leaving another 300 homeless.

In other instances, such as 1992s Hurricane ANDREW, tornadoes have appeared during a system's inland dissipation at a time when downgraded tropical storms and TROPICAL DEPRESSIONS begin to split into their base thunderstorms or merge into highly unstable extratropical fronts. On September 20, 1967, Hurricane BEULAH generated some 115 confirmed tornadoes as it disintegrated into torrential rain squalls over eastern Texas, while Hurricane Norma spawned 13 tornadoes over northeastern Texas and southern Oklahoma as it dissipated on October 13–14, 1981. Two people died in Beulah's outbreak, while another seven perished in Norma's.

Conceived by the large quantities of cool, dry air found in the upper levels of a hurricane's cumulonimbus thunderheads and nourished by rumbling clashes with warm, humid air rising from Earth's surface, pre- and post-landfall tornadoes are noted for their capriciousness as well as their destructiveness. Reaching down without warning from the underside of a tropical cyclone's cloud canopy, they have often proved fatal to those people who live in lightly constructed buildings or improperly anchored mobile homes, Between 1948 and 1986, nearly 10 percent of all deaths from North Atlantic and Gulf of Mexico hurricanes were due to storm-spun tornadoes, a tally that amply illustrates the small but constant threat posed by them.

Tropical Storm Fay spawned three tornadoes as it moved ashore in northern Texas on September 7, 2002. One damaged a mobile home near Bolig and another damaged a house in Hungerford.

Isidore spawned a tornado that touched down in Lafourche parish, Louisiana, on September 26, 2002.

As Tropical Storm BILL came ashore in Louisiana on July 1, 2003, a tornado touched down 15 miles west of New Orleans. Another tornado was reported in the marshlands of Plaquemines Parish.

In 2005, one person was killed in MISSISSIPPI when a tornado spawned by Hurricane RITA's remnants ripped through two housing divisions.

In 2005, several tornadic touchdowns were reported in central Florida as powerful Category 3 hurricane WILMA trundled ashore in late October. Earlier, Hurricane EMILY generated at least eight tornadoes in the southern Texas county of Jim Wells alone.

Toyamaru Typhoon *Japan, September 22–27, 1954* Named for an interisland ferry that it tragically capsized on September 26, the Toyamaru Typhoon (also known as Typhoon Marie) delivered 125-MPH (201-km/h) winds, abundant rains, and rocking, 20-foot (7-m) seas to northern JAPAN on September 26 and 27, 1954. Fueled by a CENTRAL BAROMETRIC PRESSURE of 28.34 inches (960 mb) at landfall near the Japanese port city of Hakodate, the Toyamaru Typhoon's associated floods, landslides and fires eventually claimed 1,567 lives, 1,172 of them aboard the ill-fated ferry. Overtaken by the Category 3 typhoon while navigating the roiling waters of the Tsugaru Strait, the 1,400-ton *Toyamaru* was driven against the volcanic crags of Hakodate Bay and holed. Quickly filling with water, the listing ferry, along with its cargo of 45 railway cars, soon rolled over completely, throwing hundreds of terror-stricken passengers and crew into the frothing sea. Of the 1,368 people aboard the *Toyamaru*, only 196 were eventually rescued, making the ferry's loss one of modern Japan's deadliest maritime incidents. A Japanese legislative committee convened to investigate the disaster subsequently ruled that the vessel's master had not only anchored his ship too close to shore but that he also failed to order its abandonment in a timely manner, thus precipitating the typhoon's final tragedy.

Tracy, Cyclone *Northwestern Australia, December 20–26, 1974* An exceptionally severe Category 5 CYCLONE, Tracy ravaged the northern Australian city of Darwin with sustained 148-MPH (238-km/h) winds, eight-inch (203-mm) PRECIPITATION counts, and raging 20-foot (7-m) seas on the morning of December 25, 1974. Possessing a frighteningly low CENTRAL BAROMETRIC PRESSURE of 27.00 inches (914 mb) at landfall, Tracy was not only one of the most intense recorded cyclones to ever strike AUSTRALIA but also that nation's worst natural disaster. Ten thousand of Darwin's 12,000 BUILDINGS—lightly built tropical houses with fero-cement walls and galvanized sheet-metal roofs—were utterly destroyed, sheared off to the floorboards. Every tree in Darwin, a resilient port city of 41,000 people on Australia's northwest coast, was stripped of its foliage, uprooted, and then split down the trunk by Tracy's 164-MPH (263-km/h) gusts. Twenty-one vessels, among them a private yacht with 15 Christmas revelers aboard, were overwhelmed by Tracy's record-breaking waves, capsized, and lost at sea. All told, 66 people—13 of them children—perished in Cyclone Tracy, while another 790 were seriously injured, some maimed for life. Damage estimates of $230 million made Tracy the single most expensive cyclone in Australian history.

The seventh cyclone of the 1974 season, Tracy originated over the western Arafura Sea, approximately 600 miles (966 km) northeast of Darwin, on the morning of December 20, 1974. Carried steadily south-southwestward at 14 MPH (23 km/h), the cyclone entered the energized 84°F (29°C) waters of the eastern Timor Sea and swiftly intensified. Upgraded to TROPICAL STORM–status on the night of December 21, a newly christened Tracy continued churning south, spreading its spiral cloud bands into Croker Island, where automated weather instruments recorded sporadic wind gusts of 561 MPH (90 km/h) and some light rain. By early afternoon of December 22, Tracy had grown into a mature Category 2 cyclone whose 105-MPH (169-km/h) EYEWALL winds and deluging downpours were clearly visible on radar screens in Darwin, 250 miles (402 km) to the south. Duly alarmed by both the cyclone's sudden period of deepening and its slow but perceptible curl toward northwestern Australia, meteorologists at the Darwin Tropical Cyclone Warning Centre began to issue terse storm bulletins to coastal towns and villages from Melville Island to Wyndham, warning of "destructive winds," pelting rains, and heavy swell activity in harbors and inlets.

As slow-moving Tracy continued to strengthen through the morning and afternoon hours of December 23, 125-MPH (201–4-km/h) winds swept the surface of the Timor Sea, sending nine-foot (3-m) swells crashing ashore at Bathurst Island and Snake Bay, 80 miles (129 km) northwest of Darwin. Although still committed to its four-MPH (5-km/h) swing southwest, the cyclone was at this point predicted to pass comfortably 150 miles (241 km) north of Darwin, remain an offshore storm as it patiently rounded the Kimberley Plateau's northwestern fingers, and crawl ashore in the vicinity of North West Cape. Residents of Darwin, many of them otherwise preoccupied with final preparations for the upcoming Christmas holiday, were advised to expect tropical storm-force winds and torrential rain showers over the city on Christmas Eve but little else.

Instead, on the afternoon of December 24, the eyewall of Cyclone Tracy—now generating wind gusts of 150 MPH (241 km/h)—was positioned just 50 miles west-northwest of Darwin, heading east at five MPH (8 km/h). Described in official weather bulletins as a "severe tropical cyclone," it was calculated that Tracy would—unless deflected by a sudden course change—howl across Darwin shortly after dawn on Christmas Day, quite possibly bringing the city its worst cyclone strike in living memory. While a number of Darwinians did indeed heed the warnings to secure their elevated tropical houses, tape closed their louvered windows, and stow away light lawn furniture and bicycles, an even greater number did not. Perhaps lulled into a mistaken sense of complacency by the capricious actions of earlier Cyclone Selma (which three weeks before had shied away from Darwin just before it was due to make landfall there), thousands of people merrily strung garlands and set out holiday lights. Those residents returning home from early evening festivities found the roads of Darwin slick with light rain and buffed to a sheen by Tracy's leading gusts but still passable.

A rare central barometric pressure of 27.00 inches (914 mb) tightly locked within its circular EYE, Cyclone Tracy vaulted ashore at Darwin just before dawn on December 25, 1974. For nearly 12 hours, Tracy's 148-

MPH (238-km/h) winds and blasting rains robbed the city's buildings of their corrugated roofs, carried away storm shutters, defaced billboards, plundered trees and telephone poles, sabotaged several moored boats, and triggered severe localized flooding. Wind-whipped waves, some as high as 20 feet (7 m), invaded some 75 miles (121 km) of coastline, occupying soaring beachheads and low-lying mangrove swamps with equally destructive ease. Darwin's civic hall was utterly demolished, its elaborate stone portal left standing as a future MONUMENT to the historic carnage then besetting the city. Taking advantage of the cyclone's furious onset, opportunistic thieves made their way from gutted store to abandoned house, taking whatever items they could carry, only to return with surprise and consternation to the ruin of their own homes. Sixteen of the 66 reported victims of the storm, forced into the maelstrom by the structural failure of their government-built houses, were struck down by flying debris, impaled by timbers, or sliced in two by skating sheets of tin roofing.

Left without electricity, water, and telephone service, it would be 10 hours before the outside world learned of Darwin's virtual extinction. In the Australian capital of Canberra, government agencies tasked with finding shelter for those 36,000 people left homeless by Tracy, scrambled to evacuate the city entirely. Within the next six days, some 35,000 people left pulverized Darwin, taking to the roads and airports in a long, exhausting exodus, the scale of which had not been witnessed on Australian soil since the Second World War. Squadrons of military and civilian aircraft, from *Boeing 727*s to *Air Force Starlifter*s, delivered tons of food, water, and medicine to the hastily cleared runways of Darwin's disabled airports and then airlifted hundreds of injured men, women, and children to hospitals and refugee camps in neighboring Queensland. In time rebuilt, Darwin emerged from the destruction of Cyclone Tracy a larger, more prosperous city. The name Tracy has been retired from the rotating list of cyclone names.

tropical cyclone This broadly defined term is used to describe those cyclonic disturbances—TROPICAL DEPRESSIONS, TROPICAL STORMS, HURRICANES, TYPHOONS, and CYCLONES—that originate over the tropical and subtropical regions of the world's oceans. By definition, a cyclone is a weather system in which winds move in a circular direction around a warm center of low BAROMETRIC PRESSURE. Unlike the larger EXTRATROPICAL CYCLONE, the tropical cyclone has no frontal systems, and its strongest winds are located near Earth's surface. Barometric pressures ranging from a fairly mild 29.25 inches (990 mb) to a devastating 27.17 inches (920 mb) or less can bring on wind speeds

of 15 MPH (24 km/h) to 200 MPH (322 km/h) or higher, making a developing or mature tropical cyclone one of the most destructive meteorological conditions on the planet. It has been estimated that a tropical cyclone of hurricane, typhoon, or cyclone strength contains the equivalent energy of 2 million atomic weapons, although only 3–4 percent of this potential is ever realized. Certain tropical cyclones, such as the titanic Hurricane GILBERT of September 1988 are so large that they synchronize up to a million cubic miles of atmospheric circulation. Despite such staggering statistics, tropical cyclones are in truth fragile meteorological conditions that require several favorable factors—such as warm water temperatures and constant upper-level winds—to ensure their formation.

Tropical cyclones are divided into three classes based upon their intensity: tropical depressions, tropical storms, and either hurricanes, typhoons, or cyclones. A tropical depression is a tropical storm in which sustained wind speeds at the surface do not exceed 38 MPH (61 km/h). A tropical storm is a tropical cyclone in which wind speeds range from 39–73 MPH (63–118 km/h), while a hurricane, typhoon, or cyclone is a tropical cyclone in which sustained surface winds are 74 MPH (119 km/h) or higher. Each class is further defined by varying degrees of barometric pressure, PRECIPITATION counts, and the radius of its cloud mass.

The life cycle of a tropical cyclone can be divided into four distinct stages: Formative, Deepening, Mature, and Decay. Each stage is defined by a host of factors, including duration, geographical position of the EYE, central barometric pressure, forward speed, and direction. For example, during the Formative stage, tropical cyclones in the Northern Hemisphere are generally located between latitudes 5 and 15 degrees North and in the Southern Hemisphere between 5 and 10 degrees South. Formative tropical cyclones have central barometric pressures above 1,000 mb and tend to move at speeds of between six to 25 MPH (10–40 km/h).

During the Deepening stage, which lasts from two to four days, a tropical cyclone experiences a steep drop in its PRESSURE GRADIENT, the degree by which barometric pressure decreases over a certain radius of the storm's mass. Still moving in either a westerly, southwesterly, northwesterly, or northeasterly direction, depending on geographical location and season, the Deepening tropical cyclone can experience forward speeds of between six and 18 MPH (10–30 km/h). At this point in its life cycle the tropical cyclone's circulation has grown more organized, and the familiar eye has become clearly visible. Barometric pressures are rapidly approaching their minimum, while wind speeds near their maximum velocity.

A large tropical depression—without doubt a downgraded hurricane—spreads across the lower half of the United States. Tropical depressions can produce considerable quantities of precipitation. *(NOAA)*

A Mature-stage tropical cyclone, which on average lasts from two to six days, has a central barometric pressure of between 28.94 and 25.69 inches (980 and 880 mb) and a forward speed of between six and 25 MPH (10–40 km/h). Now officially a hurricane, typhoon, or cyclone, the mature tropical cyclone produces sustained surface winds of 74 to 200 MPH (119–322 km/h) and seven to 17-inch (237–576-mm) rainfalls. It also carries beneath its eye a dome of seawater known as a STORM SURGE, which in the event of landfall can prove enormously destructive to low-lying shorelines. Weather satellite photographs of a Mature-stage tropical cyclone reveal a tightly wound, almost cloudless eye surrounded by vast gray spirals of cumulus, CUMULONIMBUS, and CIRROSTRATUS CLOUDS. There is also the presence of an anticyclonic "tail," or the trail of cirrus clouds that highlights a cyclonic system's OUTFLOW LAYER. The Mature tropical cyclone's ISOBARS, or the lines that denote areas of similar barometric pressure, are almost circular, and the distance between them is fairly small.

It can take a tropical cyclone several days to pass through its final stage, Decay. By this time, the tropical cyclone has either undergone RECURVATURE and moved into cooler regions above latitude 40 degrees North or else made a stormy landfall on one of the many land masses that border the tropical cyclone's breeding grounds. In the first scenario, a tropical cyclone's forward speed can continue to increase to

40 to 50 MPH (64–81 km/h), quickly whisking it into waters where surface temperatures are below 76°F (24°C). Denied the invigorating evaporation rates of the subtropical seas over which it originated, a tropical cyclone begins to break apart as barometric pressures within its eye begin to rise, collapsing the EYEWALL and redirecting its circulation. In the event of landfall, a similar process occurs but not before the tropical cyclone has unleashed a torrent of wind and rain. As it continues to move inland, the tropical cyclone undergoes a gradual filling that frequently results in a thunderous hail of TORNADOES, thunderstorms, and flash floods. It then either transforms itself into an extratropical cyclone or dissolves into an existing high-pressure system.

There are recorded instances of two Mature-stage tropical cyclones directly interacting with each other over the open expanses of the NORTH ATLANTIC and NORTH PACIFIC OCEANS. Called PINWHEEL CYCLONES, their development and characteristics were first defined in the early 1920s and are known as the FUJIWARA EFFECT.

tropical depression   This meteorological classification is given to embryonic TROPICAL CYCLONES whose maximum sustained wind speeds are less than 39 MPH (63 km/h) and that possess at least one closed ISOBAR. In a tropical depression, some of the circular motion associated with mature-stage tropical cyclones is observable, although this rotation is primarily confined to those air currents located near the ocean's surface. A precursory stage to the TROPICAL STORM, tropical depressions are not given official HURRICANE and TYPHOON NAMES unless they have been downgraded from an already named tropical cyclone, in which case they retain the designation previously given to that particular system. The symbol for a tropical depression on weather maps is rendered as T.

tropical disturbance   This meteorological term is given to a dense cluster of tropical thunderstorms that maintain their collective identity for at least 24 hours. The opening stage in the development of a TROPICAL DEPRESSION, there is no predetermined wind speed associated with tropical disturbances, nor are they awarded official HURRICANE or TYPHOON NAMES unless they should achieve TROPICAL STORM intensity. While an average HURRICANE SEASON over the NORTH ATLANTIC OCEAN and CARIBBEAN SEA will see the development of at least 20 tropical disturbances, only a fraction of these will ever reach cyclonic maturity. In the NORTH PACIFIC OCEAN, where eastern hurricanes and western TYPHOONS originate with greater frequency than North Atlantic and Caribbean systems, between 30 and 40 tropical disturbances will organize during the course of an

average season, with some 16 of them eventually becoming full-fledged cyclonic systems.

tropical storm
This meteorological designation is given to a TROPICAL CYCLONE whose sustained wind speeds measure between 39 and 73 MPH (63–118 km/h). Principally used by official weather and civil defense agencies to describe the destructive potential of a cyclonic storm system with winds of less than hurricane intensity (74 MPH or higher), the term *tropical storm* further defines that critical stage in the development of a fledgling HURRICANE, TYPHOON, or CYCLONE when the storm becomes better organized, more intense, and considerably more dangerous. Although the average tropical storm is smaller and weaker than a mature tropical cyclone, a number of them have nevertheless managed to initiate severe flood emergencies in many countries around the world. In some instances, the toll in property damage and human lives during a particular HURRICANE or TYPHOON SEASON has been much higher from tropical storms than it has been from more developed cyclonic systems.

A tropical storm is formed when a loosely organized TROPICAL DEPRESSION begins to gather its strength over the warm tropical waters of the North Atlantic, North and South Pacific, or Indian

In this photograph, a tropical disturbance lingers off Florida's east coast. A coalescing mass of thunderstorms and low barometric pressure, a tropical disturbance can often develop into a tropical depression and then into a tropical storm. *(NOAA)*

oceans. Spinning in a counterclockwise direction in the Northern Hemisphere and clockwise in the Southern Hemisphere, the tropical storm's deepening but loosely defined EYE increasingly draws low-level, moisture-laden air toward it, intensifying the burgeoning convective processes of rising warm air and cooling condensation. As the tropical storm's CENTRAL BAROMETRIC PRESSURE steadily continues to drop, sustained surface winds concomitantly rise, averaging between 39 and 52 MPH (63–84 km/h) in a small system, and 53 to 73 MPH (85–118 km/h) in a late-stage one. Less concentrated around the eye than would be a mature hurricane, typhoon, or cyclone, they tend to be spread in gusty divisions throughout the 75–200 mile-wide (121–322-km) circulation, kicking up massive quantities of foam and spray from the surface of the sea. Meteorologists believe that this fairly balanced distribution of wind allows the tropical storm to strengthen more easily. The spray and foam is moisture, and moisture, drawn up into a storm's swirling mass, energizes that storm through the heat-producing mechanisms of evaporation and condensation. In this way, the tropical storm's powerful winds serve to draw warmth from the first three feet (1 m) of the sea's surface without churning up the cold waters that lie eight to 10 feet (2–3 m) below that. Should these cold waters swell to the surface or the fragile tropical storm pass over a cold-water pool or eddy, then it will rapidly begin to lose strength and possibly even dissipate.

During its early stages, the tropical storm does not possess a clearly defined eye. Satellite photographs of young tropical storms frequently describe a wide band of CUMULONIMBUS CLOUDS shaped something like a distended crescent, with the uppermost end curling in toward the center of low pressure. But as the storm's barometric pressure continues to decrease and the cloud mass immediately adjacent to the eye grows heavier with precipitation, the system begins to assume a more elliptical pattern. The tropical storm's elongated eye becomes better defined as successive layers of spiral rain bands cluster around it. Sustained surface winds steadily begin to climb, passing the 52 MPH (84 km/h) mark in a hail of gust-driven rain. In time, the cloud cover will curl very tightly around the eye, appearing to weather satellites as a concentric chain of gray-white thunderstorms radiating out from a funnellike hub. Surface winds will begin to spike at 74 MPH (119 km/h) and higher, signifying the tropical storm's graduation to minimal hurricane, typhoon, or cyclone status. Conversely, once a hurricane, typhoon, or cyclone's winds drop below 74 MPH (119 km/h), it reverts to tropical-storm status. Such a rapid change in classification can occur quite often in the lifespan of a cyclonic system, particularly when the tropical storm in question either travels a great distance across the open sea or seesaws just above and below hurricane intensity.

Tropical storms generally contain enormous quantities of precipitation. When Tropical Storm ALBERTO passed over CUBA in July 1994, more than 10 inches (254 mm) of rain fell on the Isle of Youth in a 12-hour period. Tropical Storm Claudette dropped an estimated 43 inches (1,092 mm) of precipitation on northern TEXAS in late July 1979, shutting down astronaut training operations at the Johnson Space Center near Houston and claiming seven lives elsewhere in the state. Tropical Storm Isabel deluged FLORIDA's northeast coast with 15 inches (381 mm) of rain in October 1985. In the PHILIPPINES Tropical Storm Lucille killed more than 100 people when its torrential 16-inch (406-mm) rains flooded Manila in May 1960, while Tropical Storm Harriet claimed 911 lives in southern Thailand on October 25, 1962.

Meteorologists attribute this destructive phenomenon to the tropical storm's evolving nature. In a mature hurricane, moisture-levels are rapidly decreased by the heated process of EVAPORATION, thereby reducing the amount of rain the storm actually carries with it. But in a tropical storm, where the convective cells are still organizing, the moisture-laden air is not rising fast enough to complete the evaporative cycle. Vast amounts of rain begin to collect in the edges of the cells, darkening the storm's pale white color. This precipitation will continue to gather until the tropical storm either reaches hurricane force or makes landfall.

Because tropical storms represent a secondary stage in the development of hurricanes, typhoons, and cyclones, their frequency in various parts of the world directly follows that of the larger, more intense cyclonic systems. However, because not all tropical storms eventually reach maturity, it can be said that tropical storms are more frequent than full-grown tropical cyclones. On average, the North Atlantic Ocean sees nine to ten tropical storms per season, with five to six of them reaching hurricane status. In the southeastern North Pacific, the averages are slightly lower, with five to eight tropical storms a year, four of them achieving hurricane strength. In the North Pacific region, the average jumps to between 20 and 30 tropical storms annually, with between 17 and 26 typhoons subsequently formed. In the BAY OF BENGAL and southern Indian Ocean, between six and 11 tropical storms are generated yearly, with a majority of these reaching cyclone strength.

In many parts of the world, the graduation of a tropical depression to a tropical storm is heralded by the tradition of naming the storm. When the winds of a tropical storm reach 39 MPH (63 km/h), meteorologists consult a revolving list of alphabetized male and female HURRICANE and TYPHOON NAMES. These names, which are chosen with an eye toward regional diversity, are then assigned to the tropical storm for the duration of its existence. Such a practice allows for greater ease in identifying and defining the developing tropical storm to other meteorologists, the media, and civil defense authorities.

The symbol used to designate a tropical storm on weather and tracking charts is as illustrated above. The twin convex vanes on either side of the circle indicate the counterclockwise (or, if reversed, the clockwise) motion of the tropical storm's winds, while the circle's open center signifies a storm of less than mature strength.

tropical storm warnings   Issued by the National Weather Service (NWS), tropical storm warnings announce that a TROPICAL STORM in the NORTH ATLANTIC, CARIBBEAN SEA, or GULF OF MEXICO is predicted to make landfall on a particular section of coastline in 24 hours or less. Superseding earlier TROPICAL STORM WATCHES posted by the NWS, tropical storm warnings are designed to prompt those communities that lie in a tropical storm's path to complete their preparations immediately for the arrival of 39+ MPH (63+ km/h) winds, potentially flooding PRECIPITATION counts and dangerous STORM SURGE or STORM TIDE conditions. As is the case with HURRICANE WATCHES and HURRICANE WARNINGS, tropical storm warnings appear as advisories, extended passages within a much larger weather bulletin that not only draw attention to the respective tropical storm's size, intensity, forward speed, and direction but also to a suggested course of defensive action.

In place for the duration of a tropical storm's passage, tropical storm warnings are generally lifted when sustained wind speeds fall below 39 MPH (63 km/h). They are also posted in those instances when a former HURRICANE has been downgraded to a tropical storm but still remains a threat to inland communities.

tropical storm watches   Posted by the National Weather Service (NWS) when a TROPICAL STORM drifts to within 36 hours of landfall on a certain section of coastline, tropical storm watches are designed to alert those communities that lie in a tropical storm's probable path to immediately begin their preparations for the possible arrival of 39+ MPH (63+ km/h) winds, torrential PRECIPITATION, and minimal but potentially destructive STORM SURGE or STORM TIDE conditions. Superseded only by even more urgent TROPICAL STORM WARNINGS, HURRICANE WATCHES,

and HURRICANE WARNINGS, tropical storm watches appear in weather advisories as bulletins, highlighted passages that not only describe a particular tropical storm's size, intensity, forward speed, and direction but also suggest a timetable for the initiation of evacuation and civil defense measures. Tropical storm watches are initially announced for long stretches of coastline—100 miles (161 km) or more—in order to allow vulnerable oceanside villages, towns, and cities to empty their crowded beaches, summer cottages, and anchorages before tropical storm warnings are issued 12 hours later. Tropical storm watches are also hoisted concurrently with tropical storm warnings; they are used to alert communities on both sides of the tropical-storm-warning zone to the possibility of last-minute course changes by the system. When tropical watches are announced, residents in threatened areas should securely stow away lawn furniture, trash cans, bicycles, wheelbarrows, and any other small objects that a tropical storm's winds could turn into potentially lethal missiles. Those individuals in coastal regions should consider evacuating vulnerable barrier islands and low-lying coastal areas before limited evacuation routes become crowded or impassable because of surge conditions.

tropical wave   Sometimes referred to as an easterly or African wave, a tropical wave is a center of low barometric pressure that migrates westward across the NORTH ATLANTIC, NORTH PACIFIC, and Indian OCEANS with the easterly trade winds.

Born over the hot, dry deserts of northern Africa, most North Atlantic tropical waves are in reality embryonic thunderstorms, loosely organized clusters of nimbus and CUMULONIMBUS CLOUDS that evidence very little vorticity, or rotary motion, as they track westward at speeds of between 12–20 MPH (19–32 km/h). Appearing to weather satellites as drifting white patches of low-level condensation, the 70 or so tropical waves that skirt across the subequatorial North Atlantic each HURRICANE SEASON can either dissipate into harmless rain showers or deepen into powerful, PRECIPITATION-laden thunderstorms—the root element needed for the formation of a seedling TROPICAL DISTURBANCE.

In the western North Pacific and Indian oceans, tropical waves tend to emerge from the INTERTROPICAL CONVERGENCE ZONE (ITCZ), or that broadbanded TROUGH of low-pressure that girdles Earth between latitudes 6 and 12 degrees North and South. Far more frequent than their North Atlantic counterparts, the 120 or more tropical waves that on average develop over the North Pacific each season receive their spin from Earth's CORIOLIS EFFECT and their promotion to tropical disturbance from the combined actions of EVAPORATION, CONDENSATION, and those pumping mechanisms naturally found in upper-level wind circulations.

tropopause   The boundary zone that separates the TROPOSPHERE from the stratosphere. Although they are rare, there have been recorded instances in which the cloud tops in the EYEWALL of a TROPICAL CYCLONE have penetrated the tropopause.

troposphere   Also known as the habitation layer, the troposphere is the lowest six to 12 miles (10 km–19 km) of Earth's atmosphere. Separating Earth's surface from the stratosphere and noted for its decrease in BAROMETRIC PRESSURE and TEMPERATURE with altitude, the troposphere encompasses the climatological environments required for TROPICAL CYCLONE generation and maturation. Very much a result of the physical properties of the troposphere, the mechanical and thermal dynamics of tropical systems in part and in whole depend upon the varying temperature and pressure characteristics of the troposphere for further development. Close to Earth's surface, relatively warm air circulates around a point of low barometric pressure, collecting tons of moisture through the process of EVAPORATION. Rising into the troposphere, the warm, moist air begins to cool, undergoes CONDENSATION, and becomes a cloud. In time, increased evaporation rates will result in thicker, higher clouds that contain great quantities of PRECIPITATION. Billowing ever higher into the troposphere, the upper reaches of the CUMULONIMBUS tower cool into long, frozen tendrils of cirrus and CIRROSTRATUS CLOUDS, while rain showers wash the lower quarter of the burgeoning CONVECTION cell. Sapped of its moisture and latent heat by the cooling altitudes of the troposphere, cold air begins to descend, thereby establishing both a downdraft counterpart to the warm air updraft and a full convective cycle. Without the varying thermal characteristics of the troposphere in place, warm, moist, rising air would simply continue rising, never condense into clouds or rain or the furious RAIN BANDS of a tropical cyclone.

trough   Either close to Earth's surface or within the cooling reaches of the upper atmosphere, a trough is an elongated area of low BAROMETRIC PRESSURE that generally runs from north to south. Its meteorological opposite is the *ridge,* an elliptical zone of high barometric pressure that also extends from north to south and is likewise found near Earth's surface or within the upper atmosphere. Both low-pressure troughs and high-pressure ridges can affect a particular HURRICANE, TYPHOON, or CYCLONE by either adding potential energy to a TROPICAL CYCLONE's strengthening configuration or by influencing the strength and direction of the system's STEERING CURRENTS.

typhoon   This regional term is given to those mature TROPICAL CYCLONES that originate in the western NORTH PACIFIC OCEAN. Like its meteorological cousins, the HURRICANE and the CYCLONE, the typhoon is a highly organized storm system in which a warm center of low BAROMETRIC PRESSURE is surrounded by winds that rotate in a counterclockwise direction in the Northern Hemisphere and clockwise in the Southern Hemisphere. Characterized by sustained surface winds in excess of 74 MPH (119 km/h) and barometric pressures ranging from a fairly mild 29.25 inches to a catastrophic 27.17 inches (920 mb) or less, the typhoon generally contains enormous amounts of PRECIPITATION and may on occasion spawn TORNADOES and waterspouts. The typhoon also carries beneath its EYE a vast dome of seawater known as a STORM SURGE that in the event of a landfall can be responsible for significant property damage and loss of life.

Typhoons originate over the warm, 84–86 degree F (25–27°C) waters of the North Pacific Ocean, west of the international date line and north of the equator, every year from April to December. Although the unofficial TYPHOON SEASON extends from June to December, there are several recorded instances of mature typhoons striking CHINA, JAPAN, and the PHILIPPINES in April and May. During the early part of the season, typhoons principally form between latitudes 3 and 10 degrees North, over the energized waters that wash the eastern shores of the Philippines. While these early storms tend to have less intense winds than late-season typhoons, they carry greater amounts of rain. Relentlessly moving westward at speeds of 12–15 MPH ((20–24 km/h), a number of them have brought major flood emergencies to the southern Philippines and northern Borneo. Conversely, late-season typhoons—storms that develop primarily north of latitude 10 degrees and further out in the North Pacific—have much higher DEATH TOLLS because of their enormous size and more organized nature. For some of them, it is a 3,000-mile (4,828-km) course across sun-drenched waters before a landfall is made in Hong Kong, China, Japan, or TAIWAN—time enough for them to grow into heat-driven machines of enormous destructive potential.

North Pacific typhoons are also far more frequent than either North Atlantic hurricanes or Indian Ocean cyclones. On average, typhoons account for 30 percent of the world's annual tropical cyclone activity, more than double the North Atlantic's annual average of 12 percent or the eastern North Pacific's 15 percent. In 1949 alone, 44 typhoons formed. In 1969, one of the least active years on record, 19 typhoons developed. In terms of trajectories, early season typhoons generally move in a linear direction west, while late-season storms swing more northwest, curving in a parabolic fashion around the high-pressure ANTICY-

CLONE that dominates the North Pacific Ocean during the summer months. Like a majority of hurricanes and cyclones, typhoons do not travel in a perfectly straight line but instead wobble back and forth over an imaginary track. It is not unusual for a typhoon's eye to drift 50 or more miles in one direction and then drift back to its original position a few hours later.

Some of the lowest barometric pressures ever recorded have been observed in typhoons. On August 18, 1927, the Sapoeroea Typhoon, 460 miles off Luzon Island, the Philippines, produced a pressure reading of 26.18 inches (886 mb), far lower than the 26.35 inches (892 mb) recorded during the LABOR DAY HURRICANE of 1935 or the 26.22 inches (888 mb) measured in Hurricane GILBERT on September 9, 1988. The Nagasaki Typhoon, which struck Japan on September 17, 1828, reportedly produced a minimum pressure of 26.57 inches (900 mb), while Typhoon Tip, a late-season storm that swept through the Luzon Strait in October 1979, produced a verified instrument reading of 25.69 inches (870 mb), a world record for the time.

Such remarkably low barometric pressures are due in part to the considerable size of late-season typhoons. While a mature hurricane or cyclone rarely exceeds 300 miles (483 km) across, a routine number of SUPER-TYPHOONS—those storms with barometric pressures lower than 28.31 inches (959 mb)—have widths of between 400 and 600 miles (644–966 km). Although a large typhoon's maximum winds are found directly around its eye, its extensive size allows for a greater distribution of the PRESSURE GRADIENT, or the drop in barometric pressure over a certain distance from the eye. This brings gale-force winds, many with 74-MPH (118-km/h) gusts, to a wider area of the ocean, thereby increasing the rate of EVAPORATION through wind-driven spray. When Typhoon VERA came ashore at Nagoya, Japan, on September 26, 1959, minimum barometric pressures of 29.23 inches (990 mb) were recorded 200 miles (322 km) from the typhoon's eye, and 29.53 inches (1,000 mb) another 250 miles (402 km) from that. It was, however, within the compact, 23-mile (37-km) radius of Vera's eye that the minimum barometric pressure reached 27.31 inches (925 mb) and then gradually rose to 28.93 inches (980 mb) 100 miles (161 km) from the center. Such a steep drop in the pressure gradient over so great a distance dramatically increased Vera's intensity, making it one of the largest—and strongest—typhoons to ever strike Japan.

Those countries that ring the North Pacific basin—China, Japan, Korea, the Philippines, and Taiwan—all have long, violent histories of frequent typhoon strikes. A number of these storms have been of exceptional severity, making them of meteorological or historical interest.

A number of typhoons have also managed to either skirt, or pass over, the Philippine Islands and lay waste to the coastal regions of Vietnam. One Vietnamese typhoon in September 1953 claimed the lives of more than 1,300 people, while another in November 1956, killed 56 people and left tens of thousands of others homeless.

The intensely low barometric pressures found in many typhoons have also been responsible for creating enormous, often deadly, storm surges. Caused when low central pressures allow the sea to form a dome of water literally beneath the typhoon's eye, these storm surges have frequently crashed ashore with devastating results. During the Nagasaki Typhoon that struck Japan in September of 1828, a towering storm surge in the Ariake Sea claimed more than 10,000 lives. A similar but even more deadly instance of a storm surge occurred at Haifong, China, in 1881; an estimated 30,000 coastal residents were drowned.

As for the word itself, most scholars agree that *typhoon* is a derivation of the Cantonese phrase *t'ai fung,* or "great wind." It appears to have first become associated with tropical cyclones in 19th-century Japan, where early meteorologists began to refer to the giant storms that annually afflicted their shores as *taifu.* In ancient Japan, typhoons were generally referred to as *nowaki,* or a strong wind that "splits the grass of the field." It has also been suggested that the term originated with Aristotle's *Meteorologica,* the fourth century B.C. investigation into the causes of weather in which he referred to a cyclonic-type storm system as a *typhon.* Because the Typhon was a mythical beast with enormous destructive powers, a connection between the words *typhon* and *typhoon* and their respective subjects is not difficult to make.

INTENSITY AND SCALE OF TYPHOONS

Intensity	Pressure (Inches)	Max. Wind Spd (mph)	Scale	Radius of 1000 mb	Storm Area 25m/sec
Weak	>29.23	<56	Very Small	<100	N/A
Moderate	28.34–29.20	57–76	Small	100–200	100
Strong	28.31–27.46	77–98	Medium	200–300	200
Severe	26.57–27.43	99–121	Large	300–600	300
Catastrophic	<26.57	>123	Very Large	>600	400

To quantify and record the size and strength of approaching typhoons more easily, the region's meteorological agencies have adopted a five-point scale of typhoon intensity. Similar in design and purpose to the SAFFIR-SIMPSON SCALE (the system used to define the destructive potential of North Atlantic hurricanes), the typhoon scale details the most salient highlights of typhoon generation and construction.

typhoon alley In the broadcast sense of the definition, the name is given to that funnel-shaped region of the NORTH PACIFIC OCEAN, ranging from latitudes 10 to 30 degrees North, through which a seeming majority of westward-moving late and early season TYPHOONS travel; in its strictest sense, it is the Luzon Strait, that channel of water between the northern PHILIPPINES and southern TAIWAN. Use of the phrase *typhoon alley* is primarily found in Hong Kong, CHINA, and the PHILIPPINES, where an above-average percentage of yearly typhoons make their stormy landfall.

typhoon gun A large 48-pounder cannon positioned by the British Royal Navy near the entrance to Hong Kong harbor was used during the late-19th and early 20th centuries to signal the approach of a TYPHOON. In an age when instantaneous telephone and television communications were not available, the firing of the typhoon gun not only served as the first and only warning many Hong Kong residents received of a threatening typhoon but also was a means by which warships anchored in the harbor could be simultaneously ordered to prepare for heavy weather. During the Second Hong Kong Typhoon of September 18, 1906, the typhoon gun was fired less than an hour before the typhoon's powerful EYE swept into the harbor. The staggering 10,000-person DEATH TOLL from this storm was due more to the speed with which the typhoon came ashore than it was to the questionable effectiveness of the typhoon gun as a warning device.

Although the typhoon gun is no longer used in Hong Kong, it has been supplanted by Tropical Cyclone Warning No. 10. This signal is issued by the Hong Kong Observatory when a system with sustained winds of 73 MPH (118 km/h) or higher are expected within 62 miles (100 km) of Hong Kong. Between 1946 and 2003, some 13 typhoons necessitated the hoisting of the No. 10 signal. In 1999, Typhoon YORK prompted the signal's hoisting for the longest duration of 11 hours. On September 1, 1962, the signal was issued as Typhoon WANDA delivered a central pressure of 28.28 inches (953 mb), sustained winds of 83 MPH (133 km/h) and gusts of 161 MPH (259 km/h). An extreme gust of 176 MPH (284 km/h) was observed in Tate's Cairn during Wanda's passage over Hong Kong.

typhoon names The practice of assigning a certain name to a particular western NORTH PACIFIC OCEAN TYPHOON represents a unique cultural aspect of these often memorable but seemingly indistinguishable storms. Although scholars are not exactly certain where or when the tradition began, centuries of important typhoons have been identified though the use of geographical place names, male and female first names, a few surnames, the phonetic alphabet, and even the name of a Japanese ferry capsized by a typhoon in 1954. In this way, all classes of TROPICAL CYCLONES, from TROPICAL STORMS to typhoons, enter some aspect of the meteorological discourse, while others, particularly those storms of historical or scientific significance, become part of the common folklore.

Before the custom of assigning typhoons one-word code or female names began in the early 1940s, particularly notable storms were often named for the geographical location in which they made landfall. In CHINA, the 1881 typhoon that struck the port city of Haiphong was named the Haiphong Typhoon by contemporary sources because of the enormous amount of damage and staggering 30,000-person DEATH TOLL it wrought on the city. A majority of memorable Japanese typhoons have been given geographical names, some of them persisting in common usage despite the fact that a number of them were assigned international names by the World Meteorological Organization (WMO). The exceptionally severe Nagasaki Typhoon of 1828, the GREAT TOKYO TYPHOON (September 30, 1918), the Muroto Typhoon of September 1934, and the Makurazaki Typhoon (September 17, 1945) are all storms that were named after the principal region of their landfall. Typhoon VERA, a particularly deadly storm that struck Japan on September 26, 1959, is still referred to in Japanese chronicles as the Ise-Bay Typhoon, for the coastline southeast of Osaka. The use of geographical names in typhoon identification further extends to both TAIWAN and the PHILIPPINES, where the Taito Typhoon of September 16, 1912, and the Manila Typhoon of 1882 commemorate destructive typhoon strikes on those respective port cities. There are even instances of typhoons assuming the names of the meteorological observatories over which they passed. Perhaps christened in honor of their contributions to the study of cyclonic storms, both the Gifu Typhoon of July 9, 1903, and the Hamamatu Typhoon of September 1918 were named for the Japanese meteorological stations they assaulted on their way ashore.

One typhoon, the Nagasaki Typhoon of 1828, is also known as the Siebold Typhoon. It was named for the German meteorologist, P. F. von Siebold, who in the midst of this intense storm, dashed around rain-swept Nagasaki taking barometric pressure readings. Although von Siebold's spirited efforts yielded a treasure trove of information about 19th-century Japanese typhoons, the storm that once bore his name is more commonly known for Nagasaki, the city that it battered to the tune of 10,000 dead.

Typhoons have also been named after ships that they have overtaken at sea. The Sapoeroea Typhoon of August 1927 was named after the Dutch oil tanker *Sapoeroea,* on-board which one of the lowest BAROMETRIC PRESSURES ever recorded in a tropical cyclone—26.18 inches (886 mb)—was observed. The TOYAMARU TYPHOON, September 26, 1954, was so-named for the 4,300-ton Japanese ferry *Toyamaru,* which it drove onto the rocky shores of Hakodate and then capsized. Of the 990 passengers and crew aboard, only 196 were saved.

During World War II, typhoons were frequently given single-word code names by meteorologists who assisted the Allied forces in their strategic planning. These code names, which were based on the phonetic alphabet, represent the earliest attempt to define a means of identifying potentially hazardous typhoons systematically. By using an alphabetized system—Able, Baker, Charlie, and so on—it became possible to determine where in the TYPHOON SEASON a particular storm occurred chronologically. The shorter words also made it easier and more efficient to communicate a typhoon's position, thus ensuring that naval operations would not be disrupted by severe weather.

However, the widespread use of the phonetic alphabet for military purposes other than typhoon identification soon proved to be more confusing than no system at all, so an alternative naming system, based on that of animal species, was adopted for meteorological use. This resulted in a plethora of Typhoon Bears, Coyotes, and Wolves before the snake typhoons of Cobra and Viper gave the naming system historic but tragic notoriety. During the course of Typhoon Cobra, December 17, 1944, Admiral William "Bull" Halsey's mighty Third Fleet suffered the loss of three destroyers and nearly 800 men. Cobra's 115-MPH (185-km/h) winds and 70-foot (22-m) seas simply overwhelmed the fast but tiny vessels, capsizing and sending them to the bottom in a matter of minutes. Less than six months later, Admiral Halsey had another run-in with a named typhoon, this one christened *Viper.* In this particular instance six sailors were lost overboard, the aircraft carrier *Hornet* was badly damaged, and nearly 80 aircraft were destroyed.

By the late 1940s, the custom of assigning female names to typhoons had become the prevalent method of identifying North Pacific storms. During the 1948 typhoon season, for instance, the Philippines was struck by Typhoon Delilah in November and Typhoon Iris in December. Subsequent Filipino typhoons include 1956s Typhoon Lucille, 1958s Typhoon Lorna, and TROPICAL STORM Lucille in 1960. However, by the early 1960s, as the practice of christening typhoons after females continued to spread throughout the Pacific basin, several nations that border the region began to view the custom as a Western, or distinctly U.S. influence, and one that they desired no part of. At a time when countries like the Philippines and Japan were trying to set aside their imperial pasts in favor of a new nationalism, typhoon names began to assume a more local flavor. By the 1964 season, Filipino typhoons were being called Dading, Welming, and Sening, while Japanese typhoons were not named at all but rather were issued numbers. This custom, which had been in periodic use in Japan since the mid-1950s, first stated the year of the storm and then its particular number within the season. Thus, Typhoon VERA, or the fifteenth storm of the 1959 season, became 5915. Because Vera was the name issued to Typhoon 5915 by the World Meteorological Organization, an arm of the United Nations that had been formed to coordinate the world's diverse meteorological agencies, it found little support in Japan. In those areas like Hong Kong and the Philippines where English is spoken as a second language, the international name for the typhoon is often identified in conjunction with its native native name.

In January 2000, the World Meteorological Organization (WMO) adopted a new naming system for North Pacific Ocean tropical cyclones. Instead of using western names as had been the practice since World War II, indigenous names were substituted. Like those given to North Atlantic tropical cyclones, these names were submitted to the WMO by those countries affected by or potentially affected by tropical cyclone activity in the western North Pacific zone. Some typhoons were given indigenous names, such as Sening, etc. New names were contributed by those countries that are affected by North Pacific tropical cyclone activity, including Cambodia, China, the Koreas, Japan, Micronesia, the Philippines, Thailand, the United States, and Viet Nam. The last name used under the old naming system was Zia, while the first name used under the new system was submitted by Cambodia, and was Damrey (Elephant).

In the western North Pacific naming structure, once one column of 28 names is exhausted, the second column is implemented. It is not unusual for this to occur during the course of a single season, as witnessed in 2005, when the fifth column of names ended at Saola (Typhoon, August 2005) and the sequence resumed at the first column, Damrey (Typhoon, September 2005). It should be noted that these names are no longer alphabetical, but arranged according to the countries that submitted them, which are listed alphabetically.

Tropical cyclones in the western North Pacific are also identified through the use of numbers. This system lists the last two digits of the year (i.e., 2003 becomes 03), followed by the number of the cyclone according to that year's season. Typhoon Dujuan (Azalea, a type of flowering shrub), for instance, was identified as 0313 because it was the 13th tropical system of the 2003 western North Pacific typhoon season.

Recognizing that the naming of tropical systems is primarily designed to make it easier for citizens and emergency personnel to refer to and respond to these potentially destructive events, in 2004 and 2005, the Hong Kong Observatory sponsored a competition to suggest new names for those already submitted by the Hong Kong Typhoon Committee. A panel of experts and local luminaries judged the suggested names before submitting them to the Typhoon Committee for approval. In 2005, the winning names were Kapok and Taichi.

Other names include Haitang (Chinese for Flowering Crabapple Tree); Fengshen (Chinese for The God of Wind); Kirogi (Korean for Migrating Wild Goose); Tingting (Chinese pet name for young girls); Tokage (Japanese for Lizard); and Songda (Vietnamese for A branch of the Red River in Vietnam).

In modern practice, the naming of typhoons permits for greater ease in identifying an incoming typhoon to both civil defense authorities and the general public, thereby increasing awareness of its potential destructiveness. It also adds a persona to the nature of a typhoon, assigning to these powerful atmospheric disturbances an almost human standard. Nowhere else is this more apparent than in the media, where the movements of named typhoons are frequently identified in terms of the pronouns *he* and *she*. Even in Japan, it is not unusual for a journalist or historian to refer to a typhoon with regard for the gender of its male or female name. While it is a relatively recent practice to assign typhoons any sort of name, the manner in which the biographies of these storms is recorded indicates a long-standing cultural desire to turn them into the historical events of a civilization's history.

typhoon season Although there is no officially established typhoon season, the seven-month period extends from June until December of each year; in it, TYPHOON activity in the western NORTH PACIFIC OCEAN is at its peak. Although *typhoon season* is essentially an unofficial designation used by meteorological organizations to determine more easily the probability of a typhoon's formation in the North Pacific during a particular time span, it does have a basis in known meteorological behavior. During the summer months in the North Pacific the trade winds—winds that move from east to west just above the equator—become stronger and more frequent. Coursing across the sun-swollen waters of the ocean, they carry with them pockets of air that vary in TEMPERATURE and BAROMETRIC PRESSURE from that of the surrounding AIR MASS. Known as TROPICAL DISTURBANCES, only a fraction of these volatile pockets ever develop into mature typhoons. However, it is during the summer months that the frequency of tropical disturbances over the North Pacific greatly rises, thereby proportionally increasing the likelihood that several of them will eventually develop into powerful typhoons.

Typhoon season is also the period in which high pressure ANTICYCLONES position themselves over the North Pacific Ocean. These anticyclones—vast bodies of warm air in which the prevailing winds are moving away from the center—directly influence the course and intensity of typhoons by acting as atmospheric buffers—areas where TROPICAL CYCLONES can not penetrate because of high barometric pressure. Because anticyclones vary from month to month within the typhoon season, early and late-season typhoons will not only generate in different sections of the North Pacific but will also move across it on different trajectories as well. June typhoons tend to originate west of longitude 130 degrees and either undergo RECURVATURE north-northeast or travel in a more or less linear fashion across the northern PHILIPPINES and into central CHINA. July typhoons generally develop west of longitude 140 degrees, and, after experiencing recurvature, move in a north-northeasterly direction toward KOREA. Likewise, typhoons in August and September form between 140 and 160 degrees West and either swing around the west coast of JAPAN, where they come ashore on the island of Hokkaido, or move several hundred miles east, where they pass over the central islands of Honshu, Shikoko, and Kyushu. Those late-season typhoons that originate in October, November, and December either recurve sharply and pass back into the North Pacific or else cross the northern provinces of the Philippines.

Such a varied selection of typhoon tracks further requires those nations that are prone to

This certificate, a copy of which was kindly provided by the captain of the USS *Dewey* (DD349), was awarded to the members of the destroyer's crew who were aboard the vessel when it was nearly lost during Typhoon Cobra in December of 1944. Only the extraordinary seamanship of its captain, the courageousness of its crew, and good luck prevented the Dewey from joining three other destroyers at the bottom of the western North Pacific Ocean. *(Captain C. Raymond Calhoun, USN [Ret.])*

typhoon strikes to modify the typhoon season to fit their own requirements. In the Philippines, for instance, the typhoon season is nine months long, extending from April to December. This is because a large number of recorded typhoons have struck the central Philippines' Visayas region during April and May. In Japan, the typhoon season runs from July to October; the highest concentration of typhoons comes ashore during August and September. In Tai-

wan, the typhoon season is shortened from June to November, although there are two instances of typhoons striking the island during May and December. In China, the typhoon season begins in June, when typhoons moving between the northern Philippines and southern Taiwan crash ashore near Hong Kong, and ends in December, when the last of the Philippines typhoons has passed through the South China Sea.

U

unnamed hurricane of 1683 *Northeastern United States, August 23, 1683* Deemed an "awful intimation of Divine displeasure" by the fiery Puritan chronicler Increase Mather, the 1683 HURRICANE over NEW ENGLAND seriously inundated portions of the lower Connecticut Valley with fearsome winds and heavy rains. Rising 26 feet (10 m) above flood stage in some places, the Connecticut River submerged at least four riverside towns, blighting their corn and grain fields, and caused deep financial hardship for their weary inhabitants. Hundreds, maybe even thousands, of trees were uprooted or snapped off at ground level, and miles of fences and hedgerows were trampled into unsalvageable wreckage.

unnamed hurricane of 1727 *Northeastern United States, September 27, 1727* By all accounts one of the most severe TROPICAL CYCLONES to have struck NEW ENGLAND during the first half of the 18th century, the HURRICANE of September 27, 1727, leveled houses, prostrated trees, shrubs, and signposts, grounded ships, smashed wharves, and ruined grain and hay stocks across the region. Crashing ashore near Narragansett Bay shortly before dusk, the hurricane apparently tracked northeast, releasing tons of PRECIPITATION over Boston as it whirled into Massachusetts Bay. In Boston, extensive damage was inflicted on the waterfront by a higher-than usual tide, while a chimney collapse near the city's famed Common fatally crushed one child and injured two others. At Marblehead, an important trading and military post 50 miles (80 km) northeast of Boston, the hurricane's predatory STORM TIDE carried a number of small fishing shallops ashore, while formidable winds snapped trees, overturned barns, and blew away small farm ANIMALS.

unnamed hurricane of 1783 *Northeastern United States, October 18–19, 1783* The second documented TROPICAL CYCLONE to have transited eastern NEW ENGLAND during the 1783 HURRICANE SEASON, the unnamed HURRICANE of 1783 is principally remembered for the blanket of snow it dropped across eastern Massachusetts, New Hampshire, and Vermont. Most likely caused by the hurricane's rapid union with an encroaching cold front to the north, the snow—incongruously coupled in contemporary accounts with "excessive winds" and "great rain"—fell to a depth of two inches (51 mm) in eastern Massachusetts and to between six and eight inches (152–203 mm) over the green gorges of Vermont. Two decades later, a similar occurrence was witnessed during the infamous Snow Hurricane of October 9, 1804. In many respects a more powerful system than the 1783 storm, the Snow Hurricane unloaded close to 12 inches (305 mm) of wet, heavy snow near Litchfield, Connecticut, causing numerous bough breaks and roof collapses.

unnamed hurricane of 1991 *North Atlantic Ocean–Canada, October 28–November 2, 1991* A North Atlantic TROPICAL CYCLONE with an eccentric trajectory and without a name, the Unnamed Hurricane of 1991 originated on October 28, 1991, as an extratropical depression. Identified simply as Hurricane No. 8, it first deepened into an extratropical storm on October 29, and a TROPICAL STORM on October 31. On November 1, as its central pressure sank to 28.93 inches (980 mb) and its sustained winds topped 75 MPH (121 km/h), the system was upgraded to HURRICANE intensity. After affecting a number of mid-Atlantic loop-the-loop course changes, the system was downgraded to a tropical depression before swishing ashore in the Canadian Maritime provinces. In the modern history of tropical cyclone activity in the NORTH ATLANTIC, the Unnamed Hurricane of 1991 was the 34th tropical cyclone of tropical storm or hurricane intensity to remain unnamed. No explanation for why this occurred has been given, although it is most likely that the system was identified as possessing tropical characteristics after the 1991 hurricane season had concluded. While the system formed at about the same time and in the same location as the "Halloween Hurricane" or "The Perfect Storm" of October 1991, the two systems were not related.

This photograph shows the castlelike headquarters of the U.S. Weather Bureau following its establishment in 1870. Communications and meteorological monitoring equipment can be seen on the grounds and on the building's roof. *(NOAA)*

U.S. Weather Bureau   Established in March 1870 by an Act of Congress, the Weather Bureau was the first federally chartered meteorological service in the history of the United States. Officially renamed the National Weather Service (NWS) in 1970 and shortly thereafter placed under the operational jurisdiction of the U.S. Department of Commerce, the U.S. Weather Bureau and its 100-year history encompassed a period marked by unprecedented advances in meteorological forecasting and documentation, as well as by the development of the first weather RADAR and SATELLITE tracking systems. A number of highly publicized weather modification programs, including PROJECT CIRRUS and PROJECT STORMFURY, were conducted under its direction, while several of the world's leading authorities on TORNADOES and TROPICAL CYCLONES—among them Isaac Cline, Ivan Tannehill, GRADY NORTON, Gordon Dunn, Robert H. Simpson, Joanne Simpson, Robert C. Sheets, and Robert Burpee—have been associated with its ranks. For decades the Weather Bureau was the central clearinghouse for meteorological information in the United States (a duty presently undertaken by the National Climatic Data Center [NCDC] in Asheville, NORTH CAROLINA), making its records of invaluable use to scientists, meteorological historians, journalists, economists, and civil defense personnel.

Initially headquartered in Washington, D.C., and placed under the supervision of the Chief Signal Officer of the Army, the Weather Bureau was intended to monitor systematically meteorological conditions across the United States and then issue daily 24-hour forecasts as well as storm warnings and emergency bulletins as required. Officially designated the Division of Telegrams and Reports for the Benefit of Commerce, the Weather Bureau's founding network of 24 observatories were linked by telegraph and, as indicated by its original name, were meant to assist in the expansion of U.S. trade by providing timely, accurate information regarding the destructive onset of rain squalls, frosts, heat waves, blizzards, and tropical cyclones.

Originally housed in a four-story brick building crowned with rooftop ANEMOMETERS, BAROMETERS, and rain gauges, the Weather Bureau received wire reports twice a day concerning BAROMETRIC PRESSURE, WIND SPEED, PRECIPITATION counts, and air TEMPERATURE from across the nation. After being analyzed by meteorologists and posted on charts, the data was compiled into forecasts that were then telegraphed to more than 7,500 locations around the country, including post offices, stock and grain exchanges, shipping companies, and military installations.

Guided by the able professionalism of its first director, pioneering storm tracker Cleveland Abbe, the Weather Bureau's collection of 24 reporting stations had grown to 284 by December 1878 and to 310 by 1884. Gradually accruing a vast archive of climatological information in the process, the Weather Bureau had plotted the particulars of every known NORTH ATLANTIC OCEAN and CARIBBEAN SEA hurricane that had occurred since the bureau's inception. In late-January of 1899, after a number of violent HURRICANES had struck the United States with deadly consequence, the Weather Bureau established its first tropical cyclone forecasting center in Kingston, JAMAICA. Similar in operation to the Belen Meteorological Observatory founded in Havana, CUBA, in 1857, the Jamaica station's primary mission was to provide U.S. naval units in the GULF OF MEXICO and Caribbean Sea with accurate tracking data on hurricanes and TROPICAL STORMS. On the cessation of the Spanish-American War (April–December 1899), the

Weather Bureau—believing that closer proximity to the U.S. mainland would increase the station's forecast accuracy—transplanted the Kingston office to Havana. It remained in Cuba until November 1900, when in response to criticism that it failed to forecast accurately the GREAT GALVESTON HURRICANE, it was again moved, this time to Washington, D.C.

HIGHLIGHTS OF THE HISTORY OF THE U.S. WEATHER BUREAU

1934: A moderately powerful tropical storm lashes Corpus Christi, TEXAS with 52-MPH (84-km/h) winds and heavy rains on July 25. An early season system that had developed off the coast of SOUTH CAROLINA on July 21 before moving *southwest,* crossing northern FLORIDA, and entering the Gulf of Mexico, the July 25 storm proves a difficult one for the Weather Bureau's hurricane Washington, D.C.-based forecasting center to track accurately. Eleven people in Texas die, while damage estimates reach nearly 2 million 1934 dollars.

1935: Producing the lowest barometric pressure yet recorded in a landfalling U.S. tropical cyclone (26.35 inches), the LABOR DAY HURRICANE's sustained 160-MPH (258-km/h) winds and 25-foot (10 m) STORM SURGE visit stunning destruction on the Florida Keys on the evening of September 2. The scenic Overseas Railroad to Key West is washed out, and the town of Tavernier is flooded to the rafters. A tropical with few meteorological peers, the Labor Day Hurricane's tightly coiled fury claims the lives of over 400 people, many of them World War I veterans.

Reeling from considerable criticism in the wake of the inaccurately forecasted hurricanes of September 1926, September 1928, July 1934, and September 1935, the U.S. Congress decides to decentralize the Weather Bureau's Washington, D.C., hurricane office on December 2. New regional stations are founded in Florida, LOUISIANA, and PUERTO RICO, with a principal tracking center at Jacksonville, Florida.

1936: The chief of the Weather Bureau's marine division, Ivan Tannehill, is called to testify before a committee of the U.S. House of Representatives convened to investigate the adequacy of the Weather Bureau's HURRICANE WARNINGS in the hours leading up to the Labor Day Hurricane's landfall.

1940: In an effort to prevent a repeat performance of the catastrophic "surprise" landfall of the GREAT NEW ENGLAND HURRICANE on September 21, 1938, the Weather Bureau opens a fourth hurricane forecasting office, this one in Boston, Massachusetts, in late May.

The United States Weather Bureau's principal hurricane forecasting office is shifted from Jacksonville, Florida, to Miami, in mid-November. Placed under the guardianship of the indefatigable hurricane tracker, GRADY NORTON, the new Miami station joins the United States Army Air Corps and the Navy in planning a series of aerial reconnaissance flights into future tropical cyclones—the birth of the HURRICANE HUNTERS service.

1953: The founding of a colorful meteorological tradition occurs on June 1 when the U.S. Weather Bureau begins using female first names to identify hurricanes and tropical storms in the North Atlantic Ocean, Caribbean Sea, and Gulf of Mexico systematically. The most effective, culturally relevant method of identifying Atlantic tropical cyclones yet devised, the all-female naming practice continues until July 1979, when male first names are added to the cyclical lists.

Grady Norton, veteran hurricane tracker and director of the Weather Bureau's central hurricane-forecasting office in Miami, Florida, is counted among Hurricane HAZEL's indirect fatalities. Said to have possessed the ability to smell an approaching hurricane, Norton suffers a massive heart attack while tracking Hazel on October 11.

1955. On April 1, the U.S. Weather Bureau's hurricane-forecasting station in Miami, Florida, is given a new director and a new name: the NATIONAL HURRICANE CENTER (NHC). Supervised by the late Grady Norton's assistant, Gordon Dunn, and tasked with monitoring tropical cyclone activity in the North Atlantic, Caribbean Sea, and Gulf of Mexico, the NHC is given sophisticated data-collection technology, including several Navy and Air Force aircraft and a staff of professional meteorologists, analysts, and technicians.

1956: In part prompted by the exceptionally active, particularly expensive 1953, 1954, and 1955 HURRICANE SEASONS, Congress appropriates several million dollars for continued study into the circulatory nature and trajectories of North Atlantic tropical cyclones. In Miami, Florida, the Weather Bureau establishes the National Hurricane Research Project and appoints Dr. Robert H. Simpson its first director. Conducted with the assistance of aircraft from the Navy and Air Force, the early March formation of the National Hurricane Research Project emphasizes the ongoing alliance between meteorology and technology against the depredations of hurricanes and tropical storms.

1960: The Weather Bureau installs a network of weather radars along the North Atlantic and Gulf coasts of the United States between February and August. Able to observe precipitation and cloud cover over their respective 200-mile-radius (322 km) areas, the radar stations quickly become one of the nation's primary lines of defenses against violent tropical and extratropical storms.

The first of Earth's civilian weather satellites, the 270-pound Television and Infrared Observation Satellite (*TIROS I*) is placed in orbit by the United States on April 1. A stunning mechanical achievement that relies on photographic and infrared radiometer technology to accomplish its mission, *TIROS I* allows the Weather Bureau its first unimpeded views of the planet's weather systems from an altitude of 464 miles (747 km), thereby dramatically increasing the accuracy of four- and five-day forecasts. Powered by solar cells, *TIROS I* is able to transmit nearly 23,000

photographs before crashing out of orbit on June 17, 1960.

1961: The Weather Bureau co-hosts the International Meteorological Satellite Workshop between November 13–22, 1961. Attended by meteorologists and scientists from 27 countries, the event features lectures on the newly inaugurated *TIROS* satellite and visits to various research-related sites around the United States.

1967: After serving as the charter director of the National Hurricane Center (NHC) for 12 years, Gordon Dunn retires from the Weather Bureau in late May. He is succeeded at NHC's helm by Dr. Robert Simpson, the former director of the National Hurricane Research Project.

1970: The U.S. Weather Bureau is renamed the National Weather Service on December 1 and its operations transferred to that organization.

Vera, Typhoon *Japan, September 21–27, 1959*
More commonly known in JAPAN as the Ise Bay
Typhoon, internationally code-named Vera inflicted
catastrophic property damage and staggering DEATH
TOLLS on the central Japanese island of Honshu during the night of September 26 and 27, 1959. Whirling
around a minimum BAROMETRIC PRESSURE of 27.37
inches (927 mb) at landfall near the industrial port
city of Nagoya, 55 miles (89 km) southwest of Tokyo,
SUPERTYPHOON Vera's scathing 130-MPH (238-km/h)
gusts, 12-inch (305-mm) PRECIPITATION counts, and
capricious landslides damaged or destroyed 36,109
BUILDINGS, swept away 827 road and railway bridges,
and inundated much of Ise (pronounced EE-say) Bay
with destructive 17-foot (6-m) STORM TIDES. In the
harbor at Nagoya, seven ships—among them a 7,412-
ton British freighter—were carried ashore, while sawing 30-foot waves elsewhere along the coast sank 25
fishing boats, claiming the lives of four dozen seafarers.
In total, some 5,159 people were killed and another
10,000 injured. More than 1 million people in six of
Honshu's prefectures were rendered homeless. Damage
estimates quickly soared past the 500 billion yen mark
(about 2 billion U.S. dollars), making Vera contemporary Japan's most destructive TROPICAL CYCLONE.

The third most intense typhoon ever recorded in
Japan—only the first Muroto Typhoon of 1934 (26.93
inches) and the Makurazaki Typhoon of 1945 (27.07
inches) produced lower confirmed barometric pressures—Vera was the fifth major typhoon to strike the
island nation during the active 1959 TYPHOON SEASON.
At least 137 people perished when Typhoon Georgia slammed into central Japan on August 14; the
season closed with Typhoon Charlotte's claim of 28
lives on Okinawa island on the afternoon of October

16. Indeed, just one week before Vera came ashore,
Typhoon Sarah had addled the southern Ryukyu Islands
and the southwest coast of Kyushu with 110-MPH
(177-km/h) winds and 22-foot (7-m) seas, drowning
35 Japanese fishers and causing extensive damage to
trees, lightly built houses and shoreside installations.
Vera's landfall on September 26 is a unique meteorological coincidence in Japan; on the same date in
1954 the tragic TOYAMARU TYPHOON sailed ashore in
southern Hokkaido; Typhoon IDA decimated central
Honshu on September 26, 1958, one year to the day
before Typhoon Vera brought such widespread fear
and suffering to the densely inhabited main island.

Vietnam In modern Vietnamese history, several
powerful TYPHOONS have affected the peninsula
nation. On August 11, 2001, Typhoon Usagi savaged
Vietnam's central eastern coast with powerful winds
and deluging rains. Emergency management officials
in the country reported that one person was killed and
more than 50,000 left homeless, by Usagi's passage.

In November 2001, Typhoon Lingling hit central
Vietnam, claiming 18 lives and demolishing hundreds
of BUILDINGS. Earlier responsible for killing nearly 200
people in the PHILIPPINES, Typhoon Lingling delivered
83-MPH (137-km/h) winds to Vietnam's Yen and Binh
Dinh provinces.

The most intense typhoon to strike Vietnam since
1993, Typhoon Krovanh bashed ashore on the Vietnamese–Chinese border on August 25, 2003. One person
was killed, and nearly 1,000 buildings were damaged
or destroyed. In neighboring China, Krovanh's wrath
was more serious; the 12th typhoon to strike CHINA
during the 2003 season, Krovanh destroyed more than
11,000 homes in Guangdong and Hainan provinces.

This tinted IR image shows a strengthening Tropical Storm Vince as it scudded eastward toward the Iberian Peninsula. Eventually achieving Category 1 status, the first North Atlantic tropical cyclone to be given a "V" name under the 1953 naming system, was soon downgraded to a tropical depression. It eventually came ashore in southwestern Spain, bringing rain to the plains. *(NOAA)*

On November 2, 2005, a downgraded Typhoon Kai-Tak savaged Vinh, Vietnam, with 50-MPH (85-km/h) winds and 29 inches (737 mm) of rain. A lingering TROPICAL STORM, Kai-Tak caused widespread flooding in Quang Ngai province, killing some 15 people.

Vince, Hurricane *Eastern North Atlantic Ocean–Iberian Peninsula, October 9–11, 2005*   The first "V" named TROPICAL CYCLONE to have originated in the North Atlantic basin, Hurricane Vince formed from a nontropical low pressure system, similar to the MESOSCALE CONVECTIVE COMPLEX (MCC) that generated deadly Hurricane Mitch in 1998. On October 9, while located 140 miles (225 km/h) northwest of the Madeira archipelago. During the late-evening hours of October 9, Vince was upgraded to HURRICANE intensity with a central pressure of 29.14 inches (987 mb). In an area of characterized by unfavorable WIND SHEAR conditions, Vince quickly weakened, finally coming ashore in southwestern Spain on October 11 as a TROPICAL DEPRESSION with a central pressure of 29.59 inches (1,002 mb). A unique system on several levels, Tropical Depression Vince delivered 35 MPH (55 km/h) winds and one–two inch (25–51 mm) PRECIPITATION counts to the lower half of the Iberian Peninsula.

The name Vince has been retained on the rotating list of North Atlantic tropical cyclone identifiers.

Viñes, Benito *(1831–1893)*   Spanish-born Cuban meteorologist, director of the Belén Meteorological Observatory in Havana, CUBA, and Jesuit priest, Father Benito Viñes stands as one of the most celebrated TROPICAL CYCLONE scholars of the 19th century. A fugitive from revolutionary Spain with years of training in the atmospheric sciences, Father Viñes spent 23 years in the Caribbean, boldly defining and redefining the meteorological characteristics of West Indian HURRICANES—and, by extension, the theory and understanding of tropical cyclones around the globe. After completing his religious and scientific education in France in 1864, Viñes was appointed to succeed fellow meteorologist and priest, Antonio Cabré, as director of the prestigious meteorological observatory at the Royal College of Belén in late August of 1870. A patient, methodical man who sought to improve the accuracy of weather forecasts through daily observations of clouds and upper-level air currents, Viñes passed the following five years astutely observing the vagaries of Caribbean weather, placing a particular emphasis on hurricanes and TROPICAL STORMS and their precursory signs of approach. When in early October 1873 a powerful hurricane strike left much of southwestern Cuba in ruins, Viñes traveled to the affected area to survey the degree and nature of the destruction. On the morning of September 11, 1875, Viñes posted his first HURRICANE WARNING for Cuba's south coast. Less than 48 hours later, a fairly intense hurricane did indeed batter Havana, collapsing numerous BUILDINGS and withering acres of sugarcane fields but claiming few fatalities. This circumstance was attributed by a Havana newspaper to the timeliness and accuracy of Viñes's forecast. During the summer of 1877, Viñes published what is considered his most accomplished work, *Apuntes Relativos a los Huracanes de las Antilles.* Eventually released in the United States under the title *Practical Hints in Regard to West Indian Hurricanes,* this revolutionary treatise on the law of storms quickly became an international best seller. Viñes's uncanny ability to predict the course of individual hurricanes accurately, along with his growing reputation as the foremost authority on the circulatory patterns of tropical systems, soon earned him the fond sobriquet "Father Hurricane." In 1888, Viñes developed his "Antilles Cyclonoscope." Modeled after the famed horn cards of HENRY PIDDINGTON, the Cyclonoscope was designed, in the words of its creator, to "detect cyclone movement in the atmosphere while the vortex [eye or center of the system] is still far away." Intended for use by mariners and storm watchers, the Cyclonoscope allowed individuals to "determine the position of the observer relative to the trajectory and the direction of the trajectory," thereby providing a means by which evasive measures could be taken. Although a surviving example of the Cyclonoscope bears at its center the legend, "Ano de 1888," Viñes's explanatory treatise for the device was not published in Havana until 1902, an indication, perhaps, that its purpose and use was already widely understood. On July 23, 1893, in the midst of one of the busiest NORTH ATLANTIC HURRICANE SEASONS on record, Father Viñes died in Havana. Widely revered for his scientific intellect, devotion to duty, and lifesaving wisdom, Father Viñes's passing was announced to the world in a special newspaper bulletin headlined, "The Last Hour and Death of Father Viñes."

Virginia *Eastern United States* Even though Virginia shares 110 miles of its eastern shoreline with the storm-contested waters of the western NORTH ATLANTIC OCEAN, the Cavalier State's tidewater plains, tapered peninsulas, gentle hill country, bountiful plateaus, and dueling seaports have during the past three centuries been challenged by only a moderate number of TROPICAL CYCLONES. From 1667 to 2006, Virginia was besieged by 35 documented HURRICANES and TROPICAL STORMS, with at least 16 of these systems making landfall on the state's east coast. Of these 35 tropical cyclones, 12 were considered major hurricanes, Category 3 and 4 systems with minimum winds of 111 MPH (179 km/h), significant STORM SURGES, and sizable DEATH TOLLS.

Virginia's river-frayed eastern shores, noted for placid beaches that overlook the scenic waterways of Chesapeake Bay, are on average indirectly threatened by a tropical cyclone every six years and directly affected by a tropical storm or mature hurricane every 11 years. On at least five occasions between 1667 and 2006, Virginia sustained multiple strikes in the course of a single HURRICANE SEASON, as witnessed in 1750 when two severe tropical cyclones battled eastern portions of the state between August 18 and September 2, and in 1893 when two tropical systems caused steep losses over the Shenandoah Valley from August 29 to October 4. Both 1750 hurricanes were of probable North Atlantic origin; only the August 1893 hurricane developed to the east, in the vicinity of the Cape Verde Islands. The October 4 system first formed in the GULF OF MEXICO before bolting northward. All three Atlantic storms were systems of commanding size and severity, with the August 18, 1750, hurricane sinking two Spanish galleons as they struggled to round the Virginia capes. The September 2 hurricane was even more brutal, with 14 additional vessels either run ashore or sunk, many of them exacting a hefty loss of life.

Although many of Virginia's most destructive recorded hurricanes have in fact made landfall near the state's Atlantic ports of Norfolk and Newport News, a deadly spate of others first originated over the Gulf of Mexico or southwestern Caribbean Sea before recurving inland over LOUISIANA, MISSISSIPPI, FLORIDA,

GEORGIA, and the Carolinas. Brimming with rain, such tropical systems frequently serve gale-force winds and turbulent floods to the forest and mountain ranges of central and western Virginia, causing enormous amounts of property damage and scores of deaths.

During the remarkable 1893 hurricane season, for instance, Virginia suffered indirect strikes from two of that year's most violent tropical cyclones, the SAVANNAH-CHARLESTON HURRICANE of August 29 and the GRAND ISLE HURRICANE of October 4. Initially darting ashore between Savannah, Georgia, and Charleston, SOUTH CAROLINA, on the evening of August 27, the Savannah-Charleston Hurricane subsequently recurved northeast, passing through western NORTH CAROLINA and eastern Virginia before tracking into NEW YORK and NEW ENGLAND on August 29. Intense rains coupled with searing gales toppled trees, overturned barns, and destroyed entire tobacco fields, causing hundreds of thousands of dollars worth of property damage. Four people in Virginia perished.

Not more than one month later, the squally remains of the Grand Isle Hurricane moved eastward through Virginia before emptying into the North Atlantic Ocean near the North Carolina border. Originally slamming into Louisiana on the night of October 1–2, the Grand Isle's rapid passage through southern Virginia was not half as severe as it had been in Louisiana, where nearly 2,000 people were drowned by the hurricane's powerful storm surge. While hundreds of trees, buildings, and fences in Virginia were damaged by the Grand Isle Hurricane's 45-MPH (72-km/h) winds and whipping rains, published casualty figures recorded only one death—that of a farmer crushed beneath a tipped cow barn.

Following a similar northeasterly trajectory, the tattered remnants of Hurricane CAMILLE stitched a ribbon of record-setting devastation across western Virginia's Blue Ridge Mountains on the blustery night of August 20, 1969. A truly extraordinary tropical cyclone in any number of meteorological and historical respects, Camille dropped almost 30 inches (762 mm) of rain across the James River basin in a five-hour period, forcing the waters of the river to crest 19 feet (6 m) above flood stage. Within a matter of hours, wooded valleys were transformed into debris-mined fjords, while landslides carried entire neighborhoods into seething oblivion. Almost 4 inches (102 mm) of rain fell over the state capital at Richmond, flooding streets and temporarily closing down government offices. Before dissipating over the western North Atlantic on August 25, Camille claimed some 113 lives in central Virginia and another 42 in West Virginia and generated insured property losses in excess of $132 million.

Classified as a TROPICAL DEPRESSION for the duration of its trek across Virginia, Camille nonetheless remains that state's deadliest, most historic hurricane disaster. Very rarely does a year pass in Virginia in which the anniversary of Camille's shattering visit to the state is not recalled in newspaper articles, televised remembrances, and front-porch stories. Invariably, Camille's 20th-century tantrum is compared by storm-wise journalists and chroniclers to earlier hurricanes and tropical systems experienced in Virginia, to those pioneering occurrences that have through history, legend, and folklore shaped the state's cyclonic culture.

Indeed, the famous 1609 British expedition to settle Jamestown permanently almost ended in tragedy when an intense early season hurricane—the "most terrible and vehement storm" of July 25—overtook the nine-ship fleet as it sailed northward through the BAHAMAS Islands. Raging for nearly two days, the tempest sank one ship and scattered the rest as far north as BERMUDA. There, the flagship, *Sea Venture,* was run ashore and abandoned, its surviving passengers and crew soon planting the first permanent colony on the coral outpost. Painstakingly regrouped over a period of days, the seven remaining ships eventually landed safely in Jamestown.

Nearly six decades later, a flourishing Jamestown was itself struck by an immense midseason tropical cyclone, the "dreadful Hurry Cane" of August 27, 1667. For nearly 24 hours violent winds and a swamping, 12-foot (4-m) STORM TIDE beset the colonial settlement, uprooting trees and topping riverbanks. Hundreds of houses were overturned or collapsed, trapping their inhabitants beneath the wreckage as the dark tide gnawed far inland, devouring dozens of people and much livestock along the way. An unfinished fortress at Point Comfort, its stone foundations undermined by the hurricane's rocking waves, collapsed into the sea, taking tons of seasoned lumber intended for its completion with it. Likened by a contemporary observer to the "creation [of a] Second Chaos," the 1667 hurricane claimed at least 75 lives in Virginia and caused widespread property and agricultural losses along the east shore.

Some 80 miles (129 km) to the northeast and 39 years later, a powerful offshore hurricane grazed the northern shores of Cape Charles in August 1706, forcing its trailing storm surge into the entrance of Chesapeake Bay, where it submerged two unsuspecting fishing villages. Overwhelmed by the hurricane's sudden fury, a convoy of 14 British merchant ships was driven ashore and wrecked, drowning several sailors. Unlike a similar situation in the later hurricane of August 12, 1724 (during which numerous ships were run aground near the James River but subsequently salvaged), the entire 1706 fleet was pounded to pieces by hammering waves, its cargo scattered and eventually ruined.

One of early Virginia's most tragic hurricane disasters occurred on September 3, 1747, when an immigrant vessel en route from Ireland to Fredericksburg foundered while entering the mouth of the Rappahannock River. Quickly capsized by the tropical cyclone's rearing seas, the ship carried more than 50 people to the bottom with it, many of them indentured servants.

Known to history as the October Hurricane of 1749, the next offshore powerhouse burnished much of the Virginia coastline with fearsome winds and torrential rains on October 7 of that year. Considered by contemporaries to have been the worst hurricane to have struck southern Virginia in living memory, the October Hurricane sent a 15-foot (5-m) surge across the Chesapeake Bay and into greater Norfolk. Dozens of streets along the city's waterfront were inundated; a horde of small boats drifted inland and with the surge's retreat came to rest in cornfields. More than 50 other ships, from brigs to merchant ships, were thrown ashore along the Virginia coast, costing the lives of 22 seafarers. Damage estimates in Norfolk were substantial, totaling some 30,000 British pounds.

Three notable hurricanes affected Virginia between 1769 and 1788, all of them tropical systems of strident ferocity. The first, which lashed the colonial capital at Williamsburg with heavy winds and rains on September 8, 1769, prodded four ungainly merchant ships onto the York River delta and then swiftly reduced them to complete wrecks. Deemed "a most dreadful hurricane" by the *Virginia Gazette,* the 1769 fury destroyed two other ships on the James River and smashed a large wooden pier at Yorktown.

An even more destructive hurricane than the 1769 storm, the Equinoctial Storm of September 23–24, 1785, sent its surge crashing across the lower Chesapeake Bay, causing immense damage to the Norfolk suburb of Portsmouth. A number of warehouses filled with salt, corn, and sugar were swept into the bay. Scores of houses were unroofed as pocket floods submerged their lower stories. Some 30 ships were carried inland and stranded, many of them in thickly wooded areas. Another £30,000 pounds in losses were assessed.

The third hurricane, a powerful, early season interloper from Bermuda known to posterity as George Washington's Hurricane, shot ashore near the mouth of the Chesapeake on July 23, 1788. So dubbed because Washington himself detailed the passage of its EYE over his home at Mount Vernon in his daily weather journal, this particular system was by all accounts a very violent one. Blasting away at Norfolk and Portsmouth for nearly nine hours, the hurricane uprooted trees, toppled chimneys, breached fences, and tossed almost 50 ships onto swirling tidal banks. Entire fields of ripening corn were mowed under, while hardwood groves were pruned of their limbs and tops. Kept "at home all day" by the remarkable severity of the weather, the future president sat down to his account and soberly mourned the loss of his mill race, the "great mischief to the grain, grass, etca.," and the wind's destructive effect on nearby trees, houses, and small ships. Always a patient, methodical man, Washington further included observations regarding TEMPERATURE fluctuations ("70 in the morning, 71 at noon, and 74 at night"), wind speed ("a hurricane"), and shifts in wind direction ("No. Et. to So. Wt.").

Employed as a military and trade depot since the mid-1600s, prosperous Norfolk was in September 1821 again taxed by a great storm, the so-called NORFOLK–LONG ISLAND HURRICANE. First pressing ashore near Cape Fear, North Carolina, during the early morning hours of September 3, the Norfolk–Long Island Hurricane closely followed the eastern seaboard as it trailed north-northeast, releasing torrents of rain and winds of tremendous violence from Wilmington, North Carolina, to New York City and points north. Norfolk, west of the EYE and within the hurricane's NAVIGABLE SEMICIRCLE, sustained serious property losses; nearly every residence in the city was left either damaged or destroyed. Several large BUILDINGS—including the city courthouse and at least two banks—were partially unroofed by the hurricane's gusts, and an Episcopal church lost much of its front portico and organ loft to deluging rains. Along Norfolk's waterfront, the hurricane's bellicose seas shattered piers, inundated warehouses, swept away drawbridges, and stranded two U.S. frigates, the *Congress* and the *Guerriere*, on a sandbar near Hampton Roads. On the opposite side of Chesapeake Bay, four steamboats, struggling to raise steam and keep afloat in the hurricane's confused seas, were dragged from anchorage and pushed ashore near Virginia Beach. Hundreds of trees stood stripped of their foliage, while countless others were laid over like cordwood, crashing onto houses, barns, and fences. Many of the city's most spectacular private gardens were withered into weeds by the hurricane's salt-poisoned winds, their wooden gazebos and picturesque stone follies turned into true ruins. Judged by eyewitnesses to have been a veritable "war of the elements," the Norfolk–Long Island Hurricane did several hundred thousand dollars worth of damage to eastern Virginia and claimed at least 22 lives. Norfolk would later be struck by at least two hurricanes of similar intensity, these occurring on August 16, 1830, and September 9, 1854.

On the morning of August 7, 1827, an intense hurricane that had earlier devastated the Caribbean island of St. Kitts shuffled ashore in Savannah, Georgia. Broadly arcing inland, this hurricane passed over central Virginia, dropping enormous amounts of precipitation as it grated across the Shenandoah Mountain range. Rumbling landslides and bristling floods killed at least 54 people in the region and drowned some 5,000 cows, horses, and pigs. Reminiscent of Hurricane Camille's 1969 rampage, the 1827 hurricane would in turn be duplicated by an even deadlier rainmaker, that of September 29, 1896. A dissipating hurricane that had originally come ashore in northwestern Florida on the 28th, the 1896 storm dumped nearly seven inches (178 mm) of rain on western Virginia as it struggled over the 4,500-foot (1,500-m) Blue Ridge Mountains. Some 100 people perished, many of them inhabitants of small, isolated lumber towns. Property claims soared to more than $7 million.

Less than three years later, Virginia survived its only recorded strike from a November hurricane, an 1899 late-season surprise that appeared on November 5. Sliding inland over North Carolina shortly after dawn, the moderately severe hurricane quickly recurved into central Virginia, pounding much of the Shenandoah Valley with four-inch (102-mm) PRECIPITATION counts and tropical storm-force winds. Scores of wooden buildings were demolished, pinning dozens of victims beneath their rain-splintered carnage. Fallen trees churned down flooded ravines, scraping away log-lined embankments, finger piers, and even small buildings. At least 20 people reportedly died; damages to crops, roads, and railroad bridges were tallied at over $2 million.

During the record-setting 1933 hurricane season, the entire east coast of Virginia was savaged by the infamous Chesapeake-Potomac Hurricane of August 23. A moderately powerful Category 3 hurricane—a central pressure of 28.35 inches (960 mb) was taken at landfall near Cape Hatteras, North Carolina—this mid-season fury sharply recurved inland, passing approximately 30 miles (48 km) west of Norfolk before arcing northward into Pennsylvania. In several places along the western edge of Chesapeake Bay, the hurricane's 10-foot (3-m) storm tide dragged beachfront cottages from their foundations, stranded sleek sailboats and boxy motor yachts on tidal mud flats, and inundated much of downtown Norfolk with six to eight feet (2–3 m) of brackish water. The town of Gloucester Point, 30 miles (48 km) northwest of Norfolk, saw both its post office and drugstore leveled by the hurricane's 30-foot (10-m) waves. Fifteen people in Virginia perished; $11 million in property losses were assessed, making the Chesapeake–Potomac Hurricane one of the most expensive tropical cyclones in state history.

Quite surprisingly, considering the state's relative dearth of hurricane activity, Virginia was in the three-year period from 1953 to 1955 affected by no less than five tropical systems of hurricane intensity, with three of them—CONNIE (August 12), DIANE (August 17), and Ione (August 28)—traversing the region during the course of the 1954 season alone. This unprecedented spike in the Virginia hurricane count began in early August 1953 when Hurricane BARBARA flaunted its Category 1 winds and feeble rains over Norfolk and the

Chesapeake Bay, uprooting hundreds of trees, downing power lines, and swamping several small watercraft, but causing no fatalities. Just over one year later, on October 15, 1954, the truly exceptional Hurricane HAZEL lashed extreme western Virginia with destructive winds and torrential rains, killing three people and injuring another fifty-six. Heralding a 15-year expansion in hurricane activity along the eastern seaboard, Hazel's passage over Virginia was followed in 1958 by Hurricane Helene, in 1959 by Hurricane Gracie, in 1960 by Hurricane DONNA, in 1972 by Tropical Storm AGNES, and in 1989 by the downgraded but rain-thrashing remains of Hurricane HUGO.

During the fascinating 1996 hurricane season, Virginia was affected by two tropical cyclones, one a tornado-spiked hurricane, the other a rain-shorn killer that spawned flood emergencies across 31 Virginia counties. The first tropical system to lash the state, Hurricane Bertha, initially made landfall in North Carolina as a moderate Category 2 hurricane on the afternoon of July 12, 1996. Slowly weakening as the hurricane's EYEWALL ground across the tidal plains of eastern Virginia, Bertha's 75-MPH (121-km/h) winds and pinpoint tornadoes damaged or destroyed nearly 100 houses and mobile homes and injured 50 people—many of them seriously. Some $30 million in insured property losses are assessed in the aftermath of this early season surprise.

Not more than two months later, on September 6, 1996, the watery remnants of Hurricane Fran—a feisty Category 3 system that had initially danced ashore in North Carolina the day before—delivered the worst flooding to central and western Virginia since 1972's Hurricane AGNES. Once possessed of sustained 115-MPH (185-km/h) winds, torrential rains, and 12-foot (4-m) storm tide, Fran was downgraded to a tropical storm as it recurved through southwestern Virginia shortly after midnight on September 6. Dropping between 10 and 14 inches (254–356 mm) of rain over northern and western Virginia as it trekked northwestward across the Blue Ridge Mountains, Fran caused many of the commonwealth's rivers to top their banks. To the east, the Rappahannock River crested at seven to nine feet (2–3 m) above flood stage; the James River to the west surmounted its embankments at 15 feet over flood level. Marking the northern border between Virginia and Maryland, the Potomac River rose five feet above flood stage, turning the streets of the Washington, D.C., neighborhoods of Georgetown and Alexandria into asphalt-lined canals. Several of the capital's main through-ways were closed to traffic, and rail service in and out of Union Station was canceled. Gushing at 130 times its normal rate after nearly five inches of rain, the flooded Potomac further damaged the C&O Canal, only recently repaired since the powerful Northeastern Blizzard of January 7–8, 1996,

damaged it with its earlier flood. Claiming the lives of three people, Fran caused more than $200 million in property losses in Virginia, making it that state's third most expensive natural disaster. At the time touted as the fourth most expensive hurricane in U.S. history, Fran's overall insured losses totaled more than $1.6 billion.

The most severe tropical cyclone to have directly affected Virginia since the Category 3 Chesapeake-Norfolk Hurricane of August 1933 blistered ashore, Hurricane ISABEL directly and indirectly claimed some 30 lives as it passed northward over Virginia on September 18–19, 2003, following its initial landfall in North Carolina. Virginia was particularly hard hit by Hurricane Isabel, with precipitation counts of 20 inches (508 mm) recorded in Upper Sherando, Virginia, and a six-foot (2-m) storm surge observed at Norfolk. Langley Air Force Base was completely inundated. The most expensive tropical cyclone in Virginia in living memory, Isabel clocked damage estimates at some $1.8 billion (in 2003 U.S. dollars).

Virginia, Typhoon *North Pacific Ocean–Japan–Taiwan, June 19–27, 1957* A moderately powerful TYPHOON, Virginia delivered 110-MPH (177-km/h) winds and flooding 7-inch PRECIPITATION counts to northern TAIWAN and southern JAPAN between June 25 and 26, 1957. An early season system of unusual severity, Virginia's tropical downpours spawned flash floods and landslides that damaged or destroyed more than 1,000 BUILDINGS, swept away nearly one dozen bridges, sank a flotilla of fishing boats off the coast of Taiwan, and claimed a total of 86 lives in both nations before dissipating over the South China Sea during the night of June 27. Total losses to buildings, watercraft, and crops reportedly exceeded $20 million.

Virgin Islands *Northeastern Caribbean Sea* Tucked between PUERTO RICO to the west and the Leeward islands of Anguilla and St. Maarten to the east, the 1,600 volcanic islands and fragile limestone cays of the Virgin Islands have for centuries presented an expansive, low-lying target to incoming NORTH ATLANTIC and CARIBBEAN SEA TROPICAL CYCLONES. From 1742 to 2006, the archipelago's five major islands—St. Thomas, St. Croix, St. John, Tortola, and Virgin Gorda—were directly affected by 60 recorded HURRICANES and TROPICAL STORMS, with at least 22 of these judged to have been major hurricanes with minimum wind speeds of 111 MPH (179 km/h), 9–18-foot (3–6-m) STORM SURGES, and reeling DEATH TOLLS.

Situated on the northeastern edge of the Caribbean Sea, the Virgin Islands are most susceptible to those mid- and late-season hurricanes that sail across the North Atlantic Ocean with the trade winds. Indeed, a slight majority of the Virgin Islands' deadliest, most destructive tropical cyclones have originated well to the east, over the subequatorial waters of the Cape Verde Islands. Often allowed by a host of climatological factors to intensify rapidly within the evaporative expanses of HURRICANE ALLEY before reaching the eastern Caribbean, these tropical systems have traditionally begun their RECURVATURE, or their parabolic curve north-northwest, in the region of the Virgin Islands, thereby delivering a long, steady history of violent winds, deluging rains, and submerging storm surges to the vulnerable island chain.

One such hurricane, a midseason screamer that terrorized the island of St. Croix on the night of August 31, 1772, had earlier that morning visited great devastation on the nearby island of DOMINICA, 275 miles (443 km) southeast. Recurving northwest at nearly 17 MPH (27 km/h), the 1772 hurricane sank six merchant vessels at Fredericksted and decimated much of St. Croix's muscovado sugar and coffee crops. Two decades later, on August 1, 1792, an equally severe hurricane followed a similar trajectory through St. Croix and St. Thomas, driving two merchant ships ashore at St. Thomas and sliding a coastal packet boat onto a reef near Tortola Island. Between 1804 and 1851, nine other Cape Verde storms would trace similar paths through the islands, including the infamous Antigua-Charleston Hurricane of September 3, 1804 (see SOUTH CAROLINA), that pounded St. Thomas with winds estimated at 120 MPH (193 km/h) and torrential rains; the gargantuan hurricane of September 22, 1819, that sank more than 100 ships at St. Thomas and cast a small brig aground at Tortola; the St. Kitts Hurricane of August 18, 1827, that in brushing past the southern coast of St. Croix placed the entire island within the hurricane's DANGEROUS SEMICIRCLE; the hurricane of August 12–13, 1830, that demolished more than 100 houses on St. Thomas; and the vicious early morning whirlwind of September 12, 1846, that transited northwestern St. Thomas, overturning buildings, snapping trees, winnowing crops, and wrecking at least 12 merchant ships.

During the particularly active 1837 HURRICANE SEASON, in which 11 tropical cyclones were documented in the North Atlantic and Caribbean basins, the Virgin Islands were struck by two mature hurricanes, both viewed as storms of significant size, intensity, and duration. The first, on July 27, 1837, spent some 12 hours battering St. Croix with violent winds and downpours. A number of Dutch and English merchant ships were forced ashore near the city of Christiansted, and hundreds of BUILDING failures transformed the island's once verdant northeastern landscape into a timber-strewn mire of death and despair. No sooner had tiny St. Croix begun to rebuild

than an even greater tempest, the so-called Los Angeles Hurricane, roared out of the southeast on the evening of August 2. Said to have been of Category 3 intensity (28.00 inches) while transiting San Juan, Puerto Rico, 90 miles northwest, the Los Angeles Hurricane's 125-MPH (201-km/h) winds, flooding rains and 11-foot (4-m) storm surge unroofed houses, toppled trees, blighted acres upon acres of sugarcane, and sank scores of small boats in the harbor at Fredericksted. Survivors claimed that the storm's early evening passage over the neighboring island of St. Thomas was accompanied by a severe EARTHQUAKE and fire that left much of the island's habitation in charred ruins. At least 500 people reportedly perished, making the Los Angeles Hurricane of 1837 one of the most cataclysmic of all 19th-century tropical cyclones.

Unfortunately for the languishing trade and prosperity of the Virgin Islands, it would not be the last. Shortly before midday on October 29, 1867, an even greater whirlwind, the so-called San Narciso Hurricane, howled out of the southeast as a tightly wound Category 3 system of exceptional severity. Wheeling about a CENTRAL BAROMETRIC PRESSURE of 27.95 inches (946 MILLIBARS) at eventual landfall in Puerto Rico, San Narciso's 127-MPH (204-km/h) EYEWALL winds and strafing rains first ravaged much of the Virgin Islands, destroying nearly every structure on St. John, St. Thomas, and Tortola. Enormous waves, many estimated to have been between 20 and 25 feet (7–8 m) in height, inundated the harbor at Charlotte Amalie, undermining stone quays, splintering wooden piers, and collapsing every ware and customshouse on the waterfront. Cut from its anchorage by the hurricane's spume-sharpened gusts, a large British-registered collier careened wildly from one headland to the other before following two Danish sailing ships to the bottom. Placed under the command of its first officer by the hurricane's sudden onset, the nearby steam packet *Wye* valiantly beat to seaward in an effort to escape the roiled, debris-studded confines of the harbor. Soon pitched ashore on neighboring Buck Island, the 819-ton vessel was quickly battered to pieces by the surf, drowning 41 crew members.

Anchored four miles (6 km) south of Tortola in the Great Harbour of densely settled Peter Island, the 2,738-ton British mail steamer *Rhone* initially managed to withstand San Narciso's burgeoning fury. Its two anchors firmly entrenched in the sandy seabed and its twin paddlewheels churning at full speed ahead, the 310-foot (100-m) *Rhone* and its complement of 145 passengers and crew struggled courageously for hours against the hurricane's rocking waves and whirring 140-MPH (225-km/h) gusts. Jerking against its hawsers like a spooked racehorse, the *Rhone* skitted from side to side, its three towering masts tracing crazy arcs across the gray, rain-racked sky. Unable to endure the twisting strain and longer, the steamship's mizzen topmast finally gave way, crashing like a javelin to the deck, where it fatally crushed the vessel's first officer. Much to the relief of the *Rhone*'s terrified passengers and crew, the hurricane's winds and rains sharply abated shortly after three o'clock that afternoon, leaving the air thick and sultry, the surrounding seas confused but navigable. Surmising that the hurricane's calm EYE was now passing over his ship, the *Rhone*'s master chose to take advantage of the lull to move his deeply laden charge east-northeast beyond the supposed reach of the storm's dangerous trailing edge. Promptly weighing anchor, the *Rhone* slowly steamed eastward toward Salt Island, a tiny cay better known for its scenic salt flats than its treacherous coral lagoons. No sooner had the vessel arrived at Lee Bay than the hurricane's eye completed its passage northwest, bearing with it a period of renewed violence. Caught broadside to the storm, the iron-hull of the *Rhone* was slammed against a reef and holed. Rapidly filling, the mail ship sank in 80 feet of water, tragically claiming 123 of the 145 souls aboard, including that of its brave but unfortunate captain. Still intact after nearly a century and a half on the ocean floor, the wreck of the *Rhone* has in recent years become a popular destination for sport divers and other visitors to the Virgin Islands. All told, more than 60 vessels succumbed to the swamping seas of the San Narcisco Hurricane; 600 lives—a majority of them sailors—were lost.

On October 12, 1876, another late-season hurricane besieged St. Thomas and its surrounding islands with 105-MPH (169-km/h) winds and voluminous rains. Of lesser intensity than the notorious San Narcisco Hurricane (a pressure reading of 28.70 inches would later be observed upon landfall in CUBA), the 1876 system nonetheless stripped the roofs from scores of houses, downed hundreds of trees and fences, and sank two schooners in St. Thomas harbor. Between 75 and 125 people were reportedly killed, many by the hurricane's thudding 10-foot storm surge.

Not more than 13 years later, the same region of St. Thomas was again shaken by a large tropical cyclone, this one darting ashore on September 7, 1889. Having only the day before caused staggering devastation and loss of life on St. Kitts 160 miles (257 km) southeast, the 1889 hurricane sank another five ships in the harbor at St. Thomas and obliterated most of that season's sugar and coffee harvests on nearby Tortola. Contemporary sources place the death toll at 110 people, a number of them victims of flash floods and landslides in the highlands around Charlotte Amalie.

The first quarter of the 20th century, characterized as it was by vast improvements in the documenting of North Atlantic and Caribbean tropical cyclones, can be taken to illustrate the frequency and nature of hurricanes and tropical storms in the Virgin Islands. During the 1901 hurricane season, for instance, St. Thomas and St. Croix were affected by two tropical cyclones—the first, a weak Category 2 hurricane on September 1 and the second, a powerful tropical storm on October 7. The following year, 1902, there was no recorded cyclonic activity over the islands. In 1903, St. Thomas was struck by a July 19 hurricane, after which another hurricane did not come along until September of 1906. In 1907, nothing, but in 1908 two timid hurricanes lashed St. Croix and St. Thomas within two months' time, causing some damage and at least five deaths. During the 1909 season, the Virgin Islands were again untroubled by a tropical cyclone, but in 1910, a powerful Category 3 hurricane grazed southeastern St. Croix, demolishing buildings and sinking small boats. Then, between 1911 and 1915 and again between 1920 and 1923, and 1925 and 1927, no hurricanes or tropical storms darkened the sunny shores of the Virgin Islands, indicating a fairly common pattern of two-, three-, and four-year breaks in hurricane strikes.

Yet the end of each hiatus was signaled by the arrival of one or more hurricanes in a single season—rainmaking systems of substantial size and intensity. The four-year lag between 1911 and 1915 saw the 1916 season produce two strong hurricanes and one tropical storm; 1924 witnessed two hurricanes strikes on St. Croix and St. Thomas between August 18 and August 29, respectively. The three-year dream between 1925 and 1927 was in turn shattered by the nightmarish catastrophe of the SAN FELIPE HURRICANE of September 12–13, 1928, a colossal storm that by any standard must count as one of the Virgin Islands' most awesome natural disasters.

Enveloping the entire island of St. Croix within its Category 4 eyewall, the San Felipe Hurricane administered potent but unequal doses of destruction to the rest of the island chain. In St. Thomas, 90-MPH (145-km/h) winds and sporadic rains crumpled dozens of tin-roofed shanties. A sluggish landslide on pristine St. John, 20 miles east, carried away part of a road, hundreds of trees, and several wild donkeys. But it was on St. Croix, where San Felipe's bellicose winds blew at 140 MPH (225 km/h), that the hurricane's flooding generosity, doled out as 5-inch downpours, was most acutely suffered. Almost 90 percent of the structures on the island—from the stone plantations of East Point to the balconied townhouses of Christiansted and Frederiksted—lost roofs, windows, doors, and porches, with many residences becoming total losses. Nearly every tree on the island was either uprooted, snapped in two, or virtually defoliated. The harbor at Christiansted writhed beneath the wreckage of four battered ships, a number of shattered piers, and a fallen storehouse. On St. Croix, 62 people reportedly perished a surprisingly small figure considering the severity of San Felipe's passage over the island.

Ironically, it was the economic devastation caused by the San Felipe Hurricane that finally compelled the United States (which had all but ignored St. Thomas and St. Croix since the purchase of the Virgin Islands from Denmark in 1917) to assume a larger role in the islands' governance. Beside extending the storm-raked atoll financial assistance, the United States ensured that military rule was replaced with a more responsive civilian government in 1931, thereby allowing for more efficient development of the islands' tourism and agriculture businesses. Declared an unincorporated territory in 1936, the Virgin Islands were eligible for emergency aid from the United States when the next monster hurricane, HUGO, trod across St. Croix and St. Thomas on September 17, 1989, racking up damage estimates of $1 billion.

On the morning of September 15, 1995, both St. Croix and St. Thomas were swept by the 13th tropical cyclone of the very active 1995 season, Hurricane Marilyn. A powerful Category 2 system that had originated over the mid-Atlantic on the afternoon of September 12, Marilyn rapidly deepened as it trudged due west, became a tropical storm later that evening and a minimal, Category 1 hurricane the following day. After pelting the islands of Dominica and St. Kitts with 80-MPH (129-km/h) winds and 7-inch (178-mm) PRECIPITATION counts on September 14, Marilyn celebrated its steady course change northwest by upgrading itself to Category 2 status. Its sustained 105-MPH (169-km/h) eyewall winds punctuated by 120-MPH (193-km/h) gusts and 10-inch (254-mm) rains, Marilyn passed northeast of St. Croix but southwest of St. Thomas on September 15, thus placing the latter island firmly within its DANGEROUS SEMICIRCLE. Although a solitary gust of 127 MPH (204 km/h) was measured on St. Croix, a majority of Marilyn's fury was directed at St. Thomas, where damage was said to have rivaled that of Hurricane Hugo. As 12-foot (4-m) waves relentlessly clawed at the island's picturesque anchorages and beachheads, 60 sail- and motorboats were either run ashore or sunk, including the U.S. Coast Guard cutter *Point Ledge*. Half

of the island's residences were destroyed, while the gutted remainder sustained severe roof and wall damage. Five multistory apartment buildings in Charlotte Amalie completely crumpled, raising unfounded fears that at least 50 people were trapped beneath the rubble. (It was later determined that the complex was still under construction at the time of the hurricane and hence uninhabited.) All the windows in St. Thomas's major hospital were shattered as rumbling flash floods invaded the building's lower floors, forcing eight critically ill patients to be airlifted to safety by a helicopter. Some 28,000 of St. Thomas's 32,000 telephone lines were severed, requiring months of repair work before full service could be restored. All told, eight people died in the Virgin Islands during Hurricane Marilyn, six of them in St. Thomas and two in St. Croix. Declared a $1.5 billion disaster area on September 18, the Virgin Islands were immediately given $700 million in assistance, as well as a contingent of 100 federal marshals. Fearing a recurrence of the widespread looting that had followed Hugo's cataclysmic passage, the marshals were intended to enforce order and at the same time protect vital relief supplies then being dispatched to the island's by droves of C-130 cargo planes.

Much kinder to the Virgin Islands than the 1995 hurricane season had been, the 1996 parade of North Atlantic and Caribbean tropical cyclones only brought one slow-moving tropical storm to the islands, Hortense, on September 8. A midscale tropical system with sustained winds of 60 MPH (97 km/h), Hortense's offshore eyewall lashed St. Croix for several hours, snapping utility poles, lifting roofs, uprooting trees, and dropping 10 inches (254 mm) of rain. In Frederiksted, Hortense's 15-foot (5-m) breakers and 70-MPH (113-km/h) gusts sent sea spray dramatically cascading over the long cruise-ship pier; swiftly accumulating rains on northern elevations caused narrow roads to be paved with more than four feet (1 m) of water. A congressional primary election scheduled to be held on September 8 was postponed because of Hortense's prolonged, if ultimately mild, challenge to the territory's domestic tranquility. Although no deaths or serious injuries were reported in the aftermath of Tropical Storm Hortense, property losses topped $5 million.

W

Wanda, Typhoon *North Pacific Ocean–Southern China, August 27–September 2, 1962* One of the deadliest TROPICAL CYCLONES to affect TAIWAN and CHINA during the latter half of the 20th century, Super Typhoon Wanda (also known as Super Typhoon #17) claimed some 130 lives, destroyed more than 5,000 small craft, and left 72,000 people homeless as it glided over Hong Kong on September 1, 1962. At one point in its existence a Category 5 SUPERTYPHOON with sustained winds of 161 MPH (259 km/h), Wanda had weakened slightly as it assaulted the then British crown colony. Powered by a central pressure of 28.14 inches (953 mb), Wanda's sustained winds topped 120 MPH (193 km/h), while an extreme gust of 177 MPH (285 km/h) was observed at Tate's Cairn. A 10-foot (3-m) STORM SURGE caused widespread destruction in and around Tolo Harbor and some 12 inches (305 mm) caused flash flooding and mudslides across the colony. The destruction caused by Typhoon Wanda (said to have been the worst since a typhoon in 1937 ravaged the colony) was responsible for Hong Kong's ongoing effort to identify, monitor, and mitigate every hill in the area that possessed the ability to slide during periods of intense PRECIPITATION. Typhoon Wanda's arrival in Hong Kong garnered the raising of the No. 10 signal (the highest level) by the Hong Kong Observatory.

weather balloon Also referred to as *pibals* or *pilot balloons,* weather balloons are most frequently used to observed WIND patterns at altitudes of greater than 50,000 feet. Measuring some 40 inches wide, equipped with RADIOSONDES and DROPSONDES, and able to rise much higher into Earth's TROPOSPHERE than weather reconnaissance aircraft, weather balloons aid in the study and forecasting of TROPICAL CYCLONES by providing crucial data on the speed and direction of upper-level STEERING CURRENTS and on such individual climatological factors as air TEMPERATURE and BAROMETRIC PRESSURE. First developed in the early 20th century as a replacement for the cumbersome, ineffective kites used by the U.S. WEATHER BUREAU during the 1880s and 1890s, the efficiency of early weather balloons suffered from the lack of a viable means by which to transmit collected data back to Earth. Before advances in radio technology made creation of the radiosonde possible, weather balloons were designed to burst on achieving a certain altitude, at which point their instrument packages returned to Earth beneath a small parachute. While in 1933 there were nearly 100 weather balloon observatories in regular service around the United States, the results of their efforts were erratic in frequency and incomplete in scope as dozens of meteographs, or the recording instruments carried by weather balloons, fell into the sea or into dense undergrowth and were lost. It was not until the radiosonde was developed in the early 1940s that the weather balloon began to realize its full potential. Able to observe temperature, barometric pressure, and humidity levels as it rose through the troposphere, the radiosonde automatically relayed this information to nearby receiving stations, thereby increasing the timeliness and accuracy of the readings and their resulting forecasts.

Wendy, Typhoon *North Pacific Ocean–Japan, August 7–14, 1960* The second of two western NORTH PACIFIC OCEAN tropical systems to have been named Wendy, this moderately powerful midseason TYPHOON delivered 107-MPH (172-km/h) winds and torrential nine-inch (229-mm) PRECIPITATION counts to southern and central JAPAN on the night of August 13–14, 1960. A rainmaker of the first order, Wendy's tropical downpours triggered flash floods and mudslides that damaged or destroyed more than 1,200 BUILDINGS and claimed 18 lives. Damage estimates topped $30 million. The first Typhoon Wendy, which battered Hong Kong, CHINA, in mid-July of 1957, killed 16 people and caused property losses in the millions.

Wilma, Hurricane *Caribbean Sea–Mexico–Southern United States, October 15–25, 2005* The 21st TROPICAL CYCLONE and the 12th named hurricane of the extraordinarily active 2005 NORTH ATLANTIC HURRICANE SEASON, Wilma was a record-breaking tropical cyclone on a number of varying levels. It was the first tropical cyclone in the Atlantic basin to bear a "W" identifier. It demolished Hurricane GILBERT's 17-year

rule as the most intense tropical cyclone yet observed in the North Atlantic basin, and when upgraded to hurricane status, toppled the 1969 season's record for the most hurricanes in one season. But despite its enormous power, Wilma proved not as deadly a killer as it could have been.

Wilma originated on October 15, 2005, from a TROPICAL DEPRESSION that had formed over the extreme northwestern corner of the CARIBBEAN SEA. It remained a tropical depression for the next three days, dropping copious amounts of rain on nearby JAMAICA (where one person lost his life) and was upgraded to TROPICAL STORM intensity (and given the name Wilma) on October 17. The system steadily intensified, and became a Category 1 hurricane during the mid-morning hours of October 18, when its sustained winds reached 74 MPH (119 km/h). At this time, Wilma's CENTRAL BAROMETRIC PRESSURE of 28.85 inches (977 mb) was located about 195 miles (320 km) south-southeast of Grand Cayman Island, and about 200 miles (325 km) east-northeast

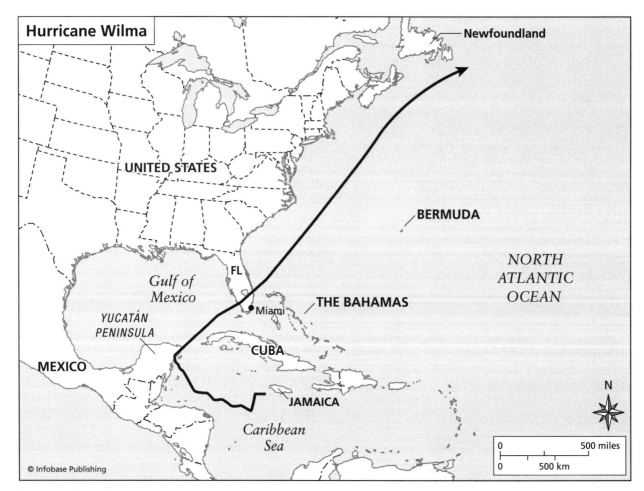

Hurricane Wilma, the most intense tropical cyclone yet observed in the North Atlantic basin, courses across the Gulf of Mexico and into Florida and the northeastern United States in October 2005.

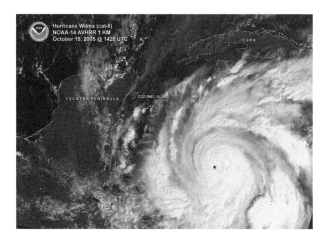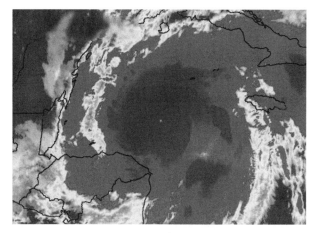

These two photographs show the most intense tropical cyclone—Hurricane Wilma—yet observed in the Atlantic basin as it lashed the waters of the Caribbean Sea and Gulf of Mexico with sustained winds of 184 MPH on October 19 and 20, 2005. Besides sporting a pinhole eye (best seen in the image [left]), Wilma's IR shot shows some of the darkest and most extensive high-altitude cloud tops ever seen in a tropical cyclone. *(NOAA)*

of Nicaragua. The hurricane moved sluggishly to the northwest at seven MPH (11 km/h), bound—as many tropical cyclones are wont to do—for the open yet narrow waters of the Yucatán Straits.

Then something interesting occurred; Wilma underwent a period of EXPLOSIVE DEEPENING. During the early evening hours of October 18, the system's central pressure stood at 28.75 inches (975 mb). Within six hours it was observed at 28.17 inches (954 mb) and within six hours of that observation, at 26.60 inches (901 mb). By noon on October 19, Wilma was generating a central barometric pressure reading of 26.02 inches (882 mb), making it the most intense tropical cyclone yet witnessed in the Atlantic.

A large tropical cyclone with a tightly coiled EYEWALL, Wilma blasted the waters of the Yucatán Straits, forcing the EVACUATION of 30,000 tourists from the coastal resort havens of Cozumel and Cancun and causing serious coastal erosion along the western coast of CUBA. Almost as quickly as it had intensified, Wilma began to weaken. Within 24 hours, its central pressure had risen to 26.39 inches (894 mb), making it a borderline Category 5 hurricane when measured by its 150-MPH (241-km/h) wind speed. Within another 24 hours, the system had further weakened to 27.25 inches (923 mb) and its wind speeds had dropped to 145 MPH (233 km/h). Meteorologists and emergency managers in MEXICO, Cuba, and the southern United States, were not, however, comforted by Wilma's slow tumble; were it to come ashore at its present intensity, it would rank as one of the top-10 most powerful hurricane landfalls in the modern history of the North Atlantic.

Wilma entered the GULF OF MEXICO on October 23. Recurving to the northeast, it had weakened to Category 2 status (with a central pressure of 28.31 inches [959 mb] and sustained winds of 98 MPH [158 km/h]) and was now bound for southern and central FLORIDA. In neighboring Cuba, Wilma's wrath was particularly severe; nearly 1 million people were evacuated from low-lying coastal areas, while waves measured at 45 feet (15 m) topped Havana's famed seawall, flooding some 10 miles (16 km/h) of the city's historic downtown.

During the early morning hours of October 23, Hurricane Wilma clubbed ashore at Cape Romano, Florida, as a Category 3 system with a central pressure of 28.20 inches (955 mb) and sustained winds of 115 MPH (161 km/h). A powerful and destructive storm, Wilma made landfall about 20 miles (32 km) west of Everglades City. Some six million people across the Sunshine State lost electrical services, while STORM SURGE flooding caused coastal erosion, destroyed shoreside property, and tossed small craft ashore. In addition, several TORNADO touchdowns were reported in central Florida. Insurance industry estimates made Wilma—at $9 billion in damages—the third most expensive hurricane-related disaster behind Hurricane KATRINA (2005) and ANDREW (1992). Key West was heavily affected; Wilma's 120 MPH (193 km/h) winds and three to five-foot (1–2-m) storm surge inundated most of the Keys. At the time, it was also said that Wilma was the most intense hurricane to strike the Fort Lauderdale area since 1950's Hurricane EASY.

A weakening Wilma traversed the Florida peninsula and by the mid-morning hours of October 25, 2005, was racing up the eastern seaboard of the United States. The warming waters of the Gulf Stream allowed Wilma to reintensify to Category 3 status with sustained winds of 115 MPH (185 km/h). The system's central barometric pressure deepened from 28.23 inches (956 mb) while the storm was exiting the Florida peninsula, to 28.20 inches (955 mb) as it moved into the Gulf Stream.

A strengthening extratropical low over the northeastern United States leached energy from Wilma's fading moisture bands, and delivered high winds, heavy rains, pounding surf, and in some places, snow, to much of the northeastern United States. Twenty-foot (7-m) waves crashed ashore on New Jersey's fragile beaches and minor coastal flooding was reported in Cape Cod, Massachusetts. Trees were toppled in Connecticut, while a wind gust of 47 MPH (76 km/h) was recorded near Boston.

As a once-weighty Wilma faded over the cooling reaches of the North Atlantic, its toll in property losses and casualties was being grimly assessed. Some 22 deaths—six in Florida, four in Mexico, one in JAMAICA, and at least 12 in Haiti—were attributed to Wilma's wrath. Damage totals reached more than $28 billion (in 2005 U.S. dollars). The name Wilma has been retired from the rotating list of North Atlantic tropical cyclone identifiers.

wind Long recognized by meteorologists, historians, and civil defense personnel as one of a TROPICAL CYCLONE's most destructive physical attributes, the horizontal transfer of energy—*wind*—assumes many forms and serves many functions in the lifecycle of a HURRICANE, TYPHOON, CYCLONE, TROPICAL STORM, TROPICAL DEPRESSION, or TROPICAL WAVE. The result of a complex interaction between air temperature, atmospheric pressure, and the topography of the land or sea surfaces over which these processes are circulating, the term *wind* can be used to define those powerful low-altitude ADVECTION streams found below a tropical cyclone's EYEWALL, the embracing upper-level movements of air that can comprise a tropical cyclone's STEERING CURRENT, or the earth-girdling circulations of the INTERTROPICAL CONVERGENCE ZONE (ITCZ) that carry embryonic tropical waves westward to the open breeding grounds of the NORTH ATLANTIC and NORTH PACIFIC OCEANS.

Within the tropical system itself, the direct influence of localized winds on the surface of the sea frequently generates enormous waves, confused pyramids of water that rock back and forth as they quickly radiate outward. Above these waves, spindrift and then spray fill the wind with moisture, merging with tons of falling PRECIPITATION to form a deep conductive layer for the tropical cyclone's continuing demand for EVAPORATION. Below these waves, long wind-pushed swells

slowly revolve around a tropical cyclone's INVERSE BAROMETER, the mount of seawater heaped up beneath the storm's center of lowest barometric pressure. Flowing downward 10–50 feet (3–25 m) and rippling outward 500–600 miles at speeds only slightly slower than those of the tropical cyclone's *sustained winds,* these undulating swells tend to raise treacherous currents and undertows on nearing a coastline, providing a secondary hazard to those residents who unwisely wade, swim, or surf at vulnerable beaches.

Sparked by the uneven heating of Earth's surface by the sun and fueled by PRESSURE GRADIENTS, or the drop in atmospheric pressure over a certain distance, winds and WIND SPEEDS within a mature-stage tropical cyclone's 1,000 MILLIBAR range are unevenly distributed across its cloudy, 300–400-mile mass. In the center of the system, where the calm EYE holds dominant sway, winds are light (five–15 MPH) to almost nonexistent in complete contrast to the immediately surrounding eyewall, where the tropical cyclone's strongest winds are located. While the eyewall's racing advection streams, driven by the horizontal movement of air along the underside of the cyclone's convection cells, do completely encompass its thickening ring, areas of faster moving winds, frequently referred to as *gust zones,* punctuate the eyewall's trailing semicircle. Briefly tearing across Earth's surface at speeds of 155 MPH (249 km/h) and higher, these gust zones are often densely positioned in the upper left corner of the eye in a westward- or northward-traveling hurricane or typhoon in the Northern Hemisphere and in the upper left of a clockwise-spinning cyclone in the Southern Hemisphere.

Elsewhere in the system, the eyewall's sustained winds gradually give way to heavy gales, falling to between 39 and 54 MPH (63–87 km/h) 50–70 miles (80–113 km) from the edge of the eye. Extending 150–250 miles (241–402 km) from the eye, these gale-force winds result from the less-intense convective activity found in a tropical cyclone's midrange RAIN BANDS. Closer to the fringe of a tropical cyclone's circulation, winds diminish into gusty rain squalls, windy downpours that lash foliage and terrorize unwary mariners with their overturning intensity. Winds at the extreme outer edge of a tropical system are generally light, with several 25+ MPH (40+ km/h) gusts present to herald the tropical cyclone's portentous approach.

A vertical cross-section of a tropical cyclone's circulation reveals a similarly layered wind pattern, only reversed. Because friction with Earth's surface decreases wind speed, a tropical cyclone's most powerful winds are situated at higher elevations, usually in the range of 3,000–15,000 feet (1,000–5,000 m). On September 28, 1966, while flying at an altitude of 8,000 feet (2,700 m), reconnaissance air-craft recorded 197-MPH (317-km/h) winds within the eyewall of Hurricane INEZ, then grinding its way across the southeastern foothills of HISPANIOLA. At the same time, Inez's sustained ground-level wind speeds tipped ANEMOMETERS at 115 MPH (185 km/h), considerably slower than those raging currents aloft. (In the interest of accuracy, it should be noted that vertically rising or descending air is referred to as a current, or CONVECTION current, not as wind.) Upon reaching the upper atmosphere, a tropical cyclone's warm convection currents cool and slow as they are absorbed into the system's OUTFLOW LAYER, the strength of which is sometimes aided by the expanding winds of a high-pressure ANTICYCLONE positioned overhead. Drawn across the top of the TROPOSPHERE, where strident west-to-east downdrafts at 40,000 feet (13,300 m) horizontally push convection currents away from the eye, the once moist, heated air is cooled and condensed to the point that it begins to grow leaden with slow-moving molecules, descending at speeds of 100 MPH (161 km/h) or more toward Earth's surface, where the mechanics of advection again turn it into wind.

The windward influence of a mature tropical cyclone's winds on the expansive waters of Earth's oceans, seas, and bays often results in the creation of giant waves. On the night of September 25, 1978, while steaming some 1,500 miles (2,414 km) off Newfoundland, CANADA, the British luxury liner *Queen Elizabeth 2* bucked into a series of towering waves spawned by distant Hurricane Ella, then centered to the south of BERMUDA. Cresting at nearly 100 feet (33 m), these oceanic behemoths battered the 55,000-ton ship for several hours, crushing bow railings, shattering windows and portholes, and crumpling part of the vessel's superstructure. Two passengers were injured, one of them seriously. Seventeen years later, on the morning of September 11, 1995, the *Queen Elizabeth 2* again encountered tropical cyclone–generated waves on the North Atlantic, these being associated with a fading Hurricane LUIS. "Significant weather and heavy waves" forced the liner to reduce its speed by almost 20 knots while passing through Luis's outer rain bands, 800 miles (1,287 km) northeast of New York. One passenger suffered a broken arm. According to the *Queen Elizabeth 2*'s officers, Luis's waves measured between 75 and 90 feet (25–30 m) in height.

The intricate array of surface and subsurface currents set in motion by a tropical cyclone's winds are fascinating to observe, though all too often deadly when personally experienced. Swirling in the same direction as the tropical cyclone's rotation, these currents move more slowly than the system but influence physical conditions over a much wider area. Running with the intensity of a river, these wind-carved currents have shattered fragile coral reefs, crunched sunken vessels, snapped sunken aircraft in two (*see* Hurricane GORDON), and aggravated dangerously high STORM SURGE and STORM TIDE emergencies along many low-lying oceanfront ecosystems. In one highly publicized incident during the 1995 North Atlantic HURRICANE SEASON, offshore currents churned up by Hurricane Erin's 85-MPH (137-km/h) winds prompted several young sand sharks to attack 13 swimmers at a resort near Vero Beach, FLORIDA. In these "nip and run" attacks, the sharks were apparently confused by the hurricane's turbulent seas and driven by an instinctive fear to slash their way through any obstacle in an effort to reach calmer feeding grounds. While no serious injuries were reported, the harrowing event only highlights the pervasive control a tropical cyclone's winds have over distant environments.

Although it is true that the wind-whipped surf caused by an approaching tropical cyclone can often be quite spectacular to watch, it is equally true that several dozen watchers have been horribly killed doing so. Capricious riptides, triggered by a collision between surface currents moving at different speeds, were blamed for at eight deaths along the east coast of the United States during Hurricane Felix's offshore passage in August 1995. Three of the victims were surfers. Either buried beneath tons of cresting water or pulled out to sea by a sudden undertow and exhausted to the point of drowning, they were as much victims of Felix's distant 120-MPH (193-km/h) winds as they were its enticing yet deadly 15-foot (5-m) seas.

As is the practice with all of Earth's wind systems, wind direction in a tropical cyclone is defined by the respective compass direction—north, northeast, northwest, south, southeast, southwest, etc.—from which it originates. In the Northern Hemisphere, a majority of landfalling hurricanes and tropical storms have winds that blow out of the north-northeast on the leading edge of the eyewall and from the south-southwest on its trailing side. In the Southern Hemisphere, clockwise-spinning cyclones in AUSTRALIA and Madagascar produce winds from the north-northeast on the advancing side of the eye and the south-southwest on the rear. In the 16th, 17th, and early 18th centuries, before it was understood that tropical cyclones were single storms that revolved around a calm center, those observers caught in their path often mistook the temporarily clearing of the eye and the subsequent onset of winds from the opposite direction as a break between two separate but equally powerful storms.

Long viewed as its most infamous meteorological by-product, the winds of the tropical cyclone have over the centuries fulfilled a number of cultural archetypes, symbolizing differing priorities to the fields of religion, science, economics and engineering. To a chastened colonist in North Carolina, the winds of a 1727 hurricane carried with them no greater a lesson than that of God's punishment; as he wrote in his journal, "Then the Lord sent a great rain and horrible wind; whereby much hurt was done. . . ." For William Dunbar, writing in his *Meteorological Observations,* the winds of the hurricane of August 18, 1779, were a perfect cause for scientific speculation, for the structuring of a belief that strangely enough foreshadows certain parallels with modern chaos theory: "It is generally believed by philosophers," he stated on the commonality of weather, "that hurricanes and perhaps the gentlest zephyrs are connected with electrical phaenomina. . . ." For a South Carolina newspaper, the *Winyaw Intelligencer,* the winds of a hurricane on the night of September 27 and 28, 1822, were none other than a ringing condemnation of the ironies of intellectual hubris of an overconfident town in which the "inhabitants apprehended no danger from the tide, as, from the violence of the gale, it was presumed that it could not continue until the period of the succeeding high water. In this expectation, however, it pleased the Almighty to disappoint them, and, by the awful result, to prove how fallacious are all human calculations." By 1869, the winds of the Great September Gale that trounced Providence, Rhode Island, on the 8th of the month had inspired an anonymous—though apparently well-versed—writer to incongruously liken them to one of William Shakespeare's most cheerful characters, the eager-to-please fairy from the aptly named play *The Tempest.* "The combined power and fury of the elements were beyond all description," he waxed. "It seemed as if nothing could withstand them. Our peaceful shores were under the wand of some mighty Ariel. . . ." In the late 20th century at a time when the preeminence of wind in the hierarchy of tropical cyclones was being surpassed by the dangers of water, insurance and engineering concerns regarded a tropical cyclone's winds not as manifestations of God's will to chastise but rather as monetary and structural hazards that technology and forewarning could overcome. Owing to this strategy, houses in hurricane-prone areas are now regularly fitted with any number of defenses against high winds, including steel storm shutters, internal wind bracings, and low-setting, hip-style roofs that are firmly secured to the building's wall by long hurricane straps.

wind shear   This term is given to that rapid change in WIND SPEED and direction that takes place over a set horizontal or vertical distance and that can cause complex WIND systems such as HURRICANES, TYPHOONS, CYCLONES, TROPICAL STORMS, and TROPICAL DEPRESSIONS to be pulled apart turbulently. Found at every altitude, wind shear can occur under a number of meteorological circumstances, including between upper- and lower-level winds that are moving in opposite directions and at different speeds; between upper- and lower-level winds that are traveling in the same direction, but at different speeds; between vertically ascending CONVECTION currents and vertically descending downdrafts; between horizontally moving winds and vertically rising updrafts; and between the different gust zones that dangerously add to the strident fury of a tropical cyclone's EYEWALL. Often forming shear lines at the point at which upper- and lower-altitude winds that are moving in the same direction but at varying speeds come together, the immediate effects of wind shear at 30,000 feet are severe air turbulence and loss of atmospheric pressure, whereas at ground level it can take the form of what the pioneering meteorologist Dr. Theodore Fujita called MINISWIRLS, TORNADOlike swaths of destruction that can often prove deadly to those BUILDINGS, people, and ANIMALS in their path.

Wind shear is no friend of the tropical cyclone. Indeed, for a TROPICAL WAVE to deepen into a tropical depression, a tropical depression to intensify into a tropical storm, and so on until reaching cyclonic maturity, wind shear between the low-altitude trade winds and the upper-altitude westerlies must be at a minimum. If it is not, the convective updrafts that form the basis of a tropical cyclone's CUMULONIMBUS CLOUD towers will be sheared off and prevented from rising into the upper TROPOSPHERE where they would normally cool and condense into descending air currents, thus establishing a mechanical convection cell.

wind speed   Wind speed, or the rate at which WIND moves horizontally across Earth's surface, remains one of the TROPICAL CYCLONE's most intriguing—and destructive—meteorological features. Tenuously measured by ANEMOMETERS, quantified by scientists, and qualified by historians, journalists, and disaster-preparedness officials, accurate wind speeds in mature-stage HURRICANES, TYPHOONS, and CYCLONES have often—if the long, mournful record of collapsed or swept-away anemometers and wind gauges is of any indication—proven difficult to obtain. At one time vaguely described in relation to any large or heavy object moved a certain distance from its original location, wind speed in hurricanes, typhoons, and other tropical systems is now observed by highly sophisticated anemometers and systematically recorded according to an internationally recognized series of standards.

Maximum sustained wind speed, for instance, is measured as the fastest mile of air sent past the three whirling cups of an anemometer.

Sustained wind speed is determined to be the average rate at which air moves past an anemometer during a one-minute period.

Gusts, those roaring tantrums caused by the drifting presence in the EYEWALL and outer RAIN BANDS of air pockets with even lower BAROMETRIC PRESSURES than the surrounding atmosphere, are measured at 1.3 times those of the maximum sustained wind speed—meaning that a tropical system that sends one mile of air past an anemometer at 120 MPH (193 km/h) can reasonably be expected to produce gusts of at least 156 MPH (251 km/h).

Some of the fastest wind speeds ever witnessed in nature have been generated by tropical cyclones. On September 13, 1928, the SAN FELIPE Hurricane produced a sustained wind speed of 150 MPH (241 km/h) for five minutes and a peak one-minute speed of 160 MPH (258 km/h). On October 18, 1944, the GREAT ATLANTIC Hurricane blasted Havana, CUBA, with sustained wind speeds of 163 MPH (262 km/h); on September 27, 1955, Hurricane JANET generated a sustained wind speed of 175 MPH (282 km/h) while besieging the city of Chetumal, MEXICO. On August 17, 1969, Hurricane CAMILLE produced

a sustained wind speed of 172 MPH (277 km/h) over the LOUISIANA delta town of Main Pass Block, and Hurricane CELIA birthed 161-MPH (259-km/h) winds at Corpus Christi, TEXAS, on August 3, 1970.

Because the SAFFIR-SIMPSON SCALE is primarily categorized according to wind speed, it is not unusual for systems to be upgraded or downgraded in category due to subsequent meteorological analysis. On August 22, 2002, for instance, nearly a decade after its devastating landfall in southeastern FLORIDA, ANDREW was upgraded from a Category 4 to a Category 5 hurricane. The decision by the NATIONAL HURRICANE CENTER (NHC) was predicated upon a number of subsequent surveys of Andrew's destruction, as well as the meteorological record. According to the NHC, "When the storm was nearing Florida, an aircraft measured the wind at 10,000 feet above the ocean. The surface wind speed was then estimated at 145 MPH (233 km/h), about 80 percent of the high-level wind. New research on the wind structure of hurricanes has determined that surface winds average 90 percent of the 10,000-foot (3,300-m) wind. The new estimated wind speed of 165 MPH (266 km/h) makes Hurricane Andrew the third Category 5 hurricane to strike the United States."

York Typhoon *North Pacific Ocean–Philippines–China, September 11–16, 1999* The 22nd tropical system of the 1999 western North Pacific typhoon season, York delivered the most violent most violent TYPHOON strike to the former British crown colony of Hong Kong since Typhoon Ellen rolled ashore in September 1983. Boasting maximum sustained winds of 93 MPH (150 km/h), Typhoon York killed two people, injured another 466, and caused considerable damage to skyscraper windows and communications antennae. The Hong Kong Observatory greeted York's arrival with the posting of a No. 10 signal, the highest level of tropical cyclone readiness in the city.

Z

Zeb, Typhoon *North Pacific Ocean–Philippines–Taiwan–Japan, October 9–17, 1998* One of the most intense TROPICAL CYCLONES yet observed in the western NORTH PACIFIC OCEAN, Supertyphoon Zeb blasted a long, deadly path of destruction across the Pacific, and into the PHILIPPINES and TAIWAN, between October 9 and 17, 1998. A system that underwent EXPLOSIVE DEEPENING while skipping toward the northern Philippines, Zeb's central pressure plummeted to an astonishingly low 25.75 inches (872 mb)—according to SATELLITE estimates—in less than 24 hours on October 12–13. Supertyphoon Zeb bashed ashore in eastern Luzon as a Category 5 system, its sustained winds of 178 MPH (287 km/h) bringing almost unprecedented devastation to the storm-prone islands. Some 39 inches (995 mm) of PRECIPITATION fell across parts of Luzon, and the resulting damage to rice harvests and other critical infrastructure totaled hundreds of millions of U.S. dollars. Weakened by its passage over Luzon, Zeb recurved to the north-northeast, passed over Taiwan on October 15 as a Category 1 system, and slid into southern JAPAN as a TROPICAL STORM on October 17. Tropical Storm Zeb's 69 MPH (111 km/h) winds and flushing rains caused some damage to trees, billboards and other lightly constructed BUILDINGS and claimed 12 lives. In total, Supertyphoon Zeb killed 99 people in the Philippines, Taiwan, and Japan, and tallied damage estimate of nearly $900 million (in 1998 U.S. dollars). Supertyphoon Zeb's harrowingly low barometric pressure has been exceeded by only one Supertyphoon—1979's Tip—and equaled by four other Supertyphoons: Gay (1992), Ivan (1997), Joan (1997) and Keith (1997).

Zoe, Cyclone *South Pacific Ocean–Solomon Islands, December 25–31, 2002* One of the most intense TROPICAL CYCLONES observed in the SOUTH PACIFIC OCEAN, Cyclone Zoe wrought widespread damage on the southeastern Solomon Islands between December 28 and December 30, 2002. The Solomon Islands are located about 1,900 miles (3,000 km) northeast of AUSTRALIA. They are small islands, but there are about 922 islands and atolls in the chain, 347 of which are inhabited.

Born over the South Pacific Ocean, south of Tuvalu, on December 25, 2002, the system underwent EXPLOSIVE DEEPENING. On December 26, Zoe's sustained wind speeds were clocked at 86 MPH (136 km/h); by December 27, those wind speeds had risen to 109 MPH (176 km/h); and later the same day, had escalated to 178 MPH (287 km/h), making the system of firm Category 5 status. Bearing a disputed central pressure reading of 26.28 inches (890 mb), the storm struck the Temotu Province, causing horrific damage on the islands of Tikopia, Anuta, and Fataka. Some media reports placed the wind speeds as high as 186 MPH (300 km/h), but in light of destroyed or nonexistent meteorological observational equipment, this figure may be inaccurate.

What is not in doubt was the ferocity of Zoe's assault. Almost every BUILDING in the Solomon Islands was either demolished or seriously damaged. Every tree was blown over or shredded. The villages of Ravenga and Namo on Tikopia Island had been virtually washed away by heavy seas. Because of limited EVACUATIONS, a relatively small population, and the use of in-ground cyclone shelters, no lives were lost to Zoe's record-setting meteorological fury.

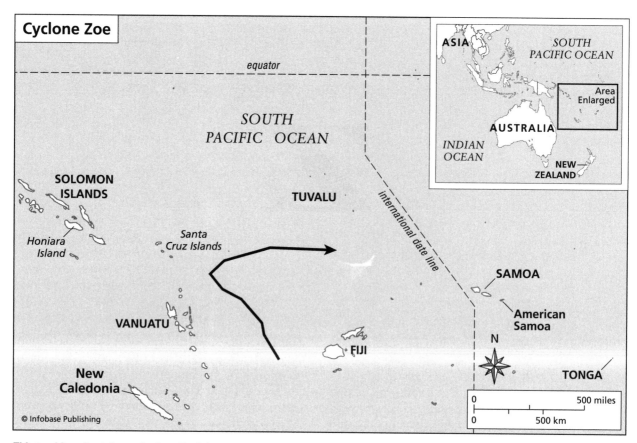

This tracking chart shows Cyclone Zoe's intense passage over the South Pacific Ocean in 2002.

Not long after Zoe's stormy passage, Cyclone Beni lashed the Solomon Islands of Rennell and Bellona on January 26–27, 2003. Of Category 4 intensity as it passed near the islands, the system did not make landfall or cause any significant damage.

APPENDIX A

HURRICANE SAFETY PROCEDURES

In order to better assist coastal residents, civil defense personnel, and local law enforcement agencies to complete efficiently the numerous preparations that precede a hurricane or tropical storm landfall in the United States, the Federal Emergency Management Agency (FEMA) has prepared a checklist of Hurricane Safety Procedures similar to the following. Because the critical, lifesaving hours before a tropical cyclone's landfall can often be marred by haste and confusion, FEMA officials recommend that residents keep a copy of the checklist in a readily accessible location.

- An understanding of the particular geographical and meteorological conditions posed by the location of your residence. These include knowledge of your home's elevation above sea level, any previous storm surge/storm tide activity, and the potential for widespread inland flooding. If uncertain, contact your local civil defense and Weather Service offices for this information.
- Familiarize yourself beforehand with the safest, most efficient evacuation routes in your area.
- Determine the location of the nearest official evacuation shelter in your neighborhood.

- If you are a boat owner or operator, know a sheltered anchorage or mooring area to which you can safely move your craft.
- Compile a complete inventory of your personal belongings, including important legal, financial, and insurance papers.
- Acquire the following emergency supplies well in advance of a tropical cyclone's approach.

  - Plywood and masking tape for securing windows.
  - Flashlights and batteries.
  - Transistor radio and batteries.
  - Candles, lamps, matches, and enough lantern fuel to last several days in the event of power loss.
  - Canned and nonperishable foods that require no refrigeration or cooking.
  - Clean, airtight containers that can be used to store sufficient quantities of drinking water for several days.
  - Tools, rope, waterproof bags, and any other materials needed to make emergency repairs to your residence.
  - A well-stocked, easily accessible first-aid kit.
- Know the location of the primary valves or switches for water, gas, and electricity within your home.

Appendix B

A CHRONOLOGY OF HURRICANES, TYPHOONS, CYCLONES, AND TROPICAL STORMS

1281   **August 23**   *Typhoon*, Southern Japan. 68,000 dead.

1494   **June 16**   *Tropical Storm*, Hispaniola.

     July 16   *Hurricane*, Virgin Islands.

     August   *Hurricane*, Hispaniola.

1495   **October**   *Hurricane*, Hispaniola.

1498   **date unknown**   *Hurricane*, Caribbean Sea–Cuba.

1500   **August**   *Hurricane*, Leeward Islands–Caribbean Sea.

1502   **July 1–2**   *Hurricane*, Hispaniola.

1504   **October 19**   *Hurricane*, North Atlantic Ocean.

1508   **August 3**   *Hurricane*, Hispaniola.

1509   **July 29**   *Hurricane*, Hispaniola.

1510   **July**   *Hurricane*, Hispaniola.

1515   **July**   *Hurricane*, Puerto Rico. Several hundred dead.

1525   **October 12**   *Hurricane*, Cuba. 70 dead.

1526   **October 4–5**   *Hurricane*, Puerto Rico–Hispaniola.

1527   **October 4**   *Hurricane*, Trinidad–Puerto Rico–Cuba. 700 dead.

1530   **July 26**   *Hurricane*, Puerto Rico.

     August 23   *Hurricane*, Puerto Rico.

     August 31   *Hurricane*, Puerto Rico. 100 dead.

1537   **August**   *Hurricane*, Puerto Rico.

     September 9   *Hurricane*, Cuba.

1557   **date unknown**   *Hurricane*, Cuba.

1574   **October 1**   *Hurricane*, Holland. 20,000 dead.

1575   **September 21**   *Hurricane*, Puerto Rico. 400 dead.

1588   **date unknown**   *Hurricane*, Cuba. 600 dead.

1591   **August 10**   *Hurricane*, North Atlantic Ocean. More than 500 dead.

     August 29   *Hurricane*, North Atlantic Ocean. More than 300 dead.

1615   **September 12**   *Hurricane*, Puerto Rico. 500 dead.

1616   **September 20**   *Hurricane*, Cuba. 70 dead.

1635   **August 20**   *Hurricane*, Windward Islands. 15 dead.

1636   **October 14**   *Hurricane*, Cuba. 200 dead.

1640   **date unknown**   *Hurricane*, Cuba. 100 dead.

1641   **October 2**   *Hurricane*, Cuba. 24 dead.

1642   **date unknown**   *Hurricane*, Guadeloupe, Martinique, St. Kitts. 600 dead.

1687   **September 15**   *Typhoon*, North Pacific Ocean. Encountered by Dampier.

1692   **October 21–24**   *Hurricane*, Puerto Rico–Cuba. San Rafael. 235 dead.

1705   **August 15**   *Hurricane*, Cuba. Nearly 100 dead.

1737   **October 7**   *Cyclone*, India. Hooghly River Cyclone. 300,000 dead.

1743   **October 22**   *Hurricane*, Northeastern United States.

1752   **September 26**   *Hurricane*, Cuba. 100 dead.

1768   **October 25**   *Hurricane*, Cuba. Nearly 250 dead.

1772   **July 20**   *Hurricane*, Cuba. 349 dead.

     September 1–5   *Hurricane*, Dominica, St. Kitts, Cuba. 900 people dead.

1775   September 9 *Hurricane*, Canada. Independence Hurricane. 4,000 dead.

1780   October 1–4 *Hurricane*, Jamaica–Cuba Savanna-La-Mar. 2,000 dead.

1787   October 4 *Cyclone*, India. 10,000 dead.

1789   December *Cyclone*, India. 20,000 dead.

1792   October 29 *Hurricane*, Cuba. 100 dead.

1794   August 27–28 *Hurricane*, Cuba.

1810   October 23–26 *Hurricane*, Cuba. 150 dead.

1812   October 12–14 *Hurricane*, Cuba.

1821   September 13 *Hurricane*, Cuba. 60 dead.

1828   September 17 *Typhoon*, Japan. 10,000 dead.

1831   August 13–14 *Hurricane*, Cuba. Great Caribbean Hurricane. 300 dead.

1839   November 23 *Cyclone*, Australia. 12 dead.

1840   May 1 *Cyclone*, India. Belle Alliance
July 21–22 *Typhoon*, Hong Kong, China. 1,000 dead.

1844   October 4–5 *Hurricane*, Cuba. Cuban Hurricane. 500 dead.

1846   October 6–14 *Hurricane*, Cuba–United States. 650 dead.

1856   August 27–28 *Hurricane*, Cuba. 200 dead.

1857   September 13 *Hurricane*, North Atlantic Ocean. 425 dead.

1862   July 27 *Typhoon*, China. 37,000 dead.

1864   October 5 *Cyclone*, India. 50,000 dead.

1870   October 7 *Hurricane*, Cuba. 20 dead.
October 20 *Hurricane*, Cuba. 12 dead.

1873   October 6 *Hurricane*, Cuba.

1875   September 12–16 *Hurricane*, Cuba–United States.

1876   October 31 *Cyclone*, India. 200,000 dead.

1881   January 6–7 *Cyclone*, Western Australia. 16 dead.
October 5 *Typhoon*, China. 30,000 dead.

1882   March 7 *Cyclone*, Western Australia. 20 dead.
June 6 *Cyclone*, India. 100,000 dead.
October 20 *Typhoon*, Philippines.

1884   January 30 *Cyclone*, Australia. Five dead.

1885   September 21 *Cyclone*, India. 10,000 dead.

1886   August 19 *Hurricane*, United States.

1888   September 4–5 *Hurricane*, Cuba. 600 dead.

1892   March 8–10 *Cyclone*, Australia. 15 dead.
November 24 *Typhoon*, Japan. 310 dead.

1893   August 27 *Hurricane*, Southeastern United States. 1,900 dead.
October 5 *Hurricane*, Southern United States. 2,000 dead.

1894   January 4 *Cyclone*, Western Australia. 50 dead.

1897   January 6–7 *Cyclone*, Australia. 30 dead.

1898   April 2 *Cyclone*, Australia.
May 5–7 *Cyclone*, Australia. 100 dead.

1899   February 17 *Cyclone*, Madagascar.
March 4–5 *Cyclone*, Australia. 300 dead.

1900   June 6–8 *Cyclone*, Australia. 12 dead.
September 8 *Hurricane*, Southern United States. 6,000–8,000 dead.

1903   January 13 *Cyclone*, South Pacific Ocean–Society Islands. 1,000 dead.
March 9 *Cyclone*, Australia. Seven dead.

1904   April 19 *Cyclone*, Western Australia. Five dead.
July 8–11 *Cyclone*, Northeastern Australia. 40 dead.

1905   June 30 *Typhoon*, North Pacific Ocean–Marshall Islands. 500 dead.

1906   September 18 *Typhoon*, Hong Kong, China. 10,000 dead.

1910   April 11 *Typhoon*, Southern Japan. 280 dead.
October 12–21 *Hurricane*, Western Cuba. 100 dead.

1911   March 16 *Cyclone*, Queensland, Australia. Two dead.

1912   September 16 *Typhoon*, Taito, Taiwan. 107 dead.
November 19 *Hurricane*, Black River, Jamaica. 124 dead.

1918   January 20–22 *Cyclone*, Queensland, Australia. 30 dead.
March 9–10 *Cyclone*, Queensland, Australia. 14 dead.

1922   August 2–3 *Typhoon*, Northeastern China. 60,000 dead.

1923   June 26   *Cyclone*, Northeastern Australia. 46 dead.

1926   September 18   *Hurricane*, Southeastern United States. 243 dead.

October 20   *Hurricane*, Cuba. 650 dead; $100 million in damages.

1928   September 9–17   *Hurricane*, Puerto Rico–United States. 3,000 dead.

September 20   *Typhoon*, China. 5,000 dead.

1931   July 5–7   *Cyclone*, Australia. 10 dead.

1932   November 9   *Hurricane*, Cuba. (Santa Cruz del Sur) 2,500 dead.

1933   August 23   *Hurricane*, Eastern United States. 17 dead.

1934   July 25   *Tropical Storm*, Southern United States. 11 dead.

September 21   *Typhoon*, Japan. 400 dead.

1935   September 2   *Hurricane*, United States. 400 dead.

1937   March 10–11   *Cyclone*, Australia. Four dead.

1938   September 21   *Hurricane*, United States. 700 dead; $700 million damages.

1939   September 29   *Tropical Storm*, California, United States. 45 dead.

1942   March 24–25   *Cyclone*, Western Australia. 15 dead.

October 16   *Cyclone*, Calcutta, India. 40,000 dead.

1944   September 13   *Hurricane*, Cuba, United States. 600 dead.

December 17–18   *Typhoon*, North Pacific Ocean. (Cobra) 790 dead.

1945   September 17   *Typhoon*, Japan. 3,000 dead.

1947   September 19   *Typhoon*, Japan. (Kathleen) 1,692 dead, 2,714 missing.

September 20   *Hurricane*, Southern United States. $5 million damages.

October 15   *Hurricane*, Southeastern United States.

December 1   *Hurricane*, North Atlantic. 165 dead.

December 26   *Typhoon*, Northern Philippines. 49 dead.

1948   September 16   *Typhoon*, Central Japan. (Ion) 838 dead.

1949   June 20   *Typhoon*, Southeastern Japan. 106 dead.

October 27   *Cyclone*, Southeastern India. 1,000 dead; 50,000 homeless.

1950   September 3   *Typhoon*, Southern Japan. (Jane) 508 dead.

September 6   *Hurricane*, United States. (Easy) No casualties.

1951   August 13–22   *Hurricane*, Jamaica–Mexico. (Charlie) 277 dead.

October 15   *Typhoon*, Japan. (Ruth) 943 dead; 34,000 buildings destroyed.

1952   February 4   *Tropical Storm*, North Atlantic Ocean–Bahamas.

1953   March 23   *Cyclone*, Western Australia.

1954   February 20–22   *Cyclone*, Eastern Australia. 26 dead.

August 31   *Hurricane*, Northeastern United States. (Carol) 60 dead.

September 10–11   *Hurricane*, Eastern United States. (Edna) 22 dead.

September 16   *Typhoon*, Japan. (Toyamaru) 1,761 dead.

October 5–16   *Hurricane*, United States–Canada. (Hazel) 300 dead.

1955   August 12–13   *Hurricane*, United States. (Connie) 41 dead; $50 million losses.

August 18–19   *Hurricane*, United States. (Diane) 191 dead.

September 19   *Hurricane*, United States. (Ione) 5 dead; $100 million losses.

September 22   *Hurricane*, Barbados–Mexico. (Janet) 450 dead.

1956   June 1   *Cyclone*, India. 480 dead; 40,000 homeless.

September 8–9   *Typhoon*, Japan–Korea. (Emma) 77 dead; damage $5 million.

1957   June 27   *Hurricane*, Southern United States. (Audrey) 400 dead.

September 21–26   *Typhoon*, Japan. (Faye) 53 dead; 79 missing.

1958   August 28   *Hurricane*, North Atlantic Ocean. (Daisy)

September 26   *Typhoon*, Japan. (Kanogawa Typhoon). 1,269 dead.

1959   September 26   *Typhoon*, Japan. (Vera) 5,129 dead; 56,000 buildings damaged.

1960   September 1–13   *Hurricane*, Puerto Rico–United States. (Donna) 168 dead.

1961   September 16 *Typhoon*, Japan. (Second Muroto Typhoon). 202 dead.

September 11 *Hurricane*, United States. (Carla) 46 dead; $400 million damage.

1963   October 4–7 *Hurricane*, Cuba. (Flora) 1,300 dead; thousands homeless.

1964   August 5–8 *Tropical Storm*, North Atlantic Ocean. (Abby)

August 25–26 *Hurricane*, Leeward Islands–Cuba–United States. (Cleo) 147 dead.

October 13 *Tropical Storm*, Cuba. (Isbell) Three dead.

1966   June 9 *Hurricane*, Central America–Cuba–United States. (Alma) 85 dead.

1968   June 1–13 *Hurricane*, Caribbean Sea–Cuba–Southeastern United States. (Abby)

October 16–17 *Hurricane*, Cuba. (Gladys) Thres dead.

1969   June–November *Statistics*, North Atlantic Ocean. The 1969 North Atlantic hurricane season generated 13 named tropical cyclones, A–M, with a record-breaking 10 of them reaching hurricane-force.

May 19 *Cyclone*, Southeastern India. 618 dead.

August 17–18 *Hurricane*, United States. (Camille) 258 dead; $1.4 billion damages.

1970   July 31–August 3 *Hurricane*, Cuba–United States. (Celia) 17 dead.

1971   October 29 *Cyclone*, Northeastern India. 10,000 dead.

1972   June 19–22 *Hurricane*, United States. (Agnes) 134 dead; $50 million damages.

1973   October 18 *Tropical Storm*, Cuba. (Gilda) No casualties reported.

1974   June–November *Statistics*, North Atlantic Ocean. The 1974 North Atlantic hurricane season produces seven named tropical cyclones, A–G, four of them reaching hurricane-intensity.

August 29–September 10 *Hurricane*, United States. (Carmen) One dead.

September 14–22 *Hurricane*, Honduras, Central America. (Fifi) 10,000 dead.

December 25 *Cyclone*, Darwin, Australia. (Tracy) 80 dead.

1975   September 27 *Hurricane*, Caribbean–United States. (Eloise) 76 dead; $200 million.

1976   August 10 *Hurricane*, United States. (Belle) 12 dead; $24 million in damages.

1977   September 2 *Hurricane*, Mexico. (Anita) No casualties reported.

November 19 *Cyclone*, Southeastern India. 20,000 dead.

1979   May 12–13 *Cyclone*, Southeastern India. 369 dead.

July 11 *Hurricane*, Southern United States. (Bob) 1 dead; $1 million in damages.

August 30–September 7 *Hurricane*, Caribbean–United States. (David) 2,068 dead.

1980   August 4–10 *Hurricane*, Caribbean Sea–United States. (Allen) 272 dead.

1982   June 4 *Cyclone*, Central India. 200 dead; 200,000 homeless; $107 million damages.

1983   August 18 *Hurricane*, Southern United States. (Alicia) 15 dead; $675 million.

1985   September 4 *Hurricane*, United States. (Elena) Three dead; $1 billion in losses.

September 27 *Hurricane*, United States. (Gloria) Three dead; $23 million damages.

1987   June–November *Statistics*, North Pacific Ocean. The 1987 eastern North Pacific hurricane season yields 18 named tropical cyclones, A–S, with nine maturing into hurricanes.

1988   September 19 *Hurricane*, Jamaica–Mexico. (Gilbert) 318 dead; $1 billion damages.

1989   June–November *Statistics*, North Atlantic Ocean. The 1989 North Atlantic hurricane season generates 11 tropical cyclones, A–K, with seven achieving hurricane status.

September 10–22 *Hurricane*, Caribbean–United States. (Hugo) 49 dead; $10 billion.

November 4–8 *Typhoon/Cyclone*, Thailand–India. (Gay) 1,124 dead.

1990   May 9–11 *Cyclone*, Central India. 450 dead; 500,000 homeless; $1 billion in losses.

October 8–10 *Tropical Depression*, Northern Cuba. (Marco) Three dead.

1991   August 18–19 *Hurricane*, United States. (Bob) 10 dead; $780 million in damages.

1992   August 23–26 *Hurricane*, United States. (Andrew) 40 dead; $20 billion in losses.

1994   November 13–14 *Hurricane*, Cuba–United States. (Gordon) Two dead.

1995 June–November *Statistics,* North Atlantic Ocean. The 1995 North Atlantic hurricane season generates 19 named tropical cyclones, A–T, 12 of them achieving hurricane-status. The second-busiest season on record, the historic 1995 season features the first North Atlantic tropical cyclones given names that begin with the letters *O* (Opal), *P* (Pablo), *R* (Roxanne), *S* (Sebastian), and *T* (Tanya).

June 5 *Hurricane,* United States. (Allison) No deaths; $1 million in damages.

June 14–22 *Cyclone,* Australia (Agnes).

June 15–21 *Hurricane,* Eastern North Pacific Ocean. (Adolph)

July 19–25 *Typhoon,* Korea. (Faye) 50 dead; $120 million in losses.

November 1 *Cyclone,* Southeastern India. Three dead; $50 million in losses.

1996 June–November *Statistics,* North Atlantic Ocean. The 1996 North Atlantic hurricane season produces 13 named tropical cyclones A (Arthur)–M (Marco), nine of them achieving hurricane status. In the eastern North Pacific Ocean, only nine tropical cyclones originate, five of which reach hurricane status. One eastern North Pacific system develops in mid-May, a weak tropical storm that remains unnamed. In the western North Pacific Ocean, 41 tropical systems form, with 30 of these reaching the mature stage. One super typhoon, Dale, produces a reported central pressure of 26.28 inches (890 mb) on November 10, one of the lowest atmospheric pressures ever observed in this region of the Earth.

June 19 *Tropical Storm,* North Atlantic Ocean–Eastern United States. (Arthur) The first tropical cyclone of the 1996 North Atlantic season, Arthur delivers 40 mph (64 kph) winds and sporadic downpours as it drifts over Cape Lookout, North Carolina. No deaths are reported, and property damage is minimal.

July 12 *Hurricane,* North Atlantic Ocean–Leeward Islands–Eastern United States. (Bertha) The first major hurricane of the 1996 season, Bertha's 105 mph (169 kph) winds (downgraded from 115 mph [185 kph] just days earlier) and flooding rains cause 12 deaths and some $270 million in property losses over large portions of North Carolina and the northeastern United States.

July 27 *Hurricane,* Caribbean Sea–Central America–Eastern North Pacific Ocean. (Cesar–

Douglas) Powered by a central barometric pressure of 29.08 inches (985 mb), Hurricane Cesar trounces ashore in Nicaragua, its Category 1 winds and heavy rains claiming 51 lives in northern Venezuela and Nicaragua. Crossing Central America to the northwest, Cesar reemerges into the eastern North Pacific Ocean, eventually restrengthening to hurricane status and earning a new name, *Douglas.*

September 5 *Hurricane,* North Atlantic Ocean–Eastern United States. (Fran) A fairly strong Category 3 hurricane at landfall at Cape Fear, North Carolina, Cape Verde–born Fran claims 24 lives and some $3.2 billion in property damage as it makes landfall on the eastern United States.

November 6 *Cyclone,* Bay of Bengal–India. (04B) A powerful cyclone with sustained 132 mph winds, Cyclone 04B drops nearly nine inches of rain near the port city of Kakinada in northern Andhra Pradesh, killing more than 2,000 people.

December 25 *Tropical Storm,* Western North Pacific Ocean–Indonesia. (Greg) Some 90 people perish as a weakening Tropical Storm Greg moves over eastern Indonesia, producing 45 mph (72 kph) winds and flooding rains.

December 22–30 *Cyclone,* South Western Pacific Ocean–Australia. (Phil).

1997 June–November *Statistics,* North Atlantic Ocean. The 1997 North Atlantic hurricane season spawns eight tropical cyclones, five of which become hurricanes. No fewer than 11 super typhoons develop in the western North Pacific Ocean, including Ivan (160 mph winds) and Joan (180 mph winds), while Cyclones Gavin and Justin stir the waters around Australia.

February 21 *Cyclone,* Western Indian Ocean. (Karlette).

March 7–11 *Cyclone,* Australia–Southwestern Pacific Ocean-Fiji. (Gavin) An intense *Cyclone,* Gavin wreaks considerable damage on the Fiji Islands, sinking a fishing trawler with 10 crew members. Several looters use coffins as boats as they move through flooded sections of Fiji, preying on shops. Twenty people perish.

March 13–21 *Cyclone,* Western South Pacific Ocean–Tonga. (Hena) One of the most powerful cyclones to ever strike the Tongan Islands (where it nearly destroys the kingdom's Parliament building), Cyclone Hena further ravages the French territories of Wallis and Futuna, destroying scores of buildings, uprooting trees, downing powerlines, and claiming the lives of 12 people.

March 24 *Cyclone,* Australia. (Justin).

July 21–29 *Typhoon,* Western North Pacific Ocean–Japan. (Rosie).

September 7 *Hurricane,* Eastern North Pacific Ocean–Mexico. (Pauline) One of the deadliest eastern North Pacific hurricanes on record, Category 3 Pauline's inundating rains kill 297 people in the mountains outside Acapulco.

October 1 *Tropical Depression,* Western North Pacific–Eastern North Pacific–Canada. (Ginger) Two weeks after coalescing to form Typhoon Ginger, the watery remnants of this long-distance, fast-moving tropical cyclone collide with a low-pressure system near Vancouver, Canada, to produce heavy downpours across much of the northwestern United States. Ginger is one of only four documented tropical cyclones to move eastward across the North Pacific Ocean.

October 14–19 *Typhoon,* Philippines–China. (Ivan) A super typhoon, Ivan produces sustained 160 mph winds as it comes ashore in the northern Philippines.

October 15–22 *Typhoon,* Philippines–China. (Joan) Following hard on the heels of Super Typhoon Ivan, Super Typhoon Joan batters the northeastern Philippines with sustained 180 mph winds. Tropical systems of remarkable intensity, both Super Typhoons Ivan and Joan break Super Typhoon Tip's 1979 endurance record for total length of time spent as a super typhoon.

1998 **June–November** *Statistics,* North Atlantic Ocean. The 1998 North Atlantic hurricane season produces 14 named tropical cyclones A (Alex)–N (Nicole), 10 of them achieving hurricane status. Two of them, Georges and Mitch, are notable for their severity. According to the National Hurricane Center (NHC), the four-year period between 1995 and 1998 produced a total of 33 hurricanes, "an all-time record." In the eastern North Pacific Ocean, 13 tropical cyclones are awarded names A (Agatha)–M (Madeline), with nine of these being mature-stage hurricanes. In the western North Pacific Ocean, 17 named typhoons and tropical storms emerge. Of these, nine were mature-stage typhoons, two of which, Zeb and Babs, were super typhoons, with sustained winds in excess of 155 mph.

August 27 *Hurricane,* North Atlantic Ocean–Eastern United States. (Bonnie) Gliding ashore near Wilmington, North Carolina, a poorly organized Bonnie delivers 110 mph (177 kph) winds and flooding rains to much of the state; three people die, $720 million in property losses are sustained.

October 24 *Hurricane,* Caribbean Sea–Central America–Florida. (Mitch) A tropical system of superlatives, the most intense October hurricane on record also proves the second-deadliest; staggering, 35-inch rainfalls claim the lives of between 9,000 and 12,000 people in Honduras and Nicaragua.

October 9–17 *Typhoon,* Philippines–Taiwan–Japan. (Zeb) A very severe super typhoon, Zeb kills 74 people in the Philippines, 31 in Taiwan, and 12 in Japan.

October 13–21 *Typhoon,* Philippines. (Babs) A powerful typhoon that leaves 132 people dead, 320,000 others homeless, and extensive flooding in central Manila, the Philippines.

November 1 *Hurricane,* North Atlantic Ocean–Puerto Rico–Hispaniola–Cuba–Florida–Mississippi. (Georges) As the last of its once-powerful cloud bands dissipate over northern Florida, the record of Hurricane Georges's violent passage through the northern Caribbean islands is compiled; 602 people are dead, while damages estimates in Puerto Rico and the United States near $6 billion.

November 4–14 *Typhoon,* Central Vietnam. (Elvis) November 4–14, 1998. The third tropical cyclone to strike central Vietnam in a period of three weeks, Elvis rocks the nation's central provinces with torrential rains, causing extensive flooding and 50 deaths. All told, Typhoons Chip, Dawn, and Elvis claim a total of 267 lives and destroy nearly 10,000 structures in Vietnam.

1999 **June–November** *Statistics,* North Atlantic Ocean. The 1999 North Atlantic tropical cyclone season yields 11 tropical systems A (Arlene)–L (Lenny), with seven of these reaching hurricane status. In the eastern North Pacific Ocean, 16 tropical systems develop, nine of them becoming mature-stage hurricanes. In the western North Pacific Ocean, a full season produces 29 tropical cyclones, 19 of which grow to typhoon intensity.

June 14–17 *Tropical Storm,* North Atlantic Ocean. (Arlene) Offshore system briefly threatens Bermuda.

June 18–23 *Hurricane,* Eastern North Pacific Ocean–Central Mexico. (Adrian) Maximum sustained winds of 100 mph, 13-foot swells.

July 30–August 5 *Typhoon,* Western North Pacific Ocean–Korea. (Olga) In South Korea, streets flooded by Typhoon Olga force mourners at a funeral to attend services in boats. All told, Olga kills 24 people in Vietnam, 30 in the Philippines (eight in one landslide), and leaves scores missing.

August 13–20 *Hurricane,* Eastern–Central–Western North Pacific Ocean. (Dora) Once boasting sustained 130 mph (209 kph) winds while positioned off the western coast of Mexico, Dora travels across the North Pacific to cross the International Date Line as a tropical storm—a feat last undertaken in 1994 by Hurricane John.

August 15 *Hurricane,* Eastern North Pacific Ocean. (Eugene) A powerful offshore system.

August 18 *Hurricane,* Caribbean Sea–Gulf of Mexico–Texas. (Bret) Strikes sparsely populated section of Texas coastline. No deaths or injuries reported.

August 22 *Typhoon,* Western North Pacific Ocean–China. (Sam) A moderately powerful typhoon, Sam lashes Hong Kong with 81 mph (130 kph) winds and heavy rains. During the typhoon's peak intensity, a jetliner crashes at Hong Kong's airport, killing two people and injuring scores of others.

September 1 *Hurricane,* North Atlantic Ocean–Eastern United States. (Dennis) System lingers off North Carolina coast, dropping intense rains.

September 7–9 *Hurricane,* Eastern North Pacific Ocean–Mexico. (Greg) Nine people perish as Hurricane Greg batters the northwestern coast of Mexico, causing damage in the resort city of Cabo San Lucas.

September 16 *Hurricane,* Bahamas–Eastern United States. (Floyd) After pummeling the Bahamas Islands as an extremely intense Category 4 hurricane, Floyd's downgraded 110 mph (177 kph) winds and swamping rains tramp ashore in North Carolina. Slogging northward, Floyd's inundating bones prompt flood emergencies from Virginia to Massachu-

setts; 72 people die, while damages estimates top $1 billion.

September 21–22 *Hurricane,* North Atlantic Ocean. (Gert) Nearly a Category 5 hurricane, with sustained winds of 150 mph.

September 22–23 *Hurricane,* Eastern North Pacific Ocean. (Hilary) A one-time Category 2 hurricane, Hilary's downgraded remains deliver strong thunderstorms to southern California.

September 23 *Typhoon,* Western North Pacific Ocean–Japan. (Bart) At one time a super typhoon with sustained winds of 160 mph, a mischievous Bart weakens considerably as it skates across Okinawa, its 120 mph (193 kph) winds leaving 110,000 of the island's residents without electricity. No deaths or injuries are reported.

September 19–24 *Tropical Storm,* Gulf of Mexico–Florida. (Harvey) Before delivering five-inch rainfalls to southern Florida, system's fringe winds batter fire-disabled cruise ship *Tropicale* in the Gulf of Mexico.

October 1–11 *Typhoon,* Western North Pacific Ocean–Philippines–China. (Dan) Bearing 111 mph (180 kph) winds, 120 mph (193 kph) gusts, and punishing rains, Typhoon Dan brushes the northern Philippines before spiraling into southeastern China.

October 7 *Hurricane,* North Atlantic Ocean. (Irene).

December 15 *Cyclone,* Australia. (John) One of the most powerful Australian cyclones on record, John pounds Australia's northwestern coast with sustained winds of 160 mph (258 kph). Despite the system's severity, no lives are lost.

2000 **June–November** *Statistics,* North Atlantic Ocean. The 2000 North Atlantic tropical cyclone season yields 18 tropical systems, with seven reaching hurricane status. In the eastern north Pacific Ocean, 21 systems develop, six of them becoming mature-stage hurricanes.

August 4–23 *Hurricane,* North Atlantic Ocean (Alberto).

October 5–7 *Tropical Storm,* North Atlantic Ocean (Leslie).

October 17–20 *Hurricane,* North Atlantic Ocean (Michael).

October 19–22 *Tropical Storm,* North Atlantic Ocean (Nadine).

2001 **June–November** *Statistics,* North Atlantic Ocean. The 2001 North Atlantic tropical cyclone season yields 17 tropical systems, with nine reaching hurricane status. In the eastern north Pacific Ocean, 20 systems develop, eight becoming mature-stage hurricanes.

October 4–9 *Hurricane,* North Atlantic Ocean (Iris).

November 5–6 *Hurricane,* North Atlantic Ocean (Noel).

2002 **June–November** *Statistics,* North Atlantic Ocean. The 2002 North Atlantic tropical cyclone season yields 14 tropical systems, with four reaching hurricane status. In the eastern North Pacific Ocean, 19 systems develop, nine of them growing into mature-stage hurricanes.

2003 **June–November** *Statistics,* North Atlantic Ocean. The 2003 North Atlantic tropical cyclone season yields 21 tropical systems, with seven reaching hurricane status. In the eastern North Pacific Ocean, 17 systems develop, seven of them developing into mature-stage hurricanes.

October 2–6 *Tropical Storm,* North Atlantic Ocean (Larry).

December 9–11 *Tropical Storm,* North Atlantic Ocean (Peter).

2004 **June–November** *Statistics,* North Atlantic Ocean. The 2004 North Atlantic tropical cyclone season yields 17 tropical systems, with seven reaching hurricane status. In the eastern North Pacific Ocean, 17 systems develop, six of them achieving mature-stage status.

August 9–15 *Hurricane,* North Atlantic Ocean–Caribbean Sea (Charley).

September 2–24 *Hurricane,* North Atlantic Ocean–Caribbean Sea–Gulf of Mexico (Ivan)

September 13–28 *Hurricane,* North Atlantic Ocean–Southern United States (Jeanne).

2005 **June–November** *Statistics,* North Atlantic Ocean. The 2005 North Atlantic tropical cyclone season generates 31 tropical systems, with 15 blossoming into hurricane status. It is the most active season on record in the North Atlantic, rivaled by the 1933 season with 21 tropical systems. In the eastern North Pacific Ocean, 17 systems develop,

with seven of them growing into mature-stage hurricanes.

July 5–13 *Hurricane,* North Atlantic Ocean–Caribbean Sea (Dennis).

July 11–21 *Hurricane,* North Atlantic Ocean–Caribbean Sea (Emily).

August 4–18 *Hurricane,* North Atlantic Ocean (Irene).

August 23–31 *Hurricane,* North Atlantic Ocean–Gulf of Mexico–United States (Katrina).

September 1–10 *Hurricane,* North Atlantic Ocean (Maria).

September 5–10 *Hurricane,* North Atlantic Ocean (Nate).

October 15–25 *Hurricane,* Caribbean Sea–Gulf of Mexico (Wilma).

December 30, 2005–January 6, 2006 *Tropical Storm,* North Atlantic Ocean (Zeta).

2006 **June–November** *Statistics,* North Atlantic Ocean. A tepid season when compared to its immediate predecessor, the 2006 North Atlantic tropical cyclone season produces only nine tropical systems, with five reaching hurricane status. In the eastern North Pacific Ocean, an active season sees 25 systems formed with 11 of them reaching mature-stage status.

September 12–24 *Hurricane,* North Atlantic Ocean (Helene).

September 27–October 2 *Hurricane,* North Atlantic Ocean (Isaac).

2007 *Statistics,* North Atlantic Ocean. The 2007 North Atlantic hurricane season featured a number of notable tropical cyclones. In August and September, two Category 5 hurricanes (Dean and Felix) came ashore in Mexico and Central America—the first Category 5 hurricanes to make landfall in the Western Hemisphere since 1992's Andrew. They were also the first two Category 5 tropical cyclones in several decades to make landfall during one season. In early September, Hurricane Humberto muscled ashore in Louisiana and Texas as a Category 1 system. At the time, meteorologists observed that Humberto underwent the fastest deepening in the Western Hemisphere on record, going from a tropical depression to a mature-stage tropical cyclone in approximately 18 hours.

APPENDIX C

TRACKING A TROPICAL CYCLONE

The deadliest tropical cyclones have always been those that have come ashore without warning. While advances in forecasting, satellite technology, and television communications have largely rendered the surprise hurricane strike a peril of the past, the National Oceanic and Atmospheric Administration (NOAA) strongly suggests that during the North Atlantic hurricane season (June 1–November 30), coastal residents, mariners, and vacationers continually monitor the proximity of dangerous tropical systems in the North Atlantic Ocean and Gulf of Mexico by plotting their daily positions on a Hurricane Tracking Chart.

The National Hurricane Center (NHC) posts hurricane and tropical cyclone advisories every six hours. Stating the location of a respective system's center, or eye, as well as its forward speed, maximum sustained wind speed, and central barometric pressure, these hurricane advisories can be easily obtained by telephone, by television and media outlets, and from the Internet. The following Internet addresses link to sites that can be used for tropical cyclone historical, tracking, and monitoring purposes:

American Meteorological Society (AMS)
http://www.ametsoc.org/

Australian Bureau of Meteorology
http://www.bom.gov.au/lam/climate/levelthree/
c20thc/cyclone.htm

Central Pacific Hurricane Center
http://www.prh.noaa.gov/hnl/cphc/

The Hong Kong Observatory
http://www.hko.gov.hk/contente.htm

Hurricanecity
http://www.hurricanecity.com/

Japan Meteorological Agency
http://www.jma.go.jp/jma/indexe.html

National Hurricane Center/Tropical Prediction Center (NHC-TPC)
http://www.nhc.noaa.gov/index.shtml
http://www.tpc.ncep.noaa.gov/pastall.shtml?text

The New York Times
http://www.nytimes.com//

NOAA Satellite and Information Services
http://www.ncdc.noaa.gov/oa/reports/weather-
events.html

UNISYS Best Track
http://weather.unisys.com/hurricane/

University of Wisconsin-Madison
http://cimss.ssec.wisc.edu/tropic/tropic.html

The Weather Channel
http://www.weather.com/

APPENDIX D

LIST OF NAMED HURRICANES, TYPHOONS, CYCLONES, AND TROPICAL STORMS

Abby (Tropical Storm)   North Atlantic Ocean, August 5–8, 1964

Abby (Hurricane)   Caribbean Sea–Cuba–SE United States, June 1–13, 1968

Adolph (Hurricane)   E North Pacific Ocean, June 15–21, 1995

Agnes (Hurricane)   SE–NE United States, June 14–22, 1972

Agnes (Cyclone)   NE Australia, April 19–21, 1995

Agnielle (Cyclone)   Indian Ocean, November 24, 1995

Alberto (Tropical Storm)   E United States–E Canada, August 5–8, 1988

Alberto (Tropical Storm)   SE United States, June 30–July 7, 1994

Aletta (Tropical Storm)   E North Pacific Ocean, May 27–29, 1974

Aletta (Tropical Storm)   E North Pacific Ocean, June 18–23, 1994

Alice (Hurricane)   Leeward Islands, December 30, 1954–January 5, 1955

Alice (Hurricane)   North Atlantic Ocean–Bermuda, July 1–7, 1973

Allen (Hurricane)   Gulf of Mexico–S United States, August 4–10, 1980

Allison (Tropical Storm)   Gulf of Mexico–United States, 1989

Allison (Hurricane)   SE United States, June 3–6, 1995

Alma (Hurricane)   Honduras–SE United States, June 7–9, 1966

Alma (Hurricane)   Honduras–SE United States, May 17–27, 1970

Alma (Tropical Storm)   Trinidad–Caribbean Sea, August 12–15, 1974

Amy (Tropical Storm)   North Atlantic–E United States, June 26–July 4, 1975

Andrew (Hurricane)   August 23–26, 1992

Angela (Typhoon)   Philippines–Vietnam, November 11, 1995

Anita (Hurricane)   Gulf of Mexico–SE Mexico, September 1, 1977

Anna (Tropical Storm)   North Atlantic Ocean, July 23–August 5, 1969

Antje (Hurricane)   Gulf of Mexico–E Mexico, August 30–September 8, 1842

Arlene (Tropical Storm)   North Atlantic Ocean, June 14–17, 1999

Audrey (Hurricane)   Gulf of Mexico–S United States, June 27, 1957

Auring (Tropical Storm)   W North Pacific Ocean–Philippines, May 30–June 2, 1995

Axel (Tropical Storm)   W North Pacific Ocean–Philippines, December 19–24, 1994

Babs (Typhoon)   Philippines, October 13–21, 1998

Barbara (Hurricane)   E North Pacific Ocean, July 7–18, 1995

Barry (Tropical Storm)   North Atlantic Ocean–Canada, July 6–10, 1995

Bart (Typhoon)   W North Pacific Ocean–Japan, September 24, 1999

Becky (Hurricane)   North Atlantic Ocean, August 26–September 2, 1974

Belle (Hurricane)   NE United States, August 9–10, 1976

Beryl (Tropical Storm)   Gulf of Mexico–S United States, August 8–10, 1988

Beryl (Tropical Storm)   SE United States, August 14–19, 1994

Beth (Hurricane)   North Atlantic–SE Canada, August 10–17, 1971

Betsy (Hurricane)   Florida Keys, East Coast, August 29–September 9, 1965

Betsy (Hurricane)   September 7–10, 1966

Betty (Hurricane)   North Atlantic Ocean, August 22–September 1, 1972

Beulah (Hurricane)   September 1963

Beulah (Hurricane)   September 16–21, 1967

Beverly (Typhoon)   North Pacific Ocean–Guam, December 2–10, 1948

Billie (Typhoon)   W North Pacific Ocean–Japan, July 9, 1967

Blanca (Tropical Storm)   E North Pacific Ocean, June 4–8, 1974

Blanche (Hurricane)   North Atlantic Ocean–SE Canada, July 23–28, 1975

Bob (Hurricane)   Gulf of Mexico–S United States, July 11, 1979

Bob (Hurricane)   North Atlantic Ocean–NE United States, August 18–19, 1991

Bobby (Cyclone)   NW Australia, February 24, 1995

Bonnie (Hurricane)   Eastern United States, August 19–30, 1998

Brenda (Tropical Storm)   North Atlantic Ocean, August 7–10, 1964

Brenda (Hurricane)   Caribbean Sea–SE Mexico, August 18–22, 1973

Bret (Hurricane)   Caribbean Sea–Gulf of Mexico–Texas, August 18–26, 1999

Bud (Tropical Storm)   E North Pacific Ocean, June 27–29, 1994

Camille (Hurricane)   Leeward Islands–Jamaica–S United States, August 5–22, 1969

Candy (Tropical Storm)   Gulf of Mexico–S United States, June 22–26, 1968

Carla (Hurricane)   Galveston Island–Texas, 1961

Carlotta (Hurricane)   E North Pacific Ocean, June 28–July 5, 1994

Carmen (Typhoon)   South Korea, August 24, 1960

Carmen (Hurricane)   NE Mexico–S United States, August 29–September 10, 1974

Carol (Hurricane)   North Atlantic Ocean, 1953

Carol (Hurricane)   Northeastern United States, August 28–31, 1954

Carolina 1713 (Hurricane)   SE United States, September 5–6 (16–17), 1713

Carolina 1728 (Hurricane)   SE United States, September 3 (14), 1728

Caroline (Hurricane)   NE Mexico–S United States, August 24–September 1, 1975

Carrie (Tropical Storm)   North Atlantic–E Canada, August 29–September 5, 1972

Celia (Hurricane)   Gulf of Mexico–S United States, July 23–August 5, 1970

Chantal (Tropical Storm)   North Atlantic Ocean, July 12–20, 1995

Charlie (Hurricane)   Jamaica–SE Mexico, August 17, 1951

Charlie (Hurricane)   SE–NE United States, August 13–20, 1986

Chesapeake-Potomac 1933 (Hurricane)   NE United States, August 17–23, 1933

Chip (Typhoon)   Central Vietnam, November 2–12, 1998

Chloe (Tropical Storm)   Caribbean Sea–Honduras, August 18–25, 1971

Chloe (Tropical Storm)   NW Australia, April 10–14, 1995

Chris (Tropical Storm)   Hispaniola–Bahamas–SE United States, August 21–29, 1988

Chris (Hurricane)   North Atlantic Ocean, August 16–23, 1994

Christine (Tropical Storm)   Leeward Islands, August 25–September 4, 1973

Chuck (Tropical Storm)   W North Pacific Ocean, May 2–5, 1995

Cindy (Hurricane)   North Atlantic Ocean, August, 1999

Claudette (Tropical Storm)   July 24–28, 1979

Cleo (Hurricane)   SE United States, August 20–September 5, 1964

Cobra (Typhoon)   Third Fleet, Philippines, December 17–18, 1944

Colleen (Tropical Storm)   W North Pacific Ocean–Wake Island, November 17, 1995

Connie (Hurricane)   NE United States, August 3–14, 1955

Connie (Hurricane)   NE North Pacific Ocean, June 6–22, 1974

Cosme (Hurricane)   E North Pacific Ocean, July 17–22, 1995

Cuban 1844 (Hurricane)   Caribbean Sea–Cuba, October 5, 1844

Cyclone   01A, Arabian Sea–India, June 2–10, 1998. 1,126 dead

Cyclone   04B, Bay of Bengal–India, November 6, 1996

Cyclone   05B, Bay of Bengal–India, October 29–November 2, 1999

Daisy (Hurricane)   August 25–28, 1958

Dale (Typhoon)   W North Pacific Ocean–Philippines–Taiwan–China, November 6–14, 1996

Dalila (Tropical Storm)   E North Pacific Ocean, July 24–August 2, 1995

Dan (Tropical Storm)   Philippines, December 27–29, 1995

Dan (Typhoon)   W North Pacific Ocean–Philippines–China, October 1–10, 1999

Daniel (Tropical Storm)   E North Pacific Ocean, July 8–14, 1994

Danny (Hurricane)   August 15, 1985

Daryl (Cyclone)   Australia–Indian Ocean, November 24, 1995

David (Hurricane)   Hispaniola–SE United States, August 30–September 7, 1979

Dawn (Hurricane)   SE United States–Bahamas, September 4–14, 1972

Dawn (Typhoon)   Central Vietnam, November 6–12, 1998

Dean (Tropical Storm)   Gulf of Mexico–S United States, July 28–August 2, 1995

Deanna (Tropical Storm)   Philippines–China, June 2–8, 1995

Debbie (Hurricane)   Leeward Islands–North Atlantic, August 13–25, 1969

Debby (Hurricane)   Gulf of Mexico–SE Mexico, August 31–September 8, 1988

Debby (Tropical Storm)   Windward Islands, September 9–11, 1994

Delia (Tropical Storm)   S United States, September 1–7, 1973

Delores (Tropical Storm)   E North Pacific Ocean–NW Mexico, June 14–16, 1974

Della (Typhoon)   Japan, August 20–28, 1960

Dennis (Hurricane)   North Atlantic Ocean–Eastern United States, September, 1999

Diana (Hurricane)   September 11–13, 1984

Diane (Hurricane)   NE United States, August 10–21, 1955

Dolly (Hurricane)   Puerto Rico–North Atlantic Ocean, September 7–10, 1953

Dolly (Tropical Storm)   North Atlantic Ocean, September 2–5, 1974

Dolly (Hurricane)   Mexico, 1996

Donna (Hurricane)   September 9–12, 1960

Dora (Hurricane)   SE United States, August 28–September 16, 1964

Dora (Hurricane)   Eastern–Central–Western North Pacific Ocean, August 11–20, 1999

Doria (Tropical Storm)   NE United States, August 20–29, 1971

Doris (Typhoon)   North Pacific Ocean–Truk Islands, May 1950

Doris (Hurricane)   North Atlantic Ocean, August 28–September 4, 1975

Dorothy (Tropical Storm)   Martinique–Caribbean Sea, August 13–23, 1970

Dot (Hurricane)   Kauai, Hawaii, 1959

Doug (Typhoon)   Philippines–China, August 8, 1994

Easy (Hurricane)   SE United States, 1950

Edith (Hurricane)   Nicaragua–Honduras–S United States, September 5–17, 1971

Edna (Hurricane)   NE United States, September 10–11, 1954

Edna (Tropical Storm)   North Atlantic Ocean, September 10–19, 1968

Eileen (Tropical Storm)   E North Pacific Ocean, July 1–4, 1974

Elaine (Tropical Storm)   North Atlantic Ocean, July 1–4, 1974

Elena (Hurricane)   S United States, September 2–3, 1985

Eli (Tropical Storm)   Mariana Islands–W North Pacific Ocean, June 2–8, 1995

Ella (Hurricane)   SE Mexico, September 7–13, 1970

Ellen (Hurricane)   North Atlantic Ocean, September 14–23, 1973

Ellie (Typhoon)   China, August 8, 1994

Eloise (Hurricane)   SE Mexico–SE United States, September 13–24, 1975

Elvis (Typhoon)   Central Vietnam, November 4–14, 1998

Emilia (Hurricane)   Central North Pacific Ocean, July 16–25, 1994

Emma (Typhoon)   Okinawa, Philippines, Korea, Japan, September 8–10, 1956

Erick (Hurricane)   E North Pacific Ocean, August 1–8, 1995

Erin (Hurricane)   SE United States, July 31–August 6, 1995

Ernesto (Tropical Storm)   North Atlantic Ocean, September 3–5, 1988

Ernesto (Tropical Storm)   North Atlantic Ocean, September 21–26, 1994

Esther (Hurricane),   1961

Ethel (Hurricane)   North Atlantic Ocean, September 4–16, 1964

Eugene (Hurricane)   E North Pacific Ocean, August 11–13, 1999

Eve (Tropical Storm)   North Atlantic Ocean, August 24–27, 1969

Eve (Tropical Storm)   W North Pacific Ocean–Vietnam, October 10–19, 1999

Fabio (Tropical Storm)   E North Pacific Ocean, July 19–24, 1994

Faith (Tropical Storm)   Central Vietnam, December 9–16, 1998

Faye (Typhoon)   W North Pacific Ocean–Okinawa, September 26, 1957

Faye (Hurricane)   North Atlantic Ocean, September 18–29, 1975

Faye (Typhoon)   W North Pacific Ocean–SE Korea, July 1995

Felice (Tropical Storm)   Gulf of Mexico–S United States, September 11–17, 1970

Felix (Hurricane)   North Atlantic Ocean, August 8–22, 1995

Fergus (Cyclone)   South Western Pacific Ocean–New Zealand, December 28–31, 1996

Fern (Hurricane)   Gulf of Mexico–S United States, September 3–13, 1971

Fernanda (Tropical Storm)   E North Pacific Ocean, August 20, 1999

Fifi (Hurricane)   Caribbean Sea–Honduras, September 14–22, 1974

Flora (Hurricane)   Cuba, Haiti, October 4–8, 1963

Florence (Tropical Storm)   North Atlantic Ocean, September 6–10, 1964

Florence (Hurricane)   Gulf of Mexico–S United States, September 7–11, 1988

Florence (Hurricane)   North Atlantic Ocean, Ocean, November 2–8, 1994

Florida 1926 (Hurricane)   SE United States, September 18, 1926

Flossie (Hurricane)   E North Pacific Ocean, August 7–14, 1995

Flossy (Hurricane)   New Orleans, Florida, Georgia, September 22–25, 1956

Fran (Typhoon)   Japan, September 8–13, 1976

Fran (Hurricane)   North Atlantic Ocean, October 8–13, 1973

Fran (Hurricane)   E United States, September 1996

Frances (Tropical Storm)   North Atlantic Ocean, September 23–30, 1968.

Francesca (Hurricane)   E North Pacific Ocean, July 14–19, 1974

Fred (Typhoon)   Taiwan–China, August 15–20, 1994

Frederic (Hurricane)   Gulf of Mexico–S United States, September 12, 1979

Gabrielle (Tropical Storm)   Gulf of Mexico–NE Mexico, August 9–12, 1995

Gavin (Cyclone)   Australia–Southwestern Pacific Ocean–Fiji, March 7–11, 1997

Gay (Typhoon)   Guam–North Pacific Ocean, 1992

Georges (Hurricane)   North Atlantic Ocean–Puerto Rico–Cuba–Florida–Mississippi, September 15–October 1, 1998

Georgia (Typhoon)   Philippines, September 14, 1970

Gertie (Tropical Storm)   Australia, December 18–22, 1995

Gertrude (Hurricane)   Windward Islands, September 28–October 3, 1974

Gil (Tropical Storm)   E North Pacific Ocean, August 20–27, 1995

Gil (Tropical Storm)   Southern Vietnam, December 11, 1998

Gilbert (Hurricane)   Leeward Islands–Jamaica–SE Mexico, September 8–19, 1988

Gilda (Typhoon)   Japan–Korea, July 7–14, 1974

Gilda (Tropical Storm)   Cuba–North Atlantic Ocean, October 16–29, 1973

Gilma (Hurricane)   E North Pacific Ocean, July 21–31, 1994

Gladys (Hurricane)   North Atlantic Ocean–E Canada, September 13–25, 1964

Gladys (Hurricane)   Bayport–Crystal River, October, 1968

Gloria (Typhoon)   North Pacific Ocean–Okinawa, July 1949

Gloria (Hurricane)   September 26–27, 1985

Ginger (Hurricane)   North Atlantic–E United States, September 5–October 5, 1971

Ginny (Hurricane)   October 21–29, 1963

Gold Coast 1949 (Hurricane)   SE United States, August, 1949

Gordon (Hurricane)   SE United States, November 8–21, 1994

Gracie (Hurricane)   South Carolina, 1959

Great Calcutta 1864 (Cyclone)   NE India, October 5, 1864

Great Coastal 1806 (Hurricane)   E United States, August 21–24, 1806

Great Galveston 1900 (Hurricane)   S United States, September 8, 1900

Great Havana 1846 (Hurricane)   Cuba–E United States, October 11, 1846

Great Hurricane 1780 (Hurricane)   Leeward Islands–Bermuda, October 9–12, 1780

Great New England 1938 (Hurricane)   NE United States, September 16–21, 1938

Great Nova Scotia 1873 (Hurricane)   SE Canada, August 24–25, 1873

Greta (Tropical Storm)   North Atlantic Ocean, September 15–October 5, 1970

Gretchen (Hurricane)   E North Pacific Ocean, July 16–21, 1974

Hallie (Tropical Storm)   North Atlantic Ocean, October 24–28, 1975

Harold (Cyclone)   S Western Pacific Ocean–Australia, February 10–23, 1997

Harriet (Typhoon)   Okinawa, Japan, September 26–27, 1956

Hattie (Hurricane)   Honduras, October 31, 1961

Hazel (Hurricane)   Haiti–E United States, October 5–16, 1954

Hector (Tropical Storm)   E North Pacific Ocean, August 7–9, 1994

Heidi (Tropical Storm)   North Atlantic–NE United States, September 10–14, 1971

Helen (Tropical Storm)   China, August 18, 1995

Helene (Hurricane)   September 23–29, 1958

Helene (Hurricane)   North Atlantic Ocean, September 19–30, 1988

Helga (Tropical Storm)   E North Pacific Ocean, August 9–13, 1974

Hena (Cyclone)   S Western Pacific Ocean–Tonga, March 13–21, 1997

Henriette (Hurricane)   E North Pacific Ocean, September 1–8, 1995

Hilary (Tropical Storm)   Eastern North Pacific Ocean, September 23, 1999

Hilda (Hurricane)   Tampico, Mexico, September 28–October 5, 1964

Hilda (Hurricane)   S United States, September 28–October 5, 1964

Hooghly River 1737 (Cyclone)   Bay of Bengal–NE India, October 7, 1737

Hope (Typhoon)   China, August 2, 1979

Hortense (Hurricane)   Puerto Rico, 1996

Hugo (Hurricane)   Puerto Rico–SE United States, September 11–22, 1989

Humberto (Hurricane)   North Atlantic Ocean, August 22–September 1, 1995

Ida (Typhoon)   Japan, September 27–28, 1958

Ike (Typhoon)   Southern Philippines, September 2, 1984

Ileana (Hurricane)   E North Pacific Ocean, August 10–14, 1994

Independence 1775 (Hurricane)   North Carolina, September 2, 1775

Indianola 1886 (Hurricane)   Indianola, Texas, August 19, 1886

Inez (Hurricane)   Dominican Republic, Haiti, Cuba, September 24–30, 1966

Iniki (Hurricane)   Kauai, Hawaii, September 11, 1992

Ione (Hurricane)   E United States, September 19, 1955

Ione (Hurricane)   E North Pacific Ocean, August 19–28, 1974

Irah (Hurricane)   E North Pacific Ocean–NW Mexico, September 20–25, 1973

Irene (Hurricane)   Caribbean Sea–Nicaragua, September 11–20, 1971

Irene (Hurricane)   North Atlantic Ocean, October 7, 1999

Iris (Typhoon)   W North Pacific Ocean–China, 1959

Iris (Hurricane)   Leeward Islands–Caribbean Sea, August 22–September 4, 1995

Irma (Typhoon)   W North Pacific Ocean–Philippines, November 24, 1985

Irving (Typhoon)   W North Pacific Ocean–China, August 25, 1995

Irwin (Tropical Storm)   E North Pacific Ocean, October 11, 1999

Isaac (Tropical Storm)   Leeward Islands, September 28–October 1, 1988

Isabel (Tropical Storm)   October 5, 1985

Isbell (Hurricane)   Cuba–SE United States, October 8–16, 1964

Ismael (Hurricane)   E North Pacific Ocean–NW Mexico, September 12–15, 1995

Iva (Hurricane)   E North Pacific Ocean, August 5–13, 1988

Ivan (Super Typhoon)   Philippines–China, October 14–19, 1997

Iwa (Hurricane)   Central Pacific–Hawaii, November 23, 1982

Janet (Hurricane)   Caribbean Sea–SE Mexico, September 22–October 4, 1955

Janice (Tropical Storm)   North Atlantic Ocean, September 21–24, 1971

Janice (Tropical Storm)   W North Pacific Ocean–China, August 25, 1995

Jérémie (Hurricane)   Haiti, October 25, 1935

Jerry (Hurricane)   S United States, October 15, 1989

Jerry (Tropical Storm)   SE United States, August 22–28, 1995

Joan (Hurricane)   Caribbean Sea–Venezuela–Nicaragua, October 11–22, 1988

Joan (Super Typhoon)   Philippines–China, October 15–22, 1997

John (Hurricane/Typhoon)   North Pacific Ocean, August 11–September 10, 1994

John (Cyclone)   Australia, December 15, 1999

Jose (Hurricane)   Caribbean Sea–Gulf of Mexico, October 17–26, 1999

Josta (Cyclone)   Madagascar–Mozambique, March 1–15, 1995

Joyce (Hurricane)   E North Pacific Ocean, August 22–27, 1974

Juan (Hurricane)   S United States, October 26, 1985

Juliette (Hurricane)   E North Pacific Ocean, September 15–26, 1995

Justin (Cyclone)   Australia, March 5–10, 1997

Karen (Tropical Storm)   North Atlantic Ocean, August 26–September 3, 1995

Kate (Hurricane)   November 19–20, 1985

Kathleen (Hurricane)   E. Pacific Ocean–SW United States, September 5–10, 1976

Keith (Tropical Storm)   E Mexico–SE United States, November 17–24, 1988

King (Hurricane)   Miami, October, 1950

Kirsten (Hurricane)   E North Pacific Ocean, August 22–29, 1974

Kit (Typhoon)   Philippines, October 7, 1960

Kit (Typhoon)   Honshu, Japan, June 28, 1966

Kristy (Tropical Storm)   North Atlantic Ocean, October 17–21, 1971

Kristy (Hurricane)   E North Pacific Ocean, August 28–September 5, 1994

Kylie (Cyclone)   Indian Ocean–Mascarene Islands, March 3–15, 1995

Labor Day 1935 (Hurricane)   Florida Keys, September 2, 1935
Lane (Hurricane)   E North Pacific Ocean, September 3–10, 1994
Laura (Tropical Storm)   Caribbean Sea–Honduras–Cuba, November 12–21, 1971
Liza (Hurricane)   October 1, 1976
Lola (Typhoon)   Philippines, October 13, 1960
Lorraine (Tropical Storm)   E North Pacific Ocean, August 23–28, 1974
Los Angeles 1837 (Hurricane)   Puerto Rico, August 2, 1837
Lucille (Typhoon)   Philippines, May 22, 1960
Luis (Hurricane)   Antigua–Virgin Islands, August 27–September 11, 1995

Maggie (Hurricane)   E North Pacific Ocean, August 26–September 1, 1974
Marge (Typhoon)   Japan, 1951
Marilyn (Hurricane)   Dominica–Martinique–Virgin Islands, September 12–22, 1995
Marlene (Tropical Storm)   Indian Ocean, April 10–15, 1995
Mary (Typhoon)   Hong Kong, June 9, 1960
Miriam (Tropical Storm)   E North Pacific Ocean, September 15–21, 1994
Muroto 1934 (Typhoon)   Osaka, Japan, September 21, 1934

Nancy (Typhoon)   Osaka, Japan, September 16, 1961
Nina (Typhoon)   Philippines, November 18–28, 1968
Nina (Typhoon)   Philippines, November 25, 1987
Nina (Tropical Storm)   Philippines, 1995
Noel (Hurricane)   North Atlantic Ocean, September 26–October 7, 1995
Norma (Hurricane)   E North Pacific Ocean–SW Mexico, September 9–10, 1974
Norman (Tropical Storm)   E North Pacific Ocean, September 19–22, 1994

Olga (Typhoon)   Philippines, May 20, 1976
Olga (Typhoon)   W North Pacific Ocean–Korea, July 30–August 5, 1999
Olive (Typhoon)   North Pacific Ocean–Wake Island, September 16, 1952
Olive (Typhoon)   Northern Philippines, June 27, 1960
Olivia (Hurricane)   October 24, 1975
Olivia (Hurricane)   E North Pacific Ocean, September 22–29, 1994

Omar (Typhoon)   Guam, 1992
Opal (Hurricane)   SE United States, September 27–October 5, 1995
Orchid (Typhoon)   Southern and Eastern Japan, September 27–30, 1994
Orlene (Hurricane)   E North Pacific Ocean–NW Mexico, September 20–24, 1974

Pablo (Tropical Storm)   North Atlantic Ocean, October 4–8, 1995
Padre Ruiz 1834 (Hurricane)   Santo Domingo, September 23, 1834
Patricia (Hurricane)   E North Pacific Ocean, October 6–17, 1974
Paul (Tropical Storm)   E North Pacific Ocean, September 24–30, 1994
Paul (Tropical Storm)   W North Pacific Ocean, August 6, 1999
Phil (Cyclone)   S Western Pacific Ocean–Australia, December 22–30, 1996
Polly (Typhoon)   Manila, Philippines, December 10, 1956

Rita (Typhoon)   Japan, August 23, 1975
Rita (Typhoon)   Philippines, October 27, 1978
Rosalie (Tropical Storm)   E North Pacific Ocean, October 20–24, 1974
Rose (Hurricane)   E North Pacific Ocean–NW Mexico, October 8–15, 1994
Rose (Typhoon)   Hong Kong, August 17, 1971
Rosie (Typhoon)   Western North Pacific Ocean–Japan, July 21–29, 1997
Roxanne (Hurricane)   Caribbean Sea–Mexico, October 7–21, 1995
Ruby (Typhoon)   Hong Kong and China, September 5, 1964

Sam (Typhoon)   W North Pacific Ocean–Philippines–Taiwan, August 11–25, 1999
San Ciprian 1932 (Hurricane)   Puerto Rico, September 26, 1932
San Ciriaco 1899 (Hurricane)   Puerto Rico, August 8, 1899
San Felipe 1876 (Hurricane)   Puerto Rico, September 13, 1876
San Felipe 1928 (Hurricane)   Puerto Rico–United States, September 13, 1928
San Narciso 1867 (Hurricane)   Virgin Islands–Puerto Rico, October 29, 1867
Santa Ana 1825 (Hurricane)   Puerto Rico, July 26, 1825
Santa Cruz del Sur 1932 (Hurricane)   Cuba, November 9, 1932
Santa Elena 1851 (Hurricane)   Puerto Rico, August 18, 1851

Sarah (Typhoon) Japan, South Korea, September 17–18, 1959

Sarah (Typhoon) China, October 1979

Sebastien (Tropical Storm) Leeward Islands, October 20–25, 1995

Sening (Typhoon) Philippines, October 14, 1970

Seth (Typhoon) South Korea, October 15–16, 1994

Shirley (Typhoon) Taiwan, July 31, 1960

Shirley (Typhoon) Japan, September 10, 1966

Sybil (Typhoon) Philippines–China, October 1–5, 1995

Tanya (Hurricane) North Atlantic Ocean, October 27–November 1, 1995

Theresa (Typhoon) Philippines, December 3, 1972

Tip (Typhoon) Luzon Island, Philippines, October 12, 1979

Titang (Typhoon) Philippines, October 15, 1970

Tracy (Cyclone) Darwin, Australia, December 25, 1974

Trix (Typhoon) W North Pacific Ocean–Japan, September 11–19, 1966

Valerian (Hurricane) Cuba–Bermuda, October 20–22, 1926

Vera (Typhoon) Honshu, Japan, September 26–27, 1959

Vera (Typhoon) Taiwan, July 31, 1977

Vera (Typhoon) Taiwan, 1979

Violet (Cyclone) SE Australia, March 8–12, 1995

Virginia (Typhoon) Japan, Formosa, June 27, 1957

Wanda (Typhoon) China, August 2, 1956

Warren (Cyclone) NE Australia, March 6–10, 1995

Wendy (Typhoon) Hong Kong, July 18, 1957

Wendy (Typhoon) Japan, August 13, 1960

Willie (Typhoon) Vietnam, September, 1996

Wilma (Typhoon) Philippines, November 1, 1952

Winnie (Typhoon) Hong Kong, Philippines, June 30, 1964

York (Typhoon) W North Pacific Ocean–China, September 16, 1999

Yuri (Typhoon) Guam, 1991

Yvette (Tropical Storm) SE Vietnam, October 21–25, 1995

Zack (Typhoon) NE Vietnam, October 28–November 2, 1995

Zeb (Typhoon) Philippines–Taiwan–Japan, October 9–17, 1998

BIBLIOGRAPHY

Addison, Doug. "Filmmaker to Storm Chasers." *Weatherwise* (June/July 1996).

Ainlay, George. "1944—Year of the Great Hurricane!" *Science Digest* (May 15, 1945).

Alexander, W. H. "Hurricanes, Especially Those of Puerto Rico and St. Kitts." *Monthly Weather Review* 33 (August 1905).

Algue, Jose. *Cyclones of the Far East*. Manila: Bureau of Public Printing, 1904.

Aubert de la Rue, Edgar. *Man and the Winds*. London: Hutchinson, 1955.

Ayoade, J. O. *Introduction to Climatology for the Tropics*. Chichester: John Wiley and Sons, 1983.

Bailey, J. F., J. L. Patterson, and J. L. H. Paulhus. "Hurricane Agnes Rainfall and Floods, June–July 1972." Professional Paper 924. Reston, Va.: U.S. Geological Survey.

Barnes, Jay. *North Carolina's Hurricane History*. Durham: University of North Carolina Press, 1995.

———. *North Carolina's Hurricane History*, third edition. Chapel Hill: University of North Carolina Press, 2001.

Battan, Louis J. *The Nature of Violent Storms*. New York: Doubleday, 1959.

Baum, Robert A. "Tropical Cyclones in the Eastern Pacific Ocean in 1974." *Weatherwise* (February 1975).

Blair, Thomas Arthur. *Weather Elements*. New York: Prentice-Hall, 1942.

Blodgett, Lorin. "List of Hurricanes on the Coast of the South Atlantic States, and on the North Coast of the Gulf of Mexico." *Climatology of the United States and the Temperate Latitudes of the North American Continent*. Philadelphia: Lippincott, 1857.

Bowie, E. H. "Formation and Movement of West Indian Hurricanes." *Monthly Weather Review* 50 (June 1922).

Breen, Henry H. *St. Lucia: Historical, Statistical and Descriptive*. London: 1844.

Brennan, J. F. *A Report on the Hurricane of Western Jamaica, October 29, 1933*. Kingston, Jamaica: 1934.

Breuer, Georg. *Weather Modification: Prospects and Problems*. Cambridge: Cambridge University Press, 1979.

Brooks, Charles F. "Hurricanes in New England." *Geographical Review* 29 (December 1939).

———. "Some Excessive Rainfalls." *Monthly Weather Review* 47 (April 1919).

———. *Why the Weather?* New York: Harcourt, Brace, 1935.

Broughner, C. C. "Hurricane Hazel." *Weather* 10 (November 1955).

Burns, Cherie. *The Great Hurricane: 1938*. New York: Atlantic Monthly Press, 2005.

Calhoun, Raymond C. *Typhoon: The Other Enemy* Annapolis, Md.: Naval Institute Press, 1981.

Canis, Wayne F., and William J. Neal. *Living with the Alabama–Mississippi Shore*. Durham, N.C.: Duke University Press, 1985.

Capper, James. *Observations on the winds and monsoons; illustrated with a chart, and accompanied with notes, geographical and meteorological*. London: 1844.

Carrier, Jim. *The Ship and the Storm*. New York: International Marine/McGraw-Hill, 2001.

Cartwright, Gary. *Galveston: A History of the Island*. New York: Atheneum, 1991.

Cerveny, Randy. "Making Weather in the Movies." *Weatherwise* (December 1996/January 1997).

Clark, Gilbert B. "The Hurricane Season of 1965." *Weatherwise* (February 1966).

Cline, Isaac Munroe. *Tropical Cyclones*. New York: Macmillan, 1926.

Cline, Joseph L. *When the Heavens Frowned*. Dallas, Tex.: M. Van Nort, 1946.

Cobb, Hugh D., III. "The Chesapeake-Potomac Hurricane of 1933." *Weatherwise* (August/September 1991).

———. "The Siege of New England." *Weatherwise* (October 1989).

Coronas, J. *The Climate and Weather of the Philippines*. Manila: Bureau of Printing, 1920.

Cry, G. W, W. H. Haggard, and H. S. White. "North Atlantic Tropical Cyclones, Technical Paper Number 35." Washington, D.C.: U.S. Weather Bureau, 1959.

Dampier, William. *Dampier's Voyages.* New York: 1906.

Daniels, George H. *American Science in the Age of Jackson.* New York: Columbia University Press, 1968.

Davis, Albert B., Jr. *Galveston's Bulwark Against the Sea: History of the Galveston Sea Wall.* Galveston: U.S. Army Corps of Engineers, 1961.

Davies, Pete. *Inside the Hurricane.* New York: Henry Holt, 2000.

Davis, W. M. *Whirlwinds, Cyclones and Tornadoes.* New York: Lee and Shepard, 1884.

Dawes, Robert A., Jr. *The Dragon's Breath, Hurricane at Sea.* Annapolis, Md.: Naval Institute Press, 1996.

DeAngelis, Dick. "The Hurricane Priest." *Weatherwise* (October 1989).

Dennis, Jerry. "The Lure of Storms." *Weatherwise* (June/July 1994).

De Wire, Elinor. "England's Great Storm." *Weatherwise* (October/November 1996).

Donn, William L. *Meteorology.* New York: McGraw Hill, 1951.

Douglas, Marjory Stoneman. *Hurricane.* New York: Rinehart, 1958.

Dunn, Gordon E., and Banner I. Miller. *Atlantic Hurricanes.* Baton Rouge: Louisiana State University Press, 1960.

Edinger, James G. *Watching for the Wind.* Garden City, N.Y.: Doubleday, 1967.

Edwards, Bryan. *The History Civil and Commercial of the British West Indies.* 5th ed. London: 1819.

Eliot, J. *Cyclones of the Bay of Bengal.* Calcutta: Superintendant of the Government Printing Office, 1888.

Emanuel, Kerry. *Divine Wind: The History and Science of Hurricanes.* Oxford: Oxford University Press, 2005.

Espy, James P. *Facts Collected by Mr. Espy, taken from newspapers of the time.* Philadelphia: Journal of the Franklin Institute, no. 23, 1839.

Esquemeling, John. *The Buccaneers of America.* New York: Dutton, 1924.

Fassig, Oliver L. *Hurricanes of the West Indies.* Washington, D.C.: U.S. Weather Bureau, 1913.

Fay, John. "The Galveston Tragedy." *Cosmopolitan* (November 1900).

Fisher, David E. *The Scariest Place on Earth.* New York: Random House, 1994.

Fishman, Jack, and Robert Kalish. *The Weather Revolution.* New York: Plenum Press, 1994.

Fleagle, Robert G., et al. *Weather Modification in the Public Interest.* Seattle: University of Washington Press, 1974.

Flintham, Victor. *Air Wars and Aircraft: A Detailed Record of Air Combat, 1945 to the Present.* New York: Facts On File, 1990.

Flora, Snowden D. *Tornadoes of the United States.* Norman: University of Oklahoma Press, 1953.

Fortier, Edouard. *Un Cyclone dans les Antilles, l'Ouragan de 1891 a la Martinique.* Paris: 1892.

Fowler, John. *A General Account of the Calamities occasioned by the late tremendous Hurricanes and Earthquakes in the West-India Islands.* London: 1781.

Fox, William Price. *Lunatic Wind: Surviving the Storm of the Century.* Chapel Hill, N.C.: Algonquin Books, 1992.

Frank, Neil L., and S. A. Husain. "The Deadliest Tropical Cyclone in History?" *Bulletin of the American Meteorological Society* 52 (June 1971).

———. "Lessons from Hurricane Eloise." *Weatherwise* (October 1976).

———. "The 1964 Hurricane Season." *Weatherwise* (February 1965).

Frazier, Kendrick. *The Violent Face of Nature.* New York: William Morrow, 1979.

Frazier, R. D. "Early Records of Tropical Hurricanes on the Texas Coast in the Vicinity of Galveston." *Monthly Weather Review* 49 (September 1921).

Froc, R. F. Louis. *Atlas of the Tracks of 620 Typhoons, 1893–1918.* Shanghai: Catholic Mission Press, 1919.

———. *Typhoon Highways in the Far East.* Shanghai: Catholic Mission Press, 1896.

Garriot, E. B. *West Indian Hurricanes.* Washington: 1900.

Gentry, R. Cecil. "Current Hurricane Research." *Weatherwise* (August 1964).

Gillispie, Charles Coulston, ed. *Dictionary of Scientific Biography.* Vol. 11. New York: Charles Scribner's Sons, 1969.

Gold, Jerome. *Hurricanes.* Seattle: Black Heron Press, 1994.

Gray, R. W. "Florida Hurricanes." *Monthly Weather Review* 61 (August 1933).

Green, Nathan C., ed. *The Story of the 1900 Galveston Hurricane.* Gretna, La.: Pelican Publishing Company, 2000.

Grenci, Lee. "The Loneliness of a Long-Distance Hurricane." *Weatherwise* (December 1994/January 1995).

Hall, Maxwell. *Notes of Hurricanes, Earthquakes.* Kingston, Jamaica: Meteorological Service, 1916.

Hearn, Lafcadio. *Chita: A Memory of Last Island.* New York: Harper, 1905.

Helm, Thomas. *Hurricane: Weather at its Worst.* New York: Dodd, Mead, 1967.

Heninger, S. K., Jr. *A Handbook of Renaissance Meteorology.* Durham, N.C.: Duke University Press, 1960.

Henry, James A., Kenneth M. Porter, and Jan Coyne. *The Climate and Weather of Florida.* Sarasota: Pineapple Press, 1994.

Henson, Robert. "Piece by Piece." *Weatherwise* (April/May 1995).

Hebert, Paul J. "The Atlantic Hurricane Season of 1973." *Weatherwise* (February 1974).

———. "The Hurricane Season of 1972." *Weatherwise* (February 1973).

Herrera, Desiderio. *Memoria sobre los huracanes en la isla de Cuba.* Havana: 1847.

Hope, John R. "Atlantic Hurricane Season of 1974." *Weatherwise* (February 1975).

House of Representatives. "Testimony Regarding the Hurricane of September 2, 1935." *Hearings before the Committee on World War Veteran's Legislation.* Washington, D.C.: Government Printing Office, 1936.

Hughes, Patrick. "The Great Galveston Hurricane." *Weatherwise* (August 1979).

Iacovelli, Debi. "Getting a Fix on Tropical Troublemakers." *Weatherwise* (June/July 1994).

———. "The New Guard." *Weatherwise* (August/September 1995).

Jennings, Gary. *The Killer Storms, Hurricanes, Typhoons and Tornadoes.* New York: Lippincott, 1970.

Jin, Fei-Fei, J. David Neilin, and Michael Ghil. "El Niño on the Devil's Staircase: Annual Subharmonic Steps to Chaos." *Science* 264 (1994).

Keen, Richard A. *Skywatch East: A Weather Guide.* Golden, Colo.: Fulcrum Publishing, 1992.

Kennedy, Betty. *Hurricane Hazel.* Toronto: Key Porter Books, 1979.

Kerr, Richard A. "Official Forecasts Pushed Out to a Year Ahead." *Science* 266 (December 23, 1994).

Kinder, Gary. *Ship of Gold in the Deep Blue Sea.* New York: Atlantic Monthly Press, 1995.

Kleinberg, Eliot. *Black Cloud: The Deadly Hurricane of 1928.* New York: Carroll & Graf, 2003.

Knox, Hugh. *Discourse . . . on the Occasion of the Hurricane which happened on the 31st day of August.* St. Croix, V.I.: 1772.

Kraft, R. H. "Great Hurricanes 1955–1965." *Mariners Weather Log* (November 1966).

Kumar, Arun, and Ants Leetmaa. "Simulations of Atmospheric Variability Induced by Sea Surface Temperatures and Implications for Global Warming." *Science* 266 (October 28, 1994).

Lamb, Hubert, and Knud Frydendahl. *Historic Storms of the North Sea, British Isles, and Northwest Europe.* Cambridge: Cambridge University Press, 1991.

Lane, Frank W. *The Elements Rage.* New York: Chilton, 1965.

Larson, Erik. *Isaac's Storm.* New York: Crown, 1999.

Laskin, David. *Braving the Elements: The Stormy History of American Weather.* New York: Doubleday, 1996.

Lawrence, Miles B. "Return of the Hurricane." *Weatherwise* (February 1989).

Lockhart, Gary. *The Weather Companion.* New York: John Wiley & Sons, 1988.

Lopez, Harvey. "Hurricane Dora Deals City Severe Blow." *Weatherwise* (October 1964).

Ludlum, David M. *The American Weather Book.* Boston: Houghton Mifflin, 1982.

———. *The Audubon Society Field Guide to North American Weather.* New York: Alfred Knopf, 1995.

———. *Early American Hurricanes 1492–1870.* Boston: American Meteorological Society, 1963.

———. *Weather Record Book.* Princeton, N.J.: Weatherwise, Inc., 1971.

Lyell, Charles. *A Second Visit to the United States of North America.* Vol. I. London: John Murray, 1849.

McDonald, W. F. "Low Barometer Readings in West Indian Disturbances of 1932 and 1933." *Monthly Weather Review* 61 (1933).

Marquis Who's Who. *Who's Who in Science and Engineering 1996–1997.* 3rd ed. New Providence, N.J.: Reed Reference, 1996.

Marx, Jennifer, and Robert Marx. *New World Shipwrecks: A Comprehensive Guide.* Dallas: Ram Books, 1994.

Mayfield, Max, and Miles Lawrence. "The Atlantic Hurricanes." *Weatherwise* (February/March 1995).

McCarthy, Joe. *Hurricane!* New York: American Heritage Publishing, 1969.

McCullough, David. *The Johnstown Flood.* New York: Simon & Schuster, 1968.

McMurray, Emily J., ed. *Notable Twentieth Century Scientists.* Vol. 2. New York: Gale Research, 1993.

Meldrum, C. *Notes on the Forms of Cyclones in the Southern Indian Ocean, and on Some of the Rules Given for Avoiding Their Centres.* London: E. Stanford, 1873.

Mitchell, C. L. *Hurricanes of the South Atlantic and Gulf States, 1879–1928.* Washington, D.C.: U.S. Weather Bureau, 1928.

———. *West Indian Hurricanes and Other Tropical Cyclones of the North Atlantic Ocean.* Washington, D.C.: U.S. Weather Bureau, 1924.

Mitchell, Valentine Edwin. *It's an Old New England Custom.* New York: Bonanza Books, 1966.

Monastersky, R. "Atlantic Current Gives Climate the Shakes." *Science News* 145 (1994).

———. "Hurricane Experts Predict Better Forecasts." *Science News* 145 (1994).

Morison, Samuel Eliot. *The Liberation of the Philippines: Luzon, Mindanao, the Visayas 1944–1945.* Boston: Little, Brown, 1989.

Morton, Robert A., and Orrin H. Pilkey, Jr. *Living with the Texas Shore.* Durham, N.C.: Duke University Press, 1983.

Mykle, Robert. *Killer 'cane: The Deadly Hurricane of 1928.* New York: Cooper Square Press, 2002.

Nash, Jay Robert. *Darkest Hours.* Chicago: Nelson Hall, 1976.

National Oceanic and Atmospheric Administration. *The Amateur Weather Forecaster.* Washington, D.C.: National Weather Service, July 1994.

———. *Flash Floods and Floods . . . the Awesome Power.* Washington, D.C.: National Weather Service, July 1992.

———. *The Naming of Hurricanes.* Washington, D.C.: National Weather Service, 1993.

———. *Spotter's Guide.* Washington, D.C.: National Weather Service, 1992.

Neal, William J., et al. *Living with the South Carolina Shore.* Durham, N.C.: Duke University Press, 1984.

Neelin, J. David, and Jochem Marotzke. "Representing Ocean Eddies in Climate Models." *Science* 262 (May 20, 1994).

Nordberg, W. "Geophysical Observations from Nimbus I." *Science* 150.

Normand, C. W. B. *Storm Tracks in the Bay of Bengal.* Calcutta: India Meteorological Department, 1925.

Norton, Grady. *Florida Hurricanes.* Washington, D.C.: U.S. Government Printing Office, 1936.

O'Connor, Neil F. "The Fujiwara Effects." *Weatherwise* (October 1964).

Okada, T. "On the Eye of the Storm." *Memoirs of the Imperial Marine Observatory.* Kobe, Japan: 1922.

Palm Beach Post. *Hurricane Andrew: Images from the Killer Storm.* Marietta, Ga.: Longstreet Press, 1992.

Pardue, L. G. "The Hurricane Season of 1970." *Weatherwise* (February 1971).

Parrott, Daniel S. *Tall Ships Down.* New York: International Marine/McGraw-Hill, 2003.

Perley, Sidney. *Historic Storms of New England.* Salem, Mass.: Salem Press, 1891.

Phillips, David. *The Day Niagara Falls Ran Dry!* Toronto: Key Porter Books, 1993.

Piddington, Henry. *The Sailors Handbook for the Law of Storms.* London: 1876.

Pilkey, Orrin H., Jr., et al. *Living with the East Florida Shore.* Durham, N.C.: Duke University Press, 1984.

Pilkey, Orrin H., Jr., and William J. Neal. *Living with the West Florida Shore.* Durham, N.C.: Duke University Press, 1984.

Poey, Andreas. *Table chronologique des quatre cents cyclones.* Paris: 1862.

Price, W. Armstrong. *Hurricanes Affecting the Coast of Texas from Galveston to Rio Grande.* Washington, D.C.: Army Corps of Engineers, 1956.

Provenzano, Eugene F., Jr., and Sandra H. Fradd. *Hurricane Andrew: The Public Schools and the Rebuilding of Community.* New York: State University of New York Press, 1995.

Rappaport, Edward N., and Lixion A. Avila. "The Atlantic Hurricanes." *Weatherwise* (February/March 1995).

Reardon, L. F. *The Florida Hurricane and Disaster.* Coral Gables, Fla.: Arva Parks and Company, 1986.

Redfield, William C. "Remarks on the prevailing storms of the Atlantic Coast of the North American States." *American Journal of Science* 20 (July 1831).

———. "On Three Hurricanes of the American Seas." *American Journal of Science* (May 1, 1846).

Reed, Mary. "Weather Talk." *Weatherwise* (June/July 1994).

———. "Weather Talk: Tall Tails." *Weatherwise* (October/November 1996).

Reid, William. *An Attempt to Develop the Law of Storms.* 3rd ed. London: 1850.

Rich, Shebnah. *Truro, Cape Cod.* Boston: D. Lothrop, 1884.

Reichelderfer, Francis W. "The World Meteorological Organization." *Weatherwise* (October 1964).

Riehl, Herbert. *Tropical Meteorology.* New York: McGraw-Hill, 1954.

Rosenfeld, Jeff. "Safety Factors." *Weatherwise* (June/July 1994).

Rufferner, James A., and Frank E. Blair. *The Weather Almanac.* 5th ed. Detroit: Gale Research Company, 1986.

Russell, Thomas H. *Flood and Cyclone Disasters.* Chicago: Thomas H. Morrison, 1913.

Saar, John. "Japanese divers discover wreckage of Mongol fleet." *Smithsonian* (October 1980).

Sadowski, Alexander. "Tornadoes with Hurricanes." *Weatherwise* (April 1966).

Schlatter, Thomas. "Weather Queries: Haze and Hurricanes, Tanning and Thunderstorms." *Weatherwise* (August 1989).

———. "Weather Queries: Hurricane Facts." *Weatherwise* (August 1988).

Schneider, Stephen H., ed. *Encyclopedia of Climate and Weather.* Oxford: Oxford University Press, 1996.

Scott, Phil. *Hemingway's Hurricane.* New York: International Marine/McGraw-Hill, 2006.

Scotti, R. A. *Sudden Sea: The Great Hurricane of 1938.* Boston: Little, Brown, 2003.

Shaw, David. "The Killer Wind." *Cruising World* (September 1996).

Sheets, Bob, and Jack Williams. *Hurricane Watch: Forecasting the Deadliest Storms on Earth.* New York: Vintage Books, 2001

Simpson, Robert H., and Herbert Riehl. *The Hurricane and Its Impact.* Baton Rouge: Louisiana State University Press, 1981.

Smith, Roger. *Catastrophes and Disasters.* New York: Chambers, 1992.

Snow, Edward Rowe. *The Fury of the Seas.* New York: Dodd, Mead, 1964.

———. *Storms and Shipwrecks of New England.* Beverly, Mass.: Commonwealth Editions, 2003.

Spencer, R. E. "Weather Proverbs." *Weekly Weather and Crop Bulletin.* Washington, D.C.: U.S. Government Printing Office, December 27, 1954.

Stewart, George R. *Storm.* New York: Random House, 1941.

Stranack, Ian. *The Andrew and the Onions*. Bermuda: The Island Press, 1975.

Strommel, Henry M., and Dennis W. Moore. *An Introduction to the Coriolis Force*. New York: Columbia University Press, 1989.

Sugg, Arnold L., and Paul J. Herbert. "The Hurricane Season of 1966." *Monthly Weather Review* (March 1967).

————. "The Hurricane Season of 1968." *Weatherwise* (February 1969).

————. "The Hurricane Season of 1969." *Weatherwise* (February 1970).

Talley, Lynne D., and Gregory C. Johnson. "Deep, Zonal Subequatorial Currents." *Science* 262 (February 25, 1994).

Tannehill, Ivan Ray. *Hurricanes*. Princeton, N.J.: Princeton University Press, 1952.

————. *The Hurricane Hunters*. New York: Dodd, Mead, 1956.

Travis, John. "NOAA's 'Arks' Sail Into a Storm." *Science* 265 (July 8, 1994).

Tufty, Barbara. *1001 Questions Answered About Hurricanes, Tornadoes, and Other Natural Air Disasters*. New York: Dover Publications, 1987.

Viñes, P. B. "The Antilles Cyclonoscope." *Weatherwise* (October 1989).

Volk, C. M. et al. "Quantifying Transport Between the Tropical and Mid-Latitude Lower Stratosphere." *Science* 272 (June 21, 1996).

Walker, George R., Joseph E. Minor, and Richard D. Marshall. "The Darwin Cyclone: Valuable Lesson in Structural Design." *Civil Engineering* 45 (December 1975).

Watson, Benjamin A. *Acts of God: The Old Farmer's Almanac—Unpredictable Guide to Weather and Natural Disasters*. New York: Random House, 1993.

Weems, John Edward. *A Weekend in September*. College Station: Texas A & M University Press, 1957, 1980, 2005.

Whipple, A. B. C. *Storm*. Alexandria, Va.: Time-Life Books, 1982.

Widger, W. K. *Meteorological Satellites*. New York: Holt, Rinehart and Winston, 1966.

Will, Lawrence E. *Okeechobee Hurricane: Killer Storms in the Everglades*. St. Petersburg, Fla.: Great Outdoors Publishing, 1967.

Williams, Jack. *The USA TODAY Weather Almanac 1995*. New York: Vintage Books, 1994.

————. *The USA TODAY Weather Book*. New York: Vintage Books, 1993.

Williams, Richard. "The Divine Wind." *Weatherwise* (October/November 1991).

Winsberg, Morton D. *Florida Weather*. Orlando, Fla.: University of Central Florida Press, 1990.

Witten, John. "Elena; She taunted us, she teased us and & then she hit us." *Weatherwise* (October 1985).

Wurman, Joshua, and Jerry M. Straka. "Fine-Scale Doppler Radar Observations of Tornadoes." *Science* 272 (June 21, 1996).

INDEX